BIRKHÄUSER

Operator Theory: Advances and Applications

Vol. 204

Editor:
I. Gohberg

Subseries
Linear Operators and Linear Systems

Operator Theory: Advances and Applications

Vol. 204

An Operator Perspective on Signals and Systems

Arthur E. Frazho
Wisuwat Bhosri

Birkhäuser
Basel · Boston · Berlin

Authors:

Arthur E. Frazho
School of Aeronautics and Astronautics
Purdue University
West Lafayette, IN 47907-2045
USA
e-mail: frazho@ecn.purdue.edu

Wisuwat Bhosri
541/15 Lamphun Road
Amphoe Muang
Chiang Mai 50000
Thailand
e-mail: wisuwat.bhosri@gmail.com

2000 Mathematics Subject Classification: 47A20, 47A57, 47B35

Library of Congress Control Number: 2009938997

Bibliographic information published by Die Deutsche Bibliothek.
Die Deutsche Bibliothek lists this publication in the Deutsche Nationalbibliografie;
detailed bibliographic data is available in the Internet at http://dnb.ddb.de

ISBN 978-3-0346-0291-4 Birkhäuser Verlag AG, Basel – Boston – Berlin

Basel · Boston · Berlin
P.O. Box 133, CH-4010 Basel, Switzerland
Part of Springer Science+Business Media
Printed on acid-free paper produced from chlorine-free pulp. TCF∞
Printed in Germany

ISBN 978-3-0346-0291 e-ISBN 978-3-0346-0292

9 8 7 6 5 4 3 2 1

www.birkhauser.ch

Preface

In this monograph, we combine operator techniques with state space methods to solve factorization, spectral estimation, and interpolation problems arising in control and signal processing. We present both the theory and algorithms with some Matlab code to solve these problems.

A classical approach to spectral factorization problems in control theory is based on Riccati equations arising in linear quadratic control theory and Kalman filtering. One advantage of this approach is that it readily leads to algorithms in the non-degenerate case. On the other hand, this approach does not easily generalize to the nonrational case, and it is not always transparent where the Riccati equations are coming from.

Operator theory has developed some elegant methods to prove the existence of a solution to some of these factorization and spectral estimation problems in a very general setting. However, these techniques are in general not used to develop computational algorithms. In this monograph, we will use operator theory with state space methods to derive computational methods to solve factorization, spectral estimation, and interpolation problems. It is emphasized that our approach is geometric and the algorithms are obtained as a special application of the theory. We will present two methods for spectral factorization. One method derives algorithms based on finite sections of a certain Toeplitz matrix. The other approach uses operator theory to develop the Riccati factorization method. Finally, we use isometric extension techniques to solve some interpolation problems.

The monograph is divided into five parts. In the first part, we present some classical results from operator theory. This includes the Wold decomposition, unilateral and bilateral shifts, the Beurling-Lax-Halmos Theorem, and the Naimark representation Theorem. Chapter 5 on the Naimark representation Theorem is one of the fundamental tools that is used throughout the monograph. The reader familiar with operator theory can skip this part and refer back to it as needed. This part also includes some results on rational functions which are not usually presented in elementary operator theory. The first part is self contained and is written for someone with a minimal background in operator theory. The other four parts are more or less independent of each other, and can be read separately. There may be some cross references. However, this should not cause any major difficulty.

In part II, we develop finite section techniques to compute the inner-outer factorization of rational functions and solve a spectral factorization problem in both the square and non-square cases. Furthermore, operator techniques are used to solve some sinusoid estimation problems in signal processing. In particular, we use geometric methods to develop the Capon-Genorimus sinusoid estimation algorithm. Finite section methods are also used to solve some sinusoid estimation problems. Many of these techniques are based on the Levinson and Kalman-Ho algorithm. Several examples using Matlab are given.

In part III, we use Riccati techniques to solve factorization and Darlington synthesis problems. These Riccati techniques are developed from the Naimark representation Theorem. Chapter 11 is devoted to the Kalman filter. This chapter can be read independently from the rest of the monograph. This is included to demonstrate where the Riccati equations in control theory originally came from, and how they can be used to solve Kalman and Wiener filtering problems.

The fourth part is an introduction to positive real and H^∞ type interpolation problems. Our approach is based on extending a contraction to an isometry. Here we do not present the set of all solutions, we just give the central solution and corresponding state space formula. The central solution is the one that is most widely used in applications.

The fifth part is the appendix which includes a short review of state space techniques used throughout the monograph. We also place a special emphasis on the Kalman-Ho algorithm, which plays a fundamental role in some of our computational techniques. The last chapter is devoted to the Levinson algorithm. Finally, the Gohberg-Semencul-Heinig inversion formula for a positive Toeplitz operator is presented.

It is assumed that the reader is familiar with linear algebra, and some elementary facts from operator theory such as the projection theorem, the adjoint of an operator and a positive operator. Our approach is geometric and we do not rely on measure theoretic techniques. It is also assumed that the reader is familiar with some elementary concepts from linear systems theory such as controllability, observability and state space realization. A review of some of these state space techniques is given in the appendix. Some sections can be skipped without any loss of continuity and the reader will be notified when this is the case. We have also used the notes at the end of the chapter to develop some technical connections between our results and some of the existing theory. Finally, it is noted that we developed our theory using Hardy spaces of functions which are analytic outside the open unit disc. This was done mainly because these Hardy spaces are more naturally suited to the fast Fourier transform algorithm in Matlab and state space methods.

We hope that this monograph is beneficial to both mathematicians and engineers. We believe that the operator theoretic foundation provides additional insight into spectral factorization and signal processing. Moreover, this framework may be useful for other engineering applications. Finally, the applications and examples may provide some additional insight into other mathematical problems.

August, 2009 The authors

Contents

Part I

Basic Operator Theory

Chapter 1

The Wold Decomposition

In this chapter we will introduce the classical Wold decomposition for an isometry U on $\mathcal{K}$. The Wold decomposition was initially used to decompose a wide sense stationary random process into its deterministic and purely nondeterministic parts. We will also study unilateral and bilateral shifts, and present an introduction to Toeplitz operators. Finally, the Wold decomposition along with Toeplitz and shift operators will play a fundamental role in our approach to signal processing and factorization theory.

1.1 The Unilateral Shift

This section is devoted to the unilateral shift and some of its properties. Recall that A is an *operator* if A is a bounded linear map acting between two Hilbert spaces. The adjoint of an operator A is denoted by A^*. If $\mathcal{H}$ is a subspace of a Hilbert space $\mathcal{K}$, then $P_{\mathcal{H}}$ denotes the orthogonal projection onto $\mathcal{H}$. Finally, the identity operator on a Hilbert space $\mathcal{X}$ is denoted by $I_{\mathcal{X}}$ or just I when the underlying space is understood.

Recall that an operator A mapping $\mathcal{X}$ into $\mathcal{Y}$ is an *isometry*, if $A^*A = I$, or equivalently, $\|Ax\| = \|x\|$ for all x in $\mathcal{X}$. In other words, an isometry is an operator which preserves the norm. In particular, an isometry is one to one. We say that an operator A is a *co-isometry*, if A^* is an isometry. The operator A mapping $\mathcal{X}$ into $\mathcal{Y}$ is *unitary*, if $A^*A = I_{\mathcal{X}}$ and $AA^* = I_{\mathcal{Y}}$. An operator A is unitary if and only if A is an isometry and A is onto $\mathcal{Y}$. Observe that A is unitary if and only if both A and A^* are isometries. Finally, for an example of an operator which is an isometry but not unitary, consider the operator A given by

$$A = \frac{1}{\sqrt{2}} \begin{bmatrix} 1 \\ 1 \end{bmatrix} : \mathbb{C} \to \mathbb{C}^2.$$

For the moment assume that U is an operator acting on a finite dimensional space $\mathcal{X}$. Then U is an isometry if and only if U is a unitary operator. If U is an isometry, then U is one to one. Because $\mathcal{X}$ is finite dimensional and U on $\mathcal{X}$ is one to one, U must be onto, and thus, U is unitary. In a moment, we will introduce the unilateral shift, which is an example of an isometry on an infinite dimensional space which is not unitary.

Let us establish some further terminology. An operator Φ mapping $\mathcal{X}$ into $\mathcal{Y}$ is *invertible* if Φ admits a bounded inverse, that is, there exists an operator Ψ mapping $\mathcal{Y}$ into $\mathcal{X}$ such that $\Phi\Psi = I_{\mathcal{Y}}$ and $\Psi\Phi = I_{\mathcal{X}}$. We say that two operators A on $\mathcal{X}$ and B on $\mathcal{Y}$ are *unitarily equivalent* (respectively *similar*), if there exists a unitary operator (respectively an invertible operator) Φ mapping $\mathcal{X}$ onto $\mathcal{Y}$ such that $\Phi A = B\Phi$. We say that an operator Q *intertwines* A on $\mathcal{X}$ with B on $\mathcal{Y}$, if Q is an operator mapping $\mathcal{X}$ into $\mathcal{Y}$ such that $QA = BQ$. In particular, A and B are unitarily equivalent (respectively similar) if and only if there exists a unitary operator (respectively an invertible operator) intertwining A with B. Finally, we say that $\mathcal{L}$ is a *cyclic* set for an operator A on $\mathcal{X}$, if $\mathcal{L} \subseteq \mathcal{X}$ and

$$\mathcal{X} = \bigvee_{n=0}^{\infty} A^n \mathcal{L}. \tag{1.1.1}$$

As expected, $\bigvee$ denotes the closed linear span.

Let U be an isometry on $\mathcal{K}$. We say that a subspace $\mathcal{L} \subseteq \mathcal{K}$ is a *wandering subspace for* U, if for all positive integers m and n, the subspace $U^m\mathcal{L}$ is orthogonal to the subspace $U^n\mathcal{L}$, when $m \neq n$. (By positive we mean greater than or equal to zero. A *subspace* is a closed linear space.) Notice that $\mathcal{L}$ is a wandering subspace for U if and only if the subspace $U^n\mathcal{L}$ is orthogonal to $\mathcal{L}$, for all integers $n \geq 1$. (The proof is left as an exercise.) If $\mathcal{L}$ is wandering for U, then $\mathcal{M}_+(\mathcal{L})$ is the invariant subspace for U defined by

$$\mathcal{M}_+(\mathcal{L}) = \bigoplus_{n=0}^{\infty} U^n \mathcal{L} = \bigvee_{n=0}^{\infty} U^n \mathcal{L}. \tag{1.1.2}$$

It is emphasized that a vector g is in $\mathcal{M}_+(\mathcal{L})$ if and only if $g = \sum_0^\infty U^n g_n$ where $g_n \in \mathcal{L}$ for all integers $n \geq 0$ and $\|g\|^2 = \sum_0^\infty \|g_n\|^2$ is finite.

If U is an isometry on $\mathcal{K}$, then $\mathcal{L} = \ker U^*$ is a wandering subspace for U. (The kernel of an operator A is denoted by $\ker A$.) To see this, let f and g be any two vectors in $\mathcal{L} = \ker U^*$. Then $(U^n f, g) = (f, U^{*n} g) = (f, 0) = 0$, for all $n \geq 1$. Hence $U^n\mathcal{L}$ is orthogonal to $\mathcal{L}$ for all $n \geq 1$. Therefore $\ker U^*$ is a wandering subspace for U.

We say that S is a *unilateral shift* or *forward shift* on $\ell_+^2(\mathcal{E})$, if S is the operator given by

$$S = \begin{bmatrix} 0 & 0 & 0 & 0 & \cdots \\ I & 0 & 0 & 0 & \cdots \\ 0 & I & 0 & 0 & \cdots \\ 0 & 0 & I & 0 & \cdots \\ \vdots & \vdots & \vdots & \vdots & \ddots \end{bmatrix} \quad \text{on } \ell^2_+(\mathcal{E}). \tag{1.1.3}$$

The identity operator I on $\mathcal{E}$ appears immediately below the main diagonal and zeros everywhere else. As expected, $\ell^2_+(\mathcal{E})$ is the Hilbert space formed by the set of all vectors of the form

$$f = \begin{bmatrix} f_0 \\ f_1 \\ f_2 \\ \vdots \end{bmatrix} \quad \text{where } \|f\|^2 = \sum_{n=0}^{\infty} \|f_n\|^2 < \infty \text{ and } f_n \in \mathcal{E} \text{ for all } n \geq 0.$$

Notice that S is an isometry. Since the first row of S is all zeros, the unilateral shift is not onto. In other words, the unilateral shift is not a unitary operator. The dimension of $\mathcal{E}$ is called the *multiplicity* of S. In a moment, we will show that two unilateral shifts with the same multiplicity are unitarily equivalent. Notice that $\mathcal{L} = \ker S^*$ is the subspace of $\ell^2_+(\mathcal{E})$ obtained by embedding $\mathcal{E}$ into the first component of $\ell^2_+(\mathcal{E})$, that is,

$$\mathcal{L} = \ker S^* = \left\{ \begin{bmatrix} f_0 & 0 & 0 & 0 & \cdots \end{bmatrix}^{tr} \in \ell^2_+(\mathcal{E}) : f_0 \in \mathcal{E} \right\}.$$

Clearly, $\ell^2_+(\mathcal{E}) = \mathcal{M}_+(\mathcal{L}) = \oplus_0^\infty S^n \mathcal{L}$. (The transpose of a vector or matrix is denoted by tr.) In other words, $\mathcal{L}$ is a cyclic and wandering subspace for S. Motivated by this analysis we define the following general or abstract version of a unilateral shift.

Definition 1.1.1. An operator U on $\mathcal{K}$ is a unilateral shift, if U is an isometry and U contains a cyclic wandering subspace $\mathcal{L}$, that is, $\mathcal{K} = \mathcal{M}_+(\mathcal{L})$. In this case, the dimension of $\mathcal{L}$ is called the multiplicity of U.

Assume that U is a (general) unilateral shift on $\mathcal{K}$. Then $\ker U^* = \mathcal{K} \ominus U\mathcal{K}$ *is the only cyclic wandering subspace for U.* Let $\mathcal{L}$ be any cyclic wandering subspace for U. It is sufficient to show that $\mathcal{L} = \ker U^*$. Since $\mathcal{L}$ is orthogonal to $U^j\mathcal{L}$ for all $j \geq 1$, and $\mathcal{K} = \oplus_0^\infty U^j \mathcal{L}$, we see that $\mathcal{L}$ is orthogonal to $U\mathcal{K}$. In other words, $\mathcal{L} \subseteq \ker U^*$. To show that $\mathcal{L} = \ker U^*$, assume that g is a vector in $\ker U^*$, which is orthogonal to $\mathcal{L}$. Since g is in $\ker U^*$, then g must be orthogonal to $U\mathcal{K} = \oplus_1^\infty U^j \mathcal{L}$. Thus g is orthogonal to $\mathcal{L}$ and $U^j\mathcal{L}$ for all $j \geq 1$. Because $\mathcal{L}$ is cyclic for U, the vector g must be zero. Therefore $\mathcal{L} = \ker U^*$, which proves our claim.

The previous analysis shows that an isometry U on $\mathcal{K}$ is a unilateral shift if and only if $\mathcal{K} \ominus U\mathcal{K}$ is cyclic for U, or equivalently, $\mathcal{K} = \mathcal{M}_+(\ker U^*)$.

We claim that two unilateral shifts are unitarily equivalent if and only if they have the same multiplicity. In particular, if U is a unilateral shift of multiplicity n, then U is unitarily equivalent to the unilateral shift S on $\ell^2_+(\mathcal{E})$ where the dimension of $\mathcal{E}$ is n.

To verify this, let U on $\mathcal{M}_+(\mathcal{L})$ and S on $\mathcal{M}_+(\mathcal{E})$ be two unilateral shifts, where $\mathcal{L} = \ker U^*$ and $\mathcal{E} = \ker S^*$ have the same dimension. Without loss of generality we can assume that $\mathcal{L} = \mathcal{E}$. Any vector f in $\mathcal{M}_+(\mathcal{E}) = \oplus_0^\infty S^k\mathcal{E}$ admits a unique representation of the form $f = \oplus_0^\infty S^k g_k$, where each g_k is in $\mathcal{E}$ and $\|f\|^2 = \sum_0^\infty \|g_k\|^2$ is finite. Let W be the linear relation mapping $\mathcal{M}_+(\mathcal{E})$ onto $\mathcal{M}_+(\mathcal{L}) = \oplus_0^\infty U^k\mathcal{L}$ defined by

$$W\sum_{k=0}^{\infty} S^k g_k = \sum_{k=0}^{\infty} U^k g_k \qquad (g_k \in \mathcal{E} \text{ and } \sum_{k=0}^{\infty} \|g_k\|^2 < \infty). \tag{1.1.4}$$

Since $\|Wf\|^2 = \|\sum_0^\infty U^k g_k\|^2 = \sum_0^\infty \|g_k\|^2 = \|f\|^2$, it follows that W is an isometry. Because W is onto, W is unitary. We claim that W intertwines S with U, that is, $WS = UW$. To show this, observe that

$$UWf = U\sum_{k=0}^{\infty} U^k g_k = \sum_{k=0}^{\infty} U^{k+1} g_k = W\sum_{k=0}^{\infty} S^{k+1} g_k = WSf.$$

Therefore $WS = UW$. In other words, U and S are unitarily equivalent.

Let U be a unilateral shift on $\mathcal{K}$ and $\mathcal{L}$ the cyclic wandering subspace for U. If $\mathcal{Y}$ is a cyclic subspace for U, then $\dim\mathcal{L} \le \dim\mathcal{Y}$. To see this, recall that $\mathcal{L} = \mathcal{K} \ominus U\mathcal{K}$. Because $\mathcal{L}$ is orthogonal to $U\mathcal{K}$, we have $P_\mathcal{L} U^n\mathcal{Y} = \{0\}$ for all integers $n \ge 1$. Since $\mathcal{Y}$ is cyclic for U, we obtain

$$\mathcal{L} = P_\mathcal{L}\mathcal{K} = P_\mathcal{L} \bigvee_{n=0}^{\infty} U^n\mathcal{Y} = \overline{P_\mathcal{L}\mathcal{Y}}.$$

In other words, $\mathcal{L} = \overline{P_\mathcal{L}\mathcal{Y}}$. Therefore $\dim\mathcal{L} \le \dim\mathcal{Y}$.

1.2 The Eigenvalues for the Backward Shift

If A is an operator on $\mathcal{X}$, then A^n converges to zero in the *strong operator topology* if $A^n x$ converges to zero as n tends to infinity for each x in $\mathcal{X}$. If S is a unilateral shift on $\mathcal{K}$, then S^* is called the *backward shift* on $\mathcal{K}$. In other words, an operator is a backward shift, if it is adjoint to a unilateral or forward shift.

As before, let S be a unilateral shift on $\mathcal{K}$. We claim that S^{*n} converges to zero in the strong operator topology. Due to unitary equivalence, without loss of generality, we can assume that S is the unilateral shift on $\ell^2_+(\mathcal{E})$ given in (1.1.3).

Then the adjoint S^* of S is given by

$$S^* = \begin{bmatrix} 0 & I & 0 & 0 & \cdots \\ 0 & 0 & I & 0 & \cdots \\ 0 & 0 & 0 & I & \cdots \\ 0 & 0 & 0 & 0 & \cdots \\ \vdots & \vdots & \vdots & \vdots & \ddots \end{bmatrix} \quad \text{on } \ell^2_+(\mathcal{E}). \tag{1.2.1}$$

In this case, S^* is called the *backward shift* on $\ell^2_+(\mathcal{E})$. To show that S^{*n} converges to zero in the strong operator topology, let f be any vector in $\ell^2_+(\mathcal{E})$. Then $S^{*n}f$ is the vector given by

$$S^{*n}f = \begin{bmatrix} f_n \\ f_{n+1} \\ f_{n+2} \\ \vdots \end{bmatrix} \quad \text{where} \quad f = \begin{bmatrix} f_0 \\ f_1 \\ f_2 \\ \vdots \end{bmatrix} \in \ell^2_+(\mathcal{E}) \qquad (n \geq 0).$$

Since $\|f\|^2 = \sum_0^\infty \|f_k\|^2$ is finite, $\|S^{*n}f\|^2 = \sum_{k=n}^\infty \|f_k\|^2$ must converge to zero as n approaches infinity. In other words, if S^* is the backward shift, then S^{*n} converges to zero in the strong operator topology.

It is easy to verify that the unilateral shift S has no eigenvalues. However, the set of all eigenvalues for S^* is the open unit disc $\mathbb{D} = \{z \in \mathbb{C} : |z| < 1\}$. In fact, for each λ in $\mathbb{D}$, a corresponding eigenvector $\psi_{\lambda f}$ is given by

$$\psi_{\lambda f} = \begin{bmatrix} f \\ \lambda f \\ \lambda^2 f \\ \vdots \end{bmatrix} \qquad (f \in \mathcal{E} \text{ and } f \neq 0). \tag{1.2.2}$$

Notice that each nonzero f in $\mathcal{E}$ determines an eigenvector for the eigenvalue λ. So the dimension of the kernel of $S^* - \lambda I$ equals the dimension of $\mathcal{E}$. In other words, the dimension of the eigenspace for S^* corresponding to the eigenvalue λ equals the dimension of $\mathcal{E}$. (Recall that the kernel of $A - \lambda I$ is called the *eigenspace* for the operator A corresponding to the eigenvalue λ.) Finally, the dimension of the kernel of $S^* - \lambda I$ equals the multiplicity of S.

To obtain the form of the eigenvectors in (1.2.2), let $g = \begin{bmatrix} g_0 & g_1 & g_2 & \cdots \end{bmatrix}^{tr}$ be a vector in $\ell^2_+(\mathcal{E})$ such that $S^*g = \lambda g$. (The transpose of a vector is denoted by ^{tr}.) Then

$$\begin{bmatrix} g_1 \\ g_2 \\ g_3 \\ \vdots \end{bmatrix} = S^*g = \lambda g = \lambda \begin{bmatrix} g_0 \\ g_1 \\ g_2 \\ \vdots \end{bmatrix}. \tag{1.2.3}$$

If $\lambda = 0$, then g_k must be zero for all $k \geq 1$. Hence $g = \psi_{0g_0}$. On the other hand, if $\lambda \neq 0$, then (1.2.3) shows that $g_1 = \lambda g_0$, $g_2 = \lambda g_1 = \lambda^2 g_0$. By continuing in this fashion, we see that $g_k = \lambda^k g_0$ for all $k \geq 0$. Therefore the eigenvector corresponding to λ is given by $g = \psi_{\lambda f}$ where $f = g_0$ is nonzero. Finally, observe that

$$\|\psi_{\lambda f}\|^2 = \|f\|^2 \sum_{n=0}^{\infty} |\lambda|^{2n} = \frac{\|f\|^2}{1 - |\lambda|^2}$$

is finite for all λ in $\mathbb{D}$. Thus $\psi_{\lambda f}$ is a nonzero vector in $\ell^2_+(\mathcal{E})$ such that $S^*\psi_{\lambda f} = \lambda\psi_{\lambda f}$. Therefore the set of all eigenvalues of S^* is given by $\mathbb{D}$. Moreover, for each fixed λ in $\mathbb{D}$, the corresponding eigenvector is $\psi_{\lambda f}$. In particular, the eigenspace for λ is given by

$$\ker(S^* - \lambda I) = \{\psi_{\lambda f} : f \in \mathcal{E}\}. \tag{1.2.4}$$

Observe that $\|f\|^2 = \|\sqrt{1 - |\lambda|^2}\psi_{\lambda f}\|^2$. Hence the mapping $f \mapsto \sqrt{1 - |\lambda|^2}\psi_{\lambda f}$ defines a unitary operator from $\mathcal{E}$ onto the eigenspace for S^* corresponding to the eigenvalue λ. Finally, it is noted that the multiplicity of the eigenvalue λ is the dimension of $\mathcal{E}$.

A *contraction* is an operator whose norm is less than or equal to one. Recall that the spectrum of an operator is closed. Because S^* is a contraction and the set of all eigenvalues for S^* is $\mathbb{D}$, it follows that the spectrum for S^* is the closed unit disc $\overline{\mathbb{D}}$. Recall also that λ is in the spectrum of an operator A if and only if $\overline{\lambda}$ is in the spectrum of A^*. Hence the spectrum for the unilateral S is also the closed unit disc $\overline{\mathbb{D}}$.

1.3 The Wold Decomposition

Let A be an operator on $\mathcal{X}$. Recall that a subspace $\mathcal{X}_1$ *reduces* A if $\mathcal{X}_1$ is an invariant subspace for both A and A^*. Observe that $\mathcal{X}_1$ reduces A if and only if its orthogonal complement $\mathcal{X}_2 = \mathcal{X} \ominus \mathcal{X}_1$ is also a reducing subspace for A^*. The notation $A = A_1 \oplus A_2$ on $\mathcal{X}_1 \oplus \mathcal{X}_2$ means that $\mathcal{X}_1$, or equivalently, $\mathcal{X}_2$ is a reducing subspace for A, and the space $\mathcal{X} = \mathcal{X}_1 \oplus \mathcal{X}_2$. Furthermore, A_1 is the operator on $\mathcal{X}_1$ defined by $A_1 = A|\mathcal{X}_1$, while A_2 is the operator on $\mathcal{X}_2$ defined by $A_2 = A|\mathcal{X}_2$. Finally, it is noted that the notation $A = A_1 \oplus A_2$ on $\mathcal{X}_1 \oplus \mathcal{X}_2$ is equivalent to saying that A admits a matrix representation of the form

$$A = \begin{bmatrix} A_1 & 0 \\ 0 & A_2 \end{bmatrix} \text{ on } \begin{bmatrix} \mathcal{X}_1 \\ \mathcal{X}_2 \end{bmatrix}.$$

The following result is known as *the Wold decomposition*, it shows that any isometry U on $\mathcal{K}$ can be uniquely decomposed into a unilateral shift and a unitary operator.

Theorem 1.3.1 (Wold decomposition). *Let U be an isometry on $\mathcal{K}$. Then U admits a unique decomposition of the form $U = S \oplus V$ on $\mathcal{K}_+ \oplus \mathcal{V}$, where S is a unilateral* *shift on* $\mathcal{K}_+$ *and* V *is a unitary operator on* $\mathcal{V}$. *Moreover, the subspace* $\mathcal{K}_+$ *and* $\mathcal{V}$ *are determined by*

$$\mathcal{K}_+ = \bigoplus_{n=0}^{\infty} U^n(\ker U^*) \qquad \textit{and} \qquad \mathcal{V} = \bigcap_{n=0}^{\infty} U^n\mathcal{K}. \tag{1.3.1}$$

In particular, U *admits a matrix representation of the form*

$$U = \begin{bmatrix} S & 0 \\ 0 & V \end{bmatrix} \textit{ on } \begin{bmatrix} \mathcal{K}_+ \\ \mathcal{V} \end{bmatrix}. \tag{1.3.2}$$

The multiplicity of S *equals the dimension of the kernel of* U^*. *Finally, it is noted that the subspaces* $\mathcal{K}_+$ *or* $\mathcal{V}$ *may be zero.*

Proof. Let $\mathcal{L} = \ker U^*$. Then $\mathcal{L}$ is a wandering subspace for U. Because

$$\ker U^* = \mathcal{K} \ominus U\mathcal{K},$$

the space $\mathcal{K} = \mathcal{L} \oplus U\mathcal{K}$. By induction,

$$\begin{aligned}
\mathcal{K} &= \mathcal{L} \oplus U\mathcal{K} = \mathcal{L} \oplus U(\mathcal{L} \oplus U\mathcal{K}) \\
&= \mathcal{L} \oplus U\mathcal{L} \oplus U^2\mathcal{K} = \mathcal{L} \oplus U\mathcal{L} \oplus U^2(\mathcal{L} \oplus U\mathcal{K}) \\
&= \mathcal{L} \oplus U\mathcal{L} \oplus U^2\mathcal{L} \oplus U^3\mathcal{K} = \cdots \\
&= \oplus_{k=0}^{n} U^k\mathcal{L} \oplus U^{n+1}\mathcal{K}.
\end{aligned}$$

This readily implies that

$$U^{n+1}\mathcal{K} = \mathcal{K} \ominus \left(\oplus_{k=0}^{n} U^k\mathcal{L}\right) \qquad (n \geq 0). \tag{1.3.3}$$

Let $\mathcal{K}_+$ be the subspace of $\mathcal{K}$ defined by $\mathcal{K}_+ = \oplus_0^\infty U^n\mathcal{L}$, and set $\mathcal{V} = \mathcal{K} \ominus \mathcal{K}_+$. We claim that the subspace $\mathcal{V} = \bigcap_{n=0}^\infty U^n\mathcal{K}$; see (1.3.1). To verify this, observe that a vector g is an element in $\mathcal{V}$ if and only if g is orthogonal to $U^k\mathcal{L}$, for all integers $k \geq 0$. By consulting (1.3.3), this vector g must be in $U^{n+1}\mathcal{K}$, for all $n \geq 0$. The subspaces $\{U^n\mathcal{K}\}_0^\infty$ are decreasing, that is, $U^{n+1}\mathcal{K} \subseteq U^n\mathcal{K}$ for all $n \geq 0$. Hence g is in $\bigcap_{n=0}^\infty U^n\mathcal{K}$. Thus g is in $\mathcal{V}$ if and only if $g \in \bigcap_0^\infty U^n\mathcal{K}$. Therefore the subspace $\mathcal{V}$ is given by (1.3.1).

Notice that both $\mathcal{K}_+$ and $\mathcal{V}$ are both invariant subspaces for U. In other words, $\mathcal{K}_+$ and $\mathcal{V}$ are reducing subspaces for U such that $\mathcal{K} = \mathcal{K}_+ \oplus \mathcal{V}$. Now let S be the operator on $\mathcal{K}_+$ defined by $S = U|\mathcal{K}_+$, and V be the operator on $\mathcal{V}$ given by $V = U|\mathcal{V}$. Then U admits a matrix representation of the form

$$U = \begin{bmatrix} S & 0 \\ 0 & V \end{bmatrix} \text{ on } \begin{bmatrix} \mathcal{K}_+ \\ \mathcal{V} \end{bmatrix}.$$

Since $\mathcal{L}$ is a cyclic wandering subspace for S, it follows that S is a unilateral shift on $\mathcal{K}_+$.

To verify that V is unitary it is sufficient to show that $\mathcal{V} = V\mathcal{V}$, or equivalently, the kernel of V^* is zero. Assume that φ is a vector in $\mathcal{V}$ such that $V^*\varphi = 0$. Since $U = S \oplus V$, we see that $U^*\varphi = 0$. Hence φ is a vector in $\mathcal{L} \subseteq \mathcal{K}_+$. In other words, φ is in $\mathcal{V} \cap \mathcal{K}_+ = \{0\}$. Therefore $\varphi = 0$, and V is unitary.

To complete the proof, it remains to show that the decomposition $U = S \oplus V$ is unique. Assume that $U = S_1 \oplus V_1$ on $\mathcal{K}_1 \oplus \mathcal{V}_1$, where S_1 is a unilateral shift and V_1 is unitary. Then the cyclic wandering subspace $\mathcal{L}_1$ for S_1 is determined by

$$\mathcal{L}_1 = \ker S_1^* = \ker(S_1^* \oplus V_1^*) = \ker U^* = \mathcal{L}.$$

Hence $\mathcal{L} = \mathcal{L}_1$. This readily implies that

$$\mathcal{K}_1 = \bigoplus_{n=0}^{\infty} S_1^n \mathcal{L}_1 = \bigoplus_{n=0}^{\infty} U^n \mathcal{L}_1 = \bigoplus_{n=0}^{\infty} U^n \mathcal{L} = \mathcal{K}_+.$$

Thus $\mathcal{K} = \mathcal{K}_+$. Because $\mathcal{V}$ is the orthogonal complement of $\mathcal{K}_+$ in $\mathcal{K}$, and $\mathcal{V}_1$ is the orthogonal complement of $\mathcal{K}_1$ in $\mathcal{K}$, it follows that $\mathcal{V} = \mathcal{V}_1$. Therefore the Wold decomposition of U is unique. □

Lemma 1.3.2. *Two unitary operators U on $\mathcal{K}$ and Z on $\mathcal{Z}$ are similar if and only if they are unitarily equivalent.*

Proof. Obviously, if U and Z are unitarily equivalent, then they are similar. Assume that U and Z are two unitary operators such that $UW = WZ$ where W is an invertible operator mapping $\mathcal{Z}$ onto $\mathcal{K}$. By taking the adjoint, we see that $W^*U^* = Z^*W^*$. Multiplying by U on the right and Z on the left, we obtain $W^*U = ZW^*$. Multiplying by W on the right with $UW = WZ$, yields $W^*WZ = ZW^*W$. This readily implies that $(W^*W)^nZ = Z(W^*W)^n$ for all integers $n \geq 0$. So for any polynomial $p(\lambda)$ this implies that

$$p(W^*W)Z = Zp(W^*W). \tag{1.3.4}$$

Recall that if A is any positive operator on $\mathcal{X}$, then there exists a sequence of polynomials $\{p_n\}_0^\infty$ such that $p_n(A)$ converges to the positive square root $A^{1/2}$ of A in the strong operator topology; see Problem 95 in Halmos [126]. So by choosing the appropriate polynomials and passing limits in (1.3.4), we see that $RZ = ZR$ where R is the positive square root of W^*W. Since W is invertible, R is also invertible. Moreover, $\Phi = WR^{-1}$ is a unitary operator from $\mathcal{Z}$ onto $\mathcal{K}$. To see this simply observe that $\Phi^*\Phi = I$ and Φ is onto. Using $W = \Phi R$, we obtain

$$U\Phi = U\Phi RR^{-1} = UWR^{-1} = WZR^{-1} = \Phi RZR^{-1} = \Phi ZRR^{-1} = \Phi Z.$$

Therefore $U\Phi = \Phi Z$. □

Theorem 1.3.3. *Two isometries U on $\mathcal{K}$ and Z on $\mathcal{Z}$ are similar if and only if they are unitarily equivalent.*

Proof. Clearly, if U and Z are unitarily equivalent, then they are similar. Assume that U and Z are two isometries such that $UW = WZ$ where W is an invertible operator mapping $\mathcal{Z}$ onto $\mathcal{K}$. Let $U = S \oplus V$ on $\mathcal{M}_+(\ker U^*) \oplus \mathcal{V}$ and $Z = \widetilde{S} \oplus \widetilde{V}$ on $\mathcal{M}_+(\ker Z^*) \oplus \widetilde{\mathcal{V}}$ be the Wold decompositions of U and Z, where S and $\widetilde{S}$ are unilateral shifts while V and $\widetilde{V}$ are unitary operators. Using $W^*U^* = Z^*W^*$ along with the fact that W^* in invertible, it follows that $\ker Z^* = W^* \ker U^*$. (This is left as an exercise.) In particular, $\ker U^*$ and $\ker Z^*$ have the same dimension. So the unilateral shifts S and $\widetilde{S}$ have the same multiplicity. In other words, S and $\widetilde{S}$ are unitarily equivalent.

It remains to show that V on $\mathcal{V} = \cap_0^\infty U^n \mathcal{K}$ and $\widetilde{V}$ on $\widetilde{\mathcal{V}} = \cap_0^\infty Z^n \mathcal{Z}$ are unitarily equivalent. For any integer $n \geq 0$, we have

$$W\widetilde{\mathcal{V}} \subseteq WZ^n\mathcal{Z} = U^n W\mathcal{Z} = U^n\mathcal{K}.$$

By taking the infinite intersections, we see that $W\widetilde{\mathcal{V}} \subseteq \mathcal{V}$. Using $W^{-1}U = ZW^{-1}$, a similar argument shows that $W^{-1}\mathcal{V} \subseteq \widetilde{\mathcal{V}}$, or equivalently, $\mathcal{V} \subseteq W\widetilde{\mathcal{V}}$. In other words, $W\widetilde{\mathcal{V}} = \mathcal{V}$. Hence W maps $\widetilde{\mathcal{V}}$ one to one and onto $\mathcal{V}$. Since $UW = WZ$, it follows that $W|\widetilde{\mathcal{V}} : \widetilde{\mathcal{V}} \to \mathcal{V}$ is an invertible operator intertwining $\widetilde{V} = Z|\widetilde{\mathcal{V}}$ with $V = Z|\mathcal{V}$. According to Lemma 1.3.2, the operators V and $\widetilde{V}$ are unitarily equivalent. □

1.4 The Bilateral Shift

This section is devoted to the bilateral shift, which plays a basic role in our approach to factorization theory. Throughout $\ell^2(\mathcal{E})$ is the Hilbert space consisting of all square summable two-sided vectors with values in $\mathcal{E}$, that is, $\ell^2(\mathcal{E})$ is the set of all vectors of the form

$$f = \begin{bmatrix} \vdots \\ f_{-1} \\ f_0 \\ f_1 \\ \vdots \end{bmatrix} \quad \text{where } \|f\|^2 = \sum_{n=-\infty}^{\infty} \|f_n\|^2 < \infty \text{ and } f_n \in \mathcal{E} \text{ for all } n.$$

Moreover, $\ell^2_-(\mathcal{E})$ is the Hilbert space formed by the set of all vectors of the form

$$f = \begin{bmatrix} \vdots \\ f_{-3} \\ f_{-2} \\ f_{-1} \end{bmatrix} \quad \text{where } \|f\|^2 = \sum_{n=-1}^{-\infty} \|f_n\|^2 < \infty \text{ and } f_n \in \mathcal{E} \text{ for all } n < 0.$$

Finally, it is noted that $\ell^2(\mathcal{E})$ admits an orthogonal decomposition of the form $\ell^2(\mathcal{E}) = \ell^2_-(\mathcal{E}) \oplus \ell^2_+(\mathcal{E})$.

Let U be a unitary operator on $\mathcal{K}$. Let us emphasize that U in this section is unitary. Using $UU^* = I$, we say that a subspace $\mathcal{L}$ is *wandering* for U if and only if $U^n\mathcal{L}$ is orthogonal to $U^k\mathcal{L}$ for all integers $n \neq k$. In this case, we define the subspace $\mathcal{M}(\mathcal{L})$ of $\mathcal{K}$ by

$$\mathcal{M}(\mathcal{L}) = \bigoplus_{n=-\infty}^{\infty} U^n\mathcal{L}. \tag{1.4.1}$$

It is emphasized that a vector g is in $\mathcal{M}(\mathcal{L})$ if and only if $g = \sum_{-\infty}^{\infty} U^n g_n$ where $g_n \in \mathcal{E}$ for all integers n and $\|g\|^2 = \sum_{-\infty}^{\infty} \|g_n\|^2$ is finite. Since $\mathcal{M}(\mathcal{L})$ is invariant for both U and U^*, the subspace $\mathcal{M}(\mathcal{L})$ is a reducing subspace for U. Moreover, the subspace $\mathcal{M}_+(\mathcal{L})$ is an invariant subspace for U. Recall that the wandering subspace $\mathcal{L}$ is uniquely determined by $\mathcal{M}_+(\mathcal{L})$, that is, $\mathcal{L} = \mathcal{M}_+(\mathcal{L}) \ominus U\mathcal{M}_+(\mathcal{L})$. However, the subspace $\mathcal{M}(\mathcal{L})$ cannot be used to determine the wandering subspace $\mathcal{L}$. For example, both $\mathcal{L}$ and $U\mathcal{L}$ are wandering for U and $\mathcal{M}(\mathcal{L}) = \mathcal{M}(U\mathcal{L})$. Clearly, $\mathcal{L}$ and $U\mathcal{L}$ have the same dimension. This is a special case of the following result.

Proposition 1.4.1. *Let $\mathcal{L}$ and $\mathcal{E}$ be two wandering subspaces for a unitary operator U on $\mathcal{K}$. Assume that $\mathcal{M}(\mathcal{E}) \subseteq \mathcal{M}(\mathcal{L})$. Then* $\dim \mathcal{E} \leq \dim \mathcal{L}$. *Finally, if $\mathcal{E}$ is finite dimensional, then $\mathcal{M}(\mathcal{E}) = \mathcal{M}(\mathcal{L})$ if and only if $\mathcal{E}$ and $\mathcal{L}$ have the same dimension.*

Proof. Recall that $\mathcal{M}(\mathcal{E}) \subseteq \mathcal{M}(\mathcal{L})$. For the moment assume that $\dim \mathcal{L} \geq \aleph$ where $\aleph$ is the cardinality of the integers. Then

$$\aleph \dim \mathcal{L} = \dim \mathcal{M}(\mathcal{L}) \geq \dim \mathcal{M}(\mathcal{E}) = \aleph \dim \mathcal{E}.$$

Hence $\dim \mathcal{L} \geq \dim \mathcal{E}$.

Now assume that the dimension of $\mathcal{E}$ equals m is finite. Let $\{e_j\}_1^m$ be an orthonormal basis for $\mathcal{E}$ and $\{\varphi_j\}_1^n$ be an orthonormal basis for $\mathcal{L}$. Because $\mathcal{E}$ and $\mathcal{L}$ are both wandering subspaces, we see that

$$\begin{aligned} &\{U^k e_j : 1 \leq j \leq m \text{ and } -\infty < k < \infty\} \\ &\{U^k \varphi_j : 1 \leq j \leq n \text{ and } -\infty < k < \infty\} \end{aligned}$$

form an orthonormal basis for $\mathcal{M}(\mathcal{E})$ and $\mathcal{M}(\mathcal{L})$, respectively.

By employing Bessel's inequality $\|\varphi_j\|^2 \geq \sum_{ki} |(\varphi_j, U^k e_i)|^2$ and Parseval's equality $\sum_{jk} |(U^{*k}\varphi_j, e_i)|^2 = \|e_i\|^2$, we obtain

$$\begin{aligned} \dim \mathcal{L} &= \sum_{j=1}^{n} \|\varphi_j\|^2 \geq \sum_{jki} |(\varphi_j, U^k e_i)|^2 \\ &= \sum_{jki} |(U^{*k}\varphi_j, e_i)|^2 = \sum_{i=1}^{m} \|e_i\|^2 \\ &= \dim \mathcal{E}. \end{aligned} \tag{1.4.2}$$

Thus $\dim \mathcal{E} \leq \dim \mathcal{L}$.

If $\dim \mathcal{E} = \dim \mathcal{L}$, then we have equality in (1.4.2). Hence

$$\|\varphi_j\|^2 = \sum_{ki} |(\varphi_j, U^k e_i)|^2 \qquad (\text{for all } 1 \le j \le n).$$

Since $\{U^k e_j\}_{jk}$ is an orthonormal basis for $\mathcal{M}(\mathcal{E})$, this implies that $\{\varphi_j\}_1^n$ is contained in the subspace $\mathcal{M}(\mathcal{E})$. Hence $\mathcal{M}(\mathcal{L}) \subseteq \mathcal{M}(\mathcal{E})$. Combining this with the hypothesis $\mathcal{M}(\mathcal{E}) \subseteq \mathcal{M}(\mathcal{L})$ yields $\mathcal{M}(\mathcal{E}) = \mathcal{M}(\mathcal{L})$. On the other hand, if $\mathcal{M}(\mathcal{E}) = \mathcal{M}(\mathcal{L})$, then $\{U^k e_j\}_{jk}$ is an orthonormal basis for $\mathcal{M}(\mathcal{L})$. In this case, we have equality in (1.4.2), that is, $\dim \mathcal{E} = \dim \mathcal{L}$. □

An operator U on $\mathcal{K}$ is called a *bilateral shift* if U is a unitary operator and $\mathcal{K} = \mathcal{M}(\mathcal{L})$ where $\mathcal{L}$ is a wandering subspace for U. The dimension of $\mathcal{L}$ is the *multiplicity* for U. The wandering subspace $\mathcal{L}$ is not uniquely determined by U. However, Proposition 1.4.1 shows that the multiplicity of U is independent of the wandering subspace satisfying $\mathcal{K} = \mathcal{M}(\mathcal{L})$. For an example of a bilateral shift consider the unitary operator U on $\ell^2(\mathcal{E})$ defined by

$$U = \begin{bmatrix} \ddots & \vdots & \vdots & \vdots & \vdots & \vdots \\ \cdots & I & 0 & 0 & 0 & \cdots \\ \cdots & 0 & I & \boxed{0} & 0 & \cdots \\ \cdots & 0 & 0 & I & 0 & \cdots \\ \cdots & 0 & 0 & 0 & I & \cdots \\ \vdots & \vdots & \vdots & \vdots & \vdots & \ddots \end{bmatrix} \text{ on } \ell^2(\mathcal{E}) = \begin{bmatrix} \vdots \\ \mathcal{E} \\ \boxed{\mathcal{E}} \\ \mathcal{E} \\ \mathcal{E} \\ \vdots \end{bmatrix}. \tag{1.4.3}$$

The box around zero represents the zero-zero entry $U_{00} = 0$ of U. Notice that the identity I on $\mathcal{E}$ appears immediately below the main diagonal and all other entries are zeros. Clearly, U in (1.4.3) is unitary. To show that this U is a bilateral shift, observe that

$$\mathcal{L} = \begin{bmatrix} \cdots & 0 & 0 & \boxed{\mathcal{E}} & 0 & 0 & \cdots \end{bmatrix}^{tr}$$

is a wandering subspace for U such that $\ell^2(\mathcal{E}) = \mathcal{M}(\mathcal{L})$. Finally, it is noted that $\dim \mathcal{E}$ is the multiplicity of U.

Two bilateral shifts with the same multiplicity are unitarily equivalent. To see this, let U on $\mathcal{M}(\mathcal{L})$ and U_1 on $\mathcal{M}(\mathcal{L}_1)$ be two bilateral shifts of the same multiplicity where $\mathcal{L}$ is wandering for U and $\mathcal{L}_1$ is wandering for U_1. Since U and U_1 have the same multiplicity, without loss of generality we can assume that $\mathcal{L} = \mathcal{L}_1$. Now let Φ be the unitary operator mapping $\mathcal{M}(\mathcal{L})$ onto $\mathcal{M}(\mathcal{L}_1)$ defined by

$$\Phi \oplus_{-\infty}^{\infty} U^n g_n = \oplus_{-\infty}^{\infty} U_1^n g_n \qquad (\oplus_{-\infty}^{\infty} U^n g_n \in \mathcal{M}(\mathcal{L})).$$

Then it is easy to verify that Φ intertwines U with U_1. In other words, U and U_1 are unitarily equivalent. Finally, it is noted that any bilateral shift with multiplicity n is unitarily equivalent to the bilateral shift on U on $\ell^2(\mathcal{E})$ in (1.4.3) where n is the dimension of $\mathcal{E}$.

Let T be an operator mapping $\mathcal{X}$ into $\mathcal{Y}$. Then we say that U is *an extension* of T if U is an operator mapping $\mathcal{H}$ into $\mathcal{K}$ such that $U|\mathcal{X} = T$ where $\mathcal{X}$ is a subspace of $\mathcal{H}$ and $\mathcal{Y}$ is a subspace of $\mathcal{K}$. Notice that U is an extension of T if and only if U admits a matrix representation of the form

$$U = \begin{bmatrix} A & 0 \\ C & T \end{bmatrix} : \begin{bmatrix} \mathcal{H} \ominus \mathcal{X} \\ \mathcal{X} \end{bmatrix} \to \begin{bmatrix} \mathcal{K} \ominus \mathcal{Y} \\ \mathcal{Y} \end{bmatrix}.$$

Now assume that T is an operator on $\mathcal{X}$. Then U on $\mathcal{K}$ is an extension of T if $U|\mathcal{X} = T$ where $\mathcal{X} \subseteq \mathcal{K}$. We say that two extensions U on $\mathcal{K}$ and U_1 on $\mathcal{K}_1$ of T are *isomorphic* if there exists a unitary operator Φ mapping $\mathcal{K}$ onto $\mathcal{K}_1$ such that $\Phi U = U_1 \Phi$ and $\Phi|\mathcal{X} = I_{\mathcal{X}}$.

For example, let S be the unilateral shift on $\ell^2_+(\mathcal{E})$ in (1.1.3) and U on $\ell^2(\mathcal{E})$ the bilateral shift in (1.4.3). Then U is an extension of S. In other words, U admits a matrix representation of the form

$$U = \begin{bmatrix} A & 0 \\ C & S \end{bmatrix} \text{ on } \ell^2(\mathcal{E}) = \begin{bmatrix} \ell^2_-(\mathcal{E}) \\ \ell^2_+(\mathcal{E}) \end{bmatrix}. \tag{1.4.4}$$

Clearly, U and S have the same multiplicity. Recall that any unilateral shift is unitarily equivalent to the unilateral shift S on $\ell^2_+(\mathcal{E})$, and any bilateral shift is unitarily equivalent to the bilateral shift U on $\ell^2(\mathcal{E})$. This readily yields the following result.

Proposition 1.4.2. *Any unilateral shift can be extended to a bilateral shift with the same multiplicity.*

As before, consider the matrix representation for the bilateral shift U on $\ell^2(\mathcal{E})$ in (1.4.4) where S on $\ell^2_+(\mathcal{E})$ is the unilateral shift. Notice that

$$\ell^2(\mathcal{E}) = \bigvee_{n=-\infty}^{\infty} U^n \ell^2_+(\mathcal{E}).$$

Because $\ell^2_+(\mathcal{E})$ is an invariant subspace for U, the previous equality is equivalent to $\ell^2(\mathcal{E}) = \bigvee_0^\infty U^{*n} \ell^2_+(\mathcal{E})$. Motivated by this, let U_+ be any isometry on $\mathcal{K}_+$. Then we say that U on $\mathcal{K}$ is a *minimal unitary extension* of U_+ if U is a unitary extension of U_+ satisfying

$$\mathcal{K} = \bigvee_{n=-\infty}^{\infty} U^n \mathcal{K}_+ \quad \text{or equivalently} \quad \mathcal{K} = \bigvee_{n=0}^{\infty} U^{*n} \mathcal{K}_+. \tag{1.4.5}$$

Clearly, the bilateral shift U on $\ell^2(\mathcal{E})$ in (1.4.4) is a minimal unitary extension of the unilateral shift S on $\ell^2_+(\mathcal{E})$. By virtue of the Wold decomposition any isometry U_+ admits a minimal unitary extension. To see this, let $U_+ = S \oplus V$ on $\mathcal{K}_+ \oplus \mathcal{V}$ be the Wold decomposition of U_+ where S is a unilateral shift and V is unitary. Without loss of generality we can assume that S is the unilateral shift on $\mathcal{K}_+ = \ell^2_+(\mathcal{E})$. Then $U \oplus V$ on $\ell^2(\mathcal{E}) \oplus \mathcal{V}$ is a minimal unitary extension of S where U is the bilateral shift on $\ell^2(\mathcal{E})$. This proves the first part of the following result.

Proposition 1.4.3. *Any isometry U_+ on $\mathcal{K}_+$ admits a minimal unitary extension. Moreover, all minimal unitary extensions of U_+ are unique up to an isomorphism.*

Proof. To complete the proof it remains to establish the uniqueness result. Let U on $\mathcal{K}$ and Z on $\mathcal{Z}$ be two minimal unitary extensions of U_+. Let $\{g_k\}_{-\infty}^{\infty}$ be a sequence of vectors in $\mathcal{K}_+$ with compact support, that is, the number of nonzero vectors in $\{g_k\}_{-\infty}^{\infty}$ is finite. Then we obtain

$$\begin{aligned}\|\sum_{k=-\infty}^{\infty} Z^k g_k\|^2 &= \sum_{k,j}(Z^k g_k, Z^j g_j) = \sum_{k\geq j}(Z^{k-j}g_k, g_j) + \sum_{k<j}(g_k, Z^{j-k}g_j)\\ &= \sum_{k\geq j}(U_+^{k-j}g_k, g_j) + \sum_{k<j}(g_k, U_+^{j-k}g_j)\\ &= \sum_{k\geq j}(U^{k-j}g_k, g_j) + \sum_{k<j}(g_k, U^{j-k}g_j) = \|\sum_{k=-\infty}^{\infty} U^k g_k\|^2.\end{aligned}$$

This and the minimality condition imposed on U and Z imply that the relation

$$\Phi \sum_{k=-\infty}^{\infty} Z^k g_k = \sum_{k=-\infty}^{\infty} U^k g_k$$

defines an isometry mapping a dense set in $\mathcal{Z}$ into a dense set in $\mathcal{K}$. So Φ can be extended by continuity to a unitary operator also denoted by Φ mapping $\mathcal{Z}$ onto $\mathcal{K}$. For any sequence of vectors $\{g_k\}_{-\infty}^{\infty}$ in $\mathcal{K}$ with compact support, the definition of Φ yields

$$\Phi Z \sum_{k=-\infty}^{\infty} Z^k g_k = \sum_{k=-\infty}^{\infty} U^{k+1} g_k = U\Phi \sum_{k=-\infty}^{\infty} Z^k g_k.$$

In other words, $\Phi Z = U\Phi$. Clearly, $\Phi|\mathcal{K}_+ = I$. □

The spectrum of the bilateral shift. Recall that the *spectrum* $\sigma(A)$ *of an operator* A *on* $\mathcal{X}$ is the set of all points λ in the complex plane such that $\lambda I - A$ is not invertible. It is well known that the spectrum for A is closed; see Halmos [126]. Recall that the unilateral shift has no eigenvalues. Moreover, the spectrum for the unilateral shift is the closed unit disc $\overline{\mathbb{D}}$. It is easy to show that the bilateral shift also has no eigenvalues. We claim that the spectrum for the bilateral shift is the unit circle $\mathbb{T} = \{z \in \mathbb{C} : |z| = 1\}$.

To see this, let U be the bilateral shift on $\ell^2(\mathcal{E})$ and λ be any complex number on the unit circle. Consider the unit vector φ_n determined by

$$\varphi_n = \frac{1}{\sqrt{n}}\begin{bmatrix}\cdots & 0 & 0 & f & \overline{\lambda}f & \cdots & \overline{\lambda}^{n-2}f & \overline{\lambda}^{n-1}f & 0 & 0 & \cdots\end{bmatrix}^{tr}$$

where f is any unit vector in $\mathcal{E}$. Notice that the vector φ_n has precisely n nonzero components. A simple calculation shows that

$$\|(\lambda I - U)\varphi_n\|^2 = \frac{\|\lambda f\|^2 + \|\overline{\lambda}^{n-1} f\|^2}{n} = \frac{2}{n}.$$

This readily implies that $\|(\lambda I - U)\varphi_n\|$ converges to zero as n tends to infinity. So for any λ on the unit circle, $\lambda I - U$ is not invertible. Hence $\mathbb{T} \subseteq \sigma(U)$. Because the spectrum of any unitary operator is contained in the unit circle, and the bilateral shift is a unitary operator, we see that $\mathbb{T} = \sigma(U)$.

1.5 Abstract Toeplitz Operators

Proposition 1.5.2 below will be used to develop some fundamental factorization results in future chapters. In order to prove Proposition 1.5.2, we needed to present the notion of an abstract Toeplitz operator and some preliminary results in Proposition 1.5.1. In Chapter 2 these results will also be used to analyze Laurent and Toeplitz block matrix operators. Finally, the proofs in this section are given for completeness. However, the techniques developed in these proofs are not used later on, and thus, may be skipped by the reader.

Let A be an operator on $\mathcal{X}$ and B an operator on $\mathcal{Y}$. The set of all operators T intertwining A with B is denoted by $\mathcal{I}(A, B)$, that is, $\mathcal{I}(A, B)$ is the set of all operators T mapping $\mathcal{X}$ into $\mathcal{Y}$ such that $TA = BT$. An *abstract "causal" Toeplitz operator* is an operator which intertwines two isometries. This is a generalization of the definition of the classical causal Toeplitz operator which intertwines two unilateral shifts on the appropriate $\ell_+^2(\cdot)$ spaces; see Chapter 2. The following result provides some insight into studying classical Toeplitz operators.

Proposition 1.5.1. *Let U on $\mathcal{K}$ be a minimal unitary extension of an isometry U_+ on $\mathcal{K}_+$ and Z on $\mathcal{Z}$ a unitary (not necessarily minimal) extension of an isometry Z_+ on $\mathcal{Z}_+$. Let T be an operator in $\mathcal{I}(U_+, Z_+)$. Then the following holds.*

(i) *There exists a unique operator L in $\mathcal{I}(U, Z)$ extending T.*

(ii) *This operator L is given by*

$$L = \textit{strong} \lim_{n\to\infty} Z^{*n} T U^n P_n \tag{1.5.1}$$

where P_n is the orthogonal projection onto $U^{*n}\mathcal{K}_+$.

(iii) *The operator L and T have the same norm, that is, $\|L\| = \|T\|$.*

(iv) *If the range of T is dense in $\mathcal{Z}_+$ and Z is a minimal unitary extension of Z_+, then the range of L is dense in $\mathcal{Z}$.*

(v) *If T is an isometry, then L is an isometry.*

(vi) *If T is unitary and Z is a minimal unitary extension of Z_+, then L is unitary.*

Proof. First let us show that L in (1.5.1) is a well-defined operator. Let g be a vector in $\mathcal{K}_+$. For any integers $n > j \geq 0$, we have

$$LU^{*j}g = Z^{*n}TU^nP_nU^{*j}g = Z^{*n}TU^{n-j}g = Z^{*n}TU_+^{n-j}g = Z^{*n}Z_+^{n-j}Tg = Z^{*j}Tg.$$

For any integer $j \geq 0$, we have

$$LU^{*j}g = Z^{*j}Tg \qquad (g \in \mathcal{K}_+ \text{ and } j \geq 0). \tag{1.5.2}$$

So L is a well-defined linear map on $U^{*j}\mathcal{K}_+$. Choosing $j = 0$, yields $L|\mathcal{K}_+ = T$. Notice that $\{U^{*j}\mathcal{K}_+\}_0^\infty$ is an increasing sequence of subspaces, that is, $U^{*j}\mathcal{K}_+ \subseteq U^{*j+1}\mathcal{K}_+$. Since U is a minimal unitary extension, $\{U^{*j}\mathcal{K}_+\}_0^\infty$ is dense in $\mathcal{K}$. In other words, L is a well-defined linear map on a dense set in $\mathcal{K}$. To show that L is bounded, notice that for any integer $j \geq 0$ and g in $\mathcal{K}_+$, we have

$$\|LU^{*j}g\| = \|Z^{*j}Tg\| \leq \|T\|\|g\| = \|T\|\|U^{*j}g\|.$$

Thus $\|L\| \leq \|T\|$. Because $L|\mathcal{K}_+ = T$, the operators L and T have the same norm. Finally, observe that L can be uniquely extended by continuity to an operator also denoted by L from $\mathcal{K}$ into $\mathcal{Z}$. In fact, the formula in (1.5.1) yields this extension of L.

We claim that L is in $\mathcal{I}(U, Z)$. For any integer $j \geq 0$ and g in $\mathcal{K}_+$, equation (1.5.2) yields

$$LU^*U^{*j}g = LU^{*j+1}g = Z^{*j+1}Tg = Z^*LU^{*j}g.$$

Since $\{U^{*j}\mathcal{K}_+\}_0^\infty$ is dense in $\mathcal{K}_+$, we see that $LU^* = Z^*L$. Multiplying by Z on the left and U on the right, shows that $ZL = LU$.

Let M be any operator in $\mathcal{I}(U, Z)$ extending T. Then (1.5.2) with $Z^*M = MU^*$ yields

$$LU^{*j}g = Z^{*j}Tg = Z^{*j}Mg = MU^{*j}g \qquad (g \in \mathcal{K}_+ \text{ and } j \geq 0).$$

Since $\{U^{*j}\mathcal{K}_+\}_0^\infty$ is dense in $\mathcal{K}$, we see that $L = M$. Therefore Parts (i) to (iii) hold.

To prove that Part (iv) holds, assume that the range of T is dense in $\mathcal{Z}_+$ and Z is a minimal unitary extension of Z_+. By consulting (1.5.2), we see that

$$\overline{\text{ran} L} \supseteq \bigvee_{j=0}^{\infty} Z^{*j}T\mathcal{K}_+ = \bigvee_{j=0}^{\infty} Z^{*j}\mathcal{Z}_+ = \mathcal{Z}.$$

In other words, the range of L is dense in $\mathcal{Z}$.

To verify that Part (v) holds, assume that T is an isometry. By consulting (1.5.2), we obtain

$$\|LU^{*j}g\| = \|Z^{*j}Tg\| = \|g\| = \|U^{*j}g\| \qquad (g \in \mathcal{K}_+ \text{ and } j \geq 0).$$

Since $\{U^{*j}\mathcal{K}_+\}_0^\infty$ is dense in $\mathcal{K}$, we see that L is an isometry.

Clearly, the range of an isometry is closed. So if an isometry L mapping $\mathcal{K}$ into $\mathcal{Z}$ is onto a dense set in $\mathcal{Z}$, then L is a unitary operator. In other words, Part (vi) is an immediate consequence of Parts (iv) and (v). □

An operator W from $\mathcal{X}$ into $\mathcal{Y}$ is a *quasi-affinity* if W is one to one and the range of W is dense in $\mathcal{Y}$. This sets the stage for the following result.

Proposition 1.5.2. *Let S be the unilateral shift on $\ell_+^2(\mathcal{E})$ where $\mathcal{E}$ is finite dimensional. Let U_+ be an isometry on $\mathcal{K}_+$ whose wandering subspace $\mathcal{L} = \mathcal{K} \ominus U_+\mathcal{K}$ has the same dimension as $\mathcal{E}$. Then S is unitarily equivalent to U_+ if and only if there is a quasi-affinity intertwining S with U_+.*

Proof. Let $U_+ = S_+ \oplus V$ on $\mathcal{M}_+(\mathcal{L}) \oplus \mathcal{V}$ be the Wold decomposition of U_+ where S_+ is the unilateral shift and V is unitary. Because S and S_+ have the same multiplicity, S and S_+ are unitarily equivalent. So without loss of generality, we assume that $S_+ = S$ is the unilateral shift on $\ell_+^2(\mathcal{E})$. To complete the proof, it remains to show that $\mathcal{V} = \{0\}$ when there exists a quasi-affinity W in $\mathcal{I}(S, U_+)$. Let U be the bilateral shift on $\ell^2(\mathcal{E})$ viewed as a minimal unitary extension of S. Let Z be the minimal unitary extension of U_+ given by $Z = U \oplus V$ on $\mathcal{Z} = \ell^2(\mathcal{E}) \oplus \mathcal{V}$. According to Proposition 1.5.1, there exists a unique operator L in $\mathcal{I}(U, Z)$ extending W. Moreover, the range of L is dense in $\mathcal{Z}$; see Part (iv) of Proposition 1.5.1.

We claim that U is unitarily equivalent to Z. By taking the adjoint of $ZL = LU$, we obtain $U^*L^* = L^*Z^*$. Because both U and Z are unitary, $L^*Z = UL^*$. Hence $ZLL^* = LL^*Z$. This implies that $Z(LL^*)^n = (LL^*)^n Z$ for all integers $n \geq 0$. So for any polynomial $p(\lambda)$ we have $Zp(LL^*) = p(LL^*)Z$. Recall that one can choose a sequence of polynomials $\{p_n\}_0^\infty$ such that $p_n(LL^*)$ converges in the strong operator topology to $(LL^*)^{1/2}$, the positive square root of LL^*; see Problem 121 in Halmos [126]. Thus $Z(LL^*)^{1/2} = (LL^*)^{1/2}Z$. Since L is onto a dense set, $\ker L^*$ is zero. So L admits a polar decomposition of the form $L = (LL^*)^{1/2}\Omega$ where Ω is a co-isometry from $\ell^2(\mathcal{E})$ onto $\mathcal{Z}$ (that is, Ω^* is an isometry). So using $ZL = LU$ once again

$$(LL^*)^{1/2}\Omega U = LU = ZL = Z(LL^*)^{1/2}\Omega = (LL^*)^{1/2}Z\Omega\,.$$

Since the kernel of $(LL^*)^{1/2}$ is zero, $\Omega U = Z\Omega$. By taking the adjoint, this readily yields $\Omega^* Z^* = U^*\Omega^*$, or equivalently, $U\Omega^* = \Omega^* Z$.

Now let us show that $\mathcal{V} = \{0\}$. Let $\{\varphi_j\}_1^n$ be an orthonormal basis for $\mathcal{E}$. Then

$$\{U^k\varphi_j : 1 \leq j \leq n \text{ and } -\infty < k < \infty\}$$

is an orthonormal basis for $\ell^2(\mathcal{E})$. Let us embed $\varphi_i \oplus 0$ in $\mathcal{E} \oplus \mathcal{V} \subseteq \ell(\mathcal{E}) \oplus \mathcal{V}$. Using Parseval's equality and Bessel's inequality along with the fact that Ω is a co-isometry, we have

$$\begin{aligned} n &= \sum_{i=1}^n \|\varphi_i \oplus 0\|^2 = \sum_{i,\,j,\,k} |(\Omega^*(\varphi_i \oplus 0), U^k\varphi_j)|^2 \\ &= \sum_{i,\,j,\,k} |(\Omega^* Z^{*k}(\varphi_i \oplus 0), \varphi_j)|^2 \leq \sum_{j=1}^n \|\varphi_j\|^2 = n. \end{aligned} \tag{1.5.3}$$

Since $n = n$, we have equality in (1.5.3). This means that $\mathcal{E}$ which equals the span of $\{\varphi_j\}_1^n$ must be contained in the span of the orthonormal set

$$\{\Omega^* Z^k(\varphi_i \oplus 0) : 1 \leq i \leq n \text{ and } -\infty < k < \infty\}.$$

In other words, $\mathcal{E} \subset \Omega^*(\ell^2(\mathcal{E}) \oplus \{0\})$ where $\mathcal{E}$ is viewed as a subspace corresponding to the zero component of $\ell^2(\mathcal{E})$. Since $\{U^n\mathcal{E}\}_{-\infty}^{\infty}$ spans $\ell^2(\mathcal{E})$ and $U^k\Omega^* = \Omega^* Z^k$, for all integers k, the subspace $\ell^2(\mathcal{E}) \subseteq \Omega^*(\ell^2(\mathcal{E}) \oplus \{0\})$. Hence the isometry Ω^* is onto and Ω^* is unitary. Therefore Ω is unitary and $\Omega\ell^2(\mathcal{E}) = \ell^2(\mathcal{E}) \oplus \{0\}$. Because Ω is unitary, $\mathcal{V} = \{0\}$ and U_+ is a unilateral shift. □

Corollary 1.5.3. *Let S be the unilateral shift on $\ell_+^2(\mathcal{E})$ where $\mathcal{E}$ is finite dimensional, and U_+ be an isometry on $\mathcal{K}_+$. Assume that there exists a quasi-affinity W in $\mathcal{I}(S, U_+)$, the subspace $\mathcal{K}_+ \ominus WS\ell_+^2(\mathcal{E})$ and $\mathcal{E}$ have the same dimension. Then S is unitarily equivalent to U_+.*

Proof. Because W is a quasi-affinity, the subspace $U_+\mathcal{K}_+$ equals the closure of $U_+W\ell_+^2(\mathcal{E}) = WS\ell_+^2(\mathcal{E})$. Hence the wandering subspace $\ker U_+^*$ for U_+ is given by

$$\mathcal{L} = \mathcal{K}_+ \ominus U_+\mathcal{K}_+ = \mathcal{K}_+ \ominus WS\ell_+^2(\mathcal{E}).$$

Since $\mathcal{L}$ and $\mathcal{E}$ have the same dimension, Proposition 1.5.2 shows that U_+ is unitarily equivalent to S. □

1.5.1 Abstract Toeplitz operators viewed as a compression

Let U_+ on $\mathcal{K}_+$ and Z_+ on $\mathcal{Z}_+$ be two isometries. Then we say that T is a *Toeplitz operator with respect to* U_+ and Z_+ if T is an operator mapping $\mathcal{K}_+$ into $\mathcal{Z}_+$ such that $T = Z_+^* T U_+$. Finally, it is noted that if T is an operator in $\mathcal{I}(U_+, Z_+)$, then T is a Toeplitz operator with respect to U_+ and Z_+. In this case, T is called a *causal Toeplitz operator* with respect to U_+ and Z_+. The results in this section will be used to prove some properties of Toeplitz and Laurent operators in Chapter 2. As mentioned earlier, the proofs in this section are given for completeness. The techniques developed in these proofs are not used later on, and thus, may be skipped by the reader.

Proposition 1.5.4. *Let U on $\mathcal{K}$ be a minimal unitary extension of an isometry U_+ on $\mathcal{K}_+$ and Z on $\mathcal{Z}$ a minimal unitary extension of an isometry Z_+ on $\mathcal{Z}_+$. Then T is a Toeplitz operator with respect to U_+ and Z_+ if and only if there exists an operator L mapping $\mathcal{K}$ into $\mathcal{Z}$ such that*

$$T = P_{\mathcal{Z}_+} L|\mathcal{K}_+ \quad \textit{where} \quad L \in \mathcal{I}(U, Z). \tag{1.5.4}$$

In this case the following holds.

(i) *There is only one operator L in $\mathcal{I}(U, Z)$ satisfying $T = P_{\mathcal{Z}_+} L|\mathcal{K}_+$.*

(ii) *This operator L is given by*

$$L = weak \lim_{n\to\infty} Z^{*n}TU^nP_n \tag{1.5.5}$$

where P_n is the orthogonal projection onto $U^{*n}\mathcal{K}_+$.

(iii) *The operators T and L have the same norm, that is, $\|L\| = \|T\|$.*

Proof. Assume that $T = P_{\mathcal{Z}_+}L|\mathcal{K}_+$ where L is in $\mathcal{I}(U,Z)$. Then for g in $\mathcal{K}_+$ and h in $\mathcal{Z}_+$, we have

$$\begin{aligned}(Z_+^*TU_+g,h) &= (P_{\mathcal{Z}_+}LU_+g,Z_+h) = (LUg,Zh)\\ &= (ZLg,Zh) = (P_{\mathcal{Z}_+}Lg,h) = (Tg,h).\end{aligned}$$

Since this holds for all g in $\mathcal{K}_+$ and h in $\mathcal{Z}_+$, we see that $T = Z_+^*TU_+$. In other words, T is a Toeplitz operator with respect to U_+ and Z_+.

On the other hand, assume that $T = Z_+^*TU_+$. For any integer $n \ge 0$, let $\mathcal{K}_n$ and $\mathcal{Z}_n$ be the subspaces defined by

$$\mathcal{K}_n = U^{*n}\mathcal{K}_+ \quad \text{and} \quad \mathcal{Z}_n = Z^{*n}\mathcal{Z}_+.$$

Observe that the subspaces $\{\mathcal{K}_n\}_0^\infty$ and $\{\mathcal{Z}_n\}_0^\infty$ are increasing, that is, $\mathcal{K}_n \subseteq \mathcal{K}_{n+1}$ and $\mathcal{Z}_n \subseteq \mathcal{Z}_{n+1}$. The minimality conditions on U and Z imply that $\{\mathcal{K}_n\}_0^\infty$ and $\{\mathcal{Z}_n\}_0^\infty$ are respectively dense in $\mathcal{K}$ and $\mathcal{Z}$. Let T_n be the operator mapping $\mathcal{K}$ into $\mathcal{Z}$ defined by

$$T_n = Z^{*n}TU^nP_n \qquad (n \ge 0). \tag{1.5.6}$$

Notice that $T_0 = TP_{\mathcal{K}_+}$. We claim that for all integers $n \ge j$:

$$P_{\mathcal{Z}_j}T_n|\mathcal{K}_j = T_j|\mathcal{K}_j \qquad (0 \le j \le n). \tag{1.5.7}$$

The equation $T = Z_+^*TU_+$ implies that $T = Z_+^{*\nu}TU_+^\nu$ for all integers $\nu \ge 0$. Moreover, $U^jP_jU^{*j}g = g$ where g is in $\mathcal{K}_+$. For g in $\mathcal{K}_+$ and h in $\mathcal{Z}_+$, we have

$$\begin{aligned}(P_{\mathcal{Z}_j}T_nU^{*j}g,Z^{*j}h) &= (T_nU^{*j}g,Z^{*j}h) = (Z^{*n}TU^nP_nU^{*j}g,Z^{*j}h)\\ &= (TU^{n-j}g,Z^{n-j}h) = (Z_+^{*n-j}TU_+^{n-j}g,h) = (Tg,h)\\ &= (Z^{*j}TU^jP_jU^{*j}g,Z^{*j}h) = (T_jU^{*j}g,Z^{*j}h).\end{aligned}$$

This yields (1.5.7). By consulting (1.5.6), we see that $\|T_n\| \le \|T\|$. Since $T = P_{\mathcal{Z}_+}T_n|\mathcal{K}_+$, we see that $\|T_n\| = \|T\|$ for all integers $n \ge 0$. Because $\{\mathcal{K}_n\}_0^\infty$ and $\{\mathcal{Z}_n\}_0^\infty$ are respectively dense in $\mathcal{K}$ and $\mathcal{Z}$, equation (1.5.7) implies that T_n converges in the weak operator topology to an operator L mapping $\mathcal{K}$ into $\mathcal{Z}$. Finally, since $\|T_n\| = \|T\|$ for all $n \ge 0$, we see that T and L have the same norm.

To show that L is in $\mathcal{I}(U,Z)$, first observe that

$$P_{\mathcal{Z}_n}L|\mathcal{K}_n = T_n|\mathcal{K}_n \qquad (n \ge 0). \tag{1.5.8}$$

For any integers $n \geq j \geq 0$ with g in $\mathcal{K}_+$ and h in $\mathcal{Z}_+$, equations (1.5.7) and (1.5.6) yield

$$\begin{aligned}(LUU^{*j+1}g, Z^{*j}h) &= (T_n U^{*j}g, Z^{*j}h) = (T_j U^{*j}g, Z^{*j}h)\\ &= (Z^{*j}TU^j P_j U^{*j}g, Z^{*j}h) = (Tg, h)\\ &= (Z^{*j+1}TU^{j+1}P_{j+1}U^{*j+1}g, Z^{*j+1}h)\\ &= (T_{j+1}U^{*j+1}g, Z^{*j+1}h) = (P_{\mathcal{Z}_{j+1}}LU^{*j+1}g, Z^{*j+1}h)\\ &= (LU^{*j+1}g, Z^{*j+1}h) = (ZLU^{*j+1}g, Z^{*j}h).\end{aligned}$$

Hence $(LUU^{*j+1}g, Z^{*j}h) = (ZLU^{*j+1}g, Z^{*j}h)$. Because $\{\mathcal{K}_n\}_0^\infty$ and $\{\mathcal{Z}_n\}_0^\infty$ are respectively dense in $\mathcal{K}$ and $\mathcal{Z}$, we arrive at $LU = ZL$.

To show that there is only one L satisfying (1.5.4), assume that M is an operator in $\mathcal{I}(U, Z)$ such that $T = P_{\mathcal{Z}_+}M|\mathcal{K}_+$. As before, choose any g in $\mathcal{K}_+$ and h in $\mathcal{Z}_+$. Then for any integer $n \geq 0$, we have

$$(LU^{*n}g, Z^{*n}h) = (Z^n LU^{*n}g, h) = (Tg, h) = (Mg, h) = (MU^{*n}g, Z^{*n}h).$$

Because $\{\mathcal{K}_n\}_0^\infty$ and $\{\mathcal{Z}_n\}_0^\infty$ are respectively dense in $\mathcal{K}$ and $\mathcal{Z}$, we arrive at $L = M$. Hence Parts (i) to (iii) hold. □

Let U_+ be an isometry on $\mathcal{K}_+$. Then we say that T is a *Toeplitz operator with respect to* U_+ if T is an operator on $\mathcal{K}_+$ satisfying $U_+^* T U_+ = T$. Recall that an operator A on $\mathcal{X}$ is the *compression* of an operator L on $\mathcal{K}$ if $\mathcal{X} \subset \mathcal{K}$ and $A = P_\mathcal{X} L|\mathcal{X}$.

Now let U on $\mathcal{K}$ be a minimal unitary extension for an isometry U_+ on $\mathcal{K}_+$, and T an operator on $\mathcal{K}_+$. Then Proposition 1.5.4 shows that T is a Toeplitz operator with respect to U_+ if and only if there exists an operator L in $\mathcal{I}(U, U)$ such that T equals the compression of L to $\mathcal{K}_+$, that is, $UL = LU$ and $T = P_{\mathcal{K}_+}L|\mathcal{K}_+$. Moreover, there is only one operator L in $\mathcal{I}(U, U)$ such that T is the compression of L to $\mathcal{K}_+$. This operator L is given by

$$L = \text{weak} \lim_{n\to\infty} U^{*n}TU^n P_n \tag{1.5.9}$$

where P_n is the orthogonal projection onto $U^{*n}\mathcal{K}_+$. Finally, L and T have the same norm.

Now assume that T is a self-adjoint Toeplitz operator with respect to U_+. Then $U^{*n}TU^nP_n$ is a sequence of self-adjoint operators. Since the weak limit of self-adjoint operators is self-adjoint, Equation (1.5.9) shows that L is also a self-adjoint operator.

Recall that an operator A on $\mathcal{X}$ is *positive* if $(Ax, x) \geq 0$ for all x in $\mathcal{X}$. We say that an operator A on $\mathcal{X}$ is *strictly positive* if there exists a positive scalar $\delta > 0$ such that $(Ax, x) \geq \delta\|x\|^2$ for all x in $\mathcal{X}$. It is noted that A is strictly positive if and only if A is positive and invertible. Finally, A is a strictly positive operator if and only if there exists a $\delta > 0$ such that

$$\delta = \inf\{(Ax, x) : \|x\| = 1\}. \tag{1.5.10}$$

In this case, $\|A^{-1}\| = 1/\delta$.

Corollary 1.5.5. *Let U on $\mathcal{K}$ be a minimal unitary extension of an isometry U_+ on $\mathcal{K}_+$, and T on $\mathcal{K}_+$ a Toeplitz operator with respect to U_+. Let L on $\mathcal{K}$ be the unique operator in $\mathcal{I}(U,U)$ such that T is the compression of L to $\mathcal{K}_+$. Then T is positive if and only if L is positive. Moreover, L is strictly positive if and only if T is strictly positive. In this case, L^{-1} and T^{-1} have the same norm.*

Proof. If L is positive, then its compression T must also be a positive operator. If T is a positive operator, then equation (1.5.6) shows that $T_n = P_n U^{*n} T U^n P_n$ is a sequence of positive operators. Since T_n converges to L in the weak operator topology, it follows that L is also a positive operator.

For any integer $n \geq 0$, we have $U^n L = L U^n$, and thus,

$$(LU^{*n}g, U^{*n}g) = (U^n L U^{*n} g, g) = (Lg, g) \qquad (g \in \mathcal{K}).$$

Recall that $\{U^{*n}\mathcal{K}_+\}_0^\infty$ is dense in $\mathcal{K}$ and $T = P_{\mathcal{K}_+} L|\mathcal{K}_+$. Using this we have

$$\begin{aligned} \inf\{(Lx,x) : \|x\| = 1\} &= \inf\{(LU^{*n}g, U^{*n}g) : \|g\| = 1, g \in \mathcal{K}_+ \text{ and } n \geq 0\} \\ &= \inf\{(Tg,g) : \|g\| = 1\}. \end{aligned}$$

Hence L is strictly positive if and only if T is strictly positive. In this case, the previous equation also shows that L^{-1} and T^{-1} have the same norm. □

1.6 Notes

The results in this chapter are classical. For an introduction to functional analysis see [59]. An introduction to operator theory with applications is given in Gohberg-Goldberg-Kaashoek [112] and Young [202]. The results concerning isometries and the Wold decomposition are also classical; see [80, 82, 126, 112, 114, 182, 198] for further results in this direction. Finally, the abstract Toeplitz theory in Section 1.5 was taken from Foias-Frazho [82].

Chapter 2

Toeplitz and Laurent Operators

Toeplitz and Laurent operators play a basic role in systems theory. For example, a lower triangular Toeplitz matrix can be viewed as an input-output map for a linear causal time invariant system. First we introduce the Fourier transform. Then we will study Toeplitz and Laurent operators. The Fourier transform will be used to turn Laurent operators into multiplication operators and visa versa.

2.1 The Fourier Transform

In this section, we will present a short review of the Fourier transform. Throughout, $L^2(\mathcal{E})$ is the Hilbert space formed by the set of all Lebesgue measurable, square integrable functions with values in $\mathcal{E}$ over the interval $[0, 2\pi)$. In all of our applications concerning the space $L^2(\mathcal{E})$ (or $L^2_+(\mathcal{E})$ or $H^2(\mathcal{E})$ introduced below), the subspace $\mathcal{E}$ is finite dimensional. So whenever we write $L^2(\mathcal{E})$ (or $L^2_+(\mathcal{E})$ or $H^2(\mathcal{E})$) it understood that $\mathcal{E}$ is finite dimensional. The inner product between two functions f and g in $L^2(\mathcal{E})$ is given by

$$(f, g) = \frac{1}{2\pi} \int_0^{2\pi} (f(e^{\imath\omega}), g(e^{\imath\omega}))_{\mathcal{E}} \, d\omega.$$

It is well known that a Lebesgue measurable function f with values in $\mathcal{E}$ is in $L^2(\mathcal{E})$ if and only if f admits a Fourier series expansion of the form

$$f(e^{\imath\omega}) = \sum_{k=-\infty}^{\infty} e^{-\imath\omega k} f_k \quad \text{where} \quad \sum_{k=-\infty}^{\infty} \|f_k\|^2 < \infty.$$

Recall that the Fourier coefficients $\{f_k\}_{-\infty}^{\infty}$ are the vectors in $\mathcal{E}$ determined by

$$f_k = \frac{1}{2\pi} \int_0^{2\pi} f(e^{\imath\omega}) e^{\imath\omega k} d\omega.$$

Finally, Parserval's theorem shows that

$$(f, g) = \sum_{k=-\infty}^{\infty} (f_k, g_k)_{\mathcal{E}},$$

where $f(e^{\imath\omega}) = \sum_{-\infty}^{\infty} e^{-\imath\omega k} f_k$, and $g(e^{\imath\omega}) = \sum_{-\infty}^{\infty} e^{-\imath\omega k} g_k$ are the Fourier series expansion for f and g respectively. In particular, $\|f\|^2 = \sum_{-\infty}^{\infty} \|f_k\|^2$.

The Fourier transform $\mathcal{F}_{\mathcal{E}}$ is the unitary operator mapping $\ell^2(\mathcal{E})$ onto $L^2(\mathcal{E})$ defined by

$$\mathcal{F}_{\mathcal{E}} \begin{bmatrix} \cdots & f_{-2} & f_{-1} & \boxed{f_0} & f_1 & f_2 & \cdots \end{bmatrix}^{tr} = \sum_{k=-\infty}^{\infty} e^{-\imath\omega k} f_k. \tag{2.1.1}$$

The box around f_0 represents the zero component of a vector f in $\ell^2(\mathcal{E})$.

We say that Z is the *bilateral shift on* $L^2(\mathcal{E})$ if Z is the unitary operator on $L^2(\mathcal{E})$ defined by

$$(Zg)(e^{\imath\omega}) = e^{-\imath\omega} g(e^{\imath\omega}) \qquad (g \in L^2(\mathcal{E})).$$

To see that Z is indeed a bilateral shift, observe that $(Z^* g)(e^{\imath\omega}) = e^{\imath\omega} g(e^{\imath\omega})$. Since Z and Z^* are both isometries, Z is unitary. Notice that $\mathcal{E}$ is a wandering subspace for Z satisfying $L^2(\mathcal{E}) = \mathcal{M}(\mathcal{E})$. Here $\mathcal{E}$ is the set of all constant functions in $L^2(\mathcal{E})$. By definition Z is a bilateral shift. Finally, it is noted that the dimension of $\mathcal{E}$ is the multiplicity of Z. Let U be the bilateral shift on $\ell^2(\mathcal{E})$ and Z the bilateral shift on $L^2(\mathcal{E})$. Since U and Z are both bilateral shifts of the same multiplicity, they are unitarily equivalent. Moreover, the Fourier transform $\mathcal{F}_{\mathcal{E}}$ is the unitary operator which intertwines U with Z, that is, $\mathcal{F}_{\mathcal{E}} U = Z \mathcal{F}_{\mathcal{E}}$.

2.1.1 The subspace $L^2_+(\mathcal{E})$

Throughout $L^2_+(\mathcal{E})$ is the subspace of $L^2(\mathcal{E})$ consisting of the set of all functions g in $L^2(\mathcal{E})$ whose Fourier coefficients are zero for all $n < 0$, that is,

$$L^2_+(\mathcal{E}) = \{g \in L^2(\mathcal{E}) : g(e^{\imath\omega}) = \sum_{k=0}^{\infty} e^{-\imath\omega k} g_k\}. \tag{2.1.2}$$

The Fourier transform $\mathcal{F}^+_{\mathcal{E}}$ is the unitary operator mapping $\ell^2_+(\mathcal{E})$ onto $L^2_+(\mathcal{E})$ given by $\mathcal{F}^+_{\mathcal{E}} = \mathcal{F}_{\mathcal{E}} | L^2_+(\mathcal{E})$, that is,

$$\mathcal{F}^+_{\mathcal{E}} \begin{bmatrix} g_0 & g_1 & g_2 & \cdots \end{bmatrix}^{tr} = \sum_{k=0}^{\infty} e^{-\imath\omega k} g_k \qquad (\begin{bmatrix} g_0 & g_1 & g_2 & \cdots \end{bmatrix}^{tr} \in \ell^2_+(\mathcal{E})).$$

We say that Z is the *unilateral shift* on $L^2_+(\mathcal{E})$ if Z is the isometry on $L^2_+(\mathcal{E})$ defined by

$$(Zg)(e^{\imath\omega}) = e^{-\imath\omega} g(e^{\imath\omega}) \qquad (g \in L^2_+(\mathcal{E})).$$

Clearly, Z is an isometry. To see that Z is indeed a unilateral shift, observe that that $\mathcal{E}$ is a wandering subspace for Z satisfying $L^2_+(\mathcal{E}) = \mathcal{M}_+(\mathcal{E})$. By definition Z is a unilateral shift. The dimension of $\mathcal{E}$ is the multiplicity of Z. Let S be the unilateral shift on $\ell^2_+(\mathcal{E})$. Recall that the multiplicity of S is also the dimension of $\mathcal{E}$. Hence S is unitarily equivalent to Z. Finally, it is noted that $\mathcal{F}^+_{\mathcal{E}}$ is the unitary operator which intertwines S with Z, that is, $\mathcal{F}^+_{\mathcal{E}} S = Z\mathcal{F}^+_{\mathcal{E}}$.

The subspaces $L^2(\mathcal{E},\mathcal{Y})$ and $L^\infty(\mathcal{E},\mathcal{Y})$. In all of our applications concerning the spaces $L^2(\mathcal{E},\mathcal{Y})$ or $L^\infty(\mathcal{E},\mathcal{Y})$, the subspace $\mathcal{E}$ and $\mathcal{Y}$ are finite dimensional. So whenever we write $L^2(\mathcal{E},\mathcal{Y})$ or $L^\infty(\mathcal{E},\mathcal{Y})$ it understood that $\mathcal{E}$ and $\mathcal{Y}$ are finite dimensional.

The set of all operators mapping $\mathcal{E}$ into $\mathcal{Y}$ is denoted by $\mathcal{L}(\mathcal{E},\mathcal{Y})$. Throughout $L^2(\mathcal{E},\mathcal{Y})$ is the set of all Lebesgue measurable functions F over the interval $[0,2\pi)$ with values in $\mathcal{L}(\mathcal{E},\mathcal{Y})$ such that

$$\|F\|_2^2 = \frac{1}{2\pi}\int_0^{2\pi} \operatorname{trace}(F(e^{\imath\omega})F(e^{\imath\omega})^*)d\omega < \infty.$$

Notice that F is in $L^2(\mathcal{E},\mathcal{Y})$ if and only if F admits a Fourier series expansion of the form $F = \sum_{-\infty}^{\infty} e^{-\imath\omega k}F_k$ and $\sum_{-\infty}^{\infty} F_k^*F_k$ is finite. In this case, $\|F\|_2^2 = \sum_{-\infty}^{\infty} \operatorname{trace}(F_kF_k^*)$. Finally, if F and G are two function in $L^2(\mathcal{E},\mathcal{Y})$, then the inner product between F and G is given by

$$(F,G) = \frac{1}{2\pi}\int_0^{2\pi} \operatorname{trace}(F(e^{\imath\omega})G(e^{\imath\omega})^*)d\omega = \frac{1}{2\pi}\int_0^{2\pi} \operatorname{trace}(G(e^{\imath\omega})^*F(e^{\imath\omega}))d\omega.$$

Throughout $L^\infty(\mathcal{E},\mathcal{Y})$ is the set of all, uniformly bounded, Lebesgue measurable functions F with values in $\mathcal{L}(\mathcal{E},\mathcal{Y})$ over the interval $[0,2\pi)$. Recall that $L^\infty(\mathcal{E},\mathcal{Y})$ is a Banach space under the norm

$$\|F\|_\infty = \text{essential-}\sup\{\|F(e^{\imath\omega})\| : 0 \le \omega < 2\pi\}. \tag{2.1.3}$$

Notice that $L^\infty(\mathcal{E},\mathcal{Y})$ is a proper subset of $L^2(\mathcal{E},\mathcal{Y})$.

2.2 Hardy Spaces

In this section we will review some standard results concerning Hardy spaces of functions which are analytic outside the closed unit disc, that is, analytic in the region

$$\mathbb{D}_+ = \{z \in \mathbb{C} : |z| > 1\}. \tag{2.2.1}$$

It is emphasized that we assume that these functions are also analytic at infinity. In other words, $f(z)$ is analytic in $\mathbb{D}_+$ if and only if $f(1/z)$ is analytic in the open unit disc $\mathbb{D} = \{z \in \mathbb{C} : |z| < 1\}$. So the function $f(z) = z$ is not analytic in $\mathbb{D}_+$.

Let $\begin{bmatrix} h_0 & h_1 & h_2 & \cdots \end{bmatrix}^{tr}$ be a vector in $\ell^2_+(\mathcal{E})$. Consider the function h defined by

$$h(z) = \sum_{k=0}^{\infty} z^{-k} h_k.$$

We claim that h is analytic in $\mathbb{D}_+$. Let z be a complex number such that $|z| > r > 1$. The Cauchy-Schwartz inequality yields

$$\left| \sum_{k=0}^{\infty} z^{-k} h_k \right|^2 \leq \sum_{k=0}^{\infty} \frac{1}{|z|^{2k}} \sum_{k=0}^{\infty} \|h_k\|^2 \leq \frac{1}{1 - r^{-2}} \sum_{k=0}^{\infty} \|h_k\|^2.$$

Hence the series $\sum_{k=0}^{\infty} z^{-k} h_k$ converges uniformly for any disc of radius $r > 1$. In other words, $h(z)$ is analytic in $\mathbb{D}_+$.

Throughout, $H^2(\mathcal{E})$ is the Hardy space consisting of the set of all $\mathcal{E}$-valued analytic functions in $\mathbb{D}_+$ such that

$$\|h\|^2 = \sum_{k=0}^{\infty} \|h_k\|^2 < \infty \quad \text{where} \quad h(z) = \sum_{k=0}^{\infty} z^{-k} h_k \tag{2.2.2}$$

is the Taylor's series expansion of h at infinity. The norm of a function h is $H^2(\mathcal{E})$ is simply the $\ell^2_+(\mathcal{E})$ norm of the Taylor coefficients $\{h_k\}_0^\infty$ of h. So $H^2(\mathcal{E})$ is a Hilbert space under this norm. If $\mathcal{E} = \mathbb{C}$ is the set of complex numbers, then we denote $H^2(\mathbb{C})$ by H^2. Finally, it is noted that not every function analytic in $\mathbb{D}_+$ is in $H^2(\mathcal{E})$. For example, the function $1/(z-1)$ is analytic in $\mathbb{D}_+$ and not in H^2 due to the pole at 1. Observe that $1/(z-1) = \sum_1^\infty z^{-n}$, and thus, $\|1/(z-1)\|^2 = \sum_1^\infty 1 = \infty$.

Let h be a function in $H^2(\mathcal{E})$ and $h(z) = \sum_0^\infty z^{-k} h_k$ its Taylor's series expansion about infinity. It is well known that as $r > 1$ tends to 1, the function $h(re^{\imath\omega})$ converges to a function $\widetilde{h}(e^{\imath\omega})$ almost everywhere with respect to the Lebesgue measure . Moreover, $\widetilde{h}$ is a function in $L^2_+(\mathcal{E})$ and admits a Fourier series expansion of the form

$$\widetilde{h}(e^{\imath\omega}) = \sum_{k=0}^{\infty} e^{-\imath\omega k} h_k.$$

It is noted that the H^2 norm of $h(z)$ equals the L^2 norm of $\widetilde{h}$, that is,

$$\|h\|_{H^2}^2 = \sum_{k=0}^{\infty} \|h_k\|^2 = \frac{1}{2\pi} \int_0^{2\pi} \|\widetilde{h}(e^{\imath\omega})\|^2 d\omega = \|\widetilde{h}\|_{L^2}^2.$$

On the other hand, assume that $\widetilde{h}(e^{\imath\omega})$ is a function in $L^2_+(\mathcal{E})$. Then $\widetilde{h}$ admits a Fourier series expansion of the form $h = \sum_0^\infty e^{-\imath\omega k} h_k$, and $h(z) = \sum_0^\infty z^{-k} h_k$ defines a function in $H^2(\mathcal{E})$. So the functions $h(z)$ in $H^2(\mathcal{E})$ and $\widetilde{h}(e^{\imath\omega})$ in $L^2_+(\mathcal{E})$

uniquely determine each other. To be precise, let Φ be the mapping from $H^2(\mathcal{E})$ into $L^2_+(\mathcal{E})$ defined by

$$\widetilde{h}(e^{\imath\omega}) = (\Phi h)(e^{\imath\omega}) = \sum_{k=0}^{\infty} e^{-\imath\omega k} h_k \quad \text{where} \quad h(z) = \sum_{k=0}^{\infty} z^{-k} h_k$$

is the Taylor's series expansion of h in $H^2(\mathcal{E})$. Then Φ defines a unitary operator from $H^2(\mathcal{E})$ onto $L^2_+(\mathcal{E})$. Due to this unitary identification we drop the tilde notation on $\widetilde{h}$ and simply write $h(e^{\imath\omega})$ for the function Φh. In other words, if h is in $H^2(\mathcal{E})$ and we write $h(e^{\imath\omega})$, then we mean that $h(e^{\imath\omega})$ is the function in $L^2_+(\mathcal{E})$ given by $h(e^{\imath\omega}) = (\Phi h)(e^{\imath\omega})$. Finally, using this identification we also view $H^2(\mathcal{E})$ as the subspace of $L^2(\mathcal{E})$ corresponding to $L^2_+(\mathcal{E})$. Motivated by this identification, we use $H^2(\mathcal{E})$ and $L^2_+(\mathcal{E})$ interchangeably.

Due to the previous identification between $H^2(\mathcal{E})$ and $L^2_+(\mathcal{E})$, we also view the Fourier transform $\mathcal{F}^+_{\mathcal{E}}$ as the unitary operator from $\ell^2_+(\mathcal{E})$ onto $H^2(\mathcal{E})$ defined by

$$(\mathcal{F}^+_{\mathcal{E}} \begin{bmatrix} f_0 \\ f_1 \\ f_2 \\ \vdots \end{bmatrix})(z) = \sum_{k=0}^{\infty} z^{-k} f_k \quad \text{where} \quad \begin{bmatrix} f_0 \\ f_1 \\ f_2 \\ \vdots \end{bmatrix} \in \ell^2_+(\mathcal{E}). \tag{2.2.3}$$

Let $\Pi_{\mathcal{E}}$ be the orthogonal projection from $\ell^2_+(\mathcal{E})$ onto $\mathcal{E}$ which picks out the first component of $\ell^2_+(\mathcal{E})$, that is,

$$\Pi_{\mathcal{E}} = \begin{bmatrix} I & 0 & 0 & \cdots \end{bmatrix} : \ell^2_+(\mathcal{E}) \to \mathcal{E}.$$

Let S be the unilateral shift on $\ell^2_+(\mathcal{E})$; see equation (1.1.3). If h is an element in $\ell^2_+(\mathcal{E})$, then the Fourier transform of h is given by the representation

$$(\mathcal{F}^+_{\mathcal{E}} h)(z) = \Pi_{\mathcal{E}}(I - z^{-1}S^*)^{-1}h = z\Pi_{\mathcal{E}}(zI - S^*)^{-1}h \qquad (h \in \ell^2_+(\mathcal{E})). \tag{2.2.4}$$

To see this, let $h = \begin{bmatrix} h_0 & h_1 & h_2 & \cdots \end{bmatrix}^{tr}$. Observe that $h_k = \Pi_{\mathcal{E}} S^{*k} h$ for all integers $k \geq 0$. Using the fact that $\|z^{-1}S^*\| < 1$, for each z in $\mathbb{D}_+$, we have

$$\begin{aligned} \Pi_{\mathcal{E}}(I - z^{-1}S^*)^{-1}h &= \Pi_{\mathcal{E}} \sum_{k=0}^{\infty} z^{-k} S^{*k} h = \sum_{k=0}^{\infty} z^{-k} \Pi_{\mathcal{E}} S^{*k} h \\ &= \sum_{k=0}^{\infty} z^{-k} h_k = (\mathcal{F}^+_{\mathcal{E}} h)(z). \end{aligned}$$

Finally, it is noted that the Fourier transform in (2.2.4) can also be written in the following "state space" form:

$$(\mathcal{F}^+_{\mathcal{E}} h)(z) = \Pi_{\mathcal{E}} h + \Pi_{\mathcal{E}}(zI - S^*)^{-1} S^* h \qquad (h \in \ell^2_+(\mathcal{E}) \text{ and } z \in \mathbb{D}_+). \tag{2.2.5}$$

Let $S_{\mathcal{E}}$ be the operator on $H^2(\mathcal{E})$ defined by multiplication by $1/z$ on $H^2(\mathcal{E})$, that is,

$$(S_{\mathcal{E}}h)(z) = \frac{h(z)}{z} \qquad (h \in H^2(\mathcal{E}) \text{ and } z \in \mathbb{D}_+). \tag{2.2.6}$$

Notice that $S_{\mathcal{E}}$ is an isometry on $H^2(\mathcal{E})$. Moreover, the set of all constant functions $\mathcal{E}$ in $H^2(\mathcal{E})$ is a cyclic wandering subspace for $S_{\mathcal{E}}$. In other words, $S_{\mathcal{E}}$ is a unilateral shift on $H^2(\mathcal{E})$. Hence $S_{\mathcal{E}}$ is unitarily equivalent to the unilateral shift S on $\ell^2_+(\mathcal{E})$. In fact, it is easy to verify that the Fourier transform $\mathcal{F}^+_{\mathcal{E}}$ intertwines the unilateral shift S on $\ell^2_+(\mathcal{E})$ with $S_{\mathcal{E}}$. To be precise, $\mathcal{F}^+_{\mathcal{E}} S = S_{\mathcal{E}} \mathcal{F}^+_{\mathcal{E}}$. Motivated by this, we will simply call the isometry $S_{\mathcal{E}}$ in (2.2.6) the *unilateral shift on* $H^2(\mathcal{E})$. The adjoint of $S_{\mathcal{E}}$ is determined by

$$(S^*_{\mathcal{E}}h)(z) = zh(z) - zh(\infty) \qquad (h \in H^2(\mathcal{E}))\,. \tag{2.2.7}$$

By a slight abuse of notation we set

$$h(\infty) = \lim_{r\to\infty} h(r) \qquad (h \in H^2(\mathcal{E})).$$

Notice that $h(\infty) = h_0$ where h_0 is the first component in the Taylor series expansion for $h(z) = \sum_0^\infty z^{-k} h_k$.

The set of all eigenvalues for the backward shift operator equals the open unit disc $\mathbb{D}$; see Section 1.2. In particular, by taking the Fourier transform of the eigenvector in (1.2.2), or by direct computation, we see that the set of all eigenvectors φ for $S^*_{\mathcal{E}}$ corresponding to the eigenvalue λ in $\mathbb{D}$ are given by

$$S^*\varphi = \lambda\varphi \quad \text{where} \quad \varphi = \frac{z}{z-\lambda} f \qquad (\lambda \in \mathbb{D} \text{ and } f \in \mathcal{E}). \tag{2.2.8}$$

Finally, it is noted that the eigenvector $\varphi = zf/(z-\lambda)$ corresponding to the eigenvalue λ has a pole at λ.

The subspaces $H^2(\mathcal{E},\mathcal{Y})$ and $H^\infty(\mathcal{E},\mathcal{Y})$. In all of our applications concerning the spaces $H^2(\mathcal{E},\mathcal{Y})$ and $H^\infty(\mathcal{E},\mathcal{Y})$, the subspace $\mathcal{E}$ and $\mathcal{Y}$ are finite dimensional. So whenever we write $H^2(\mathcal{E},\mathcal{Y})$ or $H^\infty(\mathcal{E},\mathcal{Y})$ it is understood that $\mathcal{E}$ and $\mathcal{Y}$ are finite dimensional.

We say that a function Θ is in $H^2(\mathcal{E},\mathcal{Y})$ if $\Theta(z)$ is analytic in $\mathbb{D}_+$, with values in $\mathcal{L}(\mathcal{E},\mathcal{Y})$, and

$$\|\Theta\|_2^2 = \sum_{n=0}^\infty \mathrm{trace}(\Theta_n\Theta_n^*) < \infty \quad \text{where} \quad \Theta(z) = \sum_{n=0}^\infty z^{-n}\Theta_n. \tag{2.2.9}$$

If Θ is a function in $H^2(\mathcal{E},\mathcal{Y})$, then as $r > 1$ tends to 1, the function $\Theta(re^{\imath\omega})$ converges almost everywhere to a function $\Theta(e^{\imath\omega})$ of the form $\Theta(e^{\imath\omega}) = \sum_0^\infty e^{-\imath\omega n}\Theta_n$. As expected, $\{\Theta_n\}_0^\infty$ are the Taylor coefficients for Θ in (2.2.9). In this case,

$\|\Theta\|_{H^2} = \|\Theta(e^{\imath\omega})\|_{L^2}$. So without loss of generality, we can also express Θ as a function on the unit circle, that is, $\Theta(e^{\imath\omega}) = \sum_0^\infty e^{-\imath\omega n}\Theta_n$. Finally, due to this identification $H^2(\mathcal{E},\mathcal{Y})$ can be viewed as the subspace of $L^2(\mathcal{E},\mathcal{Y})$ consisting of all functions of the form $\sum_0^\infty e^{-\imath\omega k}\Theta_k$.

Recall that $H^\infty(\mathcal{E},\mathcal{Y})$ is the Banach space formed by the set of all uniformly bounded, analytic functions in $\mathbb{D}_+$ with values in $\mathcal{L}(\mathcal{E},\mathcal{Y})$. The H^∞ norm of a function Θ in $H^\infty(\mathcal{E},\mathcal{Y})$, denoted by $\|\Theta\|_\infty$, is given by

$$\|\Theta\|_\infty = \sup\{\|\Theta(z)\| : z \in \mathbb{D}_+\}. \tag{2.2.10}$$

Notice that $H^\infty(\mathcal{E},\mathcal{Y})$ is a subset of $H^2(\mathcal{E},\mathcal{Y})$. If Θ is a function in $H^\infty(\mathcal{E},\mathcal{Y})$, then $\Theta(e^{\imath\omega})$ is a function in $L^\infty(\mathcal{E},\mathcal{Y})$ whose Fourier series expansion is of the form $\sum_0^\infty e^{-\imath\omega k}\Theta_k$. In particular, by the maximum modulus theorem, the H^∞ norm of Θ is also given by the L^∞ norm of $\Theta(e^{\imath\omega})$, that is,

$$\|\Theta\|_\infty = \text{essential-}\sup\{\|\Theta(e^{\imath\omega})\| : 0 \le \omega \le 2\pi\}. \tag{2.2.11}$$

Finally, it is noted that $H^\infty(\mathcal{E},\mathcal{Y})$ can be viewed as the subspace of $L^\infty(\mathcal{E},\mathcal{Y})$ consisting of all functions of the form $\sum_0^\infty e^{-\imath\omega k}\Theta_k$.

2.3 Multiplication Operators

Let F be a Lebesgue measurable function with values in $\mathcal{L}(\mathcal{E},\mathcal{Y})$. Throughout M_F denotes multiplication by F, that is,

$$(M_F g)(e^{\imath\omega}) = F(e^{\imath\omega})g(e^{\imath\omega}) \tag{2.3.1}$$

where g is a Lebesgue measurable function with values in $\mathcal{E}$. We say that M_F is a *multiplication operator* if $(M_F g)(e^{\imath\omega}) = F(e^{\imath\omega})g(e^{\imath\omega})$ defines an operator, that is, a bounded linear map from $L^2(\mathcal{E})$ into $L^2(\mathcal{Y})$. In this section, we will show that M_F defines an operator if and only if F is a function in $L^\infty(\mathcal{E},\mathcal{Y})$. Moreover, in this case, $\|M_F\| = \|F\|_\infty$.

Let F be a function in $L^\infty(\mathcal{E},\mathcal{Y})$. Then we claim that $\|M_F\| \le \|F\|_\infty$, and thus, M_F is a well-defined operator mapping $L^2(\mathcal{E})$ into $L^2(\mathcal{Y})$. To verify this for g in $L^2(\mathcal{E})$, we have

$$\|M_F g\|^2 = \frac{1}{2\pi}\int_0^{2\pi} \|F(e^{\imath\omega})g(e^{\imath\omega})\|^2 d\omega \le \frac{\|F\|_\infty^2}{2\pi}\int_0^{2\pi} \|g(e^{\imath\omega})\|^2 d\omega = \|F\|_\infty^2 \|g\|^2.$$

Hence $\|M_F\| \le \|F\|_\infty$. Lemma 2.3.1 below shows that $\|M_F\| = \|F\|_\infty$. Notice that M_F intertwines the bilateral shift U on $L^2(\mathcal{E})$ with the bilateral shift Z on $L^2(\mathcal{Y})$. In Section 2.4, we will show that an operator intertwines two bilateral shifts if and only if it is a multiplication operator.

We say that a function F is *rigid* if F is a function in $L^\infty(\mathcal{E},\mathcal{Y})$ and $F(e^{\imath\omega}) : \mathcal{E} \to \mathcal{Y}$ is an isometry almost everywhere with respect to the Lebesgue measure.

Lemma 2.3.1. *Let F be a Lebesgue measurable function with values in $\mathcal{L}(\mathcal{E},\mathcal{Y})$ over the interval $[0,2\pi)$. Then M_F defines an operator from $L^2(\mathcal{E})$ into $L^2(\mathcal{Y})$ if and only if F is in $L^\infty(\mathcal{E},\mathcal{Y})$. In this case, the following holds.*

(i) *The norm of M_F equals the L^∞ norm of F, that is, $\|M_F\| = \|F\|_\infty$.*

(ii) *The operator M_F is an isometry if and only if F is rigid.*

(iii) *The operator M_F is invertible if and only if F^{-1} is a function in $L^\infty(\mathcal{Y},\mathcal{E})$. In this case, $M_F^{-1} = M_{F^{-1}}$ and $\|M_F^{-1}\| = \|F^{-1}\|_\infty$.*

Proof. Assume that M_F is an operator, that is, $\|M_F\| < \infty$. Let

$$\varphi(e^{\imath\omega}) = \sum_{k=-n}^{n} \alpha_k e^{-\imath k\omega}$$

be any scalar-valued trigonometric polynomial. Then for any vector a in $\mathcal{E}$, we obtain

$$\begin{aligned}\frac{1}{2\pi}\int_0^{2\pi} |\varphi(e^{\imath\omega})|^2 \|F(e^{\imath\omega})a\|^2 d\omega &= \frac{1}{2\pi}\int_0^{2\pi} \|F(e^{\imath\omega})\varphi(e^{\imath\omega})a\|^2 d\omega \\ &= \|M_F\varphi a\|^2 \leq \|M_F\|^2\|\varphi a\|^2 \\ &= \|M_F\|^2\|a\|^2 \frac{1}{2\pi}\int_0^{2\pi} |\varphi(e^{\imath\omega})|^2 d\omega.\end{aligned}$$

(Notice that we have equality if M_F is an isometry.) Using the fact that the trigonometric polynomials are dense in the continuous functions of period 2π, we obtain

$$\frac{1}{2\pi}\int_0^{2\pi} p(\omega)\|F(e^{\imath\omega})a\|^2 d\omega \leq \|M_F\|^2\|a\|^2 \frac{1}{2\pi}\int_0^{2\pi} p(\omega)d\omega \tag{2.3.2}$$

where $p(\omega)$ is any positive continuous function. (We have equality if M_F is an isometry.) By employing Lebesgue's dominated convergence theorem, we can find an appropriate sequence of functions such that the inequality in (2.3.2) holds for all bounded positive measurable functions p. By choosing p to be the characteristic function with an arbitrary small support, we see that

$$\|F(e^{\imath\omega})a\| \leq \|M_F\|\|a\| \tag{2.3.3}$$

almost everywhere with respect to the Lebesgue measure. So $\|F\|_\infty \leq \|M_F\|$. In particular, F is in $L^\infty(\mathcal{E},\mathcal{Y})$. Recall we have already shown that $\|M_F\| \leq \|F\|_\infty$. Hence $\|M_F\| = \|F\|_\infty$ and Part (i) holds.

If M_F is an isometry, then we have equality in (2.3.3), that is, $\|F(e^{\imath\omega})a\| = \|a\|$ almost everywhere with respect to the Lebesgue measure. Hence F is rigid. On the other hand, if F is rigid, then it readily follows that $\|M_F g\| = \|g\|$ for all g in $L^2(\mathcal{E})$, and thus, M_F is an isometry. This yields Part (ii).

Recall that $(M_F g)(e^{\imath\omega}) = F(e^{\imath\omega})g(e^{\imath\omega})$ where g is in $L^2(\mathcal{E})$. So M_F is invertible if and only if $F(e^{\imath\omega})$ is almost everywhere invertible with respect to the Lebesgue measure and the mapping $y \to F(e^{\imath\omega})^{-1}y(e^{\imath\omega})$ defines a bounded linear map from $L^2(\mathcal{Y})$ into $L^2(\mathcal{E})$. In other words, M_F is invertible if and only if the multiplication map $M_{F^{-1}}$ defines an operator. In this case, $M_F^{-1} = M_{F^{-1}}$. In particular, $\|M_F^{-1}\| = \|F^{-1}\|_\infty$. □

Remark 2.3.2. Assume that F is a function in $L^\infty(\mathcal{E}, \mathcal{E})$. Then M_F is a positive operator on $L^2(\mathcal{E})$ if and only if $F(e^{\imath\omega})$ is almost everywhere a positive operator with respect to the Lebesgue measure. Moreover, M_F is strictly positive if and only if $F(e^{\imath\omega})$ is almost everywhere a positive operator with respect to the Lebesgue measure and F^{-1} is in $L^\infty(\mathcal{E}, \mathcal{E})$.

To see this observe that for any g in $L^\infty(\mathcal{E}, \mathcal{E})$, we have

$$(M_F g, g) = \frac{1}{2\pi}\int_0^{2\pi} (F(e^{\imath\omega})g(e^{\imath\omega}), g(e^{\imath\omega}))d\omega. \tag{2.3.4}$$

If $F(e^{\imath\omega})$ is almost everywhere a positive operator with respect to the Lebesgue measure, then the right-hand side of (2.3.4) is positive, and thus, M_F is a positive operator. On the other hand, if M_F is a positive operator, then choosing any function g which is constant on a set of Lebesgue measure and zero otherwise, shows that $F(e^{\imath\omega})$ is almost everywhere a positive operator. This prove the first part of Remark 2.3.2. The second half follows from Part (iii) of Lemma 2.3.1.

2.4 Laurent Operators

The set of all trigonometric polynomials with values in $\mathcal{E}$ is denoted by $\mathcal{P}_{trig}(\mathcal{E})$. In other words, $\mathcal{P}_{trig}(\mathcal{E})$ is the set of all functions of the form $\sum_{k=m}^{n} a_k e^{-\imath\omega k}$ where a_k is a vector in $\mathcal{E}$ while m and n are finite integers. Throughout $\ell^c(\mathcal{E})$ is the linear manifold consisting of the set of all vectors in $\ell^2(\mathcal{E})$ with compact support. The Fourier transform of $\ell^c(\mathcal{E})$ equals $\mathcal{P}_{trig}(\mathcal{E})$, that is, $\mathcal{F}_\mathcal{E}\ell^c(\mathcal{E}) = \mathcal{P}_{trig}(\mathcal{E})$.

We say that L is a *Laurent matrix* if L is a block matrix of the form

$$L = \begin{bmatrix} \ddots & \ddots & \vdots & \vdots & \\ \ddots & F_0 & F_{-1} & F_{-2} & \cdots \\ \cdots & F_1 & \boxed{F_0} & F_{-1} & \cdots \\ \cdots & F_2 & F_1 & F_0 & \ddots \\ & \vdots & \vdots & \ddots & \ddots \end{bmatrix}. \tag{2.4.1}$$

Here $\{F_k\}_{-\infty}^{\infty}$ is a sequence of operators mapping $\mathcal{E}$ into $\mathcal{Y}$. The box around F_0 represents the 0-0 component of the Laurent matrix. Notice that all the diagonal entries of the Laurent matrix are the same. Moreover, the $j+1$ column of the

Laurent matrix is the j column shifted down. In other words, L is a Laurent matrix if and only if the entries of $L_{j,k} = F_{j-k}$ for all integers j and k where $\{F_j\}_{-\infty}^{\infty}$ is a sequence of operators mapping $\mathcal{E}$ into $\mathcal{Y}$. Now assume that g is a vector in $\ell^c(\mathcal{E})$. Then Lg is well defined and the n-th component of Lg is given by

$$(Lg)_n = \sum_{j=-\infty}^{\infty} F_{n-j} g_j \qquad (\oplus_{-\infty}^{\infty} g_j \in \ell^c(\mathcal{E})). \tag{2.4.2}$$

Let F be the function with values in $\mathcal{L}(\mathcal{E}, \mathcal{Y})$ formally defined by

$$F(e^{\imath\omega}) = \sum_{k=-\infty}^{\infty} F_k e^{-\imath\omega k}. \tag{2.4.3}$$

Clearly, F and $\{F_k\}_{-\infty}^{\infty}$ formally determine each other. Motivated by this we say that F is the *symbol* for L. In this case, the Laurent matrix L in (2.4.1) is denoted by $L = L_F$. In systems theory terminology, a Laurent matrix corresponds to convolution between the (impulse response) sequence $\{F_j\}_{-\infty}^{\infty}$ and the (input) sequence $\{g_j\}_{-\infty}^{\infty}$.

We say that L is a *Laurent operator* mapping $\ell^2(\mathcal{E})$ into $\ell^2(\mathcal{Y})$ if L is an operator and L admits a matrix representation of the form in (2.4.1) with respect to the standard basis for $\ell^2(\mathcal{E})$ and $\ell^2(\mathcal{Y})$. Now assume that $L = L_F$ is a Laurent operator with symbol F. Since L_F is an operator all the columns in L_F are square summable, that is, $\sum_{-\infty}^{\infty} \|F_j a\|^2$ is finite for all a in $\mathcal{E}$. In particular, F must be a function in $L^2(\mathcal{E}, \mathcal{Y})$. In this case, F is the Fourier transform of $\{F_k\}_{-\infty}^{\infty}$ and the symbol F for L_F is well defined. In a moment we will see that L_F is defined as an operator from $\ell^2(\mathcal{E})$ into $\ell^2(\mathcal{Y})$ if and only if F is in $L^\infty(\mathcal{E}, \mathcal{Y})$.

Let F be a function in $L^2(\mathcal{E}, \mathcal{Y})$. Then the columns of L_F are square summable, and L_F is a well-defined linear map from $\ell^c(\mathcal{E})$ into $\ell^2(\mathcal{Y})$. If g is a vector in $\ell^c(\mathcal{E})$, then we claim that

$$\mathcal{F}_{\mathcal{Y}} L_F g = M_F \mathcal{F}_{\mathcal{E}} g \qquad (g \in \ell^c(\mathcal{E}) \text{ and } F \in L^2(\mathcal{E}, \mathcal{Y})). \tag{2.4.4}$$

In systems theory terminology, this states that convolution in the time domain corresponds to multiplication in the frequency domain. To see this observe that (2.4.2) yields

$$\begin{aligned}(\mathcal{F}_{\mathcal{Y}} L_F g)(e^{\imath\omega}) &= \sum_{n,j} e^{-\imath\omega n} F_{n-j} g_j = \sum_{n,j} e^{-\imath\omega(n-j)} F_{n-j} e^{-\imath\omega j} g_j \\ &= \sum_k e^{-\imath\omega k} F_k \sum_j e^{-\imath\omega j} g_j = F(e^{\imath\omega})(\mathcal{F}_{\mathcal{E}} g)(e^{\imath\omega}).\end{aligned}$$

Hence (2.4.4) holds. The following result shows that L_F is an operator if and only if F is a function in $L^\infty(\mathcal{E}, \mathcal{Y})$. Recall that $\mathcal{I}(A, B)$ denotes the set of all operators intertwining A on $\mathcal{X}$ and B on $\mathcal{Y}$.

Theorem 2.4.1. *Let $U_\mathcal{E}$ and $U_\mathcal{Y}$ be the bilateral shifts on $\ell^2(\mathcal{E})$ and $\ell^2(\mathcal{Y})$, respectively. Then an operator L is in $\mathcal{I}(U_\mathcal{E}, U_\mathcal{Y})$ if and only if $L = L_F$ is a Laurent operator where F is a function in $L^\infty(\mathcal{E}, \mathcal{Y})$. In this case, $\|L\| = \|F\|_\infty$. Finally, L_F is an isometry if and only if F is rigid.*

Proof. Assume that L is a Laurent operator with symbol F. As noted earlier F is a function in $L^2(\mathcal{E}, \mathcal{Y})$. Using the matrix representation for L in (2.4.1), it is easy to check that $U_\mathcal{Y} L = L U_\mathcal{E}$. Therefore L is in $\mathcal{I}(U_\mathcal{E}, U_\mathcal{Y})$.

Assume that L is an operator in $\mathcal{I}(U_\mathcal{E}, U_\mathcal{Y})$. Then we claim that L is a Laurent operator. To see this, let us identify $\mathcal{E}$ (respectively $\mathcal{Y}$) with the subspace of $\ell^2(\mathcal{E})$ corresponding to the zero component of $\ell^2(\mathcal{E})$ (respectively $\ell^2(\mathcal{Y})$). Then $\mathcal{E}$ is a wandering subspace for $U_\mathcal{E}$ such that $\ell^2(\mathcal{E}) = \mathcal{M}(\mathcal{E})$ and $\mathcal{Y}$ is a wandering subspace for $U_\mathcal{Y}$ such that $\ell^2(\mathcal{Y}) = \mathcal{M}(\mathcal{Y})$. Let $L_{j,k}$ be the j-k entry of L with respect to the basis $\{U_\mathcal{E}^n \mathcal{E}\}_{-\infty}^{\infty}$ for $\ell^2(\mathcal{E})$ and $\{U_\mathcal{Y}^n \mathcal{Y}\}_{-\infty}^{\infty}$ for $\ell^2(\mathcal{Y})$. Then $L_{j,k} = \Pi_\mathcal{Y} U_\mathcal{Y}^{*j} L U_\mathcal{E}^k \Pi_\mathcal{E}^*$ where $\Pi_\mathcal{Y}$ (respectively $\Pi_\mathcal{E}$) is the orthogonal projection from $\ell^2(\mathcal{Y})$ onto $\mathcal{Y}$ (respectively from $\ell^2(\mathcal{E})$ onto $\mathcal{E}$) which picks out the zero component $\mathcal{Y}$ in $\ell^2(\mathcal{Y})$ (respectively $\mathcal{E}$ in $\ell^2(\mathcal{E})$). Using this we have

$$L_{j,k} = \Pi_\mathcal{Y} U_\mathcal{Y}^{*j} L U_\mathcal{E}^k \Pi_\mathcal{E}^* = \Pi_\mathcal{Y} U_\mathcal{Y}^{*j} U_\mathcal{Y}^k L \Pi_\mathcal{E}^* = \Pi_\mathcal{Y} U_\mathcal{Y}^{*(j-k)} L \Pi_\mathcal{E}^* = L_{j-k,0}.$$

Hence $L_{j,k} = L_{j-k,0}$ for all integers j and k. Setting $F_n = L_{n,0}$ for all integers n, shows that L admits a matrix representation of the form in (2.4.1). Therefore L is a Laurent operator, which proves our claim. Finally, $L = L_F$ where the symbol F for L is the Fourier transform of $\{F_j\}_{-\infty}^{\infty}$.

So far we have shown that L is a Laurent operator if and only if L is an operator in $\mathcal{I}(U_\mathcal{E}, U_\mathcal{Y})$. Moreover, in this case, $L = L_F$ where F is a function in $L^2(\mathcal{E}, \mathcal{Y})$. Clearly, $\ell^c(\mathcal{E})$ is dense in $\ell^2(\mathcal{E})$. Recall that the Fourier transform $\mathcal{F}_\mathcal{V}$ is a unitary operator mapping $\ell^2(\mathcal{V})$ onto $L^2(\mathcal{V})$. According to (2.4.4), the operator M_F defines an operator mapping $L^2(\mathcal{E})$ into $L^2(\mathcal{Y})$ if and only if L_F defines an operator mapping $\ell^2(\mathcal{E})$ into $\ell^2(\mathcal{Y})$. In this case, L_F is unitarily equivalent to M_F, that is,

$$\mathcal{F}_\mathcal{Y} L_F = M_F \mathcal{F}_\mathcal{E}. \tag{2.4.5}$$

Lemma 2.3.1 implies that M_F defines an operator if and only if F is a function in $L^\infty(\mathcal{E}, \mathcal{Y})$. In other words, if L_F is a Laurent operator, then its symbol F must be a function in $L^\infty(\mathcal{E}, \mathcal{Y})$ and $\|L_F\| = \|F\|_\infty$. On the other hand, if F is a function in $L^\infty(\mathcal{E}, \mathcal{Y})$, then (2.4.5) shows that L_F is a well-defined Laurent operator. So there is a one to one correspondence between the set of all Laurent operators mapping $\ell^2(\mathcal{E})$ into $\ell^2(\mathcal{Y})$ and the set of all functions F in $L^\infty(\mathcal{E}, \mathcal{Y})$. Moreover, $\|L_F\| = \|M_F\| = \|F\|_\infty$. Finally, it is noted that Lemma 2.3.1 along with the fact that M_F and L_F are unitarily equivalent, also show that L_F is an isometry if and only if F is rigid. □

Let A be a function in $L^\infty(\mathcal{V}, \mathcal{Y})$ and B in $L^\infty(\mathcal{E}, \mathcal{V})$. Then AB is a function in $L^\infty(\mathcal{E}, \mathcal{Y})$. By consulting (2.4.5), or by performing a simple calculation we see that $L_A L_B = L_{AB}$. Finally, it is noted that if L_F is a Laurent operator with symbol F, then $L_F^* = L_{F^*}$ is a Laurent operator with symbol $F^* = F(e^{\imath\omega})^*$.

Corollary 2.4.2. *Let $Z_\mathcal{E}$ be the bilateral shift on $L^2(\mathcal{E})$ and $Z_\mathcal{Y}$ be the bilateral shift on $L^2(\mathcal{Y})$. Then an operator M is in $\mathcal{I}(Z_\mathcal{E}, Z_\mathcal{Y})$ if and only if $M = M_F$ where F is a function in $L^\infty(\mathcal{E},\mathcal{Y})$. In this case, $\|M_F\| = \|F\|_\infty$. Finally, M_F is an isometry if and only if F is rigid.*

Proof. If F is a function in $L^\infty(\mathcal{E},\mathcal{Y})$, then clearly, M_F is in $\mathcal{I}(Z_\mathcal{E}, Z_\mathcal{Y})$. Now assume that M is in $\mathcal{I}(Z_\mathcal{E}, Z_\mathcal{Y})$. Recall that $\mathcal{F}_\mathcal{E} U_\mathcal{E} = Z_\mathcal{E}\mathcal{F}_\mathcal{E}$ and $\mathcal{F}_\mathcal{Y} U_\mathcal{Y} = Z_\mathcal{Y}\mathcal{F}_\mathcal{Y}$. As before, $U_\mathcal{E}$ is the bilateral shift on $\ell^2(\mathcal{E})$ and $U_\mathcal{Y}$ the bilateral shift on $\ell^2(\mathcal{Y})$. Then $L = \mathcal{F}_\mathcal{Y}^{-1} M \mathcal{F}_\mathcal{E}$ is an operator in $\mathcal{I}(U_\mathcal{E}, U_\mathcal{Y})$. Hence $L = L_F$ is a Laurent operator where F is a function in $L^\infty(\mathcal{E},\mathcal{Y})$. According to (2.4.5), the operator $M = M_F$. □

Corollary 2.4.3. *Let $S_\mathcal{E}$ be the unilateral shift on $\ell^2_+(\mathcal{E})$ and $U_\mathcal{Y}$ the bilateral shift on $\ell^2(\mathcal{Y})$. Then an operator T is in $\mathcal{I}(S_\mathcal{E}, U_\mathcal{Y})$ if and only if $T = L_F|\ell^2_+(\mathcal{E})$ where L_F is a Laurent operator and F is a function in $L^\infty(\mathcal{E},\mathcal{Y})$. In this case, $\|T\| = \|F\|_\infty$. Finally, T is an isometry if and only if F is rigid.*

Proof. Assume that T is in $\mathcal{I}(S_\mathcal{E}, U_\mathcal{Y})$. The bilateral shift $U_\mathcal{E}$ on $\ell^2(\mathcal{E})$ is a (minimal) unitary extension of $S_\mathcal{E}$. By Proposition 1.5.1, there exists a unique operator L in $\mathcal{I}(U_\mathcal{E}, U_\mathcal{Y})$ extending T. In this case, $T = L|\ell^2_+(\mathcal{E})$ and $\|T\| = \|L\|$. Moreover, L is an isometry if and only if T is an isometry. According to Theorem 2.4.1, the operator L is Laurent, that is, $L = L_F$ where F is a function in $L^\infty(\mathcal{E},\mathcal{Y})$. Moreover, L_F is an isometry if and only if F is rigid. On the other hand, if $T = L_F|\ell^2_+(\mathcal{E})$, then T is in $\mathcal{I}(S_\mathcal{E}, U_\mathcal{Y})$ and L_F is an extension of T. Proposition 1.5.1 shows that L_F is the only extension of T in $\mathcal{I}(U_\mathcal{E}, U_\mathcal{Y})$. Hence $\|T\| = \|L_F\|$. □

By taking the appropriate Fourier transforms in Corollary 2.4.3, or by mimicking the proof of this corollary, we arrive at the following result.

Corollary 2.4.4. *Let $S_\mathcal{E}$ be the unilateral shift on $L^2_+(\mathcal{E})$ and $U_\mathcal{E}$ the bilateral shift on $L^2(\mathcal{Y})$. Then an operator T is in $\mathcal{I}(S_\mathcal{E}, U_\mathcal{E})$ if and only if $T = M_F|L^2_+(\mathcal{E})$ where M_F is a multiplication operator and F is a function in $L^\infty(\mathcal{E},\mathcal{Y})$. In this case, $\|T\| = \|F\|_\infty$. Finally, T is an isometry if and only if F is rigid.*

2.5 Toeplitz Operators and Matrices

We say that T is a *Toeplitz matrix* if T is a block matrix of the form

$$T = \begin{bmatrix} F_0 & F_{-1} & F_{-2} & \cdots \\ F_1 & F_0 & F_{-1} & \cdots \\ F_2 & F_1 & F_0 & \cdots \\ \vdots & \vdots & \vdots & \ddots \end{bmatrix}. \tag{2.5.1}$$

Here $\{F_k\}_{-\infty}^{\infty}$ is a sequence of operators mapping $\mathcal{E}$ into $\mathcal{Y}$. Let F be the function with values in $\mathcal{L}(\mathcal{E}, \mathcal{Y})$ formally defined by

$$F(e^{\imath\omega}) = \sum_{k=-\infty}^{\infty} F_k e^{-\imath\omega k}. \tag{2.5.2}$$

Clearly, F and $\{F_k\}_{-\infty}^{\infty}$ formally determine each other. Motivated by this we say that F is the *symbol* for T. In this case, the Toeplitz matrix T in (2.5.1) is denoted by $T = T_F$.

We say that the Toeplitz matrix T in (2.5.1) is an operator if it defines a bounded linear map from $\ell_+^2(\mathcal{E})$ into $\ell_+^2(\mathcal{Y})$. In other words, T is a *Toeplitz operator* mapping $\ell_+^2(\mathcal{E})$ into $\ell_+^2(\mathcal{Y})$ if T is a bounded linear map and T admits a matrix representation of the form

$$T = \begin{bmatrix} F_0 & F_{-1} & F_{-2} & \cdots \\ F_1 & F_0 & F_{-1} & \cdots \\ F_2 & F_1 & F_0 & \cdots \\ \vdots & \vdots & \vdots & \ddots \end{bmatrix} : \ell_+^2(\mathcal{E}) \to \ell_+^2(\mathcal{Y}) \tag{2.5.3}$$

with respect to the standard orthonormal basis for $\ell_+^2(\mathcal{E})$ and $\ell_+^2(\mathcal{Y})$.

Let $S_{\mathcal{E}}$ be the unilateral shift on $\ell_+^2(\mathcal{E})$ and $S_{\mathcal{Y}}$ be the unilateral shift on $\ell_+^2(\mathcal{Y})$. Let T be an operator mapping $\ell_+^2(\mathcal{E})$ into $\ell_+^2(\mathcal{Y})$. Then we claim that T is a Toeplitz operator if and only if $S_{\mathcal{Y}}^* T S_{\mathcal{E}} = T$. If T is a Toeplitz operator, then a simple calculation involving the matrix representation in (2.5.3) shows that $S_{\mathcal{Y}}^* T S_{\mathcal{E}} = T$. On the other hand, if $S_{\mathcal{Y}}^* T S_{\mathcal{E}} = T$, then the entries T_{jk} of T are determined by

$$\begin{bmatrix} T_{00} & T_{01} & T_{02} & \cdots \\ T_{10} & T_{11} & T_{12} & \cdots \\ T_{20} & T_{21} & T_{22} & \cdots \\ \vdots & \vdots & \vdots & \vdots \end{bmatrix} = T = S_{\mathcal{Y}}^* T S_{\mathcal{E}} = \begin{bmatrix} T_{11} & T_{12} & T_{13} & \cdots \\ T_{21} & T_{22} & T_{23} & \cdots \\ T_{31} & T_{31} & T_{33} & \cdots \\ \vdots & \vdots & \vdots & \vdots \end{bmatrix}.$$

So if $F_j = T_{j0}$ for $j \geq 0$ and $F_{-j} = T_{0j}$ for $j < 0$, then T admits a matrix representation of the form (2.5.3). In other words, T is a Toeplitz operator.

Recall that the bilateral shift $U_{\mathcal{E}}$ on $\ell^2(\mathcal{E})$, respectively $U_{\mathcal{Y}}$ on $\ell^2(\mathcal{Y})$ is the minimal unitary extension of $S_{\mathcal{E}}$, respectively $S_{\mathcal{Y}}$. Moreover, L is a Laurent operator mapping $\ell^2(\mathcal{E})$ into $\ell^2(\mathcal{Y})$ if and only if L is in $\mathcal{I}(U_{\mathcal{E}}, U_{\mathcal{Y}})$. Let T be an operator mapping $\ell_+^2(\mathcal{E})$ into $\ell_+^2(\mathcal{Y})$. According to Proposition 1.5.4, the operator T is Toeplitz if and only if there exists a Laurent operator L such that $T = P_+ L | \ell_+^2(\mathcal{E})$ where P_+ is the orthogonal projection onto $\ell_+^2(\mathcal{Y})$. In this case, the Laurent operator L is uniquely determined by T. Moreover, the symbol for L is a function in F in $L^\infty(\mathcal{E}, \mathcal{Y})$. Since $T = P_+ L | \ell_+^2(\mathcal{E})$, it follows that F is also the symbol for T. In other words, T and L have the same symbol F. Finally, it is noted that $\|T_F\| = \|L_F\| = \|F\|_\infty$. This analysis with Remark 2.3.2, readily yields the following result.

Proposition 2.5.1. *Let T be an operator mapping $\ell^2_+(\mathcal{E})$ into $\ell^2_+(\mathcal{Y})$ and P_+ be the orthogonal projection onto $\ell^2_+(\mathcal{Y})$. Then T is a Toeplitz operator if and only if there exists a Laurent operator L_F with symbol F such that*

$$T = P_+ L_F | \ell^2_+(\mathcal{E}) \quad \textit{where} \quad F \in L^\infty(\mathcal{E}, \mathcal{Y}). \tag{2.5.4}$$

Moreover, in this case the following holds.

(i) *There is only one Laurent operator L_F satisfying $T = P_+ L_F | \ell^2_+(\mathcal{E})$.*

(ii) *The operators $T = T_F$ and L_F have the same symbol F.*

(iii) *The operators T_F and L_F have the same norm: $\|T_F\| = \|L_F\| = \|F\|_\infty$.*

(iv) *Assume that $\mathcal{E} = \mathcal{Y}$. Then T_F is positive if and only if $F(e^{\imath\omega}) \geq 0$ almost everywhere with respect to the Lebesgue measure.*

(v) *Assume that $\mathcal{E} = \mathcal{Y}$. Then $T_F \geq \delta I$ for some scalar $\delta > 0$ if and only if $F(e^{\imath\omega}) \geq \delta I$ almost everywhere with respect to the Lebesgue measure. In this case, T is strictly positive and $\|T_F^{-1}\| = \|F^{-1}\|_\infty$.*

Finally, it is noted that if T_F is a Toeplitz operator, then its adjoint T_F^* is also a Toeplitz operator, and $(T_F)^* = T_{F^*}$. In other words, the symbol for the adjoint of T_F is F^*.

2.6 Toeplitz Matrices and H^∞ Functions

We say that L is a *lower triangular* or *causal Laurent matrix* if L is a Laurent matrix of the form

$$L = \begin{bmatrix} \ddots & \ddots & \vdots & \vdots & \\ \ddots & \Theta_0 & 0 & 0 & \cdots \\ \cdots & \Theta_1 & \boxed{\Theta_0} & 0 & \cdots \\ \cdots & \Theta_2 & \Theta_1 & \Theta_0 & \ddots \\ & \vdots & \vdots & \ddots & \ddots \end{bmatrix}. \tag{2.6.1}$$

Here $\{\Theta_k\}_0^\infty$ is a sequence of operators mapping $\mathcal{E}$ into $\mathcal{Y}$. The box around Θ_0 represents the 0-0 component of the Laurent matrix. All the entries above the main diagonal are zero, and the diagonal entries of the Laurent matrix are the same. Now assume that g is a vector in $\ell^c(\mathcal{E})$. Then Lg is well defined and the n-th component of Lg is given by

$$(Lg)_n = \sum_{j=-\infty}^{n} \Theta_{n-j} g_j \qquad (\oplus_{-\infty}^{\infty} g_j \in \ell^c(\mathcal{E})). \tag{2.6.2}$$

The symbol for this Laurent matrix is the function with values in $\mathcal{L}(\mathcal{E},\mathcal{Y})$ formally defined by

$$\Theta(e^{\imath\omega}) = \sum_{k=0}^{\infty} \Theta_k e^{-\imath\omega k}. \tag{2.6.3}$$

As before, L is also denoted by $L = L_\Theta$. The Laurent matrix L_Θ defines an operator mapping $\ell^2(\mathcal{E})$ into $\ell^2(\mathcal{Y})$ if and only if Θ is a function in $L^\infty(\mathcal{E},\mathcal{Y})$. Recall that a function $\Theta = \sum_0^\infty \Theta_k e^{-\imath\omega k}$ is in $L^\infty(\mathcal{E},\mathcal{Y})$ if and only if $\Theta(z) = \sum_0^\infty z^{-k}\Theta_k$ is in $H^\infty(\mathcal{E},\mathcal{Y})$. In this case, $\Theta(e^{\imath\omega})$ are the boundary values of $\Theta(z)$. So L_Θ defines an operator if and only if $\Theta(z)$ is in $H^\infty(\mathcal{E},\mathcal{Y})$. Finally, $\|L_\Theta\| = \|\Theta\|_\infty$.

We say that T is a *lower triangular* or *causal Toeplitz matrix* if T is a Toeplitz matrix of the form

$$T = \begin{bmatrix} \Theta_0 & 0 & 0 & \cdots \\ \Theta_1 & \Theta_0 & 0 & \cdots \\ \Theta_2 & \Theta_1 & \Theta_0 & \cdots \\ \vdots & \vdots & \vdots & \ddots \end{bmatrix}. \tag{2.6.4}$$

The *symbol* for T is the function formally defined by $\Theta(z) = \sum_0^\infty z^{-k}\Theta_k$. Finally, it is noted that T_Θ defines an operator from $\ell_+^2(\mathcal{E})$ into $\ell_+^2(\mathcal{Y})$ if and only if $\Theta(z)$ is a function in $H^\infty(\mathcal{E},\mathcal{Y})$. In this case, $\|L_\Theta\| = \|T_\Theta\| = \|\Theta\|_\infty$.

We say that T is a *lower triangular* or *causal Toeplitz operator* mapping $\ell_+^2(\mathcal{E})$ into $\ell_+^2(\mathcal{Y})$ if T admits a matrix representation of the form in (2.6.4) with respect to the standard basis for $\ell_+^2(\mathcal{E})$ and $\ell_+^2(\mathcal{Y})$. In this case, T is denoted by T_Θ where Θ is the symbol for T. Let $S_\mathcal{V}$ denote the unilateral shift on $\ell_+^2(\mathcal{V})$. Let T be an operator mapping $\ell_+^2(\mathcal{E})$ into $\ell_+^2(\mathcal{Y})$. We claim that T is a lower triangular Toeplitz operator if and only if T intertwines $S_\mathcal{E}$ with $S_\mathcal{Y}$, that is, $TS_\mathcal{E} = S_\mathcal{Y}T$. If T is a Toeplitz operator, then using the matrix representations for T, $S_\mathcal{E}$ and $S_\mathcal{Y}$, it is easy to verify that T intertwines $S_\mathcal{E}$ with $S_\mathcal{Y}$. On the other hand, if T intertwines $S_\mathcal{E}$ with $S_\mathcal{Y}$, then using $T_{j,k}$ as the j-k entry of T, we obtain

$$\begin{bmatrix} 0 & 0 & 0 & \cdots \\ T_{00} & T_{01} & T_{02} & \cdots \\ T_{10} & T_{11} & T_{12} & \cdots \\ \vdots & \vdots & \vdots & \vdots \end{bmatrix} = S_\mathcal{Y}T = TS_\mathcal{E} = \begin{bmatrix} T_{01} & T_{02} & T_{03} & \cdots \\ T_{11} & T_{12} & T_{13} & \cdots \\ T_{21} & T_{32} & T_{24} & \cdots \\ \vdots & \vdots & \vdots & \vdots \end{bmatrix}.$$

By matching the components of this matrix, we see that $T_{jk} = 0$ if $k > j$ and $T_{j,k} = T_{j-k,0}$ if $j \geq k$. So T is a lower triangular Toeplitz operator with symbol $\Theta = \sum_0^\infty z^{-j}T_{j,0}$. Summing up the previous analysis we readily obtain the following result.

Theorem 2.6.1. *The following holds for Toeplitz matrices.*

(i) *Let T_Θ be the lower triangular Toeplitz matrix given in* (2.6.4), *where $\{\Theta_k\}_0^\infty$ is a sequence of operators in $\mathcal{L}(\mathcal{E},\mathcal{Y})$. Then T_Θ is an operator mapping $\ell_+^2(\mathcal{E})$*

into $\ell^2_+(\mathcal{Y})$ *if and only if* $\Theta(z) = \sum_0^\infty z^{-k}\Theta_k$ *defines a function in* $H^\infty(\mathcal{E},\mathcal{Y})$. *In this case, the operator norm of* T_Θ *equals the* H^∞ *norm of* Θ, *that is,* $\|T_\Theta\| = \|\Theta\|_\infty$.

(ii) *Let* T *be an operator mapping* $\ell^2_+(\mathcal{E})$ *into* $\ell^2_+(\mathcal{Y})$. *Then* T *is a lower triangular Toeplitz matrix if and only if* T *intertwines the unilateral shift* $S_\mathcal{E}$ *on* $\ell^2_+(\mathcal{E})$ *with the unilateral shift* $S_\mathcal{Y}$ *on* $\ell^2_+(\mathcal{Y})$. *In this case,* $T = T_\Theta$ *where* Θ *is a function in* $H^\infty(\mathcal{E},\mathcal{Y})$.

Alternate proof of Part (i). Assume that Θ is a function in $H^\infty(\mathcal{E},\mathcal{Y})$. Then T_Θ can be written as $T_\Theta = L_\Theta|\ell^2_+(\mathcal{E})$. Since Θ is in $H^\infty(\mathcal{E},\mathcal{Y})$, we see that $L_\Theta|\ell^2_+(\mathcal{E})$ is an operator. Hence T_Θ is an operator and $\|T_\Theta\| \le \|L_\Theta\| = \|\Theta\|_\infty$. Proposition 2.5.1 shows that $\|T_\Theta\| = \|L_\Theta\| = \|\Theta\|_\infty$. However, for the moment all we need is the easy part, that is, $\|T_\Theta\| \le \|\Theta\|_\infty$.

Assume that T_Θ is an operator. The first column of T_Θ can be viewed as an operator from $\mathcal{E}$ into $\ell^2_+(\mathcal{Y})$. In particular, $\sum_0^\infty \Theta_k^*\Theta_k < \infty$. This implies that $\Theta = \sum_0^\infty z^{-k}\Theta_k$ defines a function in $H^2(\mathcal{E},\mathcal{Y})$. Fix α in $\mathbb{D}_+ = \{z \in \mathbb{C} : |z| > 1\}$ and consider the vector

$$\varphi_\alpha(f) = \begin{bmatrix} f & (\overline{\alpha})^{-1}f & (\overline{\alpha})^{-2}f & \cdots \end{bmatrix}^{tr}$$

where f is in $\mathcal{Y}$. Then

$$T_\Theta^*\varphi_\alpha(f) = \begin{bmatrix} \Theta_0^* & \Theta_1^* & \Theta_2^* & \cdots \\ 0 & \Theta_0^* & \Theta_1^* & \cdots \\ 0 & 0 & \Theta_0^* & \cdots \\ \vdots & \vdots & \vdots & \ddots \end{bmatrix} \begin{bmatrix} f \\ (\overline{\alpha})^{-1}f \\ (\overline{\alpha})^{-2}f \\ \vdots \end{bmatrix} = \begin{bmatrix} \Theta(\alpha)^*f \\ (\overline{\alpha})^{-1}\Theta(\alpha)^*f \\ (\overline{\alpha})^{-2}\Theta(\alpha)^*f \\ \vdots \end{bmatrix}.$$

Observe that the square of the norm of the vector on the right-hand side of the previous equation is given by

$$\|T_\Theta^*\varphi_\alpha(f)\|^2 = \sum_{k=0}^\infty \frac{1}{|\alpha|^{2k}}\|\Theta(\alpha)^*f\|^2 = \frac{\|\Theta(\alpha)^*f\|^2}{1-|\alpha|^{-2}}.$$

This readily implies that

$$\frac{\|\Theta(\alpha)^*f\|^2}{1-|\alpha|^{-2}} = \|T_\Theta^*\varphi_\alpha(f)\|^2 \le \|T_\Theta^*\|^2\|\varphi_\alpha(f)\|^2 = \frac{\|T_\Theta\|^2\|f\|^2}{1-|\alpha|^{-2}}.$$

Hence for each α in $\mathbb{D}_+$, we have $\|\Theta(\alpha)^*f\| \le \|T_\Theta\|\|f\|$. So $\|\Theta(\alpha)\| \le \|T_\Theta\|$. By taking the supremum over all α in $\mathbb{D}_+$, we see that Θ is a function in $H^\infty(\mathcal{E},\mathcal{Y})$ and $\|\Theta\|_\infty \le \|T_\Theta\|$. Thus Θ is in $H^\infty(\mathcal{E},\mathcal{Y})$. Because Θ is in $H^\infty(\mathcal{E},\mathcal{Y})$, our previous analysis shows that $\|T_\Theta\| \le \|\Theta\|_\infty$. Therefore $\|\Theta\|_\infty = \|T_\Theta\|$. □

Throughout $\ell^c_+(\mathcal{E})$ is the linear manifold consisting of all vectors in $\ell^2_+(\mathcal{E})$ with compact support. Clearly, $\ell^c_+(\mathcal{E})$ is dense in $\ell^2_+(\mathcal{E})$. Let $\mathcal{P}(\mathcal{E})$ be the space of all polynomials in $1/z$ with values in $\mathcal{E}$, that is, $\mathcal{P}(\mathcal{E}) = \{f : f = \sum_0^n z^{-k} f_k\}$ where each f_k is a vector in $\mathcal{E}$ and n is finite. Notice that $\mathcal{P}(\mathcal{E})$ is the Fourier transform of $\ell_+^c(\mathcal{E})$. Now let q be any vector in $\ell_+^c(\mathcal{E})$ and $p = \mathcal{F}_\mathcal{E}^+ q$ the Fourier transform of q. Let T_Θ be the lower triangular Toeplitz matrix given by (2.6.4) where Θ is in $H^2(\mathcal{E}, \mathcal{Y})$. In this case, $T_\Theta q$ is a vector in $\ell_+^2(\mathcal{Y})$, and the Fourier transform of $T_\Theta q$ is the function in $H^2(\mathcal{Y})$ determined by

$$(\mathcal{F}_\mathcal{E}^+ T_\Theta q)(z) = \Theta(z)p(z) \qquad (z \in \mathbb{D}_+).$$

In system theory terminology, $T_\Theta q$ corresponds to convolution in the discrete time domain, while $\Theta(z)p(z)$ corresponds to multiplication in the z or frequency domain. In other words, the equation $(\mathcal{F}_\mathcal{E}^+ T_\Theta q)(z) = \Theta(z)p(z)$ says that the Fourier transform of convolution in the time domain is multiplication in the frequency or z domain.

Let A be a function in $H^\infty(\mathcal{E}, \mathcal{Y})$ and B a function in $H^\infty(\mathcal{V}, \mathcal{E})$. Then it is easy to show that $T_{AB} = T_A T_B$.

We say that Θ is an *inner function* if Θ is a function in $H^\infty(\mathcal{E}, \mathcal{Y})$ and $\Theta(e^{\imath\omega})$ is almost everywhere an isometry mapping $\mathcal{E}$ into $\mathcal{Y}$ with respect to the Lebesgue measure. In other words, an inner function is a rigid function in $H^\infty(\mathcal{E}, \mathcal{Y})$. Finally, it is noted that in the engineering literature, inner functions are called *all-pass transfer functions* or *all-pass filters.*

Proposition 2.6.2. *Let F be a function in $L^\infty(\mathcal{E}, \mathcal{Y})$. Then F is an inner function if and only if T_F is an isometry. Furthermore, F is a unitary constant mapping $\mathcal{E}$ onto $\mathcal{Y}$ if and only if T_F is a unitary operator.*

Proof. Assume that F is an inner function, then F is rigid and by definition F is in $H^\infty(\mathcal{E}, \mathcal{Y})$. Hence L_F is an isometry mapping $\ell^2(\mathcal{E})$ into $\ell^2(\mathcal{Y})$. Because F is in $H^\infty(\mathcal{E}, \mathcal{Y})$, the operator L_F maps $\ell_+^2(\mathcal{E})$ into $\ell_+^2(\mathcal{Y})$, and thus, $T_F = L_F|\ell_+^2(\mathcal{E})$. Therefore T_F is an isometry.

Now assume that T_F is an isometry. Recall that T_F admits a matrix representation of the form

$$T_F = \begin{bmatrix} F_0 & F_{-1} & F_{-2} & \cdots \\ F_1 & F_0 & F_{-1} & \cdots \\ F_2 & F_1 & F_0 & \cdots \\ \vdots & \vdots & \vdots & \ddots \end{bmatrix} : \ell_+^2(\mathcal{E}) \to \ell_+^2(\mathcal{Y}) \qquad (2.6.5)$$

where $F = \sum_{-\infty}^{\infty} e^{-\imath\omega k} F_k$ is the Fourier series expansion for F. Because T_F is an isometry, all the columns of T_F are isometries. Consider the zero and n-th column of T_F. So for any a in $\mathcal{E}$ and integer $n \geq 0$, we have

$$\sum_{k=0}^{\infty} \|F_k a\|^2 = \|a\|^2 = \sum_{k=0}^{\infty} \|F_k a\|^2 + \sum_{k=-n}^{-1} \|F_k a\|^2 \qquad (a \in \mathcal{E}).$$

This readily implies that $F_k = 0$ for all integers $k < 0$. In other words, F is a function in $H^\infty(\mathcal{E}, \mathcal{Y})$. In particular, this implies that $T_F = L_F|\ell^2_+(\mathcal{E})$. Because T_F is an isometry, the operator $L_F|\ell^2_+(\mathcal{E})$ from $\ell^2_+(\mathcal{E})$ into $\ell^2_+(\mathcal{Y})$ is also an isometry. According to Corollary 2.4.3, the function F must be rigid. Therefore F is a rigid function in $H^\infty(\mathcal{E}, \mathcal{Y})$. By definition F is inner.

If F is a unitary constant mapping $\mathcal{E}$ onto $\mathcal{Y}$, then F and F^* are both inner functions. Hence T_F and $T_{F^*} = T_F^*$ are both isometries. In other words, T_F is unitary. On the other hand, if T_F is unitary, then T_F is an isometry, and thus, F is an inner function. Since $T_F^* = T_{F^*}$ is also an isometry, F^* is an inner function. Because F and F^* are both functions in the appropriate $H^\infty(\cdot, \cdot)$ spaces, F is a constant function mapping $\mathcal{E}$ into $\mathcal{Y}$. In other words, F and F^* are both constant inner functions. Therefore F is a unitary constant. □

Let Θ be a function in $H^\infty(\mathcal{E}, \mathcal{Y})$, then M_Θ^+ is the multiplication operator mapping $H^2(\mathcal{E})$ into $H^2(\mathcal{Y})$ defined by

$$(M_\Theta^+ f)(z) = \Theta(z)f(z) \qquad (z \in \mathbb{D}_+ \text{ and } f \in H^2(\mathcal{E})). \tag{2.6.6}$$

Notice that $M_\Theta^+ = M_\Theta|H^2(\mathcal{E})$. Using the Fourier transform, it follows that M_Θ^+ is unitarily equivalent to the lower triangular Toeplitz operator T_Θ, that is,

$$\mathcal{F}_\mathcal{Y}^+ T_\Theta = M_\Theta^+ \mathcal{F}_\mathcal{E}^+ \qquad (\Theta \in H^\infty(\mathcal{E}, \mathcal{Y})). \tag{2.6.7}$$

By taking the Fourier transform, Theorem 2.6.1 in the H^2 setting is given by the following result.

Theorem 2.6.3. *Let T be an operator mapping $H^2(\mathcal{E})$ into $H^2(\mathcal{Y})$. Then $T = M_\Theta^+$ for some Θ in $H^\infty(\mathcal{E}, \mathcal{Y})$ if and only if T intertwines the unilateral shift $S_\mathcal{E}$ on $H^2(\mathcal{E})$ with the unilateral shift $S_\mathcal{Y}$ on $H^2(\mathcal{Y})$. In this case, $\|T\| = \|M_\Theta^+\| = \|\Theta\|_\infty$.*

Let Θ be a function in $H^\infty(\mathcal{E}, \mathcal{Y})$, then Proposition 2.6.2 shows that M_Θ^+ is an isometry if and only if Θ is an inner function. Moreover, M_Θ^+ is unitary if and only if Θ is a unitary constant. Finally, let A be a function in $H^\infty(\mathcal{E}, \mathcal{Y})$ and B a function in $H^\infty(\mathcal{V}, \mathcal{E})$. Then it follows that $M_{AB}^+ = M_A^+ M_B^+$.

2.7 Notes

All the results in this chapter are classical; see [30, 80, 114, 168, 198] for further results on Toeplitz, Laurent and multiplication operators. In fact, some our ideas in this chapter were taken from Sz.-Nagy-Foias [198]. For some further results on Hardy spaces see Duren [76], Granett [106], Hoffman [134] and Koosis [151]. Proposition 2.5.1 is due to Brown-Halmos [45]. Finally, it is noted that we used Hardy spaces corresponding to functions which are analytic in $\mathbb{D}_+$. Many researchers in operator theory use Hardy spaces corresponding to functions which are analytic in the open unit disc $\mathbb{D}$. We used the Hardy space corresponding to functions which are analytic in $\mathbb{D}_+$, because these Hardy spaces are more naturally suited for state space realization theory in engineering and algorithms in Matlab.

Chapter 3

Inner and Outer Functions

In this chapter we will study inner and outer functions. In particular, we will show that any function in $H^2(\mathcal{E}, \mathcal{Y})$ admits a unique inner-outer factorization. Inner-outer factorizations play a fundamental role in many optimization and interpolation problems arising in systems theory and signal processing. In Chapter 4 we will study state space realizations for rational inner and outer functions. Finally, recall that throughout this monograph, we assume that the spaces $\mathcal{E}$ and $\mathcal{Y}$ in $H^2(\mathcal{E}, \mathcal{Y})$, $L^2(\mathcal{E}, \mathcal{Y})$, $H^\infty(\mathcal{E}, \mathcal{Y})$ and $L^\infty(\mathcal{E}, \mathcal{Y})$ are all finite dimensional. Many of our results hold in the case when $\mathcal{E}$ and $\mathcal{Y}$ are separable Hilbert spaces. However, they are finite dimensional in our applications. So we only work in the finite dimensional setting.

3.1 The Beurling-Lax-Halmos Theorem

This section is devoted to the Beurling-Lax-Halmos theorem, which shows that the set of all invariant subspaces for the unilateral shift are characterized by the set of all inner functions. Recall that Θ is an inner function if Θ is in $H^\infty(\mathcal{E}, \mathcal{Y})$ and $\Theta(e^{\imath\omega})$ is almost everywhere an isometry mapping $\mathcal{E}$ into $\mathcal{Y}$ with respect to the Lebesgue measure.

Theorem 3.1.1 (Beurling-Lax-Halmos). *Let S be a unilateral shift on $\ell_+^2(\mathcal{Y})$. Then $\mathcal{M}$ is an invariant subspace for S if and only if $\mathcal{M}$ admits a representation of the form $\mathcal{M} = T_\Theta \ell_+^2(\mathcal{E})$, where Θ is an inner function in $H^\infty(\mathcal{E}, \mathcal{Y})$. Moreover, this representation is unique up to a constant unitary operator on the right. To be precise, if $\mathcal{M} = T_\Psi \ell_+^2(\mathcal{D})$ where Ψ is an inner function in $H^\infty(\mathcal{D}, \mathcal{Y})$, then $\Theta(z) = \Psi(z)\Omega$ where Ω is a constant unitary operator mapping $\mathcal{E}$ onto $\mathcal{D}$.*

Proof. Let $S_\mathcal{V}$ denote the unilateral shift on $\ell_+^2(\mathcal{V})$. If Θ is an inner function in $H^\infty(\mathcal{E}, \mathcal{Y})$, then T_Θ is an isometric lower triangular Toeplitz matrix. Using

$S_\mathcal{Y} T_\Theta = T_\Theta S_\mathcal{E}$ it follows that

$$S_\mathcal{Y} T_\Theta \ell^2_+(\mathcal{E}) = T_\Theta S_\mathcal{E} \ell^2_+(\mathcal{E}) \subseteq T_\Theta \ell^2_+(\mathcal{E}).$$

Therefore $T_\Theta \ell^2_+(\mathcal{E})$ is an invariant subspace for $S_\mathcal{Y}$.

Now assume that $\mathcal{M}$ is an invariant subspace for $S_\mathcal{Y}$. Let U be the isometry on $\mathcal{M}$ defined by $U = S_\mathcal{Y}|\mathcal{M}$. Observe that

$$\bigcap_{n=0}^{\infty} U^n \mathcal{M} = \bigcap_{n=0}^{\infty} S^n_\mathcal{Y} \mathcal{M} \subseteq \bigcap_{n=0}^{\infty} S^n_\mathcal{Y} \ell^2_+(\mathcal{Y}) = \{0\}.$$

In other words, the future space in the Wold decomposition of U is $\{0\}$, that is, U is a unilateral shift. Let $\mathcal{L} = \mathcal{M} \ominus S_\mathcal{Y}\mathcal{M}$ be the cyclic wandering subspace determined by U. In particular, we have $\mathcal{M} = \oplus_0^\infty S^n_\mathcal{Y}\mathcal{L}$. Let Φ be any isometry mapping the space $\mathcal{E}$ into $\ell^2_+(\mathcal{Y})$ such that the range of Φ equals $\mathcal{L}$. By construction $\mathcal{M} = \oplus_0^\infty S^n_\mathcal{Y}\Phi\mathcal{E}$. In particular, $\{S^n_\mathcal{Y}\Phi\}_0^\infty$ forms a set of orthonormal operators mapping $\mathcal{E}$ into $\ell^2_+(\mathcal{Y})$, that is, $(S^k_\mathcal{Y}\Phi)^* S^n_\mathcal{Y}\Phi = \delta_{kn} I$ where δ_{kn} is the Kronecker delta. This readily implies that

$$T = \begin{bmatrix} \Phi & S_\mathcal{Y}\Phi & S^2_\mathcal{Y}\Phi & S^3_\mathcal{Y}\Phi & \cdots \end{bmatrix} : \ell^2_+(\mathcal{E}) \to \ell^2_+(\mathcal{Y})$$

is an isometry whose range equals $\mathcal{M}$. Notice that $S_\mathcal{Y} T = T S_\mathcal{E}$. So T is a lower triangular Toeplitz operator. Since this T is also an isometry, there exists a unique inner function Θ in $H^\infty(\mathcal{E}, \mathcal{Y})$ such that $T = T_\Theta$. In fact, Θ is the Fourier transform of Φ, that is,

$$\Theta(z) = (\mathcal{F}^+_\mathcal{Y}\Phi)(z) \qquad (z \in \mathbb{D}_+). \tag{3.1.1}$$

Therefore $\mathcal{M}$ equals the range of T_Θ.

Assume that $\mathcal{M} = T_\Theta \ell^2_+(\mathcal{E}) = T_\Psi \ell^2_+(\mathcal{D})$, where Θ is an inner function in $H^\infty(\mathcal{E}, \mathcal{Y})$ and Ψ is an inner function in $H^\infty(\mathcal{D}, \mathcal{Y})$. Because T_Θ and T_Ψ are both isometries with the same range $\mathcal{M}$, it follows that $W = T^*_\Psi T_\Theta$ is a unitary operator mapping $\ell^2_+(\mathcal{E})$ onto $\ell^2_+(\mathcal{D})$. (If $V_1 : \mathcal{V}_1 \to \mathcal{K}$ and $V_2 : \mathcal{V}_2 \to \mathcal{K}$ are two isometries with the same range $\mathcal{M}$, then $V_1 : \mathcal{V}_1 \to \mathcal{M}$ and $V_2 : \mathcal{V}_2 \to \mathcal{M}$ can be viewed as unitary operators whose ranges are onto $\mathcal{M}$. Hence $V_2^* V_1$ is unitary.) We claim that $T_\Psi W = T_\Theta$. Because $\mathcal{M}$ is the range of an isometry T_Ψ, this implies that $P_\mathcal{M} = T_\Psi T^*_\Psi$. Using this, we obtain

$$T_\Psi W = T_\Psi T^*_\Psi T_\Theta = P_\mathcal{M} T_\Theta = T_\Theta.$$

Hence $T_\Psi W = T_\Theta$. Using this along with the fact that both T_Θ and T_Ψ intertwine the appropriate unilateral shifts, we obtain

$$T_\Psi S_\mathcal{D} W = S_\mathcal{Y} T_\Psi W = S_\mathcal{Y} T_\Theta = T_\Theta S_\mathcal{E} = T_\Psi W S_\mathcal{E}.$$

Thus $T_\Psi S_\mathcal{D} W = T_\Psi W S_\mathcal{E}$. Because T_Ψ is one to one, $W S_\mathcal{E} = S_\mathcal{D} W$. Since W is a unitary operator intertwining $S_\mathcal{E}$ with $S_\mathcal{D}$, we see that $W = T_\Omega$, where Ω is a constant unitary operator mapping $\mathcal{E}$ onto $\mathcal{D}$; see Proposition 2.6.2. In other words, $T_\Theta = T_\Psi W = T_\Psi T_\Omega$. Therefore $\Theta(z) = \Psi(z)\Omega$ for all z in $\mathbb{D}_+$. $\square$

Remark 3.1.2. As before, let $\mathcal{M}$ be an invariant subspace for the unilateral shift $S_{\mathcal{Y}}$ on $\ell_+^2(\mathcal{Y})$. Let Φ be any isometry mapping a space $\mathcal{E}$ into $\ell_+^2(\mathcal{Y})$ such that the range of Φ equals $\mathcal{M} \ominus S_{\mathcal{Y}}\mathcal{M}$. The proof of the Beurling-Lax-Halmos Theorem shows that $\Theta(z) = (\mathcal{F}_{\mathcal{Y}}^+\Phi)(z)$ is an inner function in $H^\infty(\mathcal{E},\mathcal{Y})$ satisfying $\mathcal{M} = T_\Theta \ell_+^2(\mathcal{Y})$.

Recall that if Θ is any function in $H^\infty(\mathcal{E},\mathcal{Y})$, then the Fourier transform $\mathcal{F}_{\mathcal{Y}}^+ T_\Theta = M_\Theta^+ \mathcal{F}_{\mathcal{E}}^+$ where M_Θ^+ is the multiplication operator mapping $H^2(\mathcal{E})$ into $H^2(\mathcal{Y})$ determined by Θ; see (2.6.6) and (2.6.7). By taking the appropriate Fourier transforms in Theorem 3.1.1, we obtain the following H^2 version of the Beurling-Lax-Halmos theorem.

Corollary 3.1.3. *Let S be a unilateral shift on $H^2(\mathcal{Y})$, then $\mathcal{M}$ is an invariant subspace for S if and only if $\mathcal{M} = \Theta H^2(\mathcal{E})$ where Θ is an inner function in $H^\infty(\mathcal{E},\mathcal{Y})$. Moreover, this representation is unique up to a constant unitary operator on the right. To be precise, if $\mathcal{M} = \Psi H^2(\mathcal{D})$ where Ψ is an inner function in $H^\infty(\mathcal{D},\mathcal{Y})$, then $\Theta(z) = \Psi(z)\Omega$ where Ω is a constant unitary operator mapping $\mathcal{E}$ onto $\mathcal{D}$.*

3.1.1 The invariant subspaces for the backward shift

Let Θ be an inner function in $H^\infty(\mathcal{E},\mathcal{Y})$. Then $\mathcal{H}(\Theta)$ is the subspace defined by

$$\mathcal{H}(\Theta) = H^2(\mathcal{Y}) \ominus \Theta H^2(\mathcal{E}). \tag{3.1.2}$$

According to the Beurling-Lax-Halmos Theorem, $\Theta H^2(\mathcal{E})$ is an invariant subspace for the forward shift S on $H^2(\mathcal{Y})$. So $\mathcal{H}(\Theta)$ is an invariant subspace for the backward shift S^*. (Recall that $\mathcal{M}$ is an invariant subspace for an operator A on $\mathcal{X}$ if and only if $\mathcal{X} \ominus \mathcal{M}$ is an invariant subspace for A^*.) Moreover, $\mathcal{H}$ is an invariant subspace for the backward shift if and only if $\mathcal{H} = \mathcal{H}(\Theta)$ for some inner function Θ in $H^\infty(\mathcal{E},\mathcal{Y})$. Furthermore, this inner function Θ is unique up to a constant unitary operator on the right.

Let Θ be a function in $H^\infty(\mathcal{E},\mathcal{Y})$ and Ψ a function in $H^\infty(\mathcal{D},\mathcal{Y})$. Then we say that Ψ is a *left divisor of* Θ if there exists a function Φ in $H^\infty(\mathcal{E},\mathcal{D})$ such that $\Theta(z) = \Psi(z)\Phi(z)$ for all z in $\mathbb{D}_+$. In this case, if both Θ and Ψ are inner functions, then Φ must also be an inner function. To see this observe that for all a in $\mathcal{E}$, we have

$$\|\Phi(e^{\imath\omega})a\| = \|\Psi(e^{\imath\omega})\Phi(e^{\imath\omega})a\| = \|\Theta(e^{\imath\omega})\| = \|a\|$$

almost everywhere with respect to the Lebesgue measure. So Φ is a function in $H^\infty(\mathcal{E},\mathcal{D})$ whose boundary values are almost everywhere an isometry. Hence Φ is an inner function.

Assume that Ψ is an inner function in $H^\infty(\mathcal{D},\mathcal{Y})$ and Φ is an inner function in $H^\infty(\mathcal{E},\mathcal{D})$. Then we claim that

$$\mathcal{H}(\Psi\Phi) = \mathcal{H}(\Psi) \oplus \Psi\mathcal{H}(\Phi). \tag{3.1.3}$$

To see this observe that

$$\begin{aligned}\mathcal{H}(\Psi\Phi) &= H^2(\mathcal{Y}) \ominus \Psi\Phi H^2(\mathcal{E}) = H^2(\mathcal{Y}) \ominus \Psi\left(H^2(\mathcal{D}) \ominus \mathcal{H}(\Phi)\right)\\ &= H^2(\mathcal{Y}) \ominus \left(\Psi H^2(\mathcal{D}) \ominus \Psi\mathcal{H}(\Phi)\right) = \left(H^2(\mathcal{Y}) \ominus \Psi H^2(\mathcal{D})\right) \oplus \Psi\mathcal{H}(\Phi)\\ &= \mathcal{H}(\Psi) \oplus \Psi\mathcal{H}(\Phi).\end{aligned}$$

Therefore (3.1.3) holds.

Let Θ be an inner function in $H^\infty(\mathcal{E},\mathcal{Y})$ and Ψ an inner function in $H^\infty(\mathcal{D},\mathcal{Y})$. We claim that Ψ is a left divisor of Θ if and only if

$$\mathcal{H}(\Psi) \subseteq \mathcal{H}(\Theta). \tag{3.1.4}$$

To verify this, assume that $\Theta = \Psi\Phi$ where Φ is an inner function. Then (3.1.3) shows that $\mathcal{H}(\Psi) \subseteq \mathcal{H}(\Theta)$. On the other hand, assume that $\mathcal{H}(\Psi) \subseteq \mathcal{H}(\Theta)$. By taking the orthogonal complement of the subspaces $\mathcal{H}(\Psi)$ and $\mathcal{H}(\Theta)$, we see that $\Theta H^2(\mathcal{E}) \subseteq \Psi H^2(\mathcal{D})$, or equivalently, by employing the inverse Fourier transform $T_\Theta \ell^2_+(\mathcal{E}) \subseteq T_\Psi \ell^2_+(\mathcal{D})$. Since Θ and Ψ are inner, both T_Θ and T_Ψ are isometries. So the range of the isometry T_Θ is contained in the range of the isometry T_Ψ. Thus $T = (T_\Psi)^* T_\Theta$ is an isometry mapping $\ell^2_+(\mathcal{E})$ into $\ell^2_+(\mathcal{D})$. (If $V_1 : \mathcal{V}_1 \to \mathcal{K}$ and $V_2 : \mathcal{V}_2 \to \mathcal{K}$ are two isometries, then $V_1 : \mathcal{V}_1 \to V_1\mathcal{V}_1$ and $V_2 : \mathcal{V}_2 \to V_2\mathcal{V}_2$ can be viewed as unitary operators. If $V_1\mathcal{V}_1 \subseteq V_2\mathcal{V}_2$, then $V_2^*|V_1\mathcal{V}_1$ is an isometry from $V_1\mathcal{V}_1$ into $\mathcal{V}_2$. Hence $V_2^* V_1$ is an isometry.) Because $T_\Psi T_\Psi^*$ is the orthogonal projection onto the range of T_Ψ and the range of T_Θ is contained in the range of T_Ψ, we see that $T_\Psi T = T_\Theta$. Let $S_\mathcal{L}$ denote the unilateral shift on $\ell^2_+(\mathcal{L})$. Then we have

$$T_\Psi T S_\mathcal{E} = T_\Theta S_\mathcal{E} = S_\mathcal{Y} T_\Theta = S_\mathcal{Y} T_\Psi T = T_\Psi S_\mathcal{D} T.$$

Hence $T_\Psi(TS_\mathcal{E} - S_\mathcal{D}T) = 0$. Since T_Ψ is an isometry, $TS_\mathcal{E} = S_\mathcal{D}T$. In other words, T is an isometry which intertwines the unilateral shift $S_\mathcal{E}$ with $S_\mathcal{D}$. This implies that $T = T_\Phi$ where Φ is an inner function in $H^\infty(\mathcal{E},\mathcal{D})$. The identity $T_\Theta = T_\Psi T_\Phi = T_{\Psi\Phi}$, shows that $\Theta = \Psi\Phi$, which proves our claim.

Let Θ and Ψ be two inner functions acting on the appropriate $H^\infty(\cdot,\cdot)$ spaces. By taking the appropriate orthogonal complements in the Beurling-Lax-Halmos Theorem, we see that $\mathcal{H}(\Theta) = \mathcal{H}(\Psi)$ if and only if Θ equals Ψ up to a constant unitary operator on the right.

Let Θ be an inner function in $H^\infty(\mathcal{E},\mathcal{Y})$ and Ψ an inner function in $H^\infty(\mathcal{D},\mathcal{Y})$. We say that Ω is a *common left inner divisor* to Θ and Ψ if Ω is a left inner divisor for both Θ and Ψ. Moreover, Υ is the *greatest common left inner divisor* to Θ and Ψ if

- Υ is a left inner divisor for Θ and Ψ;
- if Ω is a left inner divisor for Θ and Ψ, then Ω is a left inner divisor of Υ.

The greatest common left inner divisor is unique up to a unitary constant on the right. Notice that $\mathcal{H}(\Theta) \cap \mathcal{H}(\Psi)$ is an invariant subspace for the backward shift

$S_{\mathcal{Y}}^*$. Hence there exists an inner function Υ such that $\mathcal{H}(\Theta) \cap \mathcal{H}(\Psi) = \mathcal{H}(\Upsilon)$. This inner function Υ is the greatest common left inner divisor for Θ and Ψ.

Using the fact that $\mathcal{H}(\Upsilon)$ is a subspace for both $\mathcal{H}(\Theta)$ and $\mathcal{H}(\Psi)$, it follows that Υ is a common left inner divisor for Θ and Ψ. Let Ω be any common left inner divisor for both Θ and Ψ. Because $\mathcal{H}(\Omega)$ is a subspace for both $\mathcal{H}(\Theta)$ and $\mathcal{H}(\Psi)$, we obtain

$$\mathcal{H}(\Omega) \subseteq \mathcal{H}(\Theta) \cap \mathcal{H}(\Psi) = \mathcal{H}(\Upsilon).$$

Since $\mathcal{H}(\Omega) \subseteq \mathcal{H}(\Upsilon)$, it follows that Ω is a left inner divisor for Υ. In other words, Υ the greatest common left inner divisor for Θ and Ψ. If Υ_1 is another greatest common left inner divisor for Θ and Ψ, then Υ is a left divisor for Υ_1 and Υ_1 is a left divisor for Υ. In other words, $\mathcal{H}(\Upsilon) = \mathcal{H}(\Upsilon_1)$. Therefore Υ equals Υ_1 up to a unitary constant on the right.

We say that Θ and Ψ are *prime on the left* if the only common left inner divisor between Θ and Ψ is a unitary constant. We claim that Θ and Ψ are prime on the left if and only if $\mathcal{H}(\Theta) \cap \mathcal{H}(\Psi) = \{0\}$. If Θ and Ψ are prime on the left, then the only common left inner divisor is a unitary constant. So the greatest common left inner divisor Υ is a unitary constant. In this case, $\mathcal{H}(\Upsilon) = \{0\}$. Hence $\mathcal{H}(\Theta) \cap \mathcal{H}(\Psi) = \{0\}$. On the other hand, if $\mathcal{H}(\Theta) \cap \mathcal{H}(\Psi) = \{0\}$, then $\mathcal{H}(\Upsilon) = \{0\}$ where Υ is the greatest common left inner divisor. Using the fact that $\mathcal{H}(I) = \{0\}$, it follows that Υ equals the identity I up to a unitary constant on the right. In other words, Υ is a unitary constant, and Θ and Ψ are prime on the left.

Using the fact that $\Theta H^2(\mathcal{E})$ equals the orthogonal complement of $\mathcal{H}(\Theta)$, we obtain the following result.

Remark 3.1.4. Let Θ in $H^\infty(\mathcal{E}, \mathcal{Y})$ and Ψ in $H^\infty(\mathcal{D}, \mathcal{Y})$ be inner functions. Clearly, $\Theta H^2(\mathcal{E})$ and $\Psi H^2(\mathcal{D})$ are invariant subspaces for the unilateral shift on $H^2(\mathcal{Y})$. Then $\Theta H^2(\mathcal{E}) \subseteq \Psi H^2(\mathcal{D})$ if and only if Ψ is a left inner divisor of Θ. Moreover, $\Theta H^2(\mathcal{E}) = \Psi H^2(\mathcal{D})$ if and only if Θ equals Ψ up to a constant unitary operator on the right. Finally, $H^2(\mathcal{Y}) = \Theta H^2(\mathcal{E}) \bigvee \Psi H^2(\mathcal{D})$ if and only if Θ and Ψ are prime on the left.

3.2 Inner-Outer Factorizations

In this section, we will show that any function Θ in $H^2(\mathcal{E}, \mathcal{Y})$ admits a unique inner-outer factorization.

Let $\Theta(z) = \sum_0^\infty z^{-k}\Theta_k$ be the Taylor series expansion for a function Θ in $H^2(\mathcal{E}, \mathcal{Y})$. Let T_Θ be the lower triangular Toeplitz matrix determined by Θ, that is,

$$T_\Theta = \begin{bmatrix} \Theta_0 & 0 & 0 & \cdots \\ \Theta_1 & \Theta_0 & 0 & \cdots \\ \Theta_2 & \Theta_1 & \Theta_0 & \cdots \\ \vdots & \vdots & \vdots & \ddots \end{bmatrix}. \tag{3.2.1}$$

Because Θ is in $H^2(\mathcal{E},\mathcal{Y})$, all the columns of T_Θ can be viewed as operators mapping $\mathcal{E}$ into $\ell^2_+(\mathcal{Y})$. In other words, T_Θ is a well-defined linear map from $\ell^c_+(\mathcal{E})$ into $\ell^2_+(\mathcal{Y})$. Let W be any linear map from $\ell^c_+(\mathcal{E})$ into $\ell^2_+(\mathcal{Y})$, that is, assume that W is linear and $Wx \in \ell^2_+(\mathcal{Y})$ for all x in $\ell^c_+(\mathcal{E})$. Then $W = T_\Theta$ for some function Θ in $H^2(\mathcal{E},\mathcal{Y})$ if and only if $S_\mathcal{Y} Wx = WS_\mathcal{E} x$ for all x in $\ell^c_+(\mathcal{E})$. In this case, Θ is the Fourier transform of the first column of W, that is,

$$\Theta = \mathcal{F}^+_\mathcal{Y} W \begin{bmatrix} I & 0 & 0 & \cdots \end{bmatrix}^{tr}.$$

As expected, $S_\mathcal{V}$ is the unilateral shift on $\ell^2_+(\mathcal{V})$.

We say that Θ is an *outer function* if Θ is a function in $H^2(\mathcal{E},\mathcal{Y})$ and $T_\Theta \ell^c_+(\mathcal{E})$ is dense in $\ell^2_+(\mathcal{Y})$. By taking the Fourier transform, we see that a function Θ in $H^2(\mathcal{E},\mathcal{Y})$ is outer if and only if $\overline{\Theta\mathcal{P}(\mathcal{E})} = H^2(\mathcal{Y})$. (The set of all polynomials $\sum_{n\geq 0} a_n z^{-n}$ in z^{-1} with values in $\mathcal{E}$ is denoted by $\mathcal{P}(\mathcal{E})$.) In engineering terminology, an outer function is called a *minimum phase transfer function* or *minimum phase filter*.

A function Θ in $H^2(\mathcal{E},\mathcal{Y})$ is both inner and outer if and only if Θ is a unitary constant mapping $\mathcal{E}$ onto $\mathcal{Y}$. If Θ is inner and outer, then T_Θ is a unitary operator. According to Proposition 2.6.2, the function Θ must be a unitary constant. On the other hand, if Θ is a unitary constant, then T_Θ is a unitary operator. Hence Θ is both inner and outer.

Let Θ be a function in $H^2(\mathcal{E},\mathcal{Y})$. Then we say that $\Theta = \Theta_i\Theta_o$ is an *inner-outer factorization* of Θ if Θ_i is an inner function in some $H^\infty(\mathcal{V},\mathcal{Y})$ space and Θ_o is an outer function in $H^2(\mathcal{E},\mathcal{V})$. In this case, Θ_i is called the *inner part* of Θ, and Θ_o is called the *outer part* of Θ. The following result shows that any function in $H^2(\mathcal{E},\mathcal{Y})$ admits a unique inner-outer factorization.

Theorem 3.2.1. *Let Θ be a function in $H^2(\mathcal{E},\mathcal{Y})$. Then Θ admits a unique factorization of the form $\Theta = \Theta_i\Theta_o$ where Θ_o is an outer function in $H^2(\mathcal{E},\mathcal{V})$ and Θ_i is an inner function in $H^\infty(\mathcal{V},\mathcal{Y})$. By unique we mean that if $\Theta = \Psi_i\Psi_o$ where Ψ_o is an outer function in $H^2(\mathcal{E},\mathcal{D})$ and Ψ_i is an inner function in $H^\infty(\mathcal{D},\mathcal{Y})$, then there exists a constant unitary operator Φ mapping $\mathcal{V}$ onto $\mathcal{D}$ such that $\Psi_o = \Phi\Theta_o$ and $\Psi_i\Phi = \Theta_i$.*

Proof. Set $\mathcal{M} = \overline{T_\Theta \ell^c_+(\mathcal{E})}$. Using the fact that T_Θ intertwines the forward shifts $S_\mathcal{E}$ with $S_\mathcal{Y}$, we obtain

$$S_\mathcal{Y} T_\Theta \ell^c_+(\mathcal{E}) = T_\Theta S_\mathcal{E} \ell^c_+(\mathcal{E}) \subseteq T_\Theta \ell^c_+(\mathcal{E}).$$

In other words, $\mathcal{M}$ is an invariant subspace for $S_\mathcal{Y}$. By the Beurling-Lax-Halmos theorem, $\mathcal{M} = T_{\Theta_i}\ell^2_+(\mathcal{V})$ where Θ_i is an inner function in $H^\infty(\mathcal{V},\mathcal{Y})$. Let W be the linear map from $\ell^c_+(\mathcal{E})$ into $\ell^2_+(\mathcal{V})$ defined by $W = T^*_{\Theta_i} T_\Theta$. Because $T_\Theta \ell^c_+(\mathcal{E})$ is dense in $\mathcal{M}$ and T_{Θ_i} is an isometry whose range equals $\mathcal{M}$, it follow that $W\ell^c_+(\mathcal{E})$ is dense in $\ell^2_+(\mathcal{V})$. Using $P_\mathcal{M} = T_{\Theta_i} T^*_{\Theta_i}$, we arrive at $T_{\Theta_i} Wx = T_\Theta x$ for x in $\ell^c_+(\mathcal{E})$. Moreover, we have

$$T_{\Theta_i} S_\mathcal{V} Wx = S_\mathcal{Y} T_{\Theta_i} Wx = S_\mathcal{Y} T_\Theta x = T_\Theta S_\mathcal{E} x = T_{\Theta_i} WS_\mathcal{E} x.$$

Thus $T_{\Theta_i} S_\mathcal{V} W x = T_{\Theta_i} W S_\mathcal{E} x$. Because T_{Θ_i} is one to one, we have $S_\mathcal{V} W x = W S_\mathcal{E} x$. Thus $W = T_{\Theta_o}$ where Θ_o is a function in $H^2(\mathcal{E}, \mathcal{V})$. Since $W \ell_+^c(\mathcal{E})$ is dense in $\ell_+^2(\mathcal{V})$, we see that Θ_o is an outer function. Finally, $T_{\Theta_i \Theta_o} = T_{\Theta_i} T_{\Theta_o} = T_\Theta$ implies that Θ admits an inner-outer factorization $\Theta(z) = \Theta_i(z)\Theta_o(z)$ for all z in $\mathbb{D}_+$.

Now assume that $\Theta = \Psi_i \Psi_o$ where Ψ_o is an outer function in $H^2(\mathcal{E}, \mathcal{D})$ and Ψ_i is an inner function in $H^\infty(\mathcal{D}, \mathcal{Y})$. Then we have

$$\Theta_i H^2(\mathcal{V}) = \overline{\Theta_i \Theta_o \mathcal{P}(\mathcal{E})} = \overline{\Theta \mathcal{P}(\mathcal{E})} = \overline{\Psi_i \Psi_o \mathcal{P}(\mathcal{E})} = \Psi_i H^2(\mathcal{D}).$$

In other words, $\Theta_i H^2(\mathcal{V}) = \Psi_i H^2(\mathcal{D})$. According to the Beurling-Lax-Halmos Theorem, $\Theta_i = \Psi_i \Phi$ where Φ is a constant unitary operator mapping $\mathcal{V}$ onto $\mathcal{D}$. This readily implies that

$$\Psi_i \Psi_o = \Theta = \Theta_i \Theta_o = \Psi_i \Phi \Theta_o.$$

In other words, $\Psi_i(\Psi_o - \Phi\Theta_o) = 0$. Because Ψ_i is almost everywhere an isometry on the boundary, $\Psi_o = \Phi\Theta_o$. □

Remark 3.2.2. Let Θ be a function in $H^2(\mathcal{E}, \mathcal{Y})$. Notice that $\mathcal{M} = \overline{T_\Theta \ell_+^c(\mathcal{E})}$ is an invariant subspace for the unilateral shift $S_\mathcal{Y}$ on $\ell_+^2(\mathcal{Y})$. Let Φ be any isometry mapping a space $\mathcal{E}$ into $\ell_+^2(\mathcal{Y})$ such that the range of Φ equals $\mathcal{M} \ominus S_\mathcal{Y}\mathcal{M}$. Let Θ_i be the Fourier transform of Φ, that is,

$$\Theta_i(z) = (\mathcal{F}_\mathcal{Y}^+ \Phi)(z) \qquad (z \in \mathbb{D}_+). \tag{3.2.2}$$

Then Remark 3.1.2 and the proof of Theorem 3.2.1 shows that Θ_i is the inner part of Θ. The outer part Θ_o of Θ is given by $\Theta_o(e^{\imath\omega}) = \Theta_i(e^{\imath\omega})^*\Theta(e^{\imath\omega})$.

In some future chapters, we will present algorithms to compute the inner-outer factorization for matrix-valued rational functions.

Remark 3.2.3. Assume that Θ is an outer function in $H^2(\mathcal{E}, \mathcal{Y})$ and fix α in $\mathbb{D}_+$. Then the range of $\Theta(\alpha)$ equals $\mathcal{Y}$, that is, $\Theta(\alpha)\mathcal{E} = \mathcal{Y}$. In particular, the dimension of $\mathcal{Y}$ is less than or equal to the dimension of $\mathcal{E}$. If $\mathcal{E}$ and $\mathcal{Y}$ are of the same dimension, then $\Theta(\alpha)$ is invertible for each fixed α in $\mathbb{D}_+$.

To verify our claim, first let us show that the function

$$\varphi_{\alpha y}(z) = \frac{\overline{\alpha} z}{\overline{\alpha} z - 1} y \qquad (\alpha, z \in \mathbb{D}_+ \text{ and } y \in \mathcal{Y}) \tag{3.2.3}$$

has the reproducing property

$$(f, \varphi_{\alpha y}) = (f(\alpha), y) \qquad (y \in \mathcal{Y} \text{ and } f \in H^2(\mathcal{Y})). \tag{3.2.4}$$

Let $f(z) = \sum_0^\infty f_n z^{-n}$ be the Taylor's series expansion for a function f in $H^2(\mathcal{Y})$. Thus

$$(f, \overline{\alpha} z(\overline{\alpha} z - 1)^{-1} y) = (f, \left(1 - \frac{1}{\overline{\alpha} z}\right)^{-1} y) = (\sum_{n=0}^\infty f_n z^{-n}, \sum_{n=0}^\infty (\overline{\alpha} z)^{-n} y)$$
$$= \sum_{n=0}^\infty (f_n \alpha^{-n}, y) = (f(\alpha), y).$$

This yields the reproducing property of $\overline{\alpha}z(\overline{\alpha}z-1)^{-1}y$ displayed in (3.2.4).

Now let us show that, for any fixed α in $\mathbb{D}_+$, the range of $\Theta(\alpha)$ equals $\mathcal{Y}$. Assume that there exists a vector y in $\mathcal{Y}$ which is orthogonal to $\Theta(\alpha)\mathcal{E}$. Then for any polynomial h in $\mathcal{P}(\mathcal{E})$, the reproducing property of $\overline{\alpha}z(\overline{\alpha}z-1)^{-1}y$ yields

$$(\Theta h, \varphi_{\alpha y}) = (\Theta(\alpha)h(\alpha), y) = 0 \qquad (h \in \mathcal{P}(\mathcal{E})).$$

Since Θ is an outer function in $H^2(\mathcal{E},\mathcal{Y})$, we must have $\overline{\Theta\mathcal{P}(\mathcal{E})} = H^2(\mathcal{Y})$, and thus, the vector y must be zero. Therefore the range of $\Theta(\alpha)$ equals $\mathcal{Y}$. This verifies our claim.

Remark 3.2.4. Assume that Θ is an outer function in $H^2(\mathcal{E},\mathcal{Y})$. Then $\Theta(e^{\imath\omega})\mathcal{E} = \mathcal{Y}$ almost everywhere with respect to the Lebesgue measure. In particular, if $\mathcal{E}$ and $\mathcal{Y}$ are of the same dimension, then $\Theta(e^{\imath\omega})$ is almost everywhere invertible.

Because Θ is outer we have $H^2(\mathcal{Y}) = \overline{\Theta\mathcal{P}(\mathcal{E})}$. So given any y in $\mathcal{Y}$, there exists a sequence of polynomials p_n such that Θp_n converges to y in $L^2(\mathcal{Y})$. (Here $H^2(\mathcal{Y})$ is viewed as a subspace of $L^2(\mathcal{Y})$.) This implies that $\Theta(e^{\imath\omega})p_n(e^{\imath\omega})$ converges to y almost everywhere as n tends to infinity. In other words, $y \in \Theta(e^{\imath\omega})\mathcal{E}$ except on a set of measure zero. Now let y run through a countable dense set $\mathcal{Y}_o$ in $\mathcal{Y}$. Since the countable union of sets of measure zero is also a set of measure zero, it follows that $\mathcal{Y}_o$ is contained in $\Theta(e^{\imath\omega})\mathcal{E}$ except on a set of measure zero. Therefore $\Theta(e^{\imath\omega})\mathcal{E} = \mathcal{Y}$ almost everywhere.

Co-inner and co-outer functions. Let Θ be a function which is analytic in $\mathbb{D}_+$ with values in $\mathcal{L}(\mathcal{E},\mathcal{Y})$. Then $\widetilde{\Theta}$ is the function defined by $\widetilde{\Theta}(z) = \Theta(\overline{z})^*$ where z is in $\mathbb{D}_+$. If $\Theta(z) = \sum_0^\infty z^{-n}\Theta_n$ is the Taylor series expansion for Θ, then $\widetilde{\Theta}(z) = \sum_0^\infty z^{-n}\Theta_n^*$ is the Taylor series expansion for $\widetilde{\Theta}$. Notice that $\widetilde{\Theta}$ is analytic in $\mathbb{D}_+$ with values in $\mathcal{L}(\mathcal{Y},\mathcal{E})$. Observe that $\widetilde{\widetilde{\Theta}} = \Theta$. Furthermore, $\widetilde{(AB)} = \widetilde{B}\widetilde{A}$ where A and B are analytic functions in $\mathbb{D}_+$ acting between the appropriate spaces. We say that Θ is *co-inner* if $\widetilde{\Theta}$ is inner. If Θ is inner and co-inner, then Θ is called *inner from both sides* or *a two-sided inner function.* Notice that Θ is inner from both sides if and only if Θ is an inner function and its boundary values $\Theta(e^{\imath\omega})$ are almost everywhere unitary operators with respect to the Lebesgue measure. As expected, we say that Θ is *co-outer* if $\widetilde{\Theta}$ is an outer function. We say that $\Theta = \Theta_{co}\Theta_{ci}$ is a co-outer co-inner factorization of Θ if Θ_{co} is co-outer and Θ_{ci} is co-inner. So $\Theta = \Theta_{co}\Theta_{ci}$ is a co-outer co-inner factorization of Θ if and only if $\widetilde{\Theta} = \widetilde{\Theta}_{ci}\widetilde{\Theta}_{co}$ is an inner-outer factorization of $\widetilde{\Theta}$. Since the inner-outer factorization, is unique up to a constant unitary operator, the co-outer co-inner factorization is also unique up to a constant unitary operator. Finally, it is noted that any function in $H^\infty(\mathcal{E},\mathcal{Y})$ admits a co-outer co-inner factorization. To see this, let $\widetilde{\Theta} = \Theta_i\Theta_o$ be the inner-outer factorization for $\widetilde{\Theta}$ where Θ_i is inner and Θ_o is outer. Then $\Theta = \widetilde{\Theta}_o\widetilde{\Theta}_i$ is the co-outer co-inner factorization of Θ.

3.3 Invertible Outer Functions

We say that Θ is an *invertible outer function* if Θ is a function in $H^\infty(\mathcal{E},\mathcal{Y})$, the inverse $\Theta(z)^{-1}$ exists for all z in $\mathbb{D}_+$, and Θ^{-1} is a function in $H^\infty(\mathcal{Y},\mathcal{E})$. For example, $(2z+1)/(3z+1)$ is an invertible outer function. The function $(1+z)/z$ is outer and not an invertible outer function.

Let Θ be a function in $H^\infty(\mathcal{E},\mathcal{Y})$. Then T_Θ is an invertible operator mapping $\ell^2_+(\mathcal{E})$ onto $\ell^2_+(\mathcal{Y})$ if and only if Θ is an invertible outer function. In this case, $(T_\Theta)^{-1} = T_{\Theta^{-1}}$.

Recall that if A is in $H^\infty(\mathcal{V},\mathcal{Y})$ and B is in $H^\infty(\mathcal{E},\mathcal{V})$, then $T_{AB} = T_A T_B$. So if Θ is an invertible outer function, then $T_\Theta T_{\Theta^{-1}} = T_I = I$ and $T_{\Theta^{-1}}T_\Theta = I$. In other words, $T_{\Theta^{-1}}$ is the inverse of T_Θ. Now assume that T_Θ is invertible, and let Q be the inverse of T_Θ. Let $S_\mathcal{E}$ be the unilateral shift on $\ell^2_+(\mathcal{E})$ and $S_\mathcal{Y}$ the unilateral shift on $\ell^2_+(\mathcal{Y})$. Because T_Θ intertwines $S_\mathcal{E}$ with $S_\mathcal{Y}$, it follows that Q intertwines $S_\mathcal{Y}$ with $S_\mathcal{E}$. In other words, Q is a lower triangular Toeplitz operator. Hence $Q = T_\Psi$ where Ψ is a function in $H^\infty(\mathcal{Y},\mathcal{E})$. Since $I = T_\Psi T_\Theta = T_{\Psi\Theta}$ and $I = T_\Theta T_\Psi = T_{\Theta\Psi}$, it follows that $I = \Psi(z)\Theta(z)$ and $I = \Theta(z)\Psi(z)$. Therefore $\Theta(z)^{-1} = \Psi(z)$ exists and is a function in $H^\infty(\mathcal{Y},\mathcal{E})$.

Let $\Theta = \Theta_i\Theta_o$ be the inner-outer factorization for a function Θ in $H^\infty(\mathcal{E},\mathcal{Y})$. As expected, Θ_i is an inner function in $H^\infty(\mathcal{V},\mathcal{Y})$ and Θ_o is an outer function in $H^\infty(\mathcal{E},\mathcal{V})$. We claim that $T_\Theta^* T_\Theta$ is the positive Toeplitz operator determined by

$$T_R = T_\Theta^* T_\Theta \qquad \text{where} \quad R = \Theta^*\Theta = \Theta_o^*\Theta_o \tag{3.3.1}$$

almost everywhere with respect to the Lebesgue measure on the unit circle. To verify this observe that for f and g in $\ell^2_+(\mathcal{E})$, we have

$$(T_\Theta^* T_\Theta f, g) = (T_\Theta f, T_\Theta g) = (L_\Theta f, L_\Theta g) = (L_\Theta^* L_\Theta f, g) = (L_{\Theta^*\Theta} f, g).$$

In other words, $T_\Theta^* T_\Theta$ is the compression of the Laurent operator $L_{\Theta^*\Theta}$ to $\ell^2_+(\mathcal{E})$. According to Proposition 2.5.1, the operator $T_\Theta^* T_\Theta = T_R$ where T_R is the positive Toeplitz operator with symbol $R = \Theta^*\Theta$. Since Θ_i is almost everywhere an isometry on the unit circle, $R = \Theta_o^*\Theta_o$. Because Θ is in $H^\infty(\mathcal{E},\mathcal{Y})$, it follows that R is in $L^\infty(\mathcal{E},\mathcal{E})$. Moreover, T_R is an invertible positive operator on $\ell^2_+(\mathcal{E})$ if and only if R^{-1} is in $L^\infty(\mathcal{E},\mathcal{E})$; see Part (v) of Proposition 2.5.1.

We claim that T_R is an invertible positive operator on $\ell^2_+(\mathcal{E})$ if and only if Θ_o is an invertible outer function. Using $T_\Theta = T_{\Theta_i}T_{\Theta_o}$ along with the fact that T_{Θ_i} is an isometry, $T_R = T_{\Theta_o}^* T_{\Theta_o}$. So if Θ_o is an invertible outer function, then T_{Θ_o} is invertible, and thus, T_R is invertible. On the other hand, if T_R is invertible, then there exists a $\delta > 0$ such that

$$\delta^2\|f\|^2 \le (T_R f, f) = (T_{\Theta_o}^* T_{\Theta_o} f, f) = \|T_{\Theta_o} f\|^2 \qquad (f \in \ell^2_+(\mathcal{E})).$$

In other words, the operator T_{Θ_o} is bounded below. Because the range of T_{Θ_o} is dense in $\ell^2_+(\mathcal{V})$, the operator T_{Θ_o} is onto. In other words, T_{Θ_o} is invertible. Therefore Θ_o is an invertible outer function.

Remark 3.3.1. Assume that $R = \Theta^*\Theta$ almost everywhere on the unit circle where Θ is a function in $H^\infty(\mathcal{E}, \mathcal{Y})$, then $T_R = T_\Theta^* T_\Theta = T_{\Theta_o}^* T_{\Theta_o}$ where Θ_o is the outer part of Θ. Moreover, $R = \Theta_o^*\Theta_o$ almost everywhere on the unit circle. Finally, the previous analysis shows us that the following are equivalent.

- The Toeplitz operator T_R is invertible.
- The Toeplitz operator T_{Θ_o} is invertible.
- R^{-1} is a function in $L^\infty(\mathcal{E}, \mathcal{E})$.
- Θ_o is an invertible outer function.

The following result provides us with an explicit formula to compute the inner-outer factorization when T_R is strictly positive.

Proposition 3.3.2. *Let Θ be a function in $H^\infty(\mathcal{E}, \mathcal{Y})$. Assume that $T_R = T_\Theta^* T_\Theta$ is a strictly positive operator on $\ell_+^2(\mathcal{E})$ where $R = \Theta^*\Theta$. Let Ω be the analytic function obtained by taking the Fourier transform of the first column of T_R^{-1}, that is,*

$$\Omega(z) = (\mathcal{F}_\mathcal{E}^+ T_R^{-1} \Pi_\mathcal{E}^*)(z) \quad \textit{where} \quad \Pi_\mathcal{E} = \begin{bmatrix} I & 0 & 0 & 0 & \cdots \end{bmatrix} : \ell_+^2(\mathcal{E}) \to \mathcal{E}.$$

Then the inner-outer factorization $\Theta = \Theta_i\Theta_o$ is given by

$$\begin{aligned} \Theta_i(z) &= \Theta(z)\Omega(z)N^{-1} \quad \textit{and} \quad \Theta_o(z) = N\Omega(z)^{-1}, \\ N &= \left(\Pi_\mathcal{E} T_R^{-1} \Pi_\mathcal{E}^*\right)^{1/2} \quad \textit{on } \mathcal{E}. \end{aligned} \tag{3.3.2}$$

Finally, Θ_o is an invertible outer function in $H^\infty(\mathcal{E}, \mathcal{E})$ while Θ_i is an inner function in $H^\infty(\mathcal{E}, \mathcal{Y})$.

Proof. Let $\Theta = \Theta_i\Theta_o$ be the inner-outer factorization for Θ. Because T_R is strictly positive, T_{Θ_o} is invertible. So the range of $T_\Theta = T_{\Theta_i} T_{\Theta_o}$ is closed. Let $\mathcal{M}$ be the invariant subspace for $S_\mathcal{Y}$ defined by $\mathcal{M} = T_\Theta \ell_+^2(\mathcal{E})$. Let $\mathcal{L}$ be the cyclic wandering subspace for the unilateral shift $S_\mathcal{Y}$ given by $\mathcal{L} = \mathcal{M} \ominus S_\mathcal{Y}\mathcal{M}$. We claim that $\mathcal{L} = T_\Theta T_R^{-1} \Pi_\mathcal{E}^*$. To see this first observe that

$$\mathcal{L} = \{T_\Theta f : f \in \ell_+^2(\mathcal{E}) \text{ and } T_\Theta f \perp S_\mathcal{Y} T_\Theta \ell_+^2(\mathcal{E})\}.$$

So $T_\Theta f$ is in $\mathcal{L}$ if and only if $T_\Theta f$ is orthogonal to $T_\Theta S_\mathcal{E} \ell_+^2(\mathcal{E})$, or equivalently, $T_R f = T_\Theta^* T_\Theta f$ is orthogonal to $S_\mathcal{E} \ell_+^2(\mathcal{E})$. In other words, $T_\Theta f$ is in $\mathcal{L}$ if and only if $T_R f$ is in the kernel of $S_\mathcal{E}^*$, or equivalently, $f \in T_R^{-1}\Pi_\mathcal{E}^*\mathcal{E}$. This readily implies that $\mathcal{L} = T_\Theta T_R^{-1}\Pi_\mathcal{E}^*\mathcal{E}$. We claim that

$$\Phi = T_\Theta T_R^{-1} \Pi_\mathcal{E}^* N^{-1} : \mathcal{E} \to \ell_+^2(\mathcal{E})$$

is an isometry whose range equals the wandering subspace $\mathcal{L}$. Since N is a positive invertible operator on $\mathcal{E}$, the range of Φ equals $\mathcal{L}$. Now observe that

$$\Phi^*\Phi = N^{-1}\Pi_\mathcal{E} T_R^{-1} T_\Theta^* T_\Theta T_R^{-1} \Pi_\mathcal{E}^* N^{-1} = N^{-1}\Pi_\mathcal{E} T_R^{-1} \Pi_\mathcal{E}^* N^{-1} = I.$$

Hence Φ is an isometry and $\mathcal{L} = \Phi\mathcal{E}$, which proves our claim. By Remark 3.2.2, we see that $\Theta_i = \mathcal{F}_{\mathcal{Y}}^+\Phi$ is the inner part of Θ. Using the fact that convolution in the time domain is multiplication in the frequency domain, we obtain

$$\Theta_i = \mathcal{F}_{\mathcal{Y}}^+\Phi = \mathcal{F}_{\mathcal{Y}}^+ T_\Theta T_R^{-1}\Pi_{\mathcal{E}}^* N^{-1} = \Theta\mathcal{F}_{\mathcal{E}}^+ T_R^{-1}\Pi_{\mathcal{E}}^* N^{-1} = \Theta\Omega N^{-1}.$$

In other words, $\Theta_i(z) = \Theta(z)\Omega(z)N^{-1}$. Now observe that

$$\Theta_i = \Theta\Omega N^{-1} = \Theta_i\Theta_o\Omega N^{-1}.$$

So we see that $\Theta_i(I - \Theta_o\Omega N^{-1}) = 0$. Because Θ_i is inner, $\Theta_i(e^{\imath\omega})$ is almost everywhere an isometry with respect to the Lebesgue measure. Thus $I = \Theta_o\Omega N^{-1}$. Since Θ_o is an invertible outer function, $\Theta_o = N\Omega^{-1}$. □

3.4 The Determinant of Inner and Outer Functions

Let A be an operator on a finite dimensional space $\mathcal{E}$. Then $\det[A]$ is the determinant of the matrix representation of A corresponding to any basis for $\mathcal{E}$. An operator T is a *contraction* if $\|T\| \leq 1$. Recall that Θ is an invertible outer function if Θ is a function in some $H^\infty(\mathcal{E}, \mathcal{Y})$ space and Θ^{-1} is in $H^\infty(\mathcal{Y}, \mathcal{E})$. We say that $\Omega(z)$ is a *contractive analytic function* if Ω is a function in $H^\infty(\mathcal{E}, \mathcal{Y})$ and $\|\Omega\|_\infty \leq 1$. In other words, Ω in $H^\infty(\mathcal{E}, \mathcal{Y})$ is a contractive analytic function if and only if the corresponding Toeplitz matrix T_Ω is a contraction.

Theorem 3.4.1. *Let Θ be a function in $H^\infty(\mathcal{E}, \mathcal{E})$.*

(i) *The function Θ is outer if and only if* $\det[\Theta]$ *is an outer function.*

(ii) *The function Θ is an invertible outer function if and only if* $\det[\Theta]$ *is an invertible outer function.*

(iii) *If Θ is a contractive analytic function, then Θ is an inner function if and only if* $\det[\Theta]$ *is an inner function.*

Proof. Set $\delta(z) = \det[\Theta(z)]$. Clearly, the finite product of H^∞ functions is also in H^∞. Because Θ is in $H^\infty(\mathcal{E}, \mathcal{E})$ and the determinant is formed by taking the appropriate products of the components of Θ, it follows that $\det[\Theta] = \delta$ is in H^∞. Let Ψ be the algebraic adjoint of Θ. Since Ψ is obtained by multiplying and adding the appropriate components of Θ, the function Ψ is in $H^\infty(\mathcal{E}, \mathcal{E})$ and $\Psi(z)\Theta(z) = \Theta(z)\Psi(z) = \delta(z)I$ for all z in $\mathbb{D}_+$. Now assume that δ is an outer function. Then we obtain

$$H^2(\mathcal{E}) = \overline{\delta H^2(\mathcal{E})} = \overline{\Theta\Psi H^2(\mathcal{E})} \subseteq \overline{\Theta H^2(\mathcal{E})} \subseteq H^2(\mathcal{E}).$$

This implies that $\Theta H^2(\mathcal{E})$ is dense in $H^2(\mathcal{E})$. In other words, Θ is an outer function.

On the other hand, assume that Θ is outer. Let $\delta = \delta_i \delta_o$ be the inner-outer factorization for δ where δ_i is inner and δ_o is outer. Let $\Psi = \Psi_i \Psi_o$ be the inner-outer factorization for Ψ where $\Psi_i \in H^\infty(\mathcal{V}, \mathcal{E})$ is inner and $\Psi_o \in H^\infty(\mathcal{E}, \mathcal{V})$ is outer. Thus

$$\delta_i H^2(\mathcal{E}) = \overline{\delta_i \delta_o H^2(\mathcal{E})} = \overline{\delta H^2(\mathcal{E})} = \overline{\Psi \Theta H^2(\mathcal{E})} = \overline{\Psi H^2(\mathcal{E})} = \Psi_i H^2(\mathcal{V}).$$

According to the Beurling-Lax-Halmos Theorem 3.1.1, the inner functions $\delta_i I$ and Ψ_i are equal up to a unitary constant on the right. So without loss of generality, we can assume that $\delta_i I = \Psi_i$ and $\mathcal{V} = \mathcal{E}$. This readily implies that $\delta_i \delta_o I = \Psi_i \Psi_o \Theta = \delta_i \Psi_o \Theta$. In other words, $\delta_o I = \Psi_o \Theta$. By taking the determinant, we arrive at

$$\delta_o^n = \det[\Psi_o] \det[\Theta] = \delta_i \det[\Psi_o] \delta_o$$

where n is the dimension of $\mathcal{E}$. Since the finite product of outer functions is also an outer function, we see that δ_i divides the outer function δ_o^n, that is, $\delta_o^n = \delta_i \varphi$ where φ is a function in H^∞. The only inner function which can divide an outer function is a constant of modulus 1. (Notice that $H^2 = \overline{\delta_o^n H^2} \subseteq \delta_i H^2$. So $\delta_i H^2 = H^2$, and δ_i is both inner and outer.) So without loss of generality, we can assume that $\delta_i = 1$. Therefore the determinant $\det[\Theta] = \delta = \delta_i \delta_o = \delta_o$ is an outer function. This proves Part (i).

If Θ is an invertible outer function, then Θ^{-1} is a function in $H^\infty(\mathcal{E}, \mathcal{E})$. Hence $\delta^{-1} = \det[\Theta^{-1}]$ is a function in H^∞. So the determinant of Θ is also an invertible outer function. On the other hand, if δ is an invertible outer function, then $\Psi/\delta = \Theta^{-1}$ is a well-defined function in $H^\infty(\mathcal{E}, \mathcal{E})$. This verifies Part (ii).

To prove Part (iii), recall that if A is a contraction on a finite dimensional space, then A is a unitary operator if and only if $|\det[A]| = 1$. If A is unitary, then all the eigenvalues of A are of modulus 1. Since the determinant of A is the product of all the eigenvalues of A, we must have $|\det[A]| = 1$. On the other hand, if A is a contraction such that $|\det[A]| = 1$, then A^*A is a positive contraction, such that $\det[A^*A] = 1$. Recall that all finite dimensional positive operators are unitarily equivalent to a positive diagonal matrix. So A^*A is unitarily equivalent to a positive diagonal contractive matrix Λ whose determinant is 1. So all the diagonal entries of Λ are 1. Therefore $\Lambda = I$. Because A^*A is unitarily equivalent to I, it follow that $A^*A = I$ and A is a unitary operator.

Now assume that Θ is a contractive analytic function in $H^\infty(\mathcal{E}, \mathcal{E})$. Then Θ is an inner function if and only if $\Theta(e^{\imath\omega})$ is almost everywhere a unitary operator on $\mathcal{E}$ with respect to the Lebesgue measure. In other words, Θ is an inner function if and only if $\det[\Theta(e^{\imath\omega})]$ is almost everywhere a function of modulus 1. Since $\det[\Theta]$ is in H^∞, we see that Θ is an inner function if and only if the determinant of Θ is an inner function. □

To complete this section, note that Part (iii) of Theorem 3.4.1 is not true if Θ is not a contraction. For a counter example, let Θ be the constant diagonal 2×2 matrix with 2 and $1/2$ on the diagonal and zeros elsewhere.

3.5 Notes

The scalar version of the Beurling-Lax-Halmos Theorem is due to Beurling [31]. The multivariable generalizations are due to Lax [154, 155] and Halmos [124]. The Beurling-Lax-Halmos Theorem is now a classical result in operator theory; see [80, 82, 182, 198] for further results and historical comments. Our approach to inner-outer factorization theory was motivated by Foias-Frazho [82] and Sz.-Nagy-Foias [198]. Proposition 3.3.2 is a classical method of computing the inner-outer factorization when the outer spectral factor is an invertible outer function. For further results on the determinant, inner and outer functions see Sz.-Nagy-Foias [198].

Chapter 4

Rational Inner and Outer Functions

Rational transfer functions naturally occur in many practical systems and control problems. Recall that a function admits a finite dimensional state space realization if and only if it is proper and rational. The finite dimensional state space setting is ideal for using computers to design feedback controllers and analyzing transfer functions. In this chapter, we will use state space realizations to characterize inner and outer rational functions. We will also present an algorithm to compute the inner-outer factorization for a rational transfer function. Finally, we will describe a state space method to compute the Douglas-Shapiro-Shields factorization for a rational function.

4.1 Blaschke Products

We say that a rational function Θ with values in $\mathcal{L}(\mathcal{E},\mathcal{Y})$ is *proper* if the degree of the numerator is less than or equal to the degree of the denominator, that is, $\Theta = N/d$ where $N = \sum_0^\nu z^k N_k$ is a $\mathcal{L}(\mathcal{E},\mathcal{Y})$-valued polynomial of degree ν while $d = \sum_0^\mu z^k d_k$ is a scalar-valued polynomial of degree μ and $\nu \leq \mu$. By definition, a *transfer function* is a proper rational function. A *stable* transfer function is a proper rational function whose poles are all contained in the open unit disc $\mathbb{D} = \{z : |z| < 1\}$. Finally, throughout this chapter, we assume that both $\mathcal{E}$ and $\mathcal{Y}$ are finite dimensional.

It is easy to show that a $\mathcal{L}(\mathcal{E},\mathcal{Y})$-valued rational function Θ is in $H^\infty(\mathcal{E},\mathcal{Y})$ if and only if Θ is a stable transfer function. A rational function Θ is in $H^\infty(\mathcal{E},\mathcal{Y})$ if and only if it is in $H^2(\mathcal{E},\mathcal{Y})$, or equivalently, Θ is a stable transfer function. In other words, a stable transfer function is a rational function in $H^\infty(\mathcal{E},\mathcal{Y})$, or equivalently, a rational function in $H^2(\mathcal{E},\mathcal{Y})$.

Recall that Θ is an *inner function* if Θ is a function in $H^\infty(\mathcal{E}, \mathcal{Y})$ and T_Θ is an isometry mapping $\ell_+^2(\mathcal{E})$ into $\ell_+^2(\mathcal{Y})$, or equivalently, Θ is in $H^\infty(\mathcal{E}, \mathcal{Y})$ and its boundary values $\Theta(e^{\imath\omega})$ are almost everywhere isometries mapping $\mathcal{E}$ into $\mathcal{Y}$. In control theory, an inner function is also referred to as an *all pass function.*

Let Θ be a $\mathcal{L}(\mathcal{E}, \mathcal{Y})$-valued rational transfer function. In this case, Θ is an inner function if and only if Θ is a stable transfer function and $\Theta(1/\overline{z})^*\Theta(z) = I$ for all z in $\mathbb{C}$. To see this, observe that $\Theta(1/\overline{z})^*\Theta(z) = N(z)/d(z)$ is a rational function where N and d are polynomials in z. If Θ is inner, then Θ is almost everywhere an isometry on the unit circle. Hence $N(z)/d(z) = I$ for infinitely many points on the unit circle. In particular, the two polynomials $N(z) = d(z)I$ at infinitely many points. Recall that two polynomials of degree at most ν are equal if and only if they are equal at $\nu + 1$ distinct points. Since $N(z) = d(z)I$ at infinitely many points, $N(z) = d(z)I$ for all z in $\mathbb{C}$. So $\Theta(1/\overline{z})^*\Theta(z) = N(z)/d(z) = I$ for all z. On the other hand, if $\Theta(1/\overline{z})^*\Theta(z) = I$ for all z, then clearly, $\Theta(z)^*\Theta(z) = \Theta(1/\overline{z})^*\Theta(z) = I$ on the unit circle. So Θ is an inner function.

We say that $b(z)$ is a *Blaschke product of order* n if

$$b(z) = \gamma \prod_{j=1}^{n} \frac{1 - \overline{\alpha}_j z}{z - \alpha_j}, \tag{4.1.1}$$

where $\{\alpha_j\}_1^n$ is a set of n scalars in $\mathbb{D}$, and γ is a constant of modulus 1. It is emphasized that the scalars $\{\alpha_j\}_1^n$ do not have to be distinct. For example, if $\alpha_j = 0$ for all j and $\gamma = 1$, then $1/z^n$ is a Blaschke product of order n. Finally, it is noted that $\{\alpha_j\}_1^n$ is the set of poles for the Blaschke product $b(z)$.

We claim that θ is a scalar-valued rational inner function if and only if θ is a Blaschke product of order n. To see this, first assume that $\theta(z) = b(z)$ is a Blaschke product of order n. Clearly, all the poles of $b(z)$ are in $\mathbb{D}$. Let $b_j(z)$ be the Blaschke product of order 1 given by

$$b_j(z) = \frac{1 - \overline{\alpha}_j z}{z - \alpha_j}. \tag{4.1.2}$$

A simple calculation shows that $b_j(1/\overline{z})^* b_j(z) = 1$, for all z. In other words, each b_j is an inner function. Because $b(z) = \prod_1^n b_j(z)$, and the product of inner functions is also an inner function, we see that $b(z)$ is an inner function.

Now assume that θ is a scalar-valued rational inner function. In particular, θ is a proper rational function. Let $\theta(z) = p(z)/q(z)$ where p and q are two polynomials with no common zeros of the form

$$p(z) = p_0 + p_1 z + \cdots + p_{\nu-1} z^{\nu-1} + p_\nu z^\nu \quad \text{and} \quad q(z) = q_0 + q_1 z + \cdots + q_{m-1} z^{m-1} + z^m.$$

Because θ is proper, $\nu \le m$. If $\xi(z) = \sum_0^m \xi_k z^k$ is a polynomial of degree at most m, then $\xi^\natural$ is the *reverse polynomial* defined by

$$\xi(z)^\natural = z^m \overline{\xi(1/\overline{z})}. \tag{4.1.3}$$

In other words,

$$\xi(z)^\natural = \overline{\xi}_0 z^m + \overline{\xi}_1 z^{m-1} + \overline{\xi}_2 z^{m-2} + \cdots + \overline{\xi}_{m-1} z + \overline{\xi}_m.$$

Observe that λ is a nonzero root of ξ if and only if $1/\overline{\lambda}$ is a root of $\xi^\natural$. (Zero is a root of ξ if and only if $\deg(\xi^\natural) < m$. Moreover, $\xi = (\xi^\natural)^\natural$.) Using the fact that θ is inner, we obtain

$$1 = \overline{\theta(1/\overline{z})}\theta(z) = \frac{z^m \overline{p(1/\overline{z})} p(z)}{z^m \overline{q(1/\overline{z})} q(z)} = \frac{p(z)^\natural p(z)}{q(z)^\natural q(z)}. \tag{4.1.4}$$

This readily implies that

$$\theta(z) = \frac{p(z)}{q(z)} = \frac{q(z)^\natural}{p(z)^\natural}.$$

Because p and q have no common roots, $p^\natural$ and $q^\natural$ have no common roots. Hence $p = \delta q(z)^\natural$ and $q = \delta p(z)^\natural$ where δ is a scalar. In other words, $\theta = \delta q(z)^\natural / q$. Since θ is in H^∞, it is stable and all of its poles are contained in the open unit disc $\mathbb{D}$. In other words, $q(z) = \prod_1^m (z - \alpha_j)$ where $\{\alpha_j\}_1^m$ are scalars contained in $\mathbb{D}$. This readily implies that $q(z)^\natural = \prod_1^m (1 - \overline{\alpha}_j z)$, that is,

$$\theta(z) = \delta \frac{q(z)^\natural}{q(z)} = \delta \prod_{j=1}^m \frac{1 - \overline{\alpha}_j z}{z - \alpha_j} = \delta \prod_{j=1}^m b_j(z).$$

Using $|\theta(e^{\imath\omega})| = |b_j(e^{\imath\omega})| = 1$, we arrive at $|\delta| = 1$. So θ is a Blaschke product of order m.

Remark 4.1.1. In general one can show that the set of all scalar-valued inner functions are given by $\theta(z) = \gamma b(z) s(z)$ where γ is a constant of modulus 1, b is a Blaschke product of the form

$$b(z) = \prod_{j=1}^n \frac{1 - \overline{\alpha}_j z}{z - \alpha_j} \quad \text{where} \quad \sum_{j=1}^n (1 - |\alpha_j|) < \infty,$$

and the set $\{\alpha_j\}_1^n$ is a possibly infinite set of complex numbers contained in the open unit disc $\mathbb{D}$. Finally, the singular function

$$s(z) = \exp\left(- \int_0^{2\pi} \frac{z e^{\imath\omega} + 1}{z e^{\imath\omega} - 1} d\mu \right)$$

where μ is a positive measure which is singular with respect to the Lebesgue measure. For further details on the scalar-valued inner functions see [76, 106, 134, 151, 187].

4.2 State Space Realizations for Rational Inner Functions

In this section we will use realization theory to determine when a rational transfer function is inner. A review of state space theory is given in Chapter 14. To establish some notation recall that $\{A, B, C, D\}$ is a realization for a transfer function F if

$$F(z) = D + C(zI - A)^{-1}B.$$

Here A is an operator on $\mathcal{X}$, the operator B maps $\mathcal{E}$ into $\mathcal{X}$, while the operator C maps $\mathcal{X}$ into $\mathcal{Y}$ and D is an operator from $\mathcal{E}$ into $\mathcal{Y}$. The operator A on $\mathcal{X}$ is stable, if the spectrum of A lives in some compact subset of the open unit disc. In particular, if $\mathcal{X}$ is finite dimensional, then A is stable if and only if all the eigenvalues for A are contained in the open unit disc.

Consider a pair of operators $\{C, A\}$, where A is an operator on a finite dimensional space $\mathcal{X}$ and C is an operator mapping $\mathcal{X}$ into $\mathcal{Y}$. Recall that the pair $\{C, A\}$ is *observable* if $\{A^{*k}C^*\mathcal{Y}\}_0^\infty$ spans $\mathcal{X}$, or equivalently, by the Cayley-Hamilton Theorem $\{A^{*k}C^*\mathcal{Y}\}_0^{\dim \mathcal{X}-1}$ spans $\mathcal{X}$. Moreover, we have

- If A is stable, then the *observability Gramian* P for $\{C, A\}$ is the unique solution to the Lyapunov equation

$$P = A^*PA + C^*C.$$

  In this case, $P = \sum_0^\infty A^{*n}C^*CA^n$.

- If A is stable, then P is strictly positive if and only if the pair $\{C, A\}$ is observable.

- If $\{C, A\}$ is observable, then there exists a strictly positive solution to the Lyapunov equation $P = A^*PA + C^*C$ if and only if A is stable.

Consider a pair of operators $\{A, B\}$, where A is an operator on a finite dimensional space $\mathcal{X}$ and B is an operator mapping $\mathcal{E}$ into $\mathcal{X}$. Recall that the pair $\{A, B\}$ is *controllable* if $\{A^kB\mathcal{E}\}_0^\infty$ spans $\mathcal{X}$, or equivalently, by the Cayley-Hamilton Theorem $\{A^kB\mathcal{E}\}_0^{\dim \mathcal{X}-1}$ spans $\mathcal{X}$. Furthermore, we have

- If A is stable, then the *controllability Gramian* Q for $\{A, B\}$ is the unique solution to the Lyapunov equation

$$Q = AQA^* + BB^*.$$

  In this case, $Q = \sum_0^\infty A^nBB^*A^{*n}$.

- If A is stable, then Q is strictly positive if and only if the pair $\{A, B\}$ is controllable.

- If $\{A, B\}$ is controllable, then there exists a strictly positive solution to the Lyapunov equation $Q = AQA^* + BB^*$ if and only if A is stable.

We say that $\{A \text{ on } \mathcal{X}, B, C, D\}$ is an *isometric realization* (respectively *co-isometric realization, unitary realization)* if the systems matrix

$$\Omega = \begin{bmatrix} A & B \\ C & D \end{bmatrix} : \begin{bmatrix} \mathcal{X} \\ \mathcal{E} \end{bmatrix} \to \begin{bmatrix} \mathcal{X} \\ \mathcal{Y} \end{bmatrix} \tag{4.2.1}$$

is an isometry (respectively co-isometry, unitary operator). If $\{A, B, C, D\}$ is an isometric realization, then the first column $\begin{bmatrix} A & C \end{bmatrix}^{tr}$ of Ω is also an isometry. Hence

$$I = A^*A + C^*C. \tag{4.2.2}$$

In particular, the identity I is the observability Gramian for the pair $\{C, A\}$. If $\{A, B, C, D\}$ is an observable finite dimensional isometric realization, then A is stable. This follows from the fact that $\{C, A\}$ is observable and I is a strictly positive solution to the Lyapunov equation $I = A^*A + C^*C$.

If $\{A, B, C, D\}$ is a co-isometric realization, then the first row $\begin{bmatrix} A & B \end{bmatrix}$ of Ω is also a co-isometry. Hence

$$I = AA^* + BB^*. \tag{4.2.3}$$

In particular, I is the controllability Gramian for the pair $\{A, B\}$. If $\{A, B, C, D\}$ is a controllable finite dimensional co-isometric realization, then A is stable.

Clearly, a realization is unitary if and only if it is isometric and co-isometric. So if $\{A, B, C, D\}$ is a unitary realization, then I is the solution to the Lyapunov equations in (4.2.2) and (4.2.3). Finally, it is noted that a minimal finite dimensional unitary realization is stable. For a finite dimensional system, a minimal realization is a realization of the lowest possible state dimension. A realization is minimal if and only if it is controllable and observable; see Chapter 14 for a review of state space.

Theorem 4.2.1. *Let $\{A \text{ on } \mathcal{X}, B, C, D\}$ be a minimal realization for a $\mathcal{L}(\mathcal{E}, \mathcal{Y})$-valued rational transfer function Θ. Then the following statements are equivalent.*

(i) *The function Θ is inner.*

(ii) *The operator A is stable and*

$$\begin{bmatrix} A^* & C^* \\ B^* & D^* \end{bmatrix} \begin{bmatrix} P & 0 \\ 0 & I \end{bmatrix} \begin{bmatrix} A & B \\ C & D \end{bmatrix} = \begin{bmatrix} P & 0 \\ 0 & I \end{bmatrix}, \tag{4.2.4}$$

where P is the observability Gramian for the pair $\{C, A\}$.

(iii) *The realization $\{A, B, C, D\}$ is similar to a stable isometric realization.*

In particular, a rational transfer function Θ is inner if and only if Θ admits a stable minimal isometric realization.

Proof. If Θ is an inner function, then Θ is a function in $H^\infty(\mathcal{E}, \mathcal{Y})$, and thus all poles of Θ are in $\mathbb{D}$. Since $\{A, B, C, D\}$ is a minimal (controllable and observable) realization of Θ, it follows that A is stable. So without loss of generality we can assume that $\{A, B, C, D\}$ is a stable realization for Θ. In particular, the observability Gramian P for the pair $\{C, A\}$ is strictly positive.

Let T_Θ be the lower triangular Toeplitz operator determined by Θ, that is,

$$T_\Theta = \begin{bmatrix} \Theta_0 & 0 & 0 & \cdots \\ \Theta_1 & \Theta_0 & 0 & \cdots \\ \Theta_2 & \Theta_1 & \Theta_0 & \cdots \\ \vdots & \vdots & \vdots & \ddots \end{bmatrix} : \ell_+^2(\mathcal{E}) \to \ell_+^2(\mathcal{Y}).$$

As expected, $\Theta(z) = \sum_0^\infty z^{-k}\Theta_k$ is the Taylor's series expansion of Θ. Because $\{A, B, C, D\}$ is a realization of Θ, the Taylor's coefficient of Θ can be written as

$$\Theta_0 = D \quad \text{and} \quad \Theta_k = CA^{k-1}B \quad (k \geq 1). \tag{4.2.5}$$

By definition, Θ is inner if and only if T_Θ is an isometry. Notice that T_Θ is an isometry if and only if all the columns of T_Θ are isometries mapping $\mathcal{E}$ into $\ell_+^2(\mathcal{Y})$ and the columns of T_Θ are orthogonal to each other. Due to the Toeplitz structure of T_Θ, the nth column of T_Θ is an isometry mapping $\mathcal{E}$ to $\ell_+^2(\mathcal{Y})$ if and only if $I = \sum_0^\infty \Theta_k^*\Theta_k$. By employing $P = \sum_0^\infty A^{*k}C^*CA^k$ and (4.2.5), we arrive at

$$\begin{aligned} \sum_{k=0}^\infty \Theta_k^*\Theta_k &= D^*D + \sum_{k=1}^\infty B^*A^{*(k-1)}C^*CA^{k-1}B \\ &= D^*D + B^*\left(\sum_{k=0}^\infty A^{*k}C^*CA^k\right)B \\ &= D^*D + B^*PB. \end{aligned} \tag{4.2.6}$$

In other words, the nth column of T_Θ is an isometry if and only if $I = D^*D+B^*PB$.

The Toeplitz structure of T_Θ also shows that all the columns of T_Θ are orthogonal if and only if the nth column of T_Θ is orthogonal to the first column of T_Θ for all $n \geq 1$, or equivalently, $0 = \sum_{k=0}^\infty \Theta_k^*\Theta_{k+n}$ for all $n \geq 1$. By using $P = \sum_0^\infty A^{*k}C^*CA^k$ and (4.2.5) once again, we see that

$$\begin{aligned} \sum_{k=0}^\infty \Theta_k^*\Theta_{k+n} &= D^*CA^{n-1}B + \sum_{k=1}^\infty B^*A^{*(k-1)}C^*CA^{k-1+n}B \\ &= D^*CA^{n-1}B + B^*\left(\sum_{k=0}^\infty A^{*k}C^*CA^k\right)A^nB \\ &= (D^*C + B^*PA)A^{n-1}B \qquad (n \geq 1). \end{aligned} \tag{4.2.7}$$

Clearly, if $0 = D^*C + B^*PA$, then $\sum_{k=0}^\infty \Theta_k^*\Theta_{k+n} = 0$ for all integers $n \geq 1$, and all the columns of T_Θ are orthogonal. On the other hand, if all the columns of T_Θ are orthogonal, then $(D^*C + B^*PA)A^{n-1}B = 0$ for all $n \geq 1$. Because the pair

$\{A, B\}$ is controllable, $\{A^k B\mathcal{E}\}_0^\infty$ spans $\mathcal{X}$, and thus, $D^*C + B^*PA = 0$. Therefore all columns of T_Θ are orthogonal if and only if $0 = D^*C + B^*PA$.

The above analysis shows that T_Θ is an isometry if and only if

$$I = D^*D + B^*PB \quad \text{and} \quad 0 = D^*C + B^*PA. \tag{4.2.8}$$

These equations with $P = A^*PA + C^*C$ are equivalent to (4.2.4). Hence Θ is an inner function if and only if A is stable and (4.2.4) holds. Therefore Parts (i) and (ii) are equivalent.

If $\{A, B, C, D\}$ is a stable isometric realization, then the observability Gramian $P = I$ and (4.2.4) holds. In other words, the transfer function for a stable isometric realization is inner. Recall that similar realizations have the same transfer function. So if $\{A, B, C, D\}$ is similar to a stable isometric realization, then its transfer function is inner. Hence Part (iii) implies Part (i).

Now assume that Part (ii) holds, or equivalently, $\{A, B, C, D\}$ is a minimal realization of Θ and (4.2.4) holds. By multiplying both sides of (4.2.4) by $P^{-1/2} \oplus I$, we see that

$$\begin{bmatrix} P^{1/2}AP^{-1/2} & P^{1/2}B \\ CP^{-1/2} & D \end{bmatrix} : \begin{bmatrix} \mathcal{X} \\ \mathcal{E} \end{bmatrix} \to \begin{bmatrix} \mathcal{X} \\ \mathcal{Y} \end{bmatrix} \tag{4.2.9}$$

is an isometry. In other words, $\Sigma = \{P^{1/2}AP^{-1/2}, P^{1/2}B, CP^{-1/2}, D\}$ is a stable isometric realization. Notice that $P^{1/2}$ is a similarity transformation intertwining $\{A, B, C, D\}$ with Σ. In particular, Θ is also the transfer function for Σ. Therefore $\{A, B, C, D\}$ is similar to a stable isometric realization. Hence Part (ii) implies Part (iii), and thus, all three parts are equivalent. □

Notice that $\{A, B, C, D\}$ is a realization (respectively minimal realization) for Θ if and only if $\{A^*, C^*, B^*, D^*\}$ is a realization (respectively minimal realization) for $\widetilde{\Theta}(z) = \Theta(\overline{z})^*$. By definition, Θ is co-inner if $\widetilde{\Theta}$ is inner. By employing this fact in Theorem 4.2.1, we readily obtain the following result.

Corollary 4.2.2. *Let $\{A$ on $\mathcal{X}, B, C, D\}$ be a minimal realization for a $\mathcal{L}(\mathcal{E}, \mathcal{Y})$-valued rational transfer function Θ.*

(i) *The function Θ is co-inner.*

(ii) *The operator A is stable and*

$$\begin{bmatrix} A & B \\ C & D \end{bmatrix} \begin{bmatrix} Q & 0 \\ 0 & I \end{bmatrix} \begin{bmatrix} A^* & C^* \\ B^* & D^* \end{bmatrix} = \begin{bmatrix} Q & 0 \\ 0 & I \end{bmatrix}, \tag{4.2.10}$$

where Q is the controllability Gramian for the pair $\{A, B\}$.

(iii) *The realization $\{A$ on $\mathcal{X}, B, C, D\}$ is similar to a stable co-isometric realization.*

In particular, Θ is a rational co-inner transfer function if and only if Θ admits a stable minimal co-isometric realization.

Let Θ be a function in $H^\infty(\mathcal{E},\mathcal{Y})$. Recall that Θ is a two-sided inner function if and only if $\Theta(e^{\imath\omega})$ is almost everywhere a unitary operator with respect to the Lebesgue measure. In particular, Θ is a two-sided inner function if and only if Θ is inner and $\dim\mathcal{E} = \dim\mathcal{Y}$, or equivalently, Θ is co-inner and $\dim\mathcal{E} = \dim\mathcal{Y}$. Hence Theorem 4.2.1 or Corollary 4.2.2 can be used to determine when a rational transfer function is inner from both sides. For example, assume that $\{A, B, C, D\}$ is a minimal realization for a rational transfer function in $H^\infty(\mathcal{E},\mathcal{Y})$ where $\dim\mathcal{E} = \dim\mathcal{Y}$. Then Θ is a two-sided inner function if and only if (4.2.4) holds where P is the observability Gramian, or equivalently, the realization $\{P^{1/2}AP^{-1/2},\ P^{1/2}B,\ CP^{-1/2},\ D\}$ of Θ is unitary. Likewise, Θ is a two-sided inner function if and only if (4.2.10) holds where Q is the controllability Gramian, or equivalently, the realization $\{Q^{-1/2}AQ^{1/2}, Q^{-1/2}B, CQ^{1/2}, D\}$ of Θ is unitary. Finally, Θ is a rational two-sided inner function if and only if Θ admits a stable minimal unitary realization.

Remark 4.2.3. Assume that Θ is a rational two-sided inner function in $H^\infty(\mathcal{Y},\mathcal{Y})$. Let H be the Hankel matrix generated by the Fourier coefficients $\{\Theta_n\}_1^\infty$ for Θ, that is,

$$H = \begin{bmatrix} \Theta_1 & \Theta_2 & \Theta_3 & \cdots \\ \Theta_2 & \Theta_3 & \Theta_4 & \cdots \\ \Theta_3 & \Theta_4 & \Theta_5 & \cdots \\ \vdots & \vdots & \vdots & \ddots \end{bmatrix} \quad \text{on } \ell_+^2(\mathcal{Y})$$

where $\Theta(z) = \sum_0^\infty z^{-n}\Theta_n$. We claim that there are $\delta(\Theta)$ singular values for H equal to 1 and all the other singular values are zero. (The McMillan degree of a transfer function G is denoted by $\delta(G)$.) To see this, recall that any rational square inner function Θ in $H^\infty(\mathcal{Y},\mathcal{Y})$ admits a unitary minimal stable realization $\{A \text{ on } \mathcal{X}, B, C, D\}$. Using $\Theta_n = CA^{n-1}B$, we arrive at $H = W_oW_c$ where

$$\begin{aligned} W_c &= \begin{bmatrix} B & AB & A^2B & \cdots \end{bmatrix} : \ell_+^2(\mathcal{Y}) \to \mathcal{X}, \\ W_o &= \begin{bmatrix} C & CA & CA^2 & \cdots \end{bmatrix}^{tr} : \mathcal{X} \to \ell_+^2(\mathcal{Y}). \end{aligned}$$

The rank of H equals the McMillan degree of Θ. Because $\{A, B, C, D\}$ is a stable unitary realization, the controllability and observability Gramian are both the identity. In other words, $W_cW_c^* = I$ and $W_o^*W_o = I$. Using $H = W_oW_c$, we see that $H^*H = W_c^*W_o^*W_oW_c = W_c^*W_c$. Recall that if M and N are two operators acting between the appropriate spaces, then MN and NM have the same nonzero spectrum. Hence H^*H and $W_cW_c^* = I$ have the same nonzero eigenvalues. The singular values for H are the square root of the eigenvalues for H^*H. Therefore H has $\delta(\Theta)$ nonzero singular values equal to 1 and all the other singular values are zero.

Remark 4.2.4. Let $\{A, B, C, D\}$ be a finite dimensional, stable, observable realization for a function Θ in $H^\infty(\mathcal{Y},\mathcal{Y})$. Moreover, assume that (4.2.4) holds where P is the observability Gramian for the pair $\{C, A\}$. Then $\{A, B, C, D\}$ is a minimal realization and Θ is a two-sided inner function. Finally, P^{-1} equals the controllability Gramian for $\{A, B\}$.

To verify this, let Ω be the systems matrix for $\{A, B, C, D\}$, that is,

$$\Omega = \begin{bmatrix} A & B \\ C & D \end{bmatrix} \quad \text{and set} \quad \Lambda = \begin{bmatrix} P & 0 \\ 0 & I \end{bmatrix}.$$

Notice that both Ω and Λ act on $\mathcal{X} \oplus \mathcal{Y}$. Then (4.2.4) states that $\Lambda = \Omega^*\Lambda\Omega$. Because P is invertible, Λ is invertible, and thus, Ω is also invertible. Multiplying $\Lambda = \Omega^*\Lambda\Omega$ on the left by $\Omega\Lambda^{-1}$ and on the right with $\Omega^{-1}\Lambda^{-1}$, yields $\Lambda^{-1} = \Omega\Lambda^{-1}\Omega^*$. In other words,

$$\begin{bmatrix} A & B \\ C & D \end{bmatrix} \begin{bmatrix} P^{-1} & 0 \\ 0 & I \end{bmatrix} \begin{bmatrix} A^* & C^* \\ B^* & D^* \end{bmatrix} = \begin{bmatrix} P^{-1} & 0 \\ 0 & I \end{bmatrix}.$$

In particular, $P^{-1} = AP^{-1}A^* + BB^*$. In other words, P^{-1} is the controllability Gramian for $\{A, B\}$. Because A is stable and P^{-1} is invertible, the pair $\{A, B\}$ is controllable. Hence $\{A, B, C, D\}$ is a minimal realization. According to Theorem 4.2.1, the function Θ is two-sided inner.

By replacing $\{A, B, C, D\}$ in Remark 4.2.4 with its dual $\{A^*, B^*, C^*, D^*\}$, we arrive at the following result.

Remark 4.2.5. Let $\{A, B, C, D\}$ be a finite dimensional, stable, controllable realization for a function Θ in $H^\infty(\mathcal{Y}, \mathcal{Y})$. Moreover, assume that (4.2.10) holds where Q is the controllability Gramian for the pair $\{A, B\}$. Then $\{A, B, C, D\}$ is a minimal realization and Θ is a two-sided inner function. Finally, Q^{-1} equals the observability Gramian for $\{C, A\}$.

Let us conclude this section with the following useful result.

Proposition 4.2.6. *Let F be a function in $H^2(\mathcal{E}, \mathcal{Y})$ and S the unilateral shift on $H^2(\mathcal{Y})$. Let $\mathcal{H}$ be the invariant subspace for the backward shift S^* defined by*

$$\mathcal{H} = \bigvee_{n=1}^{\infty} S^{*n} F\mathcal{E}. \tag{4.2.11}$$

Then the dimension of $\mathcal{H}$ equals the McMillan degree of F. In particular, $\mathcal{H}$ is finite dimensional if and only if F is rational. In this case, if $\{A$ on $\mathcal{X}, B, C, D\}$ is a minimal realization for F, then

$$\bigvee_{n=1}^{\infty} S^{*n} F\mathcal{E} = zC(zI - A)^{-1}\mathcal{X}. \tag{4.2.12}$$

Proof. If $F(z) = \sum_0^\infty z^{-k} F_k$ is the power series expansion for F, then $S^{*n}F = \sum_0^\infty z^{-k} F_{n+k}$. In particular, $S^{*n}F = \mathcal{F}_\mathcal{Y}^+ \begin{bmatrix} F_n & F_{n+1} & F_{n+2} & \cdots \end{bmatrix}^{tr}$ where $\mathcal{F}_\mathcal{Y}^+$ is the Fourier transform mapping $\ell_+^2(\mathcal{Y})$ onto $H^2(\mathcal{Y})$. Therefore $\mathcal{H} = \bigvee_1^\infty S^{*n} F\mathcal{Y}$ is the closure of the range of $\mathcal{F}_\mathcal{Y}^+ H$ where H is the Hankel matrix given by

$$H = \begin{bmatrix} F_1 & F_2 & F_3 & \cdots \\ F_2 & F_3 & F_4 & \cdots \\ F_3 & F_4 & F_5 & \cdots \\ \vdots & \vdots & \vdots & \ddots \end{bmatrix}. \tag{4.2.13}$$

Recall that the rank of a Hankel matrix H equals the McMillan degree of its symbol F. Moreover, the rank of H is finite if and only if its symbol F is rational; see Section 14.2. Since $\mathcal{H}$ equals the closure of the range of $\mathcal{F}_{\mathcal{Y}}^{+}H$, it follows that $\mathcal{H}$ is finite dimensional if and only if F is rational. Moreover, the dimension of $\mathcal{H}$ equals the rank of H which is the McMillan degree of F.

Assume that F is rational and $\{A \text{ on } \mathcal{X}, B, C, D\}$ is a minimal realization for F. Clearly, $F(\infty) = F_0 = D$. Recall that the backward shift S^* on $H^2(\mathcal{Y})$ is determined by

$$(Sh)(z) = \frac{1}{z}h(z) \quad \text{and} \quad (S^*h)(z) = zh(z) - zh(\infty) \qquad (h \in H^2(\mathcal{Y})). \tag{4.2.14}$$

Using $F(z) = D + C(zI - A)^{-1}B$ with (4.2.14), it follows that $S^*Fv = zC(zI - A)^{-1}Bv$ where v is in $\mathcal{E}$. Because A is stable, $z(zI - A)^{-1} = \sum_0^\infty z^{-k}A^k$. By consulting (4.2.14) once again, we arrive at

$$\begin{aligned} S^{*2}Fv &= S^*zC(zI - A)^{-1}Bv = z^2C(zI - A)^{-1}Bv - zCBv \\ &= zC\left(z(zI - A)^{-1} - I\right)Bv = zC(zI - A)^{-1}\left(zI - (zI - A)\right)Bv \\ &= zC(zI - A)^{-1}ABv. \end{aligned}$$

In other words, $S^{*2}Fv = zC(zI - A)^{-1}ABv$. By continuing in this fashion, we see that $S^{*n+1}Fv = zC(zI - A)^{-1}A^nBv$ for all integers $n \geq 0$. Hence

$$\begin{aligned} \mathcal{H} &= \bigvee_{n=1}^{\infty} S^{*n}F\mathcal{E} = \bigvee_{n=0}^{\infty} zC(zI - A)^{-1}A^nB\mathcal{E} \\ &= zC(zI - A)^{-1}\bigvee_{n=0}^{\infty} A^nB\mathcal{E} = zC(zI - A)^{-1}\mathcal{X}. \end{aligned}$$

The last equality follows from the fact that the pair $\{A, B\}$ is controllable. Therefore the subspace $\mathcal{H} = zC(zI - A)^{-1}\mathcal{X}$. □

4.3 Rational Two-Sided Inner Functions

This section is devoted to the finite dimensional invariant subspaces for the backward shift operator. Let S be the unilateral shift on $H^2(\mathcal{Y})$. Throughout we assume that $\mathcal{Y}$ is finite dimensional. The Beurling-Lax-Halmos Theorem 3.1.1 shows that the set of all invariant subspaces for the backward shift S^* on $H^2(\mathcal{Y})$ are given by $\mathcal{H}(\Theta) = H^2(\mathcal{Y}) \ominus \Theta H^2(\mathcal{E})$, where Θ is an inner function in $H^\infty(\mathcal{E}, \mathcal{Y})$. In this section, we will show that the set of all finite dimensional invariant subspaces for S^* are determined by the rational two-sided inner functions. We begin with the following result.

Lemma 4.3.1. *Let Θ be a two-sided inner function in $H^\infty(\mathcal{Y},\mathcal{Y})$. If S is the unilateral shift on $H^2(\mathcal{Y})$, then*

$$\mathcal{H}(\Theta) = H^2(\mathcal{Y}) \ominus \Theta H^2(\mathcal{Y}) = \bigvee_{n=1}^{\infty} S^{*n}\Theta\mathcal{Y}. \tag{4.3.1}$$

The invariant subspace $\mathcal{H}(\Theta)$ is finite dimensional if and only if Θ is rational. In this case, if $\{A \text{ on } \mathcal{X}, B, C, D\}$ is any minimal realization for Θ, then the subspace $\mathcal{H}(\Theta) = zC(zI - A)^{-1}\mathcal{X}$.

Proof. Let h be a function in $H^2(\mathcal{Y})$ and $y \in \mathcal{Y}$. Recall that $(Sg)(z) = z^{-1}g(z)$ where g is in $H^2(\mathcal{Y})$. Because Θ is inner, for any integer $n \geq 1$ we obtain

$$(S^{*n}\Theta y, \Theta h) = (\Theta y, S^n\Theta h) = (\Theta y, \Theta S^n h) = (y, S^n h) = 0.$$

Thus $S^{*n}\Theta\mathcal{Y}$ is orthogonal to $\Theta H^2(\mathcal{Y})$ for all $n \geq 1$. Hence $\bigvee_1^\infty S^{*n}\Theta\mathcal{Y} \subseteq \mathcal{H}(\Theta)$. Assume that g is an element in $\mathcal{H}(\Theta)$, and g is orthogonal to $S^{*n}\Theta\mathcal{Y}$ for all $n \geq 1$. This implies that

$$\begin{aligned} 0 &= (g, S^{*n}\Theta y) = (S^n g, \Theta y) = \frac{1}{2\pi}\int_0^{2\pi} (e^{-\imath\omega n} g(e^{\imath\omega}), \Theta(e^{\imath\omega})y)\, d\omega \\ &= \frac{1}{2\pi}\int_0^{2\pi} (\Theta(e^{\imath\omega})^* g(e^{\imath\omega}), e^{\imath\omega n}y)\, d\omega \qquad (n \geq 1). \end{aligned}$$

Hence $\Theta^* g$ is orthogonal to $\oplus_1^\infty e^{\imath\omega n}\mathcal{Y}$ in $L^2(\mathcal{Y})$. (Here we view $H^2(\mathcal{Y})$ as a subspace of $L^2(\mathcal{Y})$; see Section 2.2.) Because g is orthogonal to $\Theta H^2(\mathcal{Y})$, we see that $\Theta^* g$ is orthogonal to $H^2(\mathcal{Y})$. Since $H^2(\mathcal{Y}) = \oplus_0^\infty e^{-\imath\omega n}\mathcal{Y}$, the vector $\Theta^* g$ is orthogonal to $\oplus_{-\infty}^\infty e^{\imath\omega n}\mathcal{Y} = L^2(\mathcal{Y})$. In other words, $\Theta^* g = 0$, or equivalently, $\Theta(e^{\imath\omega})^* g(e^{\imath\omega}) = 0$ almost everywhere with respect to the Lebesgue measure. Because $\Theta(e^{\imath\omega})$ is almost everywhere a unitary operator, $g = 0$. So $\mathcal{H}(\Theta) = \bigvee_1^\infty S^{*n}\Theta\mathcal{Y}$ and (4.3.1) holds. The rest of the theorem follows by replacing F with Θ in Proposition 4.2.6. □

The following is the main result of this section.

Theorem 4.3.2. *Let S be the unilateral shift on $H^2(\mathcal{Y})$. Then the following statements are equivalent.*

(i) *$\mathcal{H}$ is a finite dimensional invariant subspace for S^*.*

(ii) *$\mathcal{H} = zC(zI - A)^{-1}\mathcal{X}$ where $\{C, A \text{ on } \mathcal{X}\}$ is a finite dimensional stable, observable pair.*

(iii) *$\mathcal{H} = \mathcal{H}(\Theta)$ where Θ is a rational two-sided inner function in $H^\infty(\mathcal{Y},\mathcal{Y})$.*

If $\{A, B, C, D\}$ is any minimal realization for a two-sided rational inner function Θ, then

$$\mathcal{H}(\Theta) = zC(zI - A)^{-1}\mathcal{X}. \tag{4.3.2}$$

Finally, the dimension of $\mathcal{H}(\Theta)$ equals the McMillan degree of Θ.

Recall that a scalar-valued inner function is rational if and only if it is a Blaschke product of order n where n is finite; see Section 4.1. Notice that the McMillan degree of a Blaschke product of order n is precisely n. This with Theorem 4.3.2 readily yields the following result.

Corollary 4.3.3. *Let S be the unilateral shift on H^2. Then the set of all $n(< \infty)$ dimensional invariant subspaces for the backward shift S^* are given by $\mathcal{H}(\theta)$ where θ is a Blaschke product of order n. In this case, $\mathcal{H}(\theta) = zC(zI - A)^{-1}\mathcal{X}$ where $\{A \text{ on } \mathcal{X}, B, C, D\}$ is any minimal realization for θ.*

Remark 4.3.4. Let $\{C, A \text{ on } \mathcal{X}\}$ be a stable, observable pair where $\mathcal{X}$ is finite dimensional and C maps $\mathcal{X}$ into $\mathcal{Y}$. Then one can always find operators B mapping $\mathcal{Y}$ into $\mathcal{X}$ and D on $\mathcal{Y}$ such that $\{A, B, C, D\}$ is a minimal realization for a two-sided inner function Θ and

$$\mathcal{H}(\Theta) = H^2(\mathcal{Y}) \ominus \Theta H^2(\mathcal{Y}) = zC(zI - A)^{-1}\mathcal{X}. \tag{4.3.3}$$

To find B and D, compute the singular value decomposition of the operator $\begin{bmatrix} A^*P & C^* \end{bmatrix}$ mapping $\mathcal{X} \oplus \mathcal{Y}$ into $\mathcal{X}$ to find an isometry

$$V = \begin{bmatrix} V_1 \\ V_2 \end{bmatrix} : \mathcal{Y} \to \begin{bmatrix} \mathcal{X} \\ \mathcal{Y} \end{bmatrix} \quad \text{such that} \quad \operatorname{ran} V = \ker\left(\begin{bmatrix} A^*P & C^* \end{bmatrix}\right). \tag{4.3.4}$$

(The Matlab command null can also be used to find V.) Here P is the observability Gramian for the pair $\{C, A\}$. To be precise, P is the unique solution to the Lyapunov equation $P = A^*PA + C^*C$. Then the operators B and D are given by

$$B = V_1 N^{-1/2}, \quad D = V_2 N^{-1/2} \quad \text{where} \quad N = V_1^* P V_1 + V_2^* V_2. \tag{4.3.5}$$

To verify this remark, first observe that

$$P = A^*PA + C^*C = \begin{bmatrix} A^*P & C^* \end{bmatrix} \begin{bmatrix} A \\ C \end{bmatrix}.$$

Notice that $\begin{bmatrix} A^*P & C^* \end{bmatrix}$ maps $\mathcal{X} \oplus \mathcal{Y}$ into $\mathcal{X}$. Because P is invertible, it is onto $\mathcal{X}$, and thus, the operator $\begin{bmatrix} A^*P & C^* \end{bmatrix}$ must also be onto $\mathcal{X}$. In other words, the dimension of the kernel of $\begin{bmatrix} A^*P & C^* \end{bmatrix}$ equals the dimension $\mathcal{X} \oplus \mathcal{Y}$ minus the dimension of $\mathcal{X}$. So the dimension of the kernel of $\begin{bmatrix} A^*P & C^* \end{bmatrix}$ equals the dimension of $\mathcal{Y}$. So we can construct an orthonormal basis of dimension $\dim \mathcal{Y}$ for the kernel of $\begin{bmatrix} A^*P & C^* \end{bmatrix}$. This basis readily yields an isometry V of the form (4.3.4). Using (4.3.5), it follows that (4.2.4) holds. Remark 4.2.4 shows that $\{A, B, C, D\}$ is a minimal realization and its transfer function Θ is two-sided inner. Finally, Lemma 4.3.1 yields (4.3.3).

Proof of Theorem 4.3.2. Assume that Part (i) holds. Let Φ be a unitary operator mapping a finite dimensional space $\mathcal{X}$ onto $\mathcal{H}$. Here $\mathcal{X}$ is any Hilbert space with the same dimension as $\mathcal{H}$. Let A be the operator on $\mathcal{X}$ defined by $A = \Phi^* S^* \Phi$. In other words, $S^*\Phi = \Phi A$. Notice that $S^{*n}\Phi = \Phi A^n$ for all integers $n \geq 0$. Because S^{*n} converges to zero in the strong operator topology, A^n must also converge to zero as n tends to infinity. Since $\mathcal{X}$ is finite dimensional, A is stable. The range of Φ is a subspace of $H^2(\mathcal{Y})$. Hence Φ admits a Taylor's series expansion of the form

$$(\Phi x)(z) = \sum_{k=0}^{\infty} z^{-k}\Phi_k x \qquad (x \in \mathcal{X} \text{ and } z \in \mathbb{D}_+).$$

Here Φ_k is an operator mapping $\mathcal{X}$ into $\mathcal{Y}$ for all integers $k \geq 0$. Let C be the operator mapping $\mathcal{X}$ into $\mathcal{Y}$ determined by $C = \Phi_0$. We claim that $\Phi_k = CA^k$ for all $k \geq 0$. To see this, let $\Pi_{\mathcal{Y}}$ be the orthogonal projection from $H^2(\mathcal{Y})$ onto $\mathcal{Y}$ which picks out the constant functions in $H^2(\mathcal{Y})$, that is, $\Pi_{\mathcal{Y}} h = h(\infty)$ where h is in $H^2(\mathcal{Y})$. Then for any integer $k \geq 0$, we have

$$\Phi_k = \Pi_{\mathcal{Y}} S^{*k}\Phi = \Pi_{\mathcal{Y}}\Phi A^k = \Phi_0 A^k = CA^k.$$

Thus $\Phi_k = CA^k$ for all $k \geq 0$. Hence for any x in $\mathcal{X}$, we have

$$\Phi x = \sum_{k=0}^{\infty} z^{-k}\Phi_k x = \sum_{k=0}^{\infty} z^{-k} CA^k x = C(I - z^{-1}A)^{-1}x = zC(zI - A)^{-1}x.$$

In other words, $\Phi = zC(zI - A)^{-1}$.

We claim that $\{C, A\}$ is observable. Assume $CA^n x = 0$ for all $n \geq 0$ and some x in $\mathcal{X}$. Then

$$\Phi x = zC(zI - A)^{-1}x = \sum_{n=0}^{\infty} z^{-n} CA^n x = 0.$$

Because Φ is one to one, $x = 0$. In other words, $\bigcap_0^\infty \ker CA^n = \{0\}$, and the pair $\{C, A\}$ is observable. Therefore Part (i) implies Part (ii).

Now assume Part (ii) holds. According to Remark 4.3.4, there exist operators B and D such that $\{A, B, C, D\}$ is a minimal realization for a two-sided inner function Θ and $\mathcal{H} = \mathcal{H}(\Theta)$. Hence Part (iii) holds. Lemma 4.3.1 shows that Part (iii) implies Part (i). Consulting Proposition 4.2.6 with $F = \Theta$ yields (4.3.2). □

By combining Proposition 4.2.6 with Theorem 4.3.2, we arrive at the following result which will be used in Section 4.7.

Remark 4.3.5. Let F be a rational function in $H^2(\mathcal{E}, \mathcal{Y})$. Let $\mathcal{H}$ be the invariant subspace for the backward shift S^* on $H^2(\mathcal{Y})$ defined by

$$\mathcal{H} = \bigvee_{n=1}^{\infty} S^{*n} F\mathcal{E}. \tag{4.3.6}$$

Then there exists a two-sided inner function Θ in $H^\infty(\mathcal{Y},\mathcal{Y})$ such that

$$\bigvee_{n=1}^{\infty} S^{*n}F\mathcal{E} = \mathcal{H}(\Theta) = H^2(\mathcal{Y}) \ominus \Theta H^2(\mathcal{Y}). \tag{4.3.7}$$

Let $\{A \text{ on } \mathcal{X}, B_1, C, D_1\}$ be a minimal realization for the transfer function F. Then $\mathcal{H}(\Theta) = zC(zI - A)^{-1}\mathcal{X}$. Moreover, one can use Remark 4.3.4 to compute the operators B mapping $\mathcal{Y}$ into $\mathcal{X}$ and D on $\mathcal{Y}$ such that $\{A, B, C, D\}$ is a minimal realization for Θ.

4.4 Rational Outer Functions

This section is devoted to rational outer functions and their state space realizations. A rational function Θ is in $H^2(\mathcal{E},\mathcal{Y})$ if and only if Θ is proper and all the poles of Θ are contained in the open unit disc $\mathbb{D}$. Moreover, a rational function Θ is in $H^2(\mathcal{E},\mathcal{Y})$ if and only if it is in $H^\infty(\mathcal{E},\mathcal{Y})$. Let Θ be a function in $H^2(\mathcal{E},\mathcal{Y})$, then by definition Θ is an outer function if $\overline{\Theta\mathcal{P}(\mathcal{E})} = H^2(\mathcal{Y})$. (The space of all polynomials in $1/z$ with values in $\mathcal{E}$ is denoted by $\mathcal{P}(\mathcal{E})$.) Recall that Θ is an invertible outer function if Θ is in $H^\infty(\mathcal{E},\mathcal{Y})$ and Θ^{-1} is in $H^\infty(\mathcal{Y},\mathcal{E})$. Notice that Θ is an invertible outer function if and only if Θ is in $H^\infty(\mathcal{E},\mathcal{Y})$ and its corresponding Toeplitz matrix T_Θ defines an invertible operator mapping $\ell_+^2(\mathcal{E})$ onto $\ell_+^2(\mathcal{Y})$. Finally, it is noted that a scalar-valued rational function θ is an invertible outer function if and only if θ and $1/\theta$ are rational functions in H^∞, or equivalently, $\theta = q/d$ where q and d are two polynomials of the same degree, and all the poles and zeros of θ are contained in the open unit disc $\mathbb{D}$.

Let θ be a scalar-valued rational function. Then θ is an outer function if and only if the following three conditions hold:

(i) *$\theta = q/d$ where q and d are two polynomials of the same degree;*

(ii) *the zeros of θ are contained in $\overline{\mathbb{D}}$;*

(iii) *the poles of θ are contained in $\mathbb{D}$.*

To prove this, assume that θ is a rational function in H^∞. Let $\theta = q/d$ where q and d are two polynomials with no common zeros, the zeros of d are contained in $\mathbb{D}$, and $\deg q \le \deg d$. Notice that d/z^n is an invertible outer function, where n is the degree of d. Hence

$$\theta H^2 = \frac{q}{d}H^2 = \frac{q}{d}\frac{d}{z^n}H^2 = z^{-n}qH^2. \tag{4.4.1}$$

Let $p(z) = z^n\overline{q(1/\overline{z})}$ be the reverse polynomial for q, that is, $p(z) = \overline{q}_0 z^n + \overline{q}_1 z^{n-1} + \cdots + \overline{q}_n$ where $q(z) = \sum_0^n q_j z^j$. Let S be the unilateral shift on H^2. Then we have

$$p(S^*) = \overline{q}_0 S^{*n} + \overline{q}_1 S^{*n-1} + \overline{q}_2 S^{*n-2} + \cdots + \overline{q}_{n-1}S^* + \overline{q}_n I.$$

Since $z^{-n}q(z) = \sum_0^n q_j z^{j-n}$, equation (4.4.1) implies that $\theta H^2 = p(S^*)^* H^2$. In other words, $\overline{\theta H^2}$ equals the closure of the range of $p(S^*)^*$, and thus,

$$\ker p(S^*) = H^2 \ominus \theta H^2.$$

So θ is outer if and only if the kernel of $p(S^*)$ is zero, or equivalently, zero is not an eigenvalue of $p(S^*)$. Recall that the set of all eigenvalues for S^* equals the open unit disc $\mathbb{D}$. According to the spectral mapping theorem, the set of all eigenvalues for $p(S^*)$ are given by $p(\mathbb{D})$; see [126]. Hence θ is not an outer function if and only if zero is an eigenvalue for $p(S^*)$, or equivalently, $p(\lambda) = 0$ for some $\lambda \in \mathbb{D}$. Observe that $p(0) = 0$ if and only if $q_n = 0$, or equivalently, $\deg q < n$. In particular, if $\deg q < n = \deg d$, then θ is not outer. Therefore if θ is an outer function, then $\deg q = \deg d$, and Part (i) holds.

Now assume that $\deg q = n$, or equivalently, $p(0) = \overline{q}_n \neq 0$. Notice that $\lambda \neq 0$ is a root of p if and only if $1/\overline{\lambda}$ is a root of q. So zero is an eigenvalue for $p(S^*)$ if and only if $p(\lambda) = 0$ for some $\lambda \in \mathbb{D}$, or equivalently, $q(\alpha) = 0$ for some $\alpha \in \mathbb{D}_+$. In other words, the kernel of $p(S^*)$ is zero if and only if all the zeros of q are contained in $\overline{\mathbb{D}}$. Therefore θ is outer if and only if Parts (i), (ii) and (iii) hold.

Remark 4.4.1. Let Θ be a rational function in $H^2(\mathcal{Y},\mathcal{Y})$. Let $\{A, B, C, D\}$ be a minimal realization of Θ. Since Θ is in $H^2(\mathcal{Y},\mathcal{Y})$, all the poles of Θ must be contained in $\mathbb{D}$, and A is stable. Hence any controllable and observable realization of Θ must be stable. Moreover, the following statements are equivalent.

(i) The function Θ is outer.

(ii) The determinant of Θ is an outer function.

(iii) $\Theta(\infty) = D$ is invertible, and the eigenvalues of $A - BD^{-1}C$ are all contained in $\overline{\mathbb{D}}$.

Finally, the following statements are also equivalent.

(a) Θ is an invertible outer function.

(b) The determinant of Θ is an invertible outer function.

(c) $\Theta(\infty) = D$ is invertible, and the eigenvalues of $A - BD^{-1}C$ are all contained in $\mathbb{D}$.

Theorem 3.4.1 shows that Parts (i) and (ii) are equivalent. If Θ is an outer function, then $\Theta(\infty) = D$ on $\mathcal{Y}$ must be invertible. So it remains to show that Parts (ii) and (iii) are equivalent. Recall that if N and M are two operators acting between the appropriate finite dimensional spaces, then

$$\det[I + MN] = \det[I + NM]. \tag{4.4.2}$$

Using this fact along with the state space realization $\{A, B, C, D\}$ for Θ, we obtain

$$\begin{aligned}
\delta(z) &= \det[D + C(zI - A)^{-1}B] = \det[D]\det[I + D^{-1}C(zI - A)^{-1}B] \\
&= \det[D]\det[I + (zI - A)^{-1}BD^{-1}C] \\
&= \det[D]\det[(zI - A)^{-1}\left((zI - A) + BD^{-1}C\right)] \\
&= \det[D]\frac{\det[zI - (A - BD^{-1}C)]}{\det[zI - A]}.
\end{aligned}$$

In other words, the determinant of Θ is given by

$$\det[\Theta(z)] = \det[D]\frac{\det[zI - (A - BD^{-1}C)]}{\det[zI - A]}. \tag{4.4.3}$$

Notice that $\det[zI-(A-BD^{-1}C)]$ is the characteristic polynomial for $A-BD^{-1}C$ and $\det[zI - A]$ is the characteristic polynomial for A. Without loss of generality, assume that the dimension of the state $\mathcal{X}$ is n. This readily implies that

$$\det[\Theta(z)] = \det[D]\frac{\prod_{j=1}^{n}(z - \alpha_j)}{\prod_{j=1}^{n}(z - \beta_j)} \tag{4.4.4}$$

where $\{\alpha_j\}_1^n$ are the eigenvalues of $A - BD^{-1}C$ and $\{\beta_j\}_1^n$ are the eigenvalues of A. Because A is stable, $\{\beta_j\}_1^n$ must be contained in $\mathbb{D}$. So the only possible pole zero cancellation in $\det[\Theta(z)]$ that can occur is perhaps on the eigenvalues of $A - BD^{-1}C$ contained in $\mathbb{D}$. By virtue of (4.4.3), we see that $\det[\Theta(z)]$ is outer if and only if all the eigenvalues of $A - BD^{-1}C$ are contained in $\overline{\mathbb{D}}$. Thus the equivalence of Parts (i) to (iii) is established.

By Theorem 3.4.1 and using the fact that Θ is rational, it follows that Parts (a) and (b) are equivalent. By consulting (4.4.4) once again, we see that $\det[\Theta(z)]$ is an invertible outer function if and only if all the eigenvalues of $A - BD^{-1}C$ are contained in $\mathbb{D}$. In other words, Parts (b) and (c) are equivalent.

The following result will be useful.

Lemma 4.4.2. *Let $\{A$ on $\mathcal{X}, B, C, D\}$ be a stable and controllable realization of an outer function Θ with values in $H^\infty(\mathcal{Y}, \mathcal{Y})$. Then all the eigenvalues of $A-BD^{-1}C$ are contained in $\overline{\mathbb{D}}$. Moreover, $A-BD^{-1}C$ is stable if and only if Θ is an invertible outer function.*

Proof. According to Remark 14.2.1, the inverse of Θ is given by

$$\begin{aligned}
\Theta(z)^{-1} &= D^{-1} - D^{-1}C(zI - J)^{-1}BD^{-1}, \\
J &= A - BD^{-1}C.
\end{aligned} \tag{4.4.5}$$

Let $\mathcal{X}_o$ be the observable subspace for the pair $\{C, A\}$ given by

$$\mathcal{X}_o = \operatorname{span}\{A^{*n}C^*\mathcal{Y} : 0 \le n < \dim \mathcal{X}\} = \bigvee_{n=0}^{\infty} A^{*n}C^*\mathcal{Y}.$$

Then the unobservable subspace is defined by $\mathcal{X}_{\overline{o}} = \mathcal{X} \ominus \mathcal{X}_o = \bigcap_0^\infty \ker CA^n$. Notice that $\mathcal{X}_o$ is an invariant subspace for A^*, and $\mathcal{X}_{\overline{o}}$ is invariant under A. Using this decomposition of $\mathcal{X}$, we see that $\{A, B, C, D\}$ admits matrix decompositions of the form:

$$A = \begin{bmatrix} A_o & 0 \\ \star & A_{\overline{o}} \end{bmatrix} \text{ on } \begin{bmatrix} \mathcal{X}_o \\ \mathcal{X}_{\overline{o}} \end{bmatrix}, \qquad B = \begin{bmatrix} B_o \\ B_{\overline{o}} \end{bmatrix} : \mathcal{Y} \to \begin{bmatrix} \mathcal{X}_o \\ \mathcal{X}_{\overline{o}} \end{bmatrix},$$
$$C = \begin{bmatrix} C_o & 0 \end{bmatrix} : \begin{bmatrix} \mathcal{X}_o \\ \mathcal{X}_{\overline{o}} \end{bmatrix} \to \mathcal{Y}.$$

The zero in C follows from the fact that $C|\mathcal{X}_{\overline{o}} = 0$. Here $\star$ represents an unspecified entry. By construction, $\{A_o, B_o, C_o, D\}$ is a controllable and observable realization of Θ. According to Remark 4.4.1, all the eigenvalues of $J_o = A_o - B_o D^{-1} C_o$ are contained in the closed unit disc. Moreover, J_o is stable if and only if Θ is an invertible outer function. Using the previous matrix decompositions, we see that

$$J = \begin{bmatrix} A_o & 0 \\ \star & A_{\overline{o}} \end{bmatrix} - \begin{bmatrix} B_o \\ B_{\overline{o}} \end{bmatrix} D^{-1} \begin{bmatrix} C_o & 0 \end{bmatrix} = \begin{bmatrix} J_o & 0 \\ \star & A_{\overline{o}} \end{bmatrix}.$$

Because A is stable, $A_{\overline{o}}$ must also be stable. Since all eigenvalues of J_o are contained in $\overline{\mathbb{D}}$, the previous decomposition shows that all eigenvalues of J must also be contained in $\overline{\mathbb{D}}$. Recall that Θ is an invertible outer function if and only if J_o is stable. Hence J is stable if and only if Θ is an invertible outer function. □

Example. Let g be a rational function in H^∞. In this case, g admits an inner-outer factorization of the form $g = g_i g_o$ where g_i is the Blaschke product consisting of the zeros of g in $\mathbb{D}_+$ and g_o is a rational outer function. By rearranging the poles and zeros of g, one can theoretically compute the inner-outer factorization for g. For example, consider the rational function g in H^∞ given by

$$g(z) = \frac{(z-2)(z-3)(z-1)}{(z-0.1)(z-0.4)(z-0.5)(z-0.6)}.$$

Notice that g has two zeros in $\mathbb{D}_+$ and one zero on the unit circle. Moreover, the degree of the numerator is strictly less than the degree of the denominator. Using the two zeros of g in $\mathbb{D}_+$, and the fact that the degree of the denominator minus the degree of the numerator equals 1, the inner part g_i of g is determined by

$$g_i(z) = \frac{(z-2)(z-3)}{z(1-2z)(1-3z)} = \frac{(1-z/2)(1-z/3)}{z(z-1/2)(z-1/3)}.$$

In other words, g_i is the Blaschke product of order 3 with zeros at $\{2, 3, \infty\}$. The outer part g_o for g is given by

$$g_o(z) = \frac{z(1-2z)(1-3z)(z-1)}{(z-0.1)(z-0.4)(z-0.5)(z-0.6)}.$$

Notice that the degree of the numerator and denominator for g_o are the same. Moreover, all the zeros of g_o are contained in $\overline{\mathbb{D}}$. It is well known that computing the roots of a polynomial can be numerically sensitive. So this method of computing the inner-outer factorization may not be reliable for rational functions of large order. Later we will present several other methods to compute the inner-outer factorization in the matrix case.

4.5 Inner-Outer Factorization and McMillan Degree

In this section we will show that the inner and outer factors of a rational function in $H^\infty(\cdot,\cdot)$ are also rational functions, and their corresponding McMillan degrees do not increase. The *McMillan degree* of a transfer function G is denoted by $\delta(G)$.

If $\Theta(z) = \sum_0^\infty z^{-n}\Theta_n$ is in $H^\infty(\mathcal{E},\mathcal{Y})$, then H_Θ is the Hankel operator defined by

$$H_\Theta = \begin{bmatrix} \cdots & \Theta_3 & \Theta_2 & \Theta_1 \\ \cdots & \Theta_4 & \Theta_3 & \Theta_2 \\ \cdots & \Theta_5 & \Theta_4 & \Theta_3 \\ \vdots & \vdots & \vdots & \vdots \end{bmatrix} : \ell^2_-(\mathcal{E}) \to \ell^2_+(\mathcal{Y}). \tag{4.5.1}$$

It is well known that Θ is rational if and only if the rank of H_Θ is finite. For convenience we have written the columns of H_Θ starting on the right-hand side. Moreover, the rank of H_Θ equals the McMillan degree of Θ, that is, $\operatorname{rank}(H_\Theta) = \delta(\Theta)$; see Section 14.2. Because Θ is in $H^\infty(\mathcal{E},\mathcal{Y})$, the Laurent operator L_Θ mapping $\ell^2(\mathcal{E})$ into $\ell^2(\mathcal{Y})$ is lower triangular. Furthermore, L_Θ admits a matrix decomposition of the form

$$L_\Theta = \begin{bmatrix} \Lambda_\Theta & 0 \\ H_\Theta & T_\Theta \end{bmatrix} : \begin{bmatrix} \ell^2_-(\mathcal{E}) \\ \ell^2_+(\mathcal{E}) \end{bmatrix} \to \begin{bmatrix} \ell^2_-(\mathcal{Y}) \\ \ell^2_+(\mathcal{Y}) \end{bmatrix}. \tag{4.5.2}$$

Here $\Lambda_\Theta = P_{\ell^2_-(\mathcal{Y})} L_\Theta | \ell^2_-(\mathcal{E})$. Since $\|L_\Theta\| = \|\Theta\|_\infty$, it follows that $\|H_\Theta\| \le \|\Theta\|_\infty$, and thus, H_Θ is a well-defined operator mapping $\ell^2_-(\mathcal{E})$ into $\ell^2_+(\mathcal{Y})$.

Let Θ be a rational function in $H^\infty(\mathcal{E},\mathcal{Y})$. Let $\Theta = \Theta_i\Theta_o$ be the inner-outer factorization for Θ, where Θ_i is an inner function in $H^\infty(\mathcal{V},\mathcal{Y})$ and Θ_o an outer function in $H^\infty(\mathcal{E},\mathcal{V})$. In this case, $T_\Theta = T_{\Theta_i} T_{\Theta_o}$ where T_{Θ_i} is an isometry mapping $\ell^2_+(\mathcal{V})$ into $\ell^2_+(\mathcal{Y})$, and the range of T_{Θ_o} is dense in $\ell^2_+(\mathcal{V})$.

We claim that the McMillan degrees

$$\delta(\Theta_o) \le \delta(\Theta) \quad \text{and} \quad \delta(\Theta_i) \le \delta(\Theta). \tag{4.5.3}$$

In particular, if Θ is a rational function, then its inner and outer part are both rational functions.

First let us establish the following result:

$$T^*_{\Theta_i} H_\Theta = H_{\Theta_o}. \tag{4.5.4}$$

In particular, $\operatorname{rank} H_{\Theta_o} \leq \operatorname{rank} H_\Theta$. In other words, $\delta(\Theta_o) \leq \delta(\Theta)$, and the first equation in (4.5.3) holds. If Θ is rational, then $T_{\Theta_i}^* H_\Theta = H_{\Theta_o}$ shows that the Hankel matrix H_{Θ_o} defines a bounded operator of finite rank. Therefore Θ_o admits a stable minimal realization of McMillan degree at most $\delta(\Theta)$.

Because Θ_i is an inner function, the Laurent operator L_{Θ_i} is also an isometry. Multiplying $L_\Theta = L_{\Theta_i} L_{\Theta_o}$ on the left by $L_{\Theta_i}^*$ yields $L_{\Theta_i}^* L_\Theta = L_{\Theta_o}$. Using the matrix representation for L_Θ in (4.5.2), and the corresponding matrix representations for L_{Θ_i} and L_{Θ_o}, we obtain

$$\begin{bmatrix} \Lambda_{\Theta_o} & 0 \\ H_{\Theta_o} & T_{\Theta_o} \end{bmatrix} = L_{\Theta_o} = L_{\Theta_i}^* L_\Theta = \begin{bmatrix} \Lambda_{\Theta_i}^* & H_{\Theta_i}^* \\ 0 & T_{\Theta_i}^* \end{bmatrix} \begin{bmatrix} \Lambda_\Theta & 0 \\ H_\Theta & T_\Theta \end{bmatrix}$$
$$= \begin{bmatrix} \Lambda_{\Theta_i}^* \Lambda_\Theta + H_{\Theta_i}^* H_\Theta & H_{\Theta_i}^* T_\Theta \\ T_{\Theta_i}^* H_\Theta & T_{\Theta_i}^* T_\Theta \end{bmatrix}.$$

By matching like entries, we see that $T_{\Theta_i}^* H_\Theta = H_{\Theta_o}$ which proves our claim. Finally, it is noted that $H_{\Theta_i}^* T_\Theta = 0$, and as expected, $T_{\Theta_o} = T_{\Theta_i}^* T_\Theta$. The Moore-Penrose inverse of an operator M acting between two finite dimensional spaces is denoted by M^{-r}. If M is right invertible, or equivalently onto, then $I = MM^{-r}$.

Lemma 4.5.1. *Let $\{A \text{ on } \mathcal{X}, B, C, D\}$ be a stable controllable realization for a rational function Θ in $H^\infty(\mathcal{E}, \mathcal{Y})$. Then there exists a state space realization of the form $\{A, B, C_o, D_o\}$ for Θ_o the outer part of Θ. In this case, a realization for the inner part Θ_i of Θ is given by*

$$\Theta_i(z) = DD_o^{-r} + (C - DD_o^{-r} C_o)(zI - A_i)^{-1} B D_o^{-r} \tag{4.5.5}$$

where $A_i = A - B_o D_o^{-r} C_o$. In particular, $\delta(\Theta_i) \leq \delta(\Theta)$.

Proof. Let W_o be the observability operator mapping $\mathcal{X}$ into $\ell_+^2(\mathcal{Y})$ and W_c the controllability operator mapping $\ell_+^2(\mathcal{E})$ into $\mathcal{X}$ defined by

$$W_o = \begin{bmatrix} C \\ CA \\ CA^2 \\ \vdots \end{bmatrix} : \mathcal{X} \to \ell_+^2(\mathcal{Y}),$$
$$W_c = \begin{bmatrix} \cdots & A^3B & A^2B & AB & B \end{bmatrix} : \ell_-^2(\mathcal{E}) \to \mathcal{X}. \tag{4.5.6}$$

Let $\Theta(z) = \sum_0^\infty z^{-n} \Theta_n$ be the power series expansion for Θ. Recall that $\Theta_n = CA^{n-1}B$ for all integers $n \geq 1$. Using this we obtain $H_\Theta = W_o W_c$, and thus, $H_{\Theta_o} = T_{\Theta_i}^* H_\Theta = T_{\Theta_i}^* W_o W_c$. Notice that $T_{\Theta_i}^* W_o$ admits a matrix representation of the form

$$T_{\Theta_i}^* W_o = \begin{bmatrix} C_o \\ C_1 \\ C_2 \\ \vdots \end{bmatrix} : \mathcal{X} \to \ell_+^2(\mathcal{V}).$$

We claim that $C_n = C_o A^n$ for all integers $n \geq 1$. To see this, let $S_{\mathcal{L}}$ denote the unilateral shift on $\ell^2_+(\mathcal{L})$, and $\Pi_{\mathcal{V}} = \begin{bmatrix} I & 0 & 0 & \cdots \end{bmatrix}$ the operator mapping $\ell^2_+(\mathcal{V})$ onto $\mathcal{V}$ which picks out the first component of $\ell^2_+(\mathcal{V})$. Then using $S_{\mathcal{Y}} T_{\Theta_i} = T_{\Theta_i} S_{\mathcal{V}}$ with $S^*_{\mathcal{Y}} W_o = W_o A$, we obtain

$$C_n = \Pi_{\mathcal{V}} S^{*n}_{\mathcal{V}} T^*_{\Theta_i} W_o = \Pi_{\mathcal{V}} T^*_{\Theta_i} S^{*n}_{\mathcal{Y}} W_o = \Pi_{\mathcal{V}} T^*_{\Theta_i} W_o A^n = C_0 A^n.$$

Hence $C_n = C_o A^n$ for all $n \geq 1$. Using this we have

$$\begin{bmatrix} \cdots & \Theta_{o3} & \Theta_{o2} & \Theta_{o1} \\ \cdots & \Theta_{o4} & \Theta_{o3} & \Theta_{o2} \\ \cdots & \Theta_{o5} & \Theta_{o4} & \Theta_{o3} \\ \vdots & \vdots & \vdots & \vdots \end{bmatrix} = H_{\Theta_o} = T^*_{\Theta_i} W_o W_c = \begin{bmatrix} C_o \\ C_o A \\ C_o A^2 \\ \vdots \end{bmatrix} \begin{bmatrix} \cdots & A^2 B & AB & B \end{bmatrix}.$$

Here $\Theta_o(z) = \sum_0^\infty z^{-n} \Theta_{on}$ is the power series expansion for Θ_o. This readily implies that $\Theta_{on} = C_o A^{n-1} B$ for all integers $n \geq 0$. Therefore Θ_o admits a realization of the form $\{A, B, C_o, D_o\}$.

Because Θ_o is outer, $D_o = \Theta_o(\infty)$ is onto $\mathcal{V}$, or equivalently, right invertible. Lemma 4.5.2 below yields the realization for Θ_i in (4.5.5). □

Lemma 4.5.2. *Let $\{A, B, C, D\}$ be a realization for a transfer function F with values in $\mathcal{L}(\mathcal{E}, \mathcal{Y})$. Assume that $\{A, B, C_o, D_o\}$ is a realization for a transfer function Θ_o with values in $\mathcal{L}(\mathcal{E}, \mathcal{V})$ and D_o is onto. Consider the function $Q(z)$ defined by*

$$\begin{aligned} Q(z) &= D_o^{-r} - D_o^{-r} C_o \left(zI - A_i\right)^{-1} B D_o^{-r} \\ A_i &= A - B D^{-r} C_o. \end{aligned} \tag{4.5.7}$$

Then the following holds.

(i) *A realization for the transfer function $F(z)Q(z)$ is given by*

$$F(z)Q(z) = D D_o^{-r} + (C - D D_o^{-r} C_o)(zI - A_i)^{-1} B D_o^{-r}. \tag{4.5.8}$$

(ii) *The function Q is a right inverse of Θ_o, that is, $I = \Theta_o(z)Q(z)$. In particular, if D_o is invertible, then the function Q is the inverse of Θ_o.*

(iii) *Assume that $\{A, B, C, D\}$ is a realization for Θ in $H^\infty(\mathcal{E}, \mathcal{Y})$, and $\Theta = \Theta_i \Theta_o$ is the inner-outer factorization for Θ where Θ_o is outer and Θ_i is inner. Then Θ_i admits a realization of the form*

$$\Theta_i(z) = D D_o^{-r} + (C - D D_o^{-r} C_o)(zI - A_i)^{-1} B D_o^{-r}. \tag{4.5.9}$$

In particular, $\delta(\Theta_i) \leq \delta(\Theta)$.

Proof. Using the realizations for F and Θ_o, we arrive at

$$\begin{aligned}
FQ &= (D + C(zI - A)^{-1}B)(D_o^{-r} - D_o^{-r}C_o(zI - A_i)^{-1}BD_o^{-r} \\
&= DD_o^{-r} - DD_o^{-r}C_o(zI - A_i)^{-1}BD_o^{-r} \\
&\quad + C(zI - A)^{-1}[I - BD_o^{-r}C_o(zI - A_i)^{-1}]BD_o^{-r} \\
&= DD_o^{-r} - DD_o^{-r}C_o(zI - A_i)^{-1}BD_o^{-r} \\
&\quad + C(zI - A)^{-1}[(zI - (A - BD_o^{-r}C_o) - BD_o^{-r}C_o](zI - A_i)^{-1}BD_o^{-r} \\
&= DD_o^{-r} - DD_o^{-r}C_o(zI - A_i)^{-1}BD_o^{-r} + C(zI - A_i)^{-1}BD_o^{-r} \\
&= DD_o^{-r} + (C - DD_o^{-r}C_o)(zI - (A - BD_o^{-r}C_o))^{-1}BD_o^{-r}.
\end{aligned}$$

This yields the state space formula in (4.5.8), and Part (i) holds.

To verify Part (ii), notice that if $F = \Theta_o$, then $C = C_o$ and $D = D_o$. In this case, (4.5.8) shows that $\Theta_o Q = I$, and Part (ii) holds.

For Part (iii) assume that $\Theta = \Theta_i\Theta_o$ is the inner-outer factorizations for Θ. Because Θ_o is outer, $\Theta(\infty) = D_o$ is onto the whole space $\mathcal{V}$. Since $\Theta_o Q = I$, we see that $\Theta_i = \Theta_i\Theta_o Q = \Theta Q$. Therefore the state space realization for Θ_i in (4.5.9) follows from (4.5.8). □

Let us conclude this section with the following two useful results.

Proposition 4.5.3. *Let Ψ be a function in $H^\infty(\mathcal{E},\mathcal{Y})$ and set $R = \Psi\Psi^*$. Then*

$$T_R = T_\Psi T_\Psi^* + H_\Psi H_\Psi^*. \tag{4.5.10}$$

Moreover, assume Ψ is rational and $\{A \text{ on } \mathcal{X}, B, C, D\}$ is a minimal realization for Ψ. Then

$$\begin{aligned}
T_R &= T_\Psi T_\Psi^* + W_o Q W_o^*, \\
Q &= AQA^* + BB^*, \\
W_o &= \begin{bmatrix} C & CA & CA^2 & \cdots \end{bmatrix}^{tr} : \mathcal{X} \to \ell_+^2(\mathcal{Y}).
\end{aligned} \tag{4.5.11}$$

Proof. Because Ψ is a function in $H^\infty(\mathcal{E},\mathcal{Y})$, recall that the Laurent operator L_Ψ admits a matrix representation of the form

$$L_\Psi = \begin{bmatrix} \Lambda_\Psi & 0 \\ H_\Psi & T_\Psi \end{bmatrix} : \begin{bmatrix} \ell_-^2(\mathcal{E}) \\ \ell_+^2(\mathcal{E}) \end{bmatrix} \to \begin{bmatrix} \ell_-^2(\mathcal{Y}) \\ \ell_+^2(\mathcal{Y}) \end{bmatrix}.$$

Using this decomposition, we have

$$\begin{aligned}
\begin{bmatrix} \star & \star \\ \star & T_{\Psi\Psi^*} \end{bmatrix} &= L_{\Psi\Psi^*} = L_\Psi L_\Psi^* = \begin{bmatrix} \Lambda_\Psi & 0 \\ H_\Psi & T_\Psi \end{bmatrix} \begin{bmatrix} \Lambda_\Psi^* & H_\Psi^* \\ 0 & T_\Psi^* \end{bmatrix} \\
&= \begin{bmatrix} \Lambda_\Psi\Lambda_\Psi^* & \Lambda_\Psi H_\Psi^* \\ H_\Psi\Lambda_\Psi^* & T_\Psi T_\Psi^* + H_\Psi H_\Psi^* \end{bmatrix}.
\end{aligned}$$

Here $\star$ represents an unspecified entry. By comparing the components in the lower right-hand corner, we arrive at $T_{\Psi\Psi^*} = T_\Psi T_\Psi^* + H_\Psi H_\Psi^*$.

Let $\Psi(z) = \sum_0^\infty z^{-n}\Psi_n$ be the Taylor series expansion for Ψ. Assume that $\{A, B, C, D\}$ is a minimal realization for Ψ. Recall that $\Psi_n = CA^{n-1}B$ for all integers $n \geq 1$. Then H_Ψ admits a matrix representation of the form in (4.5.1) where Ψ replaces Θ. Hence $H_\Psi = W_o W_c$ where W_o is defined in (4.5.11) and

$$W_c = \begin{bmatrix} \cdots & A^2B & AB & B \end{bmatrix} : \ell_-^2(\mathcal{E}) \to \mathcal{X}.$$

Notice that $Q = W_c W_c^* = \sum_0^\infty A^n BB^* A^{*n}$. Thus

$$H_\Psi H_\Psi^* = W_o W_c W_c^* W_o^* = W_o Q W_o^*.$$

Substituting $H_\Psi H_\Psi^* = W_o Q W_o^*$ into (4.5.10) yields (4.5.11). □

Lemma 4.5.4. *Let $\{A$ on $\mathcal{X}, B, C, D\}$ be a stable realization for a $\mathcal{L}(\mathcal{E}, \mathcal{Y})$-valued transfer function Θ. Let P be the observability Gramian for $\{C, A\}$, that is, let P be the unique solution to the Lyapunov equation*

$$P = A^*PA + C^*C. \tag{4.5.12}$$

Moreover, let

$$T_R = \begin{bmatrix} R_0 & R_{-1} & R_{-2} & \cdots \\ R_1 & R_0 & R_{-1} & \cdots \\ R_2 & R_1 & R_0 & \cdots \\ \vdots & \vdots & \vdots & \ddots \end{bmatrix} \tag{4.5.13}$$

be a Toeplitz matrix generated by a $\mathcal{L}(\mathcal{E}, \mathcal{E})$-valued sequence $\{R_n\}_{-\infty}^\infty$ where $R_{-n} = R_n^*$ for all integers $n \geq 0$. Then $T_R = T_\Theta^* T_\Theta$, or equivalently, $R = \Theta^*\Theta$ if and only if

$$\begin{aligned} R_0 &= B^*PB + D^*D \\ R_n &= (B^*PA + D^*C)\, A^{n-1}B \qquad (n \geq 1). \end{aligned} \tag{4.5.14}$$

In this case, that is, when $R = \Theta^*\Theta$, we have

$$T_R^{tr} - T_{\widetilde{\Theta}} T_{\widetilde{\Theta}}^* = W^*PW \tag{4.5.15}$$

where $\widetilde{\Theta}(z) = \Theta(\overline{z})^*$ and W is the controllability operator for $\{A, B\}$ defined by

$$W = \begin{bmatrix} B & AB & A^2B & \cdots \end{bmatrix} : \ell_+^2(\mathcal{E}) \to \mathcal{X}. \tag{4.5.16}$$

Proof. Let $\Theta(z) = \sum_0^\infty z^{-n}\Theta_n$ be the Taylor series expansion for Θ. If $T_R = T_\Theta^* T_\Theta$, then we obtain

$$\begin{bmatrix} R_0 & R_{-1} & R_{-2} & \cdots \\ R_1 & R_0 & R_{-1} & \cdots \\ R_2 & R_1 & R_0 & \cdots \\ \vdots & \vdots & \vdots & \ddots \end{bmatrix} = \begin{bmatrix} \Theta_0^* & \Theta_1^* & \Theta_2^* & \cdots \\ 0 & \Theta_0^* & \Theta_1^* & \cdots \\ 0 & 0 & \Theta_0^* & \cdots \\ \vdots & \vdots & \vdots & \ddots \end{bmatrix} \begin{bmatrix} \Theta_0 & 0 & 0 & \cdots \\ \Theta_1 & \Theta_0 & 0 & \cdots \\ \Theta_2 & \Theta_1 & \Theta_0 & \cdots \\ \vdots & \vdots & \vdots & \ddots \end{bmatrix}.$$

By performing the appropriate matrix multiplications we arrive at

$$R_n = \sum_{j=0}^{\infty} \Theta_j^* \Theta_{j+n} \qquad (n \geq 0). \tag{4.5.17}$$

Because $\{A, B, C, D\}$ is a realization for Θ, we also have

$$\Theta_0 = D \quad \text{and} \quad \Theta_n = CA^{n-1}B \qquad (n \geq 1). \tag{4.5.18}$$

Recall that the solution to the Lyapunov equation (4.5.12) is given by $P = \sum_0^\infty A^{*n}C^*CA^n$. By combining this with (4.5.17) and (4.5.18), we obtain

$$\begin{aligned} R_0 &= \Theta_0^*\Theta_0 + \sum_{j=1}^{\infty} \Theta_j^*\Theta_j = D^*D + \sum_{j=1}^{\infty}(CA^{j-1}B)^*CA^{j-1}B \\ &= D^*D + B^*(\sum_{n=0}^{\infty} A^{*n}C^*CA^n)B = D^*D + B^*PB. \end{aligned}$$

This yields the first equation in (4.5.14). To obtain the second equation in (4.5.14), observe that for any integer $n \geq 1$, we have

$$\begin{aligned} R_n &= \Theta_0^*\Theta_n + \sum_{j=1}^{\infty} \Theta_j^*\Theta_{j+n} = D^*CA^{n-1}B + \sum_{j=1}^{\infty}(CA^{j-1}B)^*CA^{j+n-1}B \\ &= D^*CA^{n-1}B + B^*(\sum_{j=1}^{\infty} A^{*j-1}C^*CA^{j-1})A^nB \\ &= D^*CA^{n-1}B + B^*PA^nB = (D^*C + B^*PA)\, A^{n-1}B. \end{aligned}$$

Therefore if $T_R = T_\Theta^* T_\Theta$, then (4.5.14) holds.

On the other hand, if (4.5.14) holds, then

$$R_0 = D^*D + B^*PB = D^*D + B^*(\sum_{j=1}^{\infty} A^{*j-1}C^*CA^{j-1})B = \Theta_0^*\Theta_0 + \sum_{j=1}^{\infty} \Theta_j^*\Theta_j.$$

Moreover, for $n \geq 1$

$$\begin{aligned} R_n &= (D^*C + B^*PA)\, A^{n-1}B = D^*CA^{n-1}B + B^*PA^nB \\ &= D^*CA^{n-1}B + B^*(\sum_{j=1}^{\infty} A^{*j-1}C^*CA^{j-1})A^nB \\ &= \Theta_0^*\Theta_n + \sum_{j=1}^{\infty} \Theta_j^*\Theta_{j+n}. \end{aligned}$$

This implies that $T_R = T_\Theta^* T_\Theta$.

To complete the proof, it remains to show that $T_R^{tr} = T_{\widetilde{\Theta}} T_{\widetilde{\Theta}}^* + W^* P W$. Recall that $\widetilde{\Theta}(z) = \Theta(\overline{z})^*$. Using $R = \Theta^*\Theta$, we obtain

$$R(e^{-\imath\omega}) = \Theta(e^{-\imath\omega})^*\Theta(e^{-\imath\omega}) = \widetilde{\Theta}(e^{\imath\omega})\widetilde{\Theta}(e^{\imath\omega})^*.$$

So $R(e^{-\imath\omega}) = \widetilde{\Theta}\widetilde{\Theta}^*$. Notice that $T_{R(e^{-\imath\omega})} = T_R^{tr}$. Moreover, $\{A^*, C^*, B^*, D^*\}$ is a realization for $\widetilde{\Theta}$. For this realization, the observability operator W_o in (4.5.11) now becomes W^* in (4.5.16), and the controllability Gramian Q becomes the observability Gramian P. By replacing R with $R(e^{-\imath\omega})$ and Ψ with $\widetilde{\Theta}$ and W_o with W^* in Proposition 4.5.3, we arrive at (4.5.15). □

4.6 Inner-Outer Factorization and Finite Sections

In this section, we will present a finite section method to compute the inner-outer factorization for a rational function. Let Θ be a rational function in $H^\infty(\mathcal{E}, \mathcal{Y})$. Let $\Theta = \Theta_i\Theta_o$ be the inner-outer factorization for Θ where Θ_i is an inner function in $H^\infty(\mathcal{V}, \mathcal{Y})$ and Θ_o is an outer function in $H^\infty(\mathcal{E}, \mathcal{V})$. Let S be the unilateral shift on $\ell_+^2(\mathcal{Y})$. Because T_{Θ_i} is an isometry, all the columns of T_{Θ_i} are orthogonal. In particular, $T_{\Theta_i}\mathcal{V}$ is orthogonal to $ST_{\Theta_i}\ell_+^2(\mathcal{V})$, where $\mathcal{V}$ denotes the subspace of $\ell_+^2(\mathcal{V})$ corresponding to the first component of $\ell_+^2(\mathcal{V})$. Using this fact along with $T_\Theta = T_{\Theta_i}T_{\Theta_o}$ and $\ell_+^2(\mathcal{V}) = \overline{T_{\Theta_o}\ell_+^2(\mathcal{E})}$, we obtain

$$\begin{aligned}\mathcal{L} &= \overline{T_\Theta\ell_+^2(\mathcal{E})} \ominus ST_\Theta\ell_+^2(\mathcal{E}) = T_{\Theta_i}\ell_+^2(\mathcal{V}) \ominus ST_{\Theta_i}\ell_+^2(\mathcal{V})\\ &= \left(T_{\Theta_i}\mathcal{V} \oplus ST_{\Theta_i}\ell_+^2(\mathcal{V})\right) \ominus ST_{\Theta_i}\ell_+^2(\mathcal{V}) = T_{\Theta_i}\mathcal{V}.\end{aligned}$$

Observe that $\mathcal{L}$ is a wandering subspace for the unilateral shift. Let $\mathcal{M}$ be the closure of $ST_\Theta\ell_+^2(\mathcal{E})$. Clearly, $\mathcal{L}$ is orthogonal to $\mathcal{M}$. Let $\Omega = \begin{bmatrix}\Theta_0 & \Theta_1 & \Theta_2 & \cdots\end{bmatrix}^{tr}$ be the first column of T_Θ where $\Theta(z) = \sum_0^\infty z^{-k}\Theta_k$ is the power series expansion for Θ. This readily implies that

$$\begin{aligned}T_{\Theta_i}\mathcal{V} &= \overline{T_\Theta\ell_+^2(\mathcal{E})} \ominus \mathcal{M} = P_{\mathcal{L}}\overline{T_\Theta\ell_+^2(\mathcal{E})} = P_{\mathcal{L}}\left(\Omega\mathcal{E}\bigvee\mathcal{M}\right)\\ &= P_{\mathcal{L}}\Omega\mathcal{E} = \{T_\Theta a - P_{\mathcal{M}}T_\Theta a : a \in \mathcal{E}\}.\end{aligned}$$

This yields the result that we have been working for,

$$T_{\Theta_i}\mathcal{V} = \{T_\Theta a - P_{\mathcal{M}}T_\Theta a : a \in \mathcal{E}\}. \tag{4.6.1}$$

From this fact we can obtain an inner-outer factorization procedure for any rational function Θ in $H^\infty(\mathcal{E}, \mathcal{Y})$. The idea is to use the finite sections of T_Θ to form an orthogonal basis for the subspace $\mathcal{L}$. Then use the Kalman-Ho algorithm to extract a minimal realization $\{A_i, B_i, C_i, D_i\}$ for the inner part Θ_i of Θ. To compute a realization for the outer part, let $\{A, B, C, D\}$ be any minimal realization for Θ. Lemma 4.6.1 below shows that a realization $\{A_o, B_o, C_o, D_o\}$ for Θ_o is given by

$$\begin{aligned} A_o &= A \quad \text{and} \quad B_o = B \\ C_o &= D_i^* C + B_i^* P A \\ D_o &= D_i^* D + B_i^* P B \\ P &= A_i^* P A + C_i^* C. \end{aligned} \tag{4.6.2}$$

Finally, it is noted that $\{A_o, B_o, C_o, D_o\}$ may not be a minimal realization for Θ_o. So one may have to extract the minimal realization from $\{A_o, B_o, C_o, D_o\}$. One can also apply the Kalman-Ho algorithm on $\{A_o, B_o, C_o, D_o\}$ to obtain a minimal realization from $\{A_o, B_o, C_o, D_o\}$. The following presents a method to compute this inner-outer factorization for a rational function Θ when the range of T_Θ is closed. Finally, it is noted that one of the disadvantages of this method is that we do not know how to pick n a priori.

(i) For n sufficiently large compute Ω_n and the Toeplitz matrix Υ_n defined by

$$\Omega_n = \begin{bmatrix} \Theta_0 \\ \Theta_1 \\ \Theta_2 \\ \vdots \\ \Theta_{n-1} \end{bmatrix} \quad \text{and} \quad \Upsilon_n = \begin{bmatrix} 0 & 0 & \cdots & 0 & 0 \\ \Theta_0 & 0 & \cdots & 0 & 0 \\ \Theta_1 & \Theta_0 & \ddots & 0 & 0 \\ \vdots & \vdots & \ddots & \ddots & 0 \\ \Theta_{n-2} & \Theta_{n-3} & \cdots & \Theta_0 & 0 \end{bmatrix}.$$

Here Ω_n maps $\mathcal{E}$ into $\mathcal{Y}^n$ and Υ_n maps $\mathcal{E}^n$ into $\mathcal{Y}^n$. It is noted that $\Theta_0 = D$ and $\Theta_k = CA^{k-1}B$ for all integers $k \geq 1$ where $\Theta(z) = \sum_0^\infty z^{-n}\Theta_n$ is the power series expansion for Θ. (One can also compute $\{\Theta_j\}_0^{n-1}$ by using the fast Fourier transform.)

- Let $U\Lambda V^*$ be the singular value decomposition of Υ_n. Let j be the number of significant singular values of Υ_n, and U_j the first j columns of U.
- Compute the singular value decomposition $\widetilde{U}\widetilde{\Lambda}\widetilde{V}^*$ for $\Omega_n - U_j U_j^* \Omega_n$. Let k be the number of significant singular values of $\Omega_n - U_j U_j^* \Omega_n$ and $\widetilde{U}_k$ the first k columns of $\widetilde{U}$. For large n, the orthogonal projection $U_j U_j^*$ approximates $P_\mathcal{M}$, and the range of $\widetilde{U}_k$ approximates $T_{\Theta_i}\mathcal{V}$.
- Now use the Kalman-Ho algorithm on $\widetilde{U}_k$ to compute a state space realization $\{A_i, B_i, C_i, D_i\}$ for the inner part Θ_i of Θ. Check to make sure that the operator A_i is stable and

$$\begin{bmatrix} A_i^* & C_i^* \\ B_i^* & D_i^* \end{bmatrix} \begin{bmatrix} Q & 0 \\ 0 & I \end{bmatrix} \begin{bmatrix} A_i & B_i \\ C_i & D_i \end{bmatrix} = \begin{bmatrix} Q & 0 \\ 0 & I \end{bmatrix}, \tag{4.6.3}$$

where Q is the observability Gramian for the pair $\{C_i, A_i\}$; see Theorem 4.2.1. If this is not the case, then one may have to increase n.

- Then use (4.6.2) to compute a realization $\{A_o, B_o, C_o, D_o\}$ for the outer part of Θ_o. Compute a minimal realization from $\{A_o, B_o, C_o, D_o\}$ of Θ_o.
- Finally, it is noted that one can also compute a minimal $\{A_o, B_o, C_o, D_o\}$ for Θ_o by using the fast Fourier transform on $\Theta_o = \Theta_i^*\Theta$ to compute the Fourier series expansion $\Theta_o(z) = \sum_0^\infty z^{-k}\Theta_{o,k}$. Then applying the Kalman-Ho algorithm to $\{\Theta_{o,k}\}$ yields a minimal realization for Θ_o.

Lemma 4.6.1. *Assume that $\Theta = \Theta_i\Theta_o$ is an inner-outer factorization for a rational transfer function Θ in $H^\infty(\mathcal{E}, \mathcal{Y})$. Assume that $\{A, B, C, D\}$ is a minimal realization for Θ and $\{A_i, B_i, C_i, D_i\}$ is a minimal realization for Θ_i. Then a realization $\{A_o, B_o, C_o, D_o\}$ for Θ_o is given by*

$$\begin{aligned} A_o &= A \quad \textit{and} \quad B = B_o \\ C_o &= D_i^*C + B_i^*PA \\ D_o &= D_i^*D + B_i^*PB \\ P &= A_i^*PA + C_i^*C. \end{aligned} \tag{4.6.4}$$

It is noted that $\{A_o, B_o, C_o, D_o\}$ may not be a minimal realization for Θ_o. Finally, one can also apply the Kalman-Ho algorithm on $\{A_o, B_o, C_o, D_o\}$ to compute a minimal realization from $\{A_o, B_o, C_o, D_o\}$.

Proof. First let us recall that the solution P to the Lyapunov equation P in (4.6.4) is determined by

$$P = \sum_{j=0}^\infty A_i^{*j}C_i^*CA^j. \tag{4.6.5}$$

For z on the unit circle, we must have

$$\begin{aligned} \Theta_o(z) &= \Theta_i(z)^*\Theta(z) = \left(D_i^* + B_i^*(\overline{z}I - A_i^*)^{-1}C_i^*\right)\left(D + C(zI - A)^{-1}B\right) \\ &= \left(D_i^* + \sum_{j=1}^\infty z^j B_i^*A_i^{*j-1}C_i^*\right)\left(D + \sum_{k=1}^\infty \overline{z}^k CA^{k-1}B\right) \\ &= D_i^*D + \sum_{j=1}^\infty B_i^*A_i^{*j-1}C_i^*CA^{j-1}B \\ &\quad + \sum_{k=1}^\infty \overline{z}^k\left(D_i^*CA^{k-1}B + \sum_{j=1}^\infty B_i^*A_i^{*j-1}C_i^*CA^{j-1}A^kB\right) \\ &= D_i^*D + B_i^*PB + \sum_{k=1}^\infty \overline{z}^k\left(D_i^*CA^{k-1}B + B_i^*PAA^{k-1}B\right) \\ &= (D_i^*D + B_i^*PB) + (D_i^*C + B_i^*PA)(zI - A)^{-1}B \\ &= D_o + C_o(zI - A_o)^{-1}B_o. \end{aligned}$$

Because Θ_o is in $H^\infty(\mathcal{E}, \mathcal{V})$, we only had to collect the Fourier coefficients of z^{-k} for all integers $k \geq 0$. Thus $\{A_o, B_o, C_o, D_o\}$ in (4.6.4) is a realization for Θ_o. □

Example. Consider the rational transfer function $\Theta = \begin{bmatrix}\theta_1 & \theta_2\end{bmatrix}$ where

$$\theta_1 = \frac{0.005855z^6 - 0.006508z^5 - 0.05401z^4 + 0.02012z^3 + 0.1136z^2 + 0.03977z + 0.002669}{d},$$
$$\theta_2 = \frac{0.0074z^5 - 0.01646z^4 - 0.04998z^3 + 0.08018z^2 + 0.05736z - 0.008991}{d},$$
$$d = z^6 - 0.0855z^5 - 0.3994z^4 + 0.1777z^3 - 0.01376z^2 - 0.03373z + 0.0101.$$

By choosing $n = 150$ in the previous algorithm, we computed the inner factor Θ_i and outer factor Θ_o for Θ. In our computations, $\Theta_o = \begin{bmatrix}\theta_{1o} & \theta_{2o}\end{bmatrix}$ where

$$\theta_{1o} = \frac{0.07838z^3 + 0.1299z^2 + 0.0408z + 0.002669}{d_o}$$
$$\theta_{2o} = \frac{0.09908z^2 + 0.0539z - 0.008991}{d_o}$$
$$d_o = z^3 + 0.2997z^2 - 0.07713z + 0.1352$$
$$\Theta_i = \frac{0.07469z^3 - 0.2068z^2 - 0.3852z + 1}{z^3 - 0.3852z^2 - 0.2068z + 0.07469}.$$

Using the fast Fourier transform, we computed: $\|\Theta - \Theta_i\Theta_o\|_\infty \approx 8.38 \times 10^{-5}$. Finally, it is noted that the outer factor is not square and this may cause problems in some inner-outer factorization techniques.

4.7 The Douglas-Shapiro-Shields Factorization

In the rest of this chapter, we will present the Douglas-Shapiro-Shields factorization for certain operator-valued functions. This allows one to factor certain functions into a causal and anti-causal part. Since these results are not used anywhere else in the monograph, an uninterested reader can move on to the next chapter.

Let K be a function in $H^\infty(\mathcal{D}, \mathcal{Y})$. Recall that Ψ is a *left inner divisor* or *left inner factor* of K if Ψ is an inner function in $H^\infty(\mathcal{E}, \mathcal{Y})$ and K admits a factorization of the form $K = \Psi K_2$ where K_2 is a function in $H^\infty(\mathcal{D}, \mathcal{E})$.

We claim that Ψ is a left inner divisor of K if and only if Ψ is a left inner divisor of K_i where $K_i \in H^\infty(\mathcal{U}, \mathcal{Y})$ is the inner part of K. If Ψ is a left inner divisor of K_i, then clearly, Ψ is a left inner divisor of K. On the other hand, if Ψ is a left inner divisor of K, then

$$K_i H^2(\mathcal{U}) = K_i\overline{K_o H^2(\mathcal{D})} = \overline{K H^2(\mathcal{D})} = \overline{\Psi K_2 H^2(\mathcal{D})} \subseteq \Psi H^2(\mathcal{E}).$$

Here $K = K_i K_o$ where K_o is the outer part of K. Since $K_i H^2(\mathcal{U}) \subseteq \Psi H^2(\mathcal{E})$, the function Ψ is a left inner divisor of K_i; see Remark 3.1.4 in Section 3.1.1.

Let Θ be an inner function in $H^\infty(\mathcal{E}, \mathcal{Y})$ and K a function in $H^\infty(\mathcal{D}, \mathcal{Y})$. Then we say that Θ and K are *prime on the left* if the only common left inner divisor between Θ and K is a unitary constant. Clearly, Θ and K are prime on the left if and only if Θ and the inner part of K are prime on the left. Finally, Remark 3.1.4 shows that Θ and K are prime on the left if and only if

$$H^2(\mathcal{Y}) = \Theta H^2(\mathcal{E}) \bigvee K H^2(\mathcal{D}).$$

Let F be a function in $L^\infty(\mathcal{E}, \mathcal{Y})$. Then we say that F admits a *Douglas-Shapiro-Shields factorization* if

$$F(e^{\imath\omega}) = \Theta(e^{\imath\omega})G(e^{\imath\omega})^* \quad \text{almost everywhere} \tag{4.7.1}$$

where Θ is a two-sided inner function in $H^\infty(\mathcal{Y}, \mathcal{Y})$ and G is a function in $H^\infty(\mathcal{Y}, \mathcal{E})$. The Douglas-Shapiro-Shields factorization $F = \Theta G^*$ is *canonical* if $\widetilde{\Theta}$ and $\widetilde{G}$ are prime on the left, that is, the only common inner factor between $\widetilde{\Theta}$ and $\widetilde{G}$ is a constant unitary operator. (If Ω is an operator-valued analytic function in $\mathbb{D}_+$, then $\widetilde{\Omega}(z) = \Omega(\overline{z})^*$.) Because Θ is a two-sided inner function, Λ is a left inner factor for $\widetilde{\Theta}$ if and only if Λ is two-sided inner and $\widetilde{\Lambda}$ is a right inner divisor for Θ, that is, $\Theta = \Upsilon\widetilde{\Lambda}$ where Υ is a two-sided inner function. So the Douglas-Shapiro-Shields factorization $F = \Theta G^*$ is canonical if and only if the only common right inner divisor between Θ and G is a unitary constant, or equivalently, Θ and G are prime on the right.

By the Beurling-Lax-Halmos Theorem 3.1.1, the set of all invariant subspaces for the backward shift operator S^* on $H^2(\mathcal{Y})$ are given by $\mathcal{H}(\Theta) = H^2(\mathcal{Y}) \ominus \Theta H^2(\mathcal{E})$ where Θ is an inner function in $H^\infty(\mathcal{E}, \mathcal{Y})$.

Theorem 4.7.1 (Douglas-Shapiro-Shields). *Consider a function F in $L^\infty(\mathcal{E}, \mathcal{Y})$. Let S be the unilateral shift on $H^2(\mathcal{Y})$ and P_+ the orthogonal projection from $L^2(\mathcal{Y})$ onto $H^2(\mathcal{Y})$. Then the following holds.*

(i) *F admits a Douglas-Shapiro-Shields factorization of the form $F = \Theta G^*$ where Θ is an inner function in $H^\infty(\mathcal{Y}, \mathcal{Y})$ if and only if $S^* P_+ F\mathcal{E} \subseteq \mathcal{H}(\Theta)$.*

(ii) *The function F admits a Douglas-Shapiro-Shields factorization if and only if the invariant subspace for the backward shift*

$$\bigvee_{n=1}^{\infty} S^{*n} P_+ F\mathcal{E} = \mathcal{H}(\Theta) \tag{4.7.2}$$

is determined by an inner function Θ in $H^\infty(\mathcal{Y}, \mathcal{Y})$. In this case, the Douglas-Shapiro-Shields factorization is of the form $F = \Theta G^*$ and this factorization is canonical.

(iii) *] All canonical Douglas-Shapiro-Shields factorizations of F are unique up to a unitary constant factor on the right, that is, if $F = \Theta G^* = \Psi Q^*$ are two* *canonical Douglas-Shapiro-Shields factorizations of F where Ψ is an inner function in $H^\infty(\mathcal{Y},\mathcal{Y})$ and Q is a function in $H^\infty(\mathcal{Y},\mathcal{E})$, then $\Psi = \Theta\Omega$ and $Q = G\Omega$ where Ω is a unitary constant.*

Proof. Let Θ be an inner function in $H^\infty(\mathcal{Y},\mathcal{Y})$. We claim that $S^*P_+F\mathcal{E} \subseteq \mathcal{H}(\Theta)$ if and only if F admits a Douglas-Shapiro-Shields factorization of the form $F = \Theta G^*$ where G is in $H^\infty(\mathcal{Y},\mathcal{E})$. To see this observe that the following statements are equivalent:

$$\begin{aligned}
S^*P_+F\mathcal{E} \subseteq \mathcal{H}(\Theta) &\Leftrightarrow S^*P_+F\mathcal{E} \perp \Theta H^2(\mathcal{Y}) \Leftrightarrow P_+F\mathcal{E} \perp S\Theta H^2(\mathcal{Y})\\
&\Leftrightarrow F\mathcal{E} \perp e^{-\imath\omega}\Theta H^2(\mathcal{Y}) \Leftrightarrow \Theta^*F\mathcal{E} \perp e^{-\imath\omega} H^2(\mathcal{Y})\\
&\Leftrightarrow \Theta(e^{\imath\omega})^*F(e^{\imath\omega}) = \sum_{n=0}^{\infty} A_n e^{\imath\omega n} \text{ where } A_n \in \mathcal{L}(\mathcal{E},\mathcal{Y})\\
&\Leftrightarrow F^*\Theta \in H^\infty(\mathcal{Y},\mathcal{E})\\
&\Leftrightarrow F^*\Theta = G \quad \text{where } G \in H^\infty(\mathcal{Y},\mathcal{E})\\
&\Leftrightarrow F^* = G\Theta^* \quad \text{where } G \in H^\infty(\mathcal{Y},\mathcal{E})\\
&\Leftrightarrow F = \Theta G^* \quad \text{where } G \in H^\infty(\mathcal{Y},\mathcal{E}).
\end{aligned}$$

Hence $S^*P_+F\mathcal{E} \subseteq \mathcal{H}(\Theta)$ if and only if $F = \Theta G^*$ where G is in $H^\infty(\mathcal{Y},\mathcal{E})$. Therefore Part (i) holds.

To prove Part (ii), assume that the invariant subspace $\bigvee_1^\infty S^{*n}P_+F\mathcal{E}$ for the backward shift is determined by an inner function Θ in $H^\infty(\mathcal{Y},\mathcal{Y})$. Clearly, $S^*P_+F\mathcal{E} \subseteq \mathcal{H}(\Theta)$. According to Part (i), the function F admits a Douglas-Shapiro-Shields factorization of the form $F = \Theta G^*$. On the other hand, assume that F admits a Douglas-Shapiro-Shields factorization $F = \Psi Q^*$ where Ψ is an inner function in $H^\infty(\mathcal{Y},\mathcal{Y})$. Part (i) implies that $S^*P_+F\mathcal{E} \subseteq \mathcal{H}(\Psi)$. Because $\mathcal{H}(\Psi)$ is an invariant subspace for the backward shift S^*, we see that

$$\mathcal{H}(\Theta) = \bigvee_{n=1}^{\infty} S^{*n}P_+F\mathcal{E} \subseteq \mathcal{H}(\Psi).$$

Here Θ is an inner function determined by the invariant subspace $\bigvee_{n=1}^\infty S^{*n}P_+F\mathcal{E}$ for S^*. Because $\mathcal{H}(\Theta) \subseteq \mathcal{H}(\Psi)$, it follows that Θ is a left inner divisor of Ψ; see Section 3.1.1. In other words, $\Psi = \Theta\Omega$. Since $\Psi(e^{\imath\omega})$ is almost everywhere a unitary operator, and $\Theta(e^{\imath\omega})$ is almost everywhere an isometry, it follows that $\Theta(e^{\imath\omega})$ is almost everywhere a unitary operator. In other words, Θ is a two-sided inner function. Multiplying Θ by the appropriate constant unitary operator on the right, without loss of generality, we can assume that Θ is an inner function in $H^\infty(\mathcal{Y},\mathcal{Y})$. Clearly, $S^*P_+F\mathcal{E} \subseteq \mathcal{H}(\Theta)$. Therefore F admits a Douglas-Shapiro-Shields factorization of the form $F = \Theta G^*$. In other words, the first sentence in Part (ii) holds.

Now assume that the invariant subspace $\bigvee_{n=1}^\infty S^{*n}P_+F\mathcal{E} = \mathcal{H}(\Theta)$ determines an inner function in $H^\infty(\mathcal{Y},\mathcal{Y})$. Then F admits a Douglas-Shapiro-Shields factorization of the form $F = \Theta G^*$. We claim that the factorization $F = \Theta G^*$ is canonical. To see this assume that $\Theta = \Upsilon\Phi$ and $G = Q\Phi$ where Φ is a two-sided inner function and Q is a function in $H^\infty(\mathcal{Y}, \mathcal{E})$. Because Υ is a left inner divisor for Θ, we have $\mathcal{H}(\Upsilon) \subseteq \mathcal{H}(\Theta)$. Moreover, $F = \Theta G^* = \Upsilon Q^*$, and thus, $\mathcal{H}(\Theta) \subseteq \mathcal{H}(\Upsilon)$. Therefore $\mathcal{H}(\Theta) = \mathcal{H}(\Upsilon)$. In other words, Θ equals Υ up to a unitary constant on the right; see Section 3.1.1. Hence Φ must be a unitary constant and the factorization $F = \Theta G^*$ is canonical. Therefore Part (ii) holds.

To complete the proof, it remains to show that two canonical factorizations are equal up to a unitary constant on the right. Assume that $F = \Psi Q^*$ is another canonical factorization and $F = \Theta G^*$ is the canonical factorization in Part (ii). Recall that $\mathcal{H}(\Theta) \subseteq \mathcal{H}(\Psi)$ and Θ is a left inner divisor of Ψ, that is, $\Psi = \Theta\Omega$ where Ω is a two-sided inner function. This readily implies that

$$\Theta G^* = F = \Psi Q^* = \Theta\Omega Q^*.$$

Since Θ is almost everywhere a unitary operator on the unit circle, $G^* = \Omega Q^*$, or equivalently, $Q = G\Omega$. Hence Ω is a right two-sided inner divisor for both Ψ and Q. Because the factorization is canonical, Ω must be a unitary constant. In other words, any canonical factorization of F is equal to the canonical factorization $F = \Theta G^*$ in Part (ii) up to a constant unitary operator on the right. Therefore any two canonical factorizations of F are equal up to a constant unitary operator on the right. □

Now assume that F is a function in $H^\infty(\mathcal{E}, \mathcal{Y})$. Then $S^*P_+F\mathcal{E} = S^*F\mathcal{E}$. Theorem 4.7.1, readily yields the following result.

Corollary 4.7.2. *Consider a function F in $H^\infty(\mathcal{E}, \mathcal{Y})$. Let S be the unilateral shift on $H^2(\mathcal{Y})$. Then the following holds.*

(i) *F admits a Douglas-Shapiro-Shields factorization of the form $F = \Theta G^*$ where Θ is an inner function in $H^\infty(\mathcal{Y}, \mathcal{Y})$ if and only if $S^*F\mathcal{E} \subseteq \mathcal{H}(\Theta)$.*

(ii) *The function F admits a Douglas-Shapiro-Shields factorization if and only if the invariant subspace for the backward shift*

$$\bigvee_{n=1}^{\infty} S^{*n}F\mathcal{E} = \mathcal{H}(\Theta) \tag{4.7.3}$$

is determined by an inner function Θ in $H^\infty(\mathcal{Y}, \mathcal{Y})$. In this case, the Douglas-Shapiro-Shields factorization is of the form $F = \Theta G^*$ and this factorization is canonical.

(iii) *All canonical Douglas-Shapiro-Shields factorizations of F are unique up to a unitary constant factor on the right.*

4.8 The Douglas-Shapiro-Shields Factorization for Rational Functions

Let F be a rational transfer function in $H^\infty(\mathcal{E},\mathcal{Y})$. Then F admits a canonical Douglas-Shapiro-Shields factorization of the form $F = \Theta G^*$ where Θ is a two-sided rational inner function in $H^\infty(\mathcal{Y},\mathcal{Y})$ and G is a rational transfer function in $H^\infty(\mathcal{Y},\mathcal{E})$.

Remark 4.3.5 shows that $\bigvee_{n=1}^\infty S^{*n}F\mathcal{E} = \mathcal{H}(\Theta)$ where Θ is an inner function in $H^\infty(\mathcal{Y},\mathcal{Y})$. Hence F admits a canonical Douglas-Shapiro-Shields factorization of the form $F = \Theta G^*$; see Part (ii) in Corollary 4.7.2. To compute the Douglas-Shapiro-Shields factorization $F = \Theta G^*$, let $\{A, B_1, C, D_1\}$ be a minimal realization for F. (Because F is in $H^\infty(\mathcal{E},\mathcal{Y})$ and the realization is minimal, it must also be stable.) Use Remark 4.3.4 to compute the operators B mapping $\mathcal{Y}$ into $\mathcal{X}$ and D on $\mathcal{Y}$ such that $\{A, B, C, D\}$ is a minimal realization for Θ; see also Remark 4.3.5. To compute a realization for G, let P be the observability Gramian for the pair $\{C, A\}$, that is, compute the unique solution P to the Lyapunov equation

$$P = A^*PA + C^*C.$$

Next compute

$$C_2 = D_1^*C + B_1^*PA \quad \text{and} \quad D_2 = D_1^*D + B_1^*PB. \tag{4.8.1}$$

Then $\{A, B, C_2, D_2\}$ is a controllable realization for G.

It is noted that $\{A, B, C_2, D_2\}$ may not be a minimal realization. For example, if $F = \Theta$ is a nontrivial two-sided inner function, then $G = I$. In this case, we can choose $B = B_1$, $D = D_1$. Then $C_2 = 0$ and $D_2 = I$; see (4.2.4) in Theorem 4.2.1. Clearly, $\{A, B, C_2, D_2\}$ is not observable.

Remark 4.3.5 shows that $\{A, B, C, D\}$ is a minimal realization for Θ. To complete our derivation, it remains to show that $\{A, B, C_2, D_2\}$ is a realization for G. Using $F = \Theta G^*$, it follows that $G(z) = F(z)^*\Theta(z)$ for all z on the unit circle $\mathbb{T}$. Since G is in $H^\infty(\mathcal{Y},\mathcal{E})$, the function G admits a Fourier series expansion of the form $G = \sum_0^\infty z^{-n}G_n$. Hence

$$\begin{aligned} G(z) &= F(z)^*\Theta(z) = \left(D_1^* + B_1^*(\overline{z}I - A^*)^{-1}C^*\right)\left(D + C(zI - A)^{-1}B\right) \\ &= \left(D_1^* + \sum_{j=1}^\infty z^j B_1^*A^{*j-1}C^*\right)\left(D + \sum_{k=1}^\infty \overline{z}^k CA^{k-1}B\right) = \sum_{j=0}^\infty z^{-n}G_n. \end{aligned}$$

By matching like coefficients of z^j we obtain

$$G_0 = D_1^*D + \sum_{j=0}^\infty B_1^*A^{*j}C^*CA^jB = D_1^*D + B_1^*PB = D_2,$$

$$\begin{aligned}G_n &= D_1^*CA^{n-1}B + \sum_{j=0}^{\infty} B_1^*A^{*j}C^*CA^jA^nB \\ &= (D_1^*C + B_1^*PA)\,A^{n-1}B = C_2A^{n-1}B,\end{aligned}$$

where $n \geq 1$. Here we used the fact that $P = \sum_0^\infty A^{*j}C^*CA^j$. In other words, $G_0 = D_2$ and $G_n = C_2A^{n-1}B$ for all integers $n \geq 1$. Therefore $\{A, B, C_2, D_2\}$ is a realization for G.

4.8.1 A Factorization for rational L^∞ functions

Assume that F is a rational function in $L^\infty(\mathcal{E}, \mathcal{Y})$. Then F admits a Douglas-Shapiro-Shields factorization of the form $F = \Theta G^*$ where Θ is a two-sided rational inner function in $H^\infty(\mathcal{Y}, \mathcal{Y})$ and G is a rational function in $H^\infty(\mathcal{Y}, \mathcal{E})$. The following steps can be used to compute Θ and G:

(i) Use fast Fourier transform (FFT) techniques to compute the Fourier coefficients $\{F_j\}_0^m$ of $F = \sum_{-\infty}^{\infty} e^{-\imath\omega n}F_n$ where m is sufficiently large.

(ii) Apply the Kalman-Ho algorithm to find a minimal state space realization $\{A, B_1, C, D_1\}$ for $\{F_j\}_0^m$ with m sufficiently large. The Kalman-Ho algorithm is classical and presented in Section 14.5.

(iii) Use Remark 4.3.4 to compute the operators B mapping $\mathcal{Y}$ into $\mathcal{X}$ and D on $\mathcal{Y}$ in (4.3.5). Then $\{A, B, C, D\}$ is a minimal realization for Θ.

(iv) Use the FFT to compute F and Θ at 2^k points around the unit circle, and then compute $F^*\Theta$ numerically. By employing the inverse FFT compute the Fourier coefficients $\{G_n\}_0^m$ where $G = F^*\Theta = \sum_0^\infty e^{-\imath\omega n}G_n$ with m sufficiently large.

(v) Apply the Kalman-Ho algorithm to find a minimal state space realization $\{A_2, B_2, C_2, D_2\}$ for $\{G_j\}_0^m$. Then $\{A_2, B_2, C_2, D_2\}$ is a minimal realization for G.

To see why this algorithm works, observe that $F = F_+ + F_-$ where $F_+ = \sum_0^\infty e^{-\imath\omega n}F_n$ and $F_- = \sum_{-\infty}^{-1} e^{-\imath\omega n}F_n$. Since F is a rational function in $L^\infty(\mathcal{E}, \mathcal{Y})$, the function F_+ is a rational function in $H^\infty(\mathcal{E}, \mathcal{Y})$, and F_- is a rational function in $L^\infty(\mathcal{E}, \mathcal{Y})$. Hence F_+ admits a minimal stable realization $\{A, B_1, C, D_1\}$. Moreover, this realization can be computed from the Kalman-Ho algorithm applied to $\{F_j\}_0^m$ for m sufficiently large. Using the decomposition $F = F_+ + F_-$, we obtain

$$\mathcal{H} = \bigvee_{n=1}^{\infty} S^{*n}P_+F\mathcal{E} = \bigvee_{n=1}^{\infty} S^{*n}F_+\mathcal{E}.$$

According to Theorem 4.7.1 and Corollary 4.7.2, the functions F and F_+ both admit canonical Douglas-Shapiro-Shields factorizations of the form $F = \Theta G^*$ and $F_+ = \Theta G_+^*$ with the same two-sided inner function Θ. Moreover, if $\{A, B_1, C, D_1\}$

is a realization for F_+, then the minimal realization $\{A, B, C, D\}$ for Θ is computed according to Remark 4.3.4; see also Remark 4.3.5. This verifies the first three parts of our algorithm. Since $F = \Theta G^*$ and Θ is two-sided inner, we have $\sum_0^\infty e^{-\imath\omega n} G_n = G = F^*\Theta$. Because F and Θ are rational, G is a rational function in $H^\infty(\mathcal{Y}, \mathcal{E})$. Furthermore, the Fourier coefficients $\{G_j\}_0^m$ can be computed using FFT techniques. Finally, one can use the Kalman-Ho algorithm to compute a minimal realization $\{A_2, B_2, C_2, D_2\}$ for G directly from $\{G_j\}_0^m$ for m sufficiently large.

Example. Let us compute the canonical Douglas-Shapiro-Shields factorization $f = \theta g^*$ for the function f in L^∞ given by

$$f(z) = \frac{z^4 + 2z^3 + 3z^2 + 4z + 5}{z^5 - 4.75z^4 + 4.625z^3 + 2.125z^2 - 0.75z}.$$

Notice that $f = \sum_{-\infty}^\infty e^{-i\omega k} f_k$ where the Fourier coefficients $\{f_k\}$ were computed by the following Matlab commands:

- $f = \text{fft}([0, 1, 2, 3, 4, 5], 2^{14})./\text{fft}([1, -4.75, 4.625, 2.125, -0.75], 2^{14})$;
- $[f_0, f_1, f_2, f_3, \ldots, f_{-3}, f_{-2}, f_{-1}] = \text{real}(\text{ifft}(f))$;

It is emphasized that $[f_0, f_1, f_2, \ldots, f_{-3}, f_{-2}, f_{-1}]$ is a row vector of length 2^{14} which approximates the Fourier coefficients $\{f_k\}$ for f. (We use real(ifft(f)) because we know that the Fourier coefficients $\{f_k\}$ are real, and we wanted to eliminate any small imaginary numbers which may numerically enter in computing $\{f_k\}$.) Then we ran the Kalman-Ho algorithm on $\{f_k\}_0^{500}$ and computed a realization $\{A, B_1, C, D_1\}$ for $f_+(z) = \sum_0^\infty z^{-k} f_k$. Now using Remark 4.3.4, we computed B and D such that $\{A, B, C, D\}$ is a realization for θ. By computing the transfer function for $\{A, B, C, D\}$ (use "ss2tf" command in Matlab), we arrived at

$$\theta(z) = \frac{-0.125z^2 + 0.25z + 1}{z^3 + 0.25z^2 - 0.125z}.$$

The poles of θ are $\{0, 1/4, -1, 2\}$. Moreover, θ is of the form $\theta = p^\natural/zp$. So θ is indeed an inner function. To compute $g \in H^\infty$ such that $f = \theta g^*$, we used the following Matlab commands:

- $\theta = \text{fft}([0, -0.125, 0.25, 1], 2^{14})./\text{fft}([1, 0.25, -0.125], 2^{14})$;
- $g = \text{conj}(f). * \theta$;
- $[g_0, g_1, g_2, g_3, \ldots, 0, 0, 0] = \text{real}(\text{ifft}(g))$;

Observe that $[g_0, g_1, g_2, g_3, \ldots, 0, 0, 0]$ is a row vector of length 2^{14} which approximates the Fourier coefficients $\{g_k\}$ for g. Next, we ran the Kalman-Ho algorithm on $\{g_k\}_0^{500}$ and computed a realization $\{A_2, B_2, C_2, D_2\}$ for $g(z) = \sum_0^\infty z^{-k} g_k$. Then computing the transfer function for this g in Matlab, we arrived at

$$g(z) = \frac{0.8333z^4 + 0.6667z^3 + 0.5z^2 + 0.3333z + 0.1667}{z^4 - 0.5833z^3 - 0.1667z^2 + 0.1458z - 0.02083}.$$

The poles for g are $\{\pm 1/2, 1/3, 1/4\}$. So indeed g is a function in H^∞. To check to see if our answer is correct, we computed

- $n = \text{fft}([0.8333, 0.6667, 0.5, 0.3333, 0.1667], 2^{14})$;
- $q = \text{fft}([1, -0.5833, -0.1667, 0.1458, -0.0208], 2^{14})$;
- $g = n./q$;
- $\text{norm}(f - \theta. * \text{conj}(g), \text{inf}) = 1.3337 \times 10^{-14}$.

In other words, $\|f - \theta g\|_\infty = 1.3337 \times 10^{-14}$. So numerically, $f = \theta g^*$ is the canonical Douglas-Shapiro-Shields factorization for f.

4.9 Notes

All the results in this section are well known. For further results on Blaschke products and H^∞ functions see Duren [76], Granett [106], Hoffman [134] and Koosis [151]. Our approach to a state space realization theory for an inner function is a special case of the theory of unitary systems in operator theory. The theory of unitary systems started with Livšic [163, 164]. Then using dilation theory Sz.-Nagy-Foias developed the characteristic function; see [198]. The characteristic function is a unitary system which plays a fundamental role in operator theory. For further details on unitary systems see Brodskii [43, 44], Chapter 28 in Gohberg-Goldberg-Kaashoek [114], Arocena [15] and Arov [19, 20]. The system matrix also plays a fundamental role in control systems; see Zhou-Doyle-Glover [204]. The results in Section 4.5 were taken from some joint work with M.A. Kaashoek and A.C.M. Ran. The results in Section 4.6 are from Bhosri-Du-Frazho [33]. Here we used finite sections, to compute the inner-outer factorization. For further results on the finite section method see Böttcher-Silbermann [36], Gohberg-Goldberg-Kaashoek [112] and Lindner [157]. In Chapter 10 we will present the classical Riccati method to compute the inner-outer factorization; see also Section 10.8. The Douglas-Shapiro-Shields factorization discussed in Section 4.7 is due to Douglas-Shapiro-Shields [73]. Our approach to the Douglas-Shapiro-Shields factorization is taken from Foias-Frazho [82]. An in depth discussion of the Douglas-Shapiro-Shields factorization is given in Cima-Ross [56]. For an application of the Douglas-Shapiro-Shields factorization to systems theory see Fuhrmann [105]. Finally, reference [94] uses the Douglas-Shapiro-Shields factorization to study stochastic realization theory.

A classical formula for outer functions. Let us present a method based on the fast Fourier transform (FFT) and the Kalman-Ho algorithm to compute the inner-outer factorization for a scalar-valued rational function. Later, we will present some state space techniques along with the Levinson algorithm to compute the inner-outer factorization for rational functions in $H^\infty(\mathcal{E}, \mathcal{Y})$.

It is well known that if g is a function in H^2, then $\ln |g(e^{\imath\omega})|$ is integrable with respect to Lebesgue measure. Moreover, an explicit formula for the outer factor g_o of g is given by

$$g_o(z) = \mu \exp\left(\frac{1}{2\pi}\int_0^{2\pi} \frac{z + e^{\imath\omega}}{z - e^{\imath\omega}} \ln|g(e^{\imath\omega})|\, d\omega\right),$$

where μ is a constant of modulus 1; see [187, 76, 134, 151] for details. We will not derive this formula. However, we will demonstrate how one can use this formula along with the fast Fourier transform and the Kalman-Ho algorithm to compute the inner-outer factorization for scalar-valued rational functions. This formula can be rewritten in the following way, which is more suitable for using the FFT to compute the outer spectral factor:

$$\begin{aligned} g_o(z) &= \mu e^{h(z)}, \\ h(z) &= \frac{a_0}{2} + \sum_{k=1}^{\infty} \frac{a_k}{z^k}, \\ a_k &= \frac{1}{2\pi}\int_0^{2\pi} e^{\imath\omega k} \ln|g(e^{\imath\omega})|^2\, d\omega \qquad (k \geq 0). \end{aligned} \tag{4.9.1}$$

Notice that $\ln|g(e^{\imath\omega})|^2 = \sum_{-\infty}^{\infty} a_k e^{-\imath\omega k}$, and $\{a_k\}_0^\infty$ are the Fourier coefficients for $\ln|g|^2$ corresponding to $\{e^{-\imath\omega k}\}_0^\infty$.

To see how one can use the fast Fourier transform (FFT) to compute the inner-outer factorization via (4.9.1), assume that

$$\begin{aligned} g &= \frac{p}{q} = \frac{p_n z^n + p_{n-1}z^{n-1} + \cdots + p_1 z + p_0}{q_n z^n + q_{n-1}z^{n-1} + \cdots + q_1 z + q_0}, \\ \text{num} &= \begin{bmatrix} p_n & p_{n-1} & \cdots & p_1 & p_0 \end{bmatrix}, \\ \text{den} &= \begin{bmatrix} q_n & q_{n-1} & \cdots & q_1 & q_0 \end{bmatrix} \end{aligned}$$

is a rational function in H^2, where p and q are polynomials of degree at most n. The method does not require p and q to be prime, that is, p and q can have common roots. (One can use the FFT to check to see if g is indeed a rational function in H^2; see Part (i) below.) The following algorithm computes the inner-outer factorization for g.

(i) Use the FFT to compute $g(e^{\imath\omega})$ at 2^j points on the unit circle. (A typical value of j is 12, 13 or 14.) In Matlab,

$$g = \text{fft}(\text{num}, 2^j)./\text{fft}(\text{den}, 2^j).$$

(Set $mg = \text{ifft}(g)$. Then g is a function in H^2 if and only if in Matlab we have that $\text{norm}(mg(2^{j/2} : 2^j))$ is numerically close to zero.)

(ii) Apply the inverse FFT to compute the Fourier coefficients $\{a_k\}$ of $\ln|g(e^{\imath\omega})|^2$. In Matlab, $a = 2 * \text{ifft}(\log(\text{abs}(g)))$. (In Matlab there is no zero index. So $a_k \approx a(k+1)$.)

(iii) Find the FFT of $\{a_k\}$, to compute h and set $g_o = e^h$. In Matlab,

$$h = \text{fft}([a(1)/2, a(2 : 2^{j/2})], 2^j) \quad \text{and} \quad g_o = \exp(h).$$

(iv) Compute the Fourier coefficients $\{g_{ok}\}_0^\infty$ by taking the inverse FFT of $g_o = \sum_0^\infty z^{-k} g_{ok}$. In Matlab, $\gamma = \text{ifft}(g_o)$, and $\gamma(k+1) \approx g_{ok}$.

(v) Apply the Kalman-Ho algorithm to $\{\gamma(k)\}_1^m$ for m sufficiently large to find a state space realization $\{A, B, C, D\}$ for g_o. Then the outer factor g_o is given by

$$g_o(z) = D + C(zI - A)^{-1}B.$$

(vi) To compute the inner factor g_i, set $g_i = g/g_o$. Now compute the Fourier coefficients $\{g_{ik}\}$ via the inverse FFT of $g_i = \sum_0^\infty z^{-k} g_{ik}$. In Matlab, $g_i = g./g_o$ and $\beta = \text{ifft}(g_i)$. Here $g_{ik} \approx \beta(k+1)$.

(vii) Apply the Kalman-Ho algorithm to $\{\beta(k)\}_1^m$ with m sufficiently large, to find a state space realization $\{A_i, B_i, C_i, D_i\}$ of g_i. Then the inner factor g_i for g is given by

$$g_i(z) = D_i + C_i(zI - A_i)^{-1}B_i.$$

This algorithm appears to work well even when g is a rational function of large order. If g has zeros on the unit circle, then any algorithm may have numerical problems.

Example. Consider the transfer function

$$g(z) = \frac{(z-2)(z-3)(z-0.9)}{(z-0.1)(z-0.4)(z-0.5)(z-0.6)}. \tag{4.9.2}$$

Let $g = g_i g_o$ be the inner-outer factorization where g_i is inner and g_o is outer. Then g_i and g_o are given by

$$\begin{aligned} g_i(z) &= \frac{(1-z/2)(1-z/3)}{z(z-1/2)(z-1/3)}, \\ g_o(z) &= \frac{6z(z-1/3)(z-0.9)}{(z-0.1)(z-0.4)(z-0.6)}. \end{aligned} \tag{4.9.3}$$

By using the previous algorithm we computed that $g \approx f_i f_o$ where f_i is the inner function and f_o is the outer function is given by

$$\begin{aligned} f_i(z) &= \frac{0.1667z^2 - 0.8333z + 1}{z^3 - 0.8333z^2 + 0.1667z}, \\ f_o(z) &= \frac{6z^3 - 7.4z^2 + 1.8z}{z^3 - 1.1z^2 + 0.34z - 0.024}. \end{aligned} \tag{4.9.4}$$

Finally, it is noted that we used a FFT of length 2^{14} which is more than enough and a Kalman-Ho of 500 data points which is also much larger than necessary.

Chapter 5

The Naimark Representation

This chapter is devoted to the Naimark representation theorem and its consequences. The Naimark dilation allows us to use geometric methods to compute inner-outer factorizations and solve signal processing problems. Let A be any matrix whose entries A_{jk} are operators mapping a Hilbert space $\mathcal{E}$ into $\mathcal{Y}$. Then $A^\sharp$ denotes the matrix obtained by taking the adjoint of the entries of A and then transposing this matrix, that is, the entries of $A^\sharp$ are given by $(A^\sharp)_{jk} = A^*_{kj}$. If A defines an operator mapping $\oplus_0^n \mathcal{E}$ into $\oplus_0^m \mathcal{Y}$, then $A^\sharp = A^*$ is the adjoint of A. Throughout $\ell_+^c(\mathcal{E})$ denotes the set of all vectors in $\ell_+^2(\mathcal{E})$ with compact support. Finally, recall that the controllability matrix W determined by the pair of operators $\{A \text{ on } \mathcal{X}, B\}$ where B maps $\mathcal{E}$ into $\mathcal{X}$ is given by

$$W = \begin{bmatrix} B & AB & A^2B & \cdots \end{bmatrix}. \tag{5.0.1}$$

In general, W is not necessarily an operator mapping $\ell_+^2(\mathcal{E})$ into $\mathcal{X}$. However, W is a well-defined linear map from $\ell_+^c(\mathcal{E})$ into $\mathcal{X}$.

Particularly in this chapter, we will be dealing with infinite matrices which do not necessarily define operators acting between Hilbert spaces. So let us note that in general, one has to be careful with the usual matrix product for infinite matrices. For instance, it may happen that this product does not exist, and when it exists it does not have to be associative. To illustrate the latter consider the infinite matrices

$$M = \begin{bmatrix} 1 & -2 & 0 & 0 & \cdots \\ 0 & 1 & -2 & 0 & \cdots \\ 0 & 0 & 1 & -2 & \cdots \\ 0 & 0 & 0 & 1 & \cdots \\ \vdots & \vdots & \vdots & \vdots & \ddots \end{bmatrix} \quad \text{and} \quad K = \begin{bmatrix} 1 \\ 1/2 \\ 1/4 \\ 1/8 \\ \vdots \end{bmatrix}$$

$$R = \begin{bmatrix} 1 & 2 & 4 & 8 & \cdots \end{bmatrix}.$$

Then the matrix products $R(MK)$ and $(RM)K$ are well defined but not equal, that is, $R(MK) = 0$ and $(RM)K = 1$. In the sequel we shall consider the product of two operator matrices only when this product exists and the product of the corresponding linear transformations makes sense.

5.1 The Naimark Representation Theorem

Let $\{R_k\}_{-\infty}^{\infty}$ be any $\mathcal{L}(\mathcal{E}, \mathcal{E})$-valued sequence of operators. Throughout we always assume that the space $\mathcal{E}$ is finite dimensional. Consider the block Toeplitz matrix given by

$$T_R = \begin{bmatrix} R_0 & R_{-1} & R_{-2} & \cdots \\ R_1 & R_0 & R_{-1} & \cdots \\ R_2 & R_1 & R_0 & \cdots \\ \vdots & \vdots & \vdots & \ddots \end{bmatrix}. \tag{5.1.1}$$

The $\{R_k\}_0^{\infty}$ determines the first column of T_R. The j-k entry of T_R is given by $(T_R)_{j,k} = R_{j-k}$. One can also view R as the function *formally defined* by $R = \sum_{-\infty}^{\infty} e^{-\imath\omega k} R_k$. In this case, R is called the *symbol* for T_R, and T_R is the Toeplitz matrix generated by R. If R is a function in $L^2(\mathcal{E}, \mathcal{E})$, then $\{R_k\}_{-\infty}^{\infty}$ are the Fourier coefficients of R.

Let S be the lower shift matrix, that is,

$$S = \begin{bmatrix} 0 & 0 & 0 & \cdots \\ I & 0 & 0 & \cdots \\ 0 & I & 0 & \cdots \\ \vdots & \vdots & \vdots & \ddots \end{bmatrix}. \tag{5.1.2}$$

Here we simply view S as a block matrix with entries in $\mathcal{L}(\mathcal{E}, \mathcal{E})$, and not necessarily as an operator on $\ell_+^2(\mathcal{E})$. Recall that a block matrix

$$A = \begin{bmatrix} A_{0,0} & A_{0,1} & A_{0,2} & \cdots \\ A_{1,0} & A_{1,1} & A_{1,2} & \cdots \\ A_{2,0} & A_{2,1} & A_{2,2} & \cdots \\ \vdots & \vdots & \vdots & \ddots \end{bmatrix}$$

with operator entries $\{A_{j,k}\}_{0,0}^{\infty,\infty}$ in $\mathcal{L}(\mathcal{E}, \mathcal{E})$ is Toeplitz if $A_{j,k} = \Lambda_{j-k}$ where $\{\Lambda_j\}_{-\infty}^{\infty}$ is a set of operators in $\mathcal{L}(\mathcal{E}, \mathcal{E})$. Clearly, the matrix T_R in (5.1.1) is Toeplitz. It is noted that A is a Toeplitz matrix if and only if $(Af, g) = (ASf, Sg)$ for all f and g in $\ell_+^c(\mathcal{E})$. Finally, A is a lower triangular Toeplitz matrix if and only if $(ASf, g) = (SAf, g)$ for all f and g in $\ell_+^c(\mathcal{E})$.

Let T_R be the block Toeplitz matrix defined in (5.1.1). By a slight abuse of terminology we say that T_R is a *self-adjoint Toeplitz matrix* if $R_{-k} = R_k^*$ for all integers $k \geq 0$, or equivalently, $T_R = (T_R)^\sharp$. If $R = \sum_{-\infty}^{\infty} e^{-\imath\omega n} R_n$ is the symbol for T_R, then T_R is self-adjoint if and only if $R_{-k} = R_k^*$ for all $k \geq 0$. Finally, if the symbol R is in $L^2(\mathcal{E}, \mathcal{E})$, then the corresponding Toeplitz matrix T_R is self-adjoint if and only if $R = R^*$ almost everywhere with respect to the Lebesgue measure. The Toeplitz matrix T_R is *positive* if

$$0 \leq (T_R g, g) = \sum_{j,k=0}^{\infty,\infty} (R_{j-k} g_k, g_j)_{\mathcal{E}} \quad \text{(for all } g = \begin{bmatrix} g_0 & g_1 & g_2 & \cdots \end{bmatrix}^{tr} \in \ell_+^c(\mathcal{E})).$$

Finally, we say that T_R is *strictly positive*, if there exists a scalar $\delta > 0$ such that $0 < \delta \|g\|^2 \leq (T_R g, g)$ for all g in $\ell_+^c(\mathcal{E})$.

Notice that T_R is positive if and only if the n by n block Toeplitz matrix contained in the upper left-hand corner of T_R is positive for all $n \geq 0$. To be precise, let $T_{R,n}$ be the n by n block Toeplitz matrix given by

$$T_{R,n} = \begin{bmatrix} R_0 & R_{-1} & \cdots & R_{1-n} \\ R_1 & R_0 & \cdots & R_{2-n} \\ \vdots & \vdots & \ddots & \vdots \\ R_{n-1} & R_{n-2} & \cdots & R_0 \end{bmatrix} \quad \text{on } \mathcal{E}^n. \tag{5.1.3}$$

Then T_R is positive if and only if $T_{R,n}$ is a positive operator on $\mathcal{E}^n = \oplus_0^{n-1} \mathcal{E}$, for all $n \geq 1$. Observe that T_R is strictly positive if and only if there exists a scalar $\delta > 0$ such that $0 < \delta I \leq T_{R,n}$ for all $n \geq 1$. Finally, it is noted that if T_R is positive, then $T_{R,n}$ must be a self-adjoint operator for all integers $n \geq 1$. In this case, $R_0 \geq 0$ and $R_{-k} = R_k^*$ for all $n \geq 0$. In other words, if T_R is a positive Toeplitz matrix, then T_R is also a self-adjoint matrix.

In this chapter, we will be mainly interested in positive Toeplitz matrices. In this case, T_R is self-adjoint and $R_{-k} = R_k^*$ for all integers $k \geq 0$. Motivated by this we say that T_R is the *self-adjoint Toeplitz matrix generated by* $\{R_k\}_0^\infty$, if T_R is the self-adjoint Toeplitz matrix given by (5.1.1) where $R_{-k} = R_k^*$ for all $k \geq 0$.

We say that $\{U \text{ on } \mathcal{K}, \Gamma\}$ is an *isometric pair* if U is an isometry on $\mathcal{K}$, and Γ is an operator mapping $\mathcal{E}$ into $\mathcal{K}$. The pair $\{U, \Gamma\}$ is *controllable*, if $\Gamma\mathcal{E}$ is cyclic for U, that is,

$$\mathcal{K} = \bigvee_{n=0}^{\infty} U^n \Gamma \mathcal{E}.$$

Let $\{R_{-n}\}_0^\infty$ be a sequence of operators on $\mathcal{E}$. We say that $\{U, \Gamma\}$ is an *isometric representation* for $\{R_{-n}\}_0^\infty$ if $R_{-n} = \Gamma^* U^n \Gamma$ for all integers $n \geq 0$. It may seem a bit odd to define an isometric representation using the negative index on R_{-n}. However, this will be useful in keeping the notation consistent. In almost all of our problems $R_n^* = R_{-n}$. In this case, $\{U, \Gamma\}$ is an *isometric representation* of $\{R_{-n}\}_0^\infty$ if and only if $R_n^* = \Gamma^* U^n \Gamma$ for all $n \geq 0$. Now let T_R be any self-adjoint Toeplitz matrix, that is, let T_R be a matrix of the form

$$T_R = \begin{bmatrix} R_0 & R_{-1} & R_{-2} & \cdots \\ R_1 & R_0 & R_{-1} & \cdots \\ R_2 & R_1 & R_0 & \cdots \\ \vdots & \vdots & \vdots & \ddots \end{bmatrix} \qquad \text{(where } R_{-n} = R_n^*\text{)}. \tag{5.1.4}$$

A self-adjoint Toeplitz matrix is uniquely determined by its first column $R_n = (T_R)_{n,0}$ for all integers $n \geq 0$, and the entries in its first row $(T_R)_{0,n} = R_n^* = R_{-n}$. In this case, $R = \sum_{-\infty}^{\infty} e^{-\imath\omega k} R_k$ is the symbol for T_R. Motivated by this we also say that $\{U, \Gamma\}$ is an *isometric representation* for a self-adjoint Toeplitz matrix T_R, if $\{U, \Gamma\}$ is an isometric pair such that

$$(T_R)_{0,n} = \Gamma^* U^n \Gamma \qquad \text{(for all integers } n \geq 0\text{)}. \tag{5.1.5}$$

Consider two operator pairs $\{U \text{ on } \mathcal{K}, \Gamma\}$ and $\{U_1 \text{ on } \mathcal{K}_1, \Gamma_1\}$ where Γ maps $\mathcal{E}$ into $\mathcal{K}$ and Γ_1 maps $\mathcal{E}$ into $\mathcal{K}_1$. Then we say that Φ *intertwines* $\{U, \Gamma\}$ with $\{U_1, \Gamma_1\}$ if Φ is an operator mapping $\mathcal{K}$ onto $\mathcal{K}_1$, such that

$$\Phi U = U_1 \Phi \qquad \text{and} \qquad \Phi\Gamma = \Gamma_1. \tag{5.1.6}$$

These two operator pairs are *unitarily equivalent*, if there exists a unitary operator Φ intertwining $\{U, \Gamma\}$ with $\{U_1, \Gamma_1\}$. Observe that if $\{U, \Gamma\}$ and $\{U_1, \Gamma_1\}$ are unitarily equivalent, then $\{U, \Gamma\}$ is controllable if and only if $\{U_1, \Gamma_1\}$ is controllable. Finally, it is noted that two unitarily equivalent isometric pairs are isometric representations of the same Toeplitz matrix. The following result is known as the *Naimark representation theorem.*

Theorem 5.1.1. *Let T_R be a self-adjoint Toeplitz matrix; see* (5.1.1). *Then T_R admits an isometric representation if and only if T_R is positive. In this case, T_R admits a controllable isometric representation. Moreover, all controllable isometric representations of T_R are unitarily equivalent.*

Proof. Assume that T_R admits an isometric representation $\{U \text{ on } \mathcal{K}, \Gamma\}$. Let W be the controllability matrix determined by $\{U, \Gamma\}$, that is,

$$W = \begin{bmatrix} \Gamma & U\Gamma & U^2\Gamma & \cdots \end{bmatrix} \qquad \text{and} \qquad W^\sharp = \begin{bmatrix} \Gamma^* \\ \Gamma^* U^* \\ \Gamma^* U^{*2} \\ \vdots \end{bmatrix}. \tag{5.1.7}$$

Observe that $W^\sharp$ is the matrix formed by transposing W and taking the adjoint of each of its entries. Let us emphasize that W and $W^\sharp$ are simply matrices, and not necessarily operators. We claim that T_R admits a factorization of the form $T_R = W^\sharp W$. By employing $(T_R)_{0,n} = R_{-n} = \Gamma^* U^n \Gamma$ and $R_n = \Gamma^* U^{*n} \Gamma$ for all integers $n \geq 0$ with $U^* U = I$, we obtain

$$W^\sharp W = \begin{bmatrix} \Gamma^*\Gamma & \Gamma^*U\Gamma & \Gamma^*U^2\Gamma & \cdots \\ \Gamma^*U^*\Gamma & \Gamma^*U^*U\Gamma & \Gamma^*U^*U^2\Gamma & \cdots \\ \Gamma^*U^{*2}\Gamma & \Gamma^*U^{*2}U\Gamma & \Gamma^*U^{*2}U^2\Gamma & \cdots \\ \vdots & \vdots & \vdots & \ddots \end{bmatrix}$$

$$= \begin{bmatrix} \Gamma^*\Gamma & \Gamma^*U\Gamma & \Gamma^*U^2\Gamma & \cdots \\ \Gamma^*U^*\Gamma & \Gamma^*\Gamma & \Gamma^*U\Gamma & \cdots \\ \Gamma^*U^{*2}\Gamma & \Gamma^*U^*\Gamma & \Gamma^*\Gamma & \cdots \\ \vdots & \vdots & \vdots & \ddots \end{bmatrix} = T_R.$$

Hence $T_R = W^\sharp W$. Now let g be any vector in $\ell_+^c(\mathcal{E})$, then the quadratic form $(T_R g, g)$ can be written as

$$(T_R g, g) = (W^\sharp W g, g) = (Wg, Wg) = \|Wg\|^2 \geq 0. \tag{5.1.8}$$

So if $\{U, \Gamma\}$ is an isometric representation of the Toeplitz matrix T_R, then T_R is positive.

Now assume that the Toeplitz matrix T_R in (5.1.1) is positive. Consider the quadratic form mapping $\ell_+^c(\mathcal{E}) \times \ell_+^c(\mathcal{E})$ into $\mathbb{C}$ determined by

$$\langle f, g \rangle \quad = \quad (T_R f, g) \qquad (f, g \in \ell_+^c(\mathcal{E})).$$

Notice that $\langle f, g \rangle$ is linear in the first variable, conjugate linear in the second variable, and satisfies $\langle f, f \rangle \geq 0$. Let $\mathcal{N}$ be the set of all vectors in $\ell_+^c(\mathcal{E})$ such that $\langle f, f \rangle = 0$. Let $\mathcal{K}_o$ be the quotient space formed by $\ell_+^c(\mathcal{E})/\mathcal{N}$. Notice that $\mathcal{K}_o$ is an inner product space with respect to the quadratic form $(T_R f, g)$. Finally, let $\mathcal{K}$ be the Hilbert space formed by completing $\mathcal{K}_o$.

To construct an isometric representation for T_R, let U be the linear map on $\mathcal{K}_o$ and Γ be the linear mapping from $\mathcal{E}$ into $\mathcal{K}_o$ defined by

$$U = \begin{bmatrix} 0 & 0 & 0 & \cdots \\ I & 0 & 0 & \cdots \\ 0 & I & 0 & \cdots \\ \vdots & \vdots & \vdots & \ddots \end{bmatrix} \quad \text{and} \quad \Gamma = \begin{bmatrix} I \\ 0 \\ 0 \\ \vdots \end{bmatrix}.$$

We claim that U is an isometry on $\mathcal{K}_o$. First, notice that U is a lower shift. Because T_R is Toeplitz, $(T_R f, f) = (T_R U f, U f)$ for all f in $\ell_+^c(\mathcal{E})$. For any vector f in $\mathcal{K}_o$, we have

$$\|Uf\|_{\mathcal{K}}^2 = (T_R U f, U f) = (T_R f, f) = \|f\|_{\mathcal{K}}^2.$$

Hence U is an isometry on $\mathcal{K}_o$. Since $\mathcal{K}_o$ is dense in $\mathcal{K}$, the isometry U admits a unique extension by continuity to an isometry on all of $\mathcal{K}$. Without loss of generality, we also denote this isometric extension by the same symbol U. For any vector a in $\mathcal{E}$, we have

$$\|\Gamma a\|_{\mathcal{K}}^2 = (T_R \Gamma a, \Gamma a) = (R_0 a, a) \leq \|R_0\| \|a\|^2.$$

Therefore Γ is bounded, and thus, we can view Γ as an operator mapping $\mathcal{E}$ into $\mathcal{K}$. Hence $\{U, \Gamma\}$ is an isometric pair. Finally, because U is the lower shift and Γ embeds $\mathcal{E}$ into the first component of $\mathcal{K}_o$, it follows that $\{U, \Gamma\}$ is controllable.

To verify that $\{U, \Gamma\}$ is an isometric representation of the Toeplitz matrix T_R, it remains to show that $R_{-n} = (T_R)_{0,n} = \Gamma^* U^n \Gamma$ for all $n \geq 0$. If a is any vector in $\mathcal{E}$, then we have

$$(U^n\Gamma a, \Gamma a)_{\mathcal{K}} = (T_R U^n \Gamma a, \Gamma a) = \left(\begin{bmatrix} R_{-n} \\ R_{1-n} \\ R_{2-n} \\ \vdots \end{bmatrix}, \begin{bmatrix} I \\ 0 \\ 0 \\ \vdots \end{bmatrix} \right) = R_{-n} = R_n^* \quad (n \geq 0).$$

Hence $\{U, \Gamma\}$ is a controllable isometric representation of T_R.

Let $\{U \text{ on } \mathcal{K}, \Gamma\}$ and $\{U_1 \text{ on } \mathcal{K}_1, \Gamma_1\}$ be two controllable isometric representations of the same positive Toeplitz matrix T_R. Let W be the controllable matrix for $\{U, \Gamma\}$ defined in (5.1.7), and W_1 be the controllable matrix for $\{U_1, \Gamma_1\}$ where U_1 replaces U and Γ_1 replaces Γ. Because $\{U, \Gamma\}$ and $\{U_1, \Gamma_1\}$ are controllable, $W\ell_+^c(\mathcal{E})$ is dense in $\mathcal{K}$ and $W_1\ell_+^c(\mathcal{E})$ is dense in $\mathcal{K}_1$. Since $\{U, \Gamma\}$ and $\{U_1, \Gamma_1\}$ are both isometric representations of T_R, we have $W^\sharp W = T_R = W_1^\sharp W_1$. Using this with g in $\ell_+^c(\mathcal{E})$, we obtain

$$\|Wg\|^2 = (T_R g, g) = \|W_1 g\|^2.$$

So there exists an isometry Φ mapping $W\ell_+^c(\mathcal{E})$ into $W_1\ell_+^c(\mathcal{E})$, such that $\Phi W = W_1$. Because $W\ell_+^c(\mathcal{E})$ is dense in $\mathcal{K}$ and $W_1\ell_+^c(\mathcal{E})$ is dense in $\mathcal{K}_1$, the isometry Φ admits a unique extension by continuity to a unitary operator mapping $\mathcal{K}$ onto $\mathcal{K}_1$, and this unitary operator is also denoted by Φ. By choosing $f = \begin{bmatrix} a & 0 & 0 & \cdots \end{bmatrix}^{tr}$ with a in $\mathcal{E}$, we have $\Gamma_1 a = W_1 f = \Phi W f = \Phi \Gamma a$. Hence $\Phi\Gamma = \Gamma_1$. Let S be the forward shift on $\ell_+^c(\mathcal{E})$ defined in (5.1.2). Notice that $UW = WS$ and $U_1 W_1 = W_1 S$. For any g in $\ell_+^c(\mathcal{E})$, we have

$$\Phi U W g = \Phi W S g = W_1 S g = U_1 W_1 g = U_1 \Phi W g.$$

Thus $\Phi U W g = U_1 \Phi W g$. Since $W\ell_+^c(\mathcal{E})$ is dense in $\mathcal{K}$, we obtain $\Phi U = U_1 \Phi$. Therefore $\{U, \Gamma\}$ and $\{U_1, \Gamma_1\}$ are unitarily equivalent. In other words, all controllable isometric representations of T_R are unitarily equivalent. □

If $\{U \text{ on } \mathcal{K}, \Gamma\}$ is any isometric representation of T_R, then we can always extract from $\{U, \Gamma\}$ a controllable isometric representation $\{U_c, \Gamma_c\}$ of T_R. To see this, let $\mathcal{K}_c$ be the invariant subspace for U defined by $\bigvee_0^\infty U^n \Gamma \mathcal{E} = \mathcal{K}_c \subset \mathcal{K}$. Let U_c on $\mathcal{K}_c$ be the isometry defined by $U_c = U|\mathcal{K}_c$, and Γ_c be the operator mapping $\mathcal{E}$ into $\mathcal{K}_c$ given by $\Gamma_c = \Gamma$. Because $\mathcal{K}_c$ is invariant under U, we have $\mathcal{K}_c = \bigvee_0^\infty U_c^n \Gamma_c \mathcal{E}$. Moreover,

$$(U_c^n \Gamma_c a, \Gamma_c a) = (U^n \Gamma a, \Gamma a) = ((T_R)_{0,n} a, a) \qquad (\text{for all } a \in \mathcal{E} \text{ and } n \geq 0).$$

Hence $\{U_c \text{ on } \mathcal{K}_c, \Gamma_c\}$ is a controllable isometric representation of T_R.

5.2 The Maximal Outer Spectral Factor

Let $\Theta(z) = \sum_0^\infty z^{-k}\Theta_k$ be the Taylor series expansion for a function Θ in $H^2(\mathcal{E},\mathcal{Y})$. Let T_Θ be the lower triangular Toeplitz matrix determined by Θ, that is,

$$T_\Theta = \begin{bmatrix} \Theta_0 & 0 & 0 & \cdots \\ \Theta_1 & \Theta_0 & 0 & \cdots \\ \Theta_2 & \Theta_1 & \Theta_0 & \cdots \\ \vdots & \vdots & \vdots & \ddots \end{bmatrix} \quad \text{and} \quad T_\Theta^\sharp = \begin{bmatrix} \Theta_0^* & \Theta_1^* & \Theta_2^* & \cdots \\ 0 & \Theta_0^* & \Theta_1^* & \cdots \\ 0 & 0 & \Theta_0^* & \cdots \\ \vdots & \vdots & \vdots & \ddots \end{bmatrix}. \tag{5.2.1}$$

Because Θ is in $H^2(\mathcal{E},\mathcal{Y})$, all the columns of T_Θ can be viewed as operators mapping $\mathcal{E}$ into $\ell_+^2(\mathcal{Y})$. In other words, T_Θ is a well-defined linear map from $\ell_+^c(\mathcal{E})$ into $\ell_+^2(\mathcal{Y})$. This also implies that the matrix $T_\Theta^\sharp T_\Theta$ is a well-defined positive Toeplitz matrix. Recall that T_Θ is an operator mapping $\ell_+^2(\mathcal{E})$ into $\ell_+^2(\mathcal{Y})$ if and only if Θ is in $H^\infty(\mathcal{E},\mathcal{Y})$. In this case, $\|T_\Theta\| = \|\Theta\|_\infty$, and $T_\Theta^\sharp = T_\Theta^*$ is the adjoint of T_Θ. Finally, Θ is an *outer function* if Θ is a function in $H^2(\mathcal{E},\mathcal{Y})$ and $T_\Theta \ell_+^c(\mathcal{E})$ is dense in $\ell_+^2(\mathcal{Y})$, or equivalently, $\overline{\Theta\mathcal{P}(\mathcal{E})} = H^2(\mathcal{Y})$. (The set of all polynomials in $1/z$ with values in $\mathcal{E}$ is denoted by $\mathcal{P}(\mathcal{E})$.)

Let T_R in (5.1.1) be a positive Toeplitz matrix determined by the symbol $R = \sum_{-\infty}^\infty e^{-\imath\omega n} R_n$ with values in $\mathcal{L}(\mathcal{E},\mathcal{E})$; see Section 5.1. We say that a function Θ in $H^2(\mathcal{E},\mathcal{Y})$ is a *maximal outer spectral factor* for a positive Toeplitz matrix T_R if the following three conditions hold:

(i) The function Θ is outer.

(ii) The inequality $T_R \geq T_\Theta^\sharp T_\Theta$ holds.

(iii) If $\Psi \in H^2(\mathcal{E},\mathcal{G})$ satisfying $T_R \geq T_\Psi^\sharp T_\Psi$, then $T_\Theta^\sharp T_\Theta \geq T_\Psi^\sharp T_\Psi$.

Finally, if there is no nonzero function in $H^2(\mathcal{E},\mathcal{Y})$ satisfying $T_R \geq T_\Theta^\sharp T_\Theta$, then $\mathcal{Y} = \{0\}$, and the maximal outer spectral factor Θ is the zero function mapping $\mathcal{E}$ into $\{0\}$. The following result shows that any positive Toeplitz matrix always admits a unique maximal outer spectral factor.

Theorem 5.2.1. *Let T_R be a positive Toeplitz matrix generated by the $\mathcal{L}(\mathcal{E},\mathcal{E})$-valued sequence $\{R_k\}_0^\infty$; see* (5.1.1) *where $R_{-k} = R_k^*$ for all $k \geq 0$. Then the following holds.*

(i) *The matrix T_R admits a maximal outer spectral factor Θ.*

(ii) *All maximal outer spectral factors of T_R are unique up to a constant unitary operator on the left. To be precise, if Θ in $H^2(\mathcal{E},\mathcal{Y})$ and Ψ in $H^2(\mathcal{E},\mathcal{G})$ are two maximal outer spectral factors for T_R, then $\Psi(z) = \Omega\Theta(z)$ for all z in $\mathbb{D}_+$ where Ω is a constant unitary operator mapping $\mathcal{Y}$ onto $\mathcal{G}$.*

(iii) *The equality $T_R = T_\Theta^\sharp T_\Theta$ holds if and only if the "future space"*

$$\bigcap_{n=0}^\infty U^n\mathcal{K} = \{0\}$$

where $\{U \text{ on } \mathcal{K}, \Gamma\}$ *is the controllable isometric representation for* T_R.

(iv) *The maximal outer spectral factor for* T_R *is given by*

$$\Theta(z) = z\Pi_{\mathcal{Y}}(zI - U^*)^{-1}\Gamma = \Pi_{\mathcal{Y}}(I - z^{-1}U^*)^{-1}\Gamma \tag{5.2.2}$$

where $\mathcal{Y} = \ker U^*$ *and* $\Pi_{\mathcal{Y}}$ *is the orthogonal projection from* $\mathcal{K}$ *onto* $\mathcal{Y}$.

(v) *The maximal outer spectral factor for* T_R *is zero if and only if* $\ker U^* = \{0\}$, *or equivalently,* U *is a unitary operator.*

Proof. Let $\{U \text{ on } \mathcal{K}, \Gamma\}$ be a controllable isometric representation for T_R. According to the Wold decomposition the isometry U admits a matrix representation of the form

$$U = \begin{bmatrix} S & 0 \\ 0 & V \end{bmatrix} \text{ on } \begin{bmatrix} \ell^2_+(\mathcal{Y}) \\ \mathcal{V} \end{bmatrix} \quad \text{and} \quad \Gamma = \begin{bmatrix} \Gamma_1 \\ \Gamma_2 \end{bmatrix} : \mathcal{E} \to \begin{bmatrix} \ell^2_+(\mathcal{Y}) \\ \mathcal{V} \end{bmatrix}, \tag{5.2.3}$$

where $\mathcal{Y} = \ker U^*$. Here S is a unilateral shift on $\ell^2_+(\mathcal{Y})$ and V is a unitary operator on $\mathcal{V}$. Notice that Γ_1 is an operator mapping $\mathcal{E}$ into $\ell^2_+(\mathcal{Y})$. To be precise, S and Γ_1 have matrix representations of the form:

$$S = \begin{bmatrix} 0 & 0 & 0 & \cdots \\ I & 0 & 0 & \cdots \\ 0 & I & 0 & \cdots \\ \vdots & \vdots & \vdots & \ddots \end{bmatrix} \text{ on } \ell^2_+(\mathcal{Y}) \quad \text{and} \quad \Gamma_1 = \begin{bmatrix} \Theta_0 \\ \Theta_1 \\ \Theta_2 \\ \vdots \end{bmatrix} : \mathcal{E} \to \ell^2_+(\mathcal{Y}). \tag{5.2.4}$$

Let Θ be the function defined by taking the Fourier transform of Γ_1, that is,

$$\Theta(z) = (\mathcal{F}^+_{\mathcal{Y}}\Gamma_1)(z) = \sum_{k=0}^{\infty} z^{-k}\Theta_k \qquad (z \in \mathbb{D}_+). \tag{5.2.5}$$

Since Γ_1 is an operator mapping $\mathcal{E}$ into $\ell^2_+(\mathcal{Y})$, the function Θ is in $H^2(\mathcal{E}, \mathcal{Y})$. In a moment, we will show that Θ is the maximal outer spectral factor for T_R.

Let W be the controllability matrix determined by $\{U, \Gamma\}$. Let W_1 be the controllability matrix determined by $\{S, \Gamma_1\}$, and W_2 be the controllability matrix determined by $\{V, \Gamma_2\}$. (The controllability matrix for the pair $\{A, B\}$ is defined in (5.0.1).) By the Wold decomposition of $\{U, \Gamma\}$ in (5.2.3), we see that

$$W = \begin{bmatrix} W_1 \\ W_2 \end{bmatrix} = \begin{bmatrix} \Gamma_1 & S\Gamma_1 & S^2\Gamma_1 & \cdots \\ \Gamma_2 & V\Gamma_2 & V^2\Gamma_2 & \cdots \end{bmatrix}. \tag{5.2.6}$$

Let us emphasize that

$$W_1 = \begin{bmatrix} \Gamma_1 & S\Gamma_1 & S^2\Gamma_1 & \cdots \end{bmatrix} = \begin{bmatrix} \Theta_0 & 0 & 0 & \cdots \\ \Theta_1 & \Theta_0 & 0 & \cdots \\ \Theta_2 & \Theta_1 & \Theta_0 & \cdots \\ \vdots & \vdots & \vdots & \ddots \end{bmatrix} = T_\Theta. \tag{5.2.7}$$

In other words, $W_1 = T_\Theta$ is the lower triangular Toeplitz matrix determined by Θ. Because the pair $\{U, \Gamma\}$ is controllable, the pair $\{S, \Gamma_1\}$ must also be controllable. By (5.2.6) the columns of $W_1 = T_\Theta$ must span a dense set in $\ell_+^2(\mathcal{Y})$. Hence $\ell_+^2(\mathcal{Y}) = \overline{T_\Theta \ell_+^c(\mathcal{E})}$, and thus, Θ is an outer function. Notice that

$$T_R = W^\sharp W = W_1^\sharp W_1 + W_2^\sharp W_2 \geq W_1^\sharp W_1 = T_\Theta^\sharp T_\Theta.$$

Therefore $T_R \geq T_\Theta^\sharp T_\Theta$.

Now assume that there exists another outer function Ψ in $H^2(\mathcal{E}, \mathcal{G})$ such that $T_R \geq T_\Psi^\sharp T_\Psi$. Recall that $T_R = W^\sharp W$. For any vector g in $\ell_+^c(\mathcal{E})$, we have

$$\|Wg\|^2 = (T_R g, g) \geq (T_\Psi^\sharp T_\Psi g, g) = \|T_\Psi g\|^2 \qquad (g \in \ell_+^c(\mathcal{E})).$$

In other words, $\|Wg\| \geq \|T_\Psi g\|$. Hence there exists a contraction Y mapping $\mathcal{K}$ into $\ell_+^2(\mathcal{G})$ such that $T_\Psi = YW$. Let $S_\mathcal{L}$ denote the forward shift on $\ell_+^2(\mathcal{L})$. By employing $UWg = WS_\mathcal{E} g$ and $T_\Psi S_\mathcal{E} g = S_\mathcal{G} T_\Psi g$ for all vectors g in $\ell_+^c(\mathcal{E})$, we obtain

$$S_\mathcal{G} YWg = S_\mathcal{G} T_\Psi g = T_\Psi S_\mathcal{E} g = YWS_\mathcal{E} g = YUWg.$$

Because $\{U, \Gamma\}$ is controllable, $S_\mathcal{G} Y = YU$. In particular, $S_\mathcal{G}^k Y = YU^k$ for all integers $k \geq 1$. By the Wold decomposition of U, the future space $\mathcal{V} = \bigcap_0^\infty U^k \mathcal{K}$. Observe that the subspaces $\{U^k \mathcal{K}\}_0^\infty$ are decreasing, that is, $U^{k+1}\mathcal{K} \subseteq U^k \mathcal{K}$ for all integers $k \geq 0$. In particular, $\mathcal{V} \subseteq U^k \mathcal{K}$ for all $k \geq 0$. Using the fact that $Y\mathcal{V} \subseteq YU^k \mathcal{K}$, we have

$$Y\mathcal{V} \subseteq \bigcap_{k=0}^{\infty} YU^k \mathcal{K} = \bigcap_{k=0}^{\infty} S_\mathcal{G}^k Y\mathcal{K} \subseteq \bigcap_{k=0}^{\infty} S_\mathcal{G}^k \ell_+^2(\mathcal{G}) = \{0\}.$$

Hence $Y|\mathcal{V} = \{0\}$, and thus, $YW = YW_1$. This implies that

$$T_\Psi = YW = YW_1 = YT_\Theta.$$

In other words, $T_\Psi = YT_\Theta$. Using the fact that Y is a contraction, $T_\Psi^\sharp T_\Psi \leq T_\Theta^\sharp T_\Theta$. By definition Θ is a maximal outer spectral factor for T_R, and Part (i) holds.

Recall that Θ is a maximal outer spectral factor for T_R. If T_R admits another maximal outer spectral factor Ψ, then $T_\Psi^\sharp T_\Psi \leq T_\Theta^\sharp T_\Theta \leq T_\Psi^\sharp T_\Psi$ implies that $T_\Theta^\sharp T_\Theta = T_\Psi^\sharp T_\Psi$. According to Lemma 5.3.1 below, there exists a constant unitary operator Ω mapping $\mathcal{Y}$ onto $\mathcal{G}$ such that $\Psi = \Omega\Theta$. Thus Part (ii) holds.

Recall that $\{U, \Gamma\}$ is a controllable isometric representation of T_R. Moreover,

$$T_R = W_1^\sharp W_1 + W_2^\sharp W_2 = T_\Theta^\sharp T_\Theta + W_2^\sharp W_2.$$

If $T_R = T_\Theta^\sharp T_\Theta$, then W_2 must be zero and there is no unitary part in the Wold decomposition of U. In other words, the future space $\mathcal{V} = \bigcap_0^\infty U^k \mathcal{K} = \{0\}$. Conversely, if the subspace $\mathcal{V} = \{0\}$, then $W_2 = 0$, and thus, $T_R = T_\Theta^\sharp T_\Theta$. Hence Part (iii) holds.

Now let us move on to Part (iv). Clearly, the unitary operator V^* has no kernel. Thus $\mathcal{Y} = \ker U^* = \ker S^*$. Because the kernel of S^* equals $\mathcal{Y} \oplus 0 \oplus 0 \oplus \cdots$, the operator $\Pi_\mathcal{Y}$ admits a matrix representation of the form:

$$\Pi_\mathcal{Y} = \begin{bmatrix} \begin{bmatrix} I & 0 & 0 & \cdots \end{bmatrix} & 0 \end{bmatrix} : \begin{bmatrix} \ell^2_+(\mathcal{Y}) \\ \mathcal{V} \end{bmatrix} \to \mathcal{Y}. \tag{5.2.8}$$

Recall that the maximal outer spectral factor Θ is given by (5.2.5). Using this with (5.2.4) and (5.2.8), for all integers $k \geq 0$, we arrive at

$$\begin{aligned} \Theta_k &= \begin{bmatrix} I & 0 & 0 & \cdots \end{bmatrix} S^{*k}\Gamma_1 \\ &= \begin{bmatrix} \begin{bmatrix} I & 0 & 0 & \cdots \end{bmatrix} & 0 \end{bmatrix} \begin{bmatrix} S^* & 0 \\ 0 & V^* \end{bmatrix}^k \begin{bmatrix} \Gamma_1 \\ \Gamma_2 \end{bmatrix} \\ &= \Pi_\mathcal{Y} U^{*k}\Gamma \qquad (k \geq 0). \end{aligned} \tag{5.2.9}$$

Hence $\Theta_k = \Pi_\mathcal{Y} U^{*k}\Gamma$ for all $k \geq 0$. Since $\|U^*\| < |z|$ for all z in $\mathbb{D}_+$, we obtain

$$\begin{aligned} \Theta(z) &= \sum_{k=0}^{\infty} z^{-k}\Theta_k = \sum_{k=0}^{\infty} z^{-k}\Pi_\mathcal{Y} U^{*k}\Gamma \\ &= \Pi_\mathcal{Y}(\sum_{k=0}^{\infty} z^{-k}U^{*k})\Gamma = \Pi_\mathcal{Y}(I - z^{-1}U^*)^{-1}\Gamma. \end{aligned}$$

Therefore Part (iv) holds.

Finally, Θ is zero if and only if the subspace $\mathcal{Y} = \ker U^* = \{0\}$. This yields Part (v). □

Let $\{U \text{ on } \mathcal{K}, \Gamma\}$ be a controllable isometric representation for T_R. Then the isometry U admits a Wold decomposition of the form $U = S \oplus V$ on $\mathcal{K}_+ \oplus \mathcal{V}$ where S is a unilateral shift and V is a unitary operator. Note that the unilateral shift S or the unitary operator V may not be present. To be precise, it can happen that $U = S$ or $U = V$.

Remark 5.2.2. Let $\{U \text{ on } \mathcal{K}, \Gamma\}$ be a controllable isometric representation for a positive Toeplitz matrix T_R with $\mathcal{L}(\mathcal{E}, \mathcal{E})$-valued symbol $R = \sum_{-\infty}^{\infty} e^{-\imath\omega n} R_n$. Let $U = S \oplus V$ and $\Gamma = \begin{bmatrix} \Gamma_1 & \Gamma_2 \end{bmatrix}^{tr}$ be the Wold decomposition for U in (5.2.3) where S is the unilateral shift on $\ell^2_+(\mathcal{Y})$ and V is unitary. Let $\mathcal{X}_1$ be the invariant subspace for S^* defined by

$$\mathcal{X}_1 = \bigvee_{n=1}^{\infty} S^{*n}\Gamma_1\mathcal{E}. \tag{5.2.10}$$

Let A on $\mathcal{X}_1$ and B mapping $\mathcal{E}$ into $\mathcal{X}_1$ and C mapping $\mathcal{X}_1$ into $\mathcal{Y}$ and D mapping $\mathcal{E}$ into $\mathcal{Y}$ be the operators defined by

$$A = S^*|\mathcal{X}_1, \quad C = S^*\Gamma_1, \quad C = \Pi_\mathcal{Y}|\mathcal{X}_1 \quad \text{and} \quad D = \Pi_\mathcal{Y}\Gamma_1. \tag{5.2.11}$$

Here $\Pi_{\mathcal{Y}} = \begin{bmatrix} I & 0 & 0 & \cdots \end{bmatrix}$ is viewed as the operator mapping $\ell_+^2(\mathcal{Y})$ onto $\mathcal{Y}$ which picks out the first component of $\ell_+^2(\mathcal{Y})$. Then $\{A, B, C, D\}$ is a controllable and observable realization for the maximal outer spectral factor Θ for T_R. In particular, Θ is rational if and only if the dimension of $\mathcal{X}_1$ is finite. The McMillan degree of Θ equals the dimension of $\mathcal{X}_1$. Finally, it is noted that $\{A, B, C, D\}$ is the restricted backward shift realization for Θ presented in Section 14.6.

To see this, first recall that Γ_1 is given by (5.2.4) where $\Theta(z) = \sum_0^\infty z^{-n}\Theta_n$ is the power series expansion for Θ. Notice that $\Theta_n = \Pi_{\mathcal{Y}} S^{*n}\Gamma_1$ for all integers $n \geq 0$. So for z in $\mathbb{D}_+$, we have

$$\begin{aligned}\Theta(z) &= \sum_{n=0}^{\infty} z^{-n}\Theta_n = \sum_{n=0}^{\infty} z^{-n}\Pi_{\mathcal{Y}} S^{*n}\Gamma_1 = \Pi_{\mathcal{Y}}(I - z^{-1}S^*)^{-1}\Gamma_1 \\ &= z\Pi_{\mathcal{Y}}(zI - S^*)^{-1}\Gamma_1 = \Pi_{\mathcal{Y}}\Gamma_1 + \Pi_{\mathcal{Y}}(zI - S^*)^{-1}S^*\Gamma_1.\end{aligned}$$

Since $D = \Pi_{\mathcal{Y}}\Gamma_1$, it follows that $\Sigma = \{S^*, S^*\Gamma_1, \Pi_{\mathcal{Y}}, D\}$ is a realization for Θ. Clearly, $\ell_+^2(\mathcal{Y}) = \oplus_0^\infty S^n \Pi_{\mathcal{Y}}^* \mathcal{Y} = \bigvee_0^\infty S^n \Pi_{\mathcal{Y}}^* \mathcal{Y}$. Hence the pair $\{\Pi_{\mathcal{Y}}, S^*\}$ is observable. So the realization Σ is observable. Notice that $\mathcal{X}_1 = \bigvee_0^\infty S^{*n}(S^*\Gamma_1)\mathcal{E}$ is the invariant subspace for S^* obtained by extracting the controllable subspace from Σ. In particular, $\{A, B, C, D\}$ is precisely the realization obtained by extracting the controllable part from the observable realization Σ for Θ. Therefore $\{A, B, C, D\}$ is a controllable and observable realization for Θ. This completes the verification of Remark 5.2.2.

Remark 5.2.3. In certain applications, it may be convenient to view the Wold decomposition for U as

$$U = \begin{bmatrix} S & 0 \\ 0 & V \end{bmatrix} \text{ on } \begin{bmatrix} H^2(\mathcal{Y}) \\ \mathcal{V} \end{bmatrix} \tag{5.2.12}$$

where $\mathcal{K} = H^2(\mathcal{Y}) \oplus \mathcal{V}$ and S is the unilateral shift on $H^2(\mathcal{Y})$. As expected, $\mathcal{Y} = \ker U^*$. If the unilateral shift is not present in the Wold decomposition of U, then we can set $\mathcal{Y} = \{0\}$ and $H^2(\mathcal{Y}) = H^2(\{0\}) = \{0\}$. Finally, it is noted that using $\mathcal{K} = H^2(\mathcal{Y}) \oplus \mathcal{V}$, it follows that Γ admits a matrix decomposition of the form:

$$\Gamma = \begin{bmatrix} \Gamma_1 \\ \Gamma_2 \end{bmatrix} : \mathcal{E} \to \begin{bmatrix} H^2(\mathcal{Y}) \\ \mathcal{V} \end{bmatrix}. \tag{5.2.13}$$

The maximal outer spectral factor Θ for T_R is uniquely determined by the relation

$$(\Gamma_1 g)(z) = \Theta(z)g \qquad (z \in \mathbb{D}_+ \text{ and } g \in \mathcal{E}); \tag{5.2.14}$$

see the proof of Theorem 5.2.1. Finally, it is noted that the maximal outer spectral factor Θ for T_R is zero if and only if $\ker U^* = \{0\}$, or equivalently, $U = V$.

5.2.1 The eigenspace for the unitary part

Throughout we assume that all of our Hilbert spaces are separable. Let $U = S \oplus V$ be the Wold decomposition for a controllable isometric pair $\{U, \Gamma\}$. Since $\{U, \Gamma\}$ is controllable, the pair $\{V, \Gamma_2\}$ is also controllable; see (5.2.3). The unitary operator V has at most a countable number of eigenvalues. Let $\{\lambda_j\}_1^n$ be the set of all distinct eigenvalues for V where n is either finite or infinite. Because the pair $\{V, \Gamma_2\}$ is controllable, each eigenvalue λ_j has multiplicity at most $\dim \mathcal{E}$. To see this, let $\mathcal{L}_{\lambda_j} = \ker(V - \lambda_j I)$ be the eigenspace for V corresponding to the eigenvalue λ_j. Since V is unitary, the eigenspace $\mathcal{L}_{\lambda_j}$ is a reducing subspace for V. In particular, V commutes with the orthogonal projection $P_{\mathcal{L}_{\lambda_j}}$ onto $\mathcal{L}_{\lambda_j}$. Using this we arrive at

$$\mathcal{L}_{\lambda_j} = P_{\mathcal{L}_{\lambda_j}} \mathcal{V} = P_{\mathcal{L}_{\lambda_j}} \bigvee_{k=0}^{\infty} V^k \Gamma_2 \mathcal{E} = \bigvee_{k=0}^{\infty} P_{\mathcal{L}_{\lambda_j}} V^k \Gamma_2 \mathcal{E} = \bigvee_{k=0}^{\infty} \lambda_j^k P_{\mathcal{L}_{\lambda_j}} \Gamma_2 \mathcal{E} = P_{\mathcal{L}_{\lambda_j}} \Gamma \mathcal{E}.$$

Hence $\mathcal{L}_{\lambda_j}$ equals $P_{\mathcal{L}_{\lambda_j}} \Gamma \mathcal{E}$, and thus, $\dim \mathcal{L}_{\lambda_j} \leq \dim \mathcal{E}$. In other words, the dimension of the eigenspace corresponding to the eigenvalue λ_j for V is at most $\dim \mathcal{E}$.

Since $\mathcal{L}_{\lambda_j}$ is a reducing subspace for V, the subspace $\oplus_1^n \mathcal{L}_{\lambda_j}$ is also a reducing subspace for V. Therefore V admits a matrix representation of the form

$$V = \begin{bmatrix} \lambda_1 I & 0 & 0 & \cdots & 0 \\ 0 & \lambda_2 I & 0 & \cdots & 0 \\ 0 & 0 & \lambda_3 I & \cdots & 0 \\ \vdots & \vdots & \vdots & \ddots & \vdots \\ 0 & 0 & 0 & \cdots & V_\circ \end{bmatrix} \text{ on } \begin{bmatrix} \mathcal{L}_{\lambda_1} \\ \mathcal{L}_{\lambda_2} \\ \mathcal{L}_{\lambda_3} \\ \vdots \\ \mathcal{V}_\circ \end{bmatrix} \tag{5.2.15}$$

where $V_\circ$ is the unitary operator on $\mathcal{V}_\circ = \mathcal{V} \ominus (\oplus_1^n \mathcal{L}_{\lambda_j})$ defined by $V_\circ = V|\mathcal{V}_\circ$. The operator $V_\circ$ is a unitary operator with no eigenvalues. Using this decomposition the operator Γ_2 admits a matrix representation of the form

$$\Gamma_2 = \begin{bmatrix} A_1 \\ A_2 \\ A_3 \\ \vdots \\ \Gamma_\circ \end{bmatrix} : \mathcal{E} \to \begin{bmatrix} \mathcal{L}_{\lambda_1} \\ \mathcal{L}_{\lambda_2} \\ \mathcal{L}_{\lambda_3} \\ \vdots \\ \mathcal{V}_\circ \end{bmatrix}. \tag{5.2.16}$$

Here $\{A_j\}_1^n$ are operators from $\mathcal{E}$ onto $\mathcal{L}_{\lambda_j}$. If $\mathcal{E} = \mathbb{C}$, then without loss of generality we can always assume that $A_j = a_j$ are scalars and $a_j > 0$ for all $j = 1, 2, \ldots, n$. Finally, $\Gamma_\circ$ is the operator mapping $\mathcal{E}$ into $\mathcal{V}_\circ$ defined by $\Gamma_\circ = P_{\mathcal{V}_\circ} \Gamma_2$.

Remark 5.2.4. Let $\{V, \Gamma_2\}$ be a controllable pair where Γ_2 is an operator from $\mathcal{E}$ into $\mathcal{V}$. Moreover, assume that V is a unitary operator on a nonzero finite dimensional space $\mathcal{V}$. Let W_2 be the controllability operator determined by $\{V, \Gamma_2\}$, that is,

$$W_2 = \begin{bmatrix} \Gamma_2 & V\Gamma_2 & V^2\Gamma_2 & \cdots \end{bmatrix}.$$

Then W_2 is an unbounded linear map from $\ell_+^2(\mathcal{E})$ into $\mathcal{V}$. In other words, W_2 does not determine an operator from $\ell_+^2(\mathcal{E})$ into $\mathcal{V}$.

To see this observe that W_2 is an operator if and only if

$$W_2^\sharp = \begin{bmatrix} \Gamma_2^* & \Gamma_2^* V^* & \Gamma_2^* V^{*2} & \cdots \end{bmatrix}^{tr}$$

defines an operator from $\mathcal{V}$ into $\ell_+^2(\mathcal{E})$. To show that $W_2^\sharp$ is unbounded, let x be an eigenvector with eigenvalue λ for V^*, that is, $V^*x = \lambda x$. Since V is unitary, λ is on the unit circle. Because the pair $\{\Gamma_2^*, V^*\}$ is observable, $\Gamma_2^* x$ is nonzero. (If $\Gamma_2^* x = 0$, then $0 = \lambda^n \Gamma_2^* x = \Gamma_2^* V^{*n} x$. Hence $\Gamma_2^* V^{*n} x = 0$ for all integers $n \geq 0$, which contradicts the observability of the pair $\{\Gamma_2^*, V^*\}$.) Using this, we obtain

$$\|W_2^\sharp x\|^2 = \sum_{n=0}^{\infty} \|\Gamma_2^* V^{*n} x\|^2 = \sum_{n=0}^{\infty} \|\lambda^n \Gamma_2^* x\|^2 = \sum_{n=0}^{\infty} \|\Gamma_2^* x\|^2 = \infty.$$

Hence $W_2^\sharp$ is always unbounded, and thus, W_2 is unbounded.

5.3 The Inner-Outer Factorization Revisited

In this section we will use Theorem 5.2.1 to show that any function Θ in $H^2(\mathcal{E}, \mathcal{Y})$ admits a unique inner-outer factorization. Recall that $\Omega(z)$ is a *contractive analytic function* if Ω is a function in $H^\infty(\mathcal{E}, \mathcal{Y})$ and $\|\Omega\|_\infty \leq 1$. In other words, Ω in $H^\infty(\mathcal{E}, \mathcal{Y})$ is a contractive analytic function if and only if T_Ω is a contraction.

Lemma 5.3.1. *Let Θ be an outer function in $H^2(\mathcal{E}, \mathcal{Y})$ and Ψ be a function in $H^2(\mathcal{E}, \mathcal{G})$. Then the following holds.*

(i) *$T_\Psi^\sharp T_\Psi \leq T_\Theta^\sharp T_\Theta$ if and only if there exists a contractive analytic function Ω in $H^\infty(\mathcal{Y}, \mathcal{G})$ such that $\Psi(z) = \Omega(z)\Theta(z)$ for all z in $\mathbb{D}_+$. In this case, Ω is the only function in $H^\infty(\mathcal{Y}, \mathcal{G})$ satisfying $\Psi = \Omega\Theta$.*

(ii) *$T_\Psi^\sharp T_\Psi = T_\Theta^\sharp T_\Theta$ if and only if there exists an inner function Ω in $H^\infty(\mathcal{Y}, \mathcal{G})$ such that $\Psi(z) = \Omega(z)\Theta(z)$ for all z in $\mathbb{D}_+$.*

(iii) *If Ψ is outer, then $T_\Psi^\sharp T_\Psi = T_\Theta^\sharp T_\Theta$ if and only if $\Psi(z) = \Omega\Theta(z)$ where Ω is a unitary constant in $H^\infty(\mathcal{Y}, \mathcal{G})$.*

(iv) *If $T_\Psi^\sharp T_\Psi \leq T_\Theta^\sharp T_\Theta$, then $\Psi(\infty)^*\Psi(\infty) \leq \Theta(\infty)^*\Theta(\infty)$.*

(v) *If $T_\Psi^\sharp T_\Psi \leq T_\Theta^\sharp T_\Theta$ where $\Psi \in H^2(\mathcal{E}, \mathcal{Y})$, then $\Psi(\infty)^*\Psi(\infty) = \Theta(\infty)^*\Theta(\infty)$ if and only if Θ equals Ψ up to a unitary constant operator on the left.*

Proof. Assume that $T_\Psi^\sharp T_\Psi \le T_\Theta^\sharp T_\Theta$. In other words, $\|T_\Psi x\|^2 \le \|T_\Theta x\|^2$ for all x in $\ell_+^c(\mathcal{E})$. This implies that there exists a contraction C mapping $T_\Theta \ell_+^c(\mathcal{E})$ into $\ell_+^2(\mathcal{G})$ such that $T_\Psi x = CT_\Theta x$. Because Θ is outer, $T_\Theta \ell_+^c(\mathcal{E})$ is dense in $\ell_+^2(\mathcal{Y})$, and it follows that C admits a unique extension by continuity to a contraction mapping $\ell_+^2(\mathcal{Y})$ into $\ell_+^2(\mathcal{G})$. We also denote this extension by C. Let $S_\mathcal{L}$ denote the unilateral shift on $\ell_+^2(\mathcal{L})$. Recall that a lower triangular Toeplitz matrix intertwines two unilateral shifts acting on the appropriate spaces. Then for any x in $\ell_+^c(\mathcal{E})$, we have

$$CS_\mathcal{Y} T_\Theta x = CT_\Theta S_\mathcal{E} x = T_\Psi S_\mathcal{E} x = S_\mathcal{G} T_\Psi x = S_\mathcal{G} CT_\Theta x.$$

Because $T_\Theta \ell_+^c(\mathcal{E})$ is dense in $\ell_+^2(\mathcal{Y})$, we obtain $CS_\mathcal{Y} = S_\mathcal{G} C$. In other words, $C = T_\Omega$ where Ω is a contractive analytic function in $H^\infty(\mathcal{Y},\mathcal{G})$. (In fact, $\|\Omega\|_\infty = \|T_\Omega\| \le 1$.) Hence $T_\Psi = T_\Omega T_\Theta$, and Θ admits a factorization of the form $\Psi = \Omega\Theta$.

On the other hand, if $\Psi = \Omega\Theta$ where Ω is a contractive analytic function, then $T_\Psi = T_\Omega T_\Theta$ and T_Ω is a contraction. So for all x in $\ell_+^c(\mathcal{E})$, we have

$$(T_\Psi^\sharp T_\Psi x, x) = \|T_\Psi x\|^2 = \|T_\Omega T_\Theta x\|^2 \le \|T_\Theta x\|^2 = (T_\Theta^\sharp T_\Theta x, x). \tag{5.3.1}$$

Therefore $T_\Psi^\sharp T_\Psi \le T_\Theta^\sharp T_\Theta$.

To complete the proof of Part (i), it remains to show that Ω is the only function in $H^\infty(\mathcal{Y},\mathcal{G})$ satisfying $\Psi = \Omega\Theta$. Assume that $\Psi = \Phi\Theta$ where Φ is a function in $H^\infty(\mathcal{Y},\mathcal{G})$. Then

$$T_\Omega T_\Theta = T_{\Omega\Theta} = T_\Psi = T_{\Phi\Theta} = T_\Phi T_\Theta.$$

Since $T_\Theta \ell_+^c(\mathcal{E})$ is dense in $\ell_+^c(\mathcal{Y})$, it follows that $T_\Omega = T_\Phi$. Thus $\Omega = \Phi$, and Part (i) holds.

To prove Part (ii), assume that $T_\Psi^\sharp T_\Psi = T_\Theta^\sharp T_\Theta$. According to Part (i), we have $\Psi = \Omega\Theta$, or equivalently, $T_\Psi = T_\Omega T_\Theta$ where Ω is a contractive analytic function. Using $T_\Psi x = T_\Omega T_\Theta x$ for all x in $\ell_+^c(\mathcal{E})$, we obtain

$$\|T_\Omega T_\Theta x\|^2 = \|T_\Psi x\|^2 = (T_\Psi^\sharp T_\Psi x, x) = (T_\Theta^\sharp T_\Theta x, x) = \|T_\Theta x\|^2.$$

In other words, $\|T_\Omega T_\Theta x\|^2 = \|T_\Theta x\|^2$. Because $T_\Theta \ell_+^c(\mathcal{E})$ is dense in $\ell_+^2(\mathcal{Y})$, it follows that T_Ω is an isometry, or equivalently, Ω is an inner function. On the other hand, if $\Psi = \Omega\Theta$ where Ω is an inner function, then $T_\Psi = T_\Omega T_\Theta$ where T_Ω is an isometry. Since $T_\Omega^\sharp T_\Omega = I$, we see that $T_\Psi^\sharp T_\Psi = T_\Theta^\sharp T_\Theta$, and thus, Part (ii) holds.

To obtain Part (iii), assume that Ψ is an outer function and $T_\Psi^\sharp T_\Psi = T_\Theta^\sharp T_\Theta$. By consulting Part (ii), we see that T_Ω is an isometry satisfying $T_\Psi x = T_\Omega T_\Theta x$ for all x in $\ell_+^c(\mathcal{E})$. Because Ψ is outer, $T_\Psi \ell_+^c(\mathcal{E})$ is dense in $\ell_+^2(\mathcal{G})$. Hence T_Ω is a unitary operator, or equivalently, Ω is a unitary constant mapping $\mathcal{Y}$ onto $\mathcal{G}$; see Proposition 2.6.2. The equation $\Psi = \Omega\Theta$, yields Part (iii).

For Part (iv), recall that $\Psi = \Omega\Theta$, where Ω is a contractive analytic function. Thus $\Omega(\infty)$ is a contraction. So using $\Psi(\infty) = \Omega(\infty)\Theta(\infty)$, with $\Omega(\infty)^*\Omega(\infty) \le I$, we obtain Part (iv).

To complete the proof assume that $T_\Psi^\sharp T_\Psi \leq T_\Theta^\sharp T_\Theta$ and $\Theta(\infty)^*\Theta(\infty) = \Psi(\infty)^*\Psi(\infty)$. Here Ψ is a function in $H^2(\mathcal{E},\mathcal{Y})$. According to Part (i), there exists a contractive analytic function Ω in $H^\infty(\mathcal{Y},\mathcal{Y})$ such that $\Omega\Theta = \Psi$. For all vectors v in $\mathcal{E}$, we have

$$\|\Theta(\infty)v\| = \|\Psi(\infty)v\| = \|\Omega(\infty)\Theta(\infty)v\|.$$

Hence $\|\Theta(\infty)v\| = \|\Omega(\infty)\Theta(\infty)v\|$ for all v in $\mathcal{E}$. Because Θ is an outer function, $\Theta(\infty)$ is onto, and thus, $\Omega(\infty)$ is an isometry on $\mathcal{Y}$. Using the fact that $\mathcal{Y}$ is finite dimensional, $\Omega(\infty)$ is unitary. We claim that $\Omega(z) = \Omega(\infty)$ for all z in $\mathbb{D}_+$, that is, Ω is a unitary constant. To see this, let $\Omega(z) = \sum_0^\infty z^{-n}\Omega_n$ be the Taylor series expansion for Ω. Since Ω is a contractive analytic function and $\Omega_0 = \Omega(\infty)$, we obtain

$$\|y\|^2 = \|\Omega(\infty)y\|^2 \leq \sum_{n=0}^{\infty} \|\Omega_n y\|^2 = \|\Omega y\|_{H^2}^2 \leq \|y\|^2 \qquad (y \in \mathcal{Y}).$$

So there is equality in the previous equation, and thus, $\Omega_n = 0$ for all $n \geq 1$. Therefore $\Omega(z) = \Omega(\infty)$ is a unitary constant. □

Previously we have used the Wold decomposition to show that any function in $H^2(\mathcal{E},\mathcal{Y})$ admits an inner-outer factorization; see Theorem 3.2.1. To complete this section we will present another proof of this fact restated here for convenience in the following theorem.

Theorem 5.3.2. *Let Θ be a function in $H^2(\mathcal{E},\mathcal{Y})$. Then Θ admits a unique factorization of the form $\Theta = \Theta_i\Theta_o$ where Θ_o is an outer function in $H^2(\mathcal{E},\mathcal{G})$ and Θ_i is an inner function in $H^\infty(\mathcal{G},\mathcal{Y})$. By unique we mean that if $\Theta = \Psi_i\Psi_o$ where Ψ_o is an outer function in $H^2(\mathcal{E},\mathcal{D})$ and Ψ_i is an inner function in $H^\infty(\mathcal{D},\mathcal{Y})$, then there exists a constant unitary operator Φ mapping $\mathcal{G}$ onto $\mathcal{D}$ such that $\Psi_o = \Phi\Theta_o$ and $\Psi_i\Phi = \Theta_i$.*

Proof. Let T_R be the positive Toeplitz matrix determined by $T_R = T_\Theta^\sharp T_\Theta$. Clearly, T_R admits a maximal outer spectral factor Θ_o in $H^2(\mathcal{E},\mathcal{G})$. Since Θ_o is a maximal outer spectral factor, $T_{\Theta_o}^\sharp T_{\Theta_o} \leq T_R$. By using the definition of a maximal outer spectral factor, we obtain

$$T_R = T_\Theta^\sharp T_\Theta \leq T_{\Theta_o}^\sharp T_{\Theta_o} \leq T_R.$$

Hence $T_\Theta^\sharp T_\Theta = T_{\Theta_o}^\sharp T_{\Theta_o}$. According to Part (ii) of Lemma 5.3.1, there exists an inner function Θ_i in $H^\infty(\mathcal{G},\mathcal{Y})$ such that $\Theta = \Theta_i\Theta_o$. In other words, Θ admits an inner-outer factorization.

Now suppose that $\Theta = \Psi_i\Psi_o$ is another inner-outer factorization of Θ. By employing $T_{\Psi_i}T_{\Psi_o} = T_\Theta$, along with the fact that T_{Ψ_i} is an isometry, we obtain

$$T_R = T_{\Theta_o}^\sharp T_{\Theta_o} = T_\Theta^\sharp T_\Theta = T_{\Psi_o}^\sharp T_{\Psi_i}^\sharp T_{\Psi_i} T_{\Psi_o} = T_{\Psi_o}^\sharp T_{\Psi_o}.$$

Hence $T^\sharp_{\Theta_o} T_{\Theta_o} = T^\sharp_{\Psi_o} T_{\Psi_o}$. According to Part (iii) of Lemma 5.3.1, there exists a unitary constant Φ such that $\Psi_o = \Phi\Theta_o$. Moreover,

$$\Theta = \Psi_i \Psi_o = \Psi_i \Phi \Theta_o = \Theta_i \Theta_o.$$

This implies that $\Psi_i(z)\Phi\Theta_o(z) = \Theta_i(z)\Theta_o(z)$ for all z in $\mathbb{D}_+$. Because Θ_o is outer, $\Psi_i \Phi = \Theta_i$; see Part (i) of Lemma 5.3.1. □

5.4 Positive Real Functions

We say that a function F is *positive real* if F is a $\mathcal{L}(\mathcal{E},\mathcal{E})$-valued analytic function in $\mathbb{D}_+$ and $\Re F(z) \geq 0$ for all z in $\mathbb{D}_+$. (If A is any operator on $\mathcal{X}$, then $\Re A = (A + A^*)/2$.) It is noted that a positive real function F is not necessarily in $H^\infty(\mathcal{E},\mathcal{E})$. For example, consider the scalar-valued function $F(z) = (z+1)/(z-1)$. Finally, F is positive real if and only if $\widetilde{F}$ is positive real. (Recall that $\widetilde{G}(z) = G(\overline{z})^*$.)

Let T_R be a self-adjoint Toeplitz matrix determined by a $\mathcal{L}(\mathcal{E},\mathcal{E})$-valued symbol $R = \sum_{-\infty}^{\infty} e^{-\imath\omega n} R_n$, that is,

$$T_R = \begin{bmatrix} R_0 & R_{-1} & R_{-2} & \cdots \\ R_1 & R_0 & R_{-1} & \cdots \\ R_2 & R_1 & R_0 & \cdots \\ \vdots & \vdots & \vdots & \ddots \end{bmatrix}. \tag{5.4.1}$$

It is emphasized that $R_{-n} = R_n^*$ for all $n \geq 0$. Recall that T_R in (5.4.1) is referred to as the Toeplitz matrix *generated* by $\{R_n\}_0^\infty$ or its symbol $R = \sum_{-\infty}^{\infty} e^{-\imath\omega n} R_n$.

Let F be a $\mathcal{L}(\mathcal{E},\mathcal{E})$-valued analytic function in $\mathbb{D}_+$, and $F(z) = \sum_0^\infty z^{-n} F_n$ be its Taylor series expansion. Let $\{R_n\}_0^\infty$ be the $\mathcal{L}(\mathcal{E},\mathcal{E})$-valued sequence of operators defined by

$$R_0 = F_0 + F_0^* \qquad \text{and} \qquad R_n = F_n \quad (\text{if } n \geq 1). \tag{5.4.2}$$

By construction, $T_R = T_F + T_F^\sharp$ where T_F is the lower triangular Toeplitz matrix determined by F. Theorem 5.4.1 below, shows that T_R is a positive Toeplitz matrix if and only if $F(z)$ is a positive real function. Finally, it is noted that R is formally given by $R = F + F^*$, where $F = \sum_0^\infty e^{-\imath\omega n} F_n$.

To gain some further insight, let $\{U \text{ on } \mathcal{K}, \Gamma\}$ be an isometric representation for a positive Toeplitz matrix T_R generated by a $\mathcal{L}(\mathcal{E},\mathcal{E})$-valued sequence $\{R_n\}_0^\infty$. To be precise, $R_{-n} = (T_R)_{0,n} = \Gamma^* U^n \Gamma$ for all integers $n \geq 0$. Let F_0 be any operator on $\mathcal{E}$ such that $R_0 = F_0 + F_0^*$. For example, since R_0 is positive, one can choose $F_0 = R_0/2$. In fact, $F_0 + F_0^* = R_0$ if and only if $F_0 = R_0/2 + \Psi$ where Ψ is an operator on $\mathcal{E}$ satisfying $\Psi = -\Psi^*$. By virtue of $R_0 = \Gamma^*\Gamma$, it follows that $F_0 + F_0^* = \Gamma^*\Gamma$. Set $F_n = R_n$ for all integers $n \geq 1$. Using $F_n = \Gamma^* U^{*n}\Gamma$ for all $n \geq 1$ and the fact that U is an isometry, we see that the operators F_n

are uniformly bounded. Hence the function F defined by $F(z) = \sum_0^\infty z^{-n} F_n$ is analytic in the open unit disc. Furthermore,

$$F(z) = F_0 + \sum_{k=1}^{\infty} z^{-k} F_k = F_0 + \sum_{k=1}^{\infty} z^{-k} \Gamma^* U^{*k-1} U^* \Gamma = F_0 + \Gamma^* (zI - U^*)^{-1} U^* \Gamma.$$

Thus F admits a state space realization of the form

$$F(z) = F_0 + \Gamma^* (zI - U^*)^{-1} U^* \Gamma \qquad (z \in \mathbb{D}_+). \tag{5.4.3}$$

In other words, $\{U^*, U^*\Gamma, \Gamma^*, F_0\}$ is a realization of F where $F_0 + F_0^* = \Gamma^*\Gamma$ and U^* is a co-isometry. Moreover, this realization is observable if and only if the isometric pair $\{U, \Gamma\}$ is controllable. Finally, it is noted that $\{U, \Gamma, \Gamma^* U, F_0^*\}$ is a realization of $\widetilde{F}$ where $F_0 + F_0^* = \Gamma^*\Gamma$ and U is an isometry.

Now let us show that F is positive real. For z in $\mathbb{D}_+$, we have

$$\begin{aligned}
F(z) + F(z)^* &= F_0 + \Gamma^*(zI - U^*)^{-1}U^*\Gamma + F_0^* + \Gamma^* U(\overline{z}I - U)^{-1}\Gamma \\
&= \Gamma^*\Gamma + \Gamma^*(zI - U^*)^{-1}U^*\Gamma + \Gamma^* U(\overline{z}I - U)^{-1}\Gamma \\
&= \Gamma^* \left[I + (zI - U^*)^{-1}U^* + U(\overline{z}I - U)^{-1}\right]\Gamma \\
&= \Gamma^*(zI - U^*)^{-1} \\
&\quad \times \left[(zI - U^*)(\overline{z}I - U) + U^*(\overline{z}I - U) + (zI - U^*)U\right] \\
&\quad \times (\overline{z}I - U)^{-1}\Gamma \\
&= (|z|^2 - 1)\Gamma^*(zI - U^*)^{-1}(\overline{z}I - U)^{-1}\Gamma.
\end{aligned}$$

In other words,

$$2\Re F(z) = (|z|^2 - 1)\Gamma^*(zI - U^*)^{-1}(\overline{z}I - U)^{-1}\Gamma \qquad (z \in \mathbb{D}_+). \tag{5.4.4}$$

So for a in $\mathcal{E}$, we obtain

$$((F(z) + F(z)^*)a, a) = (|z|^2 - 1)\|(\overline{z}I - U)^{-1}\Gamma a\|^2 \geq 0 \qquad (z \in \mathbb{D}_+).$$

Therefore F is positive real.

The previous analysis shows that if F has a state space realization of the form $\{U^*, U^*\Gamma, \Gamma^*, F_0\}$ where U is an isometry and $F_0 + F_0^* = \Gamma^*\Gamma$, then F is a positive real function. Motivated by this we say that $\{A, B, C, D\}$ is a *positive real Naimark realization* for a function G if $\{A, B, C, D\}$ is a realization of G where A is a co-isometry, $B = AC^*$ and $D + D^* = CC^*$. Assume that F has a Taylor series expansion of the form $F(z) = \sum_0^\infty z^{-n} F_n$. Then $\{A, B, C, D\}$ is a positive real Naimark realization for F if and only if:

(i) the operator A is a co-isometry, the operator $B = AC^*$;

(ii) the operator $D = F_0$ and $D + D^* = CC^*$;

(iii) the Taylor coefficient $F_n = CA^{n-1}B = CA^nC^*$ for all integers $n \geq 1$.

Two positive real Naimark realizations $\{U^*, U^*\Gamma, \Gamma^*, D\}$ and $\{U_1^*, U_1^*\Gamma_1, \Gamma_1^*, D_1\}$ are unitarily equivalent if and only if $D = D_1$ and the pairs $\{U, \Gamma\}$ and $\{U_1, \Gamma_1\}$ are unitarily equivalent. So by consulting the Naimark representation theorem, we see that two observable positive real Naimark realizations of the same function are unitarily equivalent. The following result shows that F is a positive real function if and only if F admits a positive real Naimark realization.

Theorem 5.4.1. *Let T_R be the self-adjoint Toeplitz matrix in* (5.4.1) *determined by a $\mathcal{L}(\mathcal{E}, \mathcal{E})$-valued symbol $R = \sum_{-\infty}^{\infty} e^{-\imath\omega n} R_n$. Let F be the function formally defined by $F(z) = \sum_0^\infty z^{-k} F_k$ where $F_0 + F_0^* = R_0$, and $F_k = R_k$ for all integers $k \geq 1$. Then the following statements are equivalent.*

(i) *The Toeplitz matrix T_R is positive.*

(ii) *The Toeplitz matrix T_R admits a controllable isometric representation $\{U, \Gamma\}$.*

(iii) *The function F admits an observable realization of the form*

$$F(z) = F_0 + \Gamma^*(zI - U^*)^{-1}U^*\Gamma,$$

where U is an isometry and $F_0 + F_0^* = \Gamma^*\Gamma$.

(iv) *The function $\widetilde{F}$ admits a controllable realization of the form*

$$\widetilde{F}(z) = F_0^* + \Gamma^* U(zI - U)^{-1}\Gamma,$$

where U is an isometry and $F_0 + F_0^* = \Gamma^*\Gamma$.

(v) *The function F is positive real.*

(vi) *The function $\widetilde{F}$ is positive real.*

In this case, $F_0 = R_0/2 + \Psi = \Gamma^*\Gamma/2 + \Psi$ where Ψ is an operator on $\mathcal{E}$ satisfying $\Psi = -\Psi^*$. Finally, all observable positive real Naimark realizations of F are unitarily equivalent.

Proof. The Naimark representation theorem 5.1.1 shows that Parts (i) and (ii) are equivalent. If $\{U, \Gamma\}$ is a controllable isometric representation for T_R, then (5.4.3) shows that F has an observable positive real Naimark realization. Hence Parts (i) or (ii) imply Part (iii). Clearly, Parts (iii) and (iv) are equivalent. Our previous analysis shows that if F has a positive real Naimark realization, then F is positive real. In other words, Part (iii) implies Part (v). Finally, it is noted that Parts (v) and (vi) are equivalent.

Now assume that Part (v) holds, that is, F is positive real. Let $F(z) = \sum_0^\infty z^{-k} F_k$ be the Taylor series expansion for F. Recall that $R_0 = F_0 + F_0^*$ and $R_k = F_k$ for all integers $k \geq 1$. Let $g(e^{\imath\omega}) = g_0 + g_1 e^{-\imath\omega} + \cdots + g_n e^{-\imath\omega n}$ be any polynomial with values in $\mathcal{E}$, and let $r > 1$ be a scalar. Because F is a positive real function, we obtain

$$\begin{aligned}
0 &\le \frac{1}{2\pi}\int_0^{2\pi}((F(re^{\imath\omega}) + F(re^{\imath\omega})^*)g(e^{\imath\omega}), g(e^{\imath\omega}))\,d\omega \\
&= \frac{1}{2\pi}\int_0^{2\pi}(F(re^{\imath\omega})g(e^{\imath\omega}), g(e^{\imath\omega}))\,d\omega + \frac{1}{2\pi}\int_0^{2\pi}(g(e^{\imath\omega}), F(re^{\imath\omega})g(e^{\imath\omega}))\,d\omega \\
&= \sum_{m=0}^{n}\sum_{k=0}^{m} r^{k-m}(F_{m-k}g_k, g_m) + \sum_{j=0}^{n}\sum_{\nu=0}^{j} r^{\nu-j}(g_j, F_{j-\nu}g_\nu) \\
&= \sum_{m>k\ge 0}^{n} r^{k-m}(F_{m-k}g_k, g_m) + \sum_{j>\nu\ge 0}^{n} r^{\nu-j}(g_j, F_{j-\nu}g_\nu) + \sum_{j=0}^{n}((F_0 + F_0^*)g_j, g_j) \\
&= \sum_{m>k\ge 0}^{n} r^{k-m}(R_{m-k}g_k, g_m) + \sum_{j>\nu\ge 0}^{n} r^{\nu-j}(g_j, R_{j-\nu}g_\nu) + \sum_{j=0}^{n}(R_0 g_j, g_j) \\
&= \sum_{m=0}^{n}\sum_{k=0}^{n} r^{-|m-k|}((T_R)_{m,k}g_k, g_m).
\end{aligned}$$

By letting r approach 1, we see that $\sum_{m=0}^n\sum_{k=0}^n(R_{m-k}g_k, g_m) \ge 0$. Because this holds for all integers $n \ge 0$ and $\{g_k\}_0^n$ is an arbitrary sequence, The Toeplitz matrix T_R is positive. Hence Part (v) implies (i). Therefore Parts (i) to (vi) are equivalent. □

Let us observe that if F admits a positive real Naimark realization, then F also admits an observable positive real Naimark realization. To see this, assume that $\{U^* \text{ on } \mathcal{K}, U^*\Gamma, \Gamma^*, D\}$ is a positive real Naimark realization for F. Let $\mathcal{K}_o$ be the closed linear span of $\{U^k\Gamma\mathcal{E}\}_0^\infty$. Clearly, $\mathcal{K}_o$ is an invariant subspace for U. Let U_o be the isometry on $\mathcal{K}_o$ defined by $U_o = U|\mathcal{K}_o$. Let Γ_o be the operator from $\mathcal{E}$ into $\mathcal{K}_o$ defined by $\Gamma_o = \Gamma$. By construction the pair $\{U_o, \Gamma_o\}$ is controllable, or equivalently, $\{\Gamma_o^*, U_o^*\}$ is observable. Let $F(z) = \sum_0^\infty z^{-n}F_n$ be the Taylor series expansion for F. Using $U^k|\mathcal{K}_o = U_o^k$ for all integers $k \ge 0$, it follows that

$$F_k = \Gamma^*U^{*k}\Gamma = \left(\Gamma^*U^k\Gamma\right)^* = \left(\Gamma_o^*U_o^k\Gamma_o\right)^* = \Gamma_o^*U_o^{*k}\Gamma_o \qquad (k \ge 1).$$

Thus $\{U_o^*, U_o^*\Gamma_o, \Gamma_0^*, D\}$ is an observable positive real Naimark realization for F.

Remark 5.4.2. Let $\{U, \Gamma\}$ be a controllable isometric representation for a positive Toeplitz matrix T_R. Recall that the maximal outer spectral factor for T_R is determined by

$$\Theta(z) = z\Pi_{\mathcal{Y}}(zI - U^*)^{-1}\Gamma \qquad (z \in \mathbb{D}_+)$$

where $\mathcal{Y} = \ker U^*$ and $\Pi_{\mathcal{Y}} : \mathcal{K} \to \mathcal{Y}$ is the orthogonal projection from $\mathcal{K}$ onto $\mathcal{Y}$; see Theorem 5.2.1. Let Φ be the function defined by $\Phi(z) = (zI - U^*)^{-1}$. Since U is an isometry, the orthogonal projection onto $\mathcal{Y}$ is determined by $\Pi_{\mathcal{Y}}^*\Pi_{\mathcal{Y}} = I - UU^*$. Using $z\Phi(z) = I + U^*\Phi(z)$, we obtain

$$\begin{aligned}
\Theta^*\Theta &= \Gamma^*\Phi^*\overline{z}\Pi_{\mathcal{Y}}^*\Pi_{\mathcal{Y}} z\Phi\Gamma = \Gamma^*\overline{z}\Phi^*(I - UU^*)z\Phi\Gamma \\
&= \Gamma^*\Phi^*\overline{z}z\Phi\Gamma - |z|^2\Gamma^*\Phi^*UU^*\Phi\Gamma \\
&= \Gamma^*(I + \Phi^*U)(I + U^*\Phi)\Gamma - |z|^2\Gamma^*\Phi^*UU^*\Phi\Gamma \\
&= \Gamma\Gamma^* + \Gamma^*\Phi^*U\Gamma + \Gamma^*U^*\Phi\Gamma + \Gamma^*\Phi^*UU^*\Phi\Gamma - |z|^2\Gamma^*\Phi^*UU^*\Phi\Gamma \\
&= 2\Re F(z) - (|z|^2 - 1)\Gamma^*\Phi^*UU^*\Phi\Gamma.
\end{aligned}$$

Therefore $2\Re F(z)$ is given by

$$2\Re F(z) = \Theta(z)^*\Theta(z) + (|z|^2 - 1)\Gamma^*(\overline{z}I - U)^{-1}UU^*(zI - U^*)^{-1}\Gamma \quad (z \in \mathbb{D}_+). \tag{5.4.5}$$

5.5 Minimal Isometric Liftings

Isometric liftings play an important role in operator theory. In this section we will present a brief introduction to isometric liftings, and their connection to Naimark representations. Let A be an operator on $\mathcal{X}$. We say that an operator U on $\mathcal{K}$ is a *lifting* of A if $\mathcal{X}$ is a subspace of $\mathcal{K}$ and $\Pi_{\mathcal{X}}U = A\Pi_{\mathcal{X}}$. In other words, U is a lifting of A if and only if U admits a matrix representation of the form

$$U = \begin{bmatrix} A & 0 \\ \star & \star \end{bmatrix} \text{ on } \begin{bmatrix} \mathcal{X} \\ \mathcal{M} \end{bmatrix}.$$

Here $\star$ represents an unspecified entry. Moreover, U on $\mathcal{K}$ is a lifting of A if and only if $\mathcal{X}$ is an invariant subspace for U^* and $A^* = U^*|\mathcal{X}$. Finally, if U is a lifting of A, then $\Pi_{\mathcal{X}}U^n = A^n\Pi_{\mathcal{X}}$ for all integers $n \geq 0$.

Now assume that A is a contraction on $\mathcal{X}$, that is, $\|A\| \leq 1$. We say that U is an *isometric lifting* of A if U is an isometry and U is a lifting of A. Moreover, U on $\mathcal{K}$ is a *minimal isometric lifting* of A if U is an isometric lifting of A and $\mathcal{X}$ is cyclic for U, that is,

$$\mathcal{K} = \bigvee_{n=0}^{\infty} U^n\mathcal{X}.$$

In other words, U is a minimal isometric lifting of A if U is a lifting of A and the pair $\{U, \Pi_{\mathcal{X}}^*\}$ is controllable. (Recall that $\Pi_{\mathcal{X}}^*$ is the natural embedding of $\mathcal{X}$ into $\mathcal{K}$.) Finally, we say that two isometric liftings U on $\mathcal{K}$ and U_1 on $\mathcal{K}_1$ of a contraction A on $\mathcal{X}$ are *isomorphic* if there exists a unitary operator Φ mapping $\mathcal{K}$ onto $\mathcal{K}_1$ such that $\Phi U = U_1\Phi$ and $\Phi|\mathcal{X} = I_{\mathcal{X}}$.

If A is a contraction, then A admits a minimal isometric lifting. To construct a minimal isometric lifting of A, consider the isometry U on $\mathcal{K} = \mathcal{X} \oplus \ell_+^2(\mathcal{D}_A)$

determined by

$$U = \begin{bmatrix} A & 0 & 0 & 0 & \cdots \\ D_A & 0 & 0 & 0 & \cdots \\ 0 & I & 0 & 0 & \cdots \\ 0 & 0 & I & 0 & \cdots \\ \vdots & \vdots & \vdots & \vdots & \ddots \end{bmatrix} \text{ on } \begin{bmatrix} \mathcal{X} \\ \mathcal{D}_A \\ \mathcal{D}_A \\ \mathcal{D}_A \\ \vdots \end{bmatrix}. \tag{5.5.1}$$

Here D_A equals the positive square root of $I - A^*A$, and $\mathcal{D}_A$ equals the closure of the range of D_A. By construction U is a lifting of A. It is a simple exercise to verify that U is an isometry and $\mathcal{X}$ is cyclic for U. Therefore U is a minimal isometric lifting of A. Finally, the isometry U in (5.5.1) is called the the *Sz.-Nagy-Schaffer minimal isometric lifting* of A. This proves part of the following result.

Theorem 5.5.1. *Let A be a contraction on $\mathcal{X}$. Then A admits a minimal isometric lifting. Moreover, the following statements are equivalent.*

(i) *The operator U on $\mathcal{K}$ is a minimal isometric lifting for A.*

(ii) *The pair $\{U, \Pi_{\mathcal{X}}^*\}$ is a controllable isometric representation for the Toeplitz matrix*

$$T_R = \begin{bmatrix} I & A & A^2 & \cdots \\ A^* & I & A & \cdots \\ A^{*2} & A^* & I & \cdots \\ \vdots & \vdots & \vdots & \ddots \end{bmatrix}. \tag{5.5.2}$$

The symbol $R = \sum_{-\infty}^{\infty} e^{-\imath\omega n} R_n$ where $R_{-n} = A^n$ and $R_n = A^{*n}$ for all integers $n \geq 0$. In particular, the Toeplitz matrix T_R in (5.5.2) is positive.

(iii) *The pair $\{U, \Pi_{\mathcal{X}}^*\}$ is a controllable isometric representation for $\{A^n\}_0^\infty$.*

Finally, all minimal isometric liftings of A are isomorphic.

Proof. Assume that U on $\mathcal{K}$ is a minimal isometric lifting for A. Recall that $A^n\Pi_{\mathcal{X}} = \Pi_{\mathcal{X}}U^n$ for all integers $n \geq 0$. Since $I = \Pi_{\mathcal{X}}\Pi_{\mathcal{X}}^*$, we see that $A^n = \Pi_{\mathcal{X}}U^n\Pi_{\mathcal{X}}^*$ for all $n \geq 0$. Therefore $\{U, \Pi_{\mathcal{X}}^*\}$ is a controllable isometric representation for the Toeplitz matrix T_R in (5.5.2). So Part (i) implies Parts (ii) and (iii). Finally, because all controllable isometric representations of the same Toeplitz matrix are unitarily equivalent, all minimal isometric liftings for A are isomorphic.

Assume that Part (ii) holds, that is, $\{U, \Pi_{\mathcal{X}}^*\}$ is a controllable isometric representation for T_R in (5.5.2). Now let U_1 be the Sz.-Nagy-Schaffer minimal isometric lifting of A. Then $\{U_1, \Pi_{\mathcal{X}}^*\}$ is also a controllable isometric representation for T_R. Since all controllable isometric representations of the same Toeplitz matrix are unitarily equivalent, U and U_1 must be isomorphic. Therefore U is also a minimal isometric lifting of A. So Part (ii) implies Part (i). In other words, Parts (i) and (ii) are equivalent. By definition Parts (ii) and (iii) are equivalent. □

Recall that an operator A on $\mathcal{X}$ is *strongly stable* if A^n converges to zero in the strong operator topology, that is, for each vector x in $\mathcal{X}$ the sequence $A^n x$ converges to zero as n tends to infinity.

Proposition 5.5.2. *Let U on $\mathcal{K}$ be a minimal isometric lifting for a contraction A on $\mathcal{X}$. Then A^* is strongly stable if and only if U is a unilateral shift.*

Proof. If U is a unilateral shift, then U^* is strongly stable. Since $A^* = U^*|\mathcal{X}$, we see that A^* is also strongly stable. Now assume that A^* is strongly stable. Let $U = S \oplus V$ on $\mathcal{K}_+ \oplus \mathcal{V}$ be the Wold decomposition of U where S is a unilateral shift and V is unitary. Then each vector x in $\mathcal{X}$ admits a unique decomposition of the form $x = h \oplus v$ where h is in $\mathcal{K}_+$ and v is in $\mathcal{V}$. Using the fact that A^* and S^* are strongly stable, we obtain

$$0 = \lim_{n\to\infty} \|A^{*n}x\|^2 = \lim_{n\to\infty} \|U^{*n}x\|^2 = \lim_{n\to\infty} \left(\|S^{*n}h\|^2 + \|V^{*n}v\|^2\right) = \|v\|^2.$$

Hence $v = 0$. In other words, $\mathcal{X}$ is a subspace of $\mathcal{K}_+$. Since $U = S \oplus V$ and $\mathcal{X} \subseteq \mathcal{K}_+$ is cyclic for U, we have

$$\mathcal{K} = \bigvee_{n=0}^{\infty} U^n\mathcal{X} = \bigvee_{n=0}^{\infty} S^n\mathcal{X} \subseteq \mathcal{K}_+ \subseteq \mathcal{K}.$$

Therefore $\mathcal{K}_+ = \mathcal{K}$, and $U = S$ is a unilateral shift. □

Let A be contraction on $\mathcal{X}$ and T_R the positive Toeplitz matrix in (5.5.2). Let C be the operator mapping $\mathcal{X}$ into $\mathcal{D}_{A^*}$ defined by $C = D_{A^*}$. Then

$$\Theta(z) = zC(zI - A^*)^{-1} \tag{5.5.3}$$

is the maximal outer spectral factor for T_R.

Let U on $\mathcal{K}$ be the minimal isometric lifting for A. Then $\{U, \Pi_{\mathcal{X}}^*\}$ is the controllable isometric representation for T_R. Since $\mathcal{X}$ is an invariant subspace for U^* and $A^* = U^*|\mathcal{X}$, Theorem 5.2.1 shows that

$$\Theta(z) = z\Pi_{\mathcal{Y}}(zI - U^*)^{-1}\Pi_{\mathcal{X}}^* = z\Pi_{\mathcal{Y}}(zI - A^*)^{-1} \tag{5.5.4}$$

is the outer spectral factor for T_R. Here $\mathcal{Y}$ equals the kernel of U^*. Using $\Pi_{\mathcal{Y}}^*\Pi_{\mathcal{Y}} = I - UU^*$ with x in $\mathcal{X}$, we obtain

$$\|Cx\|^2 \quad = \quad \|D_{A^*}x\|^2 = ((I - AA^*)x, x) = ((I - UU^*)x, x) = \|\Pi_{\mathcal{Y}}x\|^2.$$

This implies that there exists a unitary operator Φ such that $C = \Phi\Pi_{\mathcal{Y}}|\mathcal{X}$. Because the maximal outer spectral factor is unique up to a unitary constant operator on the left, (5.5.4) shows that $\Theta(z) = zC(zI - A^*)^{-1}$ is the maximal outer spectral factor for T_R.

Finally, it is noted that Proposition 5.5.2 with Theorem 5.2.1 show that $T_R = T_\Theta^* T_\Theta$ if and only if A^* is strongly stable.

5.6 Notes

All the results in this chapter are classical. Our approach is based on the Naimark representation theorem. The Naimark representation theorem in Section 5.1 is a standard result in operator theory; see Fillmore [80] and Sz.-Nagy-Foias [198]. For a bilinear version of the Naimark representation Theorem see Frazho [94] and Popescu [177]. The notion of a maximal outer spectral factor was developed in Chapter 5 of Sz.-Nagy-Foias [198]. Our approach to the maximal outer spectral factor was taken from Frazho-Kaashoek [98]. The results in Lemma 5.3.1 are now standard and were essentially taken from Sz.-Nagy-Foias [198] and Foias-Frazho [82]. For a generalization of the maximal outer spectral factor to a certain non-commutative case and the inner-outer factorization approach in Theorem 5.3.2; see Popescu [177]. The Naimark representation theorem and its relation to positive real functions is a classical result in operator theory; see Fillmore [80] and Frazho-Kaashoek [98]. Remark 5.4.2 was taken from Frazho-ter Horst-Kaashoek [103]. Isometric liftings play a fundamental role in operator theory; see Sz.-Nagy-Foias [198], Foias-Frazho [82] and Foias-Frazho-Gohberg-Kaashoek [84]. In general the isometric lifting is developed without using the Naimark representation theorem. In Section 5.5, we used the Naimark dilation to prove some elementary properties concerning isometric liftings.

Classical measure theoretic results. The following classical result due to Bochner shows that a Toeplitz matrix is positive if and only if its entries are the Fourier coefficients of an operator-valued positive measure.

Theorem 5.6.1 (Bochner). *Let T_R be the Toeplitz matrix determined by the self-adjoint symbol $R = \sum_{-\infty}^{\infty} e^{-\imath\omega k} R_k$ with values in $\mathcal{L}(\mathcal{E}, \mathcal{E})$. Then T_R is positive if and only if*

$$R_n = \frac{1}{2\pi} \int_0^{2\pi} e^{\imath\omega n} d\Omega \qquad (\textit{for all integers } n) \tag{5.6.1}$$

where $d\Omega$ is a positive measure with values in $\mathcal{L}(\mathcal{E}, \mathcal{E})$.

Bochner's theorem is widely used and plays a fundamental role in stochastic processes. In this monograph, we have relied more on geometric results than measure theoretic ideas to develop factorization algorithms. One can generate an isometric representation directly from Bochner's theorem. To see this, assume that $d\Omega$ is the positive measure corresponding to the Toeplitz matrix T_R. Consider the Hilbert space $\mathcal{K}$ determined by the closure of the set of all polynomials of the form $g = \sum_{n \geq 0} e^{-\imath\omega n} g_n$ under the inner product

$$\|g\|^2 = \frac{1}{2\pi} \int_0^{2\pi} (d\Omega\, g, g).$$

Let U be the isometry on $\mathcal{K}$ given by $Ug = e^{-\imath\omega} g$. Let Γ be the operator mapping $\mathcal{E}$ into $\mathcal{K}$ determined by $\Gamma a = a$ where a is in $\mathcal{E}$. Then $\{U, \Gamma\}$ is a controllable isometric representation for T_R.

The positive measure $d\Omega$ in Bochner's theorem admits a decomposition of the form

$$d\Omega = \Theta^*\Theta\, d\omega + d\Xi \tag{5.6.2}$$

where Θ is the maximal outer spectral factor for T_R. Because T_R is a positive Toeplitz matrix, it admits a unique controllable isometric representation $\{U, \Gamma\}$. Let $U = S \oplus V$ be the Wold decomposition for U where S is the unilateral shift on $H^2(\mathcal{Y})$ and V is a unitary operator on $\mathcal{V}$ while $\Gamma = \begin{bmatrix}\Gamma_1 & \Gamma_2\end{bmatrix}^{tr}$ is an operator mapping $\mathcal{E}$ into $H^2(\mathcal{Y}) \oplus \mathcal{V}$; see Remark 5.2.3. Let E_ω be the spectral measure for V, that is,

$$V = \frac{1}{2\pi}\int_0^{2\pi} e^{-\imath\omega} dE_\omega.$$

Let $d\Xi$ be the positive measure determined by $d\Xi = \Gamma_2^* dE_\omega \Gamma_2$. Without loss of generality, we can assume that $(\Gamma_1 g)(z) = \Theta(z)g$, where Θ is the maximal outer spectral factor for T_R and g is in $\mathcal{E}$; see Remark 5.2.3. So for all integers $n \geq 0$, we obtain

$$\begin{aligned}
(R_{-n}g, g) = (U^n\Gamma g, \Gamma g) &= (S^n\Gamma_1 g, \Gamma_1 g) + (V^n\Gamma_2 g, \Gamma_2 g)\\
&= \frac{1}{2\pi}\int_0^{2\pi} e^{-\imath\omega n}(\Theta g, \Theta g)d\omega + \frac{1}{2\pi}\int_0^{2\pi} e^{-\imath\omega n}(dE_\omega\Gamma_2 g, \Gamma_2 g)\\
&= \frac{1}{2\pi}\int_0^{2\pi} e^{-\imath\omega n}(d\Omega\, g, g)
\end{aligned}$$

where $d\Omega = \Theta^*\Theta\, d\omega + d\Xi$. Clearly, $d\Omega$ is a positive measure. Hence the representation for R_n in (5.6.1) holds for all integers $n \leq 0$. Recall that $R_n^* = R_{-n}$. So by taking the adjoint, we see that (5.6.1) holds for all integers.

Recall that any positive measure $d\mu$ admits a unique Lebesgue decomposition of the form $d\mu = f d\omega + d\nu$ where f is a positive Lebesgue measure function and $d\nu$ is singular with respect to the Lebesgue measure; see [183, 187]. The following result, due to Helson-Lowdenslager [130, 131], shows that the Wold decomposition yields the Lebesgue decomposition when the maximal outer spectral factor is a function in $H^2(\mathcal{E}, \mathcal{E})$.

Theorem 5.6.2. *Let T_R be the positive Toeplitz matrix determined by the symbol $R = \sum_{-\infty}^{\infty} e^{-\imath\omega k} R_k$ with values in $\mathcal{L}(\mathcal{E}, \mathcal{E})$. Assume that the maximal outer spectral factor Θ for T_R is a function in $H^2(\mathcal{E}, \mathcal{E})$. Let $d\Omega$ be the $\mathcal{L}(\mathcal{E}, \mathcal{E})$-valued positive measure determined by (5.6.1). Then the Lebesgue decomposition for $d\Omega$ is determined by*

$$d\Omega = \Theta^*\Theta\, d\omega + d\Xi \tag{5.6.3}$$

where $d\Xi$ is singular with respect to the Lebesgue measure.

Proof. Recall that the positive measure $d\Omega$ admits a decomposition of the form $d\Omega = \Theta^*\Theta\, d\omega + d\Xi$. Moreover, $d\Xi$ admits a Lebesgue decomposition of the form

$d\Xi = Qd\omega + d\nu$ where $Q(e^{\imath\omega})$ is almost everywhere a positive operator. Furthermore, $d\nu$ is a positive singular measure with respect to the Lebesgue measure. According to Lemma 5.6.4 below, there exists an outer function Ψ such that $\Psi^*\Psi = \Theta^*\Theta + Q$. This readily implies that $T_\Theta^\sharp T_\Theta \le T_\Psi^\sharp T_\Psi \le T_R$. Since Θ is the maximal outer spectral factor for T_R, we must have $T_\Psi^\sharp T_\Psi \le T_\Theta^\sharp T_\Theta$. In other words, $T_\Psi^\sharp T_\Psi = T_\Theta^\sharp T_\Theta$. Hence Θ equals Ψ up to a constant unitary operator on the left; see Lemma 5.3.1. Using $\Theta^*\Theta = \Psi^*\Psi = \Theta^*\Theta + Q$ implies that $Q = 0$. Thus $d\nu = d\Xi$ and $d\Omega = \Theta^*\Theta\, d\omega + d\Xi$ is the Lebesgue decomposition for $d\Omega$. □

In the scalar case, Theorem 5.6.2 reduces to the following result; see also Hoffman [134] for a nice proof of this Corollary.

Corollary 5.6.3. *Let T_r be the positive Toeplitz matrix determined by the self-adjoint scalar-valued symbol $r = \sum_{-\infty}^{\infty} e^{-\imath\omega k} r_k$. Assume that the maximal outer spectral factor θ for T_r is a function in H^2, that is, θ is not the zero function mapping $\mathbb{C}$ into $\{0\}$. Let $d\mu$ be the positive measure determined by*

$$r_n = \frac{1}{2\pi}\int_0^{2\pi} e^{\imath\omega n} d\mu \qquad (\text{for all integers } n). \tag{5.6.4}$$

Then the Lebesgue decomposition for $d\mu$ is given by $d\mu = |\theta|^2 d\omega + d\nu$ where $d\nu$ is singular with respect to the Lebesgue measure.

For some applications to Corollary 5.6.3 see Hoffman [134]. We say that a function F is in $L^1(\mathcal{E},\mathcal{Y})$ if F is a $\mathcal{L}(\mathcal{E},\mathcal{Y})$-valued Lebesgue measurable function and $\|F\|$ is integrable.

Lemma 5.6.4. *Let Θ be an outer function in $H^2(\mathcal{E},\mathcal{E})$. Assume Q is a function in $L^1(\mathcal{E},\mathcal{E})$ and $Q(e^{\imath\omega})$ is almost everywhere a positive operator on $\mathcal{E}$. Then $\Theta^*\Theta + Q$ admits an outer spectral factor, that is, there exists an outer function Ψ in $H^2(\mathcal{E},\mathcal{E})$ such that $\Psi^*\Psi = \Theta^*\Theta + Q$.*

Proof. Consider the positive function R in $L^1(\mathcal{E},\mathcal{E})$ defined by $R = \Theta^*\Theta + Q$. Notice that $T_\Theta^\sharp T_\Theta \le T_R$. Let Ψ in $H^2(\mathcal{E},\mathcal{Y})$ be the maximal outer spectral factor for T_R. Then $T_\Theta^\sharp T_\Theta \le T_\Psi^\sharp T_\Psi$. According to Lemma 5.3.1, there exists a contractive analytic function Φ in $H^\infty(\mathcal{Y},\mathcal{E})$ such that $\Theta = \Phi\Psi$. In particular, $\Theta(\infty) = \Phi(\infty)\Psi(\infty)$. Since $\Theta(\infty)$ is invertible, $\Phi(\infty)\Psi(\infty)$ must also be invertible. Because the space $\mathcal{E}$ is finite dimensional and Ψ is outer, $\Psi(\infty)$ is invertible. So without loss of generality we can assume that Ψ is an outer function in $H^2(\mathcal{E},\mathcal{E})$.

Let $R = \sum_{-\infty}^{\infty} R_n e^{-\imath\omega n}$ be the Fourier series expansion for R. Let $\mathcal{K}$ be the subspace of $L^2(\mathcal{E})$ defined by the closed linear span of $\{e^{-\imath\omega n}\sqrt{R}\mathcal{E}\}_0^\infty$, where $\sqrt{R}(e^{\imath\omega})$ is almost everywhere the positive square root of $R(e^{\imath\omega})$. Let U be the isometry on $L^2(\mathcal{E})$ defined by restricting the bilateral shift to $\mathcal{K}$, that is, $Uf = e^{-\imath\omega} f$ where f is in $\mathcal{K}$. Let Γ be the operator mapping $\mathcal{E}$ into $\mathcal{K}$ defined by $\Gamma a = \sqrt{R}a$ where a is in $\mathcal{E}$. Then $\{U,\Gamma\}$ is a controllable isometric representation for T_R. To see this, observe that for any integer $n \geq 0$ and a in $\mathcal{E}$, we have

$$\begin{aligned}(R_{-n}a, a) &= \frac{1}{2\pi}\int_0^{2\pi} e^{-\imath\omega n}(Ra, a)d\omega \\ &= \frac{1}{2\pi}\int_0^{2\pi} e^{-\imath\omega n}(\sqrt{R}a, \sqrt{R}a)d\omega \\ &= (U^n\Gamma a, \Gamma a).\end{aligned}$$

Hence $\{U, \Gamma\}$ is a controllable isometric representation for T_R.

By consulting the proof of Theorem 5.2.1, there exists a contraction Y mapping $\mathcal{K}$ onto $\ell_+^2(\mathcal{Y})$ such that $T_\Psi g = YWg$. Here

$$W = \begin{bmatrix}\Gamma & U\Gamma & U^2\Gamma & \cdots\end{bmatrix}$$

and g is in $\ell_+^c(\mathcal{E})$. Moreover, $SY = YU$ where S is the unilateral shift on $\ell_+^2(\mathcal{E})$ and $Y\mathcal{V} = \{0\}$ where $\mathcal{V} = \bigcap_0^\infty e^{-\imath\omega n}\mathcal{K}$. Because R is almost everywhere invertible,

$$L^2(\mathcal{E}) = \bigvee_{n=-\infty}^{\infty} e^{-\imath\omega n}\sqrt{R}\mathcal{E} = \bigvee_{n=-\infty}^{\infty} e^{-\imath\omega n}\Gamma\mathcal{E} = \bigvee_{n=-\infty}^{\infty} e^{-\imath\omega n}\mathcal{K}.$$

This readily implies that the minimal unitary extension for U is the bilateral shift on $L^2(\mathcal{E})$. Clearly, the minimal unitary extension for S is the bilateral shift on $\ell^2(\mathcal{E})$. Recall that the Fourier transform $\mathcal{F}_\mathcal{E}$ is a unitary operator which intertwines the bilateral shift on $\ell^2(\mathcal{E})$ with the bilateral shift on $L^2(\mathcal{E})$, and $SY = YU$. According to Proposition 1.5.1, the operator $\mathcal{F}_\mathcal{E}^+ Y$ admits a unique extension which commutes with the bilateral shift on $L^2(\mathcal{E})$. Therefore we can view $\mathcal{F}_\mathcal{E}^+ Y$ as the restriction of a multiplication operator $\mathcal{F}_\mathcal{E}^+ Y = M_F|\mathcal{K}$ where F is in $L^\infty(\mathcal{E}, \mathcal{E})$; see Corollary 2.4.2. For a in $\mathcal{E}$, we obtain

$$\Psi a = \mathcal{F}_\mathcal{E}^+ T_\Psi \begin{bmatrix}a & 0 & 0 & \cdots\end{bmatrix}^{tr} = M_F W \begin{bmatrix}a & 0 & 0 & \cdots\end{bmatrix}^{tr} = F\sqrt{R}a \qquad (a \in \mathcal{E}).$$

In other words, $\Psi(e^{\imath\omega}) = F(e^{\imath\omega})\sqrt{R(e^{\imath\omega})}$ almost everywhere. Because $\Psi(e^{\imath\omega})$ is almost everywhere unitary, $F(e^{\imath\omega})$ is also almost everywhere invertible. Since $\{0\} = \mathcal{F}_\mathcal{E}^+ Y\mathcal{V} = F|\mathcal{V}$, it follows that $\mathcal{V}$ is zero. So there is no unitary part in the Wold decomposition of U. Therefore $T_R = T_\Psi^\sharp T_\Psi$; see Part (iii) in Theorem 5.2.1. In other words, $\Psi^*\Psi = R = \Theta^*\Theta + Q$. □

It is noted that Helson-Lowdenslager [130, 131] showed that a positive function R in $L^1(\mathcal{E}, \mathcal{E})$, admits a square outer spectral factor if and only if

$$\int_0^{2\pi} \ln\det[R(e^{\imath\omega})]d\omega > -\infty; \tag{5.6.5}$$

see also Chapter 5 in Sz.-Nagy-Foias [198]. Using this fact one can also give another proof of Lemma 5.6.4. Here we presented a proof of this result using an isometric representation.

Chapter 6

The Rational Case

In this chapter, we will present some results on Toeplitz matrices with rational symbols. Then we will introduce the positive real lemma in systems theory. Finally, we will present some classical ergodic theorems, and show how they can be used to estimate the unitary part in the Wold decomposition of a controllable isometric representation. Throughout we assume that $\mathcal{E}$ and $\mathcal{Y}$ are finite dimensional.

6.1 Rational Symbols

In this section, we will study positive Toeplitz matrices T_R when its symbol R is a rational function. Let T_R be a positive Toeplitz matrix generated by the symbol $R = \sum_{-\infty}^{\infty} e^{-\imath\omega n} R_n$ with values in $\mathcal{L}(\mathcal{E}, \mathcal{E})$, that is,

$$T_R = \begin{bmatrix} R_0 & R_{-1} & R_{-2} & \cdots \\ R_1 & R_0 & R_{-1} & \cdots \\ R_2 & R_1 & R_0 & \cdots \\ \vdots & \vdots & \vdots & \ddots \end{bmatrix}. \tag{6.1.1}$$

We say that T_R is a *rational* Toeplitz matrix, if its symbol R defines a rational function. In this case, let F be the positive real function associated with T_R defined by

$$F(z) = \frac{R_0}{2} + \sum_{n=1}^{\infty} z^{-n} R_n. \tag{6.1.2}$$

Using $R_n^* = R_{-n}$ for all integers $n \geq 0$, it follows that $T_R = T_F + T_F^\sharp$ and its symbol $R(e^{\imath\omega}) = F(e^{\imath\omega}) + F(e^{\imath\omega})^*$. In particular, F is rational if and only if R is rational, or equivalently, T_R is a rational Toeplitz matrix. Finally, recall that F is a positive real function if F is analytic in $\mathbb{D}_+$ and the Toeplitz matrix T_R determined by $R = F + F^*$ is positive; see Section 5.4 and Theorem 5.4.1.

Assume that T_R is a positive Toeplitz matrix, or equivalently, F is positive real. Let $\{U \text{ on } \mathcal{K}, \Gamma\}$ be a controllable isometric representation for T_R. Recall that $R_n = \Gamma^* U^{*n} \Gamma$ for all integers $n \geq 0$. Using this with z in $\mathbb{D}_+$, it follows that

$$\begin{aligned} F(z) &= \frac{R_0}{2} + \sum_{n=1}^{\infty} z^{-n} R_n = \frac{1}{2}\Gamma^*\Gamma + \sum_{n=1}^{\infty} z^{-n}\Gamma^* U^{*n}\Gamma \\ &= \frac{1}{2}\Gamma^*\Gamma + z^{-1}\Gamma^*(I - z^{-1}U^*)^{-1}U^*\Gamma \\ &= \frac{1}{2}\Gamma^*\Gamma + \Gamma^*(zI - U^*)^{-1}U^*\Gamma. \end{aligned}$$

Hence $\Sigma = \{U^*, U^*\Gamma, \Gamma^*, \Gamma^*\Gamma/2\}$ is an observable realization for F. The observability follows from the fact that the pair $\{U, \Gamma\}$ is controllable. Because U^* is a contraction, $F(z)$ is analytic in $\mathbb{D}_+$. Let $\mathcal{X}$ be the invariant subspace for U^* defined by extracting the controllable subspace from Σ, that is,

$$\mathcal{X} = \bigvee_{n=0}^{\infty} U^{*n}(U^*\Gamma)\mathcal{E}. \tag{6.1.3}$$

Let $\{A, B, C\}$ be the set of operators defined by

$$A = U^*|\mathcal{X} \text{ on } \mathcal{X}, \quad B = U^*\Gamma : \mathcal{E} \to \mathcal{X} \quad \text{and} \quad C = \Gamma^*|\mathcal{X} : \mathcal{X} \to \mathcal{E}. \tag{6.1.4}$$

The realization $\{A, B, C, R_0/2\}$ was obtained by extracting the controllable part from the observable realization Σ for F. Hence $\{A, B, C, R_0/2\}$ is a controllable and observable realization for F. Therefore the dimension of $\mathcal{X}$ equals the McMillan degree of F. In other words, F is rational if and only $\mathcal{X}$ is finite dimensional. Finally, $\dim \mathcal{X} = \delta(F)$. (Recall that the McMillan degree of a transfer function G is denoted by $\delta(G)$.)

As before, let

$$U = \begin{bmatrix} S & 0 \\ 0 & V \end{bmatrix} \text{ on } \begin{bmatrix} \ell^2_+(\mathcal{Y}) \\ \mathcal{V} \end{bmatrix} \quad \text{and} \quad \Gamma = \begin{bmatrix} \Gamma_1 \\ \Gamma_2 \end{bmatrix} : \mathcal{E} \to \begin{bmatrix} \ell^2_+(\mathcal{Y}) \\ \mathcal{V} \end{bmatrix} \tag{6.1.5}$$

be the Wold decomposition for the pair $\{U, \Gamma\}$. Here S is the unilateral shift on $\ell^2_+(\mathcal{Y})$ where $\mathcal{Y} = \ker U^*$ and V is a unitary operator on $\mathcal{V}$. Now let $\mathcal{X}_1$ be the invariant subspace for S^* defined by extracting the controllable subspace from the pair $\{S^*, S^*\Gamma_1\}$, that is,

$$\mathcal{X}_1 = \bigvee_{n=0}^{\infty} S^{*n}(S^*\Gamma_1)\mathcal{E}. \tag{6.1.6}$$

Let A_1 be the operator on $\mathcal{X}_1$ and B_1 the operator mapping $\mathcal{E}$ into $\mathcal{X}_1$ defined by

$$A_1 = S^*|\mathcal{X}_1 \text{ on } \mathcal{X}_1 \quad \text{and} \quad B_1 = S^*\Gamma_1 : \mathcal{E} \to \mathcal{X}_1. \tag{6.1.7}$$

We claim that F is rational if and only if both $\mathcal{X}_1$ and $\mathcal{V}$ are finite dimensional. In this case, $\mathcal{X} = \mathcal{X}_1 \oplus \mathcal{V}$. Moreover, the minimal state space realization $\{A, B, C, R_0/2\}$ of F in (6.1.4) admits a matrix representation of the form

$$A = \begin{bmatrix} A_1 & 0 \\ 0 & V^* \end{bmatrix} \text{ on } \begin{bmatrix} \mathcal{X}_1 \\ \mathcal{V} \end{bmatrix} \quad \text{and} \quad B = \begin{bmatrix} B_1 \\ V^*\Gamma_2 \end{bmatrix} : \mathcal{E} \to \begin{bmatrix} \mathcal{X}_1 \\ \mathcal{V} \end{bmatrix},$$
$$C = \begin{bmatrix} \Gamma_1^* & \Gamma_2^* \end{bmatrix} : \begin{bmatrix} \mathcal{X}_1 \\ \mathcal{V} \end{bmatrix} \to \mathcal{E} \quad \text{and} \quad D = \frac{1}{2}(\Gamma_1^*\Gamma_1 + \Gamma_2^*\Gamma_2). \tag{6.1.8}$$

To see this, observe that the Wold decomposition $U = S \oplus V$ yields

$$\mathcal{X} = \bigvee_{n=1}^{\infty} U^{*n}\Gamma\mathcal{E} = \bigvee_{n=1}^{\infty} \{S^{*n}\Gamma_1 a \oplus V^{*n}\Gamma_2 a : a \in \mathcal{E}\} \subseteq \mathcal{X}_1 \oplus \mathcal{V}. \tag{6.1.9}$$

If $\mathcal{X}_1$ and $\mathcal{V}$ are both finite dimensional, then $\mathcal{X} \subseteq \mathcal{X}_1 \oplus \mathcal{V}$ must also be finite dimensional. Since $\{A, B, C, D\}$ is a minimal realization, F is rational.

On the other hand, if F is rational, then $\mathcal{X}$ is finite dimensional. By consulting the second equality in (6.1.9), we see that both $\mathcal{X}_1 = \bigvee_1^\infty S^{*n}\Gamma_1\mathcal{E}$ and $\mathcal{X}_2 = \bigvee_1^\infty V^{*n}\Gamma_2\mathcal{E}$ must be finite dimensional subspaces. Clearly, $\mathcal{X}_2$ is invariant under V^*, and $V^*|\mathcal{X}_2$ is an isometry on $\mathcal{X}_2$. Because the subspace $\mathcal{X}_2$ is finite dimensional, $V^*|\mathcal{X}_2$ is a unitary operator on $\mathcal{X}_2$. In particular, $\mathcal{X}_2$ is a reducing subspace for V. Using the fact that $\{V, \Gamma_2\}$ is controllable, we arrive at

$$\mathcal{V} \supseteq \mathcal{X}_2 = \bigvee_{n=0}^{\infty} V^n\mathcal{X}_2 = \bigvee_{n\geq 0, k>0} V^n V^{*k}\Gamma_2\mathcal{E} = \bigvee_{n=-\infty}^{\infty} V^n\Gamma_2\mathcal{E} \supseteq \bigvee_{n=0}^{\infty} V^n\Gamma_2\mathcal{E} = \mathcal{V}.$$

Hence $\mathcal{V} = \mathcal{X}_2$. In particular, $\mathcal{V}$ is finite dimensional. Finally, it is noted that the pair$\{V^*, V^*\Gamma_2\}$ is controllable. To see this, recall that $\mathcal{V} = \mathcal{X}_2 = \bigvee_0^\infty V^{*n}(V^*\Gamma_2)\mathcal{E}$.

It remains to show that $\mathcal{X} = \mathcal{X}_1 \oplus \mathcal{V}$, when F is rational. Recall that S^{*n} converges to zero in the strong operator topology. Since $A_1 = S^*|\mathcal{X}_1$ and $\mathcal{X}_1$ is finite dimensional, A_1^n converges to zero. In particular, A_1 is stable, and all the eigenvalues for A_1 are inside the open unit disc. Using the fact that V^* is unitary, A_1 and V^* have no common eigenvalues. Because $\{A_1, B_1\}$ and $\{V^*, V^*\Gamma_2\}$ are both controllable, Lemma 6.1.2 below shows that $\mathcal{X} = \mathcal{X}_1 \oplus \mathcal{V}$ and $A = A_1 \oplus V^*$. Using this decomposition, we obtain the matrix representation $\{A, B, C, D\}$ for F in (6.1.8). In particular, the McMillan degree of F equals the dimension of $\mathcal{X}_1$ plus the dimension of $\mathcal{V}$.

Let Θ be the maximal outer spectral factor for T_R. Then Remark 5.2.2 shows that $\{A_1, B_1, \Pi_{\mathcal{Y}}|\mathcal{X}_1, \Theta(\infty)\}$ is a controllable and observable realization for Θ. This with $\delta(F) = \dim \mathcal{X}_1 + \dim \mathcal{V}$ readily implies that

$$\delta(F) = \delta(\Theta) + \dim(\mathcal{V}). \tag{6.1.10}$$

In particular, if F is rational, then Θ is rational. Moreover, if F is rational, then $\delta(F) = \delta(\Theta)$ if and only if $U = S$ is a unilateral shift. In this case, Θ is a function in $H^\infty(\mathcal{E}, \mathcal{Y})$, and $T_R = T_\Theta^* T_\Theta$, or equivalently, $R = \Theta^*\Theta$; see Theorem 5.2.1.

Assume that F is rational, or equivalently, $\mathcal{X}$ is finite dimensional. Let $U = S \oplus V$ on $\mathcal{K}_+ \oplus \mathcal{V}$ be the Wold decomposition of U, where S is a unilateral shift and V is unitary. Recall that in the Wold decomposition, the subspace $\mathcal{K}_+$ or $\mathcal{V}$ may or may not be present. If $\mathcal{V}$ is not present, then $U = S$. On the other hand, if $\mathcal{K}_+$ is not present, then $U = V$. If $U = S$ is the unilateral shift, then $A = S^*|\mathcal{X}$ is stable. This follows from the fact that S^{*n} converges to zero in the strong operator topology. In this case, all the poles of F are inside the open unit disc $\mathbb{D}$. Finally, if $U = V$ is unitary and F is rational, then $\mathcal{V} = \mathcal{X}$, and $\{V^*, V^*\Gamma, \Gamma^*, R_0/2\}$ is a minimal realization for F. In this case, all the poles of F are on the unit circle.

Theorem 6.1.1. *Let T_R be a positive Toeplitz matrix where $R = \sum_{-\infty}^{\infty} e^{-\imath\omega n} R_n$ is rational, and its corresponding positive real function $F = R_0/2 + \sum_1^{\infty} e^{-\imath\omega n} R_n$. Let Θ be the maximal outer spectral factor for T_R. Then the following are equivalent.*

(i) *The function F admits a stable finite dimensional realization.*

(ii) *The Toeplitz matrix $T_R = T_\Theta^* T_\Theta$, or equivalently, $R = \Theta^*\Theta$.*

(iii) *The functions F and Θ have the same McMillan degree, that is, $\delta(F) = \delta(\Theta)$.*

(iv) *If $\{U, \Gamma\}$ is a controllable isometric representation for T_R, then U is a unilateral shift.*

(v) *The Toeplitz matrix T_R defines an operator on $\ell_+^2(\mathcal{E})$.*

In this case, F and Θ both admit stable minimal realizations of the form $\{A, B, \star, \star\}$ where $\star$ represents an unspecified entry.

Proof. Recall that Θ admits a minimal realization of the form $\{A_1, B_1, \star, \star\}$ where $A_1 = S^*|\mathcal{X}_1$ and $B_1 = S^*\Gamma_1$; see Remark 5.2.2. Since A_1 is stable, Θ is a rational function in $H^\infty(\mathcal{E}, \mathcal{Y})$ and T_Θ is a well-defined operator. Recall also that $\{A, B, C, R_0/2\}$ in (6.1.8) is a minimal realization for F. Because all minimal realizations for F are similar, F admits a stable realization if and only if $\mathcal{V} = \{0\}$, or equivalently, $U = S$ is a unilateral shift. Hence F admits a stable realization if and only if $T_R = T_\Theta^* T_\Theta$ where Θ is the maximal outer spectral factor for T_R; see Part (iii) of Theorem 5.2.1. Therefore Parts (i) and (ii) are equivalent. In this case, $A = A_1 = S^*|\mathcal{X}_1$ is stable and $B = B_1$. So F and Θ both admit minimal realizations of the form $\{A, B, \star, \star\}$.

Assume that Part (iii) holds, that is, F and Θ have the same McMillan degree. According to (6.1.10), the subspace $\mathcal{V} = \{0\}$. In other words, $A = A_1$ is stable and F admits a stable realization. So Part (iii) implies Part (i), or equivalently, Part (ii). If Part (ii) holds, then Theorem 5.2.1 shows that $U = S$ and V is not part of the Wold decomposition for U. Equation (6.1.10) implies that $\delta(F) = \delta(\Theta)$ and Part (iii) holds. Therefore Parts (i), (ii) and (iii) are equivalent.

Part (iii) in Theorem 5.2.1 shows that Parts (ii) and (iv) are equivalent. If F admits a finite dimensional stable realization, then F is in $H^\infty(\mathcal{E}, \mathcal{E})$. In particular, T_F defines an operator on $\ell_+^2(\mathcal{E})$. Hence $T_R = T_F + T_F^*$ is also an operator on $\ell_+^2(\mathcal{E})$. In other words, Part (i) implies Part (v). On the other hand, if T_R defines an operator on $\ell^2_+(\mathcal{E})$, then the first column $T_R|\mathcal{E}$ of T_R defines an operator mapping $\mathcal{E}$ into $\ell^2_+(\mathcal{E})$. So $F(z) = R_0/2 + \sum_1^\infty z^{-n}R_n$ is a rational function in $H^2(\mathcal{E},\mathcal{E})$. In particular, the poles of F are in the open unit disc. Therefore F admits a finite dimensional stable realization, and Part (i) holds. □

As before, assume that T_R is a rational positive Toeplitz matrix, and F is the positive real function determined by (6.1.2). The matrix representation in (6.1.8) for the minimal realization $\{A,B,C,D\}$ of $F(z)$ shows that $F = F_1 + F_2$, where F_1 and F_2 are the functions determined by

$$\begin{aligned} F_1(z) &= \frac{1}{2}\Gamma_1^*\Gamma_1 + C_1(zI - A_1)^{-1}B_1, \\ F_2(z) &= \frac{1}{2}\Gamma_2^*\Gamma_2 + \Gamma_2^*(zI - V^*)^{-1}V^*\Gamma_2. \end{aligned} \tag{6.1.11}$$

Here F_1 is the positive real function corresponding to the controllable isometric pair $\{S,\Gamma_1\}$, and F_2 is a positive real function corresponding to the controllable unitary pair $\{V,\Gamma_2\}$ in the Wold decomposition of $\{U,\Gamma\}$. As expected, C_1 is the operator from $\mathcal{X}_1$ into $\mathcal{E}$ given by $C_1 = \Gamma^*|\mathcal{X}_1$. Let $\{\widehat{A},\widehat{B},\widehat{C},\widehat{D}\}$ be any minimal realization of F. Then $\widehat{A}$ is similar to $A_1 \oplus V^*$. In particular, U is a unilateral shift if and only if $\widehat{A}$ is stable. Since F is rational and the realization is minimal, U is a unilateral shift if and only if F is in $H^\infty(\mathcal{E},\mathcal{E})$. On the other hand, U is unitary if and only if all the eigenvalues of $\widehat{A}$ are on the unit circle.

Let us present a state space method to obtain F_1 and F_2 from any minimal realization of F. By employing the Jordan decomposition of $\widehat{A}$, without loss of generality, we can assume that the minimal realization $\{\widehat{A},\widehat{B},\widehat{C},\widehat{D}\}$ of F can be written as

$$\widehat{A} = \begin{bmatrix} \widehat{A}_1 & 0 \\ 0 & \widehat{A}_2 \end{bmatrix} \text{ on } \begin{bmatrix} \widehat{\mathcal{X}}_1 \\ \widehat{\mathcal{X}}_2 \end{bmatrix}, \quad \widehat{B} = \begin{bmatrix} \widehat{B}_1 \\ \widehat{B}_2 \end{bmatrix} : \mathcal{E} \to \begin{bmatrix} \widehat{\mathcal{X}}_1 \\ \widehat{\mathcal{X}}_2 \end{bmatrix},$$
$$\widehat{C} = \begin{bmatrix} \widehat{C}_1 & \widehat{C}_2 \end{bmatrix} : \begin{bmatrix} \widehat{\mathcal{X}}_1 \\ \widehat{\mathcal{X}}_2 \end{bmatrix} \to \mathcal{E}.$$

Here $\widehat{A}_1$ is stable, and all the eigenvalues of $\widehat{A}_2$ are on the unit circle. We claim that F_1 and F_2 can be computed by the following formula:

$$\begin{aligned} F_1(z) &= (\widehat{D} - \frac{1}{2}\widehat{C}_2\widehat{A}_2^{-1}\widehat{B}_2) + \widehat{C}_1(zI - \widehat{A}_1)^{-1}\widehat{B}_1, \\ F_2(z) &= \qquad \frac{1}{2}\widehat{C}_2\widehat{A}_2^{-1}\widehat{B}_2 + \widehat{C}_2(zI - \widehat{A}_2)^{-1}\widehat{B}_2. \end{aligned} \tag{6.1.12}$$

To see this, first observe that F can be expressed as

$$F(z) = \widehat{D} + \widehat{C}_1(zI - \widehat{A}_1)^{-1}\widehat{B}_1 + \widehat{C}_2(zI - \widehat{A}_2)^{-1}\widehat{B}_2 = F_1(z) + F_2(z), \tag{6.1.13}$$

where F_1 and F_2 are given in (6.1.11). Notice that F_1 is a rational function with all its poles in $\mathbb{D}$, while F_2 is a rational function with all its poles on the unit circle. In fact, (6.1.13) can be viewed as a partial fraction expansion of F. Since $\widehat{A}_1$ is stable, and all the poles of $\widehat{A}_2$ are on the unit circle, equation (6.1.13) implies that

$$\begin{aligned}\widehat{C}_1(zI-\widehat{A}_1)^{-1}\widehat{B}_1 &= C_1(zI-A_1)^{-1}B_1\\ \widehat{C}_2(zI-\widehat{A}_2)^{-1}\widehat{B}_2 &= C_2(zI-V^*)^{-1}V^*\Gamma_2.\end{aligned}\tag{6.1.14}$$

Recall that $\widehat{D}=(\Gamma_1^*\Gamma_1+\Gamma_2^*\Gamma_2)/2$. So to compute F_1 and F_2, all we need is either $\Gamma_1^*\Gamma_1$ or $\Gamma_2^*\Gamma_2$. Since $\{\widehat{A}_2,\widehat{B}_2,\widehat{C}_2,0\}$ and $\{V^*,V^*\Gamma_2,\Gamma_2^*,0\}$ are minimal realizations of the same transfer function, there exists a similarity transformation Φ mapping $\mathcal{V}$ onto $\widehat{\mathcal{X}}_2$ such that

$$\widehat{A}_2\Phi=\Phi V^*,\quad \widehat{B}_2=\Phi V^*\Gamma_2=\widehat{A}_2\Phi\Gamma_2 \quad\text{and}\quad \widehat{C}_2\Phi=\Gamma_2^*.$$

Because $\widehat{A}_2$ is invertible, $\Phi\Gamma_2=\widehat{A}_2^{-1}\widehat{B}_2$. Using this, we obtain $\Gamma_2^*\Gamma_2=\widehat{C}_2\Phi\Gamma_2=\widehat{C}_2\widehat{A}_2^{-1}\widehat{B}_2$. Therefore F_1 and F_2 are given by (6.1.12). This proves our claim.

Lemma 6.1.2. *Let $\{A_1$ on $\mathcal{X}_1,B_1\}$ and $\{A_2$ on $\mathcal{X}_2,B_2\}$ be two finite dimensional pairs of operators such that A_1 and A_2 have no common eigenvalues. Consider the operators A on $\mathcal{X}=\mathcal{X}_1\oplus\mathcal{X}_2$ and B mapping $\mathcal{E}$ into $\mathcal{X}$ defined by*

$$A=\begin{bmatrix}A_1 & 0\\ 0 & A_2\end{bmatrix} \quad\text{and}\quad B=\begin{bmatrix}B_1\\ B_2\end{bmatrix}.$$

Then the pair $\{A,B\}$ is controllable if and only if both $\{A_1,B_1\}$ and $\{A_2,B_2\}$ are controllable.

Proof. If $\{A,B\}$ is controllable, then

$$\bigvee_{n=0}^{\infty}\{A_1^nB_1a\oplus A_2^nB_2a : a\in\mathcal{E}\}=\bigvee_{n=0}^{\infty}A^nB\mathcal{E}=\mathcal{X}_1\oplus\mathcal{X}_2.$$

Hence both $\{A_1,B_1\}$ and $\{A_2,B_2\}$ must be controllable.

On the other hand, assume that both $\{A_1,B_1\}$ and $\{A_2,B_2\}$ are controllable. Recall that A_1 and A_2 have no common eigenvalues. According to the Popov-Belevitch-Hautus test, the pair $\{A,B\}$ is controllable if and only if the range of the matrix

$$J_\lambda=\begin{bmatrix}A_1-\lambda I & 0 & B_1\\ 0 & A_2-\lambda I & B_2\end{bmatrix}$$

is onto $\mathcal{X}_1\oplus\mathcal{X}_2$ for all λ in $\mathbb{C}$. If λ is not an eigenvalue of A_1 or A_2, then clearly J_λ is onto. If λ is an eigenvalue for A_1, then λ is not an eigenvalue for A_2, and thus, $A_2-\lambda I$ is invertible. Hence $0\oplus\mathcal{X}_2$ is contained in the range of J_λ. Because $\{A_1,B_1\}$ is controllable, $\mathcal{X}_1=\operatorname{ran}[A_1-\lambda I\ B_1]$. Therefore

$$\operatorname{ran}J_\lambda\supseteq\operatorname{ran}\begin{bmatrix}A_1-\lambda I & B_1\\ 0 & B_2\end{bmatrix}\bigvee\begin{bmatrix}0\\ \mathcal{X}_2\end{bmatrix}=\begin{bmatrix}\mathcal{X}_1\\ \mathcal{X}_2\end{bmatrix},$$

and thus, J_λ is onto. A similar argument shows that if λ is an eigenvalue of A_2, then λ is not an eigenvalue of A_1 and J_λ is onto. Therefore J_λ is onto for all complex numbers λ. By the Popov-Belevitch-Hautus test, the pair $\{A, B\}$ is controllable. □

6.2 The Positive Real Lemma

This section is devoted to the positive real lemma. This lemma uses state space techniques to determine when a rational Toeplitz matrix is positive. This lemma also forms the foundation for the state space factorization techniques discussed in Chapter 10.

Lemma 6.2.1 (Positive Real). *Let $\{A \text{ on } \mathcal{X}, B, \widehat{C}, R_0/2\}$ be a stable controllable finite dimensional realization for a $\mathcal{L}(\mathcal{E},\mathcal{E})$-valued rational function F, where R_0 is positive. Then F is positive real if and only if there exists a positive operator P on $\mathcal{X}$ and C mapping $\mathcal{X}$ into $\mathcal{Y}$, and D mapping $\mathcal{E}$ onto $\mathcal{Y}$ such that the following conditions hold:*

$$\begin{aligned} P &= A^*PA + C^*C, \\ \widehat{C} &= B^*PA + D^*C, \\ D^*D &= R_0 - B^*PB. \end{aligned} \tag{6.2.1}$$

Moreover, if Θ is the transfer function for $\{A, B, C, D\}$, then Θ is a spectral factor for T_R, that is, $T_R = T_\Theta^* T_\Theta$, or equivalently, $R = \Theta^*\Theta$. Finally, the outer factor Θ_o for Θ is the outer spectral factor for T_R.

Remark 6.2.2. Let F be a positive real function in $H^\infty(\mathcal{E},\mathcal{E})$ corresponding to a positive rational Toeplitz matrix T_R, that is, $R = F + F^*$. Then the positive real lemma can be used to compute the maximal outer spectral Θ_o factor for T_R. To see this, let $\{A, B, \widehat{C}, R_0/2\}$ be a stable controllable realization for F where R_0 is positive. Let Ξ be the set of all positive operators P satisfying the three conditions in Lemma 6.2.1. Then there exists a minimal P_o in Ξ, that is, there exists a unique P_o in Ξ such that $P_o \leq P$ for all P in Ξ. Moreover, for this P_o, let D_o be any operator mapping $\mathcal{E}$ onto $\mathcal{Y}$ such that $D_o^*D_o = R_0 - B^*P_oB$. Then there exists a unique C_o mapping $\mathcal{X}$ into $\mathcal{Y}$ such that

$$P_o = A^*P_oA + C_o^*C_o \quad \text{and} \quad \widehat{C} = B^*P_oA + D_o^*C_o. \tag{6.2.2}$$

In this case, $\{A, B, C_o, D_o\}$ is a realization for the outer spectral factor Θ_o for T_R.

The proof of the Positive Real Lemma 6.2.1, uses Lemma 4.5.4 restated here for convenience as follows.

Lemma 6.2.3. *Let $\{A \text{ on } \mathcal{X}, B, C, D\}$ be a stable realization for a $\mathcal{L}(\mathcal{E},\mathcal{Y})$-valued transfer function Θ. Let P be the observability Gramian for $\{C, A\}$, that is, let P be the unique solution to the Lyapunov equation*

$$P = A^*PA + C^*C. \tag{6.2.3}$$

Moreover, let T_R be the self-adjoint Toeplitz matrix generated by a $\mathcal{L}(\mathcal{E},\mathcal{E})$ sequence $\{R_n\}_0^\infty$ where $R_{-k} = R_k^*$; see (6.1.1). Then $T_R = T_\Theta^* T_\Theta$ if and only if

$$\begin{aligned} R_0 &= B^*PB + D^*D \\ R_n &= (B^*PA + D^*C)\,A^{n-1}B \qquad (n \geq 1). \end{aligned} \tag{6.2.4}$$

In this case,

$$T_R^{tr} - T_{\widetilde{\Theta}} T_{\widetilde{\Theta}}^* = W^*PW \tag{6.2.5}$$

where $\widetilde{\Theta}(z) = \Theta(\overline{z})^*$ and W is the controllability operator for $\{A, B\}$ defined by

$$W = \begin{bmatrix} B & AB & A^2B & \cdots \end{bmatrix} : \ell_+^2(\mathcal{E}) \to \mathcal{X}. \tag{6.2.6}$$

Proof of the Positive Real Lemma 6.2.1. Let $\{A, B, \widehat{C}, R_0/2\}$ be a stable controllable realization for a function F. Let $F = R_0/2 + \sum_1^\infty z^{-n} R_n$ be the Taylor series expansion for F, and T_R the Toeplitz matrix determined by $R = F + F^*$.

Assume that T_R is a positive Toeplitz matrix. Because A is stable, $T_R = T_\Theta^* T_\Theta$ where Θ is the maximal outer spectral factor for T_R. Moreover, Θ admits a realization of the form $\{A, B, C, D\}$; see Theorem 6.1.1. Let P be the observability Gramian for the pair $\{C, A\}$. Because $\{A, B, \widehat{C}, R_0/2\}$ is a realization of F, equation (6.2.4) in Lemma 6.2.3 shows that

$$\widehat{C}A^{n-1}B = R_n = (B^*PA + D^*C)A^{n-1}B \qquad (n \geq 1).$$

Since $\{A, B\}$ is controllable, $\widehat{C} = B^*PA + D^*C$. So the first two conditions in (6.2.1) hold. The last condition $D^*D = R_0 - B^*PB$ follows from (6.2.4) in Lemma 6.2.3. In other words, all the conditions in (6.2.1) hold.

Assume that P is a positive operator satisfying (6.2.1). Lemma 6.2.3 shows that $T_R = T_\Theta^* T_\Theta$ where $\{A, B, C, D\}$ is a realization of Θ. □

Proof of Remark 6.2.2. Let P be any positive solution satisfying all three conditions in the positive real lemma. Recall that $T_R = T_\Theta^* T_\Theta$ where Θ is the transfer function determined by $\{A, B, C, D\}$. Clearly, $\Theta = \Theta_i \Theta_o$ where Θ_i is an inner function, and Θ_o is an outer function. Since $T_R = T_{\Theta_o}^* T_{\Theta_i^*} T_{\Theta_i} T_{\Theta_o} = T_{\Theta_o}^* T_{\Theta_o}$, we see that Θ_o is the maximal outer spectral factor for T_R. For a rational Toeplitz matrices, one can choose realizations of the form $\{A, B, \star, \star\}$ for both F and Θ_o; see Theorem 6.1.1. So without loss of generality, we can assume that Θ_o admits a controllable realization of the form $\{A, B, C_o, D_o\}$. If P_o is the observability Gramian for $\{C_o, A\}$, then P_o also satisfies the three conditions (6.2.1) in Lemma 6.2.1, where C_o replaces C and D_o replaces D. Since $\Theta = \Theta_i \Theta_o$, we see that $\widetilde{\Theta} = \widetilde{\Theta}_o \widetilde{\Theta}_i$. Because Θ_i is inner, $1 = \|\Theta_i\|_\infty = \|\widetilde{\Theta}_i\|_\infty$. So $\widetilde{\Theta}_i$ is a contractive analytic function, and $T_{\widetilde{\Theta}_i}$ is also a contraction. In particular, $T_{\widetilde{\Theta}_i} T_{\widetilde{\Theta}_i}^* \leq I$. By employing (6.2.5), we arrive at

$$\begin{aligned} W^*PW &= T_R^{tr} - T_{\widetilde{\Theta}} T_{\widetilde{\Theta}}^* = T_R^{tr} - T_{\widetilde{\Theta}_o} T_{\widetilde{\Theta}_i} T_{\widetilde{\Theta}_i}^* T_{\widetilde{\Theta}_o}^* \\ &\geq T_R^{tr} - T_{\widetilde{\Theta}_o} T_{\widetilde{\Theta}_o}^* = W^*P_oW. \end{aligned}$$

Hence $W^*P_oW \leq W^*PW$. Because the pair $\{A, B\}$ is controllable, this readily implies that $P_o \leq P$. In other words, P_o is the smallest solution in Ξ.

We claim that P_o is unique. If P_1 is another minimal solution in Ξ, then $P_1 \leq P_o \leq P$ for all P in Ξ. Since P_o is also a minimal solution, $P_o \leq P_1 \leq P$. Therefore $P_o = P_1$. In other words, the minimal solution P_o is unique.

To complete the proof, it remains to show that if D_o is onto and C_o is determined by (6.2.2), then $\{A, B, C_o, D_o\}$ is a realization for the maximal outer spectral factor Θ_o. If $P = P_o$, then condition $R_0 - B^*P_oB = D_o^*D_o$ implies that D_o is unique up to a unitary constant operator on the left. Without loss of generality, we can assume that $D_o = \Theta_o(\infty)$ (Recall that if Ω is an outer function, then $\Omega(\infty)$ is onto; see Remark 3.2.3.) The second condition (6.2.1) in Lemma 6.2.1, implies that $D_o^*C_o = \widehat{C} - B^*P_oA$. Since D_o^* is one to one, C_o is uniquely determined. Hence $\{A, B, C_o, D_o\}$ must be a realization for Θ_o. □

The McMillan degree of the outer factor. Let $\delta(G)$ denote the McMillan degree of a transfer function G. Let Θ be a rational transfer function in $H^\infty(\mathcal{E}, \mathcal{Y})$. Clearly, Θ admits a unique inner-outer factorization of the form $\Theta = \Theta_i\Theta_o$ where Θ_i is inner and Θ_o is outer. In this case, the following holds:

(i) Both Θ_i and Θ_o are rational functions.

(ii) $\delta(\Theta_o) \leq \delta(\Theta)$ and $\delta(\Theta_i) \leq \delta(\Theta)$.

(iii) Let $\{A \text{ on } \mathcal{X}, B, C, D\}$ be any minimal realization for Θ and P the corresponding observability Gramian. Then $\delta(\Theta_o) = \delta(\Theta)$ if and only if the pair $\{\widehat{C}, A\}$ is observable where $\widehat{C} = B^*PA + D^*C$.

Lemma 4.5.1 shows that Parts (i) and (ii) hold. Let us present an alternate proof of the fact that $\delta(\Theta_o) \leq \delta(\Theta)$. Set $T_R = T_\Theta^*T_\Theta$. According to Lemma 6.2.3, the entries $\{R_n\}_0^\infty$ in the first column of T_R are computed by (6.2.4). So $\{A \text{ on } \mathcal{X}, B, \widehat{C}, R_0/2\}$ is a controllable realization for the positive real function F corresponding to T_R. In particular, $\delta(F) \leq \dim \mathcal{X}$, with equality if and only if the pair $\{\widehat{C}, A\}$ is observable. Using $T_\Theta = T_{\Theta_i}T_{\Theta_o}$ along with the fact that T_{Θ_i} is an isometry, $T_R = T_{\Theta_o}^*T_{\Theta_o}$. Hence Θ_o is the outer spectral factor for T_R. Because F and Θ_o have the same McMillan degree (see Theorem 6.1.1), we obtain

$$\delta(\Theta_o) = \delta(F) \leq \dim \mathcal{X} = \delta(\Theta). \tag{6.2.7}$$

Hence $\delta(\Theta_o) \leq \delta(\Theta)$. Since the McMillan degree of Θ_o is finite, Θ_o must be a rational function. Finally, Part (iii) follows from the fact that there is equality in (6.2.7) if and only if $\{\widehat{C}, A\}$ is observable.

6.3 Finite Dimensional Unitary Part

In this section, we will study the case when the unitary part of the Wold decomposition acts on a finite dimensional space.

Proposition 6.3.1. *Let Θ be an outer function in $H^2(\mathcal{E},\mathcal{Y})$ and Γ_1 the operator mapping $\mathcal{E}$ into $H^2(\mathcal{Y})$ determined by $\Gamma_1 f = \Theta f$ for all f in $\mathcal{E}$. Let $\{V,\Gamma_2\}$ be a controllable isometric pair where V is a unitary operator acting on a finite dimensional space $\mathcal{V}$. Consider the isometric representation $\{U$ on $\mathcal{K},\Gamma\}$ given by*

$$U = \begin{bmatrix} S & 0 \\ 0 & V \end{bmatrix} \text{ on } \begin{bmatrix} H^2(\mathcal{Y}) \\ \mathcal{V} \end{bmatrix} \quad \text{and} \quad \Gamma = \begin{bmatrix} \Gamma_1 \\ \Gamma_2 \end{bmatrix} : \mathcal{E} \to \begin{bmatrix} H^2(\mathcal{Y}) \\ \mathcal{V} \end{bmatrix} \tag{6.3.1}$$

where S is the unilateral shift on $H^2(\mathcal{Y})$. Then $\{U,\Gamma\}$ is controllable.

Proof. Assume that $h \oplus x$ is a vector in $H^2(\mathcal{Y}) \oplus \mathcal{V}$ such that $h \oplus x$ is orthogonal to $U^n\Gamma\mathcal{E}$ for all integers $n \geq 0$. By using the Wold decomposition this implies that $\Gamma_1^* S^{*n} h + \Gamma_2^* V^{*n} x = 0$ for all n. Because the backward shift S^{*n} converges to zero in the strong operator topology, we see that $\Gamma_2^* V^{*n} x$ converges to zero as n tends to infinity. Since $\{V,\Gamma_2\}$ is controllable, the pair $\{\Gamma_2^*, V^*\}$ is observable. Lemma 6.3.2 below, shows that $x = 0$. Therefore h is orthogonal to $S^n\Gamma_1\mathcal{E}$ for all integers $n \geq 0$, or equivalently, h is orthogonal to $\Theta\mathcal{P}(\mathcal{E})$. (Recall that $\mathcal{P}(\mathcal{E})$ is the space of all polynomials in $1/z$ with values in $\mathcal{E}$.) Because Θ is an outer function, we see that $h = 0$. In other words, the only vector orthogonal to $\bigvee_0^\infty U^n\Gamma\mathcal{E}$ is the zero vector, or equivalently, $\bigvee_0^\infty U^n\Gamma\mathcal{E}$ equals $H^2(\mathcal{Y}) \oplus \mathcal{V}$. Therefore the pair $\{U,\Gamma\}$ is controllable. □

Lemma 6.3.2. *Let $\{C,A\}$ be an observable pair where A is a unitary operator on a finite dimensional space $\mathcal{X}$ and C is an operator mapping $\mathcal{X}$ into $\mathcal{Y}$. Let x be any vector in $\mathcal{X}$. Then*

$$\lim_{n\to\infty} CA^n x = 0 \tag{6.3.2}$$

if and only if $x = 0$.

Proof. Let $\mathcal{M}$ be the set of all vectors x in $\mathcal{X}$ such that $CA^n x$ converges to zero as n tends to infinity. Notice that $\mathcal{M}$ is a linear space. Moreover, if x is in $\mathcal{M}$, then CA^nAx also converges to zero. In other words, $\mathcal{M}$ is an invariant subspace for A. So $A|\mathcal{M}$ defines a unitary operator on $\mathcal{M}$. In particular, the operator $A|\mathcal{M}$ has an eigenvector. Let x be any eigenvector for A contained in $\mathcal{M}$, that is, $Ax = \lambda x$ where x is in $\mathcal{M}$. Then $CA^n x = \lambda^n Cx$ converges to zero as n tends to infinity. Because λ must be on the unit circle, $Cx = 0$. Hence $CA^n x = \lambda^n Cx = 0$ for all integers $n \geq 0$. Since the pair $\{C,A\}$ is observable, x must be zero. Therefore the subspace $\mathcal{M} = \{0\}$. □

Example. Let Θ be an outer function in $H^\infty(\mathcal{E},\mathcal{Y})$. Let A_j mapping $\mathcal{E}$ onto $\mathcal{E}_j$ where $\dim\mathcal{E}_j \leq \dim\mathcal{E}$ for $j = 1,2,\ldots,\nu$ be a finite set of operators. Consider the sequence $\{R_n\}_{-\infty}^{\infty}$ determined by

$$R_n = \frac{1}{2\pi}\int_0^{2\pi} e^{\imath\omega n}\Theta(e^{\imath\omega})^*\Theta(e^{\imath\omega})d\omega + \sum_{k=1}^{\nu} A_k^* A_k e^{-\imath\omega_k n} \tag{6.3.3}$$

where $\{\omega_k\}_1^\nu$ are distinct frequencies. Let T_R be the Toeplitz matrix determined by $R = \sum_{-\infty}^{\infty} e^{-\imath\omega n} R_n$. We claim that T_R is a positive Toeplitz matrix and Θ is the maximal outer spectral factor for T_R.

To see this, consider the isometric representation $\{U, \Gamma\}$ determined by

$$U = \begin{bmatrix} S & 0 \\ 0 & V \end{bmatrix} \text{ on } \begin{bmatrix} H^2(\mathcal{Y}) \\ \oplus_1^\nu \mathcal{E}_j \end{bmatrix} \quad \text{and} \quad \Gamma = \begin{bmatrix} \Gamma_1 \\ \Gamma_2 \end{bmatrix} : \mathcal{E} \to \begin{bmatrix} H^2(\mathcal{Y}) \\ \oplus_1^\nu \mathcal{E}_j \end{bmatrix}. \tag{6.3.4}$$

As expected, S is the unilateral shift on $H^2(\mathcal{Y})$, and Γ_1 is the operator mapping $\mathcal{E}$ into $H^2(\mathcal{Y})$ given by $\Gamma_1 \xi = \Theta \xi$ where ξ is in $\mathcal{E}$. Moreover, V is the diagonal unitary operator on $\oplus_1^\nu \mathcal{E}_j$ determined by

$$V = \begin{bmatrix} e^{\imath\omega_1} I & 0 & \cdots & 0 \\ 0 & e^{\imath\omega_2} I & \cdots & 0 \\ \vdots & \vdots & \ddots & \vdots \\ 0 & 0 & \cdots & e^{\imath\omega_\nu} I \end{bmatrix} \quad \text{and} \quad \Gamma_2 = \begin{bmatrix} A_1 \\ A_2 \\ \vdots \\ A_\nu \end{bmatrix}. \tag{6.3.5}$$

Notice that $\{e^{\imath\omega_j}\}_1^\nu$ are the eigenvalues for V. Proposition 6.3.1 guarantees that the pair $\{U, \Gamma\}$ is controllable. Finally, it is noted that in most engineering problems the eigenvalues $\{e^{\imath\omega_j}\}_1^\nu$ for V come in complex conjugate pairs. Moreover, in this case, the corresponding amplitudes $A_j^* A_j$ are the same. For example, if $e^{\imath\omega_2}$ is the complex conjugate of $e^{\imath\omega_1}$, then in applications $A_1^* A_1 = A_2^* A_2$.

A simple calculation shows that Γ_1^* is the operator mapping $H^2(\mathcal{Y})$ into $\mathcal{E}$ given by

$$\Gamma_1^* h = \frac{1}{2\pi} \int_0^{2\pi} \Theta(e^{\imath\omega})^* h(e^{\imath\omega})\, d\omega \qquad (h \in H^2(\mathcal{Y})),$$

$$\Gamma_1^* S^n \Gamma_1 = \frac{1}{2\pi} \int_0^{2\pi} e^{-\imath\omega n} \Theta(e^{\imath\omega})^* \Theta(e^{\imath\omega})\, d\omega \qquad (\text{for all integers } n \geq 0).$$

Moreover, $\Gamma_2^* V^n \Gamma_2 = \sum_{k=1}^\nu A_k^* A_k e^{\imath\omega_k n}$ for all integers n. By construction $R_{-n} = \Gamma^* U^n \Gamma$ for all integers $n \geq 0$. So $\{U, \Gamma\}$ is a controllable isometric representation for the Toeplitz matrix T_R generated by the symbol $R = \sum_{-\infty}^{\infty} e^{-\imath\omega n} R_n$. In particular, T_R is positive and Θ is the maximal outer spectral factor for T_R.

6.4 A Classical Ergodic Result

In this section, we will present a classical ergodic result involving a contraction. Then we will use the inverse fast Fourier transform on $\{R_n\}$ to find the eigenvalues for U on the unit circle, where $\{U, \Gamma\}$ is the controllable isometric representation for $\{R_n\}$. Finally, it is noted that these ergodic results may require large data sets to work effectively.

Recall that an operator C is a contraction if $\|C\| \leq 1$. Finally, it is noted that $e^{-i\theta}$ is an eigenvalue for C on $\mathcal{H}$ if and only if the kernel of $e^{-i\theta} I - C$ is nonzero. The following is a classical ergodic result for a contraction.

Theorem 6.4.1. *Let C be a contraction on $\mathcal{H}$, and set $\mathcal{L} = \ker(e^{-i\theta}I - C)$. Then the orthogonal projection $P_{\mathcal{L}}$ onto $\mathcal{L}$ is given by*

$$P_{\mathcal{L}} = \lim_{n\to\infty} \frac{1}{n}\sum_{k=0}^{n-1} e^{i\theta k}C^k \tag{6.4.1}$$

in the strong operator topology.

Proof. Set

$$C_n = \frac{1}{n}\sum_{k=0}^{n-1} e^{i\theta k}C^k.$$

Let x be a vector in $\mathcal{L}$, or equivalently, assume that $Cx = e^{-i\theta}x$. Then $C_n x = x$. In other words, $C_n x = x = P_{\mathcal{L}}x$. Notice that C_n is a contraction. To see this, observe that because C is a contraction,

$$\|C_n\| = \|\frac{1}{n}\sum_{k=0}^{n-1} e^{i\theta k}C^k\| \le \frac{1}{n}\sum_{k=0}^{n-1} \|e^{i\theta k}C^k\| \le \frac{n}{n} = 1.$$

Therefore $\|C_n\| \le 1$ and C_n is a contraction.

We claim that $\mathcal{L} = \ker(e^{i\theta}I - C^*)$. Assume that x is in $\mathcal{L}$, that is, $Cx = e^{-i\theta}x$. Using the fact that C is a contraction, we obtain

$$\begin{aligned}\|e^{i\theta}x - C^*x\|^2 &= \|e^{i\theta}x\|^2 - 2\Re(e^{i\theta}x, C^*x) + \|C^*x\|^2\\ &= \|x\|^2 + \|C^*x\|^2 - 2\Re(e^{i\theta}Cx, x)\\ &\le 2\|x\|^2 - 2\Re(e^{i\theta}e^{-i\theta}x, x) = 2\|x\|^2 - 2\|x\|^2 = 0.\end{aligned}$$

Hence $\|e^{i\theta}x - C^*x\| \le 0$, or equivalently, x is in $\ker(e^{i\theta}I - C^*)$. In other words,

$$\mathcal{L} = \ker(e^{-i\theta}I - C) \subseteq \ker(e^{i\theta}I - C^*). \tag{6.4.2}$$

Notice that this holds for any contraction C and complex number $e^{i\theta}$ on the unit circle. So replacing $e^{i\theta}$ by $e^{-i\theta}$ and C by C^* in equation (6.4.2), we see that $\ker(e^{i\theta}I - C^*) \subseteq \ker(e^{-i\theta}I - C)$. Therefore $\mathcal{L} = \ker(e^{i\theta}I - C^*) = \ker(I - e^{-i\theta}C^*)$. Recall that for any operator M, we have $\ker(M)^\perp = \overline{\mathrm{ran} M^*}$. By taking the adjoint of $I - e^{-i\theta}C^*$ this readily implies that

$$\mathcal{L}^\perp = \overline{(I - e^{i\theta}C)\mathcal{H}}. \tag{6.4.3}$$

For h in $\mathcal{H}$, we obtain

$$C_n(I - e^{i\theta}C)h = \frac{1}{n}\sum_{k=0}^{n-1} e^{i\theta k}C^k h - \frac{1}{n}\sum_{k=1}^{n} e^{i\theta k}C^k h = \frac{1}{n}\left(h - e^{i\theta n}C^n h\right).$$

Because C is a contraction, $C_n(I - e^{i\theta}C)h$ converges to zero for every h in $\mathcal{H}$. Recall that C_n is also a contraction. Since $(I - e^{i\theta}C)\mathcal{H}$ is dense in $\mathcal{L}^\perp$, it follows that $C_n h$ converges to zero for all h in $\mathcal{L}^\perp$. Recall that $C_n x = x$ for all x in $\mathcal{L}$. Therefore the sequence C_n converges to $P_{\mathcal{L}}$ in the strong operator topology. □

Let $\{U, \Gamma\}$ be any controllable isometric representation for a positive Toeplitz matrix T_R with symbol $R = \sum_{-\infty}^{\infty} e^{-\imath\omega n} R_n$. Recall that $R_{-n} = \Gamma^* U^n \Gamma$ for all integers $n \geq 0$. Moreover, U admits a Wold decomposition of the form $U = S \oplus V$ where S is a unilateral shift on $\ell_+^2(\mathcal{E})$ and V is a unitary operator on $\mathcal{V}$. Recall that S^* has no eigenvalues on the unit circle. Theorem 6.4.1 with $C = U^*$ shows that

$$\Gamma^* P_{\mathcal{L}} \Gamma = \Gamma^* \lim_{n \to \infty} \frac{1}{n} \sum_{k=0}^{n-1} e^{i\theta k} U^{*k} \Gamma = \lim_{n \to \infty} \frac{1}{n} \sum_{k=0}^{n-1} e^{i\theta k} R_k \tag{6.4.4}$$

where $P_{\mathcal{L}}$ is the orthogonal projection onto $\ker(e^{-i\theta} I - V^*) = \ker(e^{i\theta} I - V)$. If $e^{i\theta}$ is not an eigenvalue for V, then $P_{\mathcal{L}}$ equals zero.

Consider the controllable isometric pair $\{U, \Gamma\}$ presented in (6.3.4) and (6.3.5) of the example at the end of Section 6.3. Here $\{R_n\}$ is defined by (6.3.3). Then (6.4.4) reduces to

$$\begin{aligned} \lim_{n \to \infty} \frac{1}{n} \sum_{k=0}^{n-1} e^{i\theta k} R_k &= A_j^* A_j \qquad &&\text{if } \theta = \omega_j \\ &= 0 &&\text{if } \theta \notin \{\omega_j\}_1^{\nu}. \end{aligned} \tag{6.4.5}$$

The inverse fast Fourier transform with Theorem 6.4.1 can be used to compute the eigenvalues for V and sometimes even $\Gamma^* P_{\mathcal{L}} \Gamma$. To see this, recall that the inverse fast Fourier transform of a sequence $\{a_k\}$ is given by $\frac{1}{n} \sum_0^{n-1} e^{\imath\omega k} a_k$ where $e^{\imath\omega}$ is chosen at n points evenly spaced around the unit circle. For large n the inverse fast Fourier transform of $R_n = \Gamma^* U^{*n} \Gamma$, that is, $\frac{1}{n} \sum_0^{n-1} e^{\imath\omega k} R_k$ "converges to" zero if $e^{\imath\omega}$ is not an eigenvalue for V and $\Gamma^* P_{\mathcal{L}} \Gamma$ if $e^{\imath\omega}$ is an eigenvalue for V.

If one runs the inverse fast Fourier transform on the controllable isometric pair in the example in Section 6.3, then for large n equation (6.4.5) holds. Finally, it is noted that due to the partitioning of the unit circle by the fast Fourier transform, there can be a significant error in using the fast Fourier transform to estimate the amplitude $\Gamma^* P_{\mathcal{L}} \Gamma$.

Example. Consider the outer function given by

$$\theta(z) = \frac{1.1465 z^2 - 0.2850 z + 0.1125}{z^2 - 0.2802 z - 0.0585}. \tag{6.4.6}$$

Let T_R be the Toeplitz matrix with symbol $R = \sum_{-\infty}^{\infty} = r_n e^{-\imath\omega n}$ determined by

$$r_n = (e^{\imath\omega n} \theta, \theta)_{L^2} + \frac{1}{2} \cos(n\pi/2) + \frac{1}{4} \cos(n\pi/4). \tag{6.4.7}$$

In this case, r_n is real and $r_n = r_{-n}$ for all integers n. Moreover, the controllable isometric representation $\{U, \Gamma\}$ for T_R admits a Wold decomposition of the form

$$U = \begin{bmatrix} S & 0 \\ 0 & V \end{bmatrix} \text{ on } \begin{bmatrix} H^2 \\ \mathbb{C}^4 \end{bmatrix} \quad \text{and} \quad \Gamma = \begin{bmatrix} \Gamma_1 \\ \Gamma_2 \end{bmatrix} : \mathbb{C} \to \begin{bmatrix} H^2 \\ \mathbb{C}^4 \end{bmatrix}. \tag{6.4.8}$$

As expected, S is the unilateral shift on H^2, and Γ_1 is the operator mapping $\mathbb{C}$ into H^2 given by $\Gamma_1\xi = \theta\xi$ where ξ is in $\mathbb{C}$. The unitary operator V on $\mathbb{C}^4$ is determined by

$$V = \begin{bmatrix} e^{\imath\pi/2} & 0 & 0 & 0 \\ 0 & e^{-\imath\pi/2} & 0 & 0 \\ 0 & 0 & e^{\imath\pi/4} & 0 \\ 0 & 0 & 0 & e^{-\imath\pi/4} \end{bmatrix} \quad \text{and} \quad \Gamma_2 = \begin{bmatrix} 1/2 \\ 1/2 \\ 1/\sqrt{8} \\ 1/\sqrt{8} \end{bmatrix}. \tag{6.4.9}$$

Proposition 6.3.1 guarantees that the pair $\{U, \Gamma\}$ is controllable. The plot for the inverse fast Fourier transform for $\{r_n\}$ is given in Figure 6.1. The peaks in the spectrum occur at $0.785 \approx \pi/4$, $1.57 \approx \pi/2$, $4.71 \approx 3\pi/2$ and $5.5 \approx 7\pi/4$. We computed a 2^{14} point inverse fast Fourier transform of $\{r_j\}_0^{1000}$ padded with the appropriate number of zeros. Then we plotted the power spectrum which is the absolute value squared of the inverse fast Fourier transform for $\{r_j\}_0^{1000}$. The power spectrum displays the peaks better than plotting the absolute value of the inverse fast Fourier transform. The corresponding Matlab commands are

- $\text{num} = [1.1465, -0.2850, 0.1125]; \text{den} = [1, -0.2802, -0.0585];$
- $f = \text{fft}(\text{num}, 2 \wedge 14)./\text{fft}(\text{den}, 2 \wedge 14);$
- $r1 = \text{real}(\text{ifft}(\text{abs}(f). \wedge 2)); r2 = \cos((0 : 1000) * \pi/4)/4 + \cos((0 : 1000) * \pi/2)/2;$
- $a = \text{ifft}(r1(1 : 1001) + r2, 2 \wedge 14);$
- $w = \text{linspace}(0, 2 * \pi, 2 \wedge 14);$
- plot(w, (abs(a)$*2\wedge14/1001$)); Here $2^{14}/1001$ is used to normalize the padding in the fast Fourier transform and at the same time plot 2^{14} points on the "spectrum".

Finally, it is noted that the peaks in Figure 6.1 yields $1/4 = (1/2)^2$ corresponding to $e^{\pm\imath\pi/2}$ and $1/8 = (1/\sqrt{8})^2$ corresponding to $e^{\pm\imath\pi/4}$.

6.5 Another Ergodic Result

If the unitary part in the Wold decomposition is finite dimensional, then the following result can be used to compute this unitary part.

Proposition 6.5.1. *Let T_R be the positive Toeplitz matrix generated by the controllable isometric pair $\{U \text{ on } \mathcal{K}, \Gamma\}$ where the maximal outer spectral factor Θ is a function in $H^\infty(\mathcal{E}, \mathcal{Y})$ and the unitary part $\{V \text{ on } \mathcal{V}, \Gamma_2\}$ acts on a finite dimensional state space; see* (5.2.3). *Without loss of generality, assume that $\{V, \Gamma_2\}$*

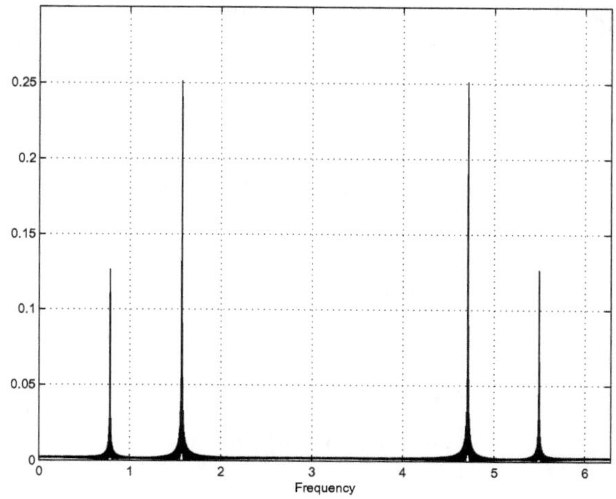

Figure 6.1: The "spectrum".

admits a matrix representation of the form:

$$V = \begin{bmatrix} \lambda_1 I & 0 & \cdots & 0 \\ 0 & \lambda_2 I & \cdots & 0 \\ \vdots & \vdots & \ddots & \vdots \\ 0 & 0 & \cdots & \lambda_\nu I \end{bmatrix} \text{ on } \begin{bmatrix} \mathcal{E}_1 \\ \mathcal{E}_2 \\ \vdots \\ \mathcal{E}_\nu \end{bmatrix} \quad \text{and} \quad \Gamma_2 = \begin{bmatrix} A_1 \\ A_2 \\ \vdots \\ A_\nu \end{bmatrix} : \mathcal{E} \to \begin{bmatrix} \mathcal{E}_1 \\ \mathcal{E}_2 \\ \vdots \\ \mathcal{E}_\nu \end{bmatrix}. \quad (6.5.1)$$

Here $\{\lambda_j\}_1^\nu$ *are* ν *distinct complex numbers on the unit circle, and* A_j *are operators from* $\mathcal{E}$ *onto* $\mathcal{E}_j$*. Let* $T_{R,n}$ *on* $\mathcal{E}^n$ *be the positive block Toeplitz matrix contained in the* $n \times n$ *upper left-hand corner of* T_R*; see* (5.1.1) *and* (5.1.3)*. Then the nonzero singular values of* $T_{R,n}/n$ *converge to the singular values of the diagonal matrix* $diag\,[\{A_j A_j^*\}_1^\nu]$.

Proof. Recall that $\{U, \Gamma\}$ admits a Wold decomposition of the form (5.2.3). Moreover, the Toeplitz matrix $T_{R,n} = W_n^* W_n$ where W_n is the controllability matrix defined by

$$\begin{aligned} W_n &= \begin{bmatrix} \Gamma & U\Gamma & U^2\Gamma & \cdots & U^{n-1}\Gamma \end{bmatrix}, \\ W_{n,1} &= \begin{bmatrix} \Gamma_1 & S\Gamma_1 & S^2\Gamma_1 & \cdots & S^{n-1}\Gamma_1 \end{bmatrix}, \\ W_{n,2} &= \begin{bmatrix} \Gamma_2 & V\Gamma_2 & V^2\Gamma_2 & \cdots & V^{n-1}\Gamma_2 \end{bmatrix}. \end{aligned} \quad (6.5.2)$$

Observe that $W_n = \begin{bmatrix} W_{n,1} & W_{n,2} \end{bmatrix}^{tr}$. Hence

$$\frac{1}{n} T_{R,n} = \frac{1}{n} W_{n,1}^* W_{n,1} + \frac{1}{n} W_{n,2}^* W_{n,2}. \quad (6.5.3)$$

Notice that $W_{n,1} = T_\Theta|\mathcal{E}^n$; see (5.2.7). Because Θ is a function in $H^\infty(\mathcal{E},\mathcal{Y})$ the Toeplitz operator T_Θ is bounded. In particular, $\|W_{n,1}\| \leq \gamma$ for some finite scalar γ and all integers $n \geq 1$. This readily implies that $\frac{1}{n}W_{n,1}^*W_{n,1}$ converges to zero as n tends to infinity. So to complete the proof, it remains to show that the nonzero singular values converge to the singular values of the diagonal matrix $\operatorname{diag}[\{A_jA_j^*\}_1^\nu]$.

Clearly, $W_{n,2}^*W_{n,2}$ and $W_{n,2}W_{n,2}^*$ have the same singular values. To finish the proof it is sufficient to show that

$$\lim_{n\to\infty} \frac{1}{n} W_{n,2}W_{n,2}^* = \operatorname{diag}[\{A_jA_j^*\}_1^\nu]. \tag{6.5.4}$$

To this end, observe that

$$W_{n,2} = \begin{bmatrix} A_1 & \lambda_1 A_1 & \lambda_1^2 A_1 & \cdots & \lambda_1^{n-1}A_1 \\ A_2 & \lambda_2 A_2 & \lambda_2^2 A_2 & \cdots & \lambda_2^{n-1}A_2 \\ \vdots & \vdots & \vdots & \cdots & \vdots \\ A_\nu & \lambda_\nu A_\nu & \lambda_\nu^2 A_\nu & \cdots & \lambda_\nu^{n-1}A_\nu \end{bmatrix}.$$

Since $\{\lambda_j\}_1^\nu$ are on the unit circle, the $\{k,k\}$ entry of $W_{n,2}W_{n,2}^*/n$ is determined by

$$\left(\frac{1}{n}W_{n,2}W_{n,2}^*\right)_{k,k} = \frac{1}{n}\sum_{j=0}^{n-1} |\lambda_k|^2 A_kA_k^* = A_kA_k^*.$$

Recall that if $r \neq 1$, then $\sum_0^{n-1} r^j = (1-r^n)/(1-r)$. Using this we see that the $\{k,m\}$ entry of $W_{n,2}W_{n,2}^*/n$ is given by

$$\left(\frac{1}{n}W_{n,2}W_{n,2}^*\right)_{k,m} = \frac{1}{n}\sum_{j=0}^{n-1} \lambda_k\overline{\lambda}_m A_kA_m^* = \frac{1-(\lambda_k\overline{\lambda}_m)^n}{n(1-\lambda_k\overline{\lambda}_m)} A_kA_m^* \to 0.$$

Combining this with $(W_{n,2}W_{n,2}^*)_{k,k}/n = A_kA_k^*$ yields (6.5.4). Finally, (6.5.3) and (6.5.4) show that the nonzero singular values of $T_{R,n}/n$ converge to the singular values of the diagonal matrix $\operatorname{diag}[\{A_jA_j^*\}_1^\nu]$. □

The Frobenius norm of a matrix M is denoted by $\|M\|_2$, that is, $\|M\|_2^2 = \operatorname{trace}(MM^*)$. Recall that $U\Lambda V^*$ is the *singular value decomposition* for a finite dimensional operator T mapping $\mathcal{X}$ into $\mathcal{Y}$, if Λ is the rectangular diagonal matrix mapping $\mathbb{C}^\nu$ into $\mathbb{C}^\mu$ consisting of the singular values of T, while V is a unitary operator mapping $\mathbb{C}^\nu$ into $\mathcal{X}$ and U is a unitary operator mapping $\mathbb{C}^\mu$ into $\mathcal{Y}$.

Theorem 6.5.2 (Procrustes problem). *Let X mapping $\mathcal{U}$ into $\mathcal{X}$ and Y mapping $\mathcal{U}$ into $\mathcal{X}$ be finite dimensional matrices. Consider the optimization problem*

$$\sigma = \inf\{\|AX - Y\|_2 : A \text{ is a unitary operator on } \mathcal{X}\}. \tag{6.5.5}$$

Then an optimal solution to (6.5.5) is given by $A = VU^*$ where $U\Lambda V^* = XY^*$ is the singular value decomposition of XY^*.

Proof. If B is a unitary operator on $\mathcal{X}$, then we have

$$\|BX - Y\|_2 = \|BX\|_2^2 - 2\Re(BX, Y)_2 + \|Y\|_2^2 = \|X\|_2^2 - 2\Re\,\mathrm{trace}(BXY^*) + \|Y\|_2^2.$$

Observe that $\|BX - Y\|_2$ is minimized by choosing B such that $\Re\,\mathrm{trace}(BU\Lambda V^*)$ is as large as possible. Since V^*BU is a contraction, the diagonal entries of V^*BU are all in the closed unit disc. Recall that $\mathrm{trace}(MN) = \mathrm{trace}(NM)$ where M and N are operators acting on the appropriate finite dimensional spaces. So by choosing $A = VU^*$, we obtain

$$\begin{aligned}
\Re\,\mathrm{trace}(BXY^*) &= \Re\,\mathrm{trace}(BU\Lambda V^*) = \Re\,\mathrm{trace}(V^*BU\Lambda) = \sum_j \Re(V^*BU)_{jj}\Lambda_{jj} \\
&\leq \sum_j \Lambda_{jj} = \mathrm{trace}(\Lambda) = \mathrm{trace}(V^*VU^*U\Lambda) \\
&= \mathrm{trace}(V^*AU\Lambda) = \mathrm{trace}(AU\Lambda V^*) \\
&= \mathrm{trace}(AXY^*) = \Re\,\mathrm{trace}(AXY^*).
\end{aligned}$$

The last equality follows from the fact that $\mathrm{trace}(AXY^*) = \mathrm{trace}(\Lambda)$ is real. Therefore $A = VU^*$ is an optimal solution to (6.5.5). □

It is noted that an optimal solution to (6.5.5) is not necessarily unique. For example, if $X = 0$ and $Y = 1$, then $A = \pm 1$ are two optimal solutions to (6.5.5).

Computing the unitary part. To motivate our algorithm to compute the unitary part, assume that $W_n = W_{n,2}$ and $n > \dim \mathcal{V}$ where $\mathcal{V} = \oplus_1^\nu \mathcal{E}_j$. In other words, due to the controllability of the pair $\{V, \Gamma_2\}$, the operator $W_{n-1,2}$ is onto $\oplus_1^\nu \mathcal{E}_j$. Now let J_n and Q_n be the matrices defined by

$$J_n = \begin{bmatrix} I \\ 0 \end{bmatrix} : \mathcal{E}^{n-1} \to \begin{bmatrix} \mathcal{E}^{n-1} \\ \mathcal{E} \end{bmatrix} \quad \text{and} \quad Q_n = \begin{bmatrix} 0 \\ I \end{bmatrix} : \mathcal{E}^{n-1} \to \begin{bmatrix} \mathcal{E} \\ \mathcal{E}^{n-1} \end{bmatrix}. \tag{6.5.6}$$

Then $W_{n,2}Q_n = VW_{n,2}J_n = VW_{n-1,2}$. Thus $V = W_{n,2}Q_n(W_{n,2}J_n)^{-r}$. (Here A^{-r} denotes the Moore-Penrose pseudo-inverse of A.) Finally, $\Gamma_2 = W_{n,2}|\mathcal{E}$ where $\mathcal{E}$ denotes the subspace of $\mathcal{E}^n$ contained in the first component of $\mathcal{E}^n$. So if one is given $W_{n,2}$ for $n > \dim \mathcal{V}$, then one can compute the unitary part $\{V, \Gamma_2\}$.

Let T_R be the positive Toeplitz matrix generated by the controllable isometric pair $\{U \text{ on } \mathcal{K}, \Gamma\}$, where the maximal outer spectral factor Θ is a function in $H^\infty(\mathcal{E}, \mathcal{Y})$ and the unitary part $\{V \text{ on } \mathcal{V}, \Gamma_2\}$ acts on a finite dimensional state space; see (5.2.3) and (6.5.1). Proposition 6.5.1 shows that for large n the singular values of $T_{R,n} = W_n^* W_n$ will contain $\dim \mathcal{V}$ large singular values approximately equal to the singular values for $n \times \mathrm{diag}\,[\{A_j A_j^*\}_1^\nu]$. In fact, the proof of Proposition 6.5.1 shows that for large n the first $\dim \mathcal{V}$ singular values of $W_n^* W_n$ are approximately equal to the nonzero singular values for $W_{n,2}^* W_{n,2}$, and the singular values contributed by $W_{n,1}^* W_{n,1}$ will be much smaller; see (6.5.3). So by keeping the large singular values for $W_n^* W_n$ we can approximate $W_{n,2}^* W_{n,2}$ and thus, $W_{n,2}$. For n large, let M_n be any operator mapping $\mathcal{E}^n$ onto $\mathcal{X}_n$ such that $T_{R,n} = M_n^* M_n$. One can use the singular value decomposition on $T_{R,n}$ to compute M_n. Then $\Phi M_n = W_n$ where Φ is a unitary operator from the range of M_n onto the range of W_n. Using this unitary equivalence, we obtain the following algorithm to compute $\{V, \Gamma_2\}$.

(i) For large n compute the singular value decomposition $U_1 \Lambda_1 V_1^*$ for M_n. One criteria for choosing n is to find n such that the significant singular values for $T_{R,n}/n$ are starting to converge. Theoretically, they should converge to the singular values for $\operatorname{diag}[\{A_j A_j^*\}_1^\nu]$.

(ii) Let μ be the number of large singular values for M_n and $\Psi = U_1|\mathbb{C}^\mu$ the first μ columns of U_1. Set $\Omega = \Psi^* M_n$ and $\Gamma_3 = \Psi^* M_n|\mathcal{E}$. Here $\mathcal{E}$ is the first component of $\mathcal{E}^n$. Due to this singular value decomposition, $\Psi^* M_n$ is approximately equal to $W_{n,2}$ up to a unitary operator on the left.

(iii) Compute the matrices J_n and Q_n defined in (6.5.6). Find a unitary operator V_3 on $\mathbb{C}^\mu$ using the Procrustes Theorem 6.5.2 such that $V_3 \Omega J_n \approx \Omega Q_n$.

(iv) To find V_3 compute the singular value decomposition $U_2 \Lambda_2 V_2^*$ for $\Omega J_n (\Omega Q_n)^*$. Then $V_3 = V_2 U_2^*$.

(v) Finally, one can compute a unitary transformation to convert $\{V_3, \Gamma_3\}$ to a controllable pair $\{V, \Gamma_2\}$ of the form in (6.5.1).

Example. Let us return to our previous example. Consider the positive Toeplitz matrix T_R with symbol $R = \sum_{-\infty}^{\infty} e^{-\imath\omega k} r_k$ determined by (6.4.6) and (6.4.7). The unitary part $\{V, \Gamma_2\}$ for T_R is given by (6.4.9). We ran the previous algorithm for $n = 200$. The six largest singular values for $T_{R,200}$ are

$$50.9631, \quad 50.9630, \quad 26.2778, \quad 26.2692, \quad 2.1686, \quad 2.1666.$$

Clearly, there are only four large singular values. In this case, the first five singular values for $T_{R,200}/200$ are $0.2548, 0.2548, 0.1314, 0.1313$ and 0.0108. As expected the first four singular values are approximately equal to $|A_1|^2 = |A_2|^2 = 1/4$ and $|A_3|^2 = |A_3|^2 = 1/8 = 0.1250$. By running the previous algorithm we obtained:

$$V = \begin{bmatrix} \imath & 0 & 0 & 0 \\ 0 & -\imath & 0 & 0 \\ 0 & 0 & 0.7072 + 0.7070\imath & 0 \\ 0 & 0 & 0 & 0.7072 - 0.7070\imath \end{bmatrix} \quad \text{and } \Gamma_2 = \begin{bmatrix} 0.5067 \\ 0.5067 \\ 0.3631 \\ 0.3631 \end{bmatrix}.$$

Finally, it is noted that our algorithm produced some complex numbers in Γ_2. However, taking the absolute value of those numbers yields our Γ_2.

The computer can easily handle $n = 200$. However, if we choose $n = 40$, then the five largest singular values for $T_{R,40}$ are 10.9779 , 10.977, 6.3067, 6.2633, 2.1541. The first four singular values for $T_{R,40}/40$ are 0.2744, 0.2744, 0.1577 and

0.1566. The corresponding computations for $\{V, \Gamma_2\}$ yield:

$$V = \begin{bmatrix} \imath & 0 & 0 & 0 \\ 0 & -\imath & 0 & 0 \\ 0 & 0 & 0.7093 + 0.7049\imath & 0 \\ 0 & 0 & 0 & 0.7093 - 0.7049\imath \end{bmatrix} \text{ and } \Gamma_2 = \begin{bmatrix} 0.5327 \\ 0.5327 \\ 0.3996 \\ 0.3996 \end{bmatrix}.$$

Finally, it is noted that for $n = 40$, we still have a good estimate for the eigenvalues for V.

6.6 Notes

The Positive Real Lemma 6.2.1 is a classical result in systems theory, and sometimes referred to as the Kalman-Popov-Yakubovich Theorem. Two fundamental papers on the positive real lemma are Anderson [8] and Hitz-Anderson [133]. For some further results and historical comments on the positive real lemma see Anderson-Moore [11], Caines [47], and Kailath-Sayed-Hassibi [143]. The positive real lemma is intimately related to the stochastic realization problem; see Caines [47], Faurre [77, 78], Foias-Frazho [81], Lindquist-Picci [158, 159, 160, 161, 162] and Ruckebusch [185, 186] for further results in this direction. The ergodic Theorem 6.4.1 is a classical result; see Halmos [125]. The fast Fourier transform method to compute the frequencies or eigenvalues for V in Section 6.4 is a classical method in signal processing. The results in Section 6.5, were motivated by Allen-Smith [5]. The algorithm in Section 6.5 to compute $\{V, \Gamma_2\}$ was taken from Bhosri [32]. Finally, it is noted that ergodic methods can be slow to converge depending on the data.

The LMI Positive Real Lemma. For completeness let us sketch a Linear Matrix Inequality (LMI) version of the positive real lemma in Faurre [77, 78]. This version of the positive real lemma includes the case when R has poles on the unit circle, and is intimately related to the Naimark representation theorem. Linear matrix inequalities play a fundamental role in systems and control theory; see Boyd-Ghaoui-Feron-Balakrishnan [37] for some nice results in this direction. Finally, the results in the rest of this section are not used anywhere in this monograph.

Recall that a rational transfer function is a proper rational function. It will be convenient to represent a rational transfer function G in the form $G(z) = zC(zI - A)^{-1}B$ where $\{A, B, C, 0\}$ is a minimal realization. Notice that any rational transfer function G admits a representation of the form $G(z) = zC(zI - A)^{-1}B$. To see this, simply observe that $z^{-1}G$ is also a rational transfer function. So $z^{-1}G$ admits a realization of the form $\{A, B, C, 0\}$, and thus, $G = zC(zI - A)^{-1}B$. In this case, all minimal realizations $\{A, B, C, 0\}$ such that $G = zC(zI - A)^{-1}B$ are unique up to a similarity transformation. Finally, observe that a rational transfer function G admits a power series expansion of the form $G(z) = \sum_0^\infty z^{-n} G_n$.

Therefore $G(z) = zC(zI - A)^{-1}B$ if and only if $G_n = CA^nB$ for all integers $n \geq 0$. The following is a LMI version of the positive real lemma.

Lemma 6.6.1 (LMI Positive Real Lemma). *Let T_R in (6.1.1) be a self-adjoint rational Toeplitz matrix determined by a $\mathcal{L}(\mathcal{E}, \mathcal{E})$-valued sequence $\{R_n\}_0^\infty$ where $R_{-n} = R_n^*$. Let G be the rational function defined by $G(z) = \sum_0^\infty z^{-n} R_n$. Let $\{A \text{ on } \mathcal{X}, B, C, 0\}$ be a minimal realization for $z^{-1}G$. Then T_R is positive if and only if there exists an operator Q on $\mathcal{X}$ such that*

$$Q > 0, \quad Q \geq A^*QA \quad \textit{and} \quad C^* = QB. \tag{6.6.1}$$

Proof. Let T_R be a positive Toeplitz matrix. Then T_R admits a controllable isometric representation $\{U \text{ on } \mathcal{K}, \Gamma\}$. Thus $R_n = \Gamma^* U^{*n} \Gamma$ for all integers $n \geq 0$, and $\{U^*, \Gamma, \Gamma^*, 0\}$ is an observable realization for $z^{-1}G$. Let $\mathcal{H}$ be the invariant subspace for U^* determined by $\mathcal{H} = \bigvee_0^\infty U^{*n}\Gamma\mathcal{E}$. Let Λ be an operator on $\mathcal{H}$ defined by $\Lambda = U^*|\mathcal{H}$, and $\Gamma_\mathcal{H}$ the operator mapping $\mathcal{E}$ into $\mathcal{H}$ given by $\Gamma_\mathcal{H} = \Gamma$. By construction the pair $\{\Lambda, \Gamma_\mathcal{H}\}$ is controllable. Moreover, for all integers $n \geq 0$, we have $\Gamma_\mathcal{H}^* \Lambda^n \Gamma_\mathcal{H} = \Gamma^* U^{*n} \Gamma = R_n$. Thus $\{\Lambda, \Gamma_\mathcal{H}, \Gamma_\mathcal{H}^*, 0\}$ is a controllable realization of $z^{-1}G$. Since $\{\Lambda, \Gamma_\mathcal{H}, \Gamma_\mathcal{H}^*, 0\}$ is the realization obtained by extracting the controllable subspace from the observable realization $\{U^*, \Gamma, \Gamma^*, 0\}$, we see that $\{\Lambda, \Gamma_\mathcal{H}, \Gamma_\mathcal{H}^*, 0\}$ is a controllable and observable realization for $z^{-1}G$.

Recall that $\{A \text{ on } \mathcal{X}, B, C, 0\}$ is also a minimal realization of $z^{-1}G$. So there exists a similarity transformation M mapping $\mathcal{X}$ onto $\mathcal{H}$ such that

$$\Lambda M = MA, \qquad \Gamma_\mathcal{H} = MB \quad \text{and} \quad \Gamma_\mathcal{H}^* M = C. \tag{6.6.2}$$

Hence $\Lambda = MAM^{-1}$. Because $\Lambda = U^*|\mathcal{H}$ and U is an isometry, it follows that Λ is a contraction. Therefore

$$0 \leq I - \Lambda^*\Lambda = I - M^{-*}A^*M^*MAM^{-1}.$$

Multiplying by M^* on the left and M on the right, yields $M^*M \geq A^*M^*MA$. Because M is invertible, M^*M is a strictly positive operator. Set $Q = M^*M$, then Q is a strictly positive operator satisfying $Q \geq A^*QA$. Furthermore,

$$C^* = M^*\Gamma_\mathcal{H} = M^*MB = QB.$$

So if $T_R \geq 0$, then there exists a strictly positive operator Q on $\mathcal{X}$ such that $Q \geq A^*QA$ and $C^* = QB$, that is, (6.6.1) holds.

Assume there exists a strictly positive operator Q such that $Q \geq A^*QA$ and $C^* = QB$. Let $\Lambda = MAM^{-1}$ where M is any operator on $\mathcal{X}$ such that $Q = M^*M$. In fact, we can choose $M = Q^{1/2}$. Then $M^*M \geq A^*M^*MA$. Multiplying the previous inequality by M^{-*} on the left M^{-1} on the right yields

$$I \geq M^{-*}A^*M^*MAM^{-1} = \Lambda^*\Lambda$$

where $\Lambda = MAM^{-1}$. In other words, $I \geq \Lambda^*\Lambda$, and thus, Λ is a contraction. Set $\Gamma_o = MB$, or equivalently, $B = M^{-1}\Gamma_o$. Since $C = B^*Q = B^*M^*M$, we have $C = \Gamma_o^*M$. Using the fact that $\{A, B, C, 0\}$ is a realization of $z^{-1}G$, we obtain

$$R_n = CA^nB = \Gamma_o^*MA^nM^{-1}\Gamma_o = \Gamma_o^*\Lambda^n\Gamma_o \quad (n \geq 0).$$

By taking the adjoint, we arrive at $R_{-n} = \Gamma_o^*\Lambda^{*n}\Gamma_o$ for all integers $n \geq 0$.

Now let us construct an isometric representation $\{U, \Gamma\}$ for T_R. Let U on $\mathcal{K}$ be an isometric lifting of the contraction Λ^*; see Section 5.5. In particular, $\mathcal{X}$ is an invariant subspace for U^* and $\Lambda = U^*|\mathcal{X}$. In other words, $\Pi_\mathcal{X}U = \Lambda^*\Pi_\mathcal{X}$. Let Γ be the operator mapping $\mathcal{E}$ into $\mathcal{K}$ given by $\Gamma = \Gamma_o$. Using the fact that $\Pi_\mathcal{X}U^n = \Lambda^{*n}\Pi_\mathcal{X}$ for all integers $n \geq 0$, we have

$$\Gamma^*U^n\Gamma = \Gamma_o^*\Pi_\mathcal{X}U^n\Gamma = \Gamma_o^*\Lambda^{*n}\Pi_\mathcal{X}\Gamma = \Gamma_o^*\Lambda^{*n}\Gamma_o = R_{-n}.$$

Therefore $\{U, \Gamma\}$ is an isometric representation for T_R. It follows that T_R is a positive Toeplitz matrix. □

Theorem 6.6.2. *Let T_R in* (6.1.1) *be a positive rational Toeplitz matrix determined by a $\mathcal{L}(\mathcal{E}, \mathcal{E})$-valued sequence $\{R_n\}_0^\infty$ where $R_{-n} = R_n^*$, and set $G(z) = \sum_0^\infty z^{-n}R_n$. Let $\{A$ on $\mathcal{X}, B, C, 0\}$ be a minimal realization for $z^{-1}G$. Let Q be an operator on $\mathcal{X}$ such that*

$$Q > 0, \quad Q \geq A^*QA \quad \text{and} \quad C^* = QB. \tag{6.6.3}$$

Let $\widehat{C}$ be an operator mapping $\mathcal{X}$ onto $\mathcal{Y}$ such that $\widehat{C}^*\widehat{C} = Q - A^*QA$, and Θ be the rational function defined by $\Theta(z) = z\widehat{C}(zI - A)^{-1}B$. Then the following holds.

(i) *The sequence $A^{*n}QA^n$ is monotonically decreasing and converges to a positive operator Ω, that is,*

$$\Omega = \lim_{n\to\infty} A^{*n}QA^n. \tag{6.6.4}$$

(ii) *The function Θ is in $H^\infty(\mathcal{E}, \mathcal{Y})$.*

(iii) *The Toeplitz matrix T_R admits a decomposition of the form*

$$T_R = T_\Theta^*T_\Theta + (W^\sharp\Omega W)^{tr} \quad \text{where} \quad W = \begin{bmatrix} B & AB & A^2B & \cdots \end{bmatrix}. \tag{6.6.5}$$

(iv) *The transpose of T_R admits a decomposition of the form*

$$T_R^{tr} = ST_{\tilde{\Theta}}T_{\tilde{\Theta}}^*S^* + W^\sharp QW, \tag{6.6.6}$$

where S is a unilateral shift on $\ell_+^2(\mathcal{E})$.

(v) *If Θ_o in $H^2(\mathcal{E}, \mathcal{L})$ is the maximal outer spectral factor for T_R, then $\Theta(z) = \Psi(z)\Theta_o(z)$ where Ψ is a contractive analytic function in $H^\infty(\mathcal{L}, \mathcal{Y})$.*

(vi) *Finally,* $T_R = T_\Theta^* T_\Theta$ *if and only if* A *is stable.*

Proof. Notice that Q is the observability Gramian for the pair $\{\widehat{C}, A\}$, that is,

$$Q = A^* Q A + \widehat{C}^* \widehat{C}. \tag{6.6.7}$$

By recursively solving for Q, we obtain

$$Q = \sum_{j=0}^{n-1} A^{*j} \widehat{C}^* \widehat{C} A^j + A^{*n} Q A^n. \tag{6.6.8}$$

Because the sum in the previous equation forms an increasing sequence of positive operators, $A^{*n} Q A^n$ is a decreasing sequence of positive operators, that is, $A^{*n} Q A \geq A^{*(n+1)} Q A^{n+1}$. Hence $A^{*n} Q A^n$ converges to a positive operator Ω. Thus Part (i) holds. Finally, it is noted that $A^* \Omega A = \Omega$.

By letting n approach infinity in (6.6.8), we obtain

$$Q = \sum_{j=0}^{\infty} A^{*j} \widehat{C}^* \widehat{C} A^j + \Omega. \tag{6.6.9}$$

Let $\Theta(z) = \sum_0^\infty z^{-n} \Theta_n$ be the Taylor's series expansion of Θ. Since $\Theta(z) = z\widehat{C}(zI - A)^{-1} B$, we see that $\Theta_n = \widehat{C} A^n B$ for all integers $n \geq 0$. By consulting (6.6.9), we have

$$B^* Q B = \sum_{n=0}^{\infty} (B^* A^{*n} \widehat{C}^*)(\widehat{C} A^n B) + B^* \Omega B = \sum_{n=0}^{\infty} \Theta_n^* \Theta_n + B^* \Omega B \geq \sum_{n=0}^{\infty} \Theta_n^* \Theta_n.$$

Hence Θ is a function in $H^2(\mathcal{E}, \mathcal{Y})$. Because Θ is a rational function, Θ must be in $H^\infty(\mathcal{E}, \mathcal{Y})$. Therefore Part (ii) holds.

To verify that Part (iii) holds, we claim that $W^\sharp \Omega W$ is a positive Toeplitz matrix. For x in $\ell_+^c(\mathcal{E})$ we have $WSx = AWx$. Using this, we obtain

$$(S^* W^\sharp \Omega W S x, x) = (\Omega A W x, W S x) = (A^* \Omega A W x, W x) = (W^\sharp \Omega W x, x).$$

So $W^\sharp \Omega W$ is Toeplitz. Because Ω is positive, $W^\sharp \Omega W$ is a positive Toeplitz matrix.

Clearly, $T_\Theta^* T_\Theta$ is a positive Toeplitz matrix. Hence the sum $T_\Theta^* T_\Theta + (W^\sharp \Omega W)^{tr}$ is also a positive Toeplitz matrix. So to verify that

$$T_R = T_\Theta^* T_\Theta + (W^\sharp \Omega W)^{tr}$$

holds, it is sufficient to show that T_R and $T_\Theta^* T_\Theta + (W^\sharp \Omega W)^{tr}$ have the same first column. Recall that $\{A, B, C, 0\}$ is a minimal realization for $z^{-1} G$ and $C = B^* Q$. By employing this with (6.6.9), the $(n+1)^{th}$ component in the first column of

$T_\Theta^* T_\Theta + (W^*\Omega W)^{tr}$ is given by

$$\begin{aligned}(T_\Theta^* T_\Theta + (W^\sharp \Omega W)^{tr})_{n,0} &= \sum_{j=0}^{\infty} \Theta_j^* \Theta_{j+n} + ((W^\sharp \Omega W)^{tr})_{n,0} \\ &= \sum_{j=0}^{\infty} (B^* A^{*j} \widehat{C}^*)(\widehat{C} A^j A^n B) + (W^\sharp \Omega W)_{0,n} \\ &= B^* (\sum_{j=0}^{\infty} A^{*j} \widehat{C}^* \widehat{C} A^j) A^n B + B^* \Omega A^n B \\ &= B^* Q A^n B = C A^n B = R_n.\end{aligned}$$

Therefore $T_R = T_\Theta^* T_\Theta + (W^*\Omega W)^{tr}$, and Part (iii) holds.

Part (iv) is left as an exercise. To verify that Part (v) holds, let Θ_o be the maximal outer spectral factor for T_R. Then Θ_o is a rational function in some $H^\infty(\mathcal{E}, \mathcal{L})$ space; see Section 6.1. Equation (6.6.5) yields $T_\Theta^* T_\Theta \leq T_R$. By the definition of the outer spectral factor, we must have $T_\Theta^* T_\Theta \leq T_{\Theta_o}^* T_{\Theta_o}$. According to Lemma 5.3.1, there exists a contractive analytic function Ψ in $H^\infty(\mathcal{L}, \mathcal{Y})$ such that $\Theta(z) = \Psi(z)\Theta_o(z)$. Therefore Part (v) holds.

To complete the proof, recall that $T_R = T_\Theta^* T_\Theta + (W^*\Omega W)^{tr}$. Hence $T_R = T_\Theta^* T_\Theta$ if and only if $\Omega = 0$. Because Q is strictly positive, $\Omega = 0$ if and only if A is stable. Therefore $T_R = T_\Theta^* T_\Theta$ if and only if A is stable. $\square$

Theorem 6.6.3. *Let T_R in* (6.1.1) *be a positive rational Toeplitz matrix determined by a $\mathcal{L}(\mathcal{E}, \mathcal{E})$-valued sequence $\{R_n\}_0^\infty$, and set $G(z) = \sum_0^\infty z^{-n} R_n$. Let $\{A$ on $\mathcal{X}, B, C, 0\}$ be a minimal realization for $z^{-1}G$. Then there exists a unique minimal solution Q_o to* (6.6.3), *that is, Q_o is a strictly positive operator on $\mathcal{X}$ such that*

$$Q_o \geq A^* Q_o A, \quad C^* = Q_o B \quad \text{and} \quad Q_o \leq Q \tag{6.6.10}$$

where Q is any other solution to (6.6.3). *In this case, the maximal outer spectral factor Θ_o for T_R is given by*

$$\Theta_o(z) = z C_o (zI - A)^{-1} B \tag{6.6.11}$$

where C_o is any operator from $\mathcal{X}$ onto $\mathcal{D}$ such that

$$C_o^* C_o = Q_o - A^* Q_o A. \tag{6.6.12}$$

Proof. Let $\{U, \Gamma\}$ be a controllable isometric representation for the Toeplitz matrix T_R. Let $\{\Lambda$ on $\mathcal{H}, \Gamma_\mathcal{H}, \Gamma_\mathcal{H}^*, 0\}$ be the controllable and observable realization for $z^{-1}G$ extracted from $\{U, \Gamma\}$ in the proof of the LMI Positive Real Lemma 6.6.1. Recall that $\mathcal{H} = \bigvee_0^\infty U^{*n} \Gamma \mathcal{E}$ and $\Lambda = U^* | \mathcal{H}$, while $\Gamma_\mathcal{H} = \Gamma$ is an operator from $\mathcal{E}$ into $\mathcal{H}$. Since $\{A, B, C, 0\}$ is also a minimal realization of $z^{-1}G$, there exists a similarity transformation M intertwining $\{A, B, C, 0\}$ with $\{\Lambda, \Gamma_\mathcal{H}, \Gamma_\mathcal{H}^*, 0\}$; see

(6.6.2). Because M is invertible, $Q_o = M^*M$ is strictly positive. The proof of Lemma 6.6.1 also shows that $Q_o = M^*M$ is a solution to (6.6.3), that is,

$$Q_o > 0, \quad Q_o \geq A^*Q_oA \quad \text{and} \quad C^* = Q_oB.$$

Let us show that the maximal outer spectral factor for T_R is determined by $\Theta_o = zC_o(zI - A)^{-1}B$ where C_o is an operator from $\mathcal{X}$ onto $\mathcal{D}$ satisfying (6.6.12). Recall that the maximal outer spectral factor Θ_o for T_R is determined by $\Theta_o(z) = z\Pi_\mathcal{Y}(zI - U^*)^{-1}\Gamma$, where $\Pi_\mathcal{Y} : \mathcal{K} \to \mathcal{Y}$ is the orthogonal projection from $\mathcal{K}$ onto $\mathcal{Y} = \ker U^*$; see Theorem 5.2.1. Using (6.6.2) with the fact that the range of Γ is contained in $\mathcal{H}$ and $\Lambda = U^*|\mathcal{H}$, we have

$$\begin{aligned}\Theta_o(z) &= z\Pi_\mathcal{Y}(zI - U^*)^{-1}\Gamma = z\Pi_\mathcal{Y}(zI - \Lambda)^{-1}\Gamma_\mathcal{H}\\ &= z\Pi_\mathcal{Y}(zI - \Lambda)^{-1}MB = z\Pi_\mathcal{Y}M(zI - A)^{-1}B.\end{aligned}$$

In other words, $\Theta_o(z) = z\Pi_\mathcal{Y}M(zI - A)^{-1}B$. Finally, because $\Theta(\infty) = \Pi_\mathcal{Y}MB$ is onto $\mathcal{Y}$, the range of $\Pi_\mathcal{Y}M$ equals $\mathcal{Y}$.

We claim that

$$(\Pi_\mathcal{Y}M)^*\Pi_\mathcal{Y}M = Q_o - A^*Q_oA. \tag{6.6.13}$$

This follows from $\Pi_\mathcal{Y}^*\Pi_\mathcal{Y} = I - UU^*$ and the calculation

$$\begin{aligned}(\Pi_\mathcal{Y}M)^*\Pi_\mathcal{Y}M &= M^*\Pi_\mathcal{H}\Pi_\mathcal{Y}^*\Pi_\mathcal{Y}M = M^*\Pi_\mathcal{H}(I - UU^*)M\\ &= Q_o - M^*\Lambda^*\Lambda M = Q_o - A^*M^*MA\\ &= Q_o - A^*Q_oA.\end{aligned}$$

Therefore (6.6.13) holds.

Let C_o be any operator mapping $\mathcal{X}$ onto $\mathcal{D}$ satisfying $C_o^*C_o = Q_o - A^*Q_oA$. Then (6.6.13) implies that $C_o^*C_o = (\Pi_\mathcal{Y}M)^*\Pi_\mathcal{Y}M$. Because both C_o and $\Pi_\mathcal{Y}M$ are onto, there exists a unitary operator Φ such that $\Phi C_o = \Pi_\mathcal{Y}M$. Recall that the maximal outer spectral factor Θ_o for T_R is given by

$$\Theta_o(z) = z\Pi_\mathcal{Y}M(zI - A)^{-1}B = z\Phi C_o(zI - A)^{-1}B.$$

Because the maximal outer spectral factor is unique up to a unitary constant on the left, $zC_o(zI - A)^{-1}B$ is the maximal outer spectral factor for T_R.

To complete the proof, it remains to show that Q_o is the smallest solution to (6.6.3). Let Q be any other solution to (6.6.3). Set $\Theta = z\widehat{C}(zI - A)^{-1}B$ where $\widehat{C}$ is any operator mapping $\mathcal{X}$ onto $\mathcal{G}$ such that $\widehat{C}\widehat{C}^* = Q - A^*QA$. By Part (v) of Theorem 6.6.2, we see that $\Theta = \Psi\Theta_o$ where Ψ is a contractive analytic function. Since T_Ψ is a contraction, Part (iv) of Theorem 6.6.2 implies that

$$\begin{aligned}W^\sharp QW &= T_R^{tr} - ST_{\tilde{\Theta}}T_{\tilde{\Theta}}^*S^* = T_R^{tr} - ST_{\tilde{\Theta}_o}T_{\tilde{\Psi}}T_{\tilde{\Psi}}^*T_{\tilde{\Theta}_o}^*S^*\\ &\geq T_R^{tr} - ST_{\tilde{\Theta}_o}T_{\tilde{\Theta}_o}^*S^* = W^\sharp Q_oW.\end{aligned}$$

Thus $W^\sharp Q_oW \leq W^\sharp QW$. Because the pair $\{A, B\}$ is controllable, $Q_o \leq Q$. $\square$

For an example, consider the positive Toeplitz matrix T_R determined by $R_0 = 2$ and $R_n = 1$ for all integers $n \neq 0$. Then $\Theta = 1$ is the maximal outer spectral factor for T_R and corresponding unitary pair $\{V, \Gamma_2\}$ is given by $V = 1$ on $\mathbb{C}$ and $\Gamma_2 = 1$. Indeed, the controllable isometric representation $\{U, \Gamma\}$ for T_R is determined by

$$U = \begin{bmatrix} S & 0 \\ 0 & 1 \end{bmatrix} \text{ on } \begin{bmatrix} \ell_+^2 \\ \mathbb{C} \end{bmatrix} \quad \text{and} \quad \Gamma = \begin{bmatrix} \Pi_{\mathbb{C}}^* \\ 1 \end{bmatrix} : \mathbb{C} \to \begin{bmatrix} \ell_+^2 \\ \mathbb{C} \end{bmatrix}.$$

Here S is the unilateral shift on ℓ_+^2 and $\Pi_{\mathbb{C}}^*$ embeds $\mathbb{C}$ in the first component of ℓ_+^2. A simple calculation shows that $R_{-n} = \Gamma^* U^n \Gamma$ for all integers $n \geq 0$. Proposition 6.3.1 guarantees that $\{U, \Gamma\}$ is controllable. So $\{U, \Gamma\}$ is a controllable isometric representation for T_R. Finally, its maximal outer spectral factor $\Theta(z) = z\Pi_{\mathbb{C}}(zI - S^*)^{-1}\Pi_{\mathbb{C}}^* = 1$; see Theorem 5.2.1.

In this setting, a minimal realization $\{A \text{ on } \mathbb{C}^2, B, C, 0\}$ for $z^{-1}G$ is given by

$$A = \begin{bmatrix} 0 & 0 \\ 0 & 1 \end{bmatrix}, \quad B = \begin{bmatrix} 1 \\ 1 \end{bmatrix},$$
$$C = \begin{bmatrix} 1 & 1 \end{bmatrix}.$$

Notice that $Q = I$ on $\mathbb{C}^2$ is the only solution to (6.6.3). In this case,

$$\Omega = \begin{bmatrix} 0 & 0 \\ 0 & 1 \end{bmatrix} \quad \text{and} \quad \widehat{C} = \begin{bmatrix} 1 & 0 \end{bmatrix}.$$

In particular, $\Theta(z) = z\widehat{C}(zI - A)^{-1}B = 1$. Finally, it is easy to verify that (6.6.5) and (6.6.6) hold.

Part II

Finite Section Techniques

Chapter 7

The Levinson Algorithm and Factorization

In this chapter we show how one can use the Levinson algorithm, along with the finite section inversion method and the Kalman-Ho algorithm to compute the inner-outer factorization for certain rational functions. We will also present some elementary results concerning contractive analytic functions and contractive realizations.

7.1 The Case when T_R is Invertible

Assume that T_R is a positive Toeplitz matrix. If Θ is an outer function in $H^2(\mathcal{E}, \mathcal{Y})$ satisfying $T_R = T_\Theta^\sharp T_\Theta$, then Θ is called the *outer spectral factor* for T_R. In this case, Θ is also the maximal outer spectral factor for T_R. According to Theorem 5.2.1, the outer spectral factor Θ for T_R is unique up to a constant unitary operator on the left. To be precise, if Ψ is an outer function in $H^2(\mathcal{E}, \mathcal{G})$ satisfying $T_R = T_\Psi^\sharp T_\Psi$, then $\Theta = \Phi\Psi$ where Φ is a unitary operator mapping $\mathcal{G}$ onto $\mathcal{Y}$.

If Θ is any function (not necessarily outer) in $H^2(\mathcal{E}, \mathcal{Y})$ satisfying $T_R = T_\Theta^\sharp T_\Theta$, then Θ is called a *spectral factor* for T_R. If T_R admits a spectral factor Θ, then the outer part Θ_o of Θ is the outer spectral factor for T_R. To see this, recall that Θ admits a unique inner-outer factorization of the form $\Theta = \Theta_i \Theta_o$ where Θ_i is an inner function and Θ_o is an outer function. In other words, $T_\Theta = T_{\Theta_i} T_{\Theta_o}$. Because T_{Θ_i} is an isometry, $T_R = T_{\Theta_o}^\sharp T_{\Theta_o}$. Therefore Θ_o is the outer spectral factor for T_R.

Let Θ be a function in $H^\infty(\mathcal{E}, \mathcal{Y})$. Recall that T_Θ is an invertible operator mapping $\ell_+^2(\mathcal{E})$ into $\ell_+^2(\mathcal{Y})$ if and only if Θ is an invertible outer function. In this case, $(T_\Theta)^{-1} = T_{\Theta^{-1}}$. Proposition 3.3.2 provides a method of computing the inner-outer factorization for a function Θ which admits an invertible outer factor. This method involved inverting the Toeplitz operator $T_\Theta^* T_\Theta$. The following is a modification of this result and shows that any strictly positive Toeplitz operator admits an invertible outer spectral factor.

Theorem 7.1.1. *Let T_R be a positive Toeplitz matrix generated a $\mathcal{L}(\mathcal{E},\mathcal{E})$-valued symbol $R = \sum_{\infty}^{\infty} e^{-\imath\omega k} R_k$; see (5.1.1). Then the following statements are equivalent.*

(i) *The Toeplitz matrix T_R defines an invertible operator on $\ell_+^2(\mathcal{E})$.*

(ii) *The function R and R^{-1} are both in $L^\infty(\mathcal{E},\mathcal{E})$.*

(iii) *The maximal outer spectral factor Θ for T_R is an invertible outer function in $H^\infty(\mathcal{E},\mathcal{E})$, and $T_R = T_\Theta^* T_\Theta$.*

In this case, the outer spectral factor Θ for R is computed by

$$\begin{aligned} \Theta(z) &= \Delta^{-1/2}\Omega(z)^{-1} \quad \text{where} \quad \Omega(z) = (\mathcal{F}_{\mathcal{E}}^+ T_R^{-1}\Pi_{\mathcal{E}}^*)(z) \qquad (z \in \mathbb{D}_+), \\ \Pi_{\mathcal{E}} &= \begin{bmatrix} I & 0 & 0 & 0 & \cdots \end{bmatrix} : \ell_+^2(\mathcal{E}) \to \mathcal{E}, \\ \Delta &= (\Pi_{\mathcal{E}} T_R^{-1}\Pi_{\mathcal{E}}^*)^{-1}. \end{aligned} \tag{7.1.1}$$

Finally, R is in $L^\infty(\mathcal{E},\mathcal{E})$ and $R = \Theta^*\Theta$.

Proof. Proposition 2.5.1 shows that Parts (i) and (ii) are equivalent. If Θ is an invertible outer function in $H^\infty(\mathcal{E},\mathcal{E})$, then T_Θ is invertible. Hence $T_R = T_\Theta^* T_\Theta$ is an invertible positive Toeplitz operator on $\ell_+^2(\mathcal{E})$, and Part (i) holds. Finally, in this case, R is in $L^\infty(\mathcal{E},\mathcal{E})$ and $R = \Theta^*\Theta$; see Proposition 2.5.1.

Now assume that T_R defines an invertible positive Toeplitz operator on $\ell_+^2(\mathcal{E})$. Let $\{U \text{ on } \mathcal{K}, \Gamma\}$ be any controllable isometric realization for T_R. Let

$$W = \begin{bmatrix} \Gamma & U\Gamma & U^2\Gamma & \cdots \end{bmatrix}$$

be the controllability matrix corresponding to $\{U,\Gamma\}$. Recall that $T_R = W^\sharp W$. For any x in $\ell_+^c(\mathcal{E})$, we have $\|Wx\|^2 = (T_R x, x) \le \|T_R\|\|x\|^2$. This implies that W defines an operator from $\ell_+^2(\mathcal{E})$ into $\mathcal{K}$. Since T_R is bounded below, there exists a constant $\delta > 0$ such that $T_R \ge \delta I$. Moreover, for any x in $\ell_+^c(\mathcal{E})$, we obtain $\|Wx\|^2 = (T_R x, x) \ge \delta\|x\|^2$. Hence W is bounded below, and thus, the range of W is closed. Because $\{U,\Gamma\}$ is controllable, W is invertible.

Observe that $UW = WS$, where S is the unilateral shift on $\ell_+^2(\mathcal{E})$. Thus $S^*W^* = W^*U^*$. Because W^* is invertible, S^* is similar to U^*. Since S^{*n} converges to zero in the strong operator topology, we see that U^{*n} also converges to zero in the strong operator topology. By virtue of the Wold decomposition, it follows that U is also a unilateral shift. Without loss of generality, we can assume that U is the unilateral shift on $\ell_+^2(\mathcal{Y})$ where $\mathcal{Y} = \ker U^*$. Since $UW = WS$, this implies that $W = T_\Theta$ where Θ is a function in $H^\infty(\mathcal{E},\mathcal{Y})$. Because W is invertible, Θ is an invertible outer function satisfying $T_R = W^*W = T_\Theta^* T_\Theta$. Finally, since $\Theta(z)$ is invertible for z in $\mathbb{D}_+$, the spaces $\mathcal{E}$ and $\mathcal{Y}$ have the same dimension. So without loss of generality we can assume that $\mathcal{E} = \mathcal{Y}$. Hence Part (iii) holds. Clearly, Part (iii) implies Part (i). Therefore Parts (i), (ii) and (iii) are equivalent.

To complete the proof, it remains to establish the inversion formula for Θ in (7.1.1). By taking the inverse of $T_R = T_\Theta^* T_\Theta$, we arrive at $T_R^{-1} = T_\Theta^{-1} T_\Theta^{-*} = T_{\Theta^{-1}} T_{\Theta^{-1}}^*$. Notice that Θ^{-1} admits a Taylor series expansion of the form

$$\Theta(z)^{-1} = \Theta(\infty)^{-1} + z^{-1}A_1 + z^{-2}A_2 + z^{-3}A_3 + \cdots .$$

(Using $I = (\Theta^{-1}\Theta)(\infty)$, it follows that $(\Theta^{-1})(\infty) = \Theta(\infty)^{-1}$.) Hence $T_{\Theta^{-1}}^*$ is an upper triangular Toeplitz matrix with the entry $\Theta(\infty)^{-*}$ on the main diagonal. Thus

$$T_R^{-1}\Pi_\mathcal{E}^* = T_\Theta^{-1}T_\Theta^{-*}\Pi_\mathcal{E}^* = T_{\Theta^{-1}}(T_{\Theta^{-1}})^*\Pi_\mathcal{E}^* = T_{\Theta^{-1}}\Pi_\mathcal{E}^*\Theta(\infty)^{-*}.$$

By taking the Fourier transform, we obtain $\mathcal{F}_\mathcal{E}^+ T_R^{-1}\Pi_\mathcal{E}^* = \Theta(z)^{-1}\Theta(\infty)^{-*}$. Using the fact that $\Pi_\mathcal{E} g = (\mathcal{F}_\mathcal{E}^+ g)(\infty)$, for any vector g in $\ell_+^2(\mathcal{E})$, we arrive at

$$\Delta^{-1} = \Pi_\mathcal{E} T_R^{-1}\Pi_\mathcal{E}^* = (\mathcal{F}_\mathcal{E}^+ T_R^{-1}\Pi_\mathcal{E}^*)(\infty) = \Theta(\infty)^{-1}\Theta(\infty)^{-*}.$$

In other words, $\Delta = \Theta(\infty)^*\Theta(\infty)$. This readily implies that there exists a unitary operator Φ on $\mathcal{E}$ such that $\Delta^{1/2} = \Phi\Theta(\infty)$. Since all maximal outer spectral factors are unique up to a constant unitary operator on the left, without loss of generality, we can assume that $\Delta^{1/2} = \Theta(\infty)$. Thus

$$\Omega(z) = (\mathcal{F}_\mathcal{E}^+ T_R^{-1}\Pi_\mathcal{E}^*)(z) = \Theta(z)^{-1}\Theta(\infty)^{-*} = \Theta(z)^{-1}\Delta^{-1/2}.$$

In other words, $\Theta(z) = \Delta^{-1/2}\Omega(z)^{-1}$, which is precisely the formula for Θ in equation (7.1.1). $\square$

Remark 7.1.2. Let R be a function in $L^\infty(\mathcal{E},\mathcal{E})$. Then Theorem 7.1.1 shows that $R = \Theta^*\Theta$ where Θ is an invertible outer function in $H^\infty(\mathcal{E},\mathcal{E})$ if and only if $\delta I \le R(e^{\imath\omega}) \le \gamma I$ almost everywhere with respect to the Lebesgue measure for some positive scalars $\delta > 0$ and $\gamma < \infty$.

Remark 7.1.3. Let T_R be a strictly positive Toeplitz operator on $\ell_+^2(\mathcal{E})$. In prediction theory one uses a slightly different formula to compute the outer spectral factor Θ for T_R. To this end, let $\{A_j\}_1^\infty$ and Δ be the operators in $\mathcal{L}(\mathcal{E},\mathcal{E})$ obtained from the unique solution to

$$T_R \begin{bmatrix} I & A_1 & A_2 & A_3 & \cdots \end{bmatrix}^{tr} = \begin{bmatrix} \Delta & 0 & 0 & 0 & \cdots \end{bmatrix}^{tr}. \tag{7.1.2}$$

Then the outer spectral factor Θ for T_R is given by

$$\Theta(z) = \Delta^{1/2}\left(I + z^{-1}A_1 + z^{-2}A_2 + z^{-3}A_3 + \cdots\right)^{-1}. \tag{7.1.3}$$

We claim that

$$\Delta = \left(\Pi_\mathcal{E} T_R^{-1}\Pi_\mathcal{E}^*\right)^{-1} \quad \text{and} \quad \begin{bmatrix} I & A_1 & A_2 & A_3 & \cdots \end{bmatrix}^{tr} = T_R^{-1}\Pi_\mathcal{E}^*\Delta \tag{7.1.4}$$

is the unique solution to (7.1.2). By construction $\Pi_\mathcal{E} T_R^{-1}\Pi_\mathcal{E}^*\Delta = I$. So the first component of $T_R^{-1}\Pi_\mathcal{E}^*\Delta$ equals I. In other words, $T_R^{-1}\Pi_\mathcal{E}^*\Delta$ is of the form presented in (7.1.4). Multiplying (7.1.4) by T_R on the left, yields $T_R \begin{bmatrix} I & A_1 & A_2 & \cdots \end{bmatrix}^{tr} = \Pi_{\mathcal{E}}^* \Delta$. Therefore the operators $\{A_j\}_1^\infty$ and Δ in (7.1.4) provide a solution to (7.1.2).

To show that this solution is unique, assume that $\{A_j\}_1^\infty$ and Δ is any solution to (7.1.2). By taking the inverse of T_R, we see that

$$\begin{bmatrix} I & A_1 & A_2 & A_3 & \cdots \end{bmatrix}^{tr} = T_R^{-1} \Pi_{\mathcal{E}}^* \Delta.$$

By applying $\Pi_{\mathcal{E}}$ to both sides, $I = \Pi_{\mathcal{E}} T_R^{-1} \Pi_{\mathcal{E}}^* \Delta$. Thus Δ equals the inverse of $\Pi_{\mathcal{E}} T_R^{-1} \Pi_{\mathcal{E}}^*$. Therefore $\{A_j\}_1^\infty$ and Δ are uniquely determined by (7.1.4).

To show that the outer spectral factor is given by (7.1.3), observe that (7.1.1) yields

$$\begin{aligned}\Theta(z)^{-1} &= (\mathcal{F}_{\mathcal{E}}^+ T_R^{-1} \Pi_{\mathcal{E}}^*)(z) \Delta^{1/2} = (\mathcal{F}_{\mathcal{E}}^+ T_R^{-1} \Pi_{\mathcal{E}}^* \Delta)(z) \Delta^{-1/2} \\ &= \left(I + z^{-1} A_1 + z^{-2} A_2 + z^{-3} A_3 + \cdots \right) \Delta^{-1/2}.\end{aligned}$$

By taking the inverse we arrive at the formula for Θ in (7.1.3).

7.2 The Classical Schur Inversion Formula

The following result is a self-adjoint version of Schur's classical matrix inversion formula.

Lemma 7.2.1. *Let T be a block self-adjoint matrix of the form*

$$T = \begin{bmatrix} A & X^* \\ X & Y \end{bmatrix} \quad on \quad \begin{bmatrix} \mathcal{U} \\ \mathcal{Y} \end{bmatrix}. \tag{7.2.1}$$

Assume that Y is strictly positive, and let $\Delta = A - X^* Y^{-1} X$ be the Schur complement of T. Then Δ is given by the error in the following optimization problem:

$$(\Delta f, f) = \min\{(Th, h) : \Pi_{\mathcal{U}} h = f\} \qquad (f \in \mathcal{U}) \tag{7.2.2}$$

where $\Pi_{\mathcal{U}} = \begin{bmatrix} I & 0 \end{bmatrix}$ maps $\mathcal{U} \oplus \mathcal{Y}$ onto $\mathcal{U}$. Moreover, T is strictly positive if and only if Δ is strictly positive. In this case,

$$T^{-1} = \begin{bmatrix} \Delta^{-1} & -\Delta^{-1} X^* Y^{-1} \\ -Y^{-1} X \Delta^{-1} & Y^{-1} + Y^{-1} X \Delta^{-1} X^* Y^{-1} \end{bmatrix} \tag{7.2.3}$$

and the Schur complement Δ is given by

$$\Delta = A - X^* Y^{-1} X = (\Pi_{\mathcal{U}} T^{-1} \Pi_{\mathcal{U}}^*)^{-1}. \tag{7.2.4}$$

Proof. A straightforward calculation shows that T admits a factorization of the form

$$T = \begin{bmatrix} I & X^* Y^{-1} \\ 0 & I \end{bmatrix} \begin{bmatrix} A - X^* Y^{-1} X & 0 \\ 0 & Y \end{bmatrix} \begin{bmatrix} I & 0 \\ Y^{-1} X & I \end{bmatrix}. \tag{7.2.5}$$

Notice that $T = L^*\Lambda L$ where

$$\Lambda = \begin{bmatrix} A - X^*Y^{-1}X & 0 \\ 0 & Y \end{bmatrix} \quad \text{and} \quad L = \begin{bmatrix} I & 0 \\ Y^{-1}X & I \end{bmatrix}.$$

Hence $(Th, h) = (\Lambda Lh, Lh)$ for h in $\mathcal{U} \oplus \mathcal{Y}$. Since L is invertible, it follows that T is strictly positive if and only if Λ is strictly positive. Because Λ is a diagonal matrix and Y is strictly positive, T is strictly positive if and only if $\Delta = A - X^*Y^{-1}X$ is strictly positive.

If Δ is strictly positive, then

$$T^{-1} = L^{-1}\Lambda^{-1}L^{-*} = \begin{bmatrix} I & 0 \\ -Y^{-1}X & I \end{bmatrix} \begin{bmatrix} \Delta^{-1} & 0 \\ 0 & Y^{-1} \end{bmatrix} \begin{bmatrix} I & -X^*Y^{-1} \\ 0 & I \end{bmatrix}.$$

By performing these matrix calculations, we obtain the form of the inverse of T in (7.2.3). Equation (7.2.4) follows from (7.2.3) and the definition of the Schur complement.

To complete the proof, it remains to establish that the Schur complement Δ is given by the error in the optimization problem (7.2.2). Let $h = f \oplus g$ be a vector in $\mathcal{U} \oplus \mathcal{Y}$. The Schur factorization of T in (7.2.5), yields

$$\begin{aligned}(Th, h) = (\Lambda Lh, Lh) &= (\Delta f, f) + (Y(Y^{-1}Xf + g), Y^{-1}Xf + g) \\ &= (\Delta f, f) + \|Y^{1/2}(Y^{-1}Xf + g)\|^2 \\ &= (\Delta f, f) + \|Y^{-1/2}Xf + Y^{1/2}g\|^2.\end{aligned}$$

So $(Th, h) \geq (\Delta f, f)$. By choosing $g = -Y^{-1}Xf$, we have $(Th, h) = (\Delta f, f)$. Hence the minimum is uniquely attained. This establishes (7.2.2). □

7.3 The Schur Complement and Toeplitz Matrices

In this section, we will use the Schur complement to gain some further insight into positive Toeplitz matrices. Let T_R be the Toeplitz matrix generated by a $\mathcal{L}(\mathcal{E}, \mathcal{E})$-valued symbol $\sum_{-\infty}^{\infty} e^{-\imath\omega n} R_n$ where $R_n = R_{-n}^*$. (Throughout $\mathcal{E}$ is finite dimensional.) Let $T_{R,n}$ be the block Toeplitz operator on $\mathcal{E}^n = \oplus_1^n \mathcal{E}$ determined by compressing T_R to $\mathcal{E}^n$, that is,

$$T_{R,n} = \begin{bmatrix} R_0 & R_1^* & R_2^* & \cdots & R_{n-1}^* \\ R_1 & R_0 & R_1^* & \cdots & R_{n-2}^* \\ R_2 & R_1 & R_0 & \cdots & R_{n-3}^* \\ \vdots & \vdots & \vdots & \ddots & \vdots \\ R_{n-1} & R_{n-2} & R_{n-3} & \cdots & R_0 \end{bmatrix} \text{ on } \begin{bmatrix} \mathcal{E} \\ \mathcal{E} \\ \mathcal{E} \\ \vdots \\ \mathcal{E} \end{bmatrix}. \tag{7.3.1}$$

Because there is only a finite number of copies of $\mathcal{E}$ in $\mathcal{E}^n$, the matrix $T_{R,n}$ is a well-defined operator. Recall that T_R is positive if and only if $T_{R,n}$ is positive for all integers $n \geq 0$. Consider the matrix decomposition of the Toeplitz matrix $T_{R,n}$ given by

$$T_{R,n} = \begin{bmatrix} R_0 & X^* \\ X & T_{R,n-1} \end{bmatrix} \tag{7.3.2}$$

where $X^* = \begin{bmatrix} R_1^* & R_2^* & \cdots & R_{n-1}^* \end{bmatrix}$. Assume that $T_{R,n-1}$ is a strictly positive operator. Then the *Schur complement* Δ_n *associated with* $T_{R,n}$ is the operator on $\mathcal{E}$ defined by

$$\Delta_n = R_0 - X^* T_{R,n-1}^{-1} X \quad \text{and} \quad \Delta_1 = R_0. \tag{7.3.3}$$

If $T_{R,n-1}^{-1}$ is not invertible, then the Schur complement Δ_n is not well defined and for convenience in this case we set $\Delta_n = -\infty$. Finally, if $T_{R,n}$ is strictly positive, then the matrix inversion Lemma 7.2.1 shows that

$$\begin{aligned} \Delta_n &= (\Pi_n T_{R,n}^{-1} \Pi_n^*)^{-1}, \\ \Pi_n &= \begin{bmatrix} I & 0 & 0 & \cdots & 0 \end{bmatrix} : \mathcal{E}^n \to \mathcal{E}. \end{aligned} \tag{7.3.4}$$

The following result allows us to determine whether or not $T_{R,n}$ is strictly positive by checking the positivity of the Schur complements.

Lemma 7.3.1. *Let $T_{R,n}$ be the $n \times n$ Toeplitz matrix on $\mathcal{E}^n$ given by* (7.3.1), *and let Δ_j on $\mathcal{E}$ be the Schur complement associated with $T_{R,j}$ for $1 \leq j \leq n$. Then $T_{R,n}$ is strictly positive if and only if Δ_j is strictly positive for all $1 \leq j \leq n$. Moreover, in this case, $\{\Delta_j\}_1^n$ forms a decreasing sequence of positive operators, that is, $\Delta_1 \geq \Delta_2 \geq \cdots \geq \Delta_n$.*

Proof. Since $\Delta_1 = R_0$, this lemma is true for $n = 1$. Now let us proceed by induction and assume that the lemma is true for $n-1$, that is, $\Delta_1, \Delta_2, \ldots, \Delta_{n-1}$ are all strictly positive if and only if $T_{R,n-1}$ is strictly positive. By applying the matrix inversion Lemma 7.2.1 to the decomposition of $T_{R,n}$ in (7.3.2), we see that $T_{R,n}$ is strictly positive if and only if $T_{R,n-1}$ and Δ_n are both strictly positive. This completes the proof of the induction.

Now assume that $\{\Delta_j\}_1^n$ are all strictly positive, or equivalently, $T_{R,n}$ is strictly positive. Notice that $T_{R,j}$ is the compression of $T_{R,n}$ to $\mathcal{E}^j$, that is, $T_{R,j}$ is contained in the $j \times j$ upper left-hand corner of $T_{R,n}$ for $j \leq n$. Hence the optimization problem in (7.2.2) shows that

$$(\Delta_j f, f) = \inf\{(T_{R,n}x, x) : x \in \mathcal{E}^j \oplus \{0\} \text{ and } \Pi_n x = f\} \quad (f \in \mathcal{E}). \tag{7.3.5}$$

Here $(\Delta_j f, f)$ is the cost of the optimization problem. Because the infimum for $(\Delta_{j+1} f, f)$ is taken over a larger set ($\mathcal{E}^{j+1} \supset \mathcal{E}^j$) than the infimum corresponding to $(\Delta_j f, f)$, it follows that $(\Delta_j f, f)$ forms a decreasing sequence. Therefore $\{\Delta_j\}$ forms a decreasing sequence of positive operators. □

Assume that $T_{R,n}$ is strictly positive for all n. The previous lemma shows that the Schur complements Δ_n form a decreasing sequence of strictly positive operators. So Δ_n converges to a positive operator Δ on $\mathcal{E}$ as n approaches infinity, that is,

$$\Delta = \lim_{n\to\infty} \Delta_n. \tag{7.3.6}$$

Moreover, by consulting the optimization problems in (7.3.5), we see that

$$(\Delta f, f) = \inf\{(T_R x, x) : x \in \ell_+^c(\mathcal{E}) \text{ and } \Pi_{\mathcal{E}} x = f\} \quad (f \in \mathcal{E}). \tag{7.3.7}$$

Here $\Pi_{\mathcal{E}}$ is the linear map from $\ell_+^c(\mathcal{E})$ onto $\mathcal{E}$ which picks out the first component of $\ell_+^c(\mathcal{E})$. If T_R defines an invertible positive operator on $\ell_+^2(\mathcal{E})$, then

$$\Delta = (\Pi T_R^{-1} \Pi^*)^{-1}. \tag{7.3.8}$$

This follows by observing that T_R admits a block matrix decomposition of the form

$$T_R = \begin{bmatrix} R_o & X^* \\ X & T_R \end{bmatrix},$$

and Δ is the Schur complement for T_R. Finally, it is noted that if T_R is invertible, then

$$(\Pi T_R^{-1} \Pi^*)^{-1} = \lim_{n\to\infty} (\Pi_n T_{R,n}^{-1} \Pi_n^*)^{-1}. \tag{7.3.9}$$

7.4 Schur Complement and Maximal Outer Factor

We say that a $\mathcal{L}(\mathcal{E},\mathcal{Y})$-valued function Θ is a *square outer* function if Θ is an outer function and $\mathcal{E}$ and $\mathcal{Y}$ have the same dimension. It is emphasized that throughout this section we assume that $\dim \mathcal{E} = \dim \mathcal{Y}$.

Let T_R be the positive Toeplitz matrix determined by a symbol $\sum_{-\infty}^{\infty} e^{-\imath\omega n} R_n$ with values in $\mathcal{L}(\mathcal{E},\mathcal{E})$. Moreover, assume that its finite sections $T_{R,n}$ are strictly positive operators for all integers $n \geq 0$. If Θ is the maximal outer spectral factor for T_R, then

$$\Theta(\infty)^*\Theta(\infty) = \lim_{n\to\infty} \Delta_n. \tag{7.4.1}$$

As expected, $\Delta_n = (\Pi_n T_{R,n}^{-1} \Pi_n^*)^{-1}$ is the Schur complement for $T_{R,n}$. To verify this, let $\{U \text{ on } \mathcal{K}, \Gamma\}$ be a controllable isometric representation for T_R and

$$W = \begin{bmatrix} \Gamma & U\Gamma & U^2\Gamma & \cdots \end{bmatrix}$$

its corresponding controllability matrix. Recall that $T_R = W^\sharp W$. Let $\mathcal{Y} = \ker U^*$. By consulting (5.2.2) in Theorem 5.2.1, we see that $\Theta(\infty) = \Pi_{\mathcal{Y}}\Gamma$. Using this with the fact that $\mathcal{K} = \overline{W\ell_+^c(\mathcal{E})}$, we obtain

$$\begin{aligned}
(\Delta f, f) &= \inf\{(T_R x, x) : x \in \ell_+^c(\mathcal{E}) \text{ and } \Pi_{\mathcal{E}} x = f\} \\
&= \inf\{\|Wx\|^2 : x \in \ell_+^c(\mathcal{E}) \text{ and } \Pi_{\mathcal{E}} x = f\} \\
&= \inf\{\|\Gamma f + UWg\|^2 : g \in \ell_+^c(\mathcal{E})\} \\
&= \inf\{\|\Gamma f + U\xi\|^2 : \xi \in \mathcal{K}\} \\
&= \|P_{\mathcal{Y}}\Gamma f\|^2 = \|\Theta(\infty) f\|^2.
\end{aligned} \tag{7.4.2}$$

The second from the last equality follows from the projection theorem. Since this holds for all f in $\mathcal{E}$, we see that $\Delta = \Theta(\infty)^*\Theta(\infty)$. Because Δ_n converges to Δ, we arrive at (7.4.1). In particular, if the maximal outer spectral factor is square, then $\Theta(\infty)$ is invertible. In this case, there exists a positive scalar δ such that $0 < \delta I \leq \Delta_n$ for all $n \geq 1$. This proves part of the following result.

Theorem 7.4.1. *Let T_R be the Toeplitz matrix determined by a $\mathcal{L}(\mathcal{E},\mathcal{E})$-valued symbol $\sum_{-\infty}^{\infty} e^{-\imath\omega n} R_n$ where $R_n = R_{-n}^*$. Let Δ_n be the Schur complements associated with the finite sections $T_{R,n}$ for T_R where $n \geq 1$. Then the Toeplitz matrix T_R is positive and admits a maximal square outer spectral factorization Θ if and only if $\Delta_n \geq \delta I$ for all n and some $\delta > 0$. In this case, the sequence $\{\Delta_n\}$ is decreasing and converges to the strictly positive operator $\Theta(\infty)^*\Theta(\infty)$. Furthermore, if G is a function in $H^2(\mathcal{E},\mathcal{E})$ such that $T_R \geq T_G^\sharp T_G$, then*

$$\Theta(\infty)^*\Theta(\infty) \geq G(\infty)^*G(\infty). \tag{7.4.3}$$

In this case, $\Theta(\infty)^*\Theta(\infty) = G(\infty)^*G(\infty)$ if and only if G is the maximal outer spectral factor of T_R.

Proof. If T_R is positive and admits a maximal square outer spectral factor Θ, then $T_{R,n}$ is strictly positive for all n. Indeed, if $(T_{R,n}x, x) = 0$ for some x in $\mathcal{E}^n$, then we have

$$0 = (T_{R,n}x, x) = (T_R(x \oplus 0), (x \oplus 0)) \geq (T_\Theta^\sharp T_\Theta(x \oplus 0), (x \oplus 0)) = \|T_\Theta(x \oplus 0)\|^2 \geq 0.$$

In other words, $T_\Theta(x \oplus 0) = 0$. In particular, this implies that

$$\begin{bmatrix} \Theta_0 & 0 & \cdots & 0 \\ \Theta_1 & \Theta_0 & \cdots & 0 \\ \vdots & \vdots & \ddots & \vdots \\ \Theta_{n-1} & \Theta_{n-2} & \cdots & \Theta_0 \end{bmatrix} x = 0$$

where $\Theta(z) = \sum_0^\infty z^{-n}\Theta_n$ is the Taylor series expansion for Θ. Because $\Theta(\infty) = \Theta_0$ is invertible, $x = 0$. Therefore $T_{R,n}$ is strictly positive for all integers $n \geq 0$. By our previous analysis, Δ_n converges to $\Theta(\infty)^*\Theta(\infty)$, and thus, $\delta I \leq \Delta_n$ for all $n \geq 1$ and some $\delta > 0$.

Now assume that $\delta I \leq \Delta_n$ for some $\delta > 0$. According to Lemma 7.3.1, the operators $T_{R,n}$ are strictly positive. In particular, T_R is a positive Toeplitz matrix. Our previous analysis shows that Δ_n converges to $\Theta(\infty)^*\Theta(\infty)$ where Θ is the maximal outer spectral factor for T_R; see (7.4.1). Since $\delta > 0$, the operator $\Theta(\infty)^*\Theta(\infty)$ must be invertible. Because Θ is outer, the range of $\Theta(\infty)$ is onto. Therefore $\Theta(\infty)$ is invertible and Θ must be a square outer function.

To complete the proof, assume that G is a function in $H^2(\mathcal{E},\mathcal{D})$ such that $T_R \geq T_G^\sharp T_G$. By the definition of the maximal outer spectral factor, $T_\Theta^\sharp T_\Theta \geq T_G^\sharp T_G$. Part (iv) of Lemma 5.3.1, yields (7.4.3). Finally, Part (v) of Lemma 5.3.1 shows that $\Theta(\infty)^*\Theta(\infty) = G(\infty)^*G(\infty)$ if and only if G is the maximal outer spectral factor of T_R. □

Remark 7.4.2. In some applications one is given a Toeplitz matrix T_R and a function G in $H^2(\mathcal{E},\mathcal{D})$ such that $T_R \geq T_G^\sharp T_G$ where $G(\infty)$ is one to one. In this case, T_R admits a square maximal outer spectral factor Θ. By definition $T_\Theta^\sharp T_\Theta \geq T_G^\sharp T_G$. Due to Part (iv) of Lemma 5.3.1, we have $\Theta(\infty)^*\Theta(\infty) \geq G(\infty)^*G(\infty)$. Because $G(\infty)$ is one to one, $\Theta(\infty)$ must be one to one. Recall that $\Theta(\infty)$ is onto. Hence Θ must be square.

Recall that G is a *spectral factor* for a Toeplitz matrix T_R if G is in $H^2(\mathcal{E},\mathcal{D})$ and $T_R = T_G^\sharp T_G$. It is noted that not all positive Toeplitz matrices admit an outer spectral factor. However, if T_R admits an outer spectral factor Θ, then Θ is also the maximal outer spectral factor for T_R. If R is a function in $L^\infty(\mathcal{E},\mathcal{E})$, then the corresponding Toeplitz matrix T_R defines an operator (by definition a bounded linear map) on $\ell_+^2(\mathcal{E})$, and $\|T_R\| = \|R\|_\infty$; see Proposition 2.5.1. In this case, T_R is positive if and only if $R \geq 0$ almost everywhere with respect to the Lebesgue measure. Finally, $G \in H^\infty(\mathcal{E},\mathcal{D})$ is a spectral factor for T_R if and only if $R = G^*G$ almost everywhere on the unit circle.

Corollary 7.4.3. *Let T_R be the Toeplitz matrix determined by a self-adjoint function R in $L^\infty(\mathcal{E},\mathcal{E})$. Let Δ_n be the Schur complements associated with the finite sections $T_{R,n}$ for T_R where $n \geq 1$. Then $T_R = T_\Theta^* T_\Theta$ where Θ is a square outer function if and only if $\Delta_n \geq \delta I$ for all n and some $\delta > 0$. In this case, the sequence $\{\Delta_n\}$ is decreasing and converges to the strictly positive operator $\Theta(\infty)^*\Theta(\infty)$. Furthermore, if $G \in L^\infty(\mathcal{E},\mathcal{E})$ is a spectral factor for T_R, that is, $T_R = T_G^* T_G$, then*

$$\Theta(\infty)^*\Theta(\infty) \geq G(\infty)^*G(\infty). \tag{7.4.4}$$

In this case, $\Theta(\infty)^*\Theta(\infty) = G(\infty)^*G(\infty)$ if and only if G is the outer spectral spectral factor for T_R.

Proof. If $T_R = T_\Theta^* T_\Theta$ where Θ is a square outer function, then Theorem 7.4.1 guarantees that $\Delta_n \geq \delta I$ for all n and some $\delta > 0$. Now assume that $\Delta_n \geq \delta I$ for all n and some $\delta > 0$. According to Theorem 7.4.1, the Toeplitz matrix T_R is positive and admits a maximal outer spectral factor Θ. It remains to show that $T_R = T_\Theta^* T_\Theta$. As before, let $\{U \text{ on } \mathcal{K},\Gamma\}$ be the controllable isometric representation for T_R, and W its corresponding controllability matrix; see (5.1.7). Because R is in $L^\infty(\mathcal{E},\mathcal{E})$, the Toeplitz matrix T_R defines an operator on $\ell_+^2(\mathcal{E})$; see Proposition 2.5.1. In particular, $\|T_R\| = \|R\|_\infty$. Since T_R is bounded and $T_R = W^\sharp W$, the controllability matrix W defines an operator from $\ell_+^2(\mathcal{E})$ onto a dense set in $\mathcal{K}$. Finally, it is noted that $W^\sharp = W^*$ and $T_R = W^*W$.

Since $0 < \delta I \leq \Delta_n$ for all integers $n \geq 1$, Lemma 7.2.1 guarantees that $T_{R,n}$ is strictly positive for all $n \geq 1$. Hence T_R is a positive operator. Moreover, the sequence Δ_n converges to a strictly positive operator Δ. We claim that $\ker T_R$ is zero. Let us proceed by contradiction, and assume that $T_R x = 0$ for some nonzero x in $\ell_+^2(\mathcal{E})$. Hence $(T_R x, x) = 0$. By using the band structure of the Toeplitz matrix T_R, we can assume without loss of generality, that $\Pi_\mathcal{E} x = f$ is nonzero.

The optimization problem in (7.3.7) shows that $0 = (T_R x, x) \geq (\Delta f, f)$. Since Δ is strictly positive f must be zero, which leads to a contradiction. Therefore $\ker T_R$ is zero. Since $T_R = W^*W$, this also implies $\ker W$ is also equal to zero.

Because the pair $\{U, \Gamma\}$ is controllable and W is one to one, W is a quasi-affinity. Notice that $UW = WS$ where S is the unilateral shift on $\ell_+^2(\mathcal{E})$. By consulting equation (5.2.2) in Theorem 5.2.1, we see that $\Theta(\infty) = \Pi_{\mathcal{Y}}\Gamma$ where Θ is the maximal outer spectral factor for T_R and $\mathcal{Y} = \ker U^*$. Theorem 7.4.1, shows that Θ is a square outer function, and thus, $\Theta(\infty)$ is invertible. So $\mathcal{Y}$ and $\mathcal{E}$ have the same dimension. According to Proposition 1.5.2, the isometries U and S are unitarily equivalent. In other words, U is a unilateral shift of multiplicity $\dim \mathcal{E}$. Therefore $T_R = T_\Theta^* T_\Theta$; see Theorem 5.2.1. The rest of this corollary follows directly from Theorem 7.4.1. □

If R is in $L^\infty(\mathcal{E}, \mathcal{E})$ and the corresponding Schur complements satisfy $\Delta_n \geq \delta I$ for all n and some $\delta > 0$, then T_R is not necessarily an invertible operator on $\ell_+^2(\mathcal{E})$. For example, consider the outer function $\theta(z) = (z-1)/z$ and set $R = |\theta|^2$. Then θ is a square outer factor for T_R. Since θ has a zero on the unit circle, T_θ is not an invertible operator on ℓ_+^2. Hence $T_R = T_\theta^* T_\theta$ is not invertible. Finally, it is noted that in this case, Δ_n converges to 1 as n tends to infinity.

If R is in $L^\infty(\mathcal{E}, \mathcal{E})$ and $T_{R,n}$ is strictly positive for all n, then T_R is positive. However, it does not necessarily follow that T_R admits a square outer spectral factorization. For example, let r be the function in L^∞ defined by $r(e^{\imath\omega}) = 1$ if $0 \leq \omega \leq \pi$ and $r = 0$ for all other ω. Since $r \geq 0$ almost everywhere, it follows that T_r defines a positive Toeplitz operator on $\ell_+^2(\mathcal{E})$; see Proposition 2.5.1. We claim that $T_{r,n}$ is strictly positive for all n and T_r does not admit an outer spectral factor θ. If $r = |\theta|^2$ for some θ in H^∞, then $\theta = 0$ almost everywhere on a set of positive Lebesgue measure, and thus, $\theta = 0$. (Recall that if f is a nonzero function in H^2, then $f(e^{\imath\omega}) \neq 0$ almost everywhere; see Chapter 5 in Hoffman [134].) So T_r does not admit an outer spectral factor. To complete the argument, it remains to show that T_r is one to one. Assume that $T_r x = 0$ for some x in ℓ_+^2. Recall that $T_r = P_+ L_r | \ell_+^2$ where L_r is the Laurent operator on ℓ^2 determined by r and P_+ is the orthogonal projection onto ℓ_+^2. This readily implies that $0 = P_+ L_r x$, and thus, $L_r x$ is in $\ell^2 \ominus \ell_+^2$. By taking the Fourier transform, $\mathcal{F} L_r x = r \mathcal{F} x$ is a function in $L^2 \ominus H^2$. The definition of r shows that $r(e^{\imath\omega})(\mathcal{F}x)(e^{\imath\omega}) = 0$ for $\pi < \omega \leq 2\pi$. Since $r(e^{-\imath\omega})(\mathcal{F}x)(e^{-\imath\omega})$ is in H^2 and equals zero on a set of positive Lebesgue measure, $r(e^{\imath\omega})(\mathcal{F}x)(e^{\imath\omega}) = 0$ almost everywhere. Recall that $r(e^{\imath\omega}) = 1$ for $0 \leq \omega \leq \pi$. So $\mathcal{F}x = 0$ on a set of positive Lebesgue measure. Since $\mathcal{F}x$ is in H^2, we must have $x = 0$. So the kernel of T_r equals zero. Therefore $T_{r,n}$ is strictly positive for all n and T_r does not admit an outer spectral factor. Finally, it is noted that the maximal outer spectral factor θ for T_r is the zero function $\theta = 0$ mapping $\mathbb{C}$ onto $\{0\}$. Moreover, the Schur complements Δ_n converge to zero as n tends to infinity.

7.5 Carathéodory Interpolation

Let T_R be a strictly positive Toeplitz operator on $\ell_+^2(\mathcal{E})$, and $R = \sum_{-\infty}^{\infty} e^{-\imath\omega n} R_n$ its symbol. Then $T_{R,n+1}$ is a strictly positive Toeplitz operator on $\mathcal{E}^{n+1}$. Here $T_{R,n+1} = \Pi_{\mathcal{E}^{n+1}} T_R | \mathcal{E}^{n+1}$ is the compression of T_R to $\mathcal{E}^{n+1}$. Now let $\{R_j\}_0^n$ be a $\mathcal{L}(\mathcal{E}, \mathcal{E})$-valued sequence of operators. Moreover, assume that the Toeplitz matrix

$$\Upsilon_{n+1} = \begin{bmatrix} R_0 & R_1^* & R_2^* & \cdots & R_n^* \\ R_1 & R_0 & R_1^* & \cdots & R_{n-1}^* \\ R_2 & R_1 & R_0 & \cdots & R_{n-2}^* \\ \vdots & \vdots & \vdots & \ddots & \vdots \\ R_n & R_{n-1} & R_{n-2} & \cdots & R_0 \end{bmatrix} \quad \text{on } \mathcal{E}^{n+1} \tag{7.5.1}$$

is strictly positive. One version of the *Carathéodory interpolation problem* is to find a strictly positive Toeplitz operator T_R on $\ell_+^2(\mathcal{E})$ such that $\Upsilon_{n+1} = T_{R,n+1}$, that is, $\Upsilon_{n+1} = \Pi_{\mathcal{E}^{n+1}} T_R | \mathcal{E}^{n+1}$ equals the compression of T_R to $\mathcal{E}^{n+1}$. In this case, T_R or its symbol R is called a *solution* to the Carathéodory interpolation problem for the data $\{R_j\}_0^n$. In this section, we will show that the Carathéodory interpolation problem always has a solution. In fact, we will construct a solution by computing a special invertible outer function Θ in $H^\infty(\mathcal{E}, \mathcal{E})$ such that $R = \Theta^*\Theta$ is a solution, that is, $\Upsilon_{n+1} = \Pi_{\mathcal{E}^{n+1}} T_{\Theta^*\Theta} | \mathcal{E}^{n+1}$.

For a *positive real formulation* of the Carathéodory interpolation problem, let $\{R_j\}_0^n$ be a $\mathcal{L}(\mathcal{E}, \mathcal{E})$-valued sequence of operators, such that Υ_{n+1} is strictly positive. Then find a $\mathcal{L}(\mathcal{E}, \mathcal{E})$-valued positive real function F of the form

$$F(z) = \frac{R_0}{2} + \sum_{k=1}^{n} z^{-k} R_k + \sum_{k=n+1}^{\infty} z^{-k} F_k.$$

In this case, $R = 2\Re F = F + F^*$ is a solution to this problem.

The solution to the Carathéodory interpolation problem is not unique. The set of all solutions is parameterized by the unit ball in the $H^\infty(\mathcal{E}, \mathcal{E})$ space; see [113] for further details. Finally, in Chapter 12 we will use isometric representations to solve a tangential Nevanlinna-Pick interpolation problem, which includes the Carathéodory interpolation problem as a special case.

Assume that the Toeplitz operator Υ_{n+1} on $\mathcal{E}^{n+1}$ is strictly positive; see (7.5.1). The *Levinson system* associated with Υ_{n+1} is defined by

$$\Upsilon_{n+1} \begin{bmatrix} I \\ A_1 \\ \vdots \\ A_n \end{bmatrix} = \begin{bmatrix} R_0 & R_1^* & \cdots & R_n^* \\ R_1 & R_0 & \cdots & R_{n-1}^* \\ \vdots & \vdots & \ddots & \vdots \\ R_n & R_{n-1} & \cdots & R_0 \end{bmatrix} \begin{bmatrix} I \\ A_1 \\ \vdots \\ A_n \end{bmatrix} = \begin{bmatrix} \Delta_{n+1} \\ 0 \\ \vdots \\ 0 \end{bmatrix}. \tag{7.5.2}$$

Here $\{A_j\}_0^n$ and Δ_{n+1} are operators on $\mathcal{E}$ with $A_0 = I$. There is a unique solution to this system of equations. In fact, the unique solution to the Levinson system

(7.5.2) is given by

$$\Delta_{n+1} = \left(\Pi_{\mathcal{E}}\Upsilon_{n+1}^{-1}\Pi_{\mathcal{E}}^*\right)^{-1} \quad \text{and} \quad \begin{bmatrix} I & A_1 & A_2 & \cdots & A_n \end{bmatrix}^{tr} = \Upsilon_{n+1}^{-1}\Pi_{\mathcal{E}}^*\Delta_{n+1}.$$

Here $\Pi_{\mathcal{E}} = \begin{bmatrix} I & 0 & 0 & \cdots & 0 \end{bmatrix}$ is the operator mapping $\mathcal{E}^{n+1}$ onto $\mathcal{E}$ which picks out the first component of $\mathcal{E}^{n+1}$. Observe that Δ_{n+1} is the Schur complement of Υ_{n+1} with respect to the operator R_0 contained in the upper left-hand corner of Υ_{n+1}. Moreover, one can use the Levinson algorithm to recursively compute the solution $\{A_j\}_0^n$ and Δ_{n+1}. The Levinson algorithm is discussed in Chapter 15. Finally, Υ_k denotes the Toeplitz matrix contained in the upper left-hand k by k corner of Υ_{n+1} when $k \leq n+1$. Motivated by Remark 7.1.3, we obtain the following solution to the Carathéodory interpolation problem.

Theorem 7.5.1. *Let Υ_{n+1} in (7.5.1) be a strictly positive Toeplitz matrix on $\mathcal{E}^{n+1}$ generated by a $\mathcal{L}(\mathcal{E},\mathcal{E})$-valued sequence of operators $\{R_j\}_0^n$. Let $\{A_j\}_0^n$ and Δ_{n+1} be the unique solution to the Levinson system in (7.5.2) with $A_0 = I$. Let Θ be the function defined by*

$$\begin{aligned}\Theta(z) &= \Delta_{n+1}^{1/2}\left(I + z^{-1}A_1 + z^{-2}A_2 + \cdots + z^{-(n-1)}A_{n-1} + z^{-n}A_n\right)^{-1} \\ &= z^n\Delta_{n+1}^{1/2}\left(A_n + A_{n-1}z + A_{n-2}z^2 + \cdots + A_1z^{n-1} + z^nI\right)^{-1}. \end{aligned} \tag{7.5.3}$$

Then Θ is a rational invertible outer function in $H^\infty(\mathcal{E},\mathcal{E})$. Moreover, $R = \Theta^*\Theta$ is a solution to the Carathéodory interpolation problem, that is, Υ_{n+1} equals the compression of T_R to $\mathcal{E}^{n+1}$. Furthermore, let A on $\mathcal{E}^n$ be the companion matrix, B the operator from $\mathcal{E}$ into $\mathcal{E}^n$ and C the operator from $\mathcal{E}^n$ into $\mathcal{E}$ be defined by

$$A = \begin{bmatrix} -A_1 & I & 0 & \cdots & 0 & 0 \\ -A_2 & 0 & I & \cdots & 0 & 0 \\ \vdots & \vdots & \vdots & \ddots & \vdots & \vdots \\ -A_{n-2} & 0 & 0 & \cdots & I & 0 \\ -A_{n-1} & 0 & 0 & \cdots & 0 & I \\ -A_n & 0 & 0 & \cdots & 0 & 0 \end{bmatrix}, \quad B = -\begin{bmatrix} A_1 \\ A_2 \\ \vdots \\ A_{n-2} \\ A_{n-1} \\ A_n \end{bmatrix},$$

$$C = \begin{bmatrix} \Delta_{n+1}^{1/2} & 0 & 0 & \cdots & 0 & 0 \end{bmatrix} \quad \text{and} \quad D = \Delta_{n+1}^{1/2}. \tag{7.5.4}$$

Then the following holds.

(i) *$\{A, B, C, D\}$ is an observable realization for $\Theta(z)$.*

(ii) *The operator Υ_n is the observability Gramian for the pair $\{C, A\}$, that is,*

$$\Upsilon_n = A^*\Upsilon_nA + C^*C. \tag{7.5.5}$$

(iii) *The operator A is stable, and*

$$\det[zI - A] = \det\left[A_n + A_{n-1}z + A_{n-2}z^2 + \cdots + A_1z^{n-1} + z^nI\right]. \tag{7.5.6}$$

(iv) *The Fourier coefficients* $R = \sum_{-\infty}^{\infty} R_k e^{-\imath\omega k}$ *are computed by*

$$\begin{aligned} R_0 &= B^*\Upsilon_n B + D^* D, \\ R_k &= (B^*\Upsilon_n A + D^* C)\, A^{k-1} B \qquad (k \geq 1). \end{aligned} \tag{7.5.7}$$

Proof. Let Ω and Γ be the $\mathcal{L}(\mathcal{E},\mathcal{E})$-valued polynomials defined by

$$\begin{aligned} \Omega(z) &= A_n + A_{n-1}z + A_{n-2}z^2 + \cdots + A_1 z^{n-1} + z^n I, \\ \Gamma(z) &= -(A_n + A_{n-1}z + A_{n-2}z^2 + \cdots + A_1 z^{n-1}). \end{aligned}$$

A simple calculation shows that

$$\Theta(z) = z^n \Delta_{n+1}^{1/2}\Omega(z)^{-1} = \Delta_{n+1}^{1/2}\Omega(z)^{-1}\left(\Omega(z) + \Gamma(z)\right) = \Delta_{n+1}^{1/2} + \Delta_{n+1}^{1/2}\Omega(z)^{-1}\Gamma(z).$$

Classical state space results in Section 14.3 show that $\{A, B, C, D\}$ is an observable realization for Θ, and (7.5.6) holds.

To show that Υ_n is the observability Gramian for $\{C, A\}$, notice that Υ_{n+1} admits a matrix decomposition of the form

$$\Upsilon_{n+1} = \begin{bmatrix} R_0 & X_n^* \\ X_n & \Upsilon_n \end{bmatrix} \text{ on } \begin{bmatrix} \mathcal{E} \\ \mathcal{E}^n \end{bmatrix} \tag{7.5.8}$$

where $X_n = \begin{bmatrix} R_1 & R_2 & \cdots & R_n \end{bmatrix}^{tr}$. In particular, the Schur complement

$$\Delta_{n+1} = R_0 - X_n^*\Upsilon_n^{-1}X_n = (\Pi_{\mathcal{E}}\Upsilon_{n+1}^{-1}\Pi_{\mathcal{E}}^*)^{-1}.$$

Using this decomposition, equation (7.5.2) implies that

$$\begin{aligned} X_n &= -\Upsilon_n \begin{bmatrix} A_1 & A_2 & \cdots & A_n \end{bmatrix}^{tr}, \\ R_0 - \Delta_{n+1} &= -X_n^* \begin{bmatrix} A_1 & A_2 & \cdots & A_n \end{bmatrix}^{tr}, \\ R_0 - \Delta_{n+1} &= \quad -\begin{bmatrix} A_1^* & A_2^* & \cdots & A_n^* \end{bmatrix} X_n. \end{aligned} \tag{7.5.9}$$

The last equation follows by taking the adjoint of the second equation. By combining the first and last equation, we arrive at

$$R_0 - \Delta_{n+1} = \begin{bmatrix} A_1^* & A_2^* & \cdots & A_n^* \end{bmatrix} \Upsilon_n \begin{bmatrix} A_1 & A_2 & \cdots & A_n \end{bmatrix}^{tr}. \tag{7.5.10}$$

Let V and Λ be the block matrices on $\mathcal{E}^n$ defined by

$$V = \begin{bmatrix} 0 & I & 0 & \cdots & 0 \\ 0 & 0 & I & \cdots & 0 \\ \vdots & \vdots & \vdots & \ddots & \vdots \\ 0 & 0 & 0 & \cdots & I \\ 0 & 0 & 0 & \cdots & 0 \end{bmatrix} \text{ and } \Lambda = \begin{bmatrix} -A_1 & 0 & \cdots & 0 \\ -A_2 & 0 & \cdots & 0 \\ \vdots & \vdots & \ddots & \vdots \\ -A_{n-1} & 0 & \cdots & 0 \\ -A_n & 0 & \cdots & 0 \end{bmatrix}. \tag{7.5.11}$$

The shift matrix V has the identity operator I immediately above the main diagonal and zeros elsewhere. The first column of Λ is $-\begin{bmatrix} A_1 & A_2 & \cdots & A_n \end{bmatrix}^{tr}$ and all the other columns of Λ are zero. By construction $A = \Lambda + V$. Observe that Υ_n admits a decomposition of the form

$$\Upsilon_n = \begin{bmatrix} R_0 & X_{n-1}^* \\ X_{n-1} & \Upsilon_{n-1} \end{bmatrix} \quad \text{and} \quad V^*\Upsilon_n V = \begin{bmatrix} 0 & 0 \\ 0 & \Upsilon_{n-1} \end{bmatrix}. \tag{7.5.12}$$

As expected, $X_{n-1} = \begin{bmatrix} R_1 & R_2 & \cdots & R_{n-1} \end{bmatrix}^{tr}$. By (7.5.9) and (7.5.10), we have

$$\begin{aligned} A^*\Upsilon_n A &= (V+\Lambda)^*\Upsilon_n(V+\Lambda) \\ &= V^*\Upsilon_n V + V^*\Upsilon_n\Lambda + \Lambda^*\Upsilon_n V + \Lambda^*\Upsilon_n\Lambda \\ &= \begin{bmatrix} 0 & 0 \\ 0 & \Upsilon_{n-1} \end{bmatrix} + \begin{bmatrix} 0 & 0 \\ X_{n-1} & 0 \end{bmatrix} \\ &\quad + \begin{bmatrix} 0 & X_{n-1}^* \\ 0 & 0 \end{bmatrix} + \begin{bmatrix} R_0 - \Delta_{n+1} & 0 \\ 0 & 0 \end{bmatrix} \\ &= \begin{bmatrix} R_0 - \Delta_{n+1} & X_{n-1}^* \\ X_{n-1} & \Upsilon_{n-1} \end{bmatrix} = \Upsilon_n - C^*C. \end{aligned}$$

Hence $\Upsilon_n = A^*\Upsilon_n A + C^*C$. In other words, Υ_n is the observability Gramian for $\{C, A\}$. Because the pair $\{C, A\}$ is observable and Υ_n is strictly positive, A is stable. Therefore Parts (i), (ii) and (iii) hold. Equation (7.5.6) follows from (14.3.5) in Chapter 14.

Since A is stable and $\Theta = D + C(zI - A)^{-1}B$, we see that Θ is a rational function in $H^\infty(\mathcal{E}, \mathcal{E})$. Because Θ^{-1} is a polynomial in $1/z$, the function Θ^{-1} is also a rational function in $H^\infty(\mathcal{E}, \mathcal{E})$. Therefore Θ is an invertible outer function.

To complete the proof, it remains to show that Υ_{n+1} is the compression of T_R to $\mathcal{E}^{n+1}$. Let $R = \sum_{-\infty}^{\infty} e^{-\imath\omega k} Q_k$ be the Fourier series expansion for R. Recall that $\{A, B, C, D\}$ is a realization for Θ. By employing Lemma 4.5.4, with $P = \Upsilon_n$, we have

$$\begin{aligned} Q_0 &= B^*\Upsilon_n B + D^*D, \\ Q_k &= (B^*\Upsilon_n A + D^*C)\, A^{k-1}B \qquad (k \geq 1). \end{aligned} \tag{7.5.13}$$

To verify that Υ_{n+1} is the compression of T_R to $\mathcal{E}^{n+1}$, it suffices to show that $Q_k = R_k$ for $0 \leq k \leq n$. According to (7.5.10), we see that

$$R_0 - \Delta_{n+1} = R_0 - D^*D = B^*\Upsilon_n B.$$

So the first equation in (7.5.7) holds. By taking the adjoint in (7.5.2) along with the definition of B, we obtain

$$\begin{bmatrix} I & -B^* \end{bmatrix} \begin{bmatrix} R_0 & X_n^* \\ X_n & \Upsilon_n \end{bmatrix} = \begin{bmatrix} \Delta_{n+1} & 0 \end{bmatrix}.$$

In particular, $B^*\Upsilon_n = X_n^*$. For $1 \le k \le n$, the equations in (7.5.2) yield

$$\begin{aligned}
(B^*\Upsilon_n A + D^*C)\, A^{k-1}B &= (X_n^* A + D^*C)\, A^{k-1}B \\
&= (X_n^*(\Lambda + V) + D^*C)\, A^{k-1}B \\
&= \begin{bmatrix} R_0 & R_1^* & R_2^* & \cdots & R_{n-1}^* \end{bmatrix} A^{k-1}B \\
&= \begin{bmatrix} R_0 & R_1^* & R_2^* & \cdots & R_{n-1}^* \end{bmatrix} (\Lambda + V)A^{k-2}B \\
&= \begin{bmatrix} R_1 & R_0 & R_1^* & \cdots & R_{n-2}^* \end{bmatrix} (\Lambda + V)A^{k-3}B \\
&= \begin{bmatrix} R_2 & R_1 & R_0 & \cdots & R_{n-3}^* \end{bmatrix} (\Lambda + V)A^{k-4}B \\
&\ \vdots \\
&= \begin{bmatrix} R_{k+1} & R_k & \cdots & R_{n-k-2}^* \end{bmatrix} B = R_k.
\end{aligned}$$

For example, to obtain the fifth equality we used the second row in (7.5.2), that is, $-R_1 = R_0 A_1 + R_1^* A_2 + \cdots R_{n-1}^* A_n$. Continuing in this fashion yields the previous equation. □

7.6 A Finite Sections Approach to Factorization

Recall that if $\mathcal{H}_1$ and $\mathcal{H}_2$ are two subspaces of $\mathcal{K}$, then $\mathcal{H}_1 \subseteq \mathcal{H}_2$ if and only if $P_{\mathcal{H}_1} \le P_{\mathcal{H}_2}$. (The orthogonal projection onto $\mathcal{H}$ is denoted by $P_{\mathcal{H}}$.) Let $\{\mathcal{H}_k\}_1^\infty$ be an increasing sequence of subspaces ($\mathcal{H}_k \subseteq \mathcal{H}_{k+1}$ for all integers $k \ge 1$) such that $\bigvee_1^\infty \mathcal{H}_k = \mathcal{K}$. Then $P_{\mathcal{H}_k}$ converges to the identity I in the strong operator topology as k tends to infinity. To be precise, $P_{\mathcal{H}_k} x$ converges to x for each vector x in $\mathcal{K}$.

The following result shows that the inverse of the finite sections of a strictly positive operator converge to the inverse in the strong operator topology.

Lemma 7.6.1. *Let T be a strictly positive operator on $\mathcal{H}$. Let $P_{\mathcal{H}_k}$ be a sequence of orthogonal projections onto a set of subspaces $\{\mathcal{H}_k\}$ for all integers $k \ge 1$. Assume that $P_{\mathcal{H}_k}$ converges to the identity I in the strong operator topology as k tends to infinity. Let T_k be the operator on $\mathcal{H}_k$ obtained by compressing T to $\mathcal{H}_k$, that is, $T_k = P_{\mathcal{H}_k} T|\mathcal{H}_k$. Then*

$$T^{-1}g = \lim_{k\to\infty} T_k^{-1} P_{\mathcal{H}_k} g \qquad (\textit{for all } g \in \mathcal{H}).$$

In other words, the inverse of T_k converges to the inverse of T is the strong operator topology.

Proof. Since T is strictly positive, there exists a scalar $\delta > 0$ such that $T \ge \delta I$. Hence $T_k \ge \delta I$ for all integers $k \ge 1$. Let $h_k \in \mathcal{H}_k$ be the unique solution to $T_k h_k = P_{\mathcal{H}_k} g$, and $h \in \mathcal{H}$ the unique solution to $Th = g$. To complete the proof it remains to show that h_k converges to h. To this end, observe that

$$\begin{aligned}\|h_k - P_{\mathcal{H}_k} h\| &= \|T_k^{-1} T_k (h_k - P_{\mathcal{H}_k} h)\| \leq \delta^{-1} \|T_k (h_k - P_{\mathcal{H}_k} h)\| \\ &= \delta^{-1} \|P_{\mathcal{H}_k} g - T_k P_{\mathcal{H}_k} h\| = \delta^{-1} \|P_{\mathcal{H}_k} g - P_{\mathcal{H}_k} T P_{\mathcal{H}_k} h\| \\ &\leq \delta^{-1} \|g - T P_{\mathcal{H}_k} h\| \to \delta^{-1} \|g - Th\| = 0.\end{aligned}$$

In other words, $\|h_k - P_{\mathcal{H}_k} h\|$ converges to zero. Since $P_{\mathcal{H}_k} h$ converges to h, the triangle inequality implies that h_k converges to h. □

7.6.1 A finite section approach to outer factorization

Let T_R be a strictly positive Toeplitz operator on $\ell_+^2(\mathcal{E})$ of the form

$$T_R = \begin{bmatrix} R_0 & R_1^* & R_2^* & \cdots \\ R_1 & R_0 & R_1^* & \cdots \\ R_2 & R_1 & R_0 & \cdots \\ \vdots & \vdots & \vdots & \ddots \end{bmatrix} \quad \text{on } \ell_+^2(\mathcal{E}). \tag{7.6.1}$$

According to Remark 7.1.3, the operator T_R admits an invertible outer spectral factor Θ. Moreover, this outer spectral factor is given by

$$\begin{aligned}\Theta(z) &= \Delta^{1/2} \left(I + z^{-1} A_1 + z^{-2} A_2 + z^{-3} A_3 + \cdots\right)^{-1}, \\ T_R \begin{bmatrix} I & A_1 & A_2 & A_3 & \cdots \end{bmatrix}^{tr} &= \begin{bmatrix} \Delta & 0 & 0 & 0 & \cdots \end{bmatrix}^{tr}.\end{aligned} \tag{7.6.2}$$

Let us use Lemma 7.6.1, to compute an approximation Ψ_k to the outer spectral factor Θ for T_R. To this end, let $T_{R,k} = \Pi_{\mathcal{E}^k} T_R | \mathcal{E}^k$ on $\mathcal{E}^k$ be the $k \times k$ Toeplitz matrix contained in the upper left-hand corner of T_R. Now consider the function defined by

$$\begin{aligned}\Psi_k(z) &= \Delta_k^{1/2} \left(I + z^{-1} A_{k,1} + z^{-2} A_{k,2} + \cdots + z^{-(k-1)} A_{k,k-1}\right)^{-1}, \\ T_{R,k} \begin{bmatrix} I & A_{k,1} & A_{k,2} & \cdots & A_{k,k-1} \end{bmatrix}^{tr} &= \begin{bmatrix} \Delta_k & 0 & 0 & \cdots & 0 \end{bmatrix}^{tr}.\end{aligned} \tag{7.6.3}$$

Recall that Δ is the Schur complement for T_R with respect to R_0 in the upper left-hand corner of T_R, while Δ_k is the Schur complement for $T_{R,k}$ with respect to R_0 in the upper left-hand corner of $T_{R,k}$; see Section 7.2. The Levinson algorithm presented in Chapter 15 can be used to recursively compute $\{A_{k,j}\}$ and Δ_k. Theorem 7.5.1 shows that Ψ_k is an invertible outer function. Recall that

$$\begin{aligned}\Delta &= (\Pi_{\mathcal{E}} T_R^{-1} \Pi_{\mathcal{E}}^*)^{-1}, \\ \Delta_k &= (\Pi_{\mathcal{E}} T_{R,k}^{-1} \Pi_{\mathcal{E}}^*)^{-1}, \\ T_R^{-1} \Pi_{\mathcal{E}}^* \Delta &= \begin{bmatrix} I & A_1 & A_2 & A_3 & \cdots \end{bmatrix}^{tr}, \\ T_{R,k}^{-1} \Pi_{\mathcal{E}}^* \Delta_k &= \begin{bmatrix} I & A_{k,1} & A_{k,2} & \cdots & A_{k,k-1} \end{bmatrix}^{tr}.\end{aligned} \tag{7.6.4}$$

By employing the finite section inversion Lemma 7.6.1, we see that Δ_k converges to Δ and $T_{R,k}^{-1}\Pi_{\mathcal{E}}^*\Delta_k$ converges to $T_R^{-1}\Pi_{\mathcal{E}}^*\Delta$. In fact, according to the results in Section 7.2, the operators Δ_k are monotonically decreasing and converge to $\Delta = \Theta(\infty)^*\Theta(\infty)$. By taking the Fourier transform, Ψ_k^{-1} converges to Θ^{-1} in the $H^2(\mathcal{E}, \mathcal{E})$ topology. Therefore $\Psi_k(z)$ converges to $\Theta(z)$ uniformly on $\{z \in \mathbb{C} : |z| > r\}$ where r is any fixed positive scalar such that $r > 1$.

Now assume that the Toeplitz matrix T_R is strictly positive and R is rational. Then T_R admits a rational invertible outer spectral factor Θ in $H^\infty(\mathcal{E}, \mathcal{E})$; see Section 6.1. Let $\Theta(z) = \sum_0^\infty z^{-n}\Theta_n$ be the Taylor series expansion for Θ. If one knows the Taylor coefficients $\{\Theta_n\}_0^{2j-1}$ where $j > \delta(\Theta)$ the McMillan degree of Θ, then one can use the Kalman-Ho algorithm to compute the minimal realization $\{A, B, C, D\}$ for Θ; see Section 14.5 for a review of the Kalman-Ho algorithm.

Remark 7.6.2. One can also compute a minimal realization for Θ directly from $\{A_n\}_0^{2j-1}$ and Δ. To see this, recall that Θ^{-1} is a rational outer function. Remark 14.2.1 shows that Θ and Θ^{-1} have the same McMillan degree. By consulting (7.6.2), we see that

$$\Theta(z)^{-1} = \left(I + \sum_{n=1}^{\infty} z^{-n}A_n\right)\Delta^{-1/2}. \tag{7.6.5}$$

Let $\{\widehat{A}, \widehat{B}, \widehat{C}, I\}$ be a minimal realization for $\sum_0^\infty z^{-n}A_n = \Theta^{-1}\Delta^{1/2}$ where $A_0 = I$. Then $\{\widehat{A}, \widehat{B}\Delta^{-1/2}, \widehat{C}, \Delta^{-1/2}\}$ is a minimal realization for Θ^{-1}. According to Remark 14.2.1, a minimal realization $\{A, B, C, D\}$ for Θ is given by

$$\begin{aligned} \Theta(z) = D + C(zI - A)^{-1}B \quad &\text{where} \quad A = \widehat{A} - \widehat{B}\widehat{C}, \\ B = \widehat{B}, \quad C = -\Delta^{1/2}\widehat{C} \quad &\text{and} \quad D = \Delta^{1/2}. \end{aligned} \tag{7.6.6}$$

So if one knows Δ and $\{A_n\}_0^{2j-1}$ where $j > \delta(\Theta)$, then one can apply the Kalman-Ho algorithm directly to $\{A_n\}_0^{2j-1}$ to compute a minimal realization $\{\widehat{A}, \widehat{B}, \widehat{C}, I\}$ for $\Theta^{-1}\Delta^{1/2}$. Then $\{A, B, C, D\}$ in (7.6.6) is a minimal realization for Θ.

According to Theorem 7.5.1, the invertible outer function Ψ_k admits an observable realization of the form $\{A \text{ on } \mathcal{X}, B, C, D\}$ where $\dim \mathcal{X} = (k-1)\times\dim(\mathcal{E})$; see (7.5.4). In general this realization is controllable and observable. So in many problems the McMillan degree of Ψ_k approaches infinity as k tends to infinity. Recall that the McMillan degree of Θ is finite. So it may appear that Ψ_k is not very useful in computing a state space realization for Θ. However, since Ψ_k converges to Θ uniformly on compact sets in $\mathbb{D}$, we can apply the Kalman-Ho algorithm to the Fourier coefficients of Ψ_k^{-1} (or even Ψ_k) to find an approximate realization for Θ. To see this, for large k one can run the Kalman-Ho algorithm on $\{A_{k,j}\}_{j=0}^{k-1}$ where $A_{k,0} = I$ and $k > 2\delta(\Theta)$. The McMillan degree of Θ does not have to be known a priori. By throwing out all the small singular values in the Hankel matrix corresponding to $\{A_{k,j}\}$, compute a minimal realization $\{\widehat{A}, \widehat{B}, \widehat{C}, I\}$ for the data

$\{A_{k,j}\}$. Then compute the realization $\{A, B, C, D\}$ in (7.6.6). For k sufficiently large, $\Theta(z) \approx D + C(zI - A)^{-1}B$.

To see why this works, observe that the Hankel matrix in the Kalman-Ho algorithm corresponding to $\{A_{k,j}\}$ is given by

$$H_k = \begin{bmatrix} A_{k,1} & A_{k,2} & \cdots & A_{k,j} \\ A_{k,2} & A_{k,3} & \cdots & A_{k,j+1} \\ \vdots & \vdots & \vdots & \vdots \\ A_{k,j} & A_{k,j+1} & \cdots & A_{k,2j-1} \end{bmatrix} \text{ and set } H = \begin{bmatrix} A_1 & A_2 & \cdots & A_j \\ A_2 & A_3 & \cdots & A_{j+1} \\ \vdots & \vdots & \vdots & \vdots \\ A_j & A_{j+1} & \cdots & A_{2j-1} \end{bmatrix}.$$

Now assume that j is fixed and $j > \delta(\Theta)$. For large k the Hankel matrix H_k will have $\delta(\Theta)$ nonzero singular values and all the other singular values will be zero. Because $T_{R,k}^{-1}\Pi_{\mathcal{E}}^*\Delta_k$ converges to $T_R^{-1}\Pi_{\mathcal{E}}^*\Delta$ in the operator topology, H_k converges to H in the operator norm; see (7.6.4). Since H_k is the Hankel matrix in the Kalman-Ho algorithm corresponding to the data $\{A_{k,\nu}\}_{\nu=0}^{2j-1}$ for $\Psi_k^{-1}\Delta_k^{1/2}$, we see that

$$\Theta(z)^{-1}\Delta^{1/2} \approx I + \widehat{C}(zI - \widehat{A})^{-1}\widehat{B}$$

where $\{\widehat{A}, \widehat{B}, \widehat{C}, I\}$ is computed by using the Kalman-Ho algorithm on the data $\{A_{k,\nu}\}_{\nu=0}^{2j-1}$. Hence for k sufficiently large, $\Theta(z) \approx D + C(zI - A)^{-1}B$.

Remark 7.6.3. One can also compute an approximate state space realization for the outer spectral factor Θ by applying the Kalman-Ho algorithm directly to the Taylor coefficients $\{\Psi_{k,\nu}\}$ in the Taylor series expansion for

$$\Psi_k(z) = \sum_{\nu=0}^{\infty} z^{-\nu}\Psi_{k,\nu}.$$

To compute the Taylor coefficients $\{\Psi_{k,\nu}\}$ by using state space methods, recall that transfer function $\Psi_k(z) = D_k + C_k(zI - A_k)^{-1}B_k$ where A_k, B_k, C_k and D_k are determined by (7.5.4). According to Theorem 7.5.1, the Taylor coefficients $\Psi_{k,0} = D_k$ and $\Psi_{k,\nu} = C_k A_k^{\nu-1}B_k$ for all integers $\nu \geq 1$. Notice that this method requires multiplying matrices with many zero entries to compute $\{\Psi_{k,\nu}\}$. If these matrices are large, this method may be inefficient. However, the fast Fourier transform is an efficient method to compute $\{\Psi_{k,\nu}\}$, especially in the scalar setting. Once one obtains $\{\Psi_{k,\nu}\}$, run the Kalman-Ho algorithm on the data $\{\Psi_{k,\nu}\}_{\nu=0}^{2j-1}$ where $j > \delta(\Theta)$ and k is sufficiently large to compute a minimal realization $\{A, B, C, D\}$ for Ψ_k. Then $\Theta(z) \approx D + C(zI - A)^{-1}B$.

7.7 An Inner-Outer Factorization Procedure

In this section, we will use the finite section inversion method in Lemma 7.6.1 to compute the inner-outer factorization for a rational transfer function G in $H^\infty(\mathcal{E}, \mathcal{Y})$. To be precise, assume that G admits an inner-outer factorization of the form $G = G_i G_o$ where G_o is an invertible outer function in $H^\infty(\mathcal{E},\mathcal{E})$ and G_i is an inner function in $H^\infty(\mathcal{E},\mathcal{Y})$. To compute G_o set $R = G^*G = G_o^*G_o$. Now use the state space techniques in Lemma 4.5.4 or the fast Fourier transform, to compute the Fourier coefficients $\{R_n\}_0^\infty$ for $R = \sum_{-\infty}^{\infty} e^{-\imath\omega n} R_n$. Then invert $T_{R,k}$ and use the Kalman-Ho algorithm to compute a minimal realization $\{A_o, B_o, C_o, D_o\}$ for Ψ_k. For k sufficiently large, $\{A_o, B_o, C_o, D_o\}$ can be used as a realization for $\Theta = G_o$ the outer spectral factor for T_R.

To compute a realization for the inner factor G_i, observe that

$$G_i = G G_o^{-1} \approx G\left(I + z^{-1}A_{k,1} + z^{-2}A_{k,2} + \cdots + z^{-(k-1)}A_{k,k-1}\right)\Delta_k^{-1/2}.$$

Now there are several different approaches to compute G_i. First one can use standard fast Fourier transform techniques to compute the Taylor coefficients $\{G_{i,n}\}$ for $G_i = \sum_{n=0}^{\infty} z^{-n} G_{i,n}$. For another method, recall that convolution in the time domain corresponds to multiplication in the z domain. Hence $\{G_{i,n}\}_{n=0}^{\infty}$ can also be computed by the matrix multiplication

$$\begin{bmatrix} G_{i,0} \\ G_{i,1} \\ G_{i,2} \\ G_{i,3} \\ \vdots \end{bmatrix} \approx T_G \begin{bmatrix} A_{k,0} \\ A_{k,1} \\ \vdots \\ A_{k,k-1} \\ 0 \\ 0 \\ \vdots \end{bmatrix} \Delta_k^{-1/2} \tag{7.7.1}$$

where $A_{k,0} = I$ and T_G is the lower triangular Toeplitz matrix determined by the Taylor coefficients $\{G_n\}_0^\infty$ for $G = \sum_0^\infty z^{-n} G_n$. To obtain $\{G_n\}_0^\infty$ by state space techniques, let $\{A, B, C, D\}$ be any realization for G, then $G_0 = D$ and $G_n = CA^{n-1}B$ for all integers $n \geq 0$. (In fact, one can use the Matlab command "dlsim" to compute $\{G_{i,n}\}$ in (7.7.1).) Now run the Kalman-Ho algorithm on $\{G_{i,n}\}_{n=0}^{2j-1}$ for j sufficiently large to compute a minimal realization $\{A_i, B_i, C_i, D_i\}$ for G_i.

One can also use state space techniques to compute a realization for G_i. To this end, let $\{A, B, C, D\}$ be a minimal realization for G, and $\{A_o, B_o, C_o, D_o\}$ be the minimal realization for its outer spectral factor G_o. Recall that

$$\{A_o - B_o D_o^{-1} C_o, B_o D_o^{-1}, -D_o^{-1} C_o, D_o^{-1}\}$$

is a realization for G_o^{-1}; see Remark 14.2.1. Using $G_i = G G_o^{-1}$ with Remark 14.2.2, we see that a realization $\{A_i, B_i, C_i, D_i\}$ for G_i is given by

$$A_i = \begin{bmatrix} A & -BD_o^{-1}C_o \\ 0 & A_o - B_o D_o^{-1} C_o \end{bmatrix}, \quad B_i = \begin{bmatrix} B \\ B_o \end{bmatrix} D_o^{-1},$$
$$C_i = \begin{bmatrix} C & -DD_o^{-1}C_o \end{bmatrix} \quad \text{and} \quad D_i = DD_o. \tag{7.7.2}$$

This realization may not be minimal. One may have to apply standard model reduction techniques on $\{A_i, B_i, C_i, D_i\}$ (the Matlab command is minreal) to find a minimal realization for G_i. Finally, one can also use the Kalman-Ho algorithm on the Taylor coefficients of G_i to compute a minimal realization for G_i.

Example. Consider the rational transfer function g in H^∞ given by

$$g = \frac{1.1909z^3 + 0.8735z^2 - 0.5210z + 0.0492}{z^7 + 0.1211z^6 - 0.3788z^5 - 0.2342z^4 + 0.0222z^3 + 0.0408z^2 + 0.0025z - 0.0011}.$$

Let $g = g_i g_o$ denote the inner-outer factorization for g where g_o is outer and g_i is inner. By keeping only three significant singular vales in the Kalman-Ho algorithm, for computing the outer part, we obtained

$$g_o(z) = \frac{1.365z^3 + 0.64z^2 - 0.2777z + 0.1551}{z^3 + 0.1308z^2 - 0.2608z - 0.1922}.$$

The singular values for the 500×500 Hankel matrix corresponding to g_o are

$$\{0.9152,\ 0.5789,\ 0.2881,\ 0.0108,\ 0.0016,\ 0,\ \cdots\}.$$

Running the Kalman-Ho algorithm on $\{g_{i,n}\}_{n=0}^{500}$ and keeping five singular values, we arrived at the inner function

$$g_i(z) = \frac{0.8722z + 1}{z^4\,(z + 0.8722)}. \tag{7.7.3}$$

In fact, the singular values for the 500×500 Hankel matrix corresponding to the g_i are $\{1, 1, 1, 1, 1, 0, \ldots\}$. Using the fast Fourier transform, $\|g\|_\infty = 2.7818$ and $\|g - g_i g_o\|_\infty = 0.014$. On the surface, it appears that $g_i g_o$ is eighth order, while g is seventh order. However, there is an approximate pole zero cancellation in $g_i g_o$. One can obtain a more accurate approximation of the inner and outer factors by keeping more singular values in the Kalman-Ho algorithm for g_o. In fact, keeping six singular values, we arrive at $\|g - g_i g_o\|_\infty = 1.5027 \times 10^{-8}$ where the McMillan degree of g_o is now six.

A typical procedure for computing the inner-outer factorization for $g = p/q$ in Matlab is given by the following steps.

(i) Set $g = \text{fft}(p, 2 \wedge 13)./\text{fft}(q, 2 \wedge 13)$. Compute $R = \text{abs}(g). \wedge 2$. Set $Rn = \text{real}(\text{ifft}(R))$. The vector $Rn(1 : 2 \wedge 12)$ contains the first 2^{12} Fourier coefficients for $R = |g|^2$. Finally, it is noted that in Matlab,

$$p = [0, 0, 0, 0, 1.1909, 0.8735, -0.5210, 0.0492];$$
$$q = [1, 0.1211, -0.3788, -0.2342, 0.0222, 0.0408, 0.0025, -0.0011];$$

One must include the zeros in p.

(ii) Now invert the Toeplitz matrix $T_{R,k}$ for k sufficiently large. In Matlab $\begin{bmatrix} a & e \end{bmatrix} = \text{levinson}(Rn(1 : 1000))$. The vector a corresponds to $\{A_{k,n}\}$ and e to Δ_k.

(iii) Compute $g_o = \sum_0^\infty z^{-n} g_{o,n}$. In Matlab $go = \mathrm{sqrt}(e)./\mathrm{fft}(a, 2 \wedge 13)$; and $gn = \mathrm{real}(\mathrm{ifft}(go))$. Then $gn(1 : 2 \wedge 12)$ contains the first 2^{12} Fourier coefficients of g_o.

(iv) Run the Kalman-Ho algorithm on $gn(1{:}500)$. Select the appropriate number of significant singular values to compute the realization $\{A, B, C, D\}$ for g_o.

(v) Compute g_i. In Matlab, compute $g_i = g./go$. Set $g_{ni} = \mathrm{real}(\mathrm{ifft}(g_i))$. Then $g_i(1 : 2 \wedge 12)$ contains the first 2^{12} Fourier coefficients of g_i.

(vi) Run the Kalman-Ho algorithm on $g_{ni}(1{:}500)$ to compute the realization $\{A_i, B_i, C_i, D_i\}$ for g_i.

There is nothing magical about 1000 for the Levinson or 500 for the Kalman-Ho. Certainly these numbers can be much smaller, or even larger depending on the problem. We choose these numbers to demonstrate that this kind of an algorithm works well for large numbers. Finally, by making minor modifications, the previous algorithm can be converted to compute the inner-outer factorization for a rational function G in $H^\infty(\mathcal{E}, \mathcal{Y})$ when the outer factor is an invertible outer function. The details are left to the reader as a simple exercise.

Let H be the Hankel matrix corresponding to a two-sided inner function Θ. Then H has $\delta(\Theta)$ singular values equal to 1 and all the other singular values are zero; see Remark 4.2.3. So it is not surprising that the singular values for the 500×500 Hankel matrix corresponding to the inner function g_i in (7.7.3) are $\{1, 1, 1, 1, 1, 0, 0, \ldots\}$ and $\delta(g_i) = 5$. In other words, numerically, there are $\delta(g_i)$ nonzero singular values for this Hankel matrix and these singular values are all one. This happens for any square rational inner function once the finite section of the corresponding Hankel matrix becomes large enough.

7.8 Rational Contractive Analytic Functions

We say that G is a *contractive analytic function* if G is a function in $H^\infty(\mathcal{E}, \mathcal{Y})$ and $\|G\|_\infty \leq 1$. In other words, a function G in $H^\infty(\mathcal{E}, \mathcal{Y})$ is a contractive analytic function if and only if its corresponding Toeplitz matrix T_G defines a contraction mapping $\ell^2_+(\mathcal{E})$ into $\ell^2_+(\mathcal{Y})$.

We say that $\{A \text{ on } \mathcal{X}, B, C, D\}$ is a *contractive realization* if its system matrix

$$\Omega = \begin{bmatrix} A & B \\ C & D \end{bmatrix} : \begin{bmatrix} \mathcal{X} \\ \mathcal{E} \end{bmatrix} \to \begin{bmatrix} \mathcal{X} \\ \mathcal{Y} \end{bmatrix} \tag{7.8.1}$$

is a contraction. An isometric realization studied in Section 4.2 is a special case of a contractive realization. In Chapter 13 we will show that G is a contractive analytic function if and only if G admits a contractive realization. It is noted that if $\{A, B, C, D\}$ is a minimal realization for a contractive analytic function, then Ω is not necessarily a contraction. For example, $\{0, 2, 1/2, 0\}$ is a minimal realization for $1/z$ and $\|\Omega\| = 2$. In this section we will concentrate on rational contractive analytic functions and their contractive realizations.

Let $\{A \text{ on } \mathcal{X}, B, C, D\}$ be a contractive realization for a function G. The state space $\mathcal{X}$ can be finite or infinite dimensional. Let T_G be the lower triangular Toeplitz matrix corresponding to the Taylor series expansion for $G(z) = \sum_0^\infty z^{-n} G_n$, and W_o the observability matrix determined by the pair $\{C, A\}$. To be precise,

$$T_G = \begin{bmatrix} G_0 & 0 & 0 & \cdots \\ G_1 & G_0 & 0 & \cdots \\ G_2 & G_1 & G_0 & \cdots \\ \vdots & \vdots & \vdots & \ddots \end{bmatrix} \quad \text{and} \quad W_o = \begin{bmatrix} C \\ CA \\ CA^2 \\ \vdots \end{bmatrix}. \tag{7.8.2}$$

Then we claim that

$$\Xi = \begin{bmatrix} W_o & T_G \end{bmatrix} : \begin{bmatrix} \mathcal{X} \\ \ell_+^2(\mathcal{E}) \end{bmatrix} \to \ell_+^2(\mathcal{Y}) \tag{7.8.3}$$

is a contraction. In particular, T_G is a contraction mapping $\ell_+^2(\mathcal{E})$ into $\ell_+^2(\mathcal{Y})$. In other words, if G admits a contractive realization, then G is a contractive analytic function.

To show that Ξ is a contraction consider the state space system

$$\begin{bmatrix} x(n+1) \\ y(n) \end{bmatrix} = \begin{bmatrix} A & B \\ C & D \end{bmatrix} \begin{bmatrix} x(n) \\ u(n) \end{bmatrix}. \tag{7.8.4}$$

Here the state $x(n)$ is in $\mathcal{X}$, the input $u(n)$ is in $\mathcal{E}$ and the output $y(n)$ is in $\mathcal{Y}$ for all integers $n \geq 0$. Using the fact that the systems matrix Ω is a contraction, we have

$$\begin{aligned} \|y(0)\|^2 + \|x(1)\|^2 &\leq \|x(0)\|^2 + \|u(0)\|^2, \\ \|y(1)\|^2 + \|x(2)\|^2 &\leq \|x(1)\|^2 + \|u(1)\|^2, \\ &\vdots \\ \|y(n)\|^2 + \|x(n+1)\|^2 &\leq \|x(n)\|^2 + \|u(n)\|^2. \end{aligned}$$

Summing up the previous inequalities yields

$$\sum_{j=0}^{n} \|y(j)\|^2 + \sum_{j=1}^{n+1} \|x(j)\|^2 \leq \sum_{j=0}^{n} \|x(j)\|^2 + \sum_{j=0}^{n} \|u(j)\|^2.$$

By eliminating the terms $\{\|x(j)\|^2\}_1^n$, we obtain

$$\sum_{j=0}^{n} \|y(j)\|^2 \leq \sum_{j=0}^{n} \|y(j)\|^2 + \|x(n+1)\|^2 \leq \|x(0)\|^2 + \sum_{j=0}^{\infty} \|u(j)\|^2.$$

If $\{u(j)\}_0^\infty$ is any square summable sequence, then we have

$$\sum_{j=0}^{\infty} \|y(j)\|^2 \leq \|x(0)\|^2 + \sum_{j=0}^{\infty} \|u(j)\|^2. \tag{7.8.5}$$

Recall that the solution to the state space system in (7.8.4) is given by

$$\begin{bmatrix} y(0) \\ y(1) \\ y(2) \\ \vdots \end{bmatrix} = \begin{bmatrix} C \\ CA \\ CA^2 \\ \vdots \end{bmatrix} x(0) + \begin{bmatrix} G_0 & 0 & 0 & \cdots \\ G_1 & G_0 & 0 & \cdots \\ G_2 & G_1 & G_0 & \cdots \\ \vdots & \vdots & \vdots & \ddots \end{bmatrix} \begin{bmatrix} u(0) \\ u(1) \\ u(2) \\ \vdots \end{bmatrix}, \tag{7.8.6}$$

where $G(z) = \sum_0^\infty z^{-n} G_n$ is the Taylor series expansion for G; see equation (14.1.5) in Section 14.1. By consulting (7.8.5), we see that

$$\|W_o x(0) + T_G(\oplus_0^\infty u(j))\|^2 = \sum_{j=0}^{\infty} \|y(j)\|^2 \leq \|x(0)\|^2 + \sum_{j=0}^{\infty} \|u(j)\|^2.$$

Therefore Ξ is a well-defined contraction mapping $\mathcal{X} \oplus \ell_+^2(\mathcal{E})$ into $\ell_+^2(\mathcal{Y})$.

In summary, if $\{A, B, C, D\}$ is a contractive realization for a transfer function G, then G is a contractive analytic function. Recall that two similar realizations have the same transfer function. So if $\{A, B, C, D\}$ is similar to a contractive realization, then its transfer function is also a contractive analytic function. This proves part of the following result known as the *bounded real lemma* in systems theory.

Theorem 7.8.1. *Let $\{A$ on $\mathcal{X}, B, C, D\}$ be a minimal realization for a $\mathcal{L}(\mathcal{E}, \mathcal{Y})$-valued rational transfer function G. Then the following statements are equivalent.*

(i) *G is a contractive analytic function.*

(ii) *The operator A is stable and*

$$\begin{bmatrix} A^* & C^* \\ B^* & D^* \end{bmatrix} \begin{bmatrix} Y & 0 \\ 0 & I \end{bmatrix} \begin{bmatrix} A & B \\ C & D \end{bmatrix} \leq \begin{bmatrix} Y & 0 \\ 0 & I \end{bmatrix}, \tag{7.8.7}$$

where Y is a strictly positive operator on $\mathcal{X}$.

(iii) *The realization $\{A, B, C, D\}$ is similar to a stable contractive realization.*

In particular, G is a rational contractive analytic function if and only if G admits a stable minimal contractive realization.

Proof. Assume that G is a contractive analytic function. Let R be the rational function in $L^\infty(\mathcal{E}, \mathcal{E})$ defined by $R(e^{\imath\omega}) = I - G(e^{\imath\omega})^* G(e^{\imath\omega})$ for all $0 \leq \omega \leq 2\pi$. Because G is a contractive analytic function, $R(e^{\imath\omega}) \geq 0$ for all $0 \leq \omega \leq 2\pi$. Hence T_R is a rational positive Toeplitz operator on $\ell_+^2(\mathcal{E})$. By consulting Lemma 4.5.4, we see that the entries $(T_R)_{j,k} = R_{j-k}$ of the Toeplitz matrix T_R are determined by

$$\begin{aligned} R_0 &= I - D^*D - B^*PB, \\ R_n &= \widehat{C}A^{n-1}B \qquad (\text{for } n \geq 1), \\ \widehat{C} &= -(D^*C + B^*PA). \end{aligned} \tag{7.8.8}$$

Here P is the observability Gramian for the pair $\{C, A\}$. The results in Section 6.1, show that T_R admits a rational outer spectral factor Θ. Moreover, according to Theorem 6.1.1, the outer spectral factor Θ admits a realization of the form $\{A, B, C_o, D_o\}$. Let Φ be the function in $H^\infty(\mathcal{E}, \mathcal{Y} \oplus \mathcal{E})$ defined by $\Phi = \begin{bmatrix} G & \Theta \end{bmatrix}^{tr}$. We claim that Φ is an inner function. Using the fact that $I - G^*G = R = \Theta^*\Theta$ on the unit circle, we have

$$\Phi^*\Phi = G^*G + \Theta^*\Theta = G^*G + I - G^*G = I$$

on the unit circle. Therefore Φ is an inner function.

Observe that $\{A, B, \begin{bmatrix} C & C_o \end{bmatrix}^{tr}, \begin{bmatrix} D & D_o \end{bmatrix}^{tr}\}$ is a minimal realization for Φ. According to Theorem 4.2.1, we have

$$\begin{bmatrix} A^* & C^* & C_o^* \\ B^* & D^* & D_o^* \end{bmatrix} \begin{bmatrix} Y & 0 & 0 \\ 0 & I_{\mathcal{Y}} & 0 \\ 0 & 0 & I_{\mathcal{E}} \end{bmatrix} \begin{bmatrix} A & B \\ C & D \\ C_o & D_o \end{bmatrix} = \begin{bmatrix} Y & 0 \\ 0 & I_{\mathcal{E}} \end{bmatrix} \tag{7.8.9}$$

where Y is the observability Gramian for $\{\begin{bmatrix} C & C_o \end{bmatrix}^{tr}, A\}$, that is,

$$Y = A^*YA + C^*C + C_o^*C_o. \tag{7.8.10}$$

So for $x \oplus v$ in $\mathcal{X} \oplus \mathcal{E}$, we obtain

$$\left\| \begin{bmatrix} Y^{1/2}A & Y^{1/2}B \\ C & D \end{bmatrix} \begin{bmatrix} x \\ v \end{bmatrix} \right\|^2 \leq \left\| \begin{bmatrix} Y^{1/2}A & Y^{1/2}B \\ C & D \\ C_o & D_o \end{bmatrix} \begin{bmatrix} x \\ v \end{bmatrix} \right\|^2 = \left\| \begin{bmatrix} Y^{1/2} & 0 \\ 0 & I \end{bmatrix} \begin{bmatrix} x \\ v \end{bmatrix} \right\|^2.$$

This yields the inequality in (7.8.7). In other words, Part (i) implies Part (ii).

Assume that Part (ii) holds. Multiplying both sides of (7.8.7) by $Y^{-1/2} \oplus I$, we see that

$$\begin{bmatrix} Y^{1/2}AY^{-1/2} & Y^{1/2}B \\ CY^{-1/2} & D \end{bmatrix} : \begin{bmatrix} \mathcal{X} \\ \mathcal{E} \end{bmatrix} \to \begin{bmatrix} \mathcal{X} \\ \mathcal{Y} \end{bmatrix} \tag{7.8.11}$$

is a contraction. In other words,

$$\Sigma = \{Y^{1/2}AY^{-1/2}, Y^{1/2}B, CY^{-1/2}, D\}$$

is a stable contractive realization of G. Notice that $Y^{1/2}$ is a similarity transformation intertwining $\{A, B, C, D\}$ with Σ. Therefore $\{A, B, C, D\}$ is similar to a stable contractive realization. So Part (ii) implies Part (iii). We have already seen that Part (iii) implies Part (i). Therefore Parts (i), (ii) and (iii) are equivalent. □

Remark 7.8.2. Let $\{A, B, C, D\}$ be a minimal realization for a rational contractive analytic function G. Let $\{A, B, C_o, D_o\}$ be a realization for the outer spectral factor Θ for $I - G^*G$. Then

$$\begin{bmatrix} A^* & C^* \\ B^* & D^* \end{bmatrix} \begin{bmatrix} Y & 0 \\ 0 & I \end{bmatrix} \begin{bmatrix} A & B \\ C & D \end{bmatrix} \leq \begin{bmatrix} Y & 0 \\ 0 & I \end{bmatrix}, \tag{7.8.12}$$

where Y is the observability Gramian for the pair $\{\begin{bmatrix} C & C_o \end{bmatrix}^{tr}, A\}$. Moreover,

$$\{Y^{1/2}AY^{-1/2}, Y^{1/2}B, CY^{-1/2}, D\}$$

is a stable contractive realization for G. Finally,

$$\{A, B, \begin{bmatrix} C & C_o \end{bmatrix}^{tr}, \begin{bmatrix} D & D_o \end{bmatrix}^{tr}\}$$

is a minimal realization for the inner function $\begin{bmatrix} G & \Theta \end{bmatrix}^{tr}$ in $H^\infty(\mathcal{E}, \mathcal{Y} \oplus \mathcal{E})$ and (7.8.9) holds.

7.8.1 A contractive realization procedure

Assume that G is a rational contractive analytic function in $H^\infty(\mathcal{E}, \mathcal{Y})$ satisfying $\|G\|_\infty < 1$. Then one can use the finite section method in Section 7.6 to compute a realization $\{A, B, C, D\}$ for G satisfying (7.8.12). To this end, compute the outer spectral factor Θ for $R = I - G^*G$. One can compute the Fourier coefficients $\{R_n\}$ for R by using the fast Fourier transform or the state space method in (7.8.8). Now let $\{A, B, \begin{bmatrix} C & C_o \end{bmatrix}^{tr}, \begin{bmatrix} D & D_o \end{bmatrix}^{tr}\}$ be a minimal realization for the inner function $\begin{bmatrix} G & \Theta \end{bmatrix}^{tr}$. Then $\{A, B, C, D\}$ satisfies the inequality in (7.8.12) where Y is the observability Gramian for $\{\begin{bmatrix} C & C_o \end{bmatrix}^{tr}, A\}$.

Example. Consider the contractive analytic function

$$g(z) = \frac{-0.7165z^2 + 0.1796z - 0.0706}{z^3 - 0.2824z^2 - 0.0580z + 0.0003}. \tag{7.8.13}$$

A simple computation shows that $\|g\|_\infty = .92$. Using the finite section method with the Kalman-algorithm, the outer spectral factor θ for $1 - |g|^2$ is given by

$$\theta(z) = \frac{0.6671z^3 - 0.2471z^2 - 0.1627z + 0.0004497}{z^3 - 0.2824z^2 - 0.058z + 0.0003}. \tag{7.8.14}$$

A minimal realization $\{A, B, C_i, D_i\}$ for the inner function $\begin{bmatrix} g & \theta \end{bmatrix}^{tr}$ is given by

$$A = \begin{bmatrix} 0.0983 & -0.3295 & -0.0116 \\ -0.4965 & 0.1097 & -0.2787 \\ 0.0178 & 0.2840 & 0.0744 \end{bmatrix}, \quad B = \begin{bmatrix} 0.8548 \\ 0.0515 \\ 0.0196 \end{bmatrix},$$

$$C_i = \begin{bmatrix} -0.8344 & -0.0757 & 0.0294 \\ -0.0882 & 0.3187 & 0.0154 \end{bmatrix} \quad \text{and} \quad D_i = \begin{bmatrix} 0 \\ 0.6671 \end{bmatrix}. \tag{7.8.15}$$

Then a realization for g is given by $\{A, B, C, 0\}$ where

$$C = \begin{bmatrix} -0.8344 & -0.0757 & 0.0294 \end{bmatrix}$$

is the first row of C_i. The observability Gramian Y for the pair $\{C_i, A\}$ is given by

$$Y = \begin{bmatrix} 0.7590 & 0 & 0 \\ 0 & 0.1934 & 0 \\ 0 & 0 & 0.0163 \end{bmatrix}. \tag{7.8.16}$$

Finally, the inequality (7.8.12) holds with $D = 0$.

A typical procedure for computing $\{A, B, C, D\}$ for the rational contractive analytic function $g = p/q$ in Matlab is given by the following steps.

(i) Compute $g = \text{fft}(p, 2 \wedge 13)./\text{fft}(q, 2 \wedge 13)$. Here

$$p = [0, -0.7165, 0.1796, -0.0706] \text{ and } q = [1, -0.2824, -0.0580, 0.0003].$$

Then $\|g\|_\infty = \text{norm}(g, \text{'inf'})$. Compute $R = 1 - \text{abs}(g). \wedge 2$. Set $Rn = \text{real}(\text{ifft}(R)$. The vector $Rn(1{:}2 \wedge 12)$ contains the first 2^{12} Fourier coefficients for $R = 1 - |g|^2$.

(ii) Invert the Toeplitz matrix $T_{R,k}$ for k sufficiently large. In Matlab $\begin{bmatrix} a & e \end{bmatrix} = \text{levinson}(Rn(1{:}1000))$. The vector a corresponds to $\{A_{k,n}\}$ and e to Δ_k.

(iii) Compute $\theta = \sum_0^\infty z^{-n}\theta_n$, the outer spectral factor for $1 - |g|^2$. In Matlab $\theta = \text{sqrt}(e)./\text{fft}(a, 2 \wedge 13)$; and $\vartheta = \text{real}(\text{ifft}(\theta))$. Then $\vartheta(1{:}2 \wedge 12)$ contains the first 2^{12} Fourier coefficients of θ.

(iv) Run the Kalman-Ho algorithm on $\vartheta(1{:}500)$. Select the appropriate number of singular values and compute the realization $\{A_o, B_o, C_o, D_o\}$ for θ.

(v) Now use standard state space techniques to find a minimal state space realization $\{A, B, C_i, D_i\}$ for the inner function $\begin{bmatrix} g & \theta \end{bmatrix}^{tr}$.

(v-a) Another method to compute $\{A, B, C_i, D_i\}$ which skips part of step (iv) is to find the Fourier expansion $g = \sum_0^\infty z^{-n} g_n$. In Matlab $gn = \text{real}(\text{ifft}(g))$; the vector gn contains $\{g_n\}_0^{2^{12}-1}$ in the vector $gb(1{:}2 \wedge 12)$. Let α be the vector defined by

$$\alpha = \begin{bmatrix} gn(1) & \vartheta(1) & gn(2) & \vartheta(2) & gn(3) & \vartheta(3) & \cdots & gn(500) & \vartheta(500) \end{bmatrix}.$$

Then run the Kalman-Ho algorithm on α to compute a minimal realization for $\begin{bmatrix} g & \theta \end{bmatrix}^{tr}$. In fact, this is how we computed $\{A, B, C_i, D_i\}$.

(vi) Finally, use the "dlyap" command to compute the observability Gramian for the pair $\{C_i, A\}$. Then $\{A, B, C, D\}$ is a realization for g where C is the first row of C_i and D is the first component of D_i.

As noted earlier there is nothing magical about 1000 for the Levinson or 500 for the Kalman-Ho. Finally, by making minor modifications, the previous algorithm can be converted to compute a minimal realization for any rational contractive analytic function G in $H^\infty(\mathcal{E}, \mathcal{Y})$ satisfying $\|G\|_\infty < 1$. The details are left as a simple exercise.

7.9 A Spectral Factorization Approach to Filtering

Let $\{A, B, C, D\}$ be a minimal stable realization for a scalar-valued rational transfer function $g(z)$ in H^∞. The corresponding state space system is given by

$$x(n+1) = Ax(n) + Bu(n) \quad \text{and} \quad y(n) = Cx(n) + Du(n). \tag{7.9.1}$$

The output $y(n)$ corresponding to the input $u(n)$ is determined by the Toeplitz matrix T_g and the observability matrix corresponding to $\{C, A\}$, that is,

$$\begin{bmatrix} y(0) \\ y(1) \\ y(2) \\ \vdots \end{bmatrix} = \begin{bmatrix} C \\ CA \\ CA^2 \\ \vdots \end{bmatrix} x(0) + \begin{bmatrix} g_0 & 0 & 0 & \cdots \\ g_1 & g_0 & 0 & \cdots \\ g_2 & g_1 & g_0 & \cdots \\ \vdots & \vdots & \vdots & \ddots \end{bmatrix} \begin{bmatrix} u(0) \\ u(1) \\ u(2) \\ \vdots \end{bmatrix}. \tag{7.9.2}$$

Here $g(z) = \sum_0^\infty z^{-k} g_k$ where $g_0 = D$ and $g_k = CA^{k-1}B$ for all integers $k \geq 1$; see Section 14.1.

The *steady state response* to an input $u(n)$ is the output $y_{ss}(n)$ after all the transients have died out. The steady state response is usually computed for a sinusoid input. For example, consider the sinusoid input $u(n) = ae^{\imath\omega_0 n}$. Here a is the amplitude of the sinusoid and ω_0 is the angular frequency. In this case, the output $y(n)$ is determined by

$$\begin{aligned} y(n) &= CA^n x(0) + \sum_{k=0}^{n} g_{n-k} a e^{\imath\omega_0 k} \\ &= CA^n x(0) + ae^{\imath\omega_0 n} \sum_{k=0}^{n} g_{n-k} e^{-\imath\omega_0 (n-k)} \\ &= CA^n x(0) + ae^{\imath\omega_0 n} \sum_{k=0}^{n} g_k e^{-\imath\omega_0 k} \\ &= CA^n x(0) + g(e^{\imath\omega_0}) ae^{\imath\omega_0 n} - ae^{\imath\omega_0 n} \sum_{k=n+1}^{\infty} g_k e^{-\imath\omega_0 k}. \end{aligned}$$

Observe $g(e^{\imath\omega_0})$ is the value of the rational transfer function $g(z)$ computed at the boundary $z = e^{\imath\omega_0}$. Because g is a rational function in H^∞, the last term converges to zero as n tends to infinity, that is,

$$\left| a e^{\imath\omega_0 n} \sum_{k=n+1}^{\infty} g_k e^{-\imath\omega_0 k} \right| \leq |a| \sum_{k=n+1}^{\infty} |g_k| \to 0.$$

Since A is stable, $CA^n x(0)$ also converges to zero as n tends to infinity. So for large n the output $y(n) \approx g(e^{\imath\omega_0}) a e^{\imath\omega_0 n}$. Recall that the steady state response is the output $y_{ss}(n)$ after all the transients have died out. So in our example $y_{ss}(n) = g(e^{\imath\omega_0}) a e^{\imath\omega_0 n}$. Notice that $g(e^{\imath\omega_0})$ admits a polar decomposition of the form $g(e^{\imath\omega_0}) = |g(e^{\imath\omega_0})| e^{\imath\phi_0}$ where ϕ_0 is the angle for $g(e^{\imath\omega_0})$. This readily implies that the steady state response is given by

$$y_{ss}(n) = a|g(e^{\imath\omega_0})| e^{\imath(\omega_0 n + \phi_0)}. \tag{7.9.3}$$

The steady state response $y_{ss}(n)$ corresponding to the input $u(n) = a e^{\imath\omega_0 n}$ is a sinusoid with the same frequency ω_0 as the input, while the amplitude had changed to $a|g(e^{\imath\omega_0 n})|$ and there has been a phase shift of ϕ_0 radians.

Now assume that the input $u(n)$ is a finite linear combination of sinusoids, that is,

$$u(n) = \sum_{k=1}^{\nu} a_k e^{\imath\omega_k n}. \tag{7.9.4}$$

Here $\{a_k\}_1^\nu$ are the amplitudes and $\{\omega_k\}_1^\nu$ are the distinct frequencies for the sinusoid input. Because the system in (7.9.1) is linear, the corresponding steady state output is given by

$$y_{ss}(n) = \sum_{k=1}^{\nu} a_k |g(e^{\imath\omega_k})| e^{\imath(\omega_k n + \phi_k)} \tag{7.9.5}$$

where $g(e^{\imath\omega_k}) = |g(e^{\imath\omega_k})| e^{\imath\phi_k}$ is the polar decomposition for $g(e^{\imath\omega_k})$.

In most applications (7.9.1) is a state space system consisting of real matrices $\{A, B, C, D\}$. In this case, the Taylor coefficients $\{g_k\}_0^\infty$ are also real. Moreover, if ϕ is the phase for $g(e^{\imath\omega})$, then we see that

$$g(e^{-\imath\omega}) = \overline{g(e^{\imath\omega})} = \overline{|g(e^{\imath\omega})| e^{\imath\phi}} = |g(e^{\imath\omega})| e^{-\imath\phi}.$$

In other words, $g(e^{\imath\omega})$ and $g(e^{-\imath\omega})$ have the same magnitude, while ϕ is the phase for $g(e^{\imath\omega})$ if and only if $-\phi$ is the phase for $g(e^{-\imath\omega})$. Recall that sines and cosines of ϕ are linear combinations of $e^{\imath\phi}$. Now consider the sinusoidal input of the form:

$$u(n) = \sum_{k=1}^{\nu} a_k \cos(\omega_k n + \varphi_k). \tag{7.9.6}$$

Here $\{a_k\}_1^\nu$ are the amplitudes and $\{\omega_k\}_1^\nu$ are the distinct frequencies and $\{\varphi_k\}_1^\nu$ are the phases. Because the system in (7.9.1) is linear and the Taylor coefficients of g are real, the steady state output is given by

$$y_{ss}(n) = \sum_{k=1}^{\nu} a_k |g(e^{\imath\omega_k})| \cos(\omega_k n + \phi_k + \varphi_k). \tag{7.9.7}$$

An ideal filter is a "causal" function θ which accepts sinusoids in a specified frequency range κ and rejects sinusoids outside the frequency range of κ, that is, if $u(n) = ae^{\imath\omega n}$, then

$$\begin{aligned} y_{ss}(n) &= ae^{\imath\omega n} && \text{if } \omega \in \kappa, \\ &= 0 && \text{otherwise.} \end{aligned}$$

A filter is called *low pass* if the filter accepts low frequencies and rejects high frequencies, that is, $\kappa = [0, \omega_0]$. A filter is *high pass* if the filter rejects low frequencies and accepts high frequencies, that is, $\kappa = [\omega_0, \pi]$. Finally, a filter is *bandpass* if it accepts frequencies in a specified band $\kappa = [\omega_0, \omega_1]$ and rejects all the other frequencies. (Filters are symmetric about π, and thus, κ is only specified in the interval $[0, \pi]$.) Filters are usually implemented as stable rational transfer functions. Because a nonzero rational transfer function cannot be zero on a set of positive measure, it is impossible to implement an ideal filter. So the practical problem is to find a rational transfer function to approximate an ideal filter. Finally, McMillan degree plays a fundamental role in filter design. In general filters with a higher McMillan degree tend to be more accurate. However, they are more expensive to implement.

There are many methods to design a filter of a specified McMillan degree; see [12, 65, 127, 172, 192]. Here we will use our outer spectral factorization theory in Section 7.6 along with the Kalman-Ho algorithm to design a filter. A major problem with this design method is that it does not allow one to specify the McMillan degree a priori. The method is really quite simple. First specify in the frequency domain a positive function $f(\omega)$ for ω in $[0, 2\pi]$ which is symmetric about π. The function should be smooth and $f(e^{\imath\omega}) \approx 1$ if ω is in some specified frequency range κ, then outside this range, the function f should decay smoothly and relatively fast to $\epsilon > 0$. The filter which we want to design will select frequencies in κ and reject frequencies outside of κ. (The faster the function decays to ϵ the larger the McMillan degree of the filter will be.) To take advantage of the fast Fourier transform, the function f is specified by 2^k points around the unit circle for some integer k say 12, 13 or 14 is sufficient. Then the idea is to find a rational outer function θ of the lowest possible McMillan degree such that $f(e^{\imath\omega}) \approx |\theta(e^{\imath\omega})|$. To accomplish this one can use the fast Fourier transform to find the Fourier coefficients $\{r_k\}_{-\infty}^{\infty}$ for $f(e^{\imath\omega})^2 = r = \sum_{-\infty}^{\infty} e^{-\imath\omega k} r_k$. Because f is symmetric about π, the Fourier coefficients r_k are real. The corresponding Toeplitz matrix

$$T_r = \begin{bmatrix} r_0 & r_1 & r_2 & \cdots \\ r_1 & r_0 & r_1 & \cdots \\ r_2 & r_1 & r_0 & \cdots \\ \vdots & \vdots & \vdots & \ddots \end{bmatrix} \quad \text{on } \ell_+^2 \tag{7.9.8}$$

is positive and invertible; see Proposition 2.5.1. Hence T_r admits an invertible outer spectral factor, or equivalently, $r = |\theta|^2$ where θ is an invertible outer function in

H^∞. Then one can use the results in Section 7.6 with the Kalman-Ho algorithm to compute a minimal realization $\{A, B, C, D\}$ for the outer spectral factor θ for $r(e^{\imath\omega})$. Finally, θ is the transfer function or filter which picks out the sinusoids in the frequency range κ.

Recall that the steady state response $y_{ss}(n)$ for a transfer function θ corresponding to the sinusoid input $u(n) = ae^{\imath\omega_0 n}$ is given by

$$y_{ss}(n) = a|\theta(e^{\imath\omega_0})|e^{\imath(\omega_0 n+\phi_0)}$$

where ϕ_0 is the phase for $\theta(e^{\imath\omega_0})$. So if the input frequency ω_0 is in κ, then $|\theta(e^{\imath\omega_0})| \approx 1$, and $y_{ss}(n) \approx ae^{\imath(\omega_0 n+\phi_0)}$. In other words, the steady state output of the filter θ is the same as the input except for a phase shift ϕ_0. In many applications, this phase shift does not pose any problems. However, one can eliminate this phase shift by adding the appropriate Blaschke product b to the transfer function θ. To see this, consider the transfer function $g = b\theta$. Recall that $|b(e^{\imath\omega})| = 1$ over all frequencies ω. Hence $r = |\theta|^2 = |g|^2$. So by choosing the appropriate Blaschke product b one can eliminate the phase shift ϕ_0 at ω_0 by implementing the filter $b\theta$. Finally, it is noted that if the input frequency ω_0 is sufficiently far from κ, then the steady state output $y_{ss}(n) \approx a|\theta(e^{\imath\omega_0})|e^{\imath(\omega_0 n+\phi_0)} \approx 0$.

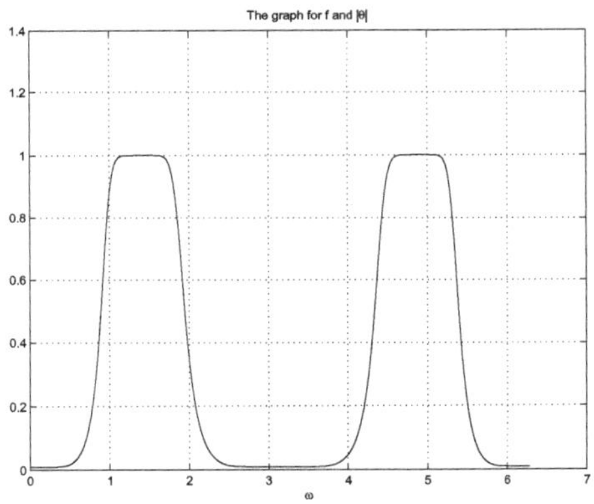

Figure 7.1: A bandpass filter.

Example. Consider the bandpass filter given in Figure 7.1. The cutoff frequencies are $3\pi/10$ and $6\pi/10$. In fact, for simplicity of presentation, we choose $f^2 = r = |\xi|^2 + 10^{-4}$ where ξ is the the fourth-order Butterworth filter with cutoff frequencies $[3\pi/10, 6\pi/10]$. (In Matlab butter(4,[3, 6]/10).) Notice that $f > 1/100$ over all frequencies. (If the function $f(e^{\imath\omega}) \approx 0$ over some interval, then the Levinson algorithm will not work properly. Moreover, since our Butterworth filter ξ has eight zeros on the unit circle, the outer spectral factor for $|\xi|^2$ is not invertible, and the Levinson algorithm may have numerical problems when inverting the corresponding Toeplitz matrix. All these problems are avoided by choosing $|\xi|^2 + 10^{-4}$.) Using the inverse fast Fourier transform on r, we computed $\{r_k\}_0^\nu$ for ν

sufficiently large. Then running this through the Levinson algorithm, we computed the transfer function $\theta = \sqrt{e}/a$. Now the fast inverse Fourier transform can be used to approximate the Fourier coefficients $\{\theta_k\}$ for $\theta = \sum_0^\infty z^{-k}\theta_k$. Finally, by running the Fourier coefficients $\{\theta_k\}$ through the Kalman-Ho algorithm, we obtained the filter

$$\theta = \frac{0.087z^8 - 0.044z^7 - 0.091z^6 + 0.024z^5 + 0.077z^4 - 0.014z^3 - 0.026z^2 + 0.002z + 0.004}{z^8 - 0.978z^7 + 1.94z^6 - 1.339z^5 + 1.627z^4 - 0.735z^3 + 0.583z^2 - 0.139z + 0.076}.$$

Figure 7.1 plots both $|\theta|$ and f on the same graph. They are so close that they appear as one graph. Our algorithm is summarized in the following steps:

(i) Compute $r = \text{ifft}(f. \wedge 2)$. The Fourier coefficients $\{r_k\}_0^\nu$ for $r = f^2$ are contained in the first ν components of r where $\nu < 2^{k-1}$ and f contains 2^k points.

(ii) Compute $[a, e] = \text{levinson}(r(1{:}\nu))$ for ν sufficiently large (we used $\nu = 1200$). Next compute $\theta = \sqrt{e}./\text{fft}(a, 2 \wedge k)$ and $\gamma = \text{real}(\text{ifft}(\theta))$. Then the first $\nu < 2^{k-1}$ Fourier coefficients $\{\theta_k\}_0^\nu$ for $\theta = \sum_0^\infty z^{-k}\theta_k$ are contained in γ.

(iii) Run the Kalman-Ho algorithm on $\gamma(1{:}\mu)$ for μ sufficiently large (we choose $\mu = 800$) to compute a state space realization $\{A, B, C, D\}$ for θ. Finally, using ss2tf we arrived at our eighth-order model for θ.

By keeping the first six singular values in the Kalman-Ho algorithm, we arrived at the following sixth-order approximation ϑ for the filter θ:

$$\vartheta = \frac{0.0874z^6 - 0.0251z^5 - 0.1158z^4 + 0.0072z^3 + 0.0972z^2 + 0.0053z - 0.0431}{z^6 - 0.7640z^5 + 1.5558z^4 - 0.8172z^3 + 1.0408z^2 - 0.2793z + 0.2230}.$$

The sixth order approximation ϑ for θ is very close. In fact, the H^∞ distance between θ and ϑ is given by $\|\theta - \vartheta\|_\infty = 0.0051$. So when $|\vartheta|$ is plotted on the same graph as $|\theta|$ or f, all the graphs are virtually indistinguishable.

Finally, by retaining only four singular values in the Kalman-Ho algorithm, we obtained the following fourth-order approximation ψ for θ:

$$\psi = \frac{0.0874z^4 - 0.0079z^3 - 0.1262z^2 - 0.0357z + 0.1666}{z^4 - 0.5406z^3 + 1.1581z^2 - 0.3621z + 0.4919}.$$

The H^∞ distance between θ and ψ is given by $\|\theta - \psi\|_\infty = 0.0886$. The plot of $|\psi|$ and f is displayed in Figure 7.2.

7.10 Notes

Theorem 7.1.1 is a classical result in inverting strictly positive Toeplitz operators; for example see Rosenblum-Rovnyak [182]. The results in Section 7.2 are well known and were taken from the Appendix in Foias-Frazho-Gohberg-Kaashoek [84]. The Carathéodory interpolation problem was solved by Carathéodory [50, 51]

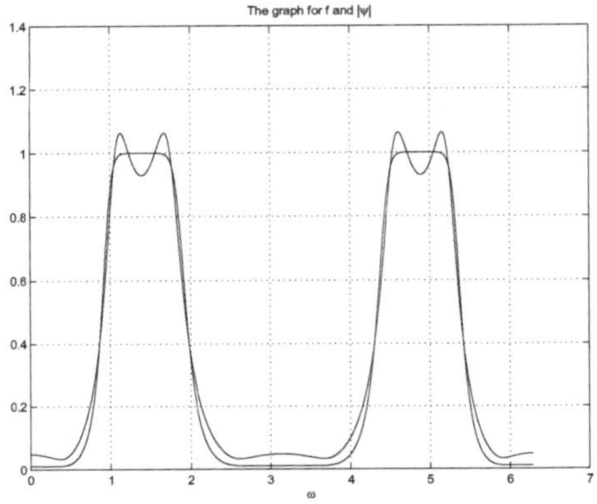

Figure 7.2: A bandpass filter approximation.

and plays a fundamental role in modern H^∞ interpolation theory. The literature on H^∞ interpolation theory is massive. For some monographs in H^∞ interpolation theory; see Agler-McCarthy [4], Bakonyi and T. Constantinescu [22], Ball-Gohberg-Rodman [24], Foias-Frazho [82], Foias-Frazho-Gohberg-Kaashoek [84], Chapter XXXIII in Gohberg-Goldberg-Kaashoek [114] and their historical comments. The finite section inversion Lemma 7.6.1 is a special case of the projection matrix inversion formulas in Section 2.17 of Gohberg-Goldberg-Kaashoek [112]. For some further results on the finite section method in operator theory see Böttcher-Silbermann [36] and Lindner [157]. Using the Levinson algorithm with the finite section inversion formulas to compute the outer spectral factor and inner-outer factorization is a straightforward application of the Kalman-Ho algorithm. The bounded real lemma (Theorem 7.8.1) is a classical result in systems theory; see Kailath-Sayed-Hassibi [143] and Zhou-Doyle-Glover [204] for historical comments and applications to systems theory. A general theory of contractive analytic functions and unitary realizations was developed independently of the bounded real lemma in operator theory. In operator theory the contractive analytic functions were not necessarily rational and the unitary realizations were in general infinite dimensional. The theory of unitary systems started with Livšic [163, 164]. Then using dilation theory, Sz.-Nagy-Foias developed the characteristic function; see [198]. The characteristic function is a unitary system which plays a fundamental role in operator theory. The Sz.-Nagy-Foias characteristic function can be used to study the spectrum and invariant subspaces of contractions. For further results on unitary systems; see Brodskii [43, 44], Chapter 28 in Gohberg-Goldberg-Kaashoek [114], Arocena [15] and Arov [19, 20]. Finally, it is noted that the results in Section 7.9 is a rather naive approach to filtering theory for sinusoids. However, it may provide a reasonable filter for certain applications. For an in-depth study of sinusoid filtering theory; see Antoniou [12], Daryanani [65], Hayes [127], Peled-Liu [172], Oppenheim-Schafer-Buck [169], Schafer-Oppenheim

[191] and Schaumann-Van Valkenburg [192].

A Wiener optimization problem. A Wiener filtering problem is discussed in Chapter 11. Motivated by Wiener filtering, a classical Wiener optimization problem is given by

$$\mu = \inf\{\|G - \Theta H\|_2 : H \in H^2(\mathcal{U}, \mathcal{E})\}. \tag{7.10.1}$$

Here Θ is a specified function in $H^\infty(\mathcal{E}, \mathcal{Y})$ and G is a function in $H^\infty(\mathcal{U}, \mathcal{Y})$. As always, we assume that the spaces $\mathcal{U}$, $\mathcal{E}$ and $\mathcal{Y}$ are all finite dimensional. Moreover the inner product between two functions P and Q in the appropriate $H^2(\cdot, \cdot)$ spaces is determined by

$$(P, Q)_2 = \frac{1}{2\pi} \int_0^{2\pi} \operatorname{trace}\left(P(e^{\imath\omega})Q(e^{\imath\omega})^*\right) d\omega.$$

Now assume that Θ admits an inner-outer factorization of the form $\Theta = \Theta_i \Theta_o$ where Θ_i is an inner function in $H^\infty(\mathcal{E}, \mathcal{Y})$ and Θ_o is an invertible outer function in $H^\infty(\mathcal{E}, \mathcal{E})$. The solution $\widehat{H}$ to the Wiener optimization problem in (7.10.1) is unique and computed by

$$\widehat{H}(z) = \Theta_o^{-1}(z)(P_+(\Theta_i^* G))(z) \quad \text{and} \quad \mu^2 = \|G\|_2^2 - \|P_+(\Theta_i^* G)\|_2^2. \tag{7.10.2}$$

Here P_+ is the orthogonal projection from $L^2(\cdot, \cdot)$ onto the appropriate $H^2(\cdot, \cdot)$ space, that is $\sum_0^\infty F_n e^{-\imath\omega n} = P_+ \sum_{-\infty}^\infty F_n e^{-\imath\omega n}$.

To verify that the solution is determined by (7.10.2), observe that

$$\Theta H^2(\mathcal{U}, \mathcal{E}) = \Theta_i \Theta_o H^2(\mathcal{U}, \mathcal{E}) = \Theta_i H^2(\mathcal{U}, \mathcal{E})$$

is a linear subspace. Because Θ_o is an invertible outer function, the projection theorem guarantees that the solution $\widehat{H}$ to the optimization problem in (7.10.1) is unique. Moreover, this solution is determined by the unique function $\widehat{H}$ in $H^2(\mathcal{U}, \mathcal{E})$ such that $\Theta_i \Theta_o \widehat{H} - G$ is orthogonal to $\Theta_i H^2(\mathcal{U}, \mathcal{E})$. So using the fact that $\Theta_i^* \Theta_i = I$ on the unit circle, $\Theta_o \widehat{H} - \Theta_i^* G$ is orthogonal to $H^2(\mathcal{U}, \mathcal{E})$ on the unit circle. Hence $\Theta_o \widehat{H} = P_+ \Theta_i^* G$, or equivalently, $\widehat{H} = \Theta_o^{-1} P_+ \Theta_i^* G$ on the unit circle. So the optimal solution $\widehat{H}$ is unique and given by the first equation in (7.10.2). To obtain the expression for the cost μ in (7.10.2), observe that

$$\begin{aligned}
\mu^2 &= \|G - \Theta_i \Theta_o \widehat{H}\|_2^2 = \|G - \Theta_i P_+ \Theta_i^* G\|_2^2 \\
&= \|G\|_2^2 - 2\Re(G, \Theta_i P_+ \Theta_i^* G) + \|\Theta_i P_+ \Theta_i^* G\|_2^2 \\
&= \|G\|_2^2 - 2\Re(\Theta_i^* G, P_+ \Theta_i^* G) + \|P_+ \Theta_i^* G\|_2^2 \\
&= \|G\|_2^2 - \|P_+ \Theta_i^* G\|_2^2.
\end{aligned}$$

Therefore the expression for the cost μ in (7.10.2) holds.

Example. Now let us try an example in Matlab. Consider the function

$$\theta = \frac{z^3 - 4.833z^2 + 3.5z - 0.6667}{z^4 - 0.7z^3 + 0.225z^2 - 0.025z},$$
$$g = \frac{z^2 + 2z + 3}{z^3 + 0.5333z^2 - 0.01667z - 0.01667}.$$

In this case, the Wiener optimization problem is $\mu = \inf\{\|g - \theta h\| : h \in H^2\}$. By using Matlab and keeping four singular values in the Kalman-Ho algorithm, we arrived at

$$\widehat{h} = \frac{-0.3976z^4 - 0.4819z^3 + 0.3073z^2 - 0.09307z + 0.001575}{z^4 - 0.1106z^3 - 0.311z^2 + 0.02745z + 0.01511}.$$

Moreover, the error $\mu = 1.4434$. The corresponding Matlab commands we used are given by

- theta=fft([0,1,−4.833,3.5,−0.6667],2^{13})./fft([1,−0.7,0.225,−0.025,0],2^{13});
- g=fft([0,1,2,3],2^{13})./fft([1,0.5333,−0.01667,−0.01667],2^{13});
- r=real(ifft(abs(theta).∧2));
- $[a, e]$ =levinson(r(1:1000));
- thetao=sqrt(e)./fft(a,2^{13});
- thetai=theta./thetao;
- q=conj(thetai).*g; dq=ifft(q);
- h=fft(dq(1:2^{12}),2^{13})./thetao;
- mh=real(ifft(h));
- $[a, b, c, d, s]$=kalho(mh(1:600));
- $[hn, hd]$=ss2tf(a,b,c,d); hopt=tf(hn,hd);
- $\mu = \text{norm}(\text{ifft}(g - \theta. * h))$;
- Finally, μ is also given by: μ^2 =norm(ifft(g)).∧2 − norm(dq(1:2^{12}))∧2.

(The Kalman-Ho algorithm is not yet in the standard Matlab package and we programmed a Kalman-Ho algorithm.) Finally, it is noted that we only kept four singular values in the Kalman-Ho algorithm.

Chapter 8

Isometric Representations and Factorization

Let T_R be a positive Toeplitz matrix with a rational symbol R. In this chapter, we will use the finite sections $T_{R,n}$ of T_R to compute the maximal outer spectral factor and the unitary part for T_R. Then we shall use the lower triangular Cholesky factorization of $T_{R,n}$ to approximate the Taylor's coefficient of the maximal outer spectral factor for T_R.

8.1 Isometric Representations and Finite Sections

Let $\{U \text{ on } \mathcal{K}, \Gamma\}$ be a controllable isometric representation for a positive Toeplitz matrix T_R with $\mathcal{L}(\mathcal{E},\mathcal{E})$-valued symbol $R = \sum_{-\infty}^{\infty} e^{-\imath\omega k} R_k$. Recall that U is an isometry and Γ is an operator mapping $\mathcal{E}$ into $\mathcal{K}$ such that $R_{-n} = \Gamma^* U^n \Gamma$ for all integers $n \geq 0$. Consider the function $G(z)$ defined by

$$G(z) = \sum_{n=0}^{\infty} \frac{1}{z^{n+1}} R_{-n}. \tag{8.1.1}$$

The function G will be used to gain some further insight into the maximal outer spectral factor for T_R. Notice that $\{U, \Gamma, \Gamma^*, 0\}$ is a controllable realization for $G(z)$. However, this realization may not be minimal. To construct a minimal realization for $G(z)$, we simply extract the observable subspace from $\{U, \Gamma, \Gamma^*, 0\}$, that is, let $\mathcal{X}$ be the observable subspace for the pair $\{\Gamma^*, U\}$, or equivalently, the controllable subspace for the pair $\{U^*, \Gamma\}$ given by

$$\mathcal{X} = \bigvee_{n=0}^{\infty} U^{*n} \Gamma \mathcal{E}.$$

Clearly, $\mathcal{X}$ is an invariant subspace for U^*. Let A be the operator on $\mathcal{X}$ and B the operator mapping $\mathcal{E}$ into $\mathcal{X}$ defined by

$$A = \Pi_{\mathcal{X}} U|\mathcal{X} \text{ on } \mathcal{X} \quad \text{and} \quad B = \Gamma : \mathcal{E} \to \mathcal{X}.$$

It is emphasized that $\{A, B, B^*, 0\}$ is obtained by extracting the observable part from the controllable realization $\{U, \Gamma, \Gamma^*, 0\}$ of G. Thus $\{A, B, B^*, 0\}$ is a controllable and observable realization for G. In particular, $\mathcal{X}$ is finite dimensional if and only if G is a rational function. Finally, the McMillan degree of G equals the dimension of $\mathcal{X}$.

Throughout we assume that G is rational, or equivalently, the symbol R for T_R is rational. To obtain the maximal outer spectral factor for T_R, recall that $\{U, \Gamma\}$ admits a Wold decomposition of the form

$$U = \begin{bmatrix} S & 0 \\ 0 & V \end{bmatrix} \text{ on } \begin{bmatrix} \ell_+^2(\mathcal{Y}) \\ \mathcal{V} \end{bmatrix} \quad \text{and} \quad \Gamma = \begin{bmatrix} \Gamma_1 \\ \Gamma_2 \end{bmatrix} : \mathcal{E} \to \begin{bmatrix} \ell_+^2(\mathcal{Y}) \\ \mathcal{V} \end{bmatrix}. \tag{8.1.2}$$

As expected, $\mathcal{Y} = \ker U^*$ and S is the unilateral shift on $\ell_+^2(\mathcal{Y})$. Moreover, V is a unitary operator on $\mathcal{V}$. We claim that $\{A, B\}$ admits a matrix representation of the form

$$A = \begin{bmatrix} A_o & 0 \\ 0 & V \end{bmatrix} \text{ on } \begin{bmatrix} \mathcal{X}_o \\ \mathcal{V} \end{bmatrix} \quad \text{and} \quad B = \begin{bmatrix} B_o \\ \Gamma_2 \end{bmatrix} : \mathcal{E} \to \begin{bmatrix} \mathcal{X}_o \\ \mathcal{V} \end{bmatrix}. \tag{8.1.3}$$

Here $B_o = \Gamma_1$ is viewed as an operator mapping $\mathcal{E}$ into $\mathcal{X}_o$, and A_o is the operator on $\mathcal{X}_o$ determined by

$$\mathcal{X}_o = \bigvee_{n=0}^{\infty} S^{*n} \Gamma_1 \mathcal{E} \quad \text{and} \quad A_o = \Pi_{\mathcal{X}_o} S|\mathcal{X}_o. \tag{8.1.4}$$

Moreover, the subspace $\mathcal{X}_o$ is a finite dimensional invariant subspace for the backward shift S^*. Finally, the operator A_o is stable.

To see this, first notice that the Wold decomposition in (8.1.2) shows that

$$\mathcal{X} = \bigvee_{n=0}^{\infty} \begin{bmatrix} S^{*n}\Gamma_1 \\ V^{*n}\Gamma_2 \end{bmatrix} \mathcal{E} \subseteq \mathcal{X}_o \oplus \mathcal{V}.$$

Because $\mathcal{X}$ is finite dimensional, $\mathcal{X}_o$ and $\mathcal{V}_o = \bigvee_0^\infty V^{*n}\Gamma_2\mathcal{E}$ are both finite dimensional. Using the fact that $\mathcal{X}_o$ is invariant for S^* and $A_o^* = S^*|\mathcal{X}_o$, it follows that $A_o^{*n}x = S^{*n}x$ converges to zero for all x in $\mathcal{X}_o$. Since the state space is finite dimensional, A_o^* is stable, and thus, A_o is stable.

Now let us show that $\mathcal{V}_o = \mathcal{V}$. Notice that $\mathcal{V}_o$ is invariant for V^*, and thus, $V^*|\mathcal{V}_o$ defines an isometry on $\mathcal{V}_o$. Since $\mathcal{V}_o$ is finite dimensional, $V^*|\mathcal{V}_o$ is unitary, and $V^*\mathcal{V}_o = \mathcal{V}_o$. Hence $V\mathcal{V}_o = \mathcal{V}_o$. In particular, $\mathcal{V}_o$ is an invariant subspace for V. Using the fact that $\Gamma_2\mathcal{E}$ is contained in $\mathcal{V}_o$ and $\mathcal{V}_o$ is invariant for V, we obtain

$\mathcal{V}_o \supseteq \bigvee_0^\infty V^n \Gamma_2 \mathcal{E} = \mathcal{V}$. The equality follows from the controllability of $\{V, \Gamma_2\}$. This readily implies that $\mathcal{V}_o = \mathcal{V}$. Finally, it is noted that the pair $\{V^*, \Gamma_2\}$ is controllable.

The above analysis shows that $\{A_o^*, B_o\}$ and $\{V^*, \Gamma_2\}$ are both controllable and $\mathcal{X} \subseteq \mathcal{X}_o \oplus \mathcal{V}$. The operator A_o^* is stable and all the eigenvalues for V^* are on the unit circle. By consulting Lemma 6.1.2, we see that $\mathcal{X} = \mathcal{X}_o \oplus \mathcal{V}$, and the pair $\{A, B\}$ admit a matrix decomposition of the form (8.1.3). This proves our claim.

Recall that the maximal outer spectral factor Θ for T_R is determined by

$$\Theta(z) = z\Pi_{\mathcal{Y}}(zI - U^*)^{-1}\Gamma = z\Pi_{\mathcal{Y}}(zI - S^*)^{-1}\Gamma_1, \tag{8.1.5}$$

where $\Pi_{\mathcal{Y}}$ is the orthogonal projection onto $\mathcal{Y} = \ker U^*$; see Theorem 5.2.1. We claim that the pair $\{\Pi_{\mathcal{Y}}|\mathcal{X}_o, A_o^*\}$ is observable. To verify this, assume that there exists a vector $x = \begin{bmatrix} x_0 & x_1 & x_2 & \cdots \end{bmatrix}^{tr}$ in $\mathcal{X}_o$ such that $0 = \Pi_{\mathcal{Y}} A_o^{*n} x$ for all integers $n \geq 0$. Since $A_o^* = S^*|\mathcal{X}_o$, we obtain

$$0 = \Pi_{\mathcal{Y}} A_o^{*n} x = \Pi_{\mathcal{Y}} S^{*n} x = x_n \qquad (n \geq 0).$$

In other words, $x_n = 0$ for all n and the vector x must be zero. Hence $\{\Pi_{\mathcal{Y}}|\mathcal{X}_o, A_o^*\}$ is observable. (Essentially, the observability of $\{\Pi_{\mathcal{Y}}|\mathcal{X}_o, A_o^*\}$ follows from the observability of $\{\Pi_{\mathcal{Y}}, S^*\}$.) By construction, the pair $\{A_o^*, B_o\}$ is controllable. So $\{A_o^*, B_o, \Pi_{\mathcal{Y}}|\mathcal{X}_o, 0\}$ is both controllable and observable. Since $\mathcal{X}$ is an invariant subspace for U^* and $A^* = U^*|\mathcal{X}$, we obtain

$$\Theta(z) = z(\Pi_{\mathcal{Y}}|\mathcal{X})(zI - A^*)^{-1}B = z(\Pi_{\mathcal{Y}}|\mathcal{X}_o)(zI - A_o^*)^{-1}B_o. \tag{8.1.6}$$

Therefore $\{A_o^*, B_o, \Pi_{\mathcal{Y}}|\mathcal{X}_o, 0\}$ is a controllable and observable realization for $z^{-1}\Theta$.

The subspace $\mathcal{X} = \mathcal{X}_o$ if and only if the unitary part V is not presented in the Wold decomposition of U. Because $\{A_o^*, B_o, \Pi_{\mathcal{Y}}|\mathcal{X}_o, 0\}$ and $\{A^*, B, B^*, 0\}$ are both controllable and observable, the McMillan degree of $z^{-1}\Theta$ equals the McMillan degree of G if and only if the unitary operator V is not presented in the Wold decomposition of U.

We claim that the maximal outer spectral factor for T_R is given by

$$\Theta(z) = zC_o(zI - A_o^*)^{-1}B_o \tag{8.1.7}$$

where C_o is any operator mapping $\mathcal{X}_o$ onto $\mathcal{Z}$ such that

$$I - A_o A_o^* = C_o^* C_o. \tag{8.1.8}$$

To see this, first observe that $\Pi_{\mathcal{Y}}^* \Pi_{\mathcal{Y}} = I - UU^*$. Let C_1 be the operator mapping $\mathcal{X}_o$ into $\mathcal{Y}$ given by $C_1 = \Pi_{\mathcal{Y}}|\mathcal{X}_o$. Using the fact that $A_o^* = S^*|\mathcal{X}_o = U^*|\mathcal{X}_o$, we arrive at

$$\begin{aligned} C_1^* C_1 &= \Pi_{\mathcal{X}_o} \Pi_{\mathcal{Y}}^* \Pi_{\mathcal{Y}} |\mathcal{X}_o = \Pi_{\mathcal{X}_o}(I - UU^*)|\mathcal{X}_o \\ &= I - \Pi_{\mathcal{X}_o} U A_o^* = I - A_o A_o^*. \end{aligned}$$

Hence $C_1^*C_1 = I - A_oA_o^*$. We claim that C_1 is onto $\mathcal{Y}$. Equation (8.1.5), shows that $\Pi_{\mathcal{Y}}\Gamma = \Theta(\infty)$. Since $\Gamma\mathcal{E}$ is a subspace of $\mathcal{X} = \mathcal{X}_o \oplus \mathcal{V}$ and $\Pi_{\mathcal{Y}}\mathcal{V} = \{0\}$, we have

$$C_1\mathcal{X}_o = \Pi_{\mathcal{Y}}\mathcal{X}_o = \Pi_{\mathcal{Y}}\mathcal{X} \supseteq \Pi_{\mathcal{Y}}\Gamma\mathcal{E} = \Theta(\infty)\mathcal{E} = \mathcal{Y}.$$

The last equality follows from the fact that Θ is outer, and thus, $\Theta(\infty)$ is onto $\mathcal{Y}$. So C_1 maps $\mathcal{X}_o$ onto $\mathcal{Y}$. Equation (8.1.6) shows that $\Theta(z) = zC_1(zI - A_o)^{-1}B_o$ is the maximal outer spectral factor for T_R.

Let C_o be any operator mapping $\mathcal{X}_o$ onto $\mathcal{Z}$ such that $I - A_oA_o^* = C_o^*C_o$. Then $C_o^*C_o = C_1^*C_1$ and there exists a unitary operator Φ such that $C_o = \Phi C_1$. Hence $zC_o(zI - A_o^*)^{-1}B_o = \Phi\Theta(z)$. Because the maximal outer spectral factor is unique up to a constant unitary operator on the left, $zC_o(zI - A_o^*)^{-1}B_o$ is also a maximal outer spectral factor for T_R. Finally, $\{A_o^*, B_o, C_o, 0\}$ is a controllable and observable realization for $z^{-1}\Theta$. This verifies our claim.

Let us construct a sequence of finite dimensional operators to approximate the pair $\{A, B\}$. This approximation will be used to develop some computational algorithms. To this end, let $\mathcal{H}_n$ be the subspace of $\mathcal{K}$ defined by

$$\mathcal{H}_n = \bigvee_{j=0}^{n-1} U^j\Gamma\mathcal{E} \qquad (n \geq 1). \tag{8.1.9}$$

Let U_n be the operator on $\mathcal{H}_n$ and B_n the operator mapping $\mathcal{E}$ into $\mathcal{H}_n$ defined by

$$U_n = UP_{\mathcal{H}_{n-1}}|\mathcal{H}_n \text{ on } \mathcal{H}_n \quad \text{and} \quad B_n = \Gamma : \mathcal{E} \to \mathcal{H}_n. \tag{8.1.10}$$

Notice that $U_nP_{\mathcal{H}_n}$ converges to U in the strong operator topology, and B_n equals Γ for all n. By employing the definition of U_n and B_n, we have $U_n^jB_n = U^j\Gamma$ for all integers $j = 0, \ldots, n-1$. Thus we arrive at

$$B_n^*U_n^jB_n = \Gamma^*U^j\Gamma = R_{-j} \qquad (\text{for } j = 0, 1, 2, \ldots, n-1). \tag{8.1.11}$$

Remark 8.1.1. Assume that $U = S$ is the unilateral shift. Then U_n is stable, that is, all eigenvalues of U_n are contained in the open unit disc $\mathbb{D}$.

Since $U_n = UP_{\mathcal{H}_{n-1}}|\mathcal{H}_n$ where $U = S$ is an isometry, $\|U_n\| \leq 1$. In other words, U_n is a contraction and all eigenvalues of U_n are contained in the closed unit disc. To prove that U_n is stable, it remains to show that U_n has no eigenvalue on the unit circle. Let us proceed by contradiction. Assume that λ is an eigenvalue of U_n on the unit circle corresponding to the eigenvector x in $\mathcal{H}_n$. Then we obtain

$$\|x\| = |\lambda|\|x\| = \|\lambda x\| = \|U_nx\| = \|SP_{\mathcal{H}_{n-1}}x\| = \|P_{\mathcal{H}_{n-1}}x\|.$$

In other words, $\|x\| = \|P_{\mathcal{H}_{n-1}}x\|$ which implies that x is a vector in $\mathcal{H}_{n-1}$. Thus

$$Sx = SP_{\mathcal{H}_{n-1}}x = U_nx = \lambda x.$$

So λ must also be an eigenvalue for the unilateral shift. Recall that the unilateral shift has no eigenvalues; see Section 1.2. This leads to a contradiction. Hence none of the eigenvalues of U_n are on the unit circle. Therefore all eigenvalues of U_n are contained in $\mathbb{D}$ and U_n is stable.

Recall that $T_R = W^\sharp W$ where W is the controllability matrix determined by

$$W = \begin{bmatrix} \Gamma & U\Gamma & U^2\Gamma & \cdots \end{bmatrix}.$$

Let $T_{R,n}$ on $\mathcal{E}^n$ be the block Toeplitz matrix determined by the n by n block upper left-hand corner of T_R, that is,

$$T_{R,n} = \begin{bmatrix} R_0 & R_1^* & \cdots & R_{n-1}^* \\ R_1 & R_0 & \cdots & R_{n-2}^* \\ \vdots & \vdots & \ddots & \vdots \\ R_{n-1} & R_{n-2} & \cdots & R_0 \end{bmatrix} \text{ on } \mathcal{E}^n. \tag{8.1.12}$$

Let W_n be the operator mapping $\mathcal{E}^n$ onto $\mathcal{H}_n$ restricted to the first n columns of W, that is,

$$W_n = \begin{bmatrix} \Gamma & U\Gamma & \cdots & U^{n-1}\Gamma \end{bmatrix} : \mathcal{E}^n \to \mathcal{H}_n \subset \mathcal{K}.$$

Then $T_{R,n}$ admits a factorization of the form $T_{R,n} = W_n^* W_n$. Let J_n and Q_n be the operators mapping $\mathcal{E}^{n-1}$ into $\mathcal{E}^n$ defined by

$$J_n = \begin{bmatrix} I \\ 0 \end{bmatrix} : \mathcal{E}^{n-1} \to \begin{bmatrix} \mathcal{E}^{n-1} \\ \mathcal{E} \end{bmatrix} \quad \text{and} \quad Q_n = \begin{bmatrix} 0 \\ I \end{bmatrix} : \mathcal{E}^{n-1} \to \begin{bmatrix} \mathcal{E} \\ \mathcal{E}^{n-1} \end{bmatrix}. \tag{8.1.13}$$

Notice that $\mathcal{H}_{n-1} = \operatorname{ran} W_n J_n$. Moreover, $U_n W_n J_n = U W_n J_n = W_n Q_n$. The contraction $U_n = W_n Q_n (W_n J_n)^{-r}$ where L^{-r} denotes the Moore-Penrose pseudo-inverse of L. Finally, $B_n = W_n|\mathcal{E}$ where $\mathcal{E}$ is identified as the subspace corresponding to the first component of $\mathcal{E}^n$.

Let M_n be any operator mapping $\mathcal{E}^n$ onto $\mathcal{X}_n$ such that $T_{R,n} = M_n^* M_n$. Then there exists a unitary operator Φ mapping $\mathcal{X}_n$ onto $\mathcal{H}_n$ such that $W_n = \Phi M_n$. Using this, we obtain

$$U_n \Phi = \Phi[M_n Q_n (M_n J_n)^{-r}] \quad \text{and} \quad B_n = \Phi M_n|\mathcal{E}.$$

Clearly, the pairs $\{U_n, B_n\}$ and $\{M_n Q_n (M_n J_n)^{-r}, M_n|\mathcal{E}\}$ are unitarily equivalent. Due to this unitary equivalence, without loss of generality we can ignore the unitary operator Φ, and thus, U_n and B_n can be computed by

$$U_n = M_n Q_n (M_n J_n)^{-r} \quad \text{and} \quad B_n = M_n|\mathcal{E}. \tag{8.1.14}$$

Notice that U_n^* converges to U^* in the strong operator topology. (This is left as an exercise.) Recall that $\mathcal{X} = \bigvee_0^\infty U^{*j}\Gamma\mathcal{E}$ and $A^* = U^*|\mathcal{X}$. For large n and fixed $k \geq \dim \mathcal{X} = \delta(G)$, the number of significant singular values of

$$\Xi = \begin{bmatrix} B_n & U_n^* B_n & U_n^{*2} B_n & \cdots & U_n^{*k} B_n \end{bmatrix} \tag{8.1.15}$$

is equal to the dimension of $\mathcal{X}$. Moreover, one can use the singular value decomposition of Ξ to find an approximation of $\{A, B\}$ and this readily leads to the outer spectral factor and the unitary part V. This computational method is described in the following algorithm. Finally, it is emphasized that this method does not require Θ to be square.

Because the pair $\{V \text{ on } \mathcal{V}, \Gamma_2\}$ acts on finite dimensional spaces and V is unitary, by performing the appropriate unitary transformation, without loss of generality we can assume that V is a diagonal operator with distinct eigenvalues $e^{\imath\omega_j}$ with multiplicity ν_j, that is,

$$V = \begin{bmatrix} e^{\imath\omega_1}I & 0 & \cdots & 0 \\ 0 & e^{\imath\omega_2}I & \cdots & 0 \\ \vdots & \vdots & \ddots & \vdots \\ 0 & 0 & \cdots & e^{\imath\omega_k}I \end{bmatrix} \text{ on } \begin{bmatrix} \mathbb{C}^{\nu_1} \\ \mathbb{C}^{\nu_2} \\ \vdots \\ \mathbb{C}^{\nu_k} \end{bmatrix} \text{ and } \Gamma_2 = \begin{bmatrix} \Gamma_{2,1} \\ \Gamma_{2,2} \\ \vdots \\ \Gamma_{2,k} \end{bmatrix}. \tag{8.1.16}$$

Here $\Gamma_{2,j}$ is an operator mapping $\mathcal{E}$ onto $\mathbb{C}^{\nu_j}$ for $j = 1, 2, \ldots, k$, and $\mathcal{V} = \oplus_1^k \mathbb{C}^{\nu_j}$.

The Wold decomposition algorithm

Assume that G in (8.1.1) is a rational function corresponding to a positive Toeplitz matrix T_R. We say that $\widetilde{U}\Lambda\widetilde{V}^*$ is the *singular value decomposition* for a finite dimensional operator T if Λ is the (rectangular) diagonal matrix consisting of the singular values for T while $\widetilde{U}$ and $\widetilde{V}$ are unitary operators acting between the appropriate spaces.

(i) For sufficiently large n, construct $T_{R,n}$ on $\mathcal{E}^n$.

(ii) Compute a factorization $T_{R,n} = M_n^* M_n$ where M_n maps $\mathcal{E}^n$ onto $\mathcal{X}_n$.

(iii) Construct J_n and Q_n in (8.1.13). Then compute $U_n = M_n Q_n (M_n J_n)^{-r}$ and $B_n = M_n|\mathcal{E}$.

(iv) Now construct the controllability matrix

$$\Xi_k = \begin{bmatrix} B_n & U_n^* B_n & U_n^{*2} B_n & \cdots & U_n^{*k} B_n \end{bmatrix}$$

for a fixed $k \geq \delta(G)$, or an integer k such that the number of significant singular values for Ξ_k and Ξ_{k+1} are the same.

(v) Let $\widetilde{U}\Lambda\widetilde{V}^*$ be the singular values decomposition of Ξ_k. By selecting the first μ largest significant singular values $\sigma_1 \geq \sigma_2 \geq \cdots \geq \sigma_\mu > 0$ where $\sigma_{\mu+1} \approx 0$, set $\widetilde{\Phi} = \widetilde{U}|\mathbb{C}^\mu$ by keeping the first μ columns of $\widetilde{U}$.

(vi) Compute $A = \widetilde{\Phi}^* U_n \widetilde{\Phi}$ and $B = \widetilde{\Phi}^* B_n$.

(vii) Compute the matrix representation of A and B of the form

$$A = \begin{bmatrix} A_o & 0 \\ 0 & V \end{bmatrix} \text{ on } \begin{bmatrix} \mathcal{X}_o \\ \mathcal{V} \end{bmatrix} \quad \text{and} \quad B = \begin{bmatrix} B_o \\ \Gamma_2 \end{bmatrix} : \mathcal{E} \to \begin{bmatrix} \mathcal{X}_o \\ \mathcal{V} \end{bmatrix}.$$

Here A_o is stable and V is unitary.

(viii) Let $\widetilde{U}_o\Lambda_o\widetilde{V}_o^*$ be the singular value decomposition of $I - A_oA_o^*$. Let Λ_q be the q by q diagonal matrix contained in the upper left-hand corner of Λ_o where q is the number of significant nonzero singular values of Λ_o, or equivalently, $I - A_oA_o^*$. Compute the matrix C_o mapping $\mathcal{X}_o$ onto $\mathbb{C}^q$ by

$$C_o = \begin{bmatrix}\Lambda_q^{1/2} & 0\end{bmatrix}\widetilde{V}_o^* : \mathcal{X}_o \to \mathbb{C}^q.$$

Observe that $C_o^*C_o = I - A_oA_o^*$.

(ix) The maximal outer spectral factor Θ for T_R is given by the state space realization

$$\Theta(z) \approx zC_o(zI - A_o^*)^{-1}B_o.$$

Here Θ is a rational function in $H^\infty(\mathcal{E}, \mathbb{C}^q)$.

(x) Finally, $\{V, \Gamma_2\}$ is our approximation for the unitary part of $\{U, \Gamma\}$. In particular, one can use a unitary transformation to convert $\{V, \Gamma_2\}$ to a diagonal representation $\{V, \Gamma_2\}$ given in (8.1.16).

It is noted that this algorithm converges faster when the unitary part is not present in the Wold decomposition. In Section 8.4, we will show how the lower triangular Cholesky factorization can also be used to compute the maximal outer spectral factor. Finally, it is noted that one disadvantage of this method is that the size n of $T_{R,n}$ is not known a priori.

Remark 8.1.2. One can also compute the maximal outer spectral factor Θ for a rational positive Toeplitz matrix T_R and its corresponding unitary pair $\{V, \Gamma_2\}$ directly from $\{U_n, B_n\}$, when n is sufficiently large. To this end, let C_n be any matrix mapping $\mathcal{H}_n$ onto $\mathbb{C}^q$ such that $C_n^*C_n = I - U_nU_n^*$. Then

$$\Theta(z) \approx zC_n(zI - U_n^*)^{-1}B_n.$$

It is noted that this realization is not necessarily minimal. One can extract a minimal realization for Θ from the system $\{U_n^*, U_n^*B_n, C_n, C_nB_n\}$. In fact, one can run the Kalman-Ho algorithm on $\{C_nU_n^{*j}B_n\}$ to compute a minimal realization $\{A, B, C, D\}$ for Θ.

To compute $\{V, \Gamma_2\}$, let Ψ be an isometry mapping $\mathbb{C}^m$ into $\mathcal{H}_n$ such that $U_n\Psi = \Psi V$ where V is a unitary matrix of the form (8.1.16) consisting of all the eigenvalues for U_n of modulus 1. Set $\Gamma_2 = \Psi^*B$. This yields $\{V, \Gamma_2\}$ in (8.1.16). Finally, it is noted that this computational method involves larger matrices.

8.2 A Basic Optimization Problem

Let $\Theta(z) = \sum_0^\infty z^{-n}\Theta_n$ in $H^2(\mathcal{E}, \mathcal{Y})$ be the maximal outer spectral factor for a positive Toeplitz matrix T_R. Recall that Θ is unique up to a unitary operator on the left. Let

$$\Omega_k = \Pi_{\mathcal{Y}^k} T_\Theta | \mathcal{E}^k = \begin{bmatrix} \Theta_0 & 0 & \cdots & 0 \\ \Theta_1 & \Theta_0 & \cdots & 0 \\ \vdots & \vdots & \ddots & \vdots \\ \Theta_{k-1} & \Theta_{k-2} & \cdots & \Theta_0 \end{bmatrix} : \mathcal{E}^k \to \mathcal{Y}^k \tag{8.2.1}$$

be the $k \times k$ block lower triangular Toeplitz matrix contained in the upper left-hand corner of T_Θ. Let $T_{R,n}$ on $\mathcal{E}^n$ be the block Toeplitz matrix contained in the upper $n \times n$ left-hand corner of T_R; see (8.1.12). For a fixed $k \le n$, the space $\mathcal{E}^n$ admits a decomposition of the form $\mathcal{E}^n = \mathcal{E}^k \oplus \mathcal{E}^{n-k}$ where $\mathcal{E}^k$ is embedded in the first k components of $\mathcal{E}^n$, while $\mathcal{E}^{n-k}$ is embedded in the last $n-k$ components of $\mathcal{E}^n$. For a fixed $k \le n$ and a specified vector f in $\mathcal{E}^k$, consider the optimization problem

$$(\Delta_{k,n} f, f) = \inf\{(T_{R,n}(f \oplus h), (f \oplus h)) : h \in \mathcal{E}^{n-k}\}. \tag{8.2.2}$$

The solution to this optimization problem is not necessarily unique. However, the cost $\rho(f) = (\Delta_{k,n} f, f)$ where $\Delta_{k,n}$ is a positive operator on $\mathcal{E}^k$.

To obtain an expression for $\Delta_{k,n}$, let M_n be any operator mapping $\mathcal{E}^n$ onto $\mathcal{H}_n$ such that $T_{R,n} = M_n^* M_n$. In fact, one could choose $M_n = T_{R,n}^{1/2}$. Now let $\mathcal{G}$ be the subspace defined by $\mathcal{G} = M_n(0 \oplus \mathcal{E}^{n-k})$. According to the projection theorem, we have

$$\begin{aligned} \rho(f) &= \inf\{(T_{R,n}(f \oplus h), (f \oplus h)) : h \in \mathcal{E}^{n-k}\} \\ &= \inf\{\|M_n(f \oplus h)\|^2 : h \in \mathcal{E}^{n-k}\} \\ &= \inf\{\|M_n(f \oplus 0) - M_n(0 \oplus h)\|^2 : h \in \mathcal{E}^{n-k}\} \\ &= \inf\{\|M_n(f \oplus 0) - \varphi\|^2 : \varphi \in M_n(0 \oplus \mathcal{E}^{n-k})\} \\ &= \|(I - P_\mathcal{G}) M_n (f \oplus 0)\|^2 = (\Pi_{\mathcal{E}^k} M_n^* (I - P_\mathcal{G}) M_n \Pi_{\mathcal{E}^k}^* f, f). \end{aligned}$$

Hence $\rho(f) = (\Delta_{k,n} f, f)$ where

$$\Delta_{k,n} = \Pi_{\mathcal{E}^k} M_n^* (I - P_\mathcal{G}) M_n \Pi_{\mathcal{E}^k}^*. \tag{8.2.3}$$

Remark 8.2.1. One can use the eigenvalue decomposition of $T_{R,n}$ to compute the cost function $\Delta_{k,n}$ in the optimization problem 8.2.2. To see this, let $\Psi \Lambda \Psi^* = T_{R,n}$ be the eigenvalue decomposition of $T_{R,n}$ where Λ on $\mathbb{C}^\nu$ is a diagonal operator consisting of the nonzero eigenvalues of $T_{R,n}$ and Ψ is an isometry mapping $\mathbb{C}^\nu$ into $\mathcal{E}^n$. Let Φ_1 be the operator mapping $\mathcal{E}^k$ into $\mathbb{C}^\nu$ defined by $\Phi_1 = \Lambda^{1/2} \Psi^* | \mathcal{E}^k$. Let Φ be any isometry mapping C^m into $\mathbb{C}^\nu$ whose range equals the range of $\Lambda^{1/2} \Psi^* (0 \oplus \mathcal{E}^{n-k})$. (One can use the singular value decomposition or the command "orth" in Matlab to compute the isometry Φ.) Then

$$\Delta_{k,n} = \Phi_1^* \left(I - \Phi \Phi^*\right) \Phi_1. \tag{8.2.4}$$

To see this simply observe that $T_{R,n} = M_n^* M_n$ where $M_n = \Lambda^{1/2}\Psi^*$. In this case, $M_n \Pi_{\mathcal{E}^k}^* = \Phi_1$ and $P_{\mathcal{G}} = \Phi\Phi^*$ is the orthogonal projection onto

$$M_n|(0 \oplus \mathcal{E}^{n-k}) = \Lambda^{1/2}\Psi^*(0 \oplus \mathcal{E}^{n-k}).$$

So equation (8.2.4) follows from (8.2.3).

Remark 8.2.2. Assume that $T_{R,n}$ is strictly positive. Consider the decomposition of $T_{R,n}$ given by

$$T_{R,n} = \begin{bmatrix} T_{R,k} & X^* \\ X & T_{R,n-k} \end{bmatrix} \text{ on } \begin{bmatrix} \mathcal{E}^k \\ \mathcal{E}^{n-k} \end{bmatrix}. \tag{8.2.5}$$

Then Lemma 7.2.1 shows that $\Delta_{k,n}$ is the Schur complement of $T_{R,n}$, that is,

$$\Delta_{k,n} = T_{R,k} - X^* T_{R,n-k}^{-1} X. \tag{8.2.6}$$

In other words, the cost $\Delta_{k,n}$ in the optimization problem 8.2.2 is precisely the Schur complement of $T_{R,k}$ in the matrix representation (8.2.5).

Theorem 8.2.3. *Let $\Theta(z) = \sum_0^\infty z^{-j}\Theta_j$ in $H^2(\mathcal{E}, \mathcal{Y})$ be the maximal outer spectral factor for a positive block Toeplitz matrix T_R with $\mathcal{L}(\mathcal{E}, \mathcal{E})$-valued symbol R. For a fixed $k \le n$ and f in $\mathcal{E}^k$, let $\Delta_{k,n}$ be the cost in the optimization problem*

$$(\Delta_{k,n} f, f) = \inf\{(T_{R,n}(f \oplus h), (f \oplus h)) : h \in \mathcal{E}^{n-k}\} \tag{8.2.7}$$

where $T_{R,n}$ on $\mathcal{E}^n$ is the block Toeplitz matrix contained in the upper $n \times n$ left-hand corner of T_R; see (8.1.12). *Then for k fixed $\{\Delta_{k,n}\}$ forms a decreasing sequence of positive operators, that is, $\Delta_{k,n+1} \le \Delta_{k,n}$. Moreover,*

$$\lim_{n\to\infty} \Delta_{k,n} = \Omega_k^* \Omega_k \tag{8.2.8}$$

where $\Omega_k = \Pi_{\mathcal{Y}^k} T_\Theta | \mathcal{E}^k$; see (8.2.1). *In particular, $\Delta_{1,n}$ converges to $\Theta(\infty)^*\Theta(\infty)$ as n tends to infinity.*

Proof. Observe that $T_{R,n}$ is contained in the $n \times n$ upper left-hand corner of $T_{R,n+1}$. To show that for k fixed $\{\Delta_{k,n}\}$ is decreasing, notice that

$$\begin{aligned}(\Delta_{k,n+1} f, f) &= \inf\{(T_{R,n+1}(f \oplus h), (f \oplus h)) : h \in \mathcal{E}^{n+1-k}\} \\ &\le \inf\{(T_{R,n+1}(f \oplus h \oplus 0), (f \oplus h \oplus 0)) : h \oplus 0 \in \mathcal{E}^{n-k} \oplus \mathcal{E}\} \\ &= \inf\{(T_{R,n}(f \oplus h), (f \oplus h)) : h \in \mathcal{E}^{n-k}\} = (\Delta_{k,n} f, f).\end{aligned}$$

Hence $\Delta_{k,n+1} \le \Delta_{k,n}$ and the sequence $\{\Delta_{k,n}\}$ is decreasing in n.

To obtain the limit in (8.2.8), let $\{U \text{ on } \mathcal{K}, \Gamma\}$ be a controllable isometric representation for T_R. Let W_n be the controllability matrix determined by

$$W_n = \begin{bmatrix} \Gamma & U\Gamma & \cdots & U^{n-1}\Gamma \end{bmatrix} : \mathcal{E}^n \to \mathcal{K}.$$

Let $U = S \oplus V$ on $\ell^2_+(\mathcal{Y}) \oplus \mathcal{V}$ and $\Gamma = \begin{bmatrix}\Gamma_1 & \Gamma_2\end{bmatrix}^{tr}$ be the Wold decomposition for $\{U, \Gamma\}$ where S is a unilateral shift and V is unitary; see equation (8.1.2). In this decomposition $\Gamma_1 = \begin{bmatrix}\Theta_0 & \Theta_1 & \Theta_2 & \cdots\end{bmatrix}^{tr}$. The Wold decomposition for $\{U, \Gamma\}$ shows that W_n admits a matrix decomposition of the form

$$W_n = \begin{bmatrix} T_\Theta \\ \star \end{bmatrix} |\mathcal{E}^n :\to \begin{bmatrix} \ell^2_+(\mathcal{Y}) \\ \mathcal{V} \end{bmatrix} \tag{8.2.9}$$

where $\star$ represents an unspecified operator; see Section 5.2. Recall that $\mathcal{Y} = \mathcal{K} \ominus U\mathcal{K}$ is the wandering subspace for U which determines S. In particular,

$$\mathcal{K} \ominus U^k\mathcal{K} = \oplus_{j=0}^{k-1} U^j \mathcal{Y} = \begin{bmatrix} \oplus_{j=0}^{k-1} S^j \mathcal{Y} \\ 0 \end{bmatrix} = \begin{bmatrix} \mathcal{Y}^k \\ 0 \end{bmatrix}$$

where $\mathcal{Y}^k$ is viewed as the first k components of $\ell^2_+(\mathcal{Y})$. This with (8.2.9) readily implies that

$$\Pi_{\mathcal{K}\ominus U^k\mathcal{K}} W_n|\mathcal{E}^k = \Pi_{\mathcal{Y}^k} T_\Theta|\mathcal{E}^k = \Omega_k \qquad (k \leq n).$$

Because the pair $\{U, \Gamma\}$ is controllable, $\bigvee_0^\infty U^j \Gamma \mathcal{E} = \mathcal{K}$, and thus, $U^k\mathcal{K} = \bigvee_k^\infty U^j \Gamma \mathcal{E}$. In particular, $U^k\mathcal{K} \supseteq \bigvee_k^{n-1} U^j \Gamma \mathcal{E}$. Using this along with the fact that $T_{R,n} = W_n^* W_n$, we obtain

$$\begin{aligned}
\|\Omega_k f\|^2 &= \|\Pi_{\mathcal{K}\ominus U^k\mathcal{K}} W_k f\|^2 \\
&= \inf\{\|W_k f - h\|^2 : h \in U^k\mathcal{K}\} \\
&\leq \inf\{\|W_k f - h\|^2 : h \in \bigvee_{j=k}^{n-1} U^j \Gamma \mathcal{E}\} \\
&= \inf\{\|W_n(f \oplus h)\|^2 : h \in \mathcal{E}^{n-k}\} \\
&= \inf\{(T_{R,n}(f \oplus h), (f \oplus h)) : h \in \mathcal{E}^{n-k}\} \\
&= (\Delta_{k,n} f, f).
\end{aligned} \tag{8.2.10}$$

Hence $\Omega_k^* \Omega_k \leq \Delta_{k,n}$. Since $\bigvee_k^\infty U^j \Gamma \mathcal{E} = U^k\mathcal{K}$, the inequality in (8.2.10) shows that $\Delta_{k,n}$ converges to $\Omega_k^* \Omega_k$ as n tends to infinity.

Finally for $k = 1$, the matrix $\Omega_1 = \Theta(\infty) = \Theta_0$. So in this case, $\Delta_{1,n}$ converges to $\Theta(\infty)^* \Theta(\infty)$ as n tends to infinity. □

8.3 The Lower Triangular Cholesky Factorization

In this section, we will develop the lower triangular Cholesky factorization for a positive matrix. In the next section, we will use this factorization to obtain an algorithm to compute the maximal outer spectral factor.

Let M be an operator mapping $\mathbb{C}^\mu$ into $\mathbb{C}^\nu$. Let $\{m_{j,k}\}_{1,1}^{\nu,\mu}$ be the entries of M. Let k_j denote the position in the j-th row of M corresponding to the last nonzero element in this row, that is, $m_{j,k_j} \neq 0$ and $m_{j,k_j+i} = 0$ for all $i \geq 1$.

Definition 8.3.1. A matrix M mapping $\mathbb{C}^\mu$ into $\mathbb{C}^\nu$ is a lower triangular Cholesky matrix if the following conditions hold:

(i) The number of rows of M is less than or equal to the number of columns, that is, $\nu \leq \mu$.

(ii) The entries $m_{j,k_j} > 0$ for all $j = 1, 2, \ldots, \nu$.

(iii) The entries $m_{j,q} = 0$ for all $q > k_j$.

(iv) $k_1 < k_2 < \cdots < k_\nu$.

A generic structure of a lower triangular Cholesky matrix is given by

$$\begin{bmatrix} \star & \star & m_{1,k_1} & & & & & & & & & \\ \star & \star & \star & \star & \star & m_{2,k_2} & & & & & & \\ \vdots & \vdots & \vdots & \vdots & \vdots & \vdots & \ddots & & & & & \\ \star & \star & \star & \star & \star & \star & \star & m_{\nu-1,k_{\nu-1}} & & & & \\ \star & \star & \star & \star & \star & \star & \star & \star & \star & \star & m_{\nu,k_\nu} & \delta \end{bmatrix}. \tag{8.3.1}$$

Here $\star$ represents unspecified entries, the blank spots represent zero elements and the last nonzero element of every row is strictly positive. Furthermore, δ is a row of zeros or δ is empty. Finally, it is noted that the rank of the lower triangular Cholesky matrix mapping $\mathbb{C}^\mu$ into $\mathbb{C}^\nu$ is ν.

We say that M is a *lower triangular Cholesky factor* for a positive operator T on $\mathbb{C}^\mu$, if M is a lower Cholesky matrix satisfying $T = M^*M$.

Theorem 8.3.2. *Let T be a positive operator on $\mathbb{C}^\mu$. Then T admits a unique lower triangular Cholesky factor.*

Proof. Let $T = \Psi\Lambda\Psi^*$ be the eigenvalue decomposition for T where Λ is a nonsingular diagonal matrix on $\mathbb{C}^\nu$ consisting of the nonzero eigenvalues for T and Ψ is an isometry from $\mathbb{C}^\nu$ into $\mathbb{C}^\mu$. Let $QM = \Lambda^{1/2}\Psi^*$ be the factorization of $\Lambda^{1/2}\Psi^*$ where Q is a unitary operator on $\mathbb{C}^\nu$ and M is a lower triangular matrix mapping $\mathbb{C}^\mu$ into $\mathbb{C}^\nu$. Because $\Lambda^{1/2}\Psi^*$ is onto $\mathbb{C}^\nu$, the matrix M is also onto $\mathbb{C}^\nu$. By multiplying each row of M by the appropriate complex number on the unit circle, with out loss of generality, we can assume that the last nonzero entry in each row of M is strictly positive. In other words, M is a lower triangular Cholesky matrix. Since Q is unitary, $M = Q^*\Lambda^{1/2}\Psi^*$ satisfies $M^*M = \Psi\Lambda\Psi^* = T$. Therefore M is a lower triangular Cholesky factor for T. (Most algorithms compute a factorization of the form $QR = A$ for a matrix A where Q is unitary and R is an upper triangular matrix. In Matlab $[Q, R] = \text{qr}(A)$. However, reversing the columns of A and then reversing the rows and columns of R yields a lower triangular matrix M.)

Assume that N is another lower triangular Cholesky factor for T. Then $N^*N = M^*M$, and thus, there exists a unitary operator Z such that $ZM = N$. Lemma 8.3.3 below shows that $Z = I$, or equivalently, $M = N$. Therefore the lower triangular Cholesky factor is uniquely determined by T. □

A Gram-Schmidt orthogonalization proof of Theorem 8.3.2. Let us present another proof based on the Gram-Schmidt orthogonalization procedure. To this end, consider the Hilbert space $\mathcal{H}$ determined by the inner product

$$(x,y)_{\mathcal{H}} = (Tx,y)_{\mathbb{C}^\mu} \qquad (x,y \in \mathbb{C}^\mu).$$

To guarantee that $\mathcal{H}$ is a Hilbert space, we say that x is zero in $\mathcal{H}$ if and only if x is in the kernel of T. In other words, $x = y$ in $\mathcal{H}$ if and only if $x - y$ is in the kernel of T. The dimension of $\mathcal{H}$ equals the rank of T. Let $\{e_j\}_1^\mu$ be the standard orthonormal basis for $\mathbb{C}^\mu$, that is, the j-th component of e_j is one and all the other components are zero. Let $\{\varphi_\nu, \varphi_{\nu-1}, \dots, \varphi_1\}$ be the orthonormal basis for $\mathcal{H}$ obtained by applying the Gram-Schmidt orthogonalization procedure to $\{e_\mu, e_{\mu-1}, \dots, e_1\}$. It is emphasized that the Gram-Schmidt procedure is done in reverse order starting with e_μ, and $\nu = \operatorname{rank} T$. Let Φ be the unitary matrix mapping $\mathbb{C}^\nu$ onto $\mathcal{H}$ defined by

$$\Phi = \begin{bmatrix}\varphi_1 & \varphi_2 & \cdots & \varphi_\nu\end{bmatrix} : \mathbb{C}^\nu \to \mathcal{H}.$$

Because the Gram-Schmidt orthogonalization procedure is done in reverse order, and $\mathcal{H}$ is composed of vectors in $\mathbb{C}^\mu$, the unitary operator Φ also defines a lower triangular matrix mapping $\mathbb{C}^\nu$ into $\mathbb{C}^\mu$. The adjoint Φ^* mapping $\mathcal{H}$ into $\mathbb{C}^\nu$ is given by $\Phi^* = \overline{\Phi}^{tr} T$ where $\overline{\Phi}^{tr}$ is the conjugate transpose of the matrix Φ. Let M be the operator mapping $\mathbb{C}^\mu$ into $\mathbb{C}^\nu$ determined by $M = \overline{\Phi}^{tr} T$. In this case, $M^* = \overline{M}^{tr}$. For x in $\mathbb{C}^\mu$, we have

$$(M^*Mx,x)_{\mathbb{C}^\mu} = \|Mx\|^2_{\mathbb{C}^\nu} = \|\Phi^*x\|^2_{\mathbb{C}^\nu} = \|x\|^2_{\mathcal{H}} = (Tx,x)_{\mathbb{C}^\mu}.$$

In other words, $(M^*Mx,x) = (Tx,x)$ for all x in $\mathbb{C}^\mu$. Therefore $T = M^*M$.

Since $T = M^*M$, to complete the proof, it remains to show that M is a lower triangular Cholesky matrix. The entries $m_{i,j}$ of M are determined by

$$m_{i,j} = (Me_j, e_i) = (\overline{\Phi}^{tr} Te_j, e_i) = (Te_j, \Phi e_i) = (Te_j, \varphi_i) = (e_j, \varphi_i)_{\mathcal{H}}.$$

Therefore the components of M are given by $m_{i,j} = (e_j, \varphi_i)_{\mathcal{H}}$.

We claim that M is a lower triangular Cholesky matrix. Starting the Gram-Schmidt orthogonalization procedure with e_μ, let φ_ν be the unit vector defined by $\varphi_\nu = e_\mu / \|e_\mu\|_{\mathcal{H}}$. Then $m_{\nu,\mu} = (e_\mu, \varphi_\nu)_{\mathcal{H}} = \|e_\mu\|_{\mathcal{H}} > 0$. (Without loss of generality, we assume that the last column of T is nonzero, and thus, e_μ is not in the kernel of T.) Let k be the largest integer strictly less than μ such that e_k is not an element in the subspace $\mathcal{G}_\nu = \operatorname{span}\{\varphi_\nu\}$. Let $\varphi_{\nu-1}$ be the next unit vector computed by the Gram-Schmidt orthogonalization procedure, that is, $\varphi_{\nu-1} = \tilde{\varphi}_{\nu-1} / \|\tilde{\varphi}_{\nu-1}\|_{\mathcal{H}}$ where $\tilde{\varphi}_{\nu-1} = e_k - P_{\mathcal{G}_\nu} e_k$. The $\nu - 1$ row of M is given by

$$\begin{bmatrix}(e_1, \varphi_{\nu-1})_{\mathcal{H}} & (e_2, \varphi_{\nu-1})_{\mathcal{H}} & \cdots & (e_k, \varphi_{\nu-1})_{\mathcal{H}} & (e_{k+1}, \varphi_{\nu-1})_{\mathcal{H}} & \cdots & (e_\mu, \varphi_{\nu-1})_{\mathcal{H}}\end{bmatrix}.$$

Since $e_{k+1}, \dots, e_\mu$ are contained in $\mathcal{G}_\nu$, the inner products $(e_j, \varphi_{\nu-1})_{\mathcal{H}}$ are all zero for $j > k$, that is, the $\nu - 1$ row of M is given by

$$\begin{bmatrix}(e_1, \varphi_{\nu-1})_{\mathcal{H}} & (e_2, \varphi_{\nu-1})_{\mathcal{H}} & \cdots & (e_k, \varphi_{\nu-1})_{\mathcal{H}} & 0 & \cdots & 0\end{bmatrix}.$$

Using the fact that $P_{\mathcal{G}_\nu} e_k$ is orthogonal to $e_k - P_{\mathcal{G}_\nu} e_k = \tilde{\varphi}_{\nu-1}$ in $\mathcal{H}$, we obtain

$$\begin{aligned}(e_k, \tilde{\varphi}_{\nu-1})_{\mathcal{H}} &= (e_k, e_k - P_{\mathcal{G}_\nu} e_k)_{\mathcal{H}} = (e_k - P_{\mathcal{G}_\nu} e_k, e_k - P_{\mathcal{G}_\nu} e_k)_{\mathcal{H}} \\ &= \|e_k - P_{\mathcal{G}_\nu} e_k\|_{\mathcal{H}}^2 = \|\tilde{\varphi}_{\nu-1}\|_{\mathcal{H}}^2 > 0.\end{aligned}$$

Since $\varphi_{\nu-1} = \tilde{\varphi}_{\nu-1}/\|\tilde{\varphi}_{\nu-1}\|_{\mathcal{H}}$, it follows that

$$m_{\nu-1,k} = (e_k, \varphi_{\nu-1})_{\mathcal{H}} = \frac{(e_k, \tilde{\varphi}_{\nu-1})_{\mathcal{H}}}{\|\tilde{\varphi}_{\nu-1}\|_{\mathcal{H}}} = \|e_k - P_{\mathcal{G}_\nu} e_k\|_{\mathcal{H}} > 0.$$

Hence the last nonzero element of the $\nu - 1$ row of M is strictly positive.

Let q be the next largest integer less than k such that e_q is not in the subspace $\mathcal{G}_{\nu-1} = \operatorname{span}\{\varphi_\nu, \varphi_{\nu-1}\}$. Then the $\nu - 2$ row of M is determined by

$$\begin{bmatrix}(e_1, \varphi_{\nu-2})_{\mathcal{H}} & (e_2, \varphi_{\nu-2})_{\mathcal{H}} & \cdots & (e_q, \varphi_{\nu-2})_{\mathcal{H}} & 0 & \cdots & 0\end{bmatrix}.$$

The zeros after $(e_q, \varphi_{\nu-2})_{\mathcal{H}}$ follow from the fact that $\{e_j\}_{q+1}^{\mu}$ are contained in $\mathcal{G}_{\nu-1}$. In this case, $\varphi_{\nu-2} = \tilde{\varphi}_{\nu-2}/\|\tilde{\varphi}_{\nu-2}\|_{\mathcal{H}}$ where $\tilde{\varphi}_{\nu-2} = e_q - P_{\mathcal{G}_{\nu-1}} e_q$. Using the fact that $P_{\mathcal{G}_{\nu-1}} e_q$ is orthogonal to $e_q - P_{\mathcal{G}_{\nu-1}} e_q$, we obtain

$$m_{\nu-2,q} = (e_q, \varphi_{\nu-2})_{\mathcal{H}} = \frac{(e_q, \tilde{\varphi}_{\nu-2})_{\mathcal{H}}}{\|\tilde{\varphi}_{\nu-2}\|_{\mathcal{H}}} = \|e_q - P_{\mathcal{G}_{\nu-1}} e_q\|_{\mathcal{H}} > 0.$$

In other words, the last nonzero element of the $\nu - 2$ row of M is also strictly positive. By continuing in this fashion, we see that M is a lower triangular Cholesky matrix and

$$\begin{aligned}m_{j,k_j} &= \|\tilde{\varphi}_j\|_{\mathcal{H}} = \|e_{k_j} - P_{\mathcal{G}_{j+1}} e_{k_j}\|_{\mathcal{H}} > 0, \\ \mathcal{G}_{j+1} &= \operatorname{span}\{e_\eta : \eta > k_j\}\end{aligned} \tag{8.3.2}$$

where k_j is the last nonzero entry in the j-th row of M. Since $T = M^*M$, the operator M is a lower triangular Cholesky factor for T. Lemma 8.3.3 below shows that the lower triangular Cholesky factor of T is unique. □

Lemma 8.3.3. *If A and B are two lower triangular Cholesky matrices such that $ZA = B$ where Z is a unitary operator, then $A = B$.*

Proof. Without loss of generality, we can assume that both A and B map $\mathbb{C}^\mu$ onto $\mathbb{C}^\nu$. Let $\{e_j\}$ be the standard orthonormal basis and let a_{ij} and b_{ij} be entries of A and B, respectively. By eliminating the appropriate number of columns, without loss of generality we can assume that the last column of A and B are nonzero. In this case, $a_{\nu\mu}$ and $b_{\nu\mu}$ are the only nonzero elements of the last column of A and B. Moreover, both $a_{\nu\mu}$ and $b_{\nu\mu}$ are strictly positive. Observe that

$$a_{\nu\mu} Z e_\mu = Z a_{\nu\mu} e_\mu = Z A e_\mu = B e_\mu = b_{\nu\mu} e_\mu.$$

In other words, $a_{\nu\mu}\|Ze_\mu\| = b_{\nu\mu}\|e_\mu\|$. Since Z is unitary and $\|e_\mu\| = 1$, it follows that $a_{\nu\mu} = b_{\nu\mu}$. Moreover, $Ze_\mu = e_\mu$. Because Z is unitary, it must admit a matrix decomposition of the form

$$Z = \begin{bmatrix} Z_{11} & 0 \\ 0 & 1 \end{bmatrix} : \begin{bmatrix} \mathbb{C}^{\nu-1} \\ \mathbb{C} \end{bmatrix} \to \begin{bmatrix} \mathbb{C}^{\nu-1} \\ \mathbb{C} \end{bmatrix}.$$

Furthermore, Z_{11} is a unitary operator on $\mathbb{C}^{\nu-1}$. Now observe that A and B admit decomposition of the form

$$A = \begin{bmatrix} A_{11} & 0 \\ A_{21} & a_{\nu\mu} \end{bmatrix} : \begin{bmatrix} \mathbb{C}^{\mu-1} \\ \mathbb{C} \end{bmatrix} \to \begin{bmatrix} \mathbb{C}^{\nu-1} \\ \mathbb{C} \end{bmatrix},$$

$$B = \begin{bmatrix} B_{11} & 0 \\ B_{21} & b_{\nu\mu} \end{bmatrix} : \begin{bmatrix} \mathbb{C}^{\mu-1} \\ \mathbb{C} \end{bmatrix} \to \begin{bmatrix} \mathbb{C}^{\nu-1} \\ \mathbb{C} \end{bmatrix}.$$

Using this decomposition with $ZA = B$, we can conclude that $A_{21} = B_{21}$ and $Z_{11}A_{11} = B_{11}$. Because A and B are lower triangular Cholesky matrices, A_{11} and B_{11} are also lower triangular Cholesky matrices. By repeating the previous analysis $\nu - 1$ times, we arrive at $Z_{11} = I$, and thus, $Z = I$. Therefore $A = B$. □

Lemma 8.3.4. *Consider a sequence $\{\Delta_n\}_1^\infty$ of positive decreasing matrices on $\mathbb{C}^\mu$, that is, $0 \le \Delta_{n+1} \le \Delta_n$ for all integers $n \ge 1$. Let T be the positive operator on $\mathbb{C}^\mu$ defined by*

$$T = \lim_{n\to\infty} \Delta_n. \tag{8.3.3}$$

Assume that all the lower triangular Cholesky factors A_n for Δ_n is a sequence of matrices mapping $\mathbb{C}^\mu$ into $\mathbb{C}^\nu$ and $\inf\{(A_n)_{j,k_j}\} > 0$ for all j and n. Then

$$\Omega = \lim_{n\to\infty} A_n \tag{8.3.4}$$

where Ω is the lower triangular Cholesky factor for T. In particular, Ω is a lower triangular Cholesky matrix mapping $\mathbb{C}^\mu$ into $\mathbb{C}^\nu$. Finally, in the special case when T is a strictly positive operator on $\mathbb{C}^\mu$, then $\{A_n\}$ are lower triangular Cholesky matrices on $\mathbb{C}^\mu$ and A_n converges to Ω.

Proof. Let $k_j(n)$ denote the position in the j-th row of A_n corresponding to the last nonzero element in this row. Let E_n be the isometry mapping $\mathbb{C}^\nu$ into $\mathbb{C}^\mu$ defined by

$$E_n = \begin{bmatrix} e_{k_1(n)} & e_{k_2(n)} & \cdots & e_{k_\nu(n)} \end{bmatrix} : \mathbb{C}^\nu \to \mathbb{C}^\mu$$

where $e_{k_j(n)}$ is the unit vector in $\mathbb{C}^\mu$ placing 1 in the $k_j(n)$ position and zeros elsewhere. Let B_n be the lower triangular Cholesky matrix on $\mathbb{C}^\nu$ defined by $B_n = A_nE_n$. Notice that $\{(A_n)_{j,k_j}\}_{j=1}^\nu$ appears on the diagonal entries of B_n. Using the hypothesis $\inf\{(A_n)_{j,k_j}\} > 0$ for all j and n, it follows that $\det[B_n] \ge \epsilon$ for some $\epsilon > 0$. Because $A_n^*A_n$ converges to T and $\|B_n\| \le \|A_n\|$, there exists a finite positive scalar γ such that $\|B_n\| \le \gamma$ for all $n \ge 1$. Since $\|B_n\|$ is the largest singular value for B_n, all the singular values for $\{B_n\}_1^\infty$ are uniformly bounded by γ. Observe that $\det[B_n]$ equals the product of all the singular values for B_n. Because all the singular values are uniformly bounded and $\det[B_n] \geq \epsilon$, the smallest singular value for B_n is greater than or equal to some $\delta > 0$. In particular, $\|B_n^* f\| \geq \delta\|f\|$ for all f in $\mathbb{C}^\nu$. Hence $\|A_n^* f\| \geq \|B_n^* f\| \geq \delta\|f\|$. Therefore A_n has ν nonzero singular values and all these singular values are greater than or equal to δ. Since $A_n^* A_n = \Delta_n$ converges to $T = \Omega^*\Omega$, it follows that T has ν nonzero eigenvalues and all these eigenvalues are greater than or equal to δ^2. So its lower triangular Cholesky factor Ω is a matrix mapping $\mathbb{C}^\mu$ into $\mathbb{C}^\nu$.

Let Ω_{ν,k_ν} be the last nonzero entry in the last row of Ω. Then for $q > k_\nu$, we have $\|A_n e_q\|^2$ converging to $\|\Omega e_q\|^2 = 0$. Since $\inf\{(A_n)_{j,k_j}\} > 0$, it follows that $(A_n)_{\nu,q} = 0$ for all $q > k_\nu$ and n sufficiently large. So without loss of generality, we can assume that the last $\mu - k_\nu$ columns of A_n are zero. Using the fact that $\|A_n e_{k_\nu}\|$ converges to $\|\Omega e_{k_\nu}\| = \Omega_{\nu,k_\nu} > 0$, we see that $(A_n)_{\nu,k_\nu}$ converges to $\Omega_{\nu,k_\nu} > 0$. By ignoring the last $\mu - k_\nu$ zero columns of A_n and Ω, without loss of generality we can assume that $\mu = k_\nu$. In this case, A_n and Ω admits a decomposition of the form

$$A_n = \begin{bmatrix} A_{11,n} & 0 \\ A_{21,n} & a_{\nu\mu,n} \end{bmatrix} \quad \text{and} \quad \Omega = \begin{bmatrix} \Omega_{11} & 0 \\ \Omega_{21} & \Omega_{\nu\mu} \end{bmatrix}. \tag{8.3.5}$$

This readily implies that

$$\begin{aligned} \Omega^*\Omega &= \begin{bmatrix} \Omega_{11}^*\Omega_{11} + \Omega_{21}^*\Omega_{21} & \Omega_{21}^*\Omega_{\nu\mu} \\ \Omega_{21}\Omega_{\nu\mu} & \Omega_{\nu\mu}^2 \end{bmatrix} = \lim_{n\to\infty} A_n^* A_n \\ &= \lim_{n\to\infty} \begin{bmatrix} A_{11,n}^* A_{11,n} + A_{21,n}^* A_{21,n} & A_{21,n}^* a_{\nu\mu,n} \\ A_{21,n} a_{\nu\mu,n} & a_{\nu\mu,n}^2 \end{bmatrix}. \end{aligned}$$

Since $(A_n)_{\nu\mu} = a_{\nu\mu,n}$ converges to $\Omega_{\nu\mu}$, the sequence $A_{21,n}$ converges to Ω_{21}. Hence $A_{21,n}^* A_{21,n}$ converges to $\Omega_{21}\Omega_{21}^*$. So $A_{11,n}^* A_{11,n}$ converges to $\Omega_{11}^*\Omega_{11}$. Notice that $A_{11,n}$ and Ω_{11} are lower triangular Cholesky matrices mapping $C^{\mu-1}$ into $C^{\nu-1}$ and $\inf\{(A_{11,n})_{j,k_j}\} > 0$. By repeating the previous argument $\nu - 1$ times, we see that A_n converges to Ω. (The fact that $A_n^* A_n$ is decreasing is not necessary in this part of the argument.)

If T is strictly positive, then $A_n^* A_n \geq T$, and thus, $\{A_n^* A_n\}_1^\infty$ is a sequence of strictly positive operators on $\mathbb{C}^\mu$. In other words, A_n is a one to one lower triangular Cholesky matrix. Because a Cholesky matrix is onto, A_n is an invertible operator on $\mathbb{C}^\mu$. Moreover, $\det[A_n]^2 = \prod (A_n)_{j,j}^2$ converges to the product of the eigenvalues for T. This, along with the fact that $A_n^* A_n$ is uniformly bounded, implies that $\inf\{(A_n)_{j,j}\} > 0$. Therefore A_n converges to Ω. □

If T is strictly positive, then the condition $\inf\{(A_n)_{j,k_j}\} > 0$ in Lemma 8.3.4 automatically holds. However, if T is not invertible, then this condition is not necessarily satisfied. For a counterexample, consider the matrix

$$A_n = \begin{bmatrix} 1 & 0 & 0 & 0 \\ 0 & -1 & \frac{1}{n} & 0 \\ 0 & 0 & 1 & 1 \end{bmatrix} \quad \text{then} \quad T = \begin{bmatrix} 1 & 0 & 0 & 0 \\ 0 & 1 & 0 & 0 \\ 0 & 0 & 1 & 1 \\ 0 & 0 & 1 & 1 \end{bmatrix} = \lim_{n \to \infty} A_n^* A_n.$$

In this case, A_n converges to A_∞ and A_∞ is not a lower triangular Cholesky matrix.

For another example of a sequence of decreasing positive matrices which violate the hypotheses of Lemma 8.3.4, consider the positive matrices Δ_n on $\mathbb{C}^2$ given by

$$\Delta_n = \begin{bmatrix} 1 & 0 \\ 0 & 1/n^2 \end{bmatrix}.$$

Clearly, the matrices Δ_n are decreasing. Notice that

$$T = \begin{bmatrix} 1 & 0 \\ 0 & 0 \end{bmatrix} = \lim_{n \to \infty} \Delta_n.$$

In this case, the lower Cholesky factors A_n for Δ_n and Ω for T are determined by

$$A_n = \begin{bmatrix} 1 & 0 \\ 0 & 1/n \end{bmatrix} \quad \text{and} \quad \Omega = \begin{bmatrix} 1 & 0 \end{bmatrix}.$$

The lower Cholesky matrices $\{A_n\}$ do not satisfy the hypotheses of Lemma 8.3.4. Obviously, $A_n^* A_n$ converges to $\Omega^*\Omega$. However, A_n does not converge to Ω. In fact, the operators A_n and Ω do not even act between the same spaces. Finally, it is noted that A_n converges to $\begin{bmatrix} \Omega & 0 \end{bmatrix}^{tr}$ as n tends to infinity.

8.4 Cholesky Factorization and Maximal Outer Spectral Factors

Let T_n be a positive block matrix on $\mathcal{E}^n$ where $\mathcal{E} = \mathbb{C}^\tau$. Let M_n from $\mathcal{E}^n$ onto $\mathbb{C}^\mu$ be the lower triangular Cholesky factor of T_n. Then M_n admits a matrix representation of the form

$$M_n = \begin{bmatrix} A_k & 0 \\ B_k & C_k \end{bmatrix} : \begin{bmatrix} \mathcal{E}^k \\ \mathcal{E}^{n-k} \end{bmatrix} \to \begin{bmatrix} \mathbb{C}^{\nu(k,n)} \\ \mathbb{C}^{\mu-\nu(k,n)} \end{bmatrix} \tag{8.4.1}$$

where A_k and C_k are both lower triangular Cholesky matrices. Here $\mu = \operatorname{rank} M_n = \operatorname{rank} T_n$. Moreover, $\nu(k,n)$ equals the rank of A_k, while $\mu - \nu(k,n)$ equals the rank of C_k. Because C_k is a lower triangular Cholesky matrix, the range of $M_n|(0 \oplus \mathcal{E}^{n-k})$ equals $0 \oplus \mathbb{C}^{\mu-\nu(k,n)}$. In particular, for a fixed f in $\mathcal{E}^k$, this implies that

$$\|A_k f\|^2 = \inf\{\|M_n(f \oplus h)\|^2 : h \in \mathcal{E}^{n-k}\}. \tag{8.4.2}$$

To see this, notice that the decomposition in (8.4.1) yields

$$\|A_k f\|^2 \leq \left\| \begin{bmatrix} A_k f \\ B_k f + C_k h \end{bmatrix} \right\|^2 = \|M_n(f \oplus h)\|^2.$$

Hence $\|A_k f\|^2 \leq \inf\{\|M_n(f \oplus h)\|^2 : h \in \mathcal{E}^{n-k}\}$. Since C_k is onto we can choose a vector h in $\mathcal{E}^{n-k}$ such that $B_k f = -C_k h$. For this h, we obtain $\|A_k f\|^2 = \|M_n(f \oplus h)\|^2$, which yields the equality in (8.4.2).

Let $\Theta(z) = \sum_0^\infty z^{-j}\Theta_j$ in $H^2(\mathcal{E}, \mathcal{Y})$ be the maximal outer spectral factor for a positive Toeplitz matrix T_R. Recall that Θ is unique up to a constant unitary operator on the left. By the appropriate choice of this unitary operator, we can always assume that $\Theta_0 = \Theta(\infty)$ is a lower triangular Cholesky matrix. In this case, we claim that the $k \times k$ block matrix

$$\Omega_k = \Pi_{\mathcal{Y}^k} T_\Theta | \mathcal{E}^k = \begin{bmatrix} \Theta_0 & 0 & \cdots & 0 \\ \Theta_1 & \Theta_0 & \cdots & 0 \\ \vdots & \vdots & \ddots & \vdots \\ \Theta_{k-1} & \Theta_{k-2} & \cdots & \Theta_0 \end{bmatrix} : \mathcal{E}^k \to \mathcal{Y}^k \tag{8.4.3}$$

contained in the upper left-hand corner of T_Θ is also a lower triangular Cholesky matrix.

To verify this, let Ψ be any maximal outer spectral factor for T_R. Then $\Psi(\infty)^*\Psi(\infty) = L^*L$ where L is a lower triangular Cholesky matrix. Since both $\Psi(\infty)$ and L are onto, there exists a unitary operator Φ such that $\Phi\Psi(\infty) = L$. Hence $\Theta = \Phi\Psi$ is also a maximal outer spectral factor for T_R, and $\Theta(\infty) = (\Phi\Psi)(\infty) = L$ is a lower triangular Cholesky matrix. Finally, Lemma 8.3.3 shows that there is only one maximal outer spectral factor Θ for T_R such that Θ_0 is a lower triangular Cholesky matrix.

Theorem 8.4.1. *Let T_R be a positive block Toeplitz matrix with $\mathcal{L}(\mathcal{E}, \mathcal{E})$-valued symbol R. Moreover, assume that*

- *$\Theta(z) = \sum_0^\infty z^{-j}\Theta_j$ in $H^2(\mathcal{E}, \mathcal{Y})$ is the maximal outer spectral factor for T_R where $\Theta_0 = \Theta(\infty)$ is a lower triangular Cholesky matrix.*
- *Let $\Omega_k = \Pi_{\mathcal{Y}^k} T_\Theta | \mathcal{E}^k$ be the block lower triangular Cholesky Toeplitz matrix contained in the $k \times k$ left-hand corner of T_Θ; see* (8.4.3).
- *Let M_n mapping $\mathcal{E}^n$ onto $\mathbb{C}^{\mu_n}$ be the lower triangular Cholesky factor of $T_{R,n}$, where $T_{R,n}$ on $\mathcal{E}^n$ is the block Toeplitz matrix contained in the upper $n \times n$ left-hand corner of T_R; see* (8.1.12).
- *For a fixed $k < n$, consider the matrix representation of M_n determined by*

$$M_n = \begin{bmatrix} A_{k,n} & 0 \\ B_{k,n} & C_{k,n} \end{bmatrix} : \begin{bmatrix} \mathcal{E}^k \\ \mathcal{E}^{n-k} \end{bmatrix} \to \begin{bmatrix} \mathbb{C}^{\nu(k,n)} \\ \mathbb{C}^{\mu_n - \nu(k,n)} \end{bmatrix} \tag{8.4.4}$$

where $A_{k,n}$ and $C_{k,n}$ are both lower triangular Cholesky matrices.

Then for any fixed f in $\mathcal{E}^k$, the operator $\Delta_{k,n} = A_{k,n}^* A_{k,n}$ determines the cost in the optimization problem

$$(\Delta_{k,n} f, f) = \inf\{(T_{R,n}(f \oplus h), (f \oplus h)) : h \in \mathcal{E}^{n-k}\}. \tag{8.4.5}$$

In particular, for fixed k the sequence $\{\Delta_{k,n}\}$ is decreasing, that is, $\Delta_{k,n+1} \leq \Delta_{k,n}$. Moreover,

$$\lim_{n\to\infty} \Delta_{k,n} = \Omega_k^* \Omega_k. \tag{8.4.6}$$

Finally, if Θ is in $H^2(\mathcal{E}, \mathcal{E})$, then $A_{k,n}$ converges to Ω_k as n tends to infinity.

Proof. By consulting (8.4.2), we obtain

$$\begin{aligned}\|A_{k,n} f\|^2 &= \inf\{\|M_n(f \oplus h)\|^2 : h \in \mathcal{E}^{n-k}\}\\ &= \inf\{(T_{R,n}(f \oplus h), (f \oplus h)) : h \in \mathcal{E}^{n-k}\}\\ &= (\Delta_{k,n} f, f).\end{aligned}$$

Hence $\|A_{k,n} f\|^2 = (\Delta_{k,n} f, f)$ for all f in $\mathcal{E}^k$. In other words, $\Delta_{k,n} = A_{k,n}^* A_{k,n}$. Theorem 8.2.3 shows that $\Delta_{k,n}$ is decreasing in n and $\Delta_{k,n}$ converges to $\Omega_k^* \Omega_k$.

If Θ is in $H^2(\mathcal{E}, \mathcal{E})$, then $\Theta(\infty) = \Theta_0$ is an invertible operator on $\mathcal{E}$. In this case, Ω_k is an invertible lower triangular Cholesky matrix on $\mathcal{E}^k$. In particular, $\Omega_k^* \Omega_k$ is strictly positive. According to Lemma 8.3.4, the lower triangular Cholesky matrices $A_{k,n}$ converge to Ω_k as n tends to infinity. □

A Cholesky Algorithm for computing the maximal outer spectral factor

Theorem 8.4.1 and Lemma 8.3.4 can be used to develop an algorithm to compute the maximal outer spectral factor Θ in $H^2(\mathcal{E}, \mathbb{C}^q)$ for a rational positive Toeplitz matrix T_R with block entries in $\mathcal{L}(\mathcal{E}, \mathcal{E})$.

(i) For sufficiently large n, construct $T_{R,n}$ on $\mathcal{E}^n$. It is assumed that for this n the cost $\Delta_{k,n}$ in the optimization problem (8.4.5) is approximately equal to $\Omega_k^* \Omega_k \approx \Delta_{k,n}$; see Theorem 8.4.1. To apply the Kalman-Ho algorithm to find a state space realization for Θ, we also assume that $k > 2\delta(\Theta)$ where $\delta(\Theta)$ is the McMillan degree of Θ.

(ii) Compute the lower triangular Cholesky factor M_n for $T_{R,n}$. Select the lower triangular Cholesky matrix $A_{k,n}$ in the upper left-hand corner of M_n. Then $A_{k,n}$ is a block lower triangular Toeplitz matrix with block entries in $\mathcal{L}(\mathcal{E}, \mathbb{C}^q)$. One can determine q from the structure of this block Toeplitz matrix. We also assume that our lower triangular Cholesky algorithm eliminated the small entries in the last row of M_n such that the conditions in Lemma 8.3.4 hold. Then $A_{k,n} \approx \Omega_k$. Finally, for perhaps a more efficient method to compute $A_{k,n}$, one can use Remark 8.2.1 (or Remark 8.2.2 in the invertible case) to compute $\Delta_{k,n}$. Then $A_{k,n}$ is the lower triangular Cholesky factor of $\Delta_{k,n}$.

(iii) Run the Kalman-Ho algorithm on the first column of $A_{k,n}$, that is, $A_{k,n}|\mathcal{E}$ to obtain a minimal realization $\{A, B, C, D\}$ for the maximal outer spectral factor Θ for T_R.

(iv) To gain some further insight, one may also run the Wold decomposition algorithm in Section 8.1 using the lower triangular Cholesky factor M_n. Compute the operators J_n and Q_n in (8.1.13), and the contraction $U_n = M_nQ_n(M_nJ_n)^{-r}$. Recall that $U_nM_nJ_n = M_nQ_n$. Because M_n is a lower triangular Cholesky matrix and $A_{k,n}$ is block Toeplitz, the upper left-hand $k \times k$ corner of U_n will be a block lower shift of multiplicity q. So one can also determine q from U_n. If the upper left hand corner of U_n is not a block lower shift, then one may have to take a larger n.

8.5 Some Examples of the Cholesky Factorization

In this section, we will present some numerical examples to demonstrate that the initial part of the first column of the lower triangular Cholesky factor converges to the Fourier coefficients of the maximal outer spectral factor $\Theta \in H^\infty(\mathcal{E}, \mathcal{Y})$ for T_R. Throughout M_n mapping $\mathcal{E}^n$ into $\mathbb{C}^\nu$ is the lower triangular Cholesky factor of $T_{R,n}$. Moreover, J_n and Q_n are the matrices defined in (8.1.13), while $U_n = M_nQ_n(M_nJ_n)^{-r}$ and $B_n = M_n|\mathcal{E}$ is the first block column of M_n. Finally, for large n, the upper left-hand corner of the matrix U_n is approximately equal to a lower shift of multiplicity $\dim \mathcal{Y}$.

8.5.1 Scalar-valued outer function

Consider the outer function θ in H^∞ given by

$$\theta(z) = 10 + \frac{9}{z} + \frac{8}{z^2} + \frac{7}{z^3} + \frac{6}{z^4} + \frac{5}{z^5} + \frac{4}{z^6} + \frac{3}{z^7} + \frac{2}{z^8} + \frac{1}{z^9}.$$

Since we know the Taylor coefficients of θ, we can construct the Toeplitz matrix $T_R = T_\theta^* T_\theta$. In fact, $R = |\theta|^2$ can be computed by using the fast Fourier transform. Let $T_{R,n}$ be the $n \times n$ matrix contained in the upper left-hand corner of T_R. In this case, $\mathcal{E} = \mathbb{C}$. We computed the lower triangular Cholesky factor M_n of $T_{R,n}$. The first fifteen elements of the first column of M_n for $n = 10$, 20, and 30 are given by the following table:

$$\begin{bmatrix} n=10 & n=20 & n=30 \\ 10.0265 & 10.0002 & 10.0000 \\ 9.0402 & 9.0007 & 9.0000 \\ 8.0259 & 8.0009 & 8.0000 \\ 6.9990 & 7.0006 & 7.0000 \\ 5.9744 & 5.9999 & 6.0000 \\ 4.9624 & 4.9990 & 5.0000 \\ 3.9678 & 3.9984 & 4.0000 \\ 2.9895 & 2.9986 & 3.0000 \\ 2.0213 & 1.9996 & 2.0000 \\ 0.5096 & 0.9979 & 1.0000 \\ & 0 & 0 \\ & 0 & 0 \\ & 0 & 0 \\ & 0 & 0 \\ & 0 & 0 \end{bmatrix}$$

(All elements with absolute value less than 0.0001 were set to zero.) Notice that M_{10} has only ten elements in its column. For $n = 30$, we practically obtain the Taylor coefficients of θ from the first column of M_{30}. Moreover, the 5×5 matrix contained in the upper left-hand corner of U_{30} is given by

$$\begin{bmatrix} 0 & 0 & 0 & 0 & 0 \\ 1.0000 & 0 & 0 & 0 & 0 \\ 0 & 1.0000 & 0 & 0 & 0 \\ 0 & 0 & 1.0000 & 0 & 0 \\ 0 & 0 & 0 & 1.0000 & 0 \end{bmatrix}.$$

The upper 5×5 left-hand corner of U_{30} is a unilateral shift of multiplicity 1. Finally, it is noted that in general U_n is not a lower shift. However, for n sufficiently large, the upper left-hand corner of U_n will be a lower shift.

8.5.2 Multi-input multi-output square outer function

Let Θ be the square outer function in $H^\infty(\mathbb{C}^2, \mathbb{C}^2)$ given by

$$\Theta(z) = \begin{bmatrix} 15 & 0 \\ 14 & 13 \end{bmatrix} + \frac{1}{z}\begin{bmatrix} 12 & 0 \\ 11 & 10 \end{bmatrix} + \frac{1}{z^2}\begin{bmatrix} 9 & 0 \\ 8 & 7 \end{bmatrix} + \frac{1}{z^3}\begin{bmatrix} 6 & 0 \\ 5 & 4 \end{bmatrix} + \frac{1}{z^4}\begin{bmatrix} 3 & 0 \\ 2 & 1 \end{bmatrix}.$$

Here each Taylor coefficient of Θ is a 2×2 matrix. In this case, $T_R = T_\Theta^* T_\Theta$ is a block Toeplitz matrix with entries in $\mathcal{L}(\mathbb{C}^2, \mathbb{C}^2)$, and thus, $\mathcal{E} = \mathbb{C}^2$. We computed $T_{R,n}$ and calculated M_n. The first fifteen elements of the first block column of M_n for $n = 5$, 10, and 15 are given by the following:

$n=5$		$n=10$		$n=15$	
15.0919	0.0000	15.0019	0	15.0000	0
14.0109	13.0288	14.0000	13.0001	14.0000	13.0000
12.0168	−0.0247	12.0012	−0.0004	12.0001	0
11.0501	10.0067	11.0007	10.0005	11.0000	10.0000
8.9979	−0.0147	8.9980	−0.0008	8.9999	0
8.0294	6.9642	8.0007	6.9997	8.0000	7.0000
6.0926	−0.0023	6.0013	−0.0006	5.9999	0
5.0893	4.0139	5.0017	3.9989	5.0000	4.0000
1.9877	−0.0161	2.9821	−0.0004	2.9998	0
1.4205	0.7103	1.9994	0.9997	2.0000	1.0000
		0	0	0	0
		0	0	0	0
		0	0	0	0
		0	0	0	0
		0	0	0	0

By inspecting the upper left-hand corner of M_{15}, we see that Θ is in $H^2(\mathbb{C}^2, \mathbb{C}^2)$. The 5×5 matrix contained in the upper left-hand corner of U_{15} is given by

$$\begin{bmatrix} 0 & 0 & 0 & 0 & 0 \\ 0 & 0 & 0 & 0 & 0 \\ 1.0000 & 0 & 0 & 0 & 0 \\ 0 & 1.0000 & 0 & 0 & 0 \\ 0 & 0 & 1.0000 & 0 & 0 \end{bmatrix}.$$

So U_n converges to the unilateral shift S of multiplicity 2. In other words, Θ must be in $H^2(\mathbb{C}^2, \mathbb{C}^2)$. Therefore at $n = 15$, we are able to obtain the Taylor coefficients of Θ.

8.5.3 Non-square outer function

Let Θ be the non-square outer function in $H^2(\mathbb{C}^2, \mathbb{C})$ given by

$$\begin{aligned} \Theta(z) = \begin{bmatrix} 20 & 19 \end{bmatrix} + \frac{1}{z}\begin{bmatrix} 18 & 17 \end{bmatrix} + \frac{1}{z^2}\begin{bmatrix} 16 & 15 \end{bmatrix} + \frac{1}{z^3}\begin{bmatrix} 14 & 13 \end{bmatrix} + \frac{1}{z^4}\begin{bmatrix} 12 & 11 \end{bmatrix} \\ + \frac{1}{z^5}\begin{bmatrix} 10 & 9 \end{bmatrix} + \frac{1}{z^6}\begin{bmatrix} 8 & 7 \end{bmatrix} + \frac{1}{z^7}\begin{bmatrix} 6 & 5 \end{bmatrix} + \frac{1}{z^8}\begin{bmatrix} 4 & 3 \end{bmatrix} + \frac{1}{z^9}\begin{bmatrix} 2 & 1 \end{bmatrix}. \quad (8.5.1) \end{aligned}$$

Let $T_R = T_\Theta^* T_\Theta$ where $R = \Theta^*\Theta$. Let M_n be the lower triangular Cholesky factor for $T_{R,n}$. The first fifteen elements of the first block column of M_n for $n = 5$, 10, and 20 are given in the following table:

$n=5$		$n=10$		$n=20$	
0.7699	0	20.0000	19.0000	20.0000	19.0000
19.9852	19.0141	−0.4296	−0.0430	18.0000	17.0000
−0.8033	−1.1705	18.0134	17.0048	16.0000	15.0000
18.0468	17.0608	−0.2689	0.0672	14.0000	13.0000
−1.7007	−1.8266	16.0144	15.0043	12.0000	11.0000
16.0844	15.0888	−0.1663	0.1164	10.0000	9.0000
−2.2335	−2.2087	14.0133	13.0033	8.0000	7.0000
14.1078	13.1058	−0.1161	0.1234	6.0000	5.0000
−4.7446	−4.5442	12.0121	11.0028	4.0000	3.0000
18.0426	15.7393	−0.1027	0.1027	2.0000	1.0000
		10.0114	9.0027	0	0
		−0.1148	0.0631	0	0
		8.0111	7.0031	0	0
		−0.1452	0.0104	0	0
		6.0112	5.0039	0	0

The matrix $T_{R,n}$ is a $2n \times 2n$ matrix. It is well known that the Gram-Schmidt orthogonalization procedure can lead to numerical errors; see Golub-Van Loan [122]. This may explain the "extra terms" in the columns for $n=5$ and $n=10$. For $n=10$, our lower triangular Cholesky factor M_{10} maps $\mathbb{C}^{20}$ into $\mathbb{C}^{19}$. However,

$$\|T_{R,10} - M_{10}^* M_{10}\| \approx 3.2767 \times 10^{-6} \quad \text{and} \quad \|T_{R,10}\| \approx 1.6492 \times 10^4.$$

So our lower triangular Cholesky factor M_{10} for $T_{R,10}$ is fairly accurate.

For $n=20$, we essentially arrive at the Taylor coefficients for Θ. In this case, $T_{R,20}$ is a 40×40 matrix. Moreover, M_{20} maps $\mathbb{C}^{40}$ into $\mathbb{C}^{29}$ while

$$\|T_{R,20} - M_{20}^* M_{20}\| \approx 3.279 \times 10^{-6} \quad \text{and} \quad \|T_{R,20}\| \approx 2.0003 \times 10^4.$$

Hence our lower triangular Cholesky factor M_{20} for $T_{R,20}$ is fairly accurate. By inspecting the upper 5×5 left-hand corner of U_{20} we arrive at

$$\begin{bmatrix} 0 & 0 & 0 & 0 & 0 \\ 1.0000 & 0 & 0 & 0 & 0 \\ 0 & 1.0000 & 0 & 0 & 0 \\ 0 & 0 & 1.0000 & 0 & 0 \\ 0 & 0 & 0 & 1.0000 & 0 \end{bmatrix}. \tag{8.5.2}$$

Clearly, the multiplicity of this matrix is 1. This also verifies that Θ is in $H^\infty(\mathbb{C}^2, \mathbb{C})$. Finally, it is noted that one may obtain much better results by using a more efficient lower triangular Cholesky factorization algorithm. To avoid the Gram-Schmidt orthogonalization one can use the QR algorithm; see the proof of Theorem 8.3.2 or Golub-Van Loan [122] for other techniques.

8.5.4 Non-square outer function and a unitary part

For our next example, consider the controllable unitary pair $\{V \text{ on } \mathbb{C}^4, \Gamma_2\}$ given by

$$V = \begin{bmatrix} e^{\imath\pi/3} & 0 & 0 & 0 \\ 0 & e^{-\imath\pi/3} & 0 & 0 \\ 0 & 0 & e^{\imath\pi/4} & 0 \\ 0 & 0 & 0 & e^{-\imath\pi/4} \end{bmatrix} \quad \text{and} \quad \Gamma_2 = \begin{bmatrix} 2 & 1 \\ 2 & 1 \\ 1 & 2 \\ 1 & 2 \end{bmatrix}. \tag{8.5.3}$$

Clearly, V is a unitary operator on $\mathbb{C}^4$. Notice that Γ_2 is an operator mapping $\mathcal{E} = \mathbb{C}^2$ into $\mathbb{C}^4$. Let $R = \sum_{-\infty}^{\infty} e^{-\imath\omega n} R_n$ be the symbol for T_R determined by

$$R_n = \frac{1}{2\pi} \int_0^{2\pi} e^{\imath\omega n} \Theta(e^{\imath\omega})^* \Theta(e^{\imath\omega})\, d\omega + \Gamma_2^* V^{*n} \Gamma_2 \tag{8.5.4}$$

where $\Theta \in H^\infty(\mathbb{C}^2, \mathbb{C})$ is the same outer function in (8.5.1) in Example 8.5.3. The results in Section 6.3 show that the Toeplitz matrix T_R determined by this R is positive. Moreover, Θ is the maximal outer spectral factor for T_R. Finally, $\{V, \Gamma_2\}$ is the unitary part in the Wold decomposition for the controllable isometric representation $\{U, \Gamma\}$ for T_R.

Because the unitary part is present in the Wold decomposition of $\{U, \Gamma\}$, we needed to choose a larger $T_{R,n}$ than the previous Example 8.5.3 in order to compute the same Θ. So for this example, we choose $n = 30$ to compute the Fourier coefficients for Θ. In this case, $T_{R,30}$ is a 60×60 matrix. Moreover, M_{30} maps $\mathbb{C}^{60}$ into $\mathbb{C}^{43}$ while

$$\|T_{R,30} - M_{30}^* M_{30}\| \approx 5.6288 \times 10^{-7} \quad \text{and} \quad \|T_{R,30}\| \approx 2.1031 \times 10^4.$$

Hence our lower triangular Cholesky factor M_{30} for $T_{R,30}$ is fairly accurate. The first 15 components in the first column $M_{30}|\mathcal{E}$ was practically the same as the first 15 components in the first column $M_{20}|\mathcal{E}$ in the previous Example 8.5.3. Finally, it is noted that the upper 5×5 left-hand corner of U_{30} was also given by the same lower shift of multiplicity 1 in (8.5.2). This also verifies that Θ is in $H^\infty(\mathbb{C}^2, \mathbb{C})$.

To compute the unitary pair $\{V, \Gamma_2\}$ using the Wold decomposition method in Remark 8.1.2, we first observed that U_{30} has four eigenvalues on the unit circle. In fact, absolute values for the eigenvalues for U_{30} in decreasing order are $1, 1, 1, 1, 0.5645, \ldots$ Then using the isometry Ψ mapping $\mathbb{C}^4$ into $\mathbb{C}^{43}$ consisting of the eigenvectors corresponding to these four eigenvalues on the unit circle, we obtained

$$\Psi^* U_{30} \Psi = \begin{bmatrix} 0.5 + 0.8660i & 0 & 0 & 0 \\ 0 & 0.5 - 0.8660i & 0 & 0 \\ 0 & 0 & 0.7071 + 0.7071i & 0 \\ 0 & 0 & 0 & 0.7071 - 0.7071i \end{bmatrix},$$

$$\Psi^* B_n = \begin{bmatrix} 2e^{\imath\varphi_1} & e^{\imath\varphi_1} \\ 2e^{\imath\varphi_2} & e^{\imath\varphi_2} \\ e^{\imath\varphi_3} & 2e^{\imath\varphi_3} \\ e^{\imath\varphi_4} & 2e^{\imath\varphi_4} \end{bmatrix}$$

where $\varphi_1 = 2.8173$, $\varphi_2 = -2.8173$, $\varphi_3 = 2.3562$ and $\varphi_4 = -2.3562$ (all in radians). So if Φ is the diagonal unitary matrix formed by $\Phi = \operatorname{diag}(\{e^{-\imath\varphi_j}\}_1^4)$, then $\Phi\Psi^* U_{30}\Psi \approx V\Phi$ and $\Phi\Psi^* B_n \approx \Gamma_2$. In other words, $\{\Psi^* U_{30}\Psi, \Psi^* B_n\}$ is unitarily equivalent to $\{V, \Gamma_2\}$, and we have essentially computed the unitary pair $\{V, \Gamma_2\}$. Finally, it is noted that the ergodic method in Section 6.5 takes much more data to pick out the unitary pair $\{V, \Gamma_2\}$. Sometimes ergodic methods converge slowly.

8.6 An Inner-Outer Factorization Algorithm

One can use the Wold decomposition algorithm in Section 8.1, or the Cholesky algorithm in Section 8.4, to compute the inner-outer factorization for a rational function G in $H^\infty(\mathcal{E}, \mathcal{D})$. To see this, for sufficiently large n compute the Toeplitz matrix $T_{R,n}$ on $\mathcal{E}^n$ where $R = G^*G$ and $T_R = T_G^* T_G$. Using the Wold decomposition algorithm or the Cholesky algorithm, compute the rational outer spectral factor Θ for R, that is, Θ is an outer function in $H^\infty(\mathcal{E}, \mathcal{Y})$ and $R = \Theta^*\Theta$ on the unit circle. This Θ is also the outer factor for G. In other words, $G = G_i\Theta$ where G_i is a rational inner function in $H^\infty(\mathcal{Y}, \mathcal{D})$. Let $\{A_o \text{ on } \mathcal{X}_o, B_o, C_o, D_o\}$ be any realization for Θ, and F the function determined by the realization

$$F(z) = D_o^{-r} - D_o^{-r}C_o(zI - (A_o - B_oD_o^{-r}C_o))^{-1}B_oD_o^{-r}.$$

(The Moore-Penrose inverse of M is denoted by M^{-r}.) According to Lemma 4.5.2, the transfer function F is the right inverse of Θ, that is, $\Theta(z)F(z) = I$. Let $\{A \text{ on } \mathcal{X}, B, C, D\}$ be a realization for G. Notice that $G_i(z) = G(z)F(z)$. Now one can use classical state space formulas for the product of two transfer functions to find a state space realization for G_i; see Remark 14.2.2. In fact, using this remark, a realization $\{A_i, B_i, C_i, D_i\}$ for G_i is given by

$$A_i = \begin{bmatrix} A & -BD_o^{-r}C_o \\ 0 & A_o - B_oD_o^{-r}C_o \end{bmatrix} \text{ on } \begin{bmatrix} \mathcal{X} \\ \mathcal{X}_o \end{bmatrix}, \quad B_i = \begin{bmatrix} BD_o^{-r} \\ B_oD_o^{-r} \end{bmatrix} : \mathcal{E} \to \begin{bmatrix} \mathcal{X} \\ \mathcal{X}_o \end{bmatrix},$$

$$C_i = \begin{bmatrix} C & -DD_o^{-r}C_o \end{bmatrix} \begin{bmatrix} \mathcal{X} \\ \mathcal{X}_o \end{bmatrix} \to \mathcal{Y} \quad \text{and} \quad D_i = DD_o^{-r}.$$

The realization $\{A_i, B_i, C_i, D_i\}$ for G_i might not be minimal. However, one can always extract a minimal realization from $\{A_i, B_i, C_i, D_i\}$ for G_i.

If Θ is scalar-valued, then one can use the fast Fourier transform to compute the Fourier series expansion $G_i(z) = \sum_0^\nu z^{-j}\gamma_j = G/\Theta$ for the inner function G_i. (In Matlab $\begin{bmatrix} \gamma_0 & \cdots & \gamma_\nu & 0 & \cdots & 0 \end{bmatrix} = \text{ifft}(G./\Theta)$; see Remark 8.6.1 below.) Applying the Kalman-Ho algorithm to $\{\gamma_j\}_0^\nu$ yields a minimal realization for G_i. This method works even when G is a column vector, that is, G is in $H^\infty(\mathbb{C}, \mathbb{C}^\nu)$.

If Θ is not square, then $G_i = G\Theta^{-r}$ on the unit circle. Now simply use the fast Fourier transform to compute the elements of the matrices in G and Θ. (Use Remark 8.6.1 to compute the fast Fourier transform of all the entries of G and Θ.) For example, in a 2^{13} fast Fourier transform, G and Θ are matrices containing 2^{13}

components. Then, by taking the Moore-Penrose inverse, compute $G_i = G\Theta^{-r}$ (this requires 2^{13} Moore-Penrose inverses). Then we obtain a matrix for G_i with 2^{13} entries. By taking the inverse fast Fourier transform of G_i, that is, each entry of G_i, we now have $G_i(z) = \sum_0^\nu z^{-j}\gamma_j$ where $\{\gamma_j\}_0^\nu$ are matrices. Finally, applying the Kalman-Ho algorithm to $\{\gamma_j\}_0^\nu$ yields a minimal realization for G_i.

Remark 8.6.1. Recall that if

$$h(z) = \frac{p}{q} = \frac{\sum_{j=0}^{\nu} z^j p_j}{\sum_{j=0}^{\nu} z^j q_j}$$

is a rational function in H^∞, then h expressed as the fast Fourier transform in Matlab is

$$h = \mathrm{fft}(\begin{bmatrix} p_\nu & p_{\nu-1} & \cdots & p_1 & p_0 \end{bmatrix}, k)./\mathrm{fft}(\begin{bmatrix} q_\nu & q_{\nu-1} & \cdots & q_1 & q_0 \end{bmatrix}, k).$$

Here k is the length of the fast Fourier transform. A typical value for k is 2^{13}. The vectors p and q are ordered from the highest value of z^ν to the lowest including zeros. If p and q are not of the same degree, then one must also include the appropriate number of zeros when calculating the fast Fourier transform. Finally,

$$\mathrm{ifft}(h) = \begin{bmatrix} \gamma_0 & \gamma_1 & \cdots & \gamma_{k/2} & 0 & 0 & \cdots & 0 \end{bmatrix}$$

yields an approximation for the Fourier coefficients $h(z) = \sum_0^{k/2} z^{-j}\gamma_j$ for h.

8.6.1 A scalar inner-outer factorization example

Consider the rational function g in H^∞ given by

$$g(z) = \frac{0.1z + 0.5}{z^3 + 0.3z^2 - 0.3z - 0.5}.$$

Notice that g has zeros at $z = \infty$ and -5. Since there are zeros of g outside the unit disc, g is not an outer function. Set $R = |g|^2 = \sum_{-\infty}^{\infty} e^{-\imath\omega n} r_n$ on the circle. (In Matlab,

$$\begin{aligned} g &= \mathrm{fft}([0, 0, 0.1, 0.5], 2^{13})./\mathrm{fft}([1, 0.3, -0.3, -0.5], 2^{13}); \\ R &= \mathrm{abs}(g). \wedge 2; \\ \mathrm{ifft}(R) &= \begin{bmatrix} r_0 & r_1 & r_2 & \cdots & r_{-2} & r_{-1} \end{bmatrix} = r; \\ T &= \mathrm{toeplitz}(r(1:20)); \end{aligned}$$

see Remark 8.6.1.) Let us now use the Cholesky algorithm in Section 8.4 to compute the outer spectral factor θ for g. By choosing $n = 20$, construct $T_{R,20}$. Then compute the lower triangular Cholesky factor M_{20} for $T_{R,20}$. By applying the Kalman-Ho algorithm to the first ten components in the first column of M_{20}, we obtain

$$\theta(z) = \frac{0.5z^3 + 0.1z^2}{z^3 + 0.3z^2 - 0.3001z - 0.5001}.$$

All the zeros and poles of $\theta(z)$ are inside the unit disc, and the McMillan degree of θ is three. Finally,

$$\||\theta|^2 - |g|^2\|_\infty \approx 1.1887 \times 10^{-12}.$$

In other words, $|\theta|^2 \approx |g|^2$ on the unit circle. Therefore θ is approximately the outer spectral factor for g.

To compute the inner part g_i for g, use the fast Fourier transform to compute $g_i = g/\theta$. (In Matlab $g_i = g./\theta$; see Remark 8.6.1.) Then using the inverse fast Fourier transform, compute the Fourier coefficients $g_i(z) = \sum_0^\nu z^{-j}\gamma_j$. Now apply the Kalman-Ho algorithm on $\{\gamma_j\}_0^\nu$ to obtain a minimal realization $\{A_i, B_i, C_i, D_i\}$ for the inner factor g_i of g. This calculation yields:

$$g_i(z) = \frac{1 + 0.2z}{z^2(z + 0.2)}.$$

As expected, g_i is an inner function whose zeros are ∞ and -5. Finally, it is noted that $\|g - g_i g_o\|_\infty \approx 2.2627 \times 10^{-14}$.

8.6.2 A non-square inner-outer factorization example

Consider the rational function G in $H^2(\mathbb{C}^2, \mathbb{C})$ given by

$$\begin{aligned} G(z) &= \frac{1}{d(z)}\left[z^3 - 5.5z^2 + 8.5z - 3, \quad z^3 - 5z^2 + 6z\right], \\ d(z) &= z^3 - 0.2684z^2 + 0.0135z - 0.1155. \end{aligned} \tag{8.6.1}$$

Let $G = G_i\Theta$ be the inner-outer factorization of G where G_i is inner and Θ is outer. In this case, Θ is a rational function in $H^2(\mathbb{C}^2, \mathbb{C})$ and G_i is a scalar-valued rational function in H^∞. Observe that G_i equals the Blaschke product formed by the common zeros of the two components of G outside the closed unit disc. In this case $\{2, 3\}$ are the corresponding common zeros outside the closed unit disc. Hence the inner part of G is determined by

$$G_i(z) = \left(\frac{z-2}{1-2z}\right)\left(\frac{z-3}{1-3z}\right) = \frac{z^2/6 - 5z/6 + 1}{z^2 - 5z/6 + 1/6}.$$

Using $G = G_i\Theta$, we see that

$$\Theta(z) = \frac{G(z)}{G_i(z)} = \frac{1}{d(z)}\left[6z^3 - 8z^2 + 3.5z - 0.5, \quad 6z^3 - 5z^2 + z\right].$$

Let us now use the Cholesky algorithm in Section 8.4 to compute G_i and Θ. First we used the fast Fourier transform to compute $R = G(e^{\imath\omega})^* G(e^{\imath\omega})$. Then using the inverse fast Fourier transform, computed the Fourier coefficients $\{R_k\}$ for $R = \sum_{-\infty}^{\infty} e^{-\imath\omega k} R_k$. Then we proceeded to form $T_{R,n}$, and compute the lower triangular Cholesky factor M_n for $T_{R,n}$. In this case, we chose $n = 15$. Hence $T_{R,15}$ is a positive Toeplitz matrix on $\mathbb{C}^{30}$. Our lower triangular Cholesky factor M_{15} maps $\mathbb{C}^{30}$ into $\mathbb{C}^{16}$. Finally,

$$\|T_{R,15} - M_{15}^* M_{15}\| \approx 1.4230 \times 10^{-13} \quad \text{and} \quad \|T_{R,15}\| \approx 249.5897.$$

Therefore our lower triangular Cholesky factorization of $T_{R,15}$ is fairly accurate.

By applying the Kalman-Ho algorithm to the first ten components of the first block column $M_{15}|\mathbb{C}^2$, we arrived at

$$\Theta(z) = \frac{1}{d(z)} \left[6z^3 - 8z^2 + 3.5z - 0.5, \quad 6z^3 - 5z^2 + z - 1.751 \times 10^{-7}\right]. \quad (8.6.2)$$

We used the fast Fourier transform to calculate $G_i(e^{\imath\omega})$ by the formula

$$G_i(e^{\imath\omega}) = \frac{G(e^{\imath\omega})\Theta(e^{\imath\omega})^*}{\Theta(e^{\imath\omega})\Theta(e^{\imath\omega})^*}.$$

By using the inverse fast Fourier transform, we computed the Taylor's coefficients of G_i. After applying the Kalman-Ho algorithm to the Taylor's coefficients of G_i, we obtained

$$G_i(z) = \frac{0.1667z^2 - 0.8333z + 1}{z^2 - 0.8333z + 0.1667}.$$

Finally, using M_{15} along with Remark 8.1.2, we computed the outer spectral factor for G, and arrived at the same Θ.

8.6.3 An outer function and a unitary part

For our next example, consider the controllable unitary pair $\{V \text{ on } \mathbb{C}^4, \Gamma_2\}$ given by (8.5.3). Let $R = \sum_{-\infty}^{\infty} e^{-\imath\omega n} R_n$ be the symbol for T_R determined by

$$R_n = \frac{1}{2\pi} \int_0^{2\pi} e^{\imath\omega n} G(e^{\imath\omega})^* G(e^{\imath\omega})\, d\omega + \Gamma_2^* V^{*n} \Gamma_2 \quad (8.6.3)$$

where $G \in H^\infty(\mathbb{C}^2, \mathbb{C})$ is the same function in (8.6.1) in Example 8.6.2. Clearly, $G^*G = \Theta^*\Theta$ where Θ is the outer part of G. The results in Section 6.3 show that the Toeplitz matrix T_R determined by this R is positive. Moreover, Θ is the maximal outer spectral factor for T_R. Finally, $\{V, \Gamma_2\}$ is the unitary part in the Wold decomposition for the controllable isometric representation $\{U, \Gamma\}$ for T_R. Now let us compute Θ using the Cholesky algorithm in Section 8.4, and $\{V, \Gamma_2\}$ by employing Remark 8.1.2.

To compute Θ and the unitary pair $\{V, \Gamma_2\}$ for T_R, we chose $n = 15$. In this case, $T_{R,15}$ is a 30×30 matrix. Moreover, our lower triangular Cholesky factor M_{15} for $T_{R,15}$ maps $\mathbb{C}^{30}$ into $\mathbb{C}^{20}$ and $\|T_{R,15} - M_{15}^* M_{15}\| \approx 3.5697 \times 10^{-10}$, where

$\|T_{R,15}\| \approx 249.6040$. Hence our lower triangular Cholesky factor M_{15} for $T_{R,15}$ is fairly accurate. By applying the Kalman-Ho algorithm to the first ten components of the first block column $M_{15}|\mathbb{C}^2$, we arrived at the same maximal outer spectral factor Θ for T_R as in (8.6.2).

To compute the unitary pair $\{V, \Gamma_2\}$ using the Wold decomposition method in Remark 8.1.2, we first observed that U_{15} has four eigenvalues on the unit circle. In fact, absolute value for the eigenvalues for U_{15} in decreasing order are $1, 1, 1, 1, 0.5647, \dots$ Then using the isometry Ψ mapping $\mathbb{C}^4$ into $\mathbb{C}^{20}$ consisting of the eigenvectors corresponding to these four eigenvalues on the unit circle, we obtained

$$\Psi^* U_{30} \Psi = \begin{bmatrix} 0.5 + 0.8660i & 0 & 0 & 0 \\ 0 & 0.5 - 0.8660i & 0 & 0 \\ 0 & 0 & 0.7071 + 0.7071i & 0 \\ 0 & 0 & 0 & 0.7071 - 0.7071i \end{bmatrix},$$

$$\Psi^* B_n = \begin{bmatrix} 2e^{\imath\varphi_1} & e^{\imath\varphi_1} \\ 2e^{\imath\varphi_2} & e^{\imath\varphi_2} \\ e^{\imath\varphi_3} & 2e^{\imath\varphi_3} \\ e^{\imath\varphi_4} & 2e^{\imath\varphi_4} \end{bmatrix}$$

where $\varphi_1 = 0.2692$, $\varphi_2 = -0.2692$, $\varphi_3 = 1.9661$ and $\varphi_4 = -1.9661$ (all in radians). So if Φ is the diagonal unitary matrix formed by $\Phi = \operatorname{diag}(\{e^{-\imath\varphi_j}\}_1^4)$, then Φ intertwines $\{\Psi^* U_{30}\Psi, \Psi^* B_n\}$ with $\{V, \Gamma_2\}$. This yields the unitary pair $\{V, \Gamma_2\}$.

8.7 Notes

The results in Section 8.1 were taken from Bhosri [32]. The Cholesky factorization of a positive matrix is classical. The Cholesky factorization plays a fundamental role in signal processing; see Kailath-Sayed-Hassibi [143] and Stoica-Moses [194]. Theorem 8.4.1 is due to Rissanen-Barbosa [180] when T_R is a rational strictly positive Toeplitz matrix. In this case, one can also derive this result using the Kalman filter; see Section 3.3 in Caines [47]. So Theorem 8.4.1 can be viewed as a mild generalization of the Cholesky factorization results in Rissanen-Barbosa [180]. The controllable isometric representation $\{U, \Gamma\}$ for T_R along with its Wold decomposition, allows us to present an elementary proof of this result; see also Bhosri [32]. Trying to compute the maximal outer spectral factor and the unitary part from a positive Toeplitz matrix can be numerically tricky depending on the data. So one should try various methods to compute Θ and $\{V, \Gamma_2\}$. For some nice results on spectrum analysis and identification techniques, see Pillai-Schim [173]. Finally in Chapter 10, we will present some Riccati equation techniques to compute the inner-outer factorization.

Sinusoid estimation in wide sense stationary processes. The problem of determining the maximal outer spectral factor or the unitary pair $\{V, \Gamma_2\}$ from a positive Toeplitz matrix T_R naturally occurs in wide sense stationary random processes;

see Caines [47], Foias-Frazho-Sherman [88, 89], Stoica-Moses [194] or Pillai-Schim [173]. (Wide sense stationary processes are discussed in Chapter 11.) To sketch how this problem arises, recall that a *wide sense stationary process* $\{y_n\}_{-\infty}^{\infty}$ with values in $\mathbb{C}^\nu$ is a sequence of random vectors $\{y_n\}_{-\infty}^{\infty}$ in $\mathbb{C}^\nu$ such that $Ey_n = \gamma$, the same constant γ for all n, and $Ey_j y_k^* = R_{k-j}$ is just a function of the difference between j and k. Here E denotes the expectation. (We choose $Ey_j y_k^* = R_{k-j}$ to be a function of $k - j$ rather than $j - k$ to fit our notation.) Moreover, it is well known that the Toeplitz matrix T_R determined by the symbol $R = \sum_{-\infty}^{\infty} e^{-\imath\omega n} R_n$ is positive. Finally, $\Xi_y = T_R$ is called the *autocorrelation Toeplitz matrix* for y_n. In this case, the entries of Ξ_y are determined by $(\Xi_y)_{kj} = Ey_j y_k^* = R_{k-j}$.

Let G be a co-outer function in $H^\infty(\mathbb{C}^\mu, \mathbb{C}^\nu)$. Let x_n be the wide sense stationary random process determined by driving a white noise process through the causal filter G, that is,

$$\begin{bmatrix} x_n \\ x_{n-1} \\ x_{n-2} \\ \vdots \end{bmatrix} = \begin{bmatrix} G_0 & G_1 & G_2 & \cdots \\ 0 & G_0 & G_1 & \cdots \\ 0 & 0 & G_0 & \cdots \\ \vdots & \vdots & \vdots & \ddots \end{bmatrix} \begin{bmatrix} w_n \\ w_{n-1} \\ w_{n-2} \\ \vdots \end{bmatrix}. \tag{8.7.1}$$

Here w_n is a white noise process with values in $\mathbb{C}^\mu$, that is, w_n is a mean zero process such that $Ew_j w_k^* = I\delta_{k-j}$ for all integers j and k where δ_j is the Kronecker delta. In particular, $\Xi_w = I$. Moreover, $G(z) = \sum_0^\infty z^{-j} G_j$ is the Taylor series expansion for G. Let Θ be the function in $H^\infty(\mathbb{C}^\nu, \mathbb{C}^\mu)$ defined by $\Theta(z) = \widetilde{G}(z) = G(\overline{z})^*$ for all z in $\mathbb{D}_+$. Notice that $\Theta(z) = \sum_0^\infty z^{-j}\Theta_j$ is the Taylor series expansion for Θ where $\Theta_j = G_j^*$ for all integers $j \geq 0$. Because G is co-outer, Θ is an outer function; see Section 3.2. Equation (8.7.1) can be rewritten as $\vec{x} = T_\Theta^* \vec{w}$. As expected, $\vec{x}$ is the column vector formed by $\{x_j\}_{-\infty}^n$ where x_n appears in the first position, x_{n-1} appears in the second position, etc. The vector $\vec{w}$ is the column vector formed by $\{w_j\}_{-\infty}^n$ in a similar fashion. Using the fact that $E\vec{w}\vec{w}^* = I$, we see that the autocorrelation matrix Ξ_x for x_n is determined by

$$\Xi_x = E\vec{x}\vec{x}^* = T_\Theta^* T_\Theta = T_{\Theta^*\Theta}. \tag{8.7.2}$$

The j, k entry of Ξ_x is given by $(\Xi_x)_{jk} = Ex_k x_j^*$. Since $\Theta^*\Theta$ is the symbol for $\Xi_x = T_\Theta^* T_\Theta$, we have

$$Ex_k x_j^* = \frac{1}{2\pi} \int_0^{2\pi} e^{\imath\omega(j-k)} \Theta(e^{\imath\omega})^* \Theta(e^{\imath\omega})\, d\omega.$$

The autocorrelation matrix Ξ_x admits a factorization of the form $\Xi_x = T_\Theta^* T_\Theta$ where Θ is an outer function. In particular, Ξ_x is a positive Toeplitz matrix with block entries in $\mathcal{L}(\mathbb{C}^\nu, \mathbb{C}^\nu)$.

Now consider the random process ξ_n determined by

$$\xi_n = \sum_{j=1}^{\kappa} \Gamma_{2,j}^* \begin{bmatrix} e^{\imath(\omega_j n + \varphi_{1j})} \\ e^{\imath(\omega_j n + \varphi_{2j})} \\ \vdots \\ e^{\imath(\omega_j n + \varphi_{\nu_j j})} \end{bmatrix}. \tag{8.7.3}$$

Here $\Gamma_{2,j}$ is a matrix mapping $\mathbb{C}^\nu$ onto $\mathbb{C}^{\nu_j}$; see also Section 11.6.1. In particular, $\nu_j \leq \nu$. Moreover, $\{\omega_j\}_1^\kappa$ are distinct constant frequencies in the interval $[0, 2\pi)$ and κ is finite. Furthermore, φ_{mj} are independent, identically distributed random variables over the interval $[0, 2\pi]$. A simple calculation shows that

$$E\xi_m \xi_n^* = \sum_{j=1}^{\kappa} \Gamma_{2,j}^* \Gamma_{2,j} \, e^{\imath\omega_j(m-n)}.$$

Let V be the diagonal unitary matrix on $\mathcal{V} = \oplus_1^\kappa \mathbb{C}^{\nu_j}$ and Γ the operator mapping $\mathbb{C}^\nu$ into $\mathcal{V}$ given by

$$V = \begin{bmatrix} e^{\imath\omega_1} I & 0 & \cdots & 0 \\ 0 & e^{\imath\omega_2} I & \cdots & 0 \\ \vdots & \vdots & \ddots & \vdots \\ 0 & 0 & \cdots & e^{\imath\omega_\kappa} I \end{bmatrix} \quad \text{and} \quad \Gamma = \begin{bmatrix} \Gamma_{2,1} \\ \Gamma_{2,2} \\ \vdots \\ \Gamma_{2,\kappa} \end{bmatrix}. \tag{8.7.4}$$

Finally, it is noted that $E\xi_m \xi_n^* = \Gamma^* V^{m-n} \Gamma$.

Let y_n be the wide sense stationary process defined by $y_n = x_n + \xi_n$. Here we assume that the white noise process w_n and φ_{jk} are all independent random variables. So y_n is a mean zero wide sense stationary process and

$$E y_j y_k^* = E x_j x_k^* + E\xi_j \xi_k^* = E x_j x_k^* + \Gamma^* V^{j-k} \Gamma.$$

In other words, if we set $R_n = E y_0 y_n^*$, then

$$R_n = \frac{1}{2\pi} \int_0^{2\pi} e^{\imath\omega n} \Theta(e^{\imath\omega})^* \Theta(e^{\imath\omega}) \, d\omega + \Gamma_2^* V^{*n} \Gamma_2.$$

By construction $\Xi_y = T_R$ where T_R is the positive Toeplitz matrix determined by the symbol $R = \sum_{-\infty}^{\infty} e^{-\imath\omega n} R_n$. The results in Section 6.3 also show that T_R is positive. Moreover, Θ is the maximal outer spectral factor for T_R. Finally, $\{V, \Gamma_2\}$ is the unitary part in the Wold decomposition for the controllable isometric representation $\{U, \Gamma\}$ for T_R. Therefore our problem of determining the maximal outer spectral factor Θ and the unitary pair $\{V, \Gamma_2\}$ from the Toeplitz matrix T_R is equivalent to finding the co-outer filter G, the frequencies $\{\omega_j\}$, and the amplitudes matrices $\{\Gamma_{2,j}^*\}$ from the wide sense stationary process y_n.

Chapter 9

Signal Processing

In this chapter, we will show how a fundamental optimization problem in prediction theory can be used to determine the maximal outer spectral factor and the eigenvalues in the unitary part corresponding to a positive Toeplitz matrix. In particular, we will present the Capon-Geronimus method for estimating these eigenvalues.

9.1 A Fundamental Optimization Problem

Let $R = \sum_{-\infty}^{\infty} e^{-\imath\omega n} R_n$ be the $\mathcal{L}(\mathcal{E},\mathcal{E})$-valued symbol for a positive Toeplitz matrix T_R. (Throughout $\mathcal{E}$ is finite dimensional.) Moreover, let $\{U \text{ on } \mathcal{K}, \Gamma\}$ be the controllable isometric representation for T_R. Recall that $\{U, \Gamma\}$ admits a Wold decomposition of the form

$$U = \begin{bmatrix} S & 0 \\ 0 & V \end{bmatrix} \text{ on } \begin{bmatrix} \ell_+^2(\mathcal{Y}) \\ \mathcal{V} \end{bmatrix} \quad \text{and} \quad \Gamma = \begin{bmatrix} \Gamma_1 \\ \Gamma_2 \end{bmatrix} : \mathcal{E} \to \begin{bmatrix} \ell_+^2(\mathcal{Y}) \\ \mathcal{V} \end{bmatrix}. \tag{9.1.1}$$

Here S is the unilateral shift on $\ell_+^2(\mathcal{Y})$ where $\mathcal{Y} = \ker U^*$, and V is a unitary operator on $\mathcal{V}$. According to the results in Section 5.2, the outer spectral factor Θ for T_R is given by the Fourier transform of Γ_1, that is, $(\mathcal{F}_{\mathcal{Y}}^+ \Gamma_1)(z) = \Theta(z)$. In other words,

$$\Theta(z) = \sum_{n=0}^{\infty} z^{-n}\Theta_n \quad \text{where} \quad \Gamma_1 = \begin{bmatrix} \Theta_0 \\ \Theta_1 \\ \Theta_2 \\ \vdots \end{bmatrix} : \mathcal{E} \to \ell_+^2(\mathcal{E}); \tag{9.1.2}$$

see equation (5.2.5) in Section 5.2. The unitary pair $\{V, \Gamma_2\}$ admits a matrix representation of the form

$$V = \begin{bmatrix} \lambda_1 I & 0 & 0 & \cdots & 0 \\ 0 & \lambda_2 I & 0 & \cdots & 0 \\ 0 & 0 & \lambda_3 I & \cdots & 0 \\ \vdots & \vdots & \vdots & \ddots & \vdots \\ 0 & 0 & 0 & \cdots & V_\circ \end{bmatrix} \text{ on } \begin{bmatrix} \mathcal{E}_1 \\ \mathcal{E}_2 \\ \mathcal{E}_3 \\ \vdots \\ \mathcal{V}_\circ \end{bmatrix},$$

$$\Gamma_2 = \begin{bmatrix} A_1 \\ A_2 \\ A_3 \\ \vdots \\ \Gamma_\circ \end{bmatrix} : \mathcal{E} \to \begin{bmatrix} \mathcal{E}_1 \\ \mathcal{E}_2 \\ \mathcal{E}_3 \\ \vdots \\ \mathcal{V}_\circ \end{bmatrix}. \tag{9.1.3}$$

Here $\{\lambda_j\}_1^\nu$ are the distinct eigenvalues for V and $\mathcal{E}_j = \ker(V - \lambda_j I)$ is identified with the eigenspace for V corresponding to the eigenvalue λ_j. The number of distinct eigenvalues ν for V can be finite or infinite. Moreover, A_j is an operator from $\mathcal{E}$ onto $\mathcal{E}_j$. In applications, $\{A_j\}_1^\nu$ are called the *amplitude matrices*. Because the pair $\{V, \Gamma_2\}$ is controllable, $\dim \mathcal{E}_j \leq \dim \mathcal{E}$ for all $j = 1, 2, \ldots, \nu$; see Lemma 9.1.2 below. If $\mathcal{E} = \mathbb{C}$, then without loss of generality, we can always assume that $A_j = a_j$ are scalars and $a_j > 0$ for all j. Furthermore, $V_\circ$ is the unitary operator on $\mathcal{V}_\circ = \mathcal{V} \ominus (\oplus_1^\nu \mathcal{E}_j)$ defined by $V_\circ = V|\mathcal{V}_\circ$. The operator $V_\circ$ is a unitary operator with no eigenvalues. Finally, it is noted that $\Gamma_\circ$ is the operator mapping $\mathcal{E}$ into $\mathcal{V}_\circ$ defined by $\Gamma_\circ = \Pi_{\mathcal{V}_\circ}\Gamma_2$.

For fixed α in $\overline{\mathbb{D}}_+ = \{z : |z| \geq 1\}$ and $f \in \mathcal{E}$, consider the optimization problem

$$\rho(\alpha, f) = \inf\{(T_R x, x) : x \in \ell_+^c(\mathcal{E}) \text{ and } (\mathcal{F}_\mathcal{E}^+ x)(\alpha) = f\}. \tag{9.1.4}$$

Here $\rho(\alpha, f)$ is called the *cost* in this optimization problem. If $\alpha = \infty$, then this optimization problem reduces to

$$\rho(\infty, f) = \inf\{(T_R x, x) : x = \begin{bmatrix} f & x_1 & x_2 & x_3 & \cdots \end{bmatrix}^{tr} \in \ell_+^c(\mathcal{E})\}. \tag{9.1.5}$$

For α in $\mathbb{D}_+$, we set

$$d_\alpha = \frac{\sqrt{|\alpha|^2 - 1}}{|\alpha|} \qquad (\alpha \in \mathbb{D}_+). \tag{9.1.6}$$

The following result shows that the cost ρ in this optimization problem is related to the maximal outer spectral factor and the amplitude matrices $\{A_j\}$. Finally, the set of all eigenvalues for V are denoted by eig (V).

Theorem 9.1.1. *Let $R = \sum_{-\infty}^{\infty} e^{-\imath\omega n} R_n$ be the $\mathcal{L}(\mathcal{E}, \mathcal{E})$-valued symbol for a positive Toeplitz matrix T_R. Consider the optimization problems in* (9.1.4) *and* (9.1.5), *where α is in $\overline{\mathbb{D}}_+$ and the vector f is in $\mathcal{E}$. Let Θ be the maximal outer spectral factor for T_R. Then*

$$\begin{aligned}
\rho(\infty, f) &= (\Theta(\infty)^*\Theta(\infty)f, f) && (\alpha = \infty),\\
\rho(\alpha, f) &= d_\alpha^2(\Theta(\alpha)^*\Theta(\alpha)f, f) && (|\alpha| > 1),\\
\rho(\alpha, f) &= (A_j^*A_j f, f) && (\overline{\alpha} = \lambda_j \textit{ and } \overline{\alpha} \in \operatorname{eig}(V)),\\
\rho(\alpha, f) &= 0 && (|\alpha| = 1 \textit{ and } \overline{\alpha} \notin \operatorname{eig}(V)).
\end{aligned} \tag{9.1.7}$$

Finally, the maximal outer spectral factor Θ is zero if and only if $\rho(\alpha, f) = 0$ for any fixed α in $\mathbb{D}_+$ and all f in $\mathcal{E}$.

Proof. Notice that $x = \begin{bmatrix} x_0 & x_1 & x_2 & \cdots \end{bmatrix}^{tr}$ is a vector in $\ell_+^c(\mathcal{E})$ satisfying the condition $(\mathcal{F}_\mathcal{E}^+ x)(\alpha) = f$ if and only if $\widehat{x}(\alpha) = f$ where $\widehat{x}$ is the polynomial in z^{-1} defined by

$$\widehat{x}(z) = \sum_{n=0}^\infty z^{-n} x_n = (\mathcal{F}_\mathcal{E}^+ x)(z).$$

In other words, x is a vector in $\ell_+^c(\mathcal{E})$ satisfying $(\mathcal{F}_\mathcal{E}^+ x)(\alpha) = f$ if and only if $\widehat{x}$ admits a representation of the form: $\widehat{x}(z) = f + (z^{-1} - \alpha^{-1})\widehat{q}$ where $\widehat{q}$ is a polynomial in z^{-1} with values in $\mathcal{E}$. By taking the inverse Fourier transform, x is in $\ell_+^c(\mathcal{E})$ and $(\mathcal{F}_\mathcal{E}^+ x)(\alpha) = f$ if and only if

$$x = \Pi_\mathcal{E}^* f + (S_\mathcal{E} - \alpha^{-1} I)q \tag{9.1.8}$$

where $\Pi_\mathcal{E} = \begin{bmatrix} I & 0 & 0 & \cdots \end{bmatrix}$ and q is a vector in $\ell_+^c(\mathcal{E})$. As expected, $S_\mathcal{E}$ is the unilateral shift on $\ell_+^2(\mathcal{E})$. Finally, since $\ell_+^c(\mathcal{E})$ is a subspace of $\ell_+^2(\mathcal{E})$, we see that $(S_\mathcal{E} - \alpha^{-1}I)q$ is a well-defined vector in $\ell_+^c(\mathcal{E})$.

Let $\{U \text{ on } \mathcal{K}, \Gamma\}$ be the controllable isometric representation for T_R, and

$$W = \begin{bmatrix} \Gamma & U\Gamma & U^2\Gamma & \cdots \end{bmatrix}$$

its corresponding controllability matrix. Notice that $UWq = WS_\mathcal{E}q$ for all q in $\ell_+^c(\mathcal{E})$ and $W\Pi_\mathcal{E}^* f = \Gamma f$. Hence if $x = \Pi_\mathcal{E}^* f + (S_\mathcal{E} - \alpha^{-1}I)q$, then we obtain $Wx = \Gamma f + (U - \alpha^{-1}I)Wq$. Recall that $T_R = W^\sharp W$. Let $\mathcal{L}_\alpha$ be the kernel of $U^* - (\overline{\alpha})^{-1}I$. By employing (9.1.8), we arrive at

$$\begin{aligned}
\rho(\alpha, f) &= \inf\{(T_R x, x) : x \in \ell_+^c(\mathcal{E}) \text{ and } (\mathcal{F}_\mathcal{E}^+ x)(\alpha) = f\}\\
&= \inf\{\|Wx\|^2 : x \in \ell_+^c(\mathcal{E}) \text{ and } (\mathcal{F}_\mathcal{E}^+ x)(\alpha) = f\}\\
&= \inf\{\|Wx\|^2 : x = \Pi_\mathcal{E}^* f + (S_\mathcal{E} - \alpha^{-1}I)q \text{ and } q \in \ell_+^c(\mathcal{E})\}\\
&= \inf\{\|\Gamma f + (U - \alpha^{-1}I)Wq\|^2 : q \in \ell_+^c(\mathcal{E})\}\\
&= \inf\{\|\Gamma f + (U - \alpha^{-1}I)h\|^2 : h \in \mathcal{K}\}\\
&= \inf\{\|\Gamma f - h\|^2 : h \in \operatorname{ran}(U - \alpha^{-1}I)\}\\
&= \inf\{\|\Gamma f - h\|^2 : h \perp \ker(U^* - (\overline{\alpha})^{-1}I)\}\\
&= \|P_{\mathcal{L}_\alpha}\Gamma f\|^2.
\end{aligned}$$

Therefore $\rho(\alpha, f) = \|P_{\mathcal{L}_\alpha}\Gamma f\|$. It is noted that $\mathcal{L}_\alpha$ is nonzero if and only if $(\overline{\alpha})^{-1}$ is an eigenvalue for U^*. In this case, $\mathcal{L}_\alpha$ is the eigenspace for U^* corresponding to the eigenvalue $(\overline{\alpha})^{-1}$. Finally, according to Lemma 9.1.2 below, $\mathcal{L}_\alpha = P_{\mathcal{L}_\alpha}\Gamma\mathcal{E}$.

By virtue of the Wold decomposition $U = S \oplus V$, we see that

$$\mathcal{L}_\alpha = \ker(S^* - (\overline{\alpha})^{-1}I) \oplus \ker(V^* - (\overline{\alpha})^{-1}I).$$

Recall that S is the unilateral shift on $\ell^2_+(\mathcal{Y})$ and V is unitary. The open unit disc equals the set of all eigenvalues for the backward shift S^*, that is, $\mathrm{eig}\,(\mathrm{S}^*) = \mathbb{D}$; see Section 1.2. Because V is unitary, the eigenvalues for V are contained in the unit circle, that is, $\mathrm{eig}\,(\mathrm{V}^*) \subset \mathbb{T}$. This readily implies that

$$\begin{aligned} \mathcal{L}_\alpha &= \ker(S^* - (\overline{\alpha})^{-1}I) \qquad (\text{if } |\alpha| > 1) \\ &= \ker(V^* - (\overline{\alpha})^{-1}I) \qquad (\text{if } |\alpha| = 1). \end{aligned} \tag{9.1.9}$$

Finally it is noted that $\mathcal{L}_\infty = \ker S^* = \Pi^*_{\mathcal{Y}}\mathcal{Y}$.

For the moment assume that α is on the unit circle $\mathbb{T}$. In this case, $\alpha = 1/\overline{\alpha}$. Because V^* is unitary, $\alpha = 1/\overline{\alpha}$ is an eigenvalue for V^* if and only if $\overline{\alpha}$ is an eigenvalue for V. Moreover, the eigenspace for V^* corresponding to α equals the eigenspace for V corresponding to $\overline{\alpha}$, that is, $\ker(V^* - \alpha I) = \ker(V - \overline{\alpha}I)$. Notice that $\overline{\alpha}$ is an eigenvalue for V if and only if $\mathcal{L}_\alpha$ is nonzero. This readily implies that

$$\rho(\alpha, f) = \|P_{\mathcal{L}_\alpha}\Gamma f\|^2 = \|\Pi_{\mathcal{L}_\alpha}\Gamma_2 f\|^2.$$

Therefore $\rho(\alpha, f) = \|\Pi_{\mathcal{L}_\alpha}\Gamma_2 f\|^2$ for all α on $\mathbb{T}$. Obviously, $P_{\mathcal{L}_\alpha} = 0$ if and only if $\overline{\alpha}$ is not an eigenvalue for V. Hence that last equation in (9.1.7) holds. In fact, according to (9.1.3), we have

$$\begin{aligned} \Pi_{\mathcal{L}_\alpha}\Gamma &= \Pi_{\mathcal{L}_\alpha}\Gamma_2 = A_j \qquad \text{if } \overline{\alpha} = \lambda_j, \\ \Pi_{\mathcal{L}_\alpha}\Gamma &= \Pi_{\mathcal{L}_\alpha}\Gamma_2 = 0 \qquad \text{if } \overline{\alpha} \notin \mathrm{eig}\,(\mathrm{V}). \end{aligned} \tag{9.1.10}$$

The second from the last equation in (9.1.7) follows from $\rho(\alpha, f) = \|\Pi_{\mathcal{L}_\alpha}\Gamma_2 f\|^2 = \|A_j f\|^2$ when $\overline{\alpha} = \lambda_j$; see (9.1.3).

It remains to derive an expression for $\rho(\alpha, f)$ when α is contained in $\mathbb{D}_+$. Consider the operator

$$\Psi_\alpha = \begin{bmatrix} I \\ (\overline{\alpha})^{-1}I \\ (\overline{\alpha})^{-2}I \\ \vdots \end{bmatrix} : \mathcal{Y} \to \ell^2_+(\mathcal{Y}) \qquad (\alpha \in \mathbb{D}_+). \tag{9.1.11}$$

Then Ψ_α is an operator whose range equals $\mathcal{L}_\alpha$ the eigenspace for the backward shift S^* corresponding to the eigenvalue $1/\overline{\alpha}$; see Section 1.2. Notice that

$$\Psi^*_\alpha\Psi_\alpha = \sum_{n=0}^{\infty} \frac{1}{|\alpha|^{2n}} I = \frac{I}{1 - |\alpha|^{-2}} = \frac{|\alpha|^2}{|\alpha|^2 - 1} I = \frac{1}{d^2_\alpha} I.$$

This readily implies that $d_\alpha \Psi_\alpha$ is an isometry from $\mathcal{Y}$ onto $\ell^2_+(\mathcal{Y})$ whose range equals $\mathcal{L}_\alpha$. So $P_{\mathcal{L}_\alpha} = d_\alpha^2 \Psi_\alpha \Psi_\alpha^*$ is the orthogonal projection onto $\mathcal{L}_\alpha$. Using the formula for Γ_1 in (9.1.2), we see that $\Psi_\alpha^* \Gamma_1 = \sum_0^\infty \alpha^{-n} \Theta_n = \Theta(\alpha)$. Hence

$$P_{\mathcal{L}_\alpha} \Gamma = \begin{bmatrix} d_\alpha^2 \Psi_\alpha \Psi_\alpha^* \Gamma_1 \\ 0 \end{bmatrix} = \begin{bmatrix} d_\alpha^2 \Psi_\alpha \Theta(\alpha) \\ 0 \end{bmatrix} \qquad (\alpha \in \mathbb{D}_+). \tag{9.1.12}$$

Using this and the fact that $d_\alpha \Psi_\alpha$ is an isometry, we obtain

$$\|P_{\mathcal{L}_\alpha} \Gamma f\|^2 = \|(d_\alpha \Psi_\alpha) d_\alpha \Theta(\alpha) f\|^2 = d_\alpha^2 \|\Theta(\alpha) f\|^2.$$

Therefore $\rho(\alpha, f) = d_\alpha^2 (\Theta(\alpha)^* \Theta(\alpha) f, f)$ when α is in $\mathbb{D}_+$. By letting $|\alpha| \to \infty$, we obtain $\rho(\infty, f) = \|\Theta(\infty) f\|^2$.

To complete the proof assume that $\rho(\alpha, f) = 0$ for all f in $\mathcal{E}$ and some fixed α in $\mathbb{D}_+$. Then $\Theta(\alpha) = 0$. Because Θ is an outer function, the range of $\Theta(\alpha)$ equals $\mathcal{Y}$. Therefore $\mathcal{Y}$ must be zero, and the maximal outer spectral factor for T_R is zero. □

Lemma 9.1.2. *Let $\{A$ on $\mathcal{X}, B\}$ be a controllable pair where B is an operator mapping a finite dimensional space $\mathcal{E}$ into $\mathcal{X}$. Fix λ in $\mathbb{C}$, and let $\mathcal{L}$ be the subspace of $\mathcal{X}$ defined by*

$$\mathcal{L} = \mathcal{X} \ominus ran(A - \lambda I) = \ker(A^* - \overline{\lambda} I).$$

Then $P_\mathcal{L} B \mathcal{E} = \mathcal{L}$. In particular, if $\overline{\lambda}$ is an eigenvalue for A^*, then $\mathcal{L}$ is the eigenspace for A^* corresponding to $\overline{\lambda}$ and the range of $P_\mathcal{L} B$ equals $\mathcal{L}$.

Proof. Without loss of generality, we can assume that $\overline{\lambda}$ is an eigenvalue for A^* and $\mathcal{L}$ is the corresponding eigenspace for A^*. Now, let x be any vector in $\mathcal{L} \ominus P_\mathcal{L} B \mathcal{E}$. Then using $A^* x = \overline{\lambda} x$, we have, for all v in $\mathcal{E}$,

$$\begin{aligned}(x, A^n B v) &= (A^{*n} x, B v) = ((\overline{\lambda})^n x, B v) = (\overline{\lambda})^n (P_\mathcal{L} x, B v) \\ &= (\overline{\lambda})^n (x, P_\mathcal{L} B v) = 0\end{aligned}$$

for all integers $n \geq 0$. Since $\{A, B\}$ is controllable, x is orthogonal to the whole space $\mathcal{X}$. Thus x must be zero and hence $\mathcal{L} \ominus P_\mathcal{L} B \mathcal{E} = \{0\}$. In other words, $\mathcal{L} = P_\mathcal{L} B \mathcal{E}$. □

9.2 Sinusoid Estimation

Let $\{U$ on $\mathcal{K}, \Gamma\}$ be the controllable isometric representation for a positive Toeplitz matrix T_R with $\mathcal{L}(\mathcal{E}, \mathcal{E})$-valued symbol $R = \sum_{-\infty}^\infty e^{-\imath \omega n} R_n$. Recall that

$$T_{R,n} = \begin{bmatrix} R_0 & R_1^* & \cdots & R_{n-1}^* \\ R_1 & R_0 & \cdots & R_{n-2}^* \\ \vdots & \vdots & \ddots & \vdots \\ R_{n-1} & R_{n-2} & \cdots & R_0 \end{bmatrix} \text{ on } \mathcal{E}^n \tag{9.2.1}$$

is the compression of T_R to $\mathcal{E}^n$. Assume that $T_{R,n}$ is invertible for all n. In this section, we will present Capon's maximal likelihood method [49] or Geronimus [107] orthogonal polynomial result to determine the eigenvalues $\{\lambda_j\}$ and $\{A_j^*A_j\}$ for the unitary part $\{V, \Gamma_2\}$ in the Wold decomposition for $\{U, \Gamma\}$; see (9.1.1), (9.1.2) and (9.1.3).

Remark 9.2.1. Let T_R be a positive Toeplitz matrix with block entries in $\mathcal{L}(\mathcal{E}, \mathcal{E})$, and $T_{R,n}$ the compression of T_R to $\mathcal{E}^n$; see (9.2.1). If the maximal spectral factor Θ for T_R is a function in $H^2(\mathcal{E}, \mathcal{E})$, then $T_{R,n}$ is invertible for all integers $n \geq 1$.

By the definition of a maximal outer spectral factor, $T_R \geq T_\Theta^\sharp T_\Theta$. So for any x in $\mathcal{E}^n$ viewed as a subspace of $\ell_+^c(\mathcal{E})$, we obtain

$$(T_{R,n}x, x) = (T_R x, x) \geq (T_\Theta^\sharp T_\Theta x, x) = \|T_\Theta x\|^2 \geq \|\Pi_{\mathcal{E}^n} T_\Theta x\|^2.$$

Because Θ is an outer function in $H^2(\mathcal{E}, \mathcal{E})$, the operator $\Theta(\infty)$ is invertible. The matrix representation for $\Pi_{\mathcal{E}^n} T_\Theta|\mathcal{E}^n$ is a lower triangular block Toeplitz matrix with $\Theta(\infty)$ on the diagonal; see (8.2.1). Hence $\Pi_{\mathcal{E}^n} T_\Theta|\mathcal{E}^n$ is invertible. Since $(T_{R,n}x, x) \geq \|\Pi_{\mathcal{E}^n} T_\Theta x\|^2$ for all x in $\mathcal{E}^n$, the operator $T_{R,n}$ is strictly positive.

Theorem 9.2.2 (Capon-Geronimus). *Let $\{U$ on $\mathcal{K}, \Gamma\}$ be the controllable isometric representation for a positive Toeplitz matrix T_R, and assume that $T_{R,n}$ is strictly positive for all integers $n \geq 1$. Let Θ be the maximal outer spectral factor for T_R. Let $\{V, \Gamma_2\}$ in (9.1.3) be the unitary part in the Wold decomposition for $\{U, \Gamma\}$. For any α and β in $\overline{\mathbb{D}}_+$ set*

$$\begin{aligned}
C_{n,\alpha} &= \begin{bmatrix} I & \alpha^{-1}I & \alpha^{-2}I & \cdots & \alpha^{1-n}I \end{bmatrix} : \mathcal{E}^n \to \mathcal{E}, \\
C_{n,\infty} &= \begin{bmatrix} I & 0 & 0 & \cdots & 0 \end{bmatrix} : \mathcal{E}^n \to \mathcal{E}, \\
K_n(\beta, \alpha) &= C_{n,\beta} T_{R,n}^{-1} C_{n,\alpha}^*.
\end{aligned} \tag{9.2.2}$$

Then $K_n(\alpha, \alpha)$ is invertible. Moreover, $K_n(\alpha, \alpha)^{-1}$ is a decreasing set of positive operators, that is, $K_n(\alpha, \alpha)^{-1} \geq K_{n+1}(\alpha, \alpha)^{-1}$. Finally,

$$\begin{aligned}
\lim_{n\to\infty} K_n(\alpha, \alpha)^{-1} &= \Theta(\infty)^*\Theta(\infty) && (\alpha = \infty) \\
&= d_\alpha^2 \Theta(\alpha)^*\Theta(\alpha) && (|\alpha| > 1) \\
&= A_j^* A_j && (\overline{\alpha} = \lambda_j \textit{ and } \overline{\alpha} \in \operatorname{eig}(V)) \\
&= 0 && (|\alpha| = 1 \textit{ and } \overline{\alpha} \notin \operatorname{eig}(V)).
\end{aligned} \tag{9.2.3}$$

In order to implement Theorem 9.2.2, observe that $T_{R,n}^{-1}$ admits a factorization of the form $T_{R,n}^{-1} = U_n U_n^*$ where U_n is an upper triangular matrix of the form:

$$U_n = \begin{bmatrix} \widehat{B}_{n,0} & \widehat{B}_{n-1,0} & \cdots & \widehat{B}_{3,0} & \widehat{B}_{2,0} & \widehat{B}_{1,0} \\ \widehat{B}_{n,1} & \widehat{B}_{n-1,1} & \cdots & \widehat{B}_{3,1} & \widehat{B}_{2,1} & 0 \\ \widehat{B}_{n,2} & \widehat{B}_{n-1,2} & \cdots & \widehat{B}_{3,2} & 0 & 0 \\ \vdots & \vdots & \vdots & \vdots & \vdots & \vdots \\ \widehat{B}_{n,n-2} & \widehat{B}_{n-1,n-2} & \cdots & \cdots & 0 & 0 \\ \widehat{B}_{n,n-1} & 0 & \cdots & \cdots & 0 & 0 \end{bmatrix}. \tag{9.2.4}$$

Here $\{\widehat{B}_{k,j}\}$ are all operators on $\mathcal{E}$. It is emphasized that $\{\widehat{B}_{k,j}\}$ can be computed recursively using the Levinson algorithm; see Remark 15.1.2. (The indices on $\widehat{B}_{k,j}$ in U_n are arranged in a nonstandard way to coincide with how they are computed from the Levinson algorithm.) Now let $\Phi_k(z)$ be the Fourier transform of $\{\widehat{B}_{k,j}\}_{j=0}^{k-1}$, that is,

$$\Phi_k(z) = \sum_{j=0}^{k-1} z^{-j}\widehat{B}_{k,j} = C_{n,z} \begin{bmatrix} \widehat{B}_{k,0} \\ \widehat{B}_{k,1} \\ \vdots \\ \widehat{B}_{k,k-1} \\ 0 \\ \vdots \\ 0 \end{bmatrix} \qquad (k \geq 1). \tag{9.2.5}$$

The equation $K_n(z,z) = C_{n,z}T_{R,n}^{-1}C_{n,z}^*$ yields

$$K_n(z,z) = \sum_{k=1}^{n} \Phi_k(z)\Phi_k(z)^*. \tag{9.2.6}$$

To verify that (9.2.6) holds, observe that

$$C_{n,z}U_n = \begin{bmatrix} \Phi_n(z) & \Phi_{n-1}(z) & \cdots & \Phi_2(z) & \Phi_1(z) \end{bmatrix},$$

$$K_n(z,z) = C_{n,z}T_{R,n}^{-1}C_{n,z}^* = C_{n,z}U_n(C_{n,z}U_n)^* = \sum_{k=1}^{n} \Phi_k(z)\Phi_k(z)^*.$$

The second from the last equality follows from $T_{R,n}^{-1} = U_nU_n^*$. Therefore (9.2.6) holds.

By combining (9.2.3) with (9.2.6), we obtain

$$\begin{aligned} \lim_{n\to\infty} \left(\sum_{k=1}^{n} \Phi_k(\alpha)\Phi_k(\alpha)^* \right)^{-1} &= \Theta(\infty)^*\Theta(\infty) && (\alpha = \infty) \\ &= d_\alpha^2\Theta(\alpha)^*\Theta(\alpha) && (\alpha \in \mathbb{D}_+) \\ &= A_j^*A_j && (\overline{\alpha} = \lambda_j \text{ and } \overline{\alpha} \in \text{eig}(V)) \\ &= 0 && (|\alpha| = 1 \text{ and } \overline{\alpha} \notin \text{eig}(V)). \end{aligned} \tag{9.2.7}$$

Finally, it is noted that Φ_k is the kth orthogonal polynomial with values in $\mathcal{L}(\mathcal{E},\mathcal{E})$ obtained in classical orthogonal polynomial theory; see Geronimus [107]. In this case, equation (9.2.7) reduces to the classical summation formulas from orthogonal polynomial theory. In particular, if $\mathcal{E}=\mathbb{C}$, then Θ is a scalar-valued outer function and A_j is a scalar. In other words, in the scalar setting

$$\begin{aligned}\lim_{n\to\infty}\left(\sum_{k=1}^{n}|\Phi_k(\alpha)|^2\right)^{-1} &= |\Theta(\infty)|^2 && (\alpha=\infty)\\ &= d_\alpha^2|\Theta(\alpha)|^2 && (\alpha\in\mathbb{D}_+)\\ &= |A_j|^2 && (\overline{\alpha}=\lambda_j \text{ and } \overline{\alpha}\in \operatorname{eig}(V))\\ &= 0 && (|\alpha|=1 \text{ and } \overline{\alpha}\notin \operatorname{eig}(V)).\end{aligned} \tag{9.2.8}$$

Remark 9.2.3. One does not need the Levinson algorithm to compute $K_n(z,z)$. As before, assume that $T_{R,n}$ is a strictly positive Toeplitz operator on $\mathcal{E}^n$. Compute any operator L_n on $\mathcal{E}^n$ such that $T_{R,n}^{-1}=L_nL_n^*$. Then L_n admits a matrix representation of the form

$$L_n=\begin{bmatrix}\psi_{1,1} & \psi_{1,2} & \cdots & \psi_{1,n}\\ \psi_{2,1} & \psi_{2,2} & \cdots & \psi_{2,n}\\ \vdots & \vdots & \cdots & \vdots\\ \psi_{n,1} & \psi_{n,2} & \cdots & \psi_{n,n}\end{bmatrix} \text{ on } \mathcal{E}^n.$$

Notice that $\{\psi_{j,k}\}$ are operators on $\mathcal{E}$. Let $\Upsilon_j(z)$ be the Fourier transform of the j-th column of L_n, that is, $\Upsilon_j(z)=\sum_{k=0}^{n-1}z^{-k}\psi_{k+1,j}$. Then

$$K_n(z,z)=\sum_{k=1}^{n}\Upsilon_k(z)\Upsilon_k(z)^*.$$

To see this, simply notice that for $z\neq 0$, we have

$$\begin{aligned}C_{n,z}L_n &= \begin{bmatrix}\Upsilon_1(z) & \Upsilon_2(z) & \cdots & \Upsilon_n(z)\end{bmatrix},\\ K_n(z,z) &= C_{n,z}T_{R,n}^{-1}C_{n,z}^* = C_{n,z}L_nL_n^*C_{n,z}^* = \sum_{k=1}^{n}\Upsilon_k(z)\Upsilon_k(z)^*.\end{aligned}$$

Hence $K_n(z,z)=\sum_1^n\Upsilon_k(z)\Upsilon_k(z)^*$. Finally, it is noted that $\{\Upsilon_k\}_1^n$ depends on the factorization $T_{R,n}^{-1}=L_nL_n^*$. So this method is not recursive. However, the Levinson algorithm in (9.2.4) and (9.2.5) is recursive.

Proof of Theorem 9.2.2. Because $C_{n,\alpha}$ is onto and $T_{R,n}$ is invertible, it follows that $K_n(\alpha,\alpha)=C_{n,\alpha}T_{R,n}^{-1}C_{n,\alpha}^*$ is invertible. For fixed α in $\overline{\mathbb{D}}_+=\{z:|z|\geq 1\}$ and $f\in\mathcal{E}$, consider the optimization problem

$$\rho_n(\alpha,f)=\inf\{(T_{R,n}x,x): x\in\mathcal{E}^n \text{ and } C_{n,\alpha}x=f\}. \tag{9.2.9}$$

We claim that $\{\rho_n(\alpha, f)\}_1^\infty$ is decreasing, that is, $\rho_{n+1}(\alpha, f) \leq \rho_n(\alpha, f)$ for all integers $n \geq 0$. To see this observe that

$$\begin{aligned}\rho_{n+1}(\alpha, f) &= \inf\{(T_{R,n+1}x, x) : x \in \mathcal{E}^{n+1} \text{ and } C_{n+1,\alpha}x = f\} \\ &\leq \inf\{(T_{R,n+1}y, y) : y = x \oplus 0 \in \mathcal{E}^n \oplus \mathcal{E} \text{ and } C_{n+1,\alpha}y = f\} \\ &= \inf\{(T_{R,n}x, x) : x \in \mathcal{E}^n \text{ and } C_{n,\alpha}x = f\} \\ &= \rho_n(\alpha, f).\end{aligned}$$

Hence $\rho_{n+1}(\alpha, f) \leq \rho_n(\alpha, f)$. By consulting Lemma 9.2.4 below with $C = C_{n,\alpha}$ and $T = T_{R,n}$, it follows that $\rho_n(\alpha, f) = (K_n(\alpha, \alpha)^{-1}f, f)$. So $K_n(\alpha, \alpha)^{-1}$ is a decreasing sequence of positive operators in n. Finally, it is noted that $K_n(\alpha, \alpha)$ is increasing.

To complete the proof, it remains to show that $\rho_n(\alpha, f)$ converges to $\rho(\alpha, f)$ as n tends to infinity. Then Theorem 9.1.1 yields (9.2.3) and completes the proof. Let us view $\mathcal{E}^n$ as the subspace of $\ell_+^c(\mathcal{E})$ contained in the first n components of $\ell_+^c(\mathcal{E})$. Observe that

$$\begin{aligned}\rho_n(\alpha, f) &= \inf\{(T_{R,n}x, x) : x \in \mathcal{E}^n \text{ and } C_{n,\alpha}x = f\} \\ &= \inf\{(T_R x, x) : x \in \mathcal{E}^n \text{ and } C_{n,\alpha}x = f\} \\ &= \inf\{(T_R x, x) : x \in \mathcal{E}^n \text{ and } (\mathcal{F}_\mathcal{E}^+ x)(\alpha) = f\} \\ &\to \inf\{(T_R x, x) : x \in \ell_+^c(\mathcal{E}) \text{ and } (\mathcal{F}_\mathcal{E}^+ x)(\alpha) = f\} \\ &= \rho(\alpha, f).\end{aligned} \tag{9.2.10}$$

Therefore $(K_n(\alpha, \alpha)^{-1}f, f) = \rho_n(\alpha, f)$ converges to $\rho(\alpha, f)$. This with Theorem 9.1.1 completes the proof. □

Lemma 9.2.4. *Let T be a strictly positive operator on $\mathcal{X}$. Let C be an operator from $\mathcal{X}$ onto $\mathcal{E}$. Consider the optimization problem*

$$\rho(f) = \inf\{(Tx, x) : x \in \mathcal{X} \textit{ and } Cx = f\}. \tag{9.2.11}$$

Then the optimal solution to this optimization problem is unique and given by

$$x_{opt} = T^{-1}C^* \left(CT^{-1}C^*\right)^{-1} f \quad \textit{and} \quad \rho(f) = (\left(CT^{-1}C^*\right)^{-1} f, f). \tag{9.2.12}$$

Proof. First let us consider an optimization problem of the form

$$\rho(f) = \inf\{\|y\|^2 : y \in \mathcal{X} \text{ and } Ay = f\}. \tag{9.2.13}$$

Here A is an operator from $\mathcal{X}$ onto $\mathcal{E}$ and f is a vector in $\mathcal{E}$. Notice that this is precisely the optimization problem in (9.2.11) with $T = I$ and $C = A$. The solution is unique and given by

$$y_{opt} = A^* \left(AA^*\right)^{-1} f \quad \text{and} \quad \rho(f) = (\left(AA^*\right)^{-1} f, f). \tag{9.2.14}$$

To see this, let y_{opt} be the unique vector in $\ker(A)^{\perp}$ such that $Ay_{opt} = f$. If y is any other vector in $\mathcal{X}$ satisfying $Ay = f$, then $y = y_{opt} + v$ where v is in $\ker(A)$. Using the fact that y_{opt} is orthogonal to v, we obtain

$$\|y\|^2 = \|y_{opt} + v\|^2 = \|y_{opt}\|^2 + 2\Re(y_{opt}, v) + \|v\|^2 = \|y_{opt}\|^2 + \|v\|^2 \geq \|y_{opt}\|^2.$$

In other words, $\|y\| \geq \|y_{opt}\|$ with equality if and only if $v = 0$, or equivalently, $y = y_{opt}$. So the optimal solution to the optimization problem in (9.2.13) is unique and determined by the unique vector y_{opt} in $\ker(A)^{\perp}$ such that $Ay_{opt} = f$. Because the range of A^* equals $\ker(A)^{\perp}$ and A^* is one to one, there exists a unique vector u in $\mathcal{E}$ such that $y_{opt} = A^*u$. Hence $f = Ay_{opt} = AA^*u$. Since A^* is one to one, AA^* is invertible, and thus, $u = (AA^*)^{-1} f$. Therefore $y_{opt} = A^*u = A^* (AA^*)^{-1} f$. Finally, the optimal cost

$$\rho(f) = \|y_{opt}\|^2 = (A^*(AA^*)^{-1}f, A^*(AA^*)^{-1}f) = ((AA^*)^{-1}f, f).$$

Thus $\rho(f) = ((AA^*)^{-1}f, f)$.

Let us return to the optimization problem in (9.2.11). Notice that $(Tx, x) = \|T^{1/2}x\|^2$ where $T^{1/2}$ is the positive square root of T. Because T is invertible, $T^{1/2}$ is also invertible. By choosing $y = T^{1/2}x$, we see that the optimization problem in (9.2.11) is equivalent to

$$\rho(f) = \inf\{\|y\|^2 : y \in \mathcal{X} \text{ and } CT^{-1/2}y = f\}. \tag{9.2.15}$$

In other words, y_{opt} is a solution to (9.2.15) if and only if x_{opt} is a solution to (9.2.11) where $y_{opt} = T^{1/2}x_{opt}$. By consulting the optimization problem in (9.2.13), with $A = CT^{-1/2}$ we see that the optimal solution y_{opt} to (9.2.15) is given by $y_{opt} = T^{-1/2}C^* \left(CT^{-1}C^*\right)^{-1} f$. Therefore

$$x_{opt} = T^{-1/2}y_{opt} = T^{-1}C^* \left(CT^{-1}C^*\right)^{-1} f.$$

Since y_{opt} is unique, x_{opt} is also unique. A simple calculation shows that

$$\rho(f) = (Tx_{opt}, x_{opt}) = \left(\left(CT^{-1}C^*\right)^{-1} f, f\right).$$

This yields (9.2.12). □

The case when T_R is invertible. Previously, we have discussed the case when T_R is positive. Now assume that T_R is an invertible positive Toeplitz operator on $\ell_+^2(\mathcal{E})$. Then T_R admits an invertible outer spectral factor Θ in $H^\infty(\mathcal{E}, \mathcal{E})$, that is, $T_R = T_\Theta^* T_\Theta$ where T_Θ is an invertible lower triangular Toeplitz operator on $\ell_+^2(\mathcal{E})$; see Theorem 7.1.1. For each α in $\mathbb{D}_+$, let $C_{\infty,\alpha}$ be the operator mapping $\ell_+^2(\mathcal{E})$ onto $\mathcal{E}$ given by

$$C_{\infty,\alpha} = \begin{bmatrix} I & \alpha^{-1}I & \alpha^{-2}I & \cdots \end{bmatrix} : \ell_+^2(\mathcal{E}) \to \mathcal{E}. \tag{9.2.16}$$

Observe that $C_{\infty,z}h = (\mathcal{F}_{\mathcal{E}}^{+}h)(z)$ for h in $\ell_+^2(\mathcal{E})$ and z in $\mathbb{D}_+$. For α and β in $\mathbb{D}_+$, let $K(\beta,\alpha)$ be the operator on $\mathcal{E}$ defined by

$$K(\beta,\alpha) = C_{\infty,\beta}T_R^{-1}C_{\infty,\alpha}^{*}. \tag{9.2.17}$$

Notice that $K(z,\alpha) = (\mathcal{F}_{\mathcal{E}}^{+}T_R^{-1}C_{\infty,\alpha}^{*})(z)$. For α in $\mathbb{D}_+$ and f in $\mathcal{E}$, consider the optimization problem in (9.1.4) rewritten as

$$\rho(\alpha,f) = \inf\{(T_R x, x) : x \in \ell_+^2(\mathcal{E}) \text{ and } C_{\infty,\alpha}x = f\}. \tag{9.2.18}$$

According to Lemma 9.2.4, the solution to this optimization problem is unique and given by

$$x_{opt} = T_R^{-1}C_{\infty,\alpha}^{*}K(\alpha,\alpha)^{-1}f \quad \text{and} \quad \rho(\alpha,f) = (K(\alpha,\alpha)^{-1}f,f). \tag{9.2.19}$$

Finally, it is noted that $(\mathcal{F}_{\mathcal{E}}^{+}x_{opt})(z) = K(z,\alpha)K(\alpha,\alpha)^{-1}f$.

Now let us show how the solution to the optimization problem in (9.2.18) can be used to compute the outer spectral factor. First we claim that

$$\begin{aligned} K(z,\alpha) &= \varphi_\alpha(z)\Theta(z)^{-1}\Theta(\alpha)^{-*} \quad \text{where} \quad \varphi_\alpha(z) = \frac{z\overline{\alpha}}{z\overline{\alpha}-1}, \\ K(\alpha,\alpha)^{-1} &= d_\alpha^2\Theta(\alpha)^{*}\Theta(\alpha) \quad \text{where} \quad d_\alpha^2 = \frac{|\alpha|^2-1}{|\alpha|^2}, \\ K(z,\infty) &= \Theta(z)^{-1}\Theta(\infty)^{-*}, \\ K(\infty,\infty)^{-1} &= \Theta(\infty)^{*}\Theta(\infty). \end{aligned} \tag{9.2.20}$$

Because the outer spectral factor is unique up to a unitary constant on the left, without loss of generality we can assume that

$$K(\alpha,\alpha)^{-1/2} = d_\alpha\Theta(\alpha) \quad \text{and} \quad K(\infty,\infty)^{-1/2} = \Theta(\infty). \tag{9.2.21}$$

By consulting the first equation in (9.2.20), we see that

$$\Theta(z) = d_\alpha\varphi_\alpha(z)K(\alpha,\alpha)^{1/2}K(z,\alpha)^{-1} \quad \text{and} \quad \Theta(z) = K(\infty,\infty)^{1/2}K(z,\infty)^{-1}. \tag{9.2.22}$$

Finally, it is noted that the two formulas in (9.2.22) do not necessarily yield the same outer spectral factor Θ. These spectral factors are equal up to a unitary constant on the left.

All of our formulas for Θ follow from the first equation in (9.2.20). So to verify that (9.2.22) holds, it remains to show that $K(z,\alpha) = \varphi_\alpha(z)\Theta(z)^{-1}\Theta(\alpha)^{-*}$. To this end, let $\Theta(z)^{-1} = \sum_0^\infty z^{-n}\Psi_n$ be the power series expansion for Θ^{-1}.

Then

$$\begin{aligned}C_{\infty,\alpha}T_{\Theta}^{-1} = C_{\infty,\alpha}T_{\Theta^{-1}} &= C_{\infty,\alpha}\begin{bmatrix}\Psi_0 & 0 & 0 & \cdots\\ \Psi_1 & \Psi_0 & 0 & \cdots\\ \Psi_2 & \Psi_1 & \Psi_0 & \cdots\\ \vdots & \vdots & \vdots & \ddots\end{bmatrix}\\ &= \begin{bmatrix}\Theta(\alpha)^{-1} & \alpha^{-1}\Theta(\alpha)^{-1} & \alpha^{-2}\Theta(\alpha)^{-1} & \cdots\end{bmatrix}\\ &= \Theta^{-1}(\alpha)C_{\infty,\alpha}.\end{aligned}$$

In other words, $C_{\infty,\alpha}T_{\Theta}^{-1} = \Theta^{-1}(\alpha)C_{\infty,\alpha}$. Using $T_R = T_{\Theta}^*T_{\Theta}$, we obtain

$$\begin{aligned}K(z,\alpha) &= C_{\infty,z}T_R^{-1}C_{\infty,\alpha}^* = C_{\infty,z}T_{\Theta}^{-1}T_{\Theta}^{-*}C_{\infty,\alpha}^* = \Theta^{-1}(z)C_{\infty,z}C_{\infty,\alpha}^*\Theta(\alpha)^{-*}\\ &= \frac{1}{1-(z\overline{\alpha})^{-1}}\Theta^{-1}(z)\Theta(\alpha)^{-*} = \varphi_\alpha(z)\Theta^{-1}(z)\Theta(\alpha)^{-*}.\end{aligned}$$

Therefore the first equation in (9.2.20) holds, and the other equations follow from the first.

Remark 9.2.5. Let T_R be an invertible positive Toeplitz operator on $\ell_+^2(\mathcal{E})$, and Θ in $H^2(\mathcal{E},\mathcal{E})$ its outer spectral factor. Let $K_n(\beta,\alpha) = C_{n,\beta}T_{R,n}^{-1}C_{n,\alpha}^*$ where $T_{R,n}$ is the Toeplitz matrix contained in the block $n\times n$ upper left-hand corner of T_R; see (9.2.1). Then for z and α in $\mathbb{D}_+$, we have

$$\begin{aligned}\Theta(z) &= d_\alpha\varphi_\alpha(z)\lim_{n\to\infty}K_n(\alpha,\alpha)^{1/2}K_n(z,\alpha)^{-1},\\ \Theta(z) &= \lim_{n\to\infty}K_n(\infty,\infty)^{1/2}K_n(z,\infty)^{-1}.\end{aligned}\tag{9.2.23}$$

Lemma 7.6.1 shows that $T_{R,n}^{-1}$ converges to T_R^{-1} in the strong operator topology. Therefore (9.2.23) follows from (9.2.22). In Section 9.5, we will use (9.2.23) with the Kalman-Ho algorithm to obtain an algorithm to compute the outer spectral factor Θ for T_R in the rational case. Finally, it is noted that one can use the Gohberg-Semencul-Heinig inversion formula in Section 15.3 to compute the inverse of $T_{R,n}$.

Remark 9.2.6. For the moment let us return to the Carathéodory interpolation problem discussed in Section 7.5. Let $T_{R,n+1} = \Upsilon_{n+1}$ on $\mathcal{E}^{n+1}$ be the strictly positive Toeplitz matrix presented in (7.5.1). Then the solution Θ to the Carathéodory interpolation problem in Theorem 7.5.1 is also determined by

$$\Theta(z) = K_{n+1}(\infty,\infty)^{1/2}K_{n+1}(z,\infty)^{-1}.$$

Finally, it is noted that in this case, the "kernel function"

$$\begin{aligned}K_{n+1}(z,\infty) &= C_{n+1,z}T_{R,n+1}^{-1}C^*,\\ K_{n+1}(\infty,\infty) &= CT_{R,n+1}^{-1}C^*,\\ C &= \begin{bmatrix}I & 0 & 0 & \cdots & 0\end{bmatrix}^*.\end{aligned}$$

Here we set $C = C_{n+1,\infty}$ mapping $\mathcal{E}^{n+1}$ onto $\mathcal{E}$. The details are left as an exercise.

9.3 Sinusoid Estimation: Capon-Geronimus

Let $R = \sum_{-\infty}^{\infty} e^{-\imath\omega n} R_n$ be the symbol for T_R determined by

$$R_n = \frac{1}{2\pi}\int_0^{2\pi} e^{\imath\omega n}\Theta(e^{\imath\omega})^*\Theta(e^{\imath\omega})\, d\omega + \sum_{k=1}^{\nu} A_k^* A_k e^{-\imath\omega_k n} \tag{9.3.1}$$

where Θ is an outer function in $H^\infty(\mathcal{E},\mathcal{E})$. Moreover, A_j is an operator mapping $\mathcal{E}$ onto $\mathcal{E}_j$ for $j = 1, 2, \ldots, \nu$. The results in Section 6.3 show that the Toeplitz matrix T_R determined by this R is positive, and Θ is the maximal outer spectral factor for T_R. The controllable isometric representation $\{U, \Gamma\}$ for T_R is determined by

$$U = \begin{bmatrix} S & 0 \\ 0 & V \end{bmatrix} \text{ on } \begin{bmatrix} H^2(\mathcal{E}) \\ \oplus_1^\nu \mathcal{E}_j \end{bmatrix} \quad \text{and} \quad \Gamma = \begin{bmatrix} \Gamma_1 \\ \Gamma_2 \end{bmatrix} : \mathcal{E} \to \begin{bmatrix} H^2(\mathcal{E}) \\ \oplus_1^\nu \mathcal{E}_j \end{bmatrix}. \tag{9.3.2}$$

As expected, S is the unilateral shift on $H^2(\mathcal{E})$, and Γ_1 is the operator mapping $\mathcal{E}$ into $H^2(\mathcal{E})$ given by $(\Gamma_1\xi)(z) = \Theta(z)\xi$ where ξ is in $\mathcal{E}$. Furthermore, V is the diagonal unitary operator on $\oplus_1^\nu \mathcal{E}_j$ determined by

$$V = \begin{bmatrix} e^{\imath\omega_1} I & 0 & \cdots & 0 \\ 0 & e^{\imath\omega_2} I & \cdots & 0 \\ \vdots & \vdots & \ddots & \vdots \\ 0 & 0 & \cdots & e^{\imath\omega_\nu} I \end{bmatrix} \quad \text{and} \quad \Gamma_2 = \begin{bmatrix} A_1 \\ A_2 \\ \vdots \\ A_\nu \end{bmatrix}. \tag{9.3.3}$$

Notice that $\{e^{\imath\omega_j}\}_1^\nu$ are the eigenvalues for V. By construction $R_{-n} = \Gamma^* U^n \Gamma$ for all integers $n \geq 0$. Proposition 6.3.1 guarantees that the pair $\{U, \Gamma\}$ is controllable, and Θ is the maximal outer spectral factor for T_R.

Because Θ is in $H^2(\mathcal{E},\mathcal{E})$, the Toeplitz matrix $T_{R,n}$ on $\mathcal{E}^n$ obtained by compressing T_R to the upper left-hand $n \times n$ corner is strictly positive; see Remark 9.2.1. Hence one can use the Levinson algorithm to compute the normalized Levinson polynomials $\{\Phi_k\}_1^n$ associated with $T_{R,n}$; see (9.2.4) and (9.2.5). Recall that $\{e^{\imath\omega_j}\}_1^\nu$ are the eigenvalues for V. In this setting, the last two equations in (9.2.7) become

$$\begin{aligned} \lim_{n\to\infty} \left(\sum_{k=1}^{n} \Phi_k(e^{\imath\omega})\Phi_k(e^{\imath\omega})^*\right)^{-1} &= A_j^* A_j && (\text{if } \omega = -\omega_j) \\ &= 0 && (\text{if } \omega \notin \{-\omega_j\}_{j=1}^\nu). \end{aligned} \tag{9.3.4}$$

So to compute the frequencies $\{\omega_j\}_1^\nu$ one simply looks at the convergence of the series

$$K_n(e^{\imath\omega}, e^{\imath\omega})^{-1} = \left(\sum_{k=1}^{n} \Phi_k(e^{\imath\omega})\Phi_k(e^{\imath\omega})^*\right)^{-1}$$

as ω varies from 0 to 2π. The positive operators $K_n(e^{\imath\omega}, e^{\imath\omega})^{-1}$ converge monotonically to $A_j^* A_j$ at $\omega = -\omega_j$ for $j = 1, 2, \ldots, \nu$ and zero otherwise. So for large n, we have $A_j^*A_j \approx K_n(e^{-\imath\omega_j}, e^{-\imath\omega_j})^{-1}$. Moreover, if $\delta = \operatorname{rank} A_j$, then $K_n(e^{-\imath\omega_j}, e^{-\imath\omega_j})^{-1}$ has δ large eigenvalues and the rest of the eigenvalues are small. Let $K_n(e^{-\imath\omega_j}, e^{-\imath\omega_j})^{-1} = \Omega\Lambda\Omega^*$ be the spectral decomposition for the operator $K_n(e^{-\imath\omega_j}, e^{-\imath\omega_j})^{-1}$ where Ω is unitary and Λ is a diagonal matrix consisting of the eigenvalues of $K_n(e^{-\imath\omega_j}, e^{-\imath\omega_j})^{-1}$. By keeping the δ significant eigenvalues values for $K_n(e^{-\imath\omega_j}, e^{-\imath\omega_j})^{-1}$, it follows that

$$K_n(e^{-\imath\omega_j}, e^{-\imath\omega_j})^{-1} \approx \Omega_n\Lambda_n\Omega_n^*$$

where Λ_n on $\mathbb{C}^\delta$ is a diagonal matrix consisting of the δ significant eigenvalues and Ω_n is an isometry mapping $\mathbb{C}^\delta$ into $\mathcal{E}$. Hence $A_j^*A_j \approx \Omega_n\Lambda_n\Omega_n^*$. Because $A_j^*A_j$ uniquely determines A_j up to a unitary operator on the left, without loss of generality we have $A_j \approx (\Lambda_n)^{1/2}\Omega_n^*$.

To compute $(\sum_{k=1}^n \Phi_k\Phi_k^*)^{-1}$ on the unit circle, one can use the fast Fourier transform to evaluate Φ_k at 2^m points around the unit circle. Then plot the norm $\|K_n(e^{\imath\omega}, e^{\imath\omega})^{-1}\|$, or equivalently, the inverse of the smallest eigenvalue of $K_n(e^{\imath\omega}, e^{\imath\omega})$ for various values of n. (In the scalar case, one simply plots $K_n(e^{\imath\omega}, e^{\imath\omega})^{-1}$.) This graph will be decreasing to $\|A_j^*A_j\|$ at $\omega = -\omega_j$ for $j = 1, 2, \ldots, \nu$ and zero elsewhere. Once the frequencies $\{\omega_j\}_1^\nu$ have been determined, then the corresponding amplitude matrices $A_j^*A_j \approx K_n(e^{-\imath\omega_j}, e^{-\imath\omega_j})^{-1}$ for n sufficiently large. Finally, it is noted that one can also use Remark 9.2.3 with the fast Fourier transform to compute $K_n(e^{\imath\omega}, e^{\imath\omega})$ and implement this algorithm.

In almost all practical problems, the eigenvalues for V come in complex conjugate pairs. In this case, $A_j^*A_j = A_k^*A_k$ when $e^{\imath\omega_j} = e^{-\imath\omega_k}$. So in practice we only have to check the frequencies in the range $0 \le \omega \le \pi$.

Example. Consider the outer function given by

$$\theta(z) = \frac{1.1465z^2 - 0.2850z + 0.1125}{z^2 - 0.2802z - 0.0585}. \tag{9.3.5}$$

Let T_R be the positive Toeplitz matrix with symbol $R = \sum_{-\infty}^{\infty} r_n\, e^{-\imath\omega n}$ determined by

$$r_n = \frac{1}{2\pi}\int_0^{2\pi} e^{\imath\omega n}|\theta(e^{\imath\omega})|^2 d\omega + \frac{1}{4}\cos(n\pi/4) + \frac{1}{2}\cos(n\pi/2). \tag{9.3.6}$$

The controllable isometric representation $\{U, \Gamma\}$ for T_R admits a Wold decomposition of the form

$$U = \begin{bmatrix} S & 0 \\ 0 & V \end{bmatrix} \text{ on } \begin{bmatrix} H^2 \\ \mathbb{C}^4 \end{bmatrix} \quad \text{and} \quad \Gamma = \begin{bmatrix} \Gamma_1 \\ \Gamma_2 \end{bmatrix} : \mathbb{C} \to \begin{bmatrix} H^2 \\ \mathbb{C}^4 \end{bmatrix}. \tag{9.3.7}$$

As expected, S is the unilateral shift on H^2, and Γ_1 is the operator mapping $\mathbb{C}$ into H^2 given by $(\Gamma_1\xi)(z) = \theta(z)\xi$ where ξ is in $\mathbb{C}$. Moreover, V is the unitary

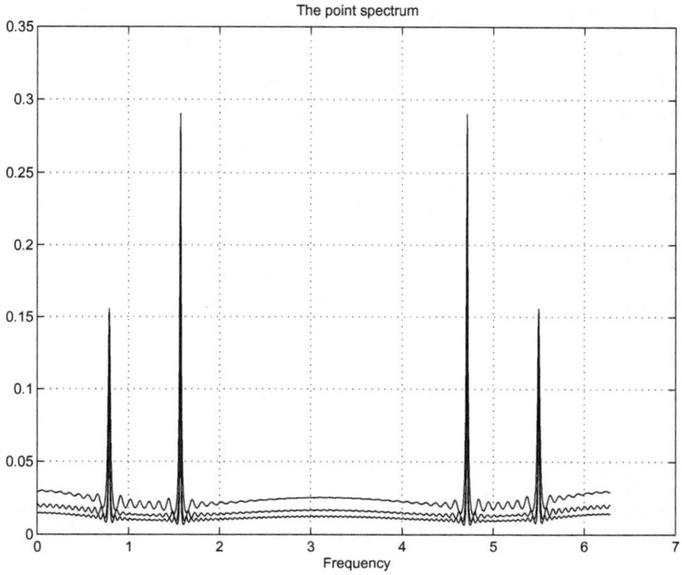

Figure 9.1: Convergence of $K_n(e^{-\imath\omega}, e^{-\imath\omega})^{-1}$

operator on $\mathbb{C}^4$ determined by

$$V = \begin{bmatrix} e^{\imath\pi/4} & 0 & 0 & 0 \\ 0 & e^{-\imath\pi/4} & 0 & 0 \\ 0 & 0 & e^{\imath\pi/2} & 0 \\ 0 & 0 & 0 & e^{-\imath\pi/2} \end{bmatrix} \quad \text{and} \quad \Gamma_2 = \begin{bmatrix} 1/\sqrt{8} \\ 1/\sqrt{8} \\ 1/2 \\ 1/2 \end{bmatrix}.$$

Proposition 6.3.1 guarantees that the pair $\{U, \Gamma\}$ is controllable and θ is the maximal outer spectral factor for T_R.

Using the Levinson algorithm with the fast Fourier transform, we plotted $\left(\sum_{k=1}^{n} |\Phi_k(e^{\imath\omega})|^2\right)^{-1}$ in Figure 9.1 for some large n. As expected, this series converges to zero if $e^{\imath\omega}$ is not one of the sinusoid frequencies, and converges to $1/8$ for $\omega = \pm\pi/4$ and $1/4$ for $\omega = \pm\pi/2$.

9.4 A Nested Optimization Problem

In this section we will present a nested set of optimization problems which will be used to approximate the maximal outer spectral factor for certain positive Toeplitz matrices.

Lemma 9.4.1. *Let $\{\mathcal{H}_n\}_1^\infty$ be an increasing sequence of finite dimensional subspaces, that is, $\mathcal{H}_n \subseteq \mathcal{H}_{n+1}$. Let C_n be a linear map from $\mathcal{H}_n$ onto a Hilbert space $\mathcal{E}$ such that $C_{n+1}|\mathcal{H}_n = C_n$ for all integers $n \geq 1$. Let E_n be a one to one operator from $\mathcal{H}_n$ into a Hilbert space $\mathcal{K}$ such that $E_{n+1}|\mathcal{H}_n = E_n$ for all integers $n \geq 1$, and set $T_n = E_n^* E_n$. Fix a vector f in $\mathcal{E}$, and consider the optimization problem*

$$\rho_n(f) = \inf\{\|E_n h\|^2 : h \in \mathcal{H}_n \text{ and } C_n h = f\}. \tag{9.4.1}$$

Then the following holds.

(i) *For each integer $n \geq 1$ an optimal solution to (9.4.1) is given by $G_n f$ where G_n is the operator mapping $\mathcal{E}$ into $\mathcal{H}_n$ determined by*

$$G_n = T_n^{-1} C_n^* \left(C_n T_n^{-1} C_n^*\right)^{-1} \quad \text{and} \quad \rho_n(f) = \left(\left(C_n T_n^{-1} C_n^*\right)^{-1} f, f\right). \tag{9.4.2}$$

(ii) *The cost functions $\{\rho_n(f)\}_1^\infty$ form a decreasing sequence of positive scalars and $\rho_n(f) = (\Delta_n f, f)$ where $\Delta_n = \left(C_n T_n^{-1} C_n^*\right)^{-1}$. The $\{\Delta_n\}_1^\infty$ forms a decreasing sequence of positive operators on $\mathcal{E}$, that is, $\Delta_{n+1} \leq \Delta_n$ for all integers $n \geq 1$. Moreover, Δ_n converges to a positive operator Δ on $\mathcal{E}$ and*

$$(\Delta f, f) = \lim_{n \to \infty} \rho_n(f). \tag{9.4.3}$$

(iii) *The operators $E_n G_n$ converge to an operator Q mapping $\mathcal{E}$ into $\mathcal{K}$ as n tends to infinity. Furthermore, $\Delta = Q^* Q$.*

(iv) *For any integer $n \geq 1$, we have $P_{\mathcal{L}} E_n G_n = Q$ where $\mathcal{L}$ is the subspace of $\mathcal{K}$ defined by*

$$\mathcal{L} = \bigcap_{n=1}^{\infty} (\mathcal{K} \ominus E_n \ker C_n) = \mathcal{K} \ominus \bigvee_{n=1}^{\infty} \{E_n \ker C_n\}. \tag{9.4.4}$$

Finally, we have

$$\|E_n G_n f - Q f\|^2 = \rho_n(f) - (\Delta f, f). \tag{9.4.5}$$

Proof. The optimization problem in (9.4.1) is equivalent to the optimization problem

$$\rho_n(f) = \inf\{(T_n x, x) : x \in \mathcal{X} \text{ and } C_n x = f\} \tag{9.4.6}$$

in (9.2.11). According to Lemma 9.2.4, the optimal solution is given by

$$T_n^{-1} C_n^* \left(C_n T_n^{-1} C_n^*\right)^{-1} f.$$

Hence Part (i) holds. Finally, it is noted that Lemma 9.2.4 also shows that $\rho_n(f) = (\Delta_n f, f)$ where $\Delta_n = \left(C_n T_n^{-1} C_n^*\right)^{-1}$.

To verify that Part (ii) holds recall that the subspace $\mathcal{H}_n$ is contained in $\mathcal{H}_{n+1}$. Moreover, $C_{n+1}|\mathcal{H}_n = C_n$ and $E_{n+1}|\mathcal{H}_n = E_n$ for all integers $n \geq 1$. So the optimization problem in (9.4.1) corresponding to $n+1$ searches over a larger set than the corresponding optimization problem for n. Thus $\rho_{n+1}(f) \le \rho_n(f)$. In other words, the sequence $\{\rho_n(f)\}$ is monotonically decreasing. Because the cost $\rho_n(f) \ge 0$, the sequence $\rho_n(f)$ converges to a positive scalar $\rho(f)$ as n tends to infinity. Recall that $\rho_n(f) = (\Delta_n f, f)$. Using the fact that $\{\rho_n(f)\}_1^\infty$ is decreasing, we obtain

$$(\Delta_{n+1} f, f) = \rho_{n+1}(f) \le \rho_n(f) = (\Delta_n f, f).$$

Thus $\{\Delta_n\}_1^\infty$ forms a decreasing sequence of positive operators on $\mathcal{E}$. Clearly, Δ_n converges to a positive operator Δ as n tends to infinity. The limit in (9.4.3) follows from the fact that

$$\rho(f) = \lim_{n\to\infty} \rho_n(f) = \lim_{n\to\infty} (\Delta_n f, f) = (\Delta f, f) \qquad (f \in \mathcal{E}).$$

Therefore Part (ii) holds.

To verify that Part (iii) holds, fix f in $\mathcal{E}$. Observe that $h_n = G_n f$ is an optimal solution to the optimization problems in (9.4.1) corresponding to the integer n. First let us show that

$$E_n h_n \perp E_n \ker C_n \quad \text{where} \quad h_n = T_n^{-1} C_n^* \Delta_n f. \tag{9.4.7}$$

For u in $\ker C_n$, we have

$$(E_n u, E_n h_n) = (T_n u, T_n^{-1} C_n^* \Delta_n f) = (C_n u, \Delta_n f) = 0.$$

Hence (9.4.7) holds. Now assume that $m \le n$. Clearly, h_m and h_n satisfy the constraints $C_m h_m = f$ and $C_n h_n = f$. Since $\mathcal{H}_m \subset \mathcal{H}_n$ and $C_n|\mathcal{H}_m = C_m$, we have $C_n h_m = C_m h_m = f = C_n h_n$. Thus $C_n(h_m - h_n) = 0$, or equivalently, $h_m - h_n$ is in $\ker C_n$. According to (9.4.7), the vector $E_n(h_m - h_n)$ is orthogonal to $E_n h_n$. This readily implies that

$$\begin{aligned}
\|E_m G_m f - E_n G_n f\|_{\mathcal{K}}^2 &= \|E_m h_m - E_n h_n\|_{\mathcal{K}}^2 \\
&= \|E_n h_m\|^2 - 2\Re(E_n h_m, E_n h_n) + \|E_n h_n\|^2 \\
&= \|E_n h_m\|^2 - 2\Re(E_n(h_m - h_n + h_n), E_n h_n) + \|E_n h_n\|^2 \\
&= \|E_n h_m\|^2 - 2\Re(E_n(h_m - h_n), E_n h_n) - \|E_n h_n\|^2 \\
&= \|E_n h_m\|^2 - \|E_n h_n\|^2 = \rho_m(f) - \rho_n(f).
\end{aligned}$$

Recall that $\rho_n(f)$ monotonically converges to $\rho(f)$ as n tends to infinity. So as both m and n tend to infinity, the difference $\rho_m(f) - \rho_n(f)$ converges to zero. In other words, $\{E_n G_n f\}_1^\infty$ is a Cauchy sequence in the Hilbert space $\mathcal{K}$. Because the space $\mathcal{E}$ is finite dimensional, the operators $E_n G_n$ converge to an operator Q mapping $\mathcal{E}$ into $\mathcal{K}$ as n approaches infinity. Since $\Delta_n = G_n^* E_n^* E_n G_n$ converges to Δ and $E_n G_n$ converges to Q, we obtain $\Delta = Q^* Q$. Therefore Part (iii) holds.

As before, assume that $m \le n$. Since $C_n|\mathcal{H}_m = C_m$, we see that $\ker C_m$ is contained in $\ker C_n$. Because $E_n|\mathcal{H}_m = E_m$, we have $\mathcal{X}_n = \mathcal{K} \ominus E_n \ker C_n$ is contained in $\mathcal{X}_m = \mathcal{K} \ominus E_m \ker C_m$. In other words, $\{\mathcal{X}_n\}_1^\infty$ forms a decreasing sequence of subspaces. So if $\mathcal{L} = \bigcap_1^\infty \mathcal{X}_n$, then $P_{\mathcal{X}_n}$ converges to $P_{\mathcal{L}}$ in the strong operator topology as n approaches infinity. Observe that $\mathcal{M}_n = E_n \ker C_n$ is a subspace of $E_n\mathcal{H}_n$. This readily implies that $\mathcal{X}_n$ admits a decomposition of the form $\mathcal{X}_n = (E_n\mathcal{H}_n)^\perp \oplus (E_n\mathcal{H}_n \ominus \mathcal{M}_n)$, and thus,

$$P_{\mathcal{X}_n} = P_{(E_n\mathcal{H}_n)^\perp} + P_{(E_n\mathcal{H}_n\ominus\mathcal{M}_n)}.$$

Recall that $h_n = G_n f$, the vector $E_n(h_m - h_n)$ is in $\mathcal{M}_n$. Equation (9.4.7) shows that $E_n h_n$ is a vector in $E_n\mathcal{H}_n \ominus \mathcal{M}_n$. This readily implies that

$$\begin{aligned}
P_{\mathcal{L}} E_m G_m f &= \lim_{n\to\infty} P_{\mathcal{X}_n} E_m h_m \\
&= \lim_{n\to\infty} P_{(E_n\mathcal{H}_n)^\perp} E_m h_m + P_{(E_n\mathcal{H}_n\ominus\mathcal{M}_n)} E_m h_m \\
&= \lim_{n\to\infty} P_{(E_n\mathcal{H}_n\ominus\mathcal{M}_n)} E_n(h_m - h_n + h_n) \\
&= \lim_{n\to\infty} E_n h_n = \lim_{n\to\infty} E_n G_n f = Qf.
\end{aligned}$$

Therefore Part (iv) holds.

To complete the proof it remains to establish (9.4.5). To this end, observe that

$$\begin{aligned}
\|E_n G_n f - Qf\|^2 &= \|E_n G_n f\|^2 - 2\Re(E_n G_n f, P_{\mathcal{L}} Qf) + \|Qf\|^2 \\
&= \rho_n(f) - 2\Re(P_{\mathcal{L}} E_n G_n f, Qf) + \|Qf\|^2 \\
&= \rho_n(f) - 2\Re(Qf, Qf) + \|Qf\|^2 \\
&= \rho_n(f) - \|Qf\|^2 \\
&= \rho_n(f) - (\Delta f, f).
\end{aligned}$$

Therefore (9.4.5) holds. □

9.5 Limit Theorems

In this section, we will use Lemma 9.4.1 to develop a limit theorem to compute the maximal outer spectral factor and the eigenvalues for the unitary part in the Wold decomposition for the isometric representation. Finally, this yields a generalization of Remark 9.2.5.

Theorem 9.5.1. *Let the pair $\{U, \Gamma\}$ be a controllable isometric representation for a positive Toeplitz matrix T_R with $\mathcal{L}(\mathcal{E},\mathcal{E})$-valued symbol R. Let $U = S \oplus V$ on $\ell_+^2(\mathcal{E}) \oplus \mathcal{V}$ and $\Gamma = \begin{bmatrix}\Gamma_1 & \Gamma_2\end{bmatrix}^{tr}$ be the Wold decomposition for $\{U, \Gamma\}$ in (9.1.1), (9.1.2) and (9.1.3) where $\Theta(z) = (\mathcal{F}_{\mathcal{Y}}^+\Gamma_1)(z)$ in $H^2(\mathcal{E},\mathcal{E})$ is the maximal outer spectral factor for T_R, and S is a unilateral shift while V is unitary. Let*

$$W_2 = \begin{bmatrix}\Gamma_2 & V\Gamma_2 & V^2\Gamma_2 & \cdots\end{bmatrix}$$

be the controllability matrix for the pair $\{V, \Gamma_2\}$. *Finally, for* α *in* $\overline{\mathbb{D}}_+$ *let* $\widehat{G}_{n,\alpha}(z)$ *be the polynomial in* $1/z$ *with values in* $\mathcal{L}(\mathcal{E},\mathcal{E})$ *be given by*

$$\begin{aligned}
\widehat{G}_{n,\alpha}(z) &= K_n(z,\alpha)K_n(\alpha,\alpha)^{-1}, \\
G_{n,\alpha} &= T_{R,n}^{-1}C_{n,\alpha}^*K_n(\alpha,\alpha)^{-1} : \mathcal{E} \to \mathcal{E}^n, \\
K_n(z,\alpha) &= C_{n,z}T_{R,n}^{-1}C_{n,\alpha}^*, \\
C_{n,\alpha} &= \begin{bmatrix} I & \alpha^{-1}I & \alpha^{-2}I & \cdots & \alpha^{1-n}I \end{bmatrix} : \mathcal{E}^n \to \mathcal{E}, \\
C_{n,\infty} &= \begin{bmatrix} I & 0 & 0 & \cdots & 0 \end{bmatrix} : \mathcal{E}^n \to \mathcal{E}, \\
\varphi_\alpha(z) &= \frac{z\overline{\alpha}}{z\overline{\alpha}-1} \quad \text{and} \quad d_\alpha = \frac{\sqrt{|\alpha|^2-1}}{|\alpha|}.
\end{aligned} \tag{9.5.1}$$

Then for α *in* $\mathbb{D}_+$, *the polynomial* $K_n(z,\alpha)$ *in* $1/z$ *is an invertible outer function in* $H^\infty(\mathcal{E},\mathcal{E})$. *Moreover, we have*

$$\begin{aligned}
&\lim_{n\to\infty} \Theta\widehat{G}_{n,\alpha} = d_\alpha^2\varphi_\alpha(z)\Theta(\alpha) \text{ in } H^2(\mathcal{E},\mathcal{E}) && (\alpha \in \mathbb{D}_+), \\
&\lim_{n\to\infty} \Theta\widehat{G}_{n,\alpha} = 0 \text{ in } H^2(\mathcal{E},\mathcal{E}) && (\alpha \in \mathbb{T}), \\
&\lim_{n\to\infty} W_2G_{n,\alpha} = \Pi_{\mathcal{E}_j}^*A_j && (\overline{\alpha} = \lambda_j \in \operatorname{eig} V), \\
&\lim_{n\to\infty} W_2G_{n,\alpha} = 0 && (\overline{\alpha} = \lambda_j \notin \operatorname{eig} V).
\end{aligned} \tag{9.5.2}$$

(In the last two equations $G_{n,\alpha}$ *is embedded in the first* $\mathcal{E}^n$ *components of* $\ell_+^c(\mathcal{E})$.*) If the maximal outer spectral factor* Θ *is an invertible outer function in* $H^\infty(\mathcal{E},\mathcal{E})$ *and* $\alpha \in \mathbb{D}_+$, *then* Θ *is determined by (up to a unitary constant on the left)*

$$\begin{aligned}
\Theta(z) &= d_\alpha\varphi_\alpha(z)\lim_{n\to\infty} K_n(\alpha,\alpha)^{1/2}K_n(z,\alpha)^{-1}, \\
\Theta(z) &= \lim_{n\to\infty} K_n(\infty,\infty)^{1/2}K_n(z,\infty)^{-1} \qquad (\text{in } H^2(\mathcal{E},\mathcal{E})).
\end{aligned} \tag{9.5.3}$$

For $\alpha \in \mathbb{D}_+$ *and* $n \geq 1$, *the rational functions* $d_\alpha\varphi_\alpha(z)K_n(\alpha,\alpha)^{1/2}K_n(z,\alpha)^{-1}$ *and* $K_n(\infty,\infty)^{1/2}K_n(z,\infty)^{-1}$ *are all invertible outer functions.*

Proof. For the moment assume that α is a fixed scalar in $\overline{\mathbb{D}}_+$. Let

$$W = \begin{bmatrix} \Gamma & U\Gamma & U^2\Gamma & \cdots \end{bmatrix}$$

be the controllability matrix determined by $\{U, \Gamma\}$. Recall that $T_R = W^\sharp W$. Consider the optimization problem

$$\rho_n(\alpha, f) = \inf\{\|E_n x\| : x \in \mathcal{E}^n \text{ and } C_{n,\alpha}x = f\} \tag{9.5.4}$$

where $E_n = W|\mathcal{E}^n$. Notice that $T_{R,n} = E_n^*E_n$. According to Lemma 9.4.1, the optimal solution to (9.5.4) is given by $G_{n,\alpha}f$ and $\rho_n(\alpha, f) = (\Delta_{n,\alpha}f, f)$ where

$G_{n,\alpha} = G_n$ and $\Delta_{n,\alpha} = K_n(\alpha,\alpha)^{-1}$. The operators $E_n G_{n,\alpha}$ converge to an operator Q in $\mathcal{L}(\mathcal{E},\mathcal{K})$ as n approaches infinity. Finally, $Q = P_{\mathcal{L}} E_m G_{m,\alpha}$ where m is any positive integer and $\mathcal{L}$ is given by (9.4.4) with $C_n = C_{n,\alpha}$.

We claim that $\ker C_{n,\alpha} = \Pi_{\mathcal{E}^n}(S - \alpha^{-1} I)\mathcal{E}^{n-1}$ where S is the unilateral shift on $\ell^2_+(\mathcal{E})$. Notice that

$$\Pi_{\mathcal{E}^n}(S-\alpha^{-1}I)|\mathcal{E}^{n-1} = \begin{bmatrix} -\alpha^{-1}I & 0 & \cdots & 0 & 0 \\ I & -\alpha^{-1}I & \cdots & 0 & 0 \\ 0 & I & \cdots & 0 & 0 \\ \vdots & \vdots & \ddots & \vdots & \vdots \\ 0 & 0 & \cdots & I & -\alpha^{-1}I \\ 0 & 0 & \cdots & 0 & I \end{bmatrix} : \mathcal{E}^{n-1} \to \mathcal{E}^n.$$

The rank of $\Pi_{\mathcal{E}^n}(S - \alpha^{-1}I)|\mathcal{E}^{n-1}$ equals the dimension of $\mathcal{E}^{n-1}$. A simple calculation shows that $C_{n,\alpha}\Pi_{\mathcal{E}^n}(S - \alpha^{-1}I)\mathcal{E}^{n-1} = 0$. So $\Pi_{\mathcal{E}^n}(S - \alpha^{-1}I)\mathcal{E}^{n-1}$ is contained in the kernel of $C_{n,\alpha}$, and the dimension of $\ker C_{n,\alpha}$ is greater than or equal to $\dim \mathcal{E}^{n-1}$. However, the rank of $C_{n,\alpha}$ equals $\dim \mathcal{E}$, and thus, the dimension of the kernel of $C_{n,\alpha}$ equals $\dim \mathcal{E}^{n-1}$. In other words, the dimension of $\Pi_{\mathcal{E}^n}(S - \alpha^{-1}I)\mathcal{E}^{n-1}$ equals the dimension of the kernel of $C_{n,\alpha}$. Therefore the kernel of $C_{n,\alpha}$ equals $\Pi_{\mathcal{E}^n}(S - \alpha^{-1}I)\mathcal{E}^{n-1}$. Finally, it is noted that

$$\mathcal{E}^n = \Pi_{\mathcal{E}^n}(S - \alpha^{-1}I)\mathcal{E}^{n-1} \oplus C^*_{n,\alpha}\mathcal{E}.$$

Using $WS = UW$ and $E_n = W|\mathcal{E}^n$, the subspace $\mathcal{L}$ in (9.4.4) is determined by

$$\begin{aligned} \mathcal{L} &= \{h \in \mathcal{K} : h \perp E_n \ker C_{n,\alpha} \text{ for all } n \geq 1\} \\ &= \{h \in \mathcal{K} : h \perp W(S - \alpha^{-1}I)\mathcal{E}^{n-1} \text{ for all } n \geq 1\} \\ &= \{h \in \mathcal{K} : h \perp (U - \alpha^{-1}I)W\mathcal{E}^{n-1} \text{ for all } n \geq 1\} \\ &= \{h \in \mathcal{K} : h \perp (U - \alpha^{-1}I)\mathcal{K}\} \\ &= \ker(U^* - (\overline{\alpha})^{-1}I) = \mathcal{L}_\alpha. \end{aligned}$$

The fourth equality follows from the fact that the pair $\{U, \Gamma\}$ is controllable. Therefore $\mathcal{L} = \mathcal{L}_\alpha = \ker(U^* - (\overline{\alpha})^{-1}I)$.

We claim that $P_{\mathcal{L}_\alpha} E_m G_{m,\alpha} = P_{\mathcal{L}_\alpha}\Gamma$ for all $m \geq 1$. Clearly, $C_{m,\alpha}G_{m,\alpha}f = f$. Since $C_{m,\alpha}\begin{bmatrix} f & 0 & 0 & \cdots & 0\end{bmatrix}^{tr} = f$, the vector $G_{m,\alpha}f$ admits a decomposition of the form

$$G_{m,\alpha}f = \begin{bmatrix} f & 0 & 0 & \cdots & 0\end{bmatrix}^{tr} + \Pi_{\mathcal{E}^m}(S - \alpha^{-1}I)q$$

where q is a vector in $\mathcal{E}^{m-1}$. This readily implies that

$$\begin{aligned} P_{\mathcal{L}_\alpha} E_m G_{m,\alpha} f &= P_{\mathcal{L}_\alpha} W(\begin{bmatrix} f & 0 & 0 & 0 & \cdots\end{bmatrix}^{tr} + (S - \alpha^{-1}I)q) \\ &= P_{\mathcal{L}_\alpha}\Gamma f + P_{\mathcal{L}_\alpha}(U - \alpha^{-1}I)Wq = P_{\mathcal{L}_\alpha}\Gamma f. \end{aligned}$$

So for any α in $\overline{\mathbb{D}}_+$, we have

$$P_{\mathcal{L}_\alpha} E_m G_{m,\alpha} = P_{\mathcal{L}_\alpha}\Gamma \qquad (\alpha \in \overline{\mathbb{D}}_+). \tag{9.5.5}$$

Now assume that α is in $\mathbb{D}_+$. By consulting (9.1.12), we obtain

$$P_{\mathcal{L}_\alpha}\Gamma = \begin{bmatrix} d_\alpha^2 \Psi_\alpha \Theta(\alpha) \\ 0 \end{bmatrix} \qquad (\alpha \in \mathbb{D}_+); \tag{9.5.6}$$

see (9.1.11) for the definition of Ψ_α. Equation (9.1.10) shows that

$$P_{\mathcal{L}_\alpha}\Gamma = \begin{bmatrix} 0 \\ \Pi_{\mathcal{E}_j}^* A_j \end{bmatrix} \qquad (\overline{\alpha} = \lambda_j \in \operatorname{eig} V). \tag{9.5.7}$$

Finally, if $\overline{\alpha} \in \mathbb{T}$ is not an eigenvalue for V, then $P_{\mathcal{L}_\alpha}\Gamma = 0$.

If x is any vector in $\ell_+^c(\mathcal{E})$, then $Wx = T_\Theta x \oplus W_2 x$ where T_Θ is the Toeplitz matrix determined by Θ and W_2 is the controllability matrix corresponding to $\{V, \Gamma_2\}$. Since $G_{n,\alpha} f$ is an optimal solution to (9.5.4), for a specified f in $\mathcal{E}$, we obtain

$$E_n G_{n,\alpha} f = W G_{n,\alpha} f = \begin{bmatrix} T_\Theta G_{n,\alpha} f \\ W_2 G_{n,\alpha} f \end{bmatrix} \to P_{\mathcal{L}_\alpha}\Gamma f. \tag{9.5.8}$$

Here $G_{n,\alpha}$ is embedded in the first $\mathcal{E}^n$ components of $\ell_+^c(\mathcal{E})$. According to Lemma 9.4.1, the operators $E_n G_{n,\alpha}$ converge to $P_{\mathcal{L}_\alpha}\Gamma$. By consulting (9.5.6) and (9.5.7), we obtain

$$\begin{aligned}
&\lim_{n\to\infty} T_\Theta G_{n,\alpha} = d_\alpha^2 \Psi_\alpha \Theta(\alpha) && (\alpha \in \mathbb{D}_+),\\
&\lim_{n\to\infty} T_\Theta G_{n,\alpha} = 0 && (\alpha \in \mathbb{T}),\\
&\lim_{n\to\infty} W_2 G_{n,\alpha} = \Pi_{\mathcal{E}_j}^* A_j && (\overline{\alpha} = \lambda_j \in \operatorname{eig} V),\\
&\lim_{n\to\infty} W_2 G_{n,\alpha} = 0 && (\overline{\alpha} \notin \operatorname{eig} V).
\end{aligned} \tag{9.5.9}$$

Notice that $(\mathcal{F}_\mathcal{E}^+ G_{n,\alpha})(z) = \widehat{G}_{n,\alpha}(z)$ and $\mathcal{F}_\mathcal{E}^+ \Psi_\alpha = \varphi_\alpha(z) I$. Because the Fourier transform of convolution is multiplication in the z domain, we obtain $\mathcal{F}_\mathcal{E}^+ T_\Theta G_{n,\alpha} = \Theta \widehat{G}_{n,\alpha}$. So by taking the Fourier transform in the first equation in (9.5.9), we see that $\Theta\widehat{G}_{n,\alpha}$ converges to $d_\alpha^2 \varphi_\alpha(z)\Theta(\alpha)$ in the $H^2(\mathcal{E},\mathcal{E})$ topology when α is in $\mathbb{D}_+$. On the other hand, if α is on the unit circle, then $\Theta\widehat{G}_{n,\alpha}$ converges to zero. Therefore the first two equation in (9.5.2) hold. The last two equations in (9.5.2) follow from (9.5.9).

Theorem 9.6.2 below shows that $K_n(z,\alpha)$ is an invertible outer function for each α in $\mathbb{D}_+$. Equation (9.2.3) in Theorem 9.2.2 shows that $K_n(\alpha,\alpha)^{-1/2}$ converges monotonically to $d_\alpha \Theta(\alpha)$ up to a unitary constant on the left. So without loss of generality, we can assume that $K_n(\alpha,\alpha)^{-1/2}$ converges to $d_\alpha\Theta(\alpha)$. If Θ is an invertible outer function, then (9.5.2) shows that

$$\begin{aligned}\Theta &= \lim_{n\to\infty} d_\alpha^2 \varphi_\alpha(z)\Theta(\alpha)\widehat{G}_{n,\alpha}^{-1}\\ &= \lim_{n\to\infty} d_\alpha^2 \varphi_\alpha(z)\Theta(\alpha)K_n(\alpha,\alpha)K_n(z,\alpha)^{-1}\\ &= \lim_{n\to\infty} d_\alpha \varphi_\alpha(z)K_n(\alpha,\alpha)^{1/2}K_n(z,\alpha)^{-1}.\end{aligned}$$

This yields (9.5.3). □

Remark 9.5.2. Let Θ in H^2 be the maximal outer spectral factor for the positive Toeplitz matrix T_R where R is a scalar-valued symbol. Let $\{U \text{ on } \mathcal{K}, \Gamma\}$ be its controllable isometric representation, and $U = S \oplus V$ its Wold decomposition in (9.1.1) and (9.1.3) where $\{A_j\}$ are all scalars on $\mathbb{C}$. Then $\widehat{G}_{n,\alpha}(z) = \sum_{j=0}^{n-1} z^{-j} g_{n,j,\alpha}$ is a scalar-valued polynomial in z^{-1}. Let $p_{n,\alpha}(\lambda)$ be the polynomial defined by $p_{n,\alpha}(\lambda) = \sum_{j=0}^{n-1} \lambda^j g_{n,j,\alpha}$. Then in the strong operator topology

$$\begin{aligned}&\lim_{n\to\infty} p_{n,\alpha}(V) = 0 && (\alpha \in \mathbb{D}_+ \text{ or } \overline{\alpha} \notin \text{eig}\,(\text{V})),\\ &\lim_{n\to\infty} p_{n,\alpha}(V)\Gamma_2 = \Pi_{\mathcal{E}_j}^* A_j && (\overline{\alpha} \in \text{eig}\,(\text{V})).\end{aligned} \tag{9.5.10}$$

If $\mathcal{E} = \mathbb{C}$, then $W_2 G_{n,\alpha} = p_{n,\alpha}(V)\Gamma_2$. Theorem 9.5.1 tells us that $p_{n,\alpha}(V)\Gamma_2$ converges to zero when α is in $\mathbb{D}_+$ or $\overline{\alpha}$ is not an eigenvalue for V. Clearly, $p_{n,\alpha}(V)V^k\Gamma$ converges to zero for all integers $k \geq 0$. Because the pair $\{V, \Gamma_2\}$ is controllable, we obtain the first equation in (9.5.10). If $\overline{\alpha} = \lambda_j$, then Theorem 9.5.1 with $W_2 G_{n,\alpha} = p_{n,\alpha}(V)\Gamma_2$ shows that $p_{n,\alpha}(V)\Gamma_2$ converges to $\Pi_{\mathcal{E}_j}^* A_j$.

Remark 9.5.3. Let Θ in $H^2(\mathcal{E},\mathcal{E})$ be the maximal outer spectral factor for the positive Toeplitz matrix T_R, and $\{U \text{ on } \mathcal{K}, \Gamma\}$ its controllable isometric representation. Remark 9.6.3 below shows that $K_n(\beta,\alpha)$ is invertible for all α and β in $\mathbb{D}_+$. We claim that

$$\lim_{n\to\infty} K_n(\beta,\alpha)^{-1} = \frac{\overline{\alpha}\beta - 1}{\overline{\alpha}\beta}\Theta(\alpha)^*\Theta(\beta) \qquad (\alpha,\beta \in \mathbb{D}_+). \tag{9.5.11}$$

This is a generalization of the fact that $K_n(\alpha,\alpha)^{-1}$ converges to $d_\alpha^2\Theta(\alpha)^*\Theta(\alpha)$ when α is in $\mathbb{D}_+$; see equation (9.2.3) in Theorem 9.2.2.

The proof of Theorem 9.5.1 shows that $WG_{n,\alpha}f$ converges to $P_{\mathcal{L}_\alpha}\Gamma f$ in $\mathcal{K}$ as n tends to infinity where f is in $\mathcal{E}$ and α is in $\overline{\mathbb{D}}_+$; see (9.5.8). In particular, this implies that

$$\lim_{n\to\infty} G_{n,\beta}^* W^* W G_{n,\alpha} = \Gamma^* P_{\mathcal{L}_\beta} P_{\mathcal{L}_\alpha}\Gamma. \tag{9.5.12}$$

By consulting the definition of $G_{n,\alpha} = T_{R,n}^{-1} C_{n,\alpha}^* K_n(\alpha,\alpha)^{-1}$, we obtain

$$\begin{aligned}G_{n,\beta}^* W^* W G_{n,\alpha} &= G_{n,\beta}^* T_{R,n} G_{n,\alpha}\\ &= K_n(\beta,\beta)^{-1} C_{n,\beta} T_{R,n}^{-1} C_{n,\alpha}^* K_n(\alpha,\alpha)^{-1}\\ &= K_n(\beta,\beta)^{-1} K_n(\beta,\alpha) K_n(\alpha,\alpha)^{-1}.\end{aligned}$$

This with (9.5.12) implies that

$$\lim_{n\to\infty} K_n(\beta,\beta)^{-1}K_n(\beta,\alpha)K_n(\alpha,\alpha)^{-1} = \Gamma^* P_{\mathcal{L}_\beta} P_{\mathcal{L}_\alpha}\Gamma. \tag{9.5.13}$$

If α is in $\mathbb{T}$ and $\beta \neq \alpha$ is in $\overline{\mathbb{D}}_+$, then $\mathcal{L}_\alpha$ is orthogonal to $\mathcal{L}_\beta$. In this case, the limit in (9.5.13) is zero. (The case when $\alpha = \beta \in \mathbb{T}$ is covered in equation (9.2.3) of Theorem 9.2.2.)

If both α and β are contained in $\mathbb{D}_+$, then (9.5.6) and (9.1.11) yield

$$\Gamma^* P_{\mathcal{L}_\beta} P_{\mathcal{L}_\alpha}\Gamma = \Theta(\beta)^*\Psi_\beta^* d_\beta^2 d_\alpha^2 \Psi_\alpha\Theta(\alpha) = \frac{\overline{\alpha}\beta d_\alpha^2 d_\beta^2}{\overline{\alpha}\beta - 1}\Theta(\beta)^*\Theta(\alpha).$$

This readily implies that

$$\lim_{n\to\infty} K_n(\beta,\beta)^{-1}K_n(\beta,\alpha)K_n(\alpha,\alpha)^{-1} = \frac{\overline{\alpha}\beta d_\alpha^2 d_\beta^2}{\overline{\alpha}\beta - 1}\Theta(\beta)^*\Theta(\alpha) \qquad (\alpha,\beta\in\mathbb{D}_+).$$

Recall that for α in $\mathbb{D}_+$, the sequence $K_n(\alpha,\alpha)^{-1}$ converges to $d_\alpha^2\Theta(\alpha)^*\Theta(\alpha)$. So $K_n(\beta,\alpha)$ must also converge as n tends to infinity. Using this in the previous limit with the fact that $\Theta(\alpha)$ and $\Theta(\beta)$ are invertible, we arrive at

$$d_\alpha^2\Theta(\beta)^*\Theta(\beta)\lim_{n\to\infty} K_n(\beta,\alpha) d_\alpha^2\Theta(\alpha)^*\Theta(\alpha) = \frac{\overline{\alpha}\beta d_\alpha^2 d_\beta^2}{\overline{\alpha}\beta - 1}\Theta(\beta)^*\Theta(\alpha),$$
$$\lim_{n\to\infty} K_n(\beta,\alpha) = \frac{\overline{\alpha}\beta}{\overline{\alpha}\beta - 1}\Theta(\beta)^{-1}\Theta(\alpha)^{-*}.$$

By taking the inverse, we arrive at the limit in (9.5.11).

Example. Let G be a rational function in $H^\infty(\mathcal{E},\mathcal{Y})$. Moreover, assume that G admits an inner-outer factorization of the form $G = G_iG_o$, where G_o is an invertible outer function in $H^\infty(\mathcal{E},\mathcal{E})$ and G_i is an inner function in $H^\infty(\mathcal{E},\mathcal{Y})$. Then the limit (9.5.3) in Theorem 9.5.1 with the Kalman-Ho algorithm can be used to compute this inner-outer factorization. To this end, compute the Toeplitz matrix $T_{R,n}$ for n sufficiently large where $R = G^*G$. One can use state space techniques (see Lemma 4.5.4) or the fast Fourier transform to compute the Fourier coefficients for the symbol $R = \sum_{-\infty}^{\infty} e^{-\imath\omega k}R_k$. Then construct $T_{R,n}$. Choose any α in $\mathbb{D}_+$ and compute

$$\Theta(z) = d_\alpha\varphi_\alpha(z)K_n(\alpha,\alpha)^{1/2}K_n(z,\alpha)^{-1} \text{ or}$$
$$\Theta(z) = K_n(\infty,\infty)^{1/2}K_n(z,\infty)^{-1}. \tag{9.5.14}$$

(Remark 9.6.3 below guarantees that Θ is an invertible outer function.) Then Θ is approximately the outer factor G_o for G. The inner factor is determined by $G_i \approx G\Theta^{-1}$. These calculations can be done by using the fast Fourier transform. One can use the fast Fourier transform to compute $K_n(z,\alpha) = (\mathcal{F}_\mathcal{E}^+ T_{R,n}^{-1} C_{n,\alpha}^*)(z)$

and $\varphi_\alpha(z)$ at 2^j points on the unit circle. Then applying (9.5.14) and $G_i \approx G\Theta^{-1}$ yields an approximation for G_o and G_i at 2^j points on the unit circle. By taking the inverse fast Fourier transform of $\Theta = \sum_0^\infty e^{-\imath\omega k}\Theta_k$ and $G_i = \sum_0^\infty e^{-\imath\omega k}G_{i,k}$, we can approximate the Fourier coefficients $\{\Theta_k\}$ for Θ and $\{G_{i,k}\}$ for G_i. Finally, applying the Kalman-Ho algorithm to these Fourier coefficients yields state space realizations for G_o and G_i. Finally, it is noted that if we use $\alpha = \infty$, then $\varphi_\infty(z) = 1$, $d_\infty = 1$ and this method is essentially the Levinson finite section method in Section 7.7 to compute the inner-outer factorization.

For example consider the rational function g in H^∞ given in Section 7.7:

$$g = \frac{1.1909z^3 + 0.8735z^2 - 0.5210z + 0.0492}{z^7 + 0.1211z^6 - 0.3788z^5 - 0.2342z^4 + 0.0222z^3 + 0.0408z^2 + 0.0025z - 0.0011}.$$

Let $g = g_i g_o$ denote the inner-outer factorization for g where g_o is outer and g_i is inner. By choosing $\alpha = 2$ we computed $C_{300,\alpha}$ and $T_{R,300}$ where $R = |g|^2$. Then we used the fast Fourier transform to compute the outer spectral factor

$$\Theta(z) \approx d_\alpha \varphi_\alpha(z) K_n(\alpha,\alpha)^{1/2} K_n(z,\alpha)^{-1}$$

for g. By taking the inverse fast Fourier transform and keeping only three significant singular vales in the Kalman-Ho algorithm, for computing the outer part, we obtained

$$g_o(z) = \frac{1.365z^3 + 0.64z^2 - 0.2777z + 0.1551}{z^3 + 0.1308z^2 - 0.2608z - 0.1922}.$$

The singular values for the 500×500 Hankel matrix corresponding to g_o are

$$\{0.9152, 0.5789, 0.2881, 0.0108, 0.0016, 0, \ldots\}.$$

Running the Kalman-Ho algorithm on $\{g_{i,n}\}_{n=0}^{500}$ and keeping five singular values, we arrived at the inner function

$$g_i(z) = \frac{0.8722z + 1}{z^4\,(z + 0.8722)}. \tag{9.5.15}$$

In fact, the singular values for the 500×500 Hankel matrix corresponding to the g_i are $\{1, 1, 1, 1, 1, 0, \ldots\}$. As expected, the McMillan degree 5 for g_i equals the number of singular values equal to 1 and all the other singular values are zero; see Remark 4.2.3. Using the fast Fourier transform, $\|g\|_\infty = 2.7818$ and $\|g - g_i g_o\|_\infty = 0.014$. One can obtain a more accurate approximation of the inner and outer factors by keeping more singular values in the Kalman-Ho algorithm for g_o.

A typical procedure for computing the inner-outer factorization for $g = p/q$ in Matlab is given by the following steps.

(i) Set $g = \text{fft}(p, 2\wedge 13)./\text{fft}(q, 2\wedge 13)$ where

$$p = [0; 0; 0; 0; 1.1909; 0.8735; -0.5210; 0.0492];$$
$$q = [1; 0.1211; -0.3788; -0.2342; 0.0222; 0.0408; 0.0025; -0.0011];$$

Compute $R = \text{abs}(g).\wedge 2$. Set $Rn = \text{real}(\text{ifft}(R))$. The vector $Rn(1{:}2 \wedge 12)$ contains the first 2^{12} Fourier coefficients for $R = |g|^2$. In Matlab set $T = \text{toeplitz}(Rn(1{:}300))$.

(ii) In Matlab $C_{n,\alpha} = C = (1/\alpha).\wedge(0{:}299)$ and $K = C * \text{inv}(T) * C'$. Moreover, in Matlab

$$\begin{aligned}
\varphi_\alpha &= \text{fft}([\text{conj}(\alpha); 0], 2 \wedge 13)./\text{fft}([\text{conj}(\alpha); -1], 2 \wedge 13);\\
d_\alpha &= \text{sqrt}(\text{abs}(\alpha)^2 - 1)/\text{abs}(\alpha);\\
Kz &= \text{fft}(\text{inv}(T) * C'_{n,\alpha}, 2 \wedge 13);\\
\Theta &= d_\alpha * \text{sqrt}(K) * \varphi_\alpha./Kz;
\end{aligned}$$

(iii) Compute $\Theta = \sum_0^\infty z^{-n} g_{o,n}$. In Matlab $gn = \text{real}(\text{ifft}(\Theta))$. Then $gn(1{:}2 \wedge 12)$ contains the first 2^{12} Fourier coefficients of Θ.

(iv) Run the Kalman-Ho algorithm on $gn(1{:}500)$. Select the appropriate number of significant singular values to compute the realization $\{A, B, C, D\}$ for Θ.

(iv) Compute g_i. In Matlab, compute $g_i = g./\Theta$. Set $g_{ni} = \text{real}(\text{ifft}(g_i))$. Then $g_i(1{:}2 \wedge 12)$ contains the first 2^{12} Fourier coefficients of g_i.

(vi) Run the Kalman-Ho algorithm on $g_{ni}(1{:}500)$ to compute the realization $\{A_i, B_i, C_i, D_i\}$ for g_i.

There is nothing magical about 300 for $T_{R,300}$ or 500 for the Kalman-Ho. Certainly these numbers can be much smaller, or even larger depending on the problem. We choose these numbers to demonstrate that this algorithm works well for large numbers. Finally, by making minor modifications, the previous algorithm can be converted to compute the inner-outer factorization for a rational function G in $H^\infty(\mathcal{E}, \mathcal{Y})$ when the outer factor is an invertible outer function. The details are left to the reader as a simple exercise.

9.6 The Outer Function $K_n(z, \alpha)$

In this section, we will complete the proof of Theorem 9.5.1, and show that the kernel function $K_n(z, \alpha) = C_{n,z} T_{R,n}^{-1} C_{n,\alpha}^*$ is an invertible outer function when α is in $\mathbb{D}_+$. Here $T_{R,n}$ is any strictly positive Toeplitz operator on $\mathcal{E}^n$; see (9.2.1). Moreover, we will present state space formulas to compute the rational functions

$$d_\alpha \varphi_\alpha(z) K_n(\alpha, \alpha)^{1/2} K_n(z, \alpha)^{-1} \quad \text{and} \quad K_n(\infty, \infty)^{1/2} K_n(z, \infty)^{-1}$$

in equation (9.5.3).

Let A be the backward shift on $\mathcal{E}^n$ and C the operator mapping $\mathcal{E}^n$ onto $\mathcal{E}$ which picks out the first component of $\mathcal{E}^n$, that is,

$$A = \begin{bmatrix} 0 & I & \cdots & 0 & 0 \\ 0 & 0 & \cdots & 0 & 0 \\ \vdots & \vdots & \ddots & \vdots & \vdots \\ 0 & 0 & \cdots & 0 & I \\ 0 & 0 & \cdots & 0 & 0 \end{bmatrix} \text{ on } \mathcal{E}^n,$$

$$C = \begin{bmatrix} I & 0 & \cdots & 0 & 0 \end{bmatrix} : \mathcal{E}^n \to \mathcal{E}. \tag{9.6.1}$$

Notice that C is onto, and the pair $\{C, A\}$ is observable. Moreover, $A^j = 0$ for $j \geq n$. Using $z(zI - A)^{-1} = \sum_0^{n-1} z^{-j} A^j$, it follows that

$$C_{n,z} = zC(zI - A)^{-1} \quad (z \neq 0) \tag{9.6.2}$$

where $C_{n,z}$ is defined in (9.2.2). In particular, this readily implies that

$$K_n(z, \alpha) = z\overline{\alpha}\, C(zI - A)^{-1} T_{R,n}^{-1} (\overline{\alpha} I - A^*)^{-1} C^*. \tag{9.6.3}$$

Finally, $T_{R,n}$ is the solution to the Lyapunov equation:

$$\begin{aligned} T_{R,n} &= A^* T_{R,n} A + \widetilde{C}^* C + C^* \widetilde{C}, \\ \widetilde{C} &= \begin{bmatrix} R_0/2 & R_1^* & R_2^* & \cdots & R_{n-1}^* \end{bmatrix}. \end{aligned} \tag{9.6.4}$$

It is noted that $\widetilde{C}$ maps $\mathcal{E}^n$ into $\mathcal{E}$. Motivated by the properties of $\{A, C, \widetilde{C}, T_{R,n}\}$, the following defines a mild generalization of the function $K_n(z, \alpha)$.

Definition 9.6.1. We say that a set of operators $\{A, C, \widetilde{C}, \Lambda\}$ define a kernel function $K(z, \alpha)$ if

$$\begin{aligned} K(z, \alpha) &= z\overline{\alpha}\, C(zI - A)^{-1} \Lambda^{-1} (\overline{\alpha} I - A^*)^{-1} C^* \qquad & (z, \alpha \in \mathbb{D}_+), \\ K(z, \infty) &= zC(zI - A)^{-1} \Lambda^{-1} C^* & (z \in \mathbb{D}_+). \end{aligned} \tag{9.6.5}$$

Here we assume that A is a stable operator on a finite dimensional space $\mathcal{X}$ and Λ is a strictly positive operator on $\mathcal{X}$. Furthermore, C maps $\mathcal{X}$ onto $\mathcal{E}$ and $\widetilde{C}$ maps $\mathcal{X}$ into $\mathcal{E}$. Finally, the pair $\{C, A\}$ is observable and Λ satisfies the Lyapunov equation

$$\Lambda = A^* \Lambda A + \widetilde{C}^* C + C^* \widetilde{C}. \tag{9.6.6}$$

We are now ready to present our invertible outer function result.

Theorem 9.6.2. *Consider the kernel function $K(z, \alpha)$ defined by $\{A, C, \widetilde{C}, \Lambda\}$.*

(i) *Let Θ be the rational function determined by*

$$\Theta(z) = K(\infty, \infty)^{1/2} K(z, \infty)^{-1} = (C\Lambda^{-1} C^*)^{1/2} K(z, \infty)^{-1}. \tag{9.6.7}$$

Then Θ is an invertible outer function in $H^\infty(\mathcal{E},\mathcal{E})$. A state space realization for Θ is given by

$$\Theta(z) = D + \widehat{C}(zI - J)^{-1}L$$

where

$$\begin{aligned} D &= \left(C\Lambda^{-1}C^*\right)^{-1/2}, \\ \widehat{C} &= -DC, \\ L &= A\Lambda^{-1}C^*D^2, \\ J &= A - LC. \end{aligned} \tag{9.6.8}$$

The McMillan degree of Θ is less than or equal to $\dim \mathcal{X}$. *Finally, the operator J is stable.*

(ii) *For α in $\mathbb{D}_+$, let Θ be the rational function determined by*

$$\Theta(z) = d_\alpha \varphi_\alpha(z) K(\alpha,\alpha)^{1/2} K(z,\alpha)^{-1},$$
$$\varphi_\alpha(z) = \frac{z\overline{\alpha}}{z\overline{\alpha} - 1} \quad \text{and} \quad d_\alpha = \frac{\sqrt{|\alpha|^2 - 1}}{|\alpha|}. \tag{9.6.9}$$

Then Θ is an invertible outer function in $H^\infty(\mathcal{E},\mathcal{E})$. A state space realization for Θ is given by

$$\Theta(z) = D + \widehat{C}(zI - J)^{-1}L$$

where

$$\begin{aligned} B &= \overline{\alpha}\Lambda^{-1}(\overline{\alpha}I - A^*)^{-1}C^*, \\ D &= d_\alpha K(\alpha,\alpha)^{1/2}(CB)^{-1}, \\ \widehat{C} &= -DC, \\ L &= (A - (\overline{\alpha})^{-1}I)B(CB)^{-1}, \\ J &= A - LC. \end{aligned} \tag{9.6.10}$$

The McMillan degree of Θ is less than or equal to $\dim \mathcal{X}$. *Finally, this operator J is also stable.*

Remark 9.6.3. Consider data set $\{A, C, \widetilde{C}, T_{R,n}\}$ where $\Lambda = T_{R,n}$ is a strictly positive Toeplitz matrix on $\mathcal{E}^n$ while A, C, and $\widetilde{C}$ are defined in (9.6.1) and (9.6.4). Theorem 9.6.2 shows that the kernel functions $K_n(z,\alpha) = C_{n,z}T_{R,n}^{-1}C_{n,\alpha}$ in Theorem 9.5.1 are invertible outer functions for all α in $\mathbb{D}_+$ and $n \geq 1$. In this case, state space realizations for

$$d_\alpha\varphi_\alpha(z)K_n(\alpha,\alpha)^{1/2}K_n(z,\alpha)^{-1} \quad \text{and} \quad K_n(\infty,\infty)^{1/2}K_n(z,\infty)^{-1}$$

in (9.5.3) are given by (9.6.10) and (9.6.8), respectively.

Proof of Theorem 9.6.2. Recall that if F admits a state space realization of the form

$$F(z) = N + C(zI - A)^{-1}E \tag{9.6.11}$$

where N is invertible, then the inverse of F exists in some neighborhood of the origin and is determined by

$$F(z)^{-1} = N^{-1} - N^{-1}C\left(zI - (A - EN^{-1}C)\right)^{-1}EN^{-1}; \tag{9.6.12}$$

see Remark 14.2.1.

Let us establish Part(i). If $\alpha = \infty$, then $K(\infty, \infty) = C\Lambda^{-1}C^* = D^{-2}$. In this case, the identity $z(zI - A)^{-1} = I + (zI - A)^{-1}A$, yields

$$\begin{aligned}\Theta(z)^{-1} &= K(z, \infty)K(\infty, \infty)^{-1/2}\\ &= zC(zI - A)^{-1}\Lambda^{-1}C^*D\\ &= C\Lambda^{-1}C^*D + C(zI - A)^{-1}A\Lambda^{-1}C^*D\\ &= D^{-1} + C(zI - A)^{-1}A\Lambda^{-1}C^*D.\end{aligned}$$

By using the previous state space method to compute the inverse, we obtain

$$\begin{aligned}\Theta(z) &= D - DC\left(zI - (A - A\Lambda^{-1}C^*D^2C)\right)^{-1}A\Lambda^{-1}C^*D^2\\ &= D - DC\left(zI - J\right)^{-1}A\Lambda^{-1}C^*D^2.\end{aligned}$$

This yields the state space formula for Θ in (9.6.8).

We claim that $J = A - A\Lambda^{-1}C^*D^2C$ satisfies the Lyapunov equation

$$\Lambda = J^*\Lambda J + C^*D^2C. \tag{9.6.13}$$

To obtain this Lyapunov equation, set

$$P = I - \Lambda^{-1}C^*D^2C \quad \text{and} \quad Q = \Lambda^{-1}C^*D^2C.$$

Using $P + Q = I$, we obtain

$$\begin{aligned}\Lambda &= (P + Q)^*\Lambda(P + Q)\\ &= P^*\Lambda P + 2\Re(P^*\Lambda Q) + Q^*\Lambda Q.\end{aligned} \tag{9.6.14}$$

(If M is any operator on $\mathcal{X}$, then $\Re M = (M + M^*)/2$.) Notice that $CP = 0$. By applying P^* to the left and P to the right of the Lyapunov equation

$$\Lambda = A^*\Lambda A + \widetilde{C}^*C + C^*\widetilde{C},$$

we obtain $P^*\Lambda P = P^*A^*\Lambda AP$. Notice that $J = AP$. This with (9.6.14) yields

$$\begin{aligned}\Lambda - J^*\Lambda J &= P^*\Lambda P + 2\Re(P^*\Lambda Q) + Q^*\Lambda Q - P^*A^*\Lambda AP\\ &= P^*\Lambda Q + Q^*\Lambda P + Q^*\Lambda Q\\ &= P^*\Lambda Q + Q^*\Lambda(P+Q)\\ &= (I-Q)^*\Lambda Q + Q^*\Lambda\\ &= \Lambda Q + Q^*\Lambda - Q^*\Lambda Q\\ &= C^*D^2C.\end{aligned}$$

Therefore the Lyapunov equation in (9.6.13) holds.

To show that J is stable, notice that $J = A - LC$ where $L = A\Lambda^{-1}C^*D^2$. Recall that the pair $\{C, A\}$ is observable. Lemma 9.6.4 below shows that the pair $\{C, J\}$ is observable. Because D is invertible, the pair $\{DC, J\}$ is also observable. Since Λ is a strictly positive solution to the Lyapunov equation in (9.6.13) and $\{DC, J\}$ is observable, J is stable; see Section 14.4. The stability of A and J imply that Θ and Θ^{-1} are both functions in $H^\infty(\mathcal{E},\mathcal{E})$. In other words, Θ is an invertible outer function. This completes the proof of Part (i).

To prove Part (ii), set $\Phi_z = (zI-A)^{-1}$. In this case, $B = \overline{\alpha}\Lambda^{-1}\Phi_\alpha^* C^*$. Using $z\Phi_z = I + \Phi_z A$, it follows that

$$\begin{aligned}d_\alpha\Theta(z)^{-1}K(\alpha,\alpha)^{1/2} &= \varphi_\alpha(z)^{-1}K(z,\alpha)\\ &= (z\overline{\alpha}-1)C\Phi_z\Lambda^{-1}\Phi_\alpha^*C^*\\ &= z\overline{\alpha}C\Phi_z\Lambda^{-1}\Phi_\alpha^*C^* - C\Phi_z\Lambda^{-1}\Phi_\alpha^*C^*\\ &= \overline{\alpha}C\left(I+\Phi_zA\right)\Lambda^{-1}\Phi_\alpha^*C^* - C\Phi_z\Lambda^{-1}\Phi_\alpha^*C^*\\ &= \overline{\alpha}C\Lambda^{-1}\Phi_\alpha^*C^* + \overline{\alpha}C\Phi_zA\Lambda^{-1}\Phi_\alpha^*C^* - C\Phi_z\Lambda^{-1}\Phi_\alpha^*C^*\\ &= CB + C\Phi_z(\overline{\alpha}A - I)\Lambda^{-1}\Phi_\alpha^*C^*\\ &= CB + C\Phi_z(A-(\overline{\alpha})^{-1}I)B.\end{aligned}$$

By using the state space technique in (9.6.11) and (9.6.12) to take the inverse, we arrive at

$$d_\alpha^{-1}K(\alpha,\alpha)^{-1/2}\Theta(z) = (CB)^{-1} - (CB)^{-1}C(zI-J)^{-1}(A-(\overline{\alpha})^{-1}I)B(CB)^{-1}.$$

In a moment we will show that CB is invertible. Using $D = d_\alpha K(\alpha,\alpha)^{1/2}(CB)^{-1}$, we obtain the state space realization for Θ in (9.6.10).

We claim that J satisfies the Lyapunov equation

$$\Lambda = J^*\Lambda J + C^*D^*DC. \tag{9.6.15}$$

To obtain this Lyapunov equation, set

$$P = I - B(CB)^{-1}C \quad \text{and} \quad Q = B(CB)^{-1}C.$$

Using $P+Q=I$ with x in $\mathcal{X}$, we obtain

$$\begin{aligned}(\Lambda x, x) &= (\Lambda(P+Q)x, (P+Q)x)\\ &= (\Lambda Px, Px) + 2\Re(\Lambda Px, Qx) + (\Lambda Qx, Qx).\end{aligned}\tag{9.6.16}$$

By employing $J = AP + (\overline{\alpha})^{-1}Q$, we obtain

$$\begin{aligned}\|\Lambda^{1/2}APx\|^2 &= \|\Lambda^{1/2}Jx - (\overline{\alpha})^{-1}\Lambda^{1/2}Qx\|^2\\ &= \|\Lambda^{1/2}Jx\|^2 - 2\Re(\Lambda Jx, (\overline{\alpha})^{-1}Qx) + |\alpha|^{-2}\|\Lambda^{1/2}Qx\|^2\\ &= \|\Lambda^{1/2}Jx\|^2 - 2\Re(\Lambda(APx + (\overline{\alpha})^{-1}Qx), (\overline{\alpha})^{-1}Qx)\\ &\quad + |\alpha|^{-2}\|\Lambda^{1/2}Qx\|^2\\ &= (\Lambda Jx, Jx) - 2\Re(Px, (\overline{\alpha})^{-1}A^*\Lambda Qx) - |\alpha|^{-2}(\Lambda Qx, Qx).\end{aligned}$$

In other words,

$$(\Lambda Jx, Jx) = (\Lambda APx, APx) + 2\Re(Px, (\overline{\alpha})^{-1}A^*\Lambda Qx) + |\alpha|^{-2}(\Lambda Qx, Qx). \tag{9.6.17}$$

Notice that $CP = 0$. By applying P^* to the left and P to the right of the Lyapunov equation

$$\Lambda = A^*\Lambda A + \widetilde{C}^*C + C^*\widetilde{C},$$

we obtain $P^*\Lambda P = P^*A^*\Lambda AP$. This with (9.6.16) and (9.6.17) yields

$$\begin{aligned}(\Lambda x, x) - (\Lambda Jx, Jx) &= (\Lambda Px, Px) + 2\Re(Px, \Lambda Qx) + (\Lambda Qx, Qx)\\ &\quad - (\Lambda APx, APx) - 2\Re(Px, (\overline{\alpha})^{-1}A^*\Lambda Qx) - |\alpha|^{-2}(\Lambda Qx, Qx)\\ &= 2\Re(Px, (I - (\overline{\alpha})^{-1}A^*)\Lambda Qx) + d_\alpha^2(\Lambda Qx, Qx)\\ &= 2\Re(Px, (\overline{\alpha}I - A^*)(\overline{\alpha})^{-1}\Lambda B(CB)^{-1}Cx) + d_\alpha^2(\Lambda Qx, Qx)\\ &= 2\Re(CPx, (CB)^{-1}Cx) + d_\alpha^2(\Lambda Qx, Qx)\\ &= d_\alpha^2(B^*\Lambda B(CB)^{-1}Cx, (CB)^{-1}Cx)\\ &= d_\alpha^2(K(\alpha,\alpha)(CB)^{-1}Cx, (CB)^{-1}Cx)\\ &= (C^*D^*DCx, x).\end{aligned}$$

Therefore the Lyapunov equation in (9.6.15) holds.

We claim that J is stable. Set $L = (A - (\overline{\alpha})^{-1}I)B(CB)^{-1}$. Recall that the pair $\{C, A\}$ is observable. Lemma 9.6.4 below shows that the pair $\{C, J\}$ is observable. Because D is invertible, the pair $\{DC, J\}$ is also observable. Since Λ is a strictly positive solution to the Lyapunov equation in (9.6.15) and $\{DC, J\}$ is observable, J is stable; see Section 14.4. The stability of A and J imply that Θ and Θ^{-1} are both function in $H^\infty(\mathcal{E},\mathcal{E})$. In other words, Θ is an invertible outer function.

To complete the proof, it remains to show that CB is invertible. In Part (i), we have proven that $K(z,\infty) = zC(zI - A)^{-1}\Lambda^{-1}C^*$ is an invertible outer function. So for any α in $\mathbb{D}_+$, the operator $K(\alpha,\infty) = \alpha C(\alpha I - A)^{-1}\Lambda^{-1}C^* = B^*C^*$ is invertible. Therefore CB is invertible. □

Lemma 9.6.4. *Let $\{C, A\}$ be an observable pair, where A is an operator acting on a finite dimensional state space $\mathcal{X}$ and C maps $\mathcal{X}$ into $\mathcal{E}$. Let J be the operator on $\mathcal{X}$ given by $J = A - LC$, where L is an operator mapping $\mathcal{E}$ into $\mathcal{X}$. Then the pair $\{C, J\}$ is observable.*

Proof. This lemma follows from the Popov-Belevitch-Hautus test; see Section 14.2. Let us present a direct proof of this result. Let k be any positive integer. Using $J = A - LC$, we obtain

$$\begin{bmatrix} C \\ CA \\ CA^2 \\ \vdots \\ CA^k \end{bmatrix} = \begin{bmatrix} I & 0 & 0 & \cdots & 0 \\ CL & I & 0 & \cdots & 0 \\ CAL & CL & I & \cdots & 0 \\ \vdots & \vdots & \vdots & \ddots & \vdots \\ CA^{k-1}L & CA^{k-2}L & CA^{k-3}L & \cdots & I \end{bmatrix} \begin{bmatrix} C \\ CJ \\ CJ^2 \\ \vdots \\ CJ^k \end{bmatrix}. \tag{9.6.18}$$

The square operator matrix in (9.6.18) is a lower triangular Toeplitz matrix with the identity on the main diagonal. In particular, this matrix is invertible. Because $\{C, A\}$ is observable, equation (9.6.18) implies that $\{C, J\}$ is observable. □

9.7 Notes

Most of the results in this chapter are classical, and the literature in this area is massive. So we will only mention a few results which we used to develop this chapter. The optimization problems in (9.1.4) and (9.1.5) are standard problems in prediction theory; see Caines [47], Kailath [138] and Helson-Lowdenslager [130, 131]. Theorem 9.1.1 is a modification of some of the results in Helson-Lowdenslager [130, 131]. The algorithm to estimate the eigenvalues for V in the unitary pair $\{V, \Gamma_2\}$ in (9.3.4) is known as the Capon [49] sinusoid estimation algorithm. This algorithm was also discovered earlier using the theory of orthogonal polynomials; see Geronimus [110]. Our approach to Section 9.2 and the Capon-Geronimus algorithm was taken from Foias-Frazho-Sherman [88, 89] and Frazho-Sherman [101]; see also Frazho-Yagci-Sumali [104] and Georgiou [107, 108, 109] for further results in this direction. The Capon-Geronimus algorithm is robust in a noisy environment, and plays a fundamental role in signal processing; see Stoica-Moses [194]. For a brief discussion on how these kinds of estimation problems naturally arise in random processes see the notes in Section 8.7. The optimization problems in Section 9.4 are standard least squares optimization problems in Hilbert space; see Corless-Frazho [60], Gohberg-Goldberg-Kaashoek [112] and Luenberger [166] for a further discussion of Hilbert space optimization problems. The results in Section 9.5 are essentially a Naimark reformulation of some of the prediction theory results in Helson-Lowdenslager [130, 131]. Finally, Theorem 9.6.2 is a standard result for reproducing kernels. Our proof of Theorem 9.6.2 was taken from Bhosri-Frazho-Yagci [34], where they used the kernel functions $K_n(z, \alpha)$ to solve a special Nevanlinna-Pick interpolation problem.

It is noted that the method in the example in Section 9.5 does not necessarily lead to an efficient computational algorithm to compute the maximal outer spectral factor Θ for a positive Toeplitz matrix T_R, when the unitary part $\{V, \Gamma_2\}$ is present in the Wold decomposition. To see what can happen assume that R is a rational scalar-valued symbol determined by (9.3.1) where Θ is an invertible rational function in H^∞ and $\{A_j\}$ are a finite set of scalars. In this case, the unitary pair $\{V, \Gamma_2\}$ is determined by (9.3.3). Assume that all the eigenvalues of V come in complex conjugate pairs. Theorem 9.5.1 guarantees that for any α in $\mathbb{D}_+$, then

$$\Theta = \lim_{n\to\infty} d_\alpha \varphi_\alpha(z) K_n(\alpha,\alpha)^{1/2} K_n(z,\alpha)^{-1}$$

in the H^2 norm. (All this means is that the square of the area in the difference must converge to zero.) Equation (9.5.10) in Remark 9.5.2 shows that the poles of $K_n(z,\alpha)^{-1}$ have to annihilate the eigenvalues of V for large n. In other words, for large n the rational function $K_n(z,\alpha)^{-1}$ blows up near the eigenvalues for V. This is a good way to determine the eigenvalues for V. Simply use the fast Fourier transform to plot the absolute value of $K_n(z,\alpha)^{-1}$ on the unit circle and look for the peaks in the spectrum. However, in applying the Kalman-Ho algorithm to $d_\alpha \varphi_\alpha(z) K_n(\alpha,\alpha)^{1/2} K_n(z,\alpha)^{-1}$ these poles or eigenvalues for V interfere with finding a realization for Θ.

A classical optimization problem in prediction theory. The optimization problem in (9.1.5) was motivated by prediction theory. To see how this problem arises in prediction theory, let $\{y_j\}_{-\infty}^{\infty}$ be a mean zero wide sense stationary random process with values in $\mathbb{C}^\nu$; see Section 8.7 and Chapter 11. If ξ is a mean zero random vector with values in $\mathbb{C}^\nu$, then the covariance of ξ is denoted by $\mathrm{cov}(\xi) = E\xi\xi^*$ where E denotes the expectation. The classical prediction problem is to find the best estimate of the future y_n given the past $\{y_j\}_{-\infty}^{n-1}$. To convert this to an optimization problem, let $\{A_j\}_1^\infty$ be a set of matrices on $\mathbb{C}^\nu$ with compact support. Then the prediction problem leads to the optimization problem

$$\rho = \inf\{\|\mathrm{cov}(y_n - \sum_{j=1}^{\infty} A_j y_{n-j})\|_2 : \{A_j\}_1^\infty \text{ has compact support}\}. \tag{9.7.1}$$

Here $\|\cdot\|_2$ denotes the Frobenius or the trace norm. Let $\vec{y}$ be the vector and Φ the matrix defined by

$$\vec{y} = \begin{bmatrix} y_n \\ y_{n-1} \\ y_{n-2} \\ \vdots \end{bmatrix} \quad \text{and} \quad \Phi = \begin{bmatrix} I & -A_1 & -A_2 & \cdots \end{bmatrix}.$$

Let T_R be the positive Toeplitz matrix defined by $T_R = E\vec{y}\vec{y}^*$. It is noted that the entries of T_R are given by $(T_R)_{jk} = Ey_k y_j^* = Ey_{k-j} y_0^*$. Using this notation, we arrive at

$$\operatorname{cov}(y_n - \sum_{j=1}^{\infty} A_j y_{n-j}) = E(\Phi\vec{y})(\Phi\vec{y})^* = \Phi E\vec{y}\vec{y}^{\,*}\Phi^* = \Phi T_R \Phi^*.$$

So the optimization problem in (9.7.1) is equivalent to the optimization problem

$$\rho = \inf\{\|\Phi T_R \Phi^*\|_2 : \Phi = \begin{bmatrix} I & -A_1 & -A_2 & \cdots \end{bmatrix} \text{ has compact support}\}.$$

Hence the optimization problem in (9.1.5) is a minor modification of the classical optimization problem occurring in prediction theory. For some further results on prediction theory; see Caines [47] and Helson-Lowdenslager [130, 131].

Reproducing Kernel Hilbert spaces. As expected, $K_n(z, \alpha)$ is a reproducing kernel. A classical reference on reproducing kernel spaces is Aronszajn [16]; see also Cucker-Smale [62]. Here we only used the notation $K_n(z, \alpha)$ and did not exploit any properties of reproducing kernel Hilbert spaces. For a reproducing kernel approach to time series and random processes see Parzen [171]. Finally, to see how reproducing kernels play a role in interpolation problems see Agler-McCarthy [3, 4].

Let us sketch how reproducing kernels naturally arise in our problem. To recall the definition of a reproducing kernel Hilbert space, let $\mathcal{H}$ be a Hilbert space consisting of $\mathcal{E}$-valued functions defined on some subset $\mathcal{D}$ of $\mathbb{C}$. For each α in $\mathcal{D}$, consider the linear map C_α from $\mathcal{H}$ into $\mathcal{E}$ defined by

$$C_\alpha h = h(\alpha) \qquad (h \in \mathcal{H}). \tag{9.7.2}$$

We say that $\mathcal{H}$ is a *reproducing kernel Hilbert space* if the linear map C_α is bounded for every α in $\mathcal{D}$, that is, $\|C_\alpha\| < \infty$ for each α in $\mathcal{D}$ (but not necessarily uniformly bounded). In other words, $\mathcal{H}$ is a reproducing kernel Hilbert space if C_α defines an operator mapping $\mathcal{H}$ into $\mathcal{E}$ for every α in $\mathcal{D}$. If $\mathcal{H}$ is finite dimensional, then $\mathcal{H}$ is always a reproducing kernel Hilbert space.

Assume $\mathcal{H}$ is a reproducing kernel Hilbert space. We say that $K(z, \alpha)$ is a *reproducing kernel* for $\mathcal{H}$ if $K(z, \alpha)$ is an operator-valued function such that the following three conditions hold:

(i) For every α and z in $\mathcal{D}$, the function $K(z, \alpha)$ is in $\mathcal{L}(\mathcal{E}, \mathcal{E})$.

(ii) For each y in $\mathcal{E}$ and α in $\mathcal{D}$, the function $K(\cdot, \alpha)y$ is in $\mathcal{H}$.

(iii) The operator-valued function K has the reproducing property

$$(h(\alpha), y)_{\mathcal{E}} = (h, K(\cdot, \alpha)y)_{\mathcal{H}} \qquad (h \in \mathcal{H}, y \in \mathcal{E}, \alpha \in \mathcal{D}). \tag{9.7.3}$$

If $\mathcal{H}$ is a reproducing kernel Hilbert space, then $\mathcal{H}$ admits a unique reproducing kernel. In fact, the reproducing kernel K is given by

$$K(\beta, \alpha) = C_\beta C_\alpha^* \qquad (\alpha, \beta \in \mathcal{D}). \tag{9.7.4}$$

To see this simply observe that for h in $\mathcal{H}$, we have

$$(h(\alpha), y)_{\mathcal{E}} = (C_\alpha h, y)_{\mathcal{E}} = (h, C_\alpha^* y)_{\mathcal{H}}.$$

Hence $K(z,\alpha)y = (C_\alpha^* y)(z)$ defines a reproducing K kernel for $\mathcal{H}$. This readily implies that

$$K(\beta,\alpha)y = C_\beta K(\cdot,\alpha)y = C_\beta C_\alpha^* y.$$

In other words, $K(\beta,\alpha) = C_\beta C_\alpha^*$. Because $K(\alpha,\alpha) = C_\alpha C_\alpha^*$, it follow that $K(\alpha,\alpha)$ is always a positive operator on $\mathcal{E}$. Finally, $K(\alpha,\alpha)$ is strictly positive if and only if C_α is onto.

To show that the reproducing kernel is unique, let $\widetilde{K}$ be another reproducing kernel for $\mathcal{H}$. Then for h in $\mathcal{H}$, we have

$$(h, K(\cdot,\alpha)y)_{\mathcal{H}} = (h(\alpha), y)_{\mathcal{E}} = (h, \widetilde{K}(\cdot,\alpha)y)_{\mathcal{H}}.$$

Since this holds for all h in $\mathcal{H}$, we must have $K(z,\alpha)y = \widetilde{K}(z,\alpha)y$ for all z,α in $\mathcal{D}$ and y in $\mathcal{E}$. Therefore the reproducing kernel K is uniquely determined by $\mathcal{H}$.

Let Φ_j for $j = 1, 2, \ldots, \nu$ be a sequence of isometries mapping a Hilbert space $\mathcal{E}_j$ into a reproducing kernel Hilbert space $\mathcal{H}$ such that $\mathcal{H} = \oplus_1^\nu \Phi_j \mathcal{E}_j$. The integer ν can be finite or infinite. We claim that the reproducing kernel K for $\mathcal{H}$ is also determined by

$$K(\beta,\alpha) = \sum_{j=1}^{\nu} \Phi_j(\beta)\Phi_j(\alpha)^*. \tag{9.7.5}$$

Because Φ_j is an isometry, $\Phi_j\Phi_j^*$ is the orthogonal projection onto $\Phi_j\mathcal{E}_j$ the range of Φ_j. Since $\mathcal{H} = \oplus_1^\nu \Phi_j \mathcal{E}_j$, we see that $I = \sum_1^\nu \Phi_j\Phi_j^*$. Using $C_\alpha \Phi_j = \Phi_j(\alpha)$, we obtain

$$\begin{aligned} K(\beta,\alpha) = C_\beta I C_\alpha^* &= \sum_{j=1}^{\nu} C_\beta \Phi_j \Phi_j^* C_\alpha^* \\ &= \sum_{j=1}^{\nu} \Phi_j(\beta)(C_\alpha \Phi_j)^* = \sum_{j=1}^{\nu} \Phi_j(\beta)\Phi_j(\alpha)^*. \end{aligned}$$

Therefore (9.7.5) holds.

Let $\mathcal{H}$ be a reproducing kernel Hilbert space. Moreover, let us assume that the evaluation operator C_α is onto $\mathcal{E}$ for all α in $\mathcal{D}$. Consider the optimization problem

$$\rho(f) = \inf\{\|h\|^2 : h \in \mathcal{H} \text{ and } C_\alpha h = f\}. \tag{9.7.6}$$

By employing Lemma 9.2.4, we see that the optimal solution $\widehat{h}$ is given by

$$\widehat{h} = C_\alpha^*(C_\alpha C_\alpha^*)^{-1} f \quad \text{and} \quad \rho(f) = ((C_\alpha C_\alpha^*)^{-1} f, f). \tag{9.7.7}$$

By consulting (9.7.4), the optimal solution in terms of the reproducing kernel K for $\mathcal{H}$ is determined by

$$\widehat{h}(z) = K(z,\alpha)K(\alpha,\alpha)^{-1}f \quad \text{and} \quad \rho(f) = (K(\alpha,\alpha)^{-1}f, f). \tag{9.7.8}$$

This readily implies that the solution to the optimization problem in (9.7.6) is given by $\widehat{h} = \widehat{G}f$ where $\widehat{G}(z) = K(z,\alpha)K(\alpha,\alpha)^{-1}$. Finally, it is noted that a similar $\widehat{G}$ plays a fundamental role in Theorem 9.5.1.

Now let us show how the reproducing kernel arises in our problem. As before, let $\{U \text{ on } \mathcal{K}, \Gamma\}$ be a controllable isometric representation for a positive Toeplitz matrix T_R, and

$$W = \begin{bmatrix} \Gamma & U\Gamma & U^2\Gamma & \cdots \end{bmatrix}$$

its controllability matrix. Throughout this section we assume that the maximal outer spectral factor Θ for T_R is in $H^2(\mathcal{E},\mathcal{E})$. Let $\mathcal{H}_n = \mathcal{F}_{\mathcal{E}}^{+}\mathcal{E}^n$ be the subspace determined by taking the Fourier transform of $\mathcal{E}^n$. (Here $\mathcal{E}^n$ is viewed as the subspace of $\ell_+^2(\mathcal{E})$ contained in the first n components of $\ell_+^2(\mathcal{E})$.) Notice that $\mathcal{H}_n$ is the space consisting of all $\mathcal{E}$-valued polynomials in z^{-1} of degree at most $n-1$. Consider the norm on $\mathcal{H}_n$ determined by $\|h\|_{\mathcal{H}_n} = \|Wx\|_{\mathcal{K}}$, where $h = \mathcal{F}_{\mathcal{E}}^{+}x$ and x is in $\mathcal{E}^n$, that is,

$$\|h\|_{\mathcal{H}_n} = \|Wx\|_{\mathcal{K}} \text{ where } x = \begin{bmatrix} x_0 & x_1 & \cdots & x_{n-1} \end{bmatrix}^{tr} \text{ and } h(z) = \sum_{k=0}^{n-1} z^{-k}x_k.$$

It is noted that

$$\|h\|_{\mathcal{H}_n}^2 = \|Wx\|^2 = \|W(x \oplus 0)\|^2 = (T_{R,n}x, x) \quad (x \in \mathcal{E}^n),$$

where $T_{R,n}$ is the $n \times n$ block Toeplitz matrix contained in the upper left-hand corner of T_R; see (9.2.1). Because Θ is square, $\|h\|_{\mathcal{H}_n} = 0$ if and only if h is zero; see Remark 9.2.1. So

$$(h, g)_{\mathcal{H}_n} = (T_{R,n}x, \xi) \qquad (h = \mathcal{F}_{\mathcal{E}}^{+}x \text{ and } g = \mathcal{F}_{\mathcal{E}}^{+}\xi)$$

where x and ξ are in $\mathcal{E}^n$ defines an inner product on $\mathcal{H}_n$. Finally, since $\mathcal{H}_n$ is a finite dimensional function space, $\mathcal{H}_n$ is also a reproducing kernel Hilbert space.

We claim that the reproducing kernel K_n for $\mathcal{H}_n$ is given by

$$K_n(\beta,\alpha) = C_{n,\beta}T_{R,n}^{-1}C_{n,\alpha}^* \qquad (\alpha, \beta \in \overline{\mathbb{D}}_+). \tag{9.7.9}$$

As expected, $C_{n,\alpha}$ is the operator mapping $\mathcal{E}^n$ onto $\mathcal{E}$ defined in (9.5.1). It is emphasized that in (9.7.9), the adjoint $C_{n,\alpha}^* = C_{n,\overline{\alpha}}^{tr}$ where tr denotes the transpose.

To obtain this formula for $K_n(z,\alpha)$, let C_α be the operator mapping $\mathcal{H}_n$ onto $\mathcal{E}$ given by $C_\alpha h = h(\alpha)$ where $h = \mathcal{F}_{\mathcal{E}}^{+}x$ is in $\mathcal{H}_n$. Notice that C_α and $C_{n,\alpha}$ are two different operators. In fact, $C_\alpha h = C_{n,\alpha}x = C_{n,\alpha}(\mathcal{F}_{\mathcal{E}}^{+})^{-1}h$. We claim that

$$C_\alpha^* = \mathcal{F}_{\mathcal{E}}^{+}T_{R,n}^{-1}C_{n,\alpha}^*.$$

To see this, observe that for $h = \mathcal{F}_{\mathcal{E}}^{+} x$ and f in $\mathcal{E}$, we have

$$\begin{aligned}(h, C_\alpha^* f)_{\mathcal{H}_n} &= (C_\alpha h, f)_{\mathcal{E}} = (C_{n,\alpha} x, f)_{\mathcal{E}} \\ &= (T_{R,n} x, T_{R,n}^{-1} C_{n,\alpha}^* f)_{\mathcal{E}^n} \\ &= (h, \mathcal{F}_{\mathcal{E}}^{+} T_{R,n}^{-1} C_{n,\alpha}^* f)_{\mathcal{H}_n}.\end{aligned}$$

Hence $C_\alpha^* = \mathcal{F}_{\mathcal{E}}^{+} T_{R,n}^{-1} C_{n,\alpha}^*$. Using this, we obtain

$$K_n(\beta, \alpha) = C_\beta C_\alpha^* = C_\beta \mathcal{F}_{\mathcal{E}}^{+} T_{R,n}^{-1} C_{n,\alpha}^* = C_{n,\beta} T_{R,n}^{-1} C_{n,\alpha}^*.$$

Therefore (9.7.9) holds.

Using $(T_{R,n} x, x) = \|h\|_{\mathcal{H}_n}^2$ where $h = \mathcal{F}_{\mathcal{E}}^{+} x$, we see that the optimization problem in (9.2.9) is equivalent to the optimization problem

$$\rho_n(\alpha, f) = \inf\{\|h\|_{\mathcal{H}_n} : h \in \mathcal{H}_n \text{ and } h(\alpha) = f\} \tag{9.7.10}$$

where f is in $\mathcal{E}$. For each integer n the optimal solution to this optimization problem is unique and given by

$$\widehat{h}_n(z) = K_n(z, \alpha) K_n(\alpha, \alpha)^{-1} f \quad \text{and} \quad \rho_n(\alpha, f) = (K_n(\alpha, \alpha)^{-1} f, f). \tag{9.7.11}$$

Here K_n is the reproducing kernel for $\mathcal{H}_n$. In this setting, the $\mathcal{L}(\mathcal{E}, \mathcal{E})$-valued function $\widehat{G}$ in (9.5.1) is given by

$$\widehat{G}_{n,\alpha}(z) = K_n(z, \alpha) K_n(\alpha, \alpha)^{-1} \quad \text{and} \quad \rho_n(\alpha, f) = \|\widehat{G}_n f\|^2. \tag{9.7.12}$$

To obtain another expression for K_n, let $\{\Phi_k\}_1^\infty$ be a sequence of isometries vectors such that the range of Φ_k equals $\mathcal{H}_k \ominus \mathcal{H}_{k-1}$ for all integers $k \geq 1$. Here Φ_k maps $\mathcal{E}_k$ onto $\mathcal{H}_k \ominus \mathcal{H}_{k-1}$ and $\mathcal{H}_0 = \{0\}$. Notice that

$$\mathcal{H}_n = \oplus_{k=1}^n (\mathcal{H}_k \ominus \mathcal{H}_{k-1}) = \oplus_{k=1}^n \Phi_k \mathcal{E}_k.$$

By consulting (9.7.5), we see that

$$K_n(\beta, \alpha) = \sum_{k=1}^n \Phi_k(\beta) \Phi_k(\alpha)^*. \tag{9.7.13}$$

In particular, these polynomials $\{\Phi_k\}$ can be used in the limit in (9.2.7). Finally, it is noted that Φ_k is the kth orthogonal polynomial with values in $\mathcal{L}(\mathcal{E}, \mathcal{E})$ obtained in classical orthogonal polynomial theory; see Geronimus [107]. In this case, equation (9.2.7) reduces to the classical summation formulas from orthogonal polynomial theory.

Part III

Riccati Methods

Chapter 10

Riccati Equations and Factorization

In this chapter we will show how one can use Riccati equations to solve spectral factorization problems.

10.1 Algebraic Riccati Equations

In this section, we will derive an algebraic Riccati equation corresponding to the case when the rational Toeplitz matrix T_R admits a square outer spectral factor. This derivation is based in part on Lemma 6.2.3 restated here for convenience as follows.

Lemma 10.1.1. *Let $\{A$ on $\mathcal{X}, B, C, D\}$ be a stable realization for a rational function Θ in $H^\infty(\mathcal{E}, \mathcal{Y})$. Let P be the observability Gramian for $\{C, A\}$, that is, let P be the unique solution to the Lyapunov equation*

$$P = A^*PA + C^*C. \tag{10.1.1}$$

Let T_R be the self-adjoint Toeplitz matrix generated by a $\mathcal{L}(\mathcal{E}, \mathcal{E})$ symbol $R = \sum_{-\infty}^{\infty} e^{-\imath\omega n} R_n$. Then $T_R = T_\Theta^* T_\Theta$ if and only if

$$\begin{aligned} R_0 &= B^*PB + D^*D, \\ R_n &= (B^*PA + D^*C)\, A^{n-1}B \qquad (n \geq 1). \end{aligned} \tag{10.1.2}$$

Let $\{A, B, C, D\}$ be a stable, controllable, finite dimensional realization for a rational outer function Θ in $H^\infty(\mathcal{E}, \mathcal{E})$. Let T_R be the positive Toeplitz matrix determined by $T_R = T_\Theta^* T_\Theta$. Because A is stable, the function R is in $L^\infty(\mathcal{E}, \mathcal{E})$ and T_R is a well-defined operator on $\ell_+^2(\mathcal{E})$; see Proposition 2.5.1. Let F be the positive real function corresponding to T_R, that is, $F(z) = R_0/2 + \sum_1^\infty z^{-n} R_n$

where $\{R_n\}_0^\infty$ forms the first column of T_R. According to Lemma 10.1.1, we can always construct a realization for F of the form $\{A, B, \widehat{C}, R_0/2\}$. To obtain $\widehat{C}$ from $\{A, B, C, D\}$, let P be the observability Gramian for the pair $\{C, A\}$. By consulting equation (10.1.2), we have

$$R_0 = D^*D + B^*PB \quad \text{and} \quad R_n = \widehat{C}A^{n-1}B \qquad (n \geq 1),$$
$$\widehat{C} = B^*PA + D^*C.$$

Since $\{A, B\}$ is controllable, it follows that $\widehat{C}$ is the only operator such that $\{A, B, \widehat{C}, R_0/2\}$ is a realization for F. Because Θ is in $H^\infty(\mathcal{E}, \mathcal{E})$ and outer, $D = \Theta(\infty)$ must be invertible. Thus $C = D^{-*}(\widehat{C} - B^*PA)$, and $D^*D = R_0 - B^*PB$ is strictly positive. Substituting this into $P = A^*PA + C^*C$ yields the algebraic Riccati equation

$$P = A^*PA + (\widehat{C} - B^*PA)^*(R_0 - B^*PB)^{-1}(\widehat{C} - B^*PA). \tag{10.1.3}$$

Moreover, since Θ is an outer function, the corresponding feedback operator

$$J = A - BD^{-1}C = A - B(R_0 - B^*PB)^{-1}(\widehat{C} - B^*PA) \tag{10.1.4}$$

has all its eigenvalues in the closed unit disc $\overline{\mathbb{D}}$; see Lemma 4.4.2. Finally, Θ is an invertible outer function if and only if J is stable, or equivalently, T_R defines a strictly positive operator on $\ell_+^2(\mathcal{E})$; see Lemma 4.4.2.

Definition 10.1.2. We say that P is a positive solution to the algebraic Riccati equation in (10.1.3), if P is a positive operator on $\mathcal{X}$ satisfying (10.1.3) and $R_0 - B^*PB$ is strictly positive. Moreover, P is a stabilizing solution (respectively marginally stabilizing solution) to (10.1.3), if P is a positive solution to (10.1.3) and the feedback operator

$$J = A - B(R_0 - B^*PB)^{-1}(\widehat{C} - B^*PA) \tag{10.1.5}$$

is stable (respectively J has all its eigenvalues in $\overline{\mathbb{D}}$).

In a moment, we will see that the marginal stabilizing solution is unique. In particular, the stabilizing solution is also unique. Finally, it is noted that one can use Matlab to compute a stabilizing solution for the algebraic Riccati equation in (10.1.3).

Assume that P is a self-adjoint operator solving the algebraic Riccati equation in (10.1.3) where $R_0 - B^*PB$ is strictly positive. Then P is also a positive solution to this algebraic Riccati equation. To see this observe that $P \geq A^*PA$ where A is stable. According to Lemma 10.1.3 below, P is positive.

Lemma 10.1.3. *Let P be a self-adjoint operator on $\mathcal{X}$ satisfying $P \geq A^*PA$ where A is a stable operator on $\mathcal{X}$. Then P is a positive operator.*

Proof. Multiplying $P \geq A^*PA$ by A^* on the left and A on the right, we obtain $P \geq A^*PA \geq A^{*2}PA^2$. By continuing in this fashion, $P \geq A^{*n}PA^n$ for all integers $n \geq 0$. Because A is stable, $A^{*n}PA^n$ converges to zero. Therefore $P \geq 0$. □

Recall that Θ is a spectral factor for a function R, or its corresponding Toeplitz matrix T_R, if Θ is a function in $H^2(\mathcal{E}, \mathcal{Y})$ such that $R = \Theta^*\Theta$, or equivalently, $T_R = T_\Theta^\sharp T_\Theta$. If Θ is a rational spectral factor for R, then Θ is in $H^\infty(\mathcal{E}, \mathcal{Y})$, the function $R = \Theta^*\Theta$ is also a rational function and $T_R = T_\Theta^* T_\Theta$. The following result allows us to determine whether or not a rational Toeplitz matrix T_R admits a two-sided outer spectral factor.

Theorem 10.1.4. *Let $F = R_0/2 + \sum_1^\infty z^{-n}R_n$ be the $\mathcal{L}(\mathcal{E}, \mathcal{E})$-valued function corresponding to a rational Toeplitz matrix T_R with symbol $R = \sum_{-\infty}^{\infty} e^{-\imath\omega n}R_n$ where $R_{-n} = R_n^*$ and $R_0 > 0$. Assume that $\{A, B, \widehat{C}, R_0/2\}$ is a stable, controllable, finite dimensional realization of F. Then the following are equivalent.*

(i) *The Toeplitz matrix T_R is positive and admits an outer spectral factor Θ_o in $H^\infty(\mathcal{E}, \mathcal{E})$.*

(ii) *There exists a positive solution P to the algebraic Riccati equation in* (10.1.3). *In this case, a spectral factor Θ for T_R is given by*

$$\begin{aligned} \Theta(z) &= D + C(zI - A)^{-1}B, \\ D &= (R_0 - B^*PB)^{1/2}, \\ C &= D^{-1}(\widehat{C} - B^*PA). \end{aligned} \tag{10.1.6}$$

(iii) *There exists a unique marginally stabilizing solution P_o to the algebraic Riccati equation in* (10.1.3). *In this case, the outer spectral Θ_o for T_R is given by*

$$\begin{aligned} \Theta_o(z) &= D_o + C_o(zI - A)^{-1}B, \\ D_o &= (R_0 - B^*P_oB)^{1/2}, \\ C_o &= D_o^{-1}(\widehat{C} - B^*P_oA). \end{aligned} \tag{10.1.7}$$

If Θ is determined by (10.1.6), *then $\Theta = \Theta_i\Theta_o$ where Θ_i is an inner function in $H^\infty(\mathcal{E}, \mathcal{E})$. Finally, T_R defines a strictly positive operator on $\ell_+^2(\mathcal{E})$ if and only if there exists a stabilizing solution to the algebraic Riccati equation in* (10.1.3). *In this case, Θ_o is the invertible outer spectral factor for T_R.*

Proof. Assume T_R is positive and admits a two-sided outer spectral factor $\Theta = \Theta_o$. In this case, F is a positive real rational function. Moreover, we can always choose a stable controllable realization for Θ of the form $\{A, B, C, D\}$, that is,

$$\Theta(z) = D + C(zI - A)^{-1}B;$$

see Theorem 6.1.1. Because Θ is a two-sided outer function, $D = \Theta(\infty)$ is invertible. Furthermore, all the eigenvalues of $A - BD^{-1}C$ are contained in the $\overline{\mathbb{D}}$; see Lemma 4.4.2. According to Lemma 10.1.1, we see that $R_0 = B^*PB + D^*D$ where P is the observability Gramian for the pair $\{C, A\}$. Hence $D^*D = R_0 - B^*PB$ is strictly positive. Because $\{A, B, \widehat{C}, R_0/2\}$ is a realization of F, equation (10.1.2) yields

$$\widehat{C}A^{n-1}B = R_n = (B^*PA + D^*C)A^{n-1}B \qquad (n \geq 1). \tag{10.1.8}$$

Since the pair $\{A, B\}$ is controllable, we obtain $\widehat{C} = B^*PA + D^*C$. Hence $C = D^{-*}(\widehat{C} - B^*PA)$. Substituting this with $D^*D = R_0 - B^*PB$ into the Lyapunov equation $P = A^*PA + C^*C$ yields the algebraic Riccati equation in (10.1.3). Finally, observe that

$$A - BD^{-1}C = A - B(R_0 - B^*PB)^{-1}(\widehat{C} - B^*PA).$$

Since all the eigenvalues of $A - BD^{-1}C$ are contained in $\overline{\mathbb{D}}$, we see that P is the marginally stabilizing solution to (10.1.3). Hence Part (i) implies that there exists a marginally stabilizing solution to (10.1.3). In particular, there exists a positive solution P to the algebraic Riccati equation in (10.1.3).

To complete the proof of Part (i) implies Part (ii), assume that P is a positive solution to the algebraic Riccati equation in (10.1.3). Set

$$D = (R_0 - B^*PB)^{1/2} \quad \text{and} \quad C = D^{-1}(\widehat{C} - B^*PA).$$

Clearly, $R_0 = B^*PB + D^*D$. Using the fact $\{A, B, \widehat{C}, R_0/2\}$ is a realization of F, we have

$$(B^*PA + D^*C)A^{n-1}B = \widehat{C}A^{n-1}B = R_n \qquad (n \geq 1).$$

By substituting C into the algebraic Riccati equation in (10.1.3), we see that P is a unique solution to the Lyapunov equation $P = A^*PA + C^*C$. Lemma 10.1.1 implies that $\Theta = D + C(zI - A)^{-1}B$ is a spectral factor for T_R. Therefore Part (ii) holds.

We have previously shown that if Part (i) holds, then there exists a marginally stabilizing solution P to (10.1.3). So Part (i) also implies all of Part (iii), except for the fact that the marginally stabilizing solution is unique.

Assume that Part (ii) holds. In particular, P is a positive solution to (10.1.3). We have just seen in the previous paragraph that because P is a positive solution, the function $\Theta = D + C(zI - A)^{-1}B$ is a spectral factor for T_R. Notice that Θ is a function in $H^\infty(\mathcal{E}, \mathcal{E})$. Hence Θ admits an inner-outer factorization of the form $\Theta = \Theta_i\Theta_o$ where Θ_i is an inner function in $H^\infty(\mathcal{V}, \mathcal{E})$ and Θ_o is an outer function in $H^\infty(\mathcal{E}, \mathcal{V})$. Since $D = \Theta_i(\infty)\Theta_o(\infty)$ is invertible and $\Theta_o(\infty)$ is onto $\mathcal{V}$, it follows that both $\Theta_i(\infty)$ and $\Theta_o(\infty)$ are invertible. So without loss of generality, we can assume that the intermediate space $\mathcal{V} = \mathcal{E}$, and Θ_i and Θ_o are functions in $H^\infty(\mathcal{E}, \mathcal{E})$. Using the fact that T_{Θ_i} is an isometry, $T_R = T_{\Theta_o}^*T_{\Theta_o}$. In other words,

Θ_o is a two-sided outer spectral factor for T_R. Thus Part (ii) implies Part (i). Therefore Parts (i) and (ii) are equivalent. Clearly, Part (iii) implies Part (ii).

To complete the proof, it remains to show that the marginally stabilizing solution to the algebraic Riccati equation in (10.1.3) is unique. Assume that P_1 and P_2 are two marginally stabilizing solutions to (10.1.3). For $j = 1, 2$, set

$$D_j = (R_0 - B^* P_j B)^{1/2}, \quad C_j = D_j^{-1}(\widehat{C} - B^* P_j A), \quad \text{and}$$
$$\Theta_j(z) = D_j + C_j(zI - A)^{-1}B.$$

Since P_1 and P_2 are both marginally stabilizing solutions, Θ_1 and Θ_2 are two outer spectral factors of T_R. So there exists a unitary operator Φ on $\mathcal{E}$ such that $\Theta_1 = \Phi\Theta_2$. In other words, $D_1 = \Phi D_2$ and $C_1 A^{n-1} B = \Phi C_2 A^{n-1} B$, for all $n \geq 1$. Because $\{A, B\}$ is controllable, we must have $C_1 = \Phi C_2$. By substituting C_1 into (10.1.3), we see that P_1 is the observability Gramian for $\{C_1, A\}$. Thus

$$P_1 = A^* P_1 A + C_1^* C_1 = A^* P_1 A + C_2^* \Phi^* \Phi C_2 = A^* P_1 A + C_2^* C_2.$$

In other words, $P_1 = A^* P_1 A + C_2^* C_2$. Hence P_1 is also the observability Gramian for $\{C_2, A\}$. However, we know that the observability Gramian for $\{C_2, A\}$ is unique and is given by P_2. Therefore $P_1 = P_2$ and the marginally stabilizing solution to algebraic Riccati equation (10.1.3) is unique. □

Assume that $\{A, B, \widehat{C}, R_0/2\}$ is a stable minimal realization for a positive real function F in Theorem 10.1.4. Then the realization $\{A, B, C_o, D_o\}$ in (10.1.7) for its corresponding outer spectral factor Θ_o is also minimal. This follows from the fact that F and Θ_o have the same McMillan degree; see Theorem 6.1.1.

Remark 10.1.5. Let $\{A, B, \widehat{C}, R_0/2\}$ be a stable, controllable finite dimensional realization for a $\mathcal{L}(\mathcal{E}, \mathcal{E})$-valued positive real function F. Moreover, assume that the corresponding algebraic Riccati equation (10.1.3) admits a positive solution, or equivalently, the corresponding Toeplitz operator has a square outer spectral factor. Let P be any positive solution to the algebraic Riccati equation (10.1.3), and set

$$D = (R_0 - B^* P B)^{1/2} \quad \text{and} \quad C = D^{-1}(\widehat{C} - B^* P A).$$

It follows that P,C, and D satisfy all three conditions in the Positive Real Lemma 6.2.1. In other words, let Ξ_{ARE} be the set of all positive solutions to the algebraic Riccati equation (10.1.3), and Ξ be the set of all positive operators satisfying all three conditions in the Positive Real Lemma 6.2.1. Then $\Xi_{ARE} \subseteq \Xi$. In general, $\Xi_{ARE} \neq \Xi$. For example, consider the case when $T_R = I$ on ℓ_+^2 and $\{0, 1, 0, 1/2\}$ is a realization for F. Then $P = 0$ is the only positive solution to (10.1.3), and it satisfies all three conditions in Lemma 6.2.1. However, $P = 1$, $C = 1$, and $D = 0$ also satisfies all three conditions in Lemma 6.2.1, and $P = 1$ is not a positive solution to (10.1.3). Finally, observe that if P_o is the marginally stabilizing solution to (10.1.3), then the corresponding spectral factor Θ is outer. According to Remark 6.2.2, we have $P_o \leq P$ for all P in Ξ. Since $\Xi_{ARE} \subseteq \Xi$, it follows that

$P_o \le P$ for all P in Ξ_{ARE}. In other words, the marginally stabilizing solution is the unique minimal solution to the algebraic Riccati equation (10.1.3).

Remark 10.1.6. If the algebraic Riccati equation in (10.1.3) admits a marginally stabilizing solution, then this solution is unique, and one does not need to assume that the pair $\{A, B\}$ is controllable.

To see this, assume that P is a marginally stabilizing solution. Then Θ in (10.1.6) defines an outer spectral factor for T_R where $\Theta(\infty)$ is strictly positive. Moreover, the inverse of Θ is given by

$$\begin{aligned}\Theta(z)^{-1} &= D^{-1} - D^{-1}C(zI - J)^{-1}BD^{-1},\\ J &= A - BD^{-1}C = A - B(R_0 - B^*PB)^{-1}(\widehat{C} - B^*PA);\end{aligned}$$

see Remark 14.2.1. In particular, $\Theta(\overline{z})^{-*} = \sum_{n=0}^{\infty} z^{-n}\Psi_n$ where $\{\Psi_n\}_0^\infty$ is the sequence of operators defined by

$$\Psi_0 = D^{-1} \quad \text{and} \quad \Psi_n = -D^{-1}B^*J^{*n-1}C^*D^{-1} \qquad (n \ge 1).$$

Consider the Toeplitz matrix T_Ψ and the observability operator W_o defined by

$$T_\Psi = \begin{bmatrix} \Psi_0 & \Psi_1 & \Psi_2 & \cdots \\ 0 & \Psi_0 & \Psi_1 & \cdots \\ 0 & 0 & \Psi_0 & \cdots \\ \vdots & \vdots & \vdots & \ddots \end{bmatrix} \quad \text{and} \quad W_o = \begin{bmatrix} \widehat{C} \\ \widehat{C}A \\ \widehat{C}A^2 \\ \vdots \end{bmatrix}.$$

It is emphasized that T_Ψ may be an unbounded operator. (If J is stable, then T_Ψ is a well-defined operator and $T_\Psi = T^*_{\Theta^{-1}}$.) However, W_o is a well-defined operator mapping $\mathcal{X}$ into $\ell^2_+(\mathcal{Y})$. Now observe that

$$T_\Psi W_o = \begin{bmatrix} \widetilde{C} \\ \widetilde{C}A \\ \widetilde{C}A^2 \\ \vdots \end{bmatrix},$$

$$\widetilde{C} = D^{-1}\widehat{C} - \sum_{n=0}^{\infty} D^{-1}B^*J^{*n}C^*D^{-1}\widehat{C}A^nA.$$

Because A is stable and the eigenvalues for J are contained in the closed unit disc, the multiplication $T_\Psi W_o$ is well defined.

We claim that $\widetilde{C} = C$. To see this, notice that the algebraic Riccati equation in (10.1.3) can be rewritten as

$$\begin{aligned}P &= A^*PA + (\widehat{C} - B^*PA)^*(R_0 - B^*PB)^{-1}(\widehat{C} - B^*PA)\\ &= \left(A^* - (\widehat{C} - B^*PA)^*(R_0 - B^*PB)^{-1}B^*\right)PA + C^*D^{-1}\widehat{C}\\ &= J^*PA + C^*D^{-1}\widehat{C}.\end{aligned}$$

In other words, $P = \sum_0^\infty J^{*n} C^* D^{-1} \widehat{C} A^n$. Using this in the definition of $\widetilde{C}$, we see that

$$\widetilde{C} = D^{-1}\widehat{C} - D^{-1}B^*PA = C.$$

Therefore $\widetilde{C} = C$.

The algebraic Riccati equation in (10.1.3) is equivalent to the Lyapunov equation $P = A^*PA + C^*C$, and thus, $P = \sum_0^\infty A^{*n}C^*CA^n$. For x in $\mathcal{X}$, we have

$$\|T_\Psi W_o x\|^2 = \sum_{n=0}^{\infty} \|CA^n x\|^2 = (Px, x).$$

Each marginally stabilizing solution P uniquely determines an outer spectral factor Θ for T_R. Because the outer spectral factor is unique up to a unitary constant on the left and $\Theta(\infty)$ is strictly positive, each marginally stabilizing solution yields the same Θ, and the same Ψ. So P is uniquely determined by $(Px, x) = \|T_\Psi W_o x\|^2$. In other words, the marginally stabilizing solution to the algebraic Riccati equation (10.1.3) is unique.

10.2 The Case when T_R is Strictly Positive

To gain some further insight into the algebraic Riccati equation, we will study the case when T_R is strictly positive. In particular, we will present an explicit formula for the stabilizing solution. To this end, recall that if $\mathcal{E}$ is a subspace of $\mathcal{K}$, then $\Pi_\mathcal{E}$ is the operator from $\mathcal{K}$ onto $\mathcal{E}$ given by $\Pi_\mathcal{E} = P_\mathcal{E}$ where $P_\mathcal{E}$ on $\mathcal{K}$ is the orthogonal projection onto $\mathcal{E}$.

Proposition 10.2.1. *Let T_R be a positive Toeplitz matrix generated by a $\mathcal{L}(\mathcal{E},\mathcal{E})$-valued sequence $\{R_n\}_0^\infty$. Then T_R is a strictly positive operator on $\ell_+^2(\mathcal{E})$ if and only if $T_R = T_\Theta^* T_\Theta$ where Θ is an invertible outer function in $H^\infty(\mathcal{E},\mathcal{E})$. In this case, the outer spectral factor Θ for T_R is given by*

$$\Theta(z) = \Delta^{1/2} + \Delta^{1/2}\Pi_\mathcal{E} T_R^{-1}(zI - S^*)^{-1} S^* T_R \Pi_\mathcal{E}^*,$$
$$\Pi_\mathcal{E} = \begin{bmatrix} I & 0 & 0 & 0 & \cdots \end{bmatrix} : \ell_+^2(\mathcal{E}) \to \mathcal{E}. \tag{10.2.1}$$

Here S is the unilateral shift on $\ell_+^2(\mathcal{E})$, while $\Pi_\mathcal{E}$ is the operator which picks out the first component of $\ell_+^2(\mathcal{E})$. Finally, Δ is the strictly positive operator on $\mathcal{E}$ determined by

$$\Delta = (\Pi_\mathcal{E} T_R^{-1} \Pi_\mathcal{E}^*)^{-1}. \tag{10.2.2}$$

Proof. Theorem 7.1.1 shows that T_R is a strictly positive operator on $\ell_+^2(\mathcal{E})$ if and only if $T_R = T_\Theta^* T_\Theta$ where Θ is an invertible outer function in $H^\infty(\mathcal{E},\mathcal{E})$.

Now assume that T_R is a strictly positive Toeplitz operator on $\ell_+^2(\mathcal{E})$. Let $\{U \text{ on } \mathcal{K}, \Gamma\}$ be any controllable isometric representation for T_R, and W its corresponding controllability matrix, that is,

$$W = \begin{bmatrix} \Gamma & U\Gamma & U^2\Gamma & \cdots \end{bmatrix}.$$

Recall that $T_R = W^\sharp W$. Because T_R is invertible and $\{U, \Gamma\}$ is controllable, it follows that W defines an invertible operator from $\ell_+^2(\mathcal{E})$ onto $\mathcal{K}$; see the proof of Theorem 7.1.1. Moreover, $T_R = W^*W$ and $UW = WS$. (Because U is similar to S, the isometries U and S are unitarily equivalent; see Theorem 1.3.3. In particular, U is a unilateral shift.)

To obtain the formula for the outer spectral factor Θ for T_R in (10.2.1), recall

$$\Theta(z) = z\Pi_{\mathcal{Y}}(zI - U^*)^{-1}\Gamma \qquad (z \in \mathbb{D}_+) \tag{10.2.3}$$

where $\mathcal{Y} = \ker U^*$; see equation (5.2.2) in Theorem 5.2.1. By taking the adjoint of $U = WSW^{-1}$, we obtain $U^* = W^{-*}S^*W^*$. Since $\Pi_{\mathcal{E}}^*\mathcal{E} = \ker S^*$, we see that

$$\ker U^* = \{y \in \mathcal{K} : W^*y \in \Pi_{\mathcal{E}}^*\mathcal{E}\}.$$

In other words, $\mathcal{Y} = \ker U^* = W^{-*}\Pi_{\mathcal{E}}^*\mathcal{E}$. Let E be the operator from $\mathcal{E}$ into $\ell_+^2(\mathcal{Y})$ given by $E = W^{-*}\Pi_{\mathcal{E}}^*$. The operator E is one to one and the range of E equals $\mathcal{Y}$. Hence the orthogonal projection $P_{\mathcal{Y}}$ onto the subspace $\mathcal{Y}$ can be computed by

$$\begin{aligned} P_{\mathcal{Y}} &= E\left(E^*E\right)^{-1}E^* = W^{-*}\Pi_{\mathcal{E}}^*(\Pi_{\mathcal{E}}W^{-1}W^{-*}\Pi_{\mathcal{E}}^*)^{-1}\Pi_{\mathcal{E}}W^{-1} \\ &= W^{-*}\Pi_{\mathcal{E}}^*(\Pi_{\mathcal{E}}T_R^{-1}\Pi_{\mathcal{E}}^*)^{-1}\Pi_{\mathcal{E}}W^{-1} = W^{-*}\Pi_{\mathcal{E}}^*\Delta\Pi_{\mathcal{E}}W^{-1}. \end{aligned}$$

For x in $\ell_+^2(\mathcal{Y})$, we have

$$\|\Pi_{\mathcal{Y}}x\|^2 = \|P_{\mathcal{Y}}x\|^2 = \|\Delta^{1/2}\Pi_{\mathcal{E}}W^{-1}x\|^2.$$

This implies that there exists a unitary operator Φ mapping $\mathcal{Y}$ onto $\mathcal{E}$ such that $\Phi\Pi_{\mathcal{Y}} = \Delta^{1/2}\Pi_{\mathcal{E}}W^{-1}$. Since we do not distinguish between two outer spectral factors which are equal up to a constant unitary operator on the left, the outer spectral factor Θ for T_R is given by

$$\begin{aligned} \Theta(z) &= z\Phi\Pi_{\mathcal{Y}}(zI - U^*)^{-1}\Gamma \\ &= z\Delta^{1/2}\Pi_{\mathcal{E}}W^{-1}(zI - W^{-*}S^*W^*)^{-1}\Gamma \\ &= z\Delta^{1/2}\Pi_{\mathcal{E}}W^{-1}W^{-*}(zI - S^*)^{-1}W^*W\Pi_{\mathcal{E}}^* \\ &= z\Delta^{1/2}\Pi_{\mathcal{E}}T_R^{-1}(zI - S^*)^{-1}T_R\Pi_{\mathcal{E}}^* \\ &= \Delta^{1/2} + \Delta^{1/2}\Pi_{\mathcal{E}}T_R^{-1}(zI - S^*)^{-1}S^*T_R\Pi_{\mathcal{E}}^*. \end{aligned}$$

This yields the formula for Θ in (10.2.1). □

Remark 10.2.2. Assume that T_R is a strictly positive operator on $\ell_+^2(\mathcal{E})$ and Θ is its outer spectral factor. Then one can use the formula for Θ in (10.2.1) to show that the inverse of Θ is determined by

$$\Theta(z)^{-1} = z\Pi_{\mathcal{E}}(zI - S^*)^{-1}T_R^{-1}\Pi_{\mathcal{E}}^*\Delta^{1/2} = (\mathcal{F}_{\mathcal{E}}^+T_R^{-1}\Pi_{\mathcal{E}}^*)(z)\Delta^{1/2}. \tag{10.2.4}$$

This is precisely the formula for Θ^{-1} presented in Theorem 7.1.1.

Theorem 10.2.3. *Let T_R be a rational strictly positive Toeplitz operator on $\ell_+^2(\mathcal{E})$. Let $\{A, B, \widehat{C}, R_0/2\}$ be a stable controllable realization for the positive real function F corresponding to T_R, that is, $\{R_n\}_0^\infty$ forms the first column of T_R and $F = R_0/2 + \sum_1^\infty z^{-n} R_n$. Then the outer spectral factor Θ for T_R is given by*

$$\begin{aligned} \Theta(z) &= D + C(zI - A)^{-1}B, \\ D &= (R_0 - B^*PB)^{1/2}, \\ C &= D^{-1}(\widehat{C} - B^*PA), \\ P &= W_o^* T_R^{-1} W_o. \end{aligned} \tag{10.2.5}$$

Here W_o is the observability operator determined by

$$W_o = \begin{bmatrix} \widehat{C} \\ \widehat{C}A \\ \widehat{C}A^2 \\ \vdots \end{bmatrix} : \mathcal{X} \to \ell_+^2(\mathcal{E}). \tag{10.2.6}$$

This P is the unique stabilizing solution to the algebraic Riccati equation (10.1.3).

Proof. Recall that $R_n = \widehat{C}A^{n-1}B$ for all integers $n \geq 1$. Hence T_R admits a matrix representation of the form

$$T_R = \begin{bmatrix} R_0 & B^*W_o^* \\ W_oB & T_R \end{bmatrix} \text{ on } \begin{bmatrix} \mathcal{E} \\ \ell_+^2(\mathcal{E}) \end{bmatrix}. \tag{10.2.7}$$

Using $P = W_o^* T_R^{-1} W_o$, the Schur compliment $\triangle$ for T_R with respect to R_0 is determined by

$$\triangle = R_0 - B^*W_o^*T_R^{-1}W_oB = R_0 - B^*PB.$$

Because T_R is strictly positive, $\triangle$ is also strictly positive. By employing the matrix inversion formula in Lemma 7.2.1, we obtain

$$T_R^{-1} = \begin{bmatrix} \triangle^{-1} & -\triangle^{-1}B^*W_o^*T_R^{-1} \\ -T_R^{-1}W_oB\triangle^{-1} & T_R^{-1} + T_R^{-1}W_oB\triangle^{-1}B^*W_o^*T_R^{-1} \end{bmatrix} \text{ on } \begin{bmatrix} \mathcal{E} \\ \ell_+^2(\mathcal{E}) \end{bmatrix}.$$

Notice that $\triangle^{-1} = \Pi_{\mathcal{E}} T_R^{-1} \Pi_{\mathcal{E}}^*$. Moreover, $S^*W_o = W_oA$, where S is the unilateral shift on $\ell_+^2(\mathcal{E})$. So for any z in $\mathbb{D}_+$, we have $(zI - S^*)W_o = W_o(zI - A)$. By taking appropriate inverses, we arrive at

$$(zI - S^*)^{-1}W_o = W_o(zI - A)^{-1}.$$

The matrix decomposition for T_R in (10.2.7) yields $S^*T_R\Pi_{\mathcal{E}}^* = W_oB$. Observe that W_o admits a matrix decomposition of the form

$$W_o = \begin{bmatrix} \widehat{C} \\ W_oA \end{bmatrix} : \mathcal{X} \to \begin{bmatrix} \mathcal{E} \\ \ell_+^2(\mathcal{E}) \end{bmatrix}. \tag{10.2.8}$$

By employing this in the previous 2×2 matrix decomposition of T_R^{-1}, we have

$$\Pi_{\mathcal{E}} T_R^{-1} W_o = \Delta^{-1}(\widehat{C} - B^* P A).$$

The state space representation for Θ in (10.2.1) with $\Delta^{1/2} = (R_0 - B^* P B)^{1/2}$, yields

$$\begin{aligned}\Theta(z) &= \Delta^{1/2} + \Delta^{1/2}\Pi_{\mathcal{E}} T_R^{-1}(zI - S^*)^{-1} S^* T_R \Pi_{\mathcal{E}}^* \\ &= \Delta^{1/2} + \Delta^{1/2}\Pi_{\mathcal{E}} T_R^{-1}(zI - S^*)^{-1} W_o B \\ &= \Delta^{1/2} + \Delta^{1/2}\Pi_{\mathcal{E}} T_R^{-1} W_o (zI - A)^{-1} B \\ &= \Delta^{1/2} + \Delta^{-1/2}(\widehat{C} - B^* P A)(zI - A)^{-1} B.\end{aligned}$$

Therefore the outer spectral factor Θ for T_R is given by (10.2.5).

To complete the proof it remains to show that $P = W_o^* T_R^{-1} W_o$ is a stabilizing solution to the algebraic Riccati equation (10.1.3). To this end, observe that

$$\begin{aligned}P &= W_o^* T_R^{-1} W_o \\ &= \begin{bmatrix} \widehat{C}^* & A^* W_o^* \end{bmatrix} \\ &\quad \times \begin{bmatrix} \Delta^{-1} & -\Delta^{-1} B^* W_o^* T_R^{-1} \\ -T_R^{-1} W_o B \Delta^{-1} & T_R^{-1} + T_R^{-1} W_o B \Delta^{-1} B^* W_o^* T_R^{-1} \end{bmatrix} \begin{bmatrix} \widehat{C} \\ W_o A \end{bmatrix} \\ &= \begin{bmatrix} \widehat{C}^* & A^* W_o^* \end{bmatrix} \begin{bmatrix} \Delta^{-1}\widehat{C} - \Delta^{-1} B^* P A \\ -T_R^{-1} W_o B \Delta^{-1} \widehat{C} + T_R^{-1} W_o A + T_R^{-1} W_o B \Delta^{-1} B^* P A \end{bmatrix} \\ &= \widehat{C}^* \Delta^{-1} \widehat{C} - \widehat{C}^* \Delta^{-1} B^* P A - A^* P B \Delta^{-1} \widehat{C} + A^* P A + A^* P B \Delta^{-1} B^* P A \\ &= A^* P A + \widehat{C}^* \Delta^{-1}(\widehat{C} - B^* P A) - A^* P B \Delta^{-1}(\widehat{C} - B^* P A) \\ &= A^* P A + (\widehat{C}^* - A^* P B)\Delta^{-1}(\widehat{C} - B^* P A) \\ &= A^* P A + (\widehat{C} - B^* P A)^*(R_0 - B^* P B)^{-1}(\widehat{C} - B^* P A).\end{aligned}$$

Therefore $P = W_o^* T_R^{-1} W_o$ is a positive solution to the algebraic Riccati equation in (10.1.3).

By construction $\{A, B, C, D\}$ is a stable controllable realization for the invertible outer function Θ. According to Lemma 4.4.2, the feedback operator

$$J = A - BD^{-1}C = A - B(R_0 - B^* P B)^{-1}(\widehat{C} - B^* P A) \tag{10.2.9}$$

is stable. So P is a stabilizing solution to the Riccati equation (10.1.3). □

10.3 The Riccati Difference Equation

Let $\{A \text{ on } \mathcal{X}, B, \widehat{C}, R_0/2\}$ be a controllable, stable, finite dimensional realization for a $\mathcal{L}(\mathcal{E}, \mathcal{E})$-valued function F, where R_0 is strictly positive, and set $R = F + F^*$

on the unit circle. By Theorem 10.1.4, the Toeplitz matrix $T_R = T_F + T_F^*$ admits a square outer factor if and only if there exists a positive solution P to the algebraic Riccati equation

$$P = A^*PA + (\widehat{C} - B^*PA)^*(R_0 - B^*PB)^{-1}(\widehat{C} - B^*PA). \qquad (10.3.1)$$

Recall that P is a positive solution to (10.3.1) if P is a positive operator solving (10.3.1) and $R_0 - B^*PB$ is strictly positive. Consider the Riccati difference equation

$$P_{n+1} = A^*P_nA + (\widehat{C} - B^*P_nA)^*(R_0 - B^*P_nB)^{-1}(\widehat{C} - B^*P_nA) \qquad (P_0 = 0). \qquad (10.3.2)$$

For P_{n+1} to be well defined $R_0 - B^*P_jB$ must be invertible for all $0 \le j \le n$. In fact, in our applications we will require $R_0 - B^*P_jB$ to be strictly positive for all j. So if $R_0 - B^*P_nB$ is singular, then we simply say that P_{n+1} is undefined. We say that P_∞ is a *positive steady state solution* to the Riccati difference equation in (10.3.2) if P_n converges to a positive operator P_∞ and the operator $R_0 - B^*P_\infty B$ is strictly positive. (Of course, we assume that $R_0 - B^*P_nB$ are invertible for all $n \ge 0$.) If there exists a positive steady state solution P_∞ to the Riccati difference equation (10.3.2), then P_∞ is a positive solution to the algebraic Riccati equation in (10.3.1). So the Toeplitz matrix T_R is positive and admits an outer spectral factor; see Theorem 10.1.4. In this case, P_∞ is the marginally stabilizing solution to the algebraic Riccati equation (10.3.1).

Theorem 10.3.1. *Assume that* $\{A, B, \widehat{C}, R_0/2\}$ *is a stable, controllable, finite dimensional realization of a* $\mathcal{L}(\mathcal{E},\mathcal{E})$*-valued function* F *where* $R_0 > 0$. *Let* $T_R = T_F + T_F^*$ *be the Toeplitz matrix corresponding to* $R = F + F^*$. *Let* P_n *be the solution to the Riccati difference equation in* (10.3.2). *Then the following are equivalent.*

(i) *The Toeplitz matrix* T_R *is positive and admits a square outer spectral factor* Θ.

(ii) *The algebraic Riccati equation* (10.3.1) *admits a positive marginally stabilizing solution.*

(iii) *There exists a positive scalar* $\delta > 0$ *such that*

$$R_0 - B^*P_nB \ge \delta I > 0 \qquad (\textit{for all } n \ge 0). \qquad (10.3.3)$$

(iv) *The solution set* $\{P_n\}_0^\infty$ *to the Riccati difference equation in (10.3.2) is increasing, uniformly bounded and*

$$R_0 - B^*P_\infty B > 0 \quad \textit{where} \quad P_\infty = \lim_{n\to\infty} P_n. \qquad (10.3.4)$$

If (i), (ii), (iii) *or* (iv) *hold, then the outer spectral factor* Θ *for* T_R *is given by*

$$\begin{aligned} \Theta(z) &= D + C(zI - A)^{-1}B, \\ D &= (R_0 - B^*P_\infty B)^{1/2}, \\ C &= D^{-1}(\widehat{C} - B^*P_\infty A). \end{aligned} \qquad (10.3.5)$$

Finally, P_∞ is the marginally stabilizing solution for the Riccati equation (10.3.1).

Proof. Theorem 10.1.4 shows that Parts (i) and (ii) are equivalent. Let $W_{o,n}$ be the operator from $\mathcal{X}$ into $\mathcal{E}^n = \oplus_1^n \mathcal{E}$ defined by

$$W_{o,n} = \begin{bmatrix} \widehat{C} & \widehat{C}A & \widehat{C}A^2 & \cdots & \widehat{C}A^{n-1} \end{bmatrix}^{tr}. \tag{10.3.6}$$

Here $\mathcal{X}$ is the state space for A. As before, let $T_{R,n}$ be the $n \times n$ Toeplitz matrix contained in the upper left-hand corner of T_R, that is, $R = \sum_{-\infty}^{\infty} e^{-\imath\omega n} R_n$ and

$$T_{R,n} = \begin{bmatrix} R_0 & R_1^* & \cdots & R_{n-1}^* \\ R_1 & R_0 & \cdots & R_{n-2}^* \\ \vdots & \vdots & \ddots & \vdots \\ R_{n-1} & R_{n-2} & \cdots & R_0 \end{bmatrix} \text{ on } \mathcal{E}^n. \tag{10.3.7}$$

Recall that $R_j = F_j$ for $j \geq 1$ is the j-th Taylor coefficient of z^{-j} for F. Using $R_j = \widehat{C}A^{j-1}B$, it follows that $T_{R,\,n+1}$ admits a matrix partition of the form

$$T_{R,\,n+1} = \begin{bmatrix} R_0 & B^* W_{o,n}^* \\ W_{o,n}B & T_{R,n} \end{bmatrix} \text{ on } \begin{bmatrix} \mathcal{E} \\ \mathcal{E}^n \end{bmatrix}. \tag{10.3.8}$$

For the moment assume that $T_{R,n}$ is strictly positive and set $P_n = W_{o,n}^* T_{R,n}^{-1} W_{o,n}$. Then the Schur complement Δ_{n+1} for $T_{R,n+1}$ is given by

$$\Delta_{n+1} = R_0 - B^* P_n B \quad \text{where} \quad P_n = W_{o,n}^* T_{R,n}^{-1} W_{o,n}; \tag{10.3.9}$$

see Lemma 7.2.1. So according to Corollary 7.4.3, the Toeplitz operator T_R admits a square outer spectral factor Θ if and only if

$$\Delta_{n+1} = R_0 - B^* P_n B \geq \delta I \qquad (\text{for all } n \geq 0 \text{ and some } \delta > 0). \tag{10.3.10}$$

In this case, $\{R_0 - B^* P_n B\}$ forms a decreasing sequence of positive operators such that

$$\Theta(\infty)^* \Theta(\infty) = \lim_{n \to \infty} (R_0 - B^* P_n B) \tag{10.3.11}$$

where Θ is the outer spectral factor for T_R. Moreover, $T_{R,n}$ is strictly positive for all n.

If Part (i) holds, then the Schur complements Δ_n satisfy (10.3.10). On the other hand, if Part (iii) holds, then $T_{R,n}$ is strictly positive for all n; see Lemma 7.3.1. So if either Part (i) or (iii) hold, then $T_{R,n}$ is strictly positive for all $n \geq 1$. To show that Parts (i) and (iii) are equivalent, it remains to show that $P_n = W_{o,n}^* T_{R,n}^{-1} W_{o,n}$ is the solution to the Riccati difference equation (10.3.2) for all $n \geq 0$. Clearly, this holds for $n = 1$. Now let us use induction and assume that $P_n = W_{o,n}^* T_{R,n}^{-1} W_{o,n}$ is the n-th solution to the Riccati difference equation (10.3.2).

By employing the matrix inversion Lemma 7.2.1 on the partition for $T_{R,n+1}$ in (10.3.8), we obtain

$$\begin{aligned}P_{n+1} &= W_{o,n+1}^* T_{R,n+1}^{-1} W_{o,n+1}\\ &= \begin{bmatrix} \widehat{C}^* & A^* W_{o,n}^* \end{bmatrix}\\ &\times \begin{bmatrix} \Delta^{-1} & -\Delta^{-1} B^* W_{o,n}^* T_{R,n}^{-1} \\ -T_{R,n}^{-1} W_{o,n} B \Delta^{-1} & T_{R,n}^{-1} + T_{R,n}^{-1} W_{o,n} B \Delta^{-1} B^* W_{o,n}^* T_{R,n}^{-1} \end{bmatrix} \begin{bmatrix} \widehat{C} \\ W_{o,n} A \end{bmatrix}\\ &= A^* P_n A + (\widehat{C} - B^* P_n A)^* (R_0 - B^* P_n B)^{-1} (\widehat{C} - B^* P_n A)\end{aligned}$$

where $\Delta = \Delta_{n+1} = R_0 - B^* P_n B$. This is precisely the Riccati difference equation in (10.3.2). Therefore Parts (i) and (iii) are equivalent.

Now assume that Part (iii) holds. We claim that $\{P_n\}$ forms an increasing sequence of operators. To see this, observe that $T_{R,n+1}$ admits a decomposition of the form

$$T_{R,n+1} = \begin{bmatrix} T_{R,n} & X^* \\ X & R_0 \end{bmatrix} \text{ on } \begin{bmatrix} \mathcal{E}^n \\ \mathcal{E} \end{bmatrix} \qquad (10.3.12)$$

where $X = \begin{bmatrix} R_n & R_{n-1} & \cdots & R_1 \end{bmatrix}$. Because $T_{R,n}$ is invertible, $T_{R,n+1}$ admits a Schur factorization of the form

$$T_{R,n+1} = \begin{bmatrix} I & 0 \\ X T_{R,n}^{-1} & I \end{bmatrix} \begin{bmatrix} T_{R,n} & 0 \\ 0 & R_0 - X T_{R,n}^{-1} X^* \end{bmatrix} \begin{bmatrix} I & T_{R,n}^{-1} X^* \\ 0 & I \end{bmatrix}. \qquad (10.3.13)$$

Since $T_{R,n+1}$ is strictly positive, the Schur complement $\Lambda = R_0 - X T_{R,n}^{-1} X^*$ is also strictly positive. In particular, the inverse of $T_{R,n+1}$ is given by

$$T_{R,n+1}^{-1} = \begin{bmatrix} I & -T_{R,n}^{-1} X^* \\ 0 & I \end{bmatrix} \begin{bmatrix} T_{R,n}^{-1} & 0 \\ 0 & \Lambda^{-1} \end{bmatrix} \begin{bmatrix} I & 0 \\ -X T_{R,n}^{-1} & I \end{bmatrix}. \qquad (10.3.14)$$

So using $W_{o,n+1} = [W_{o,n}, \widehat{C} A^n]^{tr}$ along with x in $\mathcal{X}$, we have

$$\begin{aligned}(P_{n+1} x, x) &= (T_{R,n+1}^{-1} W_{o,n+1} x, W_{o,n+1} x)\\ &= (T_{R,n+1}^{-1} (W_{o,n} x \oplus \widehat{C} A^n x), W_{o,n} x \oplus \widehat{C} A^n x) \qquad (10.3.15)\\ &= (T_{R,n}^{-1} W_{o,n} x, W_{o,n} x) + \|\Lambda^{-1/2} (-X T_{R,n}^{-1} W_{o,n} + \widehat{C} A^n) x\|^2\\ &\geq (P_n x, x).\end{aligned}$$

Therefore $P_n \leq P_{n+1}$, and $\{P_n\}$ is increasing.

Now let us show that the sequence $\{P_n\}$ is uniformly bounded. Recall that $\{R_0 - B^* P_n B\}$ forms a decreasing sequence which converge to the strictly positive operator $\Theta(\infty)^* \Theta(\infty)$. Hence $\{B^* P_n B\}$ forms an increasing sequence of positive operators which converge to a bounded operator M. By applying B^* to the left and B on the right of the Riccati difference equation in (10.3.2), we see that

$$\begin{aligned}B^* P_{n+1} B &= B^* A^* P_n A B + B^* (\widehat{C} - B^* P_n A)^* (R_0 - B^* P_n B)^{-1} (\widehat{C} - B^* P_n A) B\\ &\geq B^* A^* P_n A B.\end{aligned}$$

Hence $B^*A^*P_nAB \leq B^*P_{n+1}B$. So $B^*A^*P_nAB \leq M$ for all n. Multiplying (10.3.2) by B^*A^* on the left and AB on the right, we obtain $B^*A^*P_{n+1}AB \geq B^*A^{*2}P_nA^2B$. Since $B^*A^*P_{n+1}AB \leq B^*P_{n+2}B$, we see that $B^*A^{*2}P_nA^2B \leq M$. By continuing in this fashion $B^*A^{*k}P_nA^kB \leq M$ for all k and n. This readily implies that the diagonal entries of the positive operator

$$Q_n = \begin{bmatrix} B & AB & \cdots & A^{\nu-1}B \end{bmatrix}^* P_n \begin{bmatrix} B & AB & \cdots & A^{\nu-1}B \end{bmatrix} \qquad (10.3.16)$$

are all bounded by M. Here ν is the dimension of the state space $\mathcal{X}$. Since Q_n is positive and its diagonal entries are bounded by M, it follows that

$$\|Q_n\| \leq \text{trace } Q_n \leq \nu \times \text{trace } M < \infty.$$

Hence $\|Q_n\| \leq \gamma < \infty$ for all n. Because the pair $\{A, B\}$ is controllable, the operator $[B, AB, \cdots, A^{\nu-1}B]$ in (10.3.16) is onto $\mathcal{X}$. Thus $P_n \leq \alpha I$ for some finite α. In fact, $P_n = \Omega^* Q_n \Omega$ where Ω is the Moore-Penrose or right inverse of $\begin{bmatrix} B & AB & \cdots & A^{\nu-1}B \end{bmatrix}$. Hence P_n is uniformly bounded. Since the sequence $\{P_n\}$ is increasing and uniformly bounded, P_n converges to a positive operator P_∞. Equation (10.3.11) yields (10.3.4). Therefore Part (iv) holds.

If Part (iv) holds, then P_∞ is a positive solution to the algebraic Riccati equation in (10.3.1). According to Theorem 10.1.4, the Toeplitz operator T_R admits a square outer spectral factor. In other words, Part (i) hold. Therefore Parts (i) through (iv) are equivalent.

If any one of (i) through (iv) hold, then P_∞ is a positive solution to the algebraic Riccati equation in (10.3.1). Theorem 10.1.4 shows that Θ in (10.3.5) is a spectral factor for T_R. Corollary 7.4.3 with (10.3.11) guarantees that this Θ is the outer spectral factor for T_R. □

By consulting the finite section inversion Lemma 7.6.1, we obtain the following result.

Remark 10.3.2. If T_R defines a strictly positive operator on $\ell^2_+(\mathcal{E})$, then $P_n = W^*_{o,n}T^{-1}_{R,n}W_{o,n}$ converges to $P_\infty = W^*_o T_R^{-1} W_o$.

10.4 A Riccati Approach to Inner-Outer Factorizations

In this section, we will use the Riccati equation to compute the inner-outer factorization for rational functions which admit a square outer factor. Let G be a rational function in $H^\infty(\mathcal{E}, \mathcal{Y})$. Let us remind the reader that we always assume that $\mathcal{E}$ and $\mathcal{Y}$ are both finite dimensional. For the moment assume that G admits an inner-outer factorization of the form $G = G_i G_o$ where G_i is inner and G_o is a square outer function. Since $G_o(\infty)$ is invertible, without loss of generality, we can assume that G_o is a rational function in $H^\infty(\mathcal{E}, \mathcal{E})$ and G_i is an inner function in $H^\infty(\mathcal{E}, \mathcal{Y})$. Because $G_i(e^{\imath\omega})$ is an isometry, $\dim \mathcal{E} \leq \dim \mathcal{Y}$. So if we are looking for a square outer factor, we must assume that $\dim \mathcal{E} \leq \dim \mathcal{Y}$. Since G is a rational function in $H^\infty(\mathcal{E},\mathcal{Y})$, the function $G(e^{\imath\omega})$ is well defined for all $0 \leq \omega \leq 2\pi$. Furthermore,

$$G(e^{\imath\omega})^*G(e^{\imath\omega}) = G_o(e^{\imath\omega})^*G_o(e^{\imath\omega}) \qquad (0 \leq \omega \leq 2\pi).$$

The outer factor G_o is square if and only if $G_o(e^{\imath\omega})$ is invertible everywhere except possibility at a finite number of points on the unit circle. In other words, G admits a square outer factor if and only if $\det[G(e^{\imath\omega})^*G(e^{\imath\omega})]$ is nonzero everywhere except at a finite number of points.

Let $\{A \text{ on } \mathcal{X}, B, C, D\}$ be a stable, controllable, finite dimensional realization for G in $H^\infty(\mathcal{E},\mathcal{Y})$ where $\dim \mathcal{E} \leq \dim \mathcal{Y}$. Assume that G admits a square outer factor. Here we shall first present a special algebraic Riccati equation and its corresponding state space formula for computing G_i and G_o. Then we will derive this algebraic Riccati equation. To compute a state space realization for G_i and G_o, let X be the marginally stabilizing, positive solution to the algebraic Riccati equation

$$X = A^*XA-(D^*C+B^*XA)^*(D^*D+B^*XB)^{-1}(D^*C+B^*XA)+C^*C\,. \qquad (10.4.1)$$

As expected, X is a *positive solution* to (10.4.1) if X is a positive operator satisfying (10.4.1) where $D^*D + B^*XB$ is strictly positive. Moreover, X is a *marginally stabilizing, positive solution* if X is a positive solution to (10.4.1) and the feedback operator

$$J = A - B(D^*D + B^*XB)^{-1}(D^*C + B^*XA) \qquad (10.4.2)$$

has all its eigenvalues in the closed unit disc $\overline{\mathbb{D}}$. An operator X is a positive stabilizing solution if X is a positive solution and J is stable. Finally, it is noted that the algebraic Riccati equation in (10.4.1) can be rewritten as

$$\begin{aligned} X &= (A - B\Lambda)^*X(A - B\Lambda) + (C - D\Lambda)^*(C - D\Lambda),\\ \Lambda &= (D^*D + B^*XB)^{-1}\,(D^*C + B^*XA)\,. \end{aligned} \qquad (10.4.3)$$

The marginally stabilizing solution X can also be obtained by taking the limit of a certain Riccati difference equation. To be precise, consider the Riccati difference equation

$$\begin{aligned} X_{n+1} &= (A - B\Lambda_n)^*X_n(A - B\Lambda_n) + (C - D\Lambda_n)^*(C - D\Lambda_n),\\ \Lambda_n &= (D^*D + B^*X_nB)^{-1}\,(D^*C + B^*X_nA)\,,\\ X_0 &= Q \quad \text{where} \quad Q = A^*QA + C^*C. \end{aligned} \qquad (10.4.4)$$

The set of all $\{X_n\}_0^\infty$ are decreasing, that is, $X_{n+1} \leq X_n$. Notice that the initial condition $X_0 = Q$, the observability Gramian for the pair $\{C, A\}$. (If we choose $X_0 = 0$ for the initial condition, then $D^*D + B^*0B = D^*D$ may not be invertible. Moreover, if D is invertible, then $X_n = 0$ for all n when $X_0 = 0$.) Finally, it is emphasized that the following are equivalent:

(i) The rational function G in $H^\infty(\mathcal{E},\mathcal{Y})$ admits an outer factor G_o in $H^\infty(\mathcal{E},\mathcal{E})$.

(ii) The algebraic Riccati equation (10.4.1) has a positive marginally stabilizing solution.

(iii) There exists a $\delta > 0$ such that

$$D^*D + B^*X_nB \geq \delta I \quad (n \geq 0). \tag{10.4.5}$$

In this case, the marginally stabilizing positive solution to (10.4.1) or (10.4.3) is given by

$$X = \lim_{n\to\infty} X_n. \tag{10.4.6}$$

Now assume that (i), (ii) or (iii) holds. To compute the inner-outer factorization for G, recall that $\{A, B, C, D\}$ is a stable, controllable, finite dimensional realization for G. Let X be the marginally stabilizing positive solution for the algebraic Riccati equation in (10.4.1). Let C_o, D_o, A_i, B_i, C_i and D_i be the operators defined by

$$\begin{aligned}
D_o &= (D^*D + B^*XB)^{1/2},\\
C_o &= D_o^{-1}(D^*C + B^*XA),\\
A_i &= A - BD_o^{-1}C_o \quad\text{and}\quad B_i = BD_o^{-1},\\
C_i &= C - DD_o^{-1}C_o \quad\text{and}\quad D_i = DD_o^{-1}.
\end{aligned} \tag{10.4.7}$$

Then $\Sigma_i = \{A_i, B_i, C_i, D_i\}$ and $\Sigma_o = \{A, B, C_o, D_o\}$ are controllable realizations for the inner factor G_i and square outer factor G_o for G, respectively. Notice that the realizations Σ_i and Σ_o may not be observable. For example, if $G = G_o$ is outer, then $G_i = D_i$ is a unitary constant, $C_i = 0$, and thus, Σ_i is not observable. However, one can extract the minimal realizations from Σ_i and Σ_o by standard state space techniques.

Let us show that the state space realizations for G_i and G_o are indeed given by (10.4.7). According to Lemma 6.2.3, the function G is a spectral factor for the positive Toeplitz operator T_R with symbol $G^*G = R = \sum_{-\infty}^{\infty} e^{-\imath\omega n}R_n$ generated by the sequence

$$\begin{aligned}
R_0 &= B^*QB + D^*D,\\
R_n &= (B^*QA + D^*C)\,A^{n-1}B \qquad (n \geq 1).
\end{aligned} \tag{10.4.8}$$

In this case, Q is the observability Gramian for $\{C, A\}$. Moreover, if we set

$$\widehat{C} = D^*C + B^*QA \quad\text{and}\quad R_0 = D^*D + B^*QB, \tag{10.4.9}$$

then $\widehat{C}A^{n-1}B = R_n$ for all integers $n \geq 1$ and R_0 is specified in terms of D, B and Q. By consulting Theorem 10.1.4, we see that T_R admits a square outer spectral factor if and only if there exists a marginally stabilizing solution P to the following algebraic Riccati equation:

$$\begin{aligned} P &= A^*PA + (\widehat{C} - B^*PA)^*(R_0 - B^*PB)^{-1}(\widehat{C} - B^*PA) \qquad (10.4.10)\\ &= A^*PA \\ &\quad + (D^*C + B^*(Q-P)A)^*(DD^* + B^*(Q-P)B)^{-1}(D^*C + B^*(Q-P)A). \end{aligned}$$

By subtracting this equation from $Q = A^*QA + C^*C$, and setting $X = Q - P$, we arrive at the algebraic Riccati equation in (10.4.1). If P is a marginally stabilizing solution to the previous algebraic Riccati equation, then the feedback operator

$$\begin{aligned} J &= A - B(R_0 - B^*PB)^{-1}(\widehat{C} - B^*PA) \\ &= A - B(D^*D + B^*XB)^{-1}(D^*C + B^*XA) \qquad (10.4.11) \end{aligned}$$

has all its eigenvalues in the closed unit disc $\overline{\mathbb{D}}$. In other words, $X = Q - P$ is a marginally stabilizing solution for the algebraic Riccati equation in (10.4.1). Let C_o and D_o be the operators defined in (10.4.7). In this case, Theorem 10.1.4 shows that $\Sigma_o = \{A, B, C_o, D_o\}$ is a state space realization for the outer spectral factor for T_R. Since $R = G^*G$, it follows that Σ_o is a state space realization for the outer spectral factor for G.

Because $\{A, B, C_o, D_o\}$ is a realization of G_o, the inverse of $G_o(z)$ is given by

$$G_o(z)^{-1} = D_o^{-1} - D_o^{-1}C_o\left(zI - (A - BD_o^{-1}C_o)\right)^{-1}BD_o^{-1}.$$

Using this with $\{A_i, B_i, C_i, D_i\}$ defined in (10.4.7), we arrive at

$$\begin{aligned} G_i(z) &= G(z)G_o(z)^{-1} \\ &= (D + C(zI - A)^{-1}B) \\ &\quad \times (D_o^{-1} - D_o^{-1}C_o(zI - (A - BD_o^{-1}C_o))^{-1}BD_o^{-1} \\ &= DD_o^{-1} - DD_o^{-1}C_o(zI - (A - BD_o^{-1}C_o))^{-1}BD_o^{-1} \\ &\quad + C(zI - A)^{-1}[I - BD_o^{-1}C_o(zI - (A - BD_o^{-1}C_o))^{-1}]BD_o^{-1} \\ &= DD_o^{-1} - DD_o^{-1}C_o(zI - A_i)^{-1}BD_o^{-1} \\ &\quad + C(zI - A)^{-1}[(zI - (A - BD_o^{-1}C_o) - BD_o^{-1}C_o](zI - A_i)^{-1}BD_o^{-1} \\ &= DD_o^{-1} - DD_o^{-1}C_o(zI - A_i)^{-1}BD_o^{-1} + C(zI - A_i)^{-1}BD_o^{-1} \\ &= DD_o^{-1} + (C - DD_o^{-1}C_o)(zI - A_i)^{-1}BD_o^{-1} \\ &= D_i + C_i(zI - A_i)^{-1}B_i. \end{aligned}$$

Therefore $\{A_i, B_i, C_i, D_i\}$ is state space realization for the inner part G_i of G.

To obtain the Riccati difference equation in (10.4.4), observe that the Riccati difference equation in (10.3.2) corresponding to our data $\{A, B, \widehat{C}, R_0/2\}$ in

(10.4.9) is given by

$$P_{n+1} = A^* P_n A + (D^*C + B^*(Q - P_n)A)^*(D^*D + B^*(Q - P_n)B)^{-1}(D^*C + B^*(Q - P_n)A). \tag{10.4.12}$$

Recall that $\{P_n\}_0^\infty$ is an increasing sequence of positive operators which converge to P, the marginally stabilizing solution to (10.4.10), where $P_0 = 0$. By setting $X_n = Q - P_n$ and employing $Q = A^*QA + C^*C$, we arrive at the Riccati difference equation

$$X_{n+1} = A^* X_n A + C^*C - (D^*C + B^* X_n A)^*(D^*D + B^* X_n B)^{-1}(D^*C + B^* X_n A). \tag{10.4.13}$$

Theorem 10.3.1 guarantees that Parts (i) to (iii) are equivalent. So we see that $\{X_n\}$ is a decreasing sequence of operators which converge to the marginally stabilizing solution for the algebraic Riccati equation in (10.4.1).

To obtain the form of the Riccati difference equation in (10.4.4), observe that

$$\begin{aligned}
X_{n+1} &= A^* X_n A - (C^*D + A^* X_n B)\Lambda_n + C^*C \\
&= A^* X_n (A - B\Lambda_n) - C^*D\Lambda_n + C^*C \\
&= (A - B\Lambda_n)^* X_n (A - B\Lambda_n) + \Lambda_n^* B^* X_n (A - B\Lambda_n) - C^*D\Lambda_n + C^*C \\
&= (A - B\Lambda_n)^* X_n (A - B\Lambda_n) - \Lambda_n^*(B^* X_n B + D^*D)\Lambda_n + \Lambda_n^* D^*D\Lambda_n \\
&\quad + \Lambda_n^* B^* X_n A - C^*D\Lambda_n + C^*C \\
&= (A - B\Lambda_n)^* X_n (A - B\Lambda_n) - \Lambda_n^*(D^*C + B^* X_n A) + \Lambda_n^* D^*D\Lambda_n \\
&\quad + \Lambda_n^* B^* X_n A - C^*D\Lambda_n + C^*C \\
&= (A - B\Lambda_n)^* X_n (A - B\Lambda_n) + C^*C + \Lambda_n^* D^*D\Lambda_n - \Lambda_n^* D^*C - C^*D\Lambda_n \\
&= (A - B\Lambda_n)^* X_n (A - B\Lambda_n) + (C - D\Lambda_n)^*(C - D\Lambda_n).
\end{aligned} \tag{10.4.14}$$

This yields the Riccati difference equation in (10.4.4). The form of the Riccati difference equation in (10.4.4) guarantees that X_n is positive for all $n \geq 0$. By dropping the subscript n and $n+1$ in (10.4.13) and mimicking the calculation in (10.4.14), we obtain the form of the algebraic Riccati equation in (10.4.3).

To complete this section, let us observe that there is only one marginally stabilizing positive solution to the algebraic Riccati equation in (10.4.1). To see this, assume that X is a marginally stabilizing positive solution (10.4.1). Then $P = Q - X$ is a self-adjoint solution to the algebraic Riccati equation in (10.4.10). Since $R_0 - B^*PB = D^*D + B^*XB$ is strictly positive, by consulting (10.4.10), we obtain $P \geq A^*PA$. According to Lemma 10.1.3, the operator P is positive. So P is a positive solution to the algebraic Riccati equation in (10.4.10). Because the feedback operator J in (10.4.11) is marginally stable, P is the unique marginally stabilizing positive solution to (10.4.10). Since the marginally stabilizing solution is unique, $X = Q - P$ is also uniquely determined.

10.5 The Outer Factor for $\gamma^2 I - G^*G$

Assume that G is a rational function in $H^\infty(\mathcal{E}, \mathcal{Y})$ satisfying $\|G\|_\infty < \gamma$. Let R be the function in $L^\infty(\mathcal{E}, \mathcal{E})$ defined by $R(e^{\imath\omega}) = \gamma^2 I - G(e^{\imath\omega})^* G(e^{\imath\omega})$ for all $0 \leq \omega \leq 2\pi$. Then R and R^{-1} are both positive invertible functions in $L^\infty(\mathcal{E}, \mathcal{E})$. So the Toeplitz matrix T_R defines a strictly positive operator on $\ell_+^2(\mathcal{E})$; see Proposition 2.5.1. In particular, R admits an invertible rational outer spectral factor Θ. To compute this outer spectral factor, let $\{A, B, C, D\}$ be a stable, controllable, finite dimensional realization for G. Consider the algebraic Riccati equation

$$
\begin{aligned}
Y &= A^*YA + C^*C \\
&\quad + (D^*C + B^*YA)^*(\gamma^2 I - D^*D - B^*YB)^{-1}(D^*C + B^*YA).
\end{aligned}
\tag{10.5.1}
$$

We say that Y is a *positive solution* to this Riccati equation if Y is a positive operator satisfying (10.5.1) and $\gamma^2 I - D^*D - B^*YB$ is strictly positive. Moreover, Y is a *stabilizing solution* if Y is a positive solution to (10.5.1) and the corresponding feedback operator

$$
J = A + B(\gamma^2 I - D^*D - B^*YB)^{-1}(D^*C + B^*YA) \tag{10.5.2}
$$

is stable. Finally, it is noted that if a stabilizing solution exists, then it is unique.

If Y is any self-adjoint solution to the algebraic Riccati equation in (10.5.1) such that $\gamma^2 I - D^*D - B^*YB$ is strictly positive, then Y is a positive solution. To see this simply observe that $Y \geq A^*YA$ where A is stable. Lemma 10.1.3 shows that Y is positive.

The Riccati difference equation corresponding to (10.5.1) is determined by

$$
\begin{aligned}
Y_{n+1} &= A^*Y_nA + C^*C \\
&\quad + (D^*C + B^*Y_nA)^*(\gamma^2 I - D^*D - B^*Y_nB)^{-1}(D^*C + B^*Y_nA) \\
Y_0 &= Q \quad \text{where} \quad Q = A^*QA + C^*C.
\end{aligned}
\tag{10.5.3}
$$

The initial condition $Y_0 = Q$ is the observability Gramian for the pair $\{C, A\}$. In this setting, the set $\{Y_n\}_0^\infty$ forms an increasing sequence of positive operators. Moreover, Y_n converges to a positive operator Y if and only if there exists a $\delta > 0$ such that

$$
\gamma^2 I - D^*D - B^*Y_nB \geq \delta I \qquad (n \geq 0). \tag{10.5.4}
$$

In this case, Y is a positive solution to the algebraic Riccati equation in (10.5.1). Moreover, the following are equivalent:

(i) The algebraic Riccati equation in (10.5.1) admits a positive stabilizing solution.

(ii) There exists a $\delta > 0$ such that (10.5.4) holds and the feedback operator J is stable when $Y = \lim_{n\to\infty} Y_n$.

In this case, Y is the unique stabilizing solution to the algebraic Riccati equation in (10.5.1).

Proposition 10.5.1. *Let $\{A, B, C, D\}$ be a stable, controllable realization for a rational function G in $H^\infty(\mathcal{E}, \mathcal{Y})$. Then $\|G\|_\infty < \gamma$ if and only if there exists a stabilizing solution Y to the algebraic Riccati equation in (10.5.1). In this case, the stabilizing solution is unique. Moreover, the invertible outer spectral factor Θ for $\gamma^2 I - G^*G$ is given by*

$$\begin{aligned}\Theta(z) &= D_o + C_o(zI - A)^{-1}B,\\ D_o &= (\gamma^2 I - D^*D - B^*YB)^{1/2},\\ C_o &= -D_o^{-1}(D^*C + B^*YA).\end{aligned} \tag{10.5.5}$$

Finally, $\|G\|_\infty$ equals the infimum over the set of all $\gamma > 0$ such that the algebraic Riccati equation in (10.5.1) admits a stabilizing solution.

Proof. Set $R = \gamma^2 I - G^*G$. Since R is in $L^\infty(\mathcal{E}, \mathcal{E})$, the Toeplitz matrix T_R is a well-defined self-adjoint operator on $\ell_+^2(\mathcal{E})$. Moreover, $\|G\|_\infty < \gamma$ if and only if T_R is strictly positive, or equivalently, T_R admits an invertible outer spectral factor. Notice that $T_R = \gamma^2 I - T_{G^*G}$. By consulting Lemma 6.2.3, we see that the components of the Toeplitz matrix $(T_R)_{j,k} = R_{j-k}$ are determined by

$$\begin{aligned}R_0 &= \gamma^2 I - D^*D - B^*QB,\\ R_n &= \widehat{C}A^{n-1}B \qquad (\text{for } n \geq 1),\\ \widehat{C} &= -(D^*C + B^*QA).\end{aligned} \tag{10.5.6}$$

As before, Q is the observability Gramian for the pair $\{C, A\}$.

Now we can use Theorem 10.1.4, to determine whether or not T_R admits a square outer spectral factor. Using the expressions for R_0 and $\widehat{C}$ in (10.5.6), the algebraic Riccati equation for P in (10.1.3) becomes

$$\begin{aligned}P &= A^*PA + (\widehat{C} - B^*PA)^*(R_0 - B^*PB)^{-1}(\widehat{C} - B^*PA) \qquad (10.5.7)\\ &= A^*PA\\ &\quad + (D^*C + B^*(Q+P)A)^*(\gamma^2 I - D^*D - B^*(Q+P)B)^{-1}(D^*C + B^*(Q+P)A).\end{aligned}$$

The Toeplitz operator T_R admits a square outer spectral factor if and only if the algebraic Riccati equation (10.5.7) admits a positive solution. Moreover, T_R admits an invertible outer spectral factor if and only if (10.5.7) admits a stabilizing solution. In other words, $\|G\|_\infty < \gamma$ if and only if (10.5.7) admits a stabilizing solution. In particular, if $\|G\|_\infty < \gamma$, then (10.5.7) admits a stabilizing solution. On the other hand, if (10.5.7) admits a positive solution, then $T_R = \gamma^2 I - T_G^* T_G$ is positive, and thus, $\|G\|_\infty \leq \gamma$. Therefore $\|G\|_\infty$ equals the infimum over the set of all γ such that the Riccati equation in (10.5.7) admits a stabilizing solution.

If we add the observability Gramian $Q = A^*QA + C^*C$ to (10.5.7) and set $Y = Q + P$, we arrive at the algebraic Riccati equation for Y in (10.5.1).

We claim that P is a positive solution to (10.5.7) if and only if $Y = Q + P$ is a positive solution to (10.5.1). Obviously, if P is a positive solution to (10.5.7), then $Y = Q + P$ is a positive solution to (10.5.1). On the other hand, if Y is a positive solution to (10.5.1), then $P = Y - Q$ is a self-adjoint operator solving the algebraic Riccati equation (10.5.7) and $R_0 - B^*PB = \gamma^2 I - D^*D - B^*YB$ is strictly positive. Therefore P is a positive solution to (10.5.7), which proves our claim. So $\|G\|_\infty$ equals the infimum over the set of all $\gamma > 0$ such that the algebraic Riccati equation for Y in (10.5.1) admits a stabilizing solution. This yields the last part of the proposition.

Recall that P is a stabilizing solution to (10.5.7) if P is a positive solution to (10.5.7) and the feedback operator

$$\begin{aligned} J &= A - B(R_0 - B^*PB)^{-1}(\widehat{C} - B^*PA) \\ &= A + B(\gamma^2 I - D^*D - B^*YB)^{-1}(D^*C + B^*YA) \end{aligned}$$

is stable. By consulting (10.5.2), we see that this is the same feedback operator for the algebraic Riccati equation (10.5.1). Thus Y is a stabilizing solution for (10.5.1) if and only if $P = Y - Q$ is a stabilizing solution for (10.5.7). Because the stabilizing solution P for (10.5.7) is unique, there is a unique stabilizing solution Y for (10.5.1).

To complete the proof, assume that $\|G\|_\infty < \gamma$ and Y is the stabilizing solution to the algebraic Riccati equation (10.5.1). Then $P = Y - Q$ is the stabilizing solution to the algebraic Riccati equation (10.5.7). According to Theorem 10.1.4, the outer spectral factor Θ for $\gamma^2 I - G^*G$ is given by (10.1.7) where R_0 and $\widehat{C}$ are now specified by (10.5.6). Using this R_0 and $\widehat{C}$ in (10.1.7) along with $P = P_o = Y - Q$, we arrive at the state space realization for Θ in (10.5.5). □

Now let us verify our comments concerning the limit of Riccati difference equations in (10.5.3). The Riccati difference equation corresponding to the algebraic Riccati equation in (10.5.7) is given by

$$\begin{aligned} P_{n+1} &= A^*P_nA \\ &\quad + (D^*C + B^*(Q + P_n)A)^*(\gamma^2 I - D^*D - B^*(Q + P_n)B)^{-1}(D^*C + B^*(Q + P_n)A). \end{aligned}$$

The initial condition is $P_0 = 0$. In this setting the bounds in (10.3.3) and (10.5.4) are equivalent. According to Theorem 10.3.1, the $\{P_n\}$ forms an increasing sequence of positive operators. Moreover, P_n converge to an operator P if and only if P is the marginally stabilizing solution of the algebraic Riccati equation in (10.5.7). Observe that the difference equation in (10.5.3) is obtained by setting $Y_n = Q + P_n$ subject to the initial condition $Y_0 = Q$. Therefore $\{Y_n\}_0^\infty$ forms an increasing sequence of positive operators. Finally, Y_n converges to a positive operator Y and J is stable if and only if Y is a stabilizing solution to the algebraic Riccati equation in (10.5.1).

The previous proposition shows that $\|G\|_\infty$ is the infimum over the set of all γ such that the algebraic Riccati equation (10.5.1) admits a stabilizing solution. So one can compute $\|G\|_\infty$ by iterating on γ to find the smallest γ such that (10.5.1) admits a stabilizing solution. Finally, it is noted that in the scalar case, one can also use fast Fourier transform techniques to efficiently compute the L^∞ norm of any rational function in H^∞.

Remark 10.5.2. Recall that any rational contractive analytic function admits a contractive realization; see Section 7.8. Let us compute a contractive realization for any rational contractive analytic function G such that $\|G\|_\infty < 1$. To this end, let $\{A, B, C, D\}$ be any minimal realization for a rational function G in $H^\infty(\mathcal{E}, \mathcal{Y})$, and assume that $\|G\|_\infty < 1$. Let Y be the stabilizing solution to the algebraic Riccati equation

$$Y = A^*YA + (D^*C + B^*YA)^*(I - D^*D - B^*YB)^{-1}(D^*C + B^*YA) + C^*C. \quad (10.5.8)$$

Then

$$\begin{bmatrix} A^* & C^* \\ B^* & D^* \end{bmatrix} \begin{bmatrix} Y & 0 \\ 0 & I \end{bmatrix} \begin{bmatrix} A & B \\ C & D \end{bmatrix} \leq \begin{bmatrix} Y & 0 \\ 0 & I \end{bmatrix}. \quad (10.5.9)$$

Finally,

$$\Sigma = \{Y^{1/2}AY^{-1/2}, Y^{1/2}B, CY^{-1/2}, D\}$$

is a minimal, contractive realization for G.

To verify this, let C_o and D_o be the operators defined by equation (10.5.5) where $\gamma = 1$. Proposition 10.5.1 shows that $\{A, B, C_o, D_o\}$ is a controllable realization for the outer spectral factor Θ for $R = I - G^*G$. Using

$$C_o = -(I - D^*D - B^*YB)^{-1/2}(D^*C + B^*YA)$$

in (10.5.8), we obtain

$$Y = A^*YA + C_o^*C_o + C^*C.$$

Hence Y is the observability Gramian for the pair $\{\begin{bmatrix} C & C_o \end{bmatrix}^{tr}, A\}$. Remark 7.8.2 shows that (10.5.9) holds, and Σ is a contractive realization for G.

10.6 Darlington Synthesis

Let G be a rational function in $H^\infty(\mathcal{E}, \mathcal{Y})$ and assume that $\|G\|_\infty \leq 1$. The Darlington synthesis problem is to find a two-sided inner function $\Psi(z)$ in the Hardy space $H^\infty(\mathcal{E} \oplus \mathcal{Y}, \mathcal{Y} \oplus \mathcal{E})$ of the form

$$\Psi(z) = \begin{bmatrix} G(z) & G_{12}(z) \\ G_{21}(z) & G_{22}(z) \end{bmatrix}. \quad (10.6.1)$$

In this case, Ψ is called a *Darlington extension* of G.

To solve the Darlington synthesis problem, let $\{A \text{ on } \mathcal{X}, B, C, D\}$ be any minimal realization for a rational function G. Let Θ be the outer spectral factor for $I - G^*G$. Recall that Θ admits a controllable realization of the form

$\{A, B, C_o, D_o\}$; see Proposition 10.5.1. Let Y be the observability Gramian for the pair $\{\begin{bmatrix} C & C_o \end{bmatrix}^{tr}, A\}$. If $\|G\|_\infty < 1$, then one can use $\gamma = 1$ in Proposition 10.5.1 to compute $\{A, B, C_o, D_o\}$. Moreover, in this case Y is the stabilizing solution to the algebraic Riccati equation in (10.5.1). Using the singular value decomposition (or "null" in Matlab), we can compute an isometry Γ mapping $\mathcal{Y}$ into $\mathcal{X} \oplus \mathcal{Y} \oplus \mathcal{E}$ of the form

$$\Gamma = \begin{bmatrix} \Gamma_1 \\ \Gamma_2 \\ \Gamma_3 \end{bmatrix} : \mathcal{Y} \to \begin{bmatrix} \mathcal{X} \\ \mathcal{Y} \\ \mathcal{E} \end{bmatrix} \quad \text{such that} \quad \Gamma\mathcal{Y} = \ker \begin{bmatrix} A^*Y & C^* & C_o^* \\ B^*Y & D^* & D_o^* \end{bmatrix}. \qquad (10.6.2)$$

Now let B_1 mapping $\mathcal{Y}$ into $\mathcal{X}$, and D_{12} on $\mathcal{Y}$, and D_{22} mapping $\mathcal{Y}$ into $\mathcal{E}$ be the operators defined by

$$\begin{bmatrix} B_1 \\ D_{12} \\ D_{22} \end{bmatrix} = \begin{bmatrix} \Gamma_1 \\ \Gamma_2 \\ \Gamma_3 \end{bmatrix} (\Gamma_1^* Y \Gamma_1 + \Gamma_2^* \Gamma_2 + \Gamma_3^* \Gamma_3)^{-1/2}. \qquad (10.6.3)$$

Then a state space realization for a solution to the Darlington synthesis problem is given by

$$\Psi = \begin{bmatrix} G(z) & G_{12}(z) \\ \Theta(z) & G_{22}(z) \end{bmatrix} = \begin{bmatrix} D & D_{12} \\ D_o & D_{22} \end{bmatrix} + \begin{bmatrix} C \\ C_o \end{bmatrix} (zI - A)^{-1} \begin{bmatrix} B & B_1 \end{bmatrix}. \qquad (10.6.4)$$

In other words, Ψ is a Darlington extension of G. Finally, it is noted that Ψ and G have the same McMillan degree.

To prove that Ψ in (10.6.4) is a Darlington extension of G, let us first recall equation (7.8.9) in the proof of Theorem 7.8.1, that is,

$$\begin{bmatrix} A^* & C^* & C_o^* \\ B^* & D^* & D_o^* \end{bmatrix} \begin{bmatrix} Y & 0 & 0 \\ 0 & I_{\mathcal{Y}} & 0 \\ 0 & 0 & I_{\mathcal{E}} \end{bmatrix} \begin{bmatrix} A & B \\ C & D \\ C_o & D_o \end{bmatrix} = \begin{bmatrix} Y & 0 \\ 0 & I_{\mathcal{E}} \end{bmatrix}. \qquad (10.6.5)$$

Since Y is strictly positive, $\operatorname{rank}(Y \oplus I_{\mathcal{E}}) = \dim \mathcal{X} + \dim \mathcal{E}$. Using this and the fact that the rank of M^*M equals the rank of M, equation (10.6.5) implies that

$$\dim \mathcal{X} + \dim \mathcal{E} = \operatorname{rank} \begin{bmatrix} Y^{1/2}A & Y^{1/2}B \\ C & D \\ C_o & D_o \end{bmatrix} = \operatorname{rank} \begin{bmatrix} A & B \\ C & D \\ C_o & D_o \end{bmatrix}.$$

According to the rank nullity theorem, we obtain

$$\dim \mathcal{Y} = \dim \ker \begin{bmatrix} A^*Y & C^* & C_o^* \\ B^*Y & D^* & D_o^* \end{bmatrix}.$$

So there exists an isometry Γ from $\mathcal{Y}$ into $\mathcal{X} \oplus \mathcal{Y} \oplus \mathcal{E}$ whose range equals the kernel of the previous 2×3 block matrix, and this isometry admits a matrix representation of the form (10.6.2). By construction, $B_1^* Y B_1 + D_{12}^* D_{12} + D_{22}^* D_{22} = I$. Moreover, (10.6.2) and (10.6.3) yield

$$\begin{bmatrix} A^* & C^* & C_o^* \\ B^* & D^* & D_o^* \\ B_1^* & D_{12}^* & D_{22}^* \end{bmatrix} \begin{bmatrix} Y & 0 & 0 \\ 0 & I_{\mathcal{Y}} & 0 \\ 0 & 0 & I_{\mathcal{E}} \end{bmatrix} \begin{bmatrix} A & B & B_1 \\ C & D & D_{12} \\ C_o & D_o & D_{22} \end{bmatrix} = \begin{bmatrix} Y & 0 & 0 \\ 0 & I_{\mathcal{E}} & 0 \\ 0 & 0 & I_{\mathcal{Y}} \end{bmatrix}.$$

Recall that A is stable and Y is the observability Gramian for $\{\begin{bmatrix} C & C_o \end{bmatrix}^{tr}, A\}$. By applying Theorem 4.2.1 to the systems matrix

$$\begin{bmatrix} A & \begin{bmatrix} B & B_1 \end{bmatrix} \\ \begin{bmatrix} C \\ C_o \end{bmatrix} & \begin{bmatrix} D & D_{12} \\ D_o & D_{22} \end{bmatrix} \end{bmatrix},$$

we see that the transfer function Ψ in (10.6.4) is a two-sided inner function in $H^\infty(\mathcal{E} \oplus \mathcal{Y}, \mathcal{Y} \oplus \mathcal{E})$. Since $\{A, B, C, D\}$ is a minimal realization for G, it follows that G is contained in the upper left-hand corner of Ψ. In other words, Ψ in (10.6.4) is a Darlington extension of G.

Example. Consider the contractive analytic function g in Section 7.8.1 given by

$$\begin{aligned} g &= \frac{-0.7165z^2 + 0.1796z - 0.0706}{z^3 - 0.2824z^2 - 0.0580z + 0.0003}, \\ \theta &= \frac{0.6671z^3 - 0.2471z^2 - 0.1627z + 0.0004497}{z^3 - 0.2824z^2 - 0.058z + 0.0003}. \end{aligned} \tag{10.6.6}$$

Recall that $\|g\|_\infty = .92$, and θ is the outer spectral factor for $1 - |g|^2$. Moreover, a minimal realization $\{A, B, C_i, D_i\}$ for the inner function $\begin{bmatrix} g & \theta \end{bmatrix}^{tr}$ is given by

$$\begin{aligned} A &= \begin{bmatrix} 0.0983 & -0.3295 & -0.0116 \\ -0.4965 & 0.1097 & -0.2787 \\ 0.0178 & 0.2840 & 0.0744 \end{bmatrix}, \quad B = \begin{bmatrix} 0.8548 \\ 0.0515 \\ 0.0196 \end{bmatrix}, \\ C_i &= \begin{bmatrix} -0.8344 & -0.0757 & 0.0294 \\ -0.0882 & 0.3187 & 0.0154 \end{bmatrix} \quad \text{and} \quad D_i = \begin{bmatrix} 0 \\ 0.6671 \end{bmatrix}. \end{aligned} \tag{10.6.7}$$

In this case, $D = 0$ and $D_o = 0.6671$, while

$$C = \begin{bmatrix} -0.8344 & -0.0757 & 0.0294 \end{bmatrix} \quad \text{and} \quad C_o = \begin{bmatrix} -0.0882 & 0.3187 & 0.0154 \end{bmatrix}.$$

Hence $\{A, B, C, 0\}$ is a realization for g, and $\{A, B, C_o, D_o\}$ is a realization for θ. Finally, the observability Gramian Y for the pair $\{C_i, A\}$ is given by

$$Y = \begin{bmatrix} 0.7590 & 0 & 0 \\ 0 & 0.1934 & 0 \\ 0 & 0 & 0.0163 \end{bmatrix}. \tag{10.6.8}$$

In this case, Γ is an isometry mapping $\mathbb{C}$ into $\mathbb{C}^5 = \mathbb{C}^3 \oplus \mathbb{C} \oplus \mathbb{C}$; see (10.6.2). By using the "null" command in Matlab to compute Γ with (10.6.3), we arrived at

$$B_1 = \begin{bmatrix} -0.0663 \\ -0.1433 \\ -7.7815 \end{bmatrix}, \quad D_{12} = 0.0004 \quad \text{and} \quad D_{22} = 0.0706.$$

By computing the transfer function for the last two components of Ψ, we obtained

$$g_{12} = \frac{0.0004447z^3 - 0.1627z^2 - 0.2471z + 0.667}{z^3 - 0.2824z^2 - 0.0580z + 0.0003},$$
$$g_{22} = \frac{0.07058z^3 - 0.1796z^2 + 0.7165z - 1.128 \times 10^{-5}}{z^3 - 0.2824z^2 - 0.0580z + 0.0003}. \tag{10.6.9}$$

In other words,

$$\Phi = \begin{bmatrix} g & g_{12} \\ \theta & g_{22} \end{bmatrix} = \begin{bmatrix} D & D_{12} \\ D_o & D_{22} \end{bmatrix} + \begin{bmatrix} C \\ C_o \end{bmatrix} (zI - A)^{-1} \begin{bmatrix} B & B_1 \end{bmatrix} \tag{10.6.10}$$

is a Darlington extension of g.

To complete this section let us observe that one can use the finite section method in Section 7.7 to compute the outer spectral factor θ for $1 - |g|^2$. Then the Kalman-Ho algorithm can be used to find a minimal realization $\{A, B, C_i, D_i\}$ for $\begin{bmatrix} g & \theta \end{bmatrix}^{tr}$. Then using $\{A, B, C_i, D_i\}$ one can follow the procedure in (10.6.2) and (10.6.3) to compute a state space realization for the Darlington extension of g in (10.6.10).

10.7 Riccati Equations

In this section, we will present some classical results concerning certain algebraic Riccati equations arising in control theory. These results are used in the stability analysis of the steady state Kalman filter; see Chapter 11. Consider the algebraic Riccati equation determined by

$$P = A^*PA + R_{11} - (B^*PA + R_{21})^* (B^*PB + R_{22})^{-1} (B^*PA + R_{21}). \tag{10.7.1}$$

Here A is an operator on a finite dimensional space $\mathcal{X}$ and B maps the finite dimensional space $\mathcal{E}$ into $\mathcal{X}$. Throughout we assume that

$$R = \begin{bmatrix} R_{11} & R_{12} \\ R_{21} & R_{22} \end{bmatrix} \quad \text{on} \quad \begin{bmatrix} \mathcal{X} \\ \mathcal{E} \end{bmatrix} \tag{10.7.2}$$

is positive and R_{22} is strictly positive. Another form of the algebraic Riccati equation in (10.7.1) is given by

$$P = (A - BK_P)^*P(A - BK_P) + \begin{bmatrix} I & -K_P^* \end{bmatrix} \begin{bmatrix} R_{11} & R_{12} \\ R_{21} & R_{22} \end{bmatrix} \begin{bmatrix} I \\ -K_P \end{bmatrix},$$
$$K_P = (B^*PB + R_{22})^{-1} (B^*PA + R_{21}). \tag{10.7.3}$$

To verify that the algebraic Riccati equations in (10.7.3) and (10.7.1) are equivalent observe that

$$\begin{aligned}
P &= (A - BK_P)^*P(A - BK_P) + \begin{bmatrix} I & -K_P^* \end{bmatrix} \begin{bmatrix} R_{11} & R_{12} \\ R_{21} & R_{22} \end{bmatrix} \begin{bmatrix} I \\ -K_P \end{bmatrix} \\
&= A^*PA - 2\Re(A^*PBK_P) + K_P^*B^*PBK_P + R_{11} - 2\Re(R_{12}K_P) + K_P^*R_{22}K_P \\
&= A^*PA + R_{11} - 2\Re((B^*PA + R_{21})^*K_P) + K_P^*(B^*PB + R_{22})K_P \\
&= A^*PA + R_{11} - 2\Re(K_P^*(B^*PB + R_{22})K_P) + K_P^*(B^*PB + R_{22})K_P \\
&= A^*PA + R_{11} - K_P^*(B^*PB + R_{22})K_P \\
&= A^*PA + R_{11} - (B^*PA + R_{21})^* (B^*PB + R_{22})^{-1} (B^*PA + R_{21}) .
\end{aligned}$$

Thus the algebraic Riccati equations in (10.7.3) and (10.7.1) are identical. The following result shows that the solution to the algebraic Riccati equation in (10.7.3) can be obtained by passing to the limit in a corresponding Riccati difference equation. The proof of this result will be given in Section 10.7.2.

Theorem 10.7.1. *Consider the pair $\{A \text{ on } \mathcal{X}, B\}$ where B maps $\mathcal{E}$ into $\mathcal{X}$. Let Q_n be the solution for the Riccati difference equation*

$$\begin{aligned}
Q_{n+1} = {} & A^*Q_nA + R_{11} \\
& - (B^*Q_nA + R_{21})^* (B^*Q_nB + R_{22})^{-1} (B^*Q_nA + R_{21})
\end{aligned} \tag{10.7.4}$$

subject to the initial condition $Q_0 = 0$. Moreover, assume that R in (10.7.2) is positive and R_{22} is strictly positive. Finally, let $\Delta = R_{11} - R_{12}R_{22}^{-1}R_{21}$ be the Schur complement for R with respect to R_{11}. Then the following holds.

(i) *The solution $\{Q_n\}_0^\infty$ forms an increasing sequence of positive operators. To be precise, $Q_n \le Q_{n+1}$ for all integers $n \ge 0$.*

(ii) *If the pair $\{A, B\}$ is controllable, then Q_n converges to a positive operator P as n tends to infinity, that is,*

$$P = \lim_{n\to\infty} Q_n. \tag{10.7.5}$$

In this case, P is a positive solution for the algebraic Riccati equation in (10.7.1) or equivalently, (10.7.3).

(iii) *If $\{\Delta, A - BR_{22}^{-1}R_{21}\}$ is observable and $\{A, B\}$ is controllable and P is any positive solution to the algebraic Riccati equation in (10.7.1), then P is strictly positive. In this case, $A - BK_P$ is stable where K_P is the feedback operator defined in (10.7.3).*

(iv) *If $\{\Delta, A - BR_{22}^{-1}R_{21}\}$ is observable and $\{A, B\}$ is controllable, then there is only one positive solution P to the algebraic Riccati equation in (10.7.1). In this case, P is strictly positive. This solution is given by $P = \lim_{n\to\infty} Q_n$.*

Consider the positive operator R determined by

$$R = \begin{bmatrix} C^*C & 0 \\ 0 & D^*D \end{bmatrix} \text{ on } \begin{bmatrix} \mathcal{X} \\ \mathcal{E} \end{bmatrix}. \tag{10.7.6}$$

Substituting this R into (10.7.1), we obtain the classical algebraic Riccati equation

$$P = A^*PA + C^*C - A^*PB\,(B^*PB + D^*D)^{-1}\,B^*PA. \tag{10.7.7}$$

Here $\{A \text{ on } \mathcal{X}, B, C, D\}$ is a state space system where B maps $\mathcal{E}$ into $\mathcal{X}$, the operator C maps $\mathcal{X}$ into $\mathcal{Y}$ and D is a one to one operator mapping $\mathcal{E}$ into $\mathcal{Y}$. Another form for the algebraic Riccati equation in (10.7.7) is determined by

$$\begin{aligned} P &= (A - BK_P)^*P(A - BK_P) + C^*C + K_P^*D^*DK_P, \\ K_P &= (B^*PB + D^*D)^{-1}\,B^*PA. \end{aligned} \tag{10.7.8}$$

By applying Theorem 10.7.1 with R in (10.7.6), we arrive at the following result.

Corollary 10.7.2. *Consider the finite dimensional system $\{A \text{ on } \mathcal{X}, B, C, D\}$ where D^*D is invertible. Let Q_n be the solution for the Riccati difference equation*

$$Q_{n+1} = A^*Q_nA + C^*C - A^*Q_nB\,(B^*Q_nB + D^*D)^{-1}\,B^*Q_nA \tag{10.7.9}$$

subject to the initial condition $Q_0 = 0$. Then the following holds.

(i) *The solution $\{Q_n\}_0^\infty$ forms an increasing sequence of positive operators. To be precise, $Q_n \le Q_{n+1}$ for all integers $n \ge 0$.*

(ii) *If the pair $\{A, B\}$ is controllable, then Q_n converges to a positive operator P as n tends to infinity, that is,*

$$P = \lim_{n\to\infty} Q_n. \tag{10.7.10}$$

In this case, P is a positive solution for the algebraic Riccati equation in (10.7.7) or equivalently, (10.7.8).

(iii) *If $\{A, B, C, D\}$ is controllable and observable and P is any positive solution to the algebraic Riccati equation in (10.7.7), then P is strictly positive. In this case, $A - BK_P$ is stable where $K_P = (B^*PB + D^*D)^{-1}B^*PA$.*

(iv) *If $\{A, B, C, D\}$ is controllable and observable, then there is only one positive solution P to the algebraic Riccati equation in (10.7.7). In this case, P is strictly positive. Finally, this solution is given by $P = \lim_{n\to\infty} Q_n$.*

Remark 10.7.3. For the moment, let Q_n be the solution to the Riccati difference equation in (10.7.4). Another form of this Riccati difference equation is given by

$$\begin{aligned} Q_{n+1} &= (A - BK_{Q_n})^*Q_n(A - BK_{Q_n}) + \begin{bmatrix} I & -K_{Q_n}^* \end{bmatrix} \begin{bmatrix} R_{11} & R_{12} \\ R_{21} & R_{22} \end{bmatrix} \begin{bmatrix} I \\ -K_{Q_n} \end{bmatrix}, \\ K_{Q_n} &= (B^*Q_nB + R_{22})^{-1}\,(B^*Q_nA + R_{21})\,. \end{aligned} \tag{10.7.11}$$

Moreover, an equivalent form for the Riccati difference equation in (10.7.9) is determined by

$$Q_{n+1} = (A - BK_{Q_n})^* Q_n (A - BK_{Q_n}) + C^*C + K_{Q_n}^* D^*DK_{Q_n},$$
$$K_{Q_n} = (B^*Q_nB + D^*D)^{-1} B^*Q_nA. \tag{10.7.12}$$

In applications it may be better to use the Riccati difference equation in (10.7.11), respectively (10.7.12), rather than the Riccati difference equation in (10.7.4), respectively (10.7.9). The solution Q_n subject to the initial condition $Q_0 = 0$ must be positive. Observe that the Riccati difference equation in (10.7.11) or (10.7.12) is the sum of positive terms, while the Riccati difference equation in (10.7.4) or (10.7.9) is the sum of two positive terms and a negative term. In some applications this negative term may numerically lead to a solution Q_n which is not positive. To avoid this problem one may want to use the Riccati difference equation in (10.7.11) or (10.7.12). One can also numerically guarantee that Q_n is positive by forming the spectral decomposition of Q_n and setting all the negative eigenvalues for Q_n equal to zero. For even better methods to numerically guarantee that Q_n is positive, see Kailath-Sayed-Hassibi [143].

10.7.1 The minimum principle

Now let us return to Theorem 10.7.1. If the pair $\{A, B\}$ is not controllable, then the solution Q_n to the Riccati difference equation (10.7.4) may diverge. For example, consider the system $A = 2$, $B = 0$ and $R = I$ on $\mathbb{C}^2$. Then $Q_{n+1} = 4Q_n + 1$ subject to the initial condition $Q_0 = 0$. In this case, the solution $Q_n = \sum_{k=0}^{n-1} 4^k = (4^n - 1)/3$ for $n \geq 1$. Clearly, Q_n approaches infinity as n tends to infinity.

To prove Theorem 10.7.1, we use the following result known as the minimum principle.

Lemma 10.7.4 (Minimum principle.). *Consider the pair $\{A \text{ on } \mathcal{X}, B\}$. Assume that R in* (10.7.2) *is positive and R_{22} is strictly positive. Let Q_n be the solution to the Riccati difference equation* (10.7.11) *subject to the initial condition Q_0. Let V_n be the solution to the Riccati difference equation*

$$\begin{aligned} V_{n+1} &= (A - B\Phi_n)^* V_n (A - B\Phi_n) \\ &\quad + \begin{bmatrix} I & -\Phi_n^* \end{bmatrix} \begin{bmatrix} R_{11} & R_{12} \\ R_{21} & R_{22} \end{bmatrix} \begin{bmatrix} I \\ -\Phi_n \end{bmatrix} \end{aligned} \tag{10.7.13}$$

where Φ_n is any operator mapping $\mathcal{X}$ into $\mathcal{E}$, and V_0 is the initial condition. If $Q_0 \leq V_0$, then $Q_n \leq V_n$ for all integers $n \geq 0$.

Our proof of this minimum principle is based on the following result.

Lemma 10.7.5. *Consider the pair $\{A, B\}$. Assume that R in (10.7.2) is positive and R_{22} is strictly positive. Let V be any positive operator on $\mathcal{X}$ and Φ an operator* *mapping $\mathcal{X}$ into $\mathcal{E}$. Then we have*

$$(A - BK_V)^*V(A - BK_V) + \begin{bmatrix} I & -K_V^* \end{bmatrix} \begin{bmatrix} R_{11} & R_{12} \\ R_{21} & R_{22} \end{bmatrix} \begin{bmatrix} I \\ -K_V \end{bmatrix}$$
$$\leq (A - B\Phi)^*V(A - B\Phi) + \begin{bmatrix} I & -\Phi^* \end{bmatrix} \begin{bmatrix} R_{11} & R_{12} \\ R_{21} & R_{22} \end{bmatrix} \begin{bmatrix} I \\ -\Phi \end{bmatrix} \tag{10.7.14}$$

where the operator

$$K_V = (B^*VB + R_{22})^{-1} (B^*VA + R_{21}) .$$

Finally, we have equality in (10.7.14) if and only if $\Phi = K_V$.

Proof. Let $V^{1/2}$ be the positive square root of V and $R^{1/2}$ be the positive square root of R. Let Ω_Φ be the operator defined by the right-hand side of (10.7.14), that is,

$$\Omega_\Phi = (A - B\Phi)^*V(A - B\Phi) + \begin{bmatrix} I & -\Phi^* \end{bmatrix} \begin{bmatrix} R_{11} & R_{12} \\ R_{21} & R_{22} \end{bmatrix} \begin{bmatrix} I \\ -\Phi \end{bmatrix}. \tag{10.7.15}$$

For x in $\mathcal{X}$, we have

$$\begin{aligned} (\Omega_\Phi x, x) &= \|V^{1/2}(A - B\Phi)x\|^2 + \|R^{1/2} (x \oplus -\Phi x) \|^2 \\ &= \left\| \begin{bmatrix} V^{1/2}Ax \\ R^{1/2} \begin{bmatrix} I \\ 0 \end{bmatrix} x \end{bmatrix} - \begin{bmatrix} V^{1/2}B \\ R^{1/2} \begin{bmatrix} 0 \\ I \end{bmatrix} \end{bmatrix} \Phi x \right\|^2 . \end{aligned} \tag{10.7.16}$$

Now let T be the operator from $\mathcal{E}$ into $\mathcal{X} \oplus \mathcal{X} \oplus \mathcal{E}$ and g the vector in $\mathcal{X} \oplus \mathcal{X} \oplus \mathcal{E}$ defined by

$$T = \begin{bmatrix} V^{1/2}B \\ R^{1/2} \begin{bmatrix} 0 \\ I \end{bmatrix} \end{bmatrix} \quad \text{and} \quad g = \begin{bmatrix} V^{1/2}A \\ R^{1/2} \begin{bmatrix} I \\ 0 \end{bmatrix} \end{bmatrix} x. \tag{10.7.17}$$

Clearly, $g = g_x$ is a linear function in x. By consulting (10.7.16), we have $(\Omega_\Phi x, x) = \|g - T\Phi x\|^2$. Hence

$$(\Omega_\Phi x, x) = \|g - T\Phi x\|^2 \geq \inf\{\|g - Tu\|^2 : u \in \mathcal{E}\}. \tag{10.7.18}$$

Notice that $T^*T = B^*VB + R_{22}$. Since R_{22} is strictly positive, T^*T is invertible. By consulting standard least squares optimization theory, the solution to the optimization problem in (10.7.18) is unique and given by

$$\widehat{u} = (T^*T)^{-1}T^*g_x = (B^*VB + R_{22})^{-1} (B^*VA + R_{21})\, x = K_V x.$$

So the optimal solution $\widehat{u} = K_V x$. By employing $\widehat{u} = K_V x$ in (10.7.18), we obtain

$$\begin{aligned} (\Omega_\Phi x, x) &= \|g - T\Phi x\|^2 \geq \inf\{\|g - Tu\|^2 : u \in \mathcal{E}\} \\ &= \|g - TK_V x\|^2 = (\Omega_{K_V} x, x). \end{aligned} \tag{10.7.19}$$

Therefore $\Omega_\Phi \geq \Omega_{K_V}$. Notice that $\Omega_\Phi = \Omega_{K_V}$ if and only if $\|g_x - T\Phi x\| = \|g_x - TK_V x\|$ for all x in $\mathcal{X}$. Because $\widehat{u} = K_V x$ is the unique solution to the optimization problem in (10.7.18), we see that $\Omega_\Phi = \Omega_{K_V}$ if and only if $\Phi x = K_V x$ for all x, or equivalently, $\Phi = K_V$. □

Proof of the minimum principle Lemma 10.7.4. Recall that

$$K_{Q_n} = (B^*Q_nB + R_{22})^{-1}(B^*Q_nA + R_{21}),$$
$$K_{V_n} = (B^*V_nB + R_{22})^{-1}(B^*V_nA + R_{21}).$$

According to the hypothesis $Q_0 \leq V_0$. Now let us apply induction and assume that $Q_n \leq V_n$. By using Lemma 10.7.5 twice with K_{Q_n} and K_{V_n}, we have

$$\begin{aligned} Q_{n+1} &= (A - BK_{Q_n})^*Q_n(A - BK_{Q_n}) + \begin{bmatrix} I & -K_{Q_n}^* \end{bmatrix} \begin{bmatrix} R_{11} & R_{12} \\ R_{21} & R_{22} \end{bmatrix} \begin{bmatrix} I \\ -K_{Q_n} \end{bmatrix} \\ &\leq (A - BK_{V_n})^*Q_n(A - BK_{V_n}) + \begin{bmatrix} I & -K_{V_n}^* \end{bmatrix} \begin{bmatrix} R_{11} & R_{12} \\ R_{21} & R_{22} \end{bmatrix} \begin{bmatrix} I \\ -K_{V_n} \end{bmatrix} \\ &\leq (A - BK_{V_n})^*V_n(A - BK_{V_n}) + \begin{bmatrix} I & -K_{V_n}^* \end{bmatrix} \begin{bmatrix} R_{11} & R_{12} \\ R_{21} & R_{22} \end{bmatrix} \begin{bmatrix} I \\ -K_{V_n} \end{bmatrix} \\ &\leq (A - B\Phi_n)^*V_n(A - B\Phi_n) + \begin{bmatrix} I & -\Phi_n^* \end{bmatrix} \begin{bmatrix} R_{11} & R_{12} \\ R_{21} & R_{22} \end{bmatrix} \begin{bmatrix} I \\ -\Phi_n \end{bmatrix} \\ &= V_{n+1}. \end{aligned}$$

Therefore $Q_{n+1} \leq V_{n+1}$. □

10.7.2 Proof of Theorem 10.7.1

Proof of Part (i). By assumption the initial condition $Q_0 = 0$. (If Q_0 is nonzero, then the corresponding solution sequence $\{Q_n\}$ is not necessarily increasing.) We claim that $Q_{n+1} \geq V_n$ where V_n is the solution to the Riccati difference equation in (10.7.13), subject to the initial condition $V_0 = 0$, and $\Phi_{k-1} = K_{Q_k}$ for $k \geq 1$ where $K_{Q_k} = (B^*Q_kB + R_{22})^{-1}(B^*Q_kA + R_{21})$. To prove this we use induction. Since $R_{12} = R_{21}^*$, we obtain

$$Q_1 = \left(R_{11} - R_{12}R_{22}^{-1}R_{21}\right) \geq 0 = V_0.$$

Because R is positive, its Schur complement Q_1 is positive. Hence $Q_1 \geq V_0$.

Now let us proceed by induction and assume that $Q_k \geq V_{k-1}$ for some integer $k \geq 2$. By choosing $\Phi_{k-1} = K_{Q_k}$ in (10.7.13), we have

$$\begin{aligned} V_k &= (A - BK_{Q_k})^*V_{k-1}(A - BK_{Q_k}) \\ &\quad + \begin{bmatrix} I & -K_{Q_k}^* \end{bmatrix} \begin{bmatrix} R_{11} & R_{12} \\ R_{21} & R_{22} \end{bmatrix} \begin{bmatrix} I \\ -K_{Q_k} \end{bmatrix}. \end{aligned} \tag{10.7.20}$$

Subtracting (10.7.20) from (10.7.11) yields

$$Q_{k+1} - V_k = (A - BK_{Q_k})^* (Q_k - V_{k-1})(A - BK_{Q_k}) \geq 0.$$

Thus $Q_{k+1} \geq V_k$. By induction it follows that $Q_{n+1} \geq V_n$ for all integers $n \geq 0$. Using the minimum principle, $Q_n \leq V_n \leq Q_{n+1}$. Therefore $\{Q_n\}_0^\infty$ is an increasing sequence of positive operators. □

Proof of Part (ii). Now assume that the pair $\{A, B\}$ is controllable. Then we claim that the sequence $\{Q_n\}_0^\infty$ is uniformly bounded, that is, there exists a finite constant γ such that $Q_n \leq \gamma I$ for all integers $n \geq 0$. Because the pair $\{A, B\}$ is controllable, there exists a feedback operator L from $\mathcal{X}$ into $\mathcal{E}$ such that $A - BL$ is stable, that is, all the eigenvalues of $A - BL$ are contained in the open unit disc; see [60, 140, 189]. By setting $\Phi_n = L$ for all n in (10.7.13), we arrive at the following Riccati difference equation

$$V_{n+1} = (A - BL)^* V_n (A - BL) + \begin{bmatrix} I & -L^* \end{bmatrix} \begin{bmatrix} R_{11} & R_{12} \\ R_{21} & R_{22} \end{bmatrix} \begin{bmatrix} I \\ -L \end{bmatrix}.$$

By recursively solving for V_n with $V_0 = 0$, we see that

$$V_n = \sum_{j=0}^{n-1} (A - BL)^{*j} \begin{bmatrix} I & -L^* \end{bmatrix} \begin{bmatrix} R_{11} & R_{12} \\ R_{21} & R_{22} \end{bmatrix} \begin{bmatrix} I \\ -L \end{bmatrix} (A - BL)^j.$$

Since the term between $(A-BL)^{*j}$ and $(A-BL)^j$ is positive, this readily implies that

$$V_n \leq \sum_{j=0}^{\infty} (A - BL)^{*j} \begin{bmatrix} I & -L^* \end{bmatrix} \begin{bmatrix} R_{11} & R_{12} \\ R_{21} & R_{22} \end{bmatrix} \begin{bmatrix} I \\ -L \end{bmatrix} (A - BL)^j = V_\infty.$$

Here we set V_∞ equal to the infinite sum. Because $A - BL$ is stable, V_∞ is a bounded positive operator. In particular, $V_n \leq V_\infty$ where V_∞ is the solution to the Lyapunov equation

$$V_\infty = (A - BL)^* V_\infty (A - BL) + \begin{bmatrix} I & -L^* \end{bmatrix} \begin{bmatrix} R_{11} & R_{12} \\ R_{21} & R_{22} \end{bmatrix} \begin{bmatrix} I \\ -L \end{bmatrix}.$$

By employing the minimum principle, $Q_n \leq V_n \leq V_\infty$. Hence $\{Q_n\}_0^\infty$ is uniformly bounded. Because $\{Q_n\}_0^\infty$ is a uniformly bounded, increasing sequence of positive operators, Q_n converges to a positive operator P as n tends to infinity; see Halmos [126]. So by passing to limits in the Riccati difference equation (10.7.11), we arrive at a positive solution P to the algebraic Riccati equation in (10.7.1), or equivalently, (10.7.3). □

Proof of Part (iii). Assume that $\{\Delta, A - BR_{22}^{-1}R_{21}\}$ is observable and $\{A, B\}$ is controllable. Moreover, assume that P is any positive solution to the algebraic

Riccati equation in (10.7.1). Then we claim that P is strictly positive and $A-BK_P$ is stable. (Notice that we did not assume that P is given by the limit in (10.7.5). The conclusions in Part (iii) follow from the hypothesis that P is a positive solution to the algebraic Riccati equation and the corresponding pairs are controllable and observable.) Recall that P is a solution to the algebraic Riccati equation

$$\begin{aligned} P &= (A-BK_P)^*P(A-BK_P) + \begin{bmatrix} I & -K_P^* \end{bmatrix} \begin{bmatrix} R_{11} & R_{12} \\ R_{21} & R_{22} \end{bmatrix} \begin{bmatrix} I \\ -K_P \end{bmatrix}, \\ K_P &= (B^*PB + R_{22})^{-1} (B^*PA + R_{21}). \end{aligned} \tag{10.7.21}$$

Let $\Delta^{1/2}$ on $\mathcal{X}$ be the positive square root of the Schur complement $\Delta = R_{11}-R_{12}R_{22}^{-1}R_{21}$ for R. Motivated by the Schur decomposition in (7.2.5) in Section 7.2, let Ω be the lower triangular operator defined by

$$\Omega = \begin{bmatrix} \Delta^{1/2} & 0 \\ R_{22}^{-1/2}R_{21} & R_{22}^{1/2} \end{bmatrix} \text{ on } \begin{bmatrix} \mathcal{X} \\ \mathcal{E} \end{bmatrix}.$$

Because $\{\Delta, A - BR_{22}^{-1}R_{21}\}$, is observable, the pair $\{\Delta^{1/2}, A - BR_{22}^{-1}R_{21}\}$ is observable. A direct calculation shows that $R = \Omega^*\Omega$, and thus,

$$\begin{bmatrix} I & -K_P^* \end{bmatrix} R \begin{bmatrix} I \\ -K_P \end{bmatrix} = \begin{bmatrix} I & -K_P^* \end{bmatrix} \Omega^*\Omega \begin{bmatrix} I \\ -K_P \end{bmatrix}.$$

By consulting (10.7.21), we see that P satisfies the Lyapunov equation

$$P = (A-BK_P)^*P(A-BK_P) + \begin{bmatrix} I & -K_P^* \end{bmatrix} \Omega^*\Omega \begin{bmatrix} I \\ -K_P \end{bmatrix}. \tag{10.7.22}$$

We claim that the pair

$$\Sigma = \left\{ \Omega \begin{bmatrix} I \\ -K_P \end{bmatrix}, A - BK_P \right\} \tag{10.7.23}$$

is observable. According to the Popov-Belevitch-Hautus observability test, a pair $\{\Gamma, \Lambda \text{ on } \mathcal{X}\}$ is observable if and only if

$$\operatorname{rank} \begin{bmatrix} \Lambda - \lambda I \\ \Gamma \end{bmatrix} = \dim \mathcal{X} \qquad (\text{for all } \lambda \in \mathbb{C});$$

see Section 14.2. Because the pair $\{\Delta^{1/2}, A - BR_{22}^{-1}R_{21}\}$ is observable, we have

$$\operatorname{rank}\begin{bmatrix} A - BK_P - \lambda I \\ \Omega \begin{bmatrix} I \\ -K_P \end{bmatrix} \end{bmatrix} = \operatorname{rank}\begin{bmatrix} A - BK_P - \lambda I \\ \Delta^{1/2} \\ R_{22}^{-1/2}R_{21} - R_{22}^{1/2}K_P \end{bmatrix}$$

$$= \operatorname{rank}\begin{bmatrix} A - BK_P - \lambda I \\ \Delta^{1/2} \\ R_{22}^{-1}R_{21} - K_P \end{bmatrix}$$

$$= \operatorname{rank}\begin{bmatrix} A - BK_P - B(R_{22}^{-1}R_{21} - K_P) - \lambda I \\ \Delta^{1/2} \\ R_{22}^{-1}R_{21} - K_P \end{bmatrix}$$

$$= \operatorname{rank}\begin{bmatrix} A - BR_{22}^{-1}R_{21} - \lambda I \\ \Delta^{1/2} \\ R_{22}^{-1}R_{21} - K_P \end{bmatrix} = \dim \mathcal{X}$$

for all λ in $\mathbb{C}$. The second equality follows by multiplying the last row by $R_{22}^{-1/2}$ on the left, and the third equality by multiplying the last row by B and subtracting this from the first row. Clearly, these elementary operations do not change the rank. Hence the pair Σ in (10.7.23) is observable.

We claim that P is strictly positive. To see this first let us show that the kernel of P is an invariant subspace for $A - BR_{22}^{-1}R_{21}$. If $Px = 0$ for some x in $\mathcal{X}$, then the Lyapunov equation in (10.7.22) implies that

$$0 = (Px, x) = \|P^{1/2}(A - BK_P)x\|^2 + \|\Omega \begin{bmatrix} x \\ -K_P x \end{bmatrix}\|^2.$$

Thus $P^{1/2}(A - BK_P)x$ and $\Delta^{1/2}x$ and $R_{22}^{1/2}(R_{22}^{-1}R_{21} - K_P)x$ are all zero. This implies that $(R_{22}^{-1}R_{21} - K_P)x$ is also zero and

$$P^{1/2}(A - BR_{22}^{-1}R_{21})x = P^{1/2}(A - BR_{22}^{-1}R_{21} + B(R_{22}^{-1}R_{21} - K_P))x = 0.$$

In particular, $P(A - BR_{22}^{-1}R_{21})x = 0$. So the kernel of P is an invariant subspace for $A - BR_{22}^{-1}R_{21}$. Let x be any eigenvector for $A - BR_{22}^{-1}R_{21}$ in the kernel of P, that is, assume that $(A - BR_{22}^{-1}R_{21})x = \lambda x$ for some nonzero vector x in the kernel of P. Since $Px = 0$, our previous analysis shows that $\Delta^{1/2}x = 0$. Hence for all integers $k \geq 0$, we obtain

$$\Delta^{1/2}(A - BR_{22}^{-1}R_{21})^k x = \Delta^{1/2}\lambda^k x = \lambda^k \Delta^{1/2} x = 0.$$

Because $\{\Delta^{1/2}, A - BR_{22}^{-1}R_{21}\}$ is observable, x must be zero. This contradicts the fact that an eigenvector is nonzero. In other words, the kernel of P is zero. Therefore P is strictly positive.

The above analysis shows that the pair Σ in (10.7.23) is an observable, and P is a strictly positive, solution to the Lyapunov equation in (10.7.22). By consulting Section 14.4, we see that $A - BK_P$ is stable. □

Proof of Part (iv). To this end, assume that P and V are two arbitrary positive solutions to the algebraic Riccati equation in (10.7.21). By implementing Lemma 10.7.5, we have

$$\begin{aligned} -P &= -(A - BK_P)^* P(A - BK_P) - \begin{bmatrix} I & -K_P^* \end{bmatrix} \begin{bmatrix} R_{11} & R_{12} \\ R_{21} & R_{22} \end{bmatrix} \begin{bmatrix} I \\ -K_P \end{bmatrix} \\ &\geq -(A - BK_V)^* P(A - BK_V) - \begin{bmatrix} I & -K_V^* \end{bmatrix} \begin{bmatrix} R_{11} & R_{12} \\ R_{21} & R_{22} \end{bmatrix} \begin{bmatrix} I \\ -K_V \end{bmatrix}. \end{aligned}$$

Adding this $-P$ to

$$V = (A - BK_V)^* V(A - BK_V) + \begin{bmatrix} I & -K_V^* \end{bmatrix} \begin{bmatrix} R_{11} & R_{12} \\ R_{21} & R_{22} \end{bmatrix} \begin{bmatrix} I \\ -K_V \end{bmatrix},$$

we obtain

$$V - P \geq (A - BK_V)^* (V - P)(A - BK_V). \tag{10.7.24}$$

Since V is a positive solution to the algebraic Riccati equation (10.7.21), the operator $A - BK_V$ is stable; see Part (iii). By consulting Lemma 10.1.3, we see that $V - P \geq 0$. In other words, $V \geq P$. Since V and P are two arbitrary positive solutions to the algebraic Riccati equation in (10.7.21), we can interchange the roles of V and P, which yield $P \geq V$. Therefore $P = V$. □

Let $\{A, B, C, D\}$ be a stable system where D^*D is invertible. Recall that the algebraic Riccati equation in (10.4.1) to compute the inner-outer factorization is of the form

$$P = A^*PA + C^*C - (B^*PA + D^*C)^*(B^*PB + D^*D)^{-1}(B^*PA + D^*C). \tag{10.7.25}$$

The corresponding algebraic Riccati difference equation is given by

$$Q_{n+1} = A^*Q_nA + C^*C - (B^*Q_nA + D^*C)^*(B^*Q_nB + D^*D)^{-1}(B^*Q_nA + D^*C). \tag{10.7.26}$$

Now set

$$R = \begin{bmatrix} C^*C & C^*D \\ D^*C & D^*D \end{bmatrix} = \begin{bmatrix} C & D \end{bmatrix}^* \begin{bmatrix} C & D \end{bmatrix}. \tag{10.7.27}$$

By choosing $R_{11} = C^*C$, $R_{21} = D^*C$ and $R_{22} = D^*D$, we see that the algebraic Riccati equation in (10.7.25) is a special case of the algebraic Riccati equation in (10.7.1). If D is invertible and the initial condition $Q_0 = 0$ for the Riccati difference equation (10.7.26), then $Q_n = 0$ for all integers $n \geq 0$. The Schur complement for R is given by

$$\Delta = R_{11} - R_{12}R_{22}^{-1}R_{21} = C^*C - C^*D(D^*D)^{-1}D^*C.$$

So if D is invertible, then the Schur complement $\Delta = 0$. In this case, we cannot implement Theorem 10.7.1 to find a nontrivial solution to the algebraic Riccati equation in (10.7.25).

To get around this problem, we assumed that A is stable and the initial condition Q_0 is the observability Gramian for the pair $\{C, A\}$, that is, $Q_0 = A^*Q_0A + C^*C$. In this case, $\{Q_n\}$ forms a decreasing set of positive operators; see also (10.4.13) and the paragraph immediately following this equation. Hence Q_n converges to a positive solution P to the algebraic Riccati equation (10.7.25).

10.8 Notes

The Riccati equation in control theory started with the Kalman filter [144] and Kalman's solution to the linear quadratic regular problem [145]. The Riccati equation is now a standard tool in solving optimal control, filtering and H^∞ control problems; see Anderson-Moore [10, 11], Caines [47], Corless-Frazho [60], Davis [66], Green-Limebeer [123], Ionescu-Oară-Weiss [135], Kailath-Sayed-Hassibi [143], Kwakernaak-Sivan [152], Lancaster-Rodman [153] and Zhou-Doyle-Glover [204] for further results on Riccati equations and a history of the subject area. Ionescu-Oară-Weiss [135] also presents Riccati techniques for computing the inner-outer factorization when the outer part is not square. Our method for obtaining the Riccati difference equation in Section 10.3 is classical, and due to Faurre [78]. In fact, our approach to the discrete time Riccati equation was highly motivated by the stochastic realization problem; see Caines [47], Faurre [77, 78], Foias-Frazho [81], Lindquist-Picci [158, 159, 160, 161, 162] and Ruckebusch [185, 186] for further results in this direction. Proposition 10.2.1 uses the Naimark representation to give an explicit state space formula for the invertible outer spectral factor. It is also noted that the Naimark dilation also plays a fundamental role in Foias-Frazho [81] and Ruckebusch's [185, 186] solution to the stochastic realization problem.

The results in Section 10.1 to 10.6 were taken from Foias-Frazho-Gohberg-Kaashoek [84]. Remark 10.1.6 was taken from Frazho-Kaashoek-Ran [100]. Computing the inner-outer factorization and contractive realizations using state space methods is now standard; see Bart-Gohberg-Kaashoek-Ran [28], Ionescu-Oară-Weiss [135], Lancaster-Rodman [153] and Zhou-Doyle-Glover [204] for further results in this direction. Our approach to solving the Darlington synthesis problem is well known. Darlington synthesis [64] started with a problem in network theory. For some further results on Darlington synthesis and its relation to network theory, see Belevitch [29], Carlin[52], and Hazony [128]. For some theoretical results on Darlington synthesis, see Arov [17, 18], Dewilde [70], Douglas-Helton [72], and Feintuch [79].

The results in Section 10.7 were taken from Section 3.5 in Caines [47]. It is noted that passing to limits on the Riccati difference equation is not the most efficient method of solving an algebraic Riccati equation. A more numerically efficient way to compute the solution to the algebraic Riccati equation is to use pencils and invariant subspaces; see Bart-Gohberg-Kaashoek-Ran [28] and Lancaster-Rodman [153]. Developing this theory would take us too far from the main emphasis of this book. Finally, it is noted that computers are becoming very fast, and for many problems one can simply pass to limits on the Riccati difference equation to compute the stabilizing solution to the algebraic Riccati equation.

Chapter 11

Kalman and Wiener Filtering

In this chapter we will present a brief introduction to Kalman and Wiener filtering. The main emphasis is to develop a connection between filtering theory and the Riccati equation.

Let us briefly review the notion of random variable. Consider a probability space $(\Omega, \mathcal{A}, P)$ where Ω is the universal set, $\mathcal{A}$ is a σ-algebra, and P is the probability measure. Recall that a *random variable* x is a measurable function mapping $\mathcal{A}$ into $\mathbb{C}$. The *mean* of x is given by

$$Ex = \int_{\Omega} x dP$$

where E denotes the expectation. The *variance* of x is determined by

$$\sigma_x^2 = E|x - \mu_x|^2 = \int_{\Omega} |x - \mu_x|^2 dP.$$

Moreover, σ_x is the *standard deviation* for x. Let f and g be two measurable functions of random variables x and y respectively. If x and y are independent, then it follows from probability theory that $Ef(x)g(y) = Ef(x)Eg(y)$.

We say that a sequence $\{y(n)\}_{-\infty}^{\infty}$ is a *stochastic process* or *random process* if each $y(n)$ is a random variable. Let $L^2(\Omega, \mathcal{A}, P)$ be the Hilbert space of all square integrable random variables with respect to probability measure dP. The inner product is given by $(x, y) = Ex\overline{y}$ for all x and y in $L^2(\Omega, \mathcal{A}, P)$. Throughout we assume that all random variables are in $L^2(\Omega, \mathcal{A}, P)$. Moreover, let $\{u(n)\}$ and $\{v(n)\}$ be random processes where each $u(n)$ and $v(n)$ is an element in $L^2(\Omega, \mathcal{A}, P)$. We say that the random processes $u(n)$ and $v(n)$ are *orthogonal*, if for all integers i and j, the random variable $u(i)$ is orthogonal to $v(j)$, or equivalently, the inner product $(u(i), v(j)) = Eu(i)\overline{v(j)} = 0$. In this case, let $y(n)$ be a random process given by $y(n) = u(n) + v(n)$. Then we have

$$(y(i), y(j)) = (u(i), u(j)) + (v(i), v(j)).$$

Finally, if $u(n)$ and $v(n)$ are independent random processes and $u(n)$ or $v(n)$ has zero mean for all n, then $u(n)$ and $v(n)$ are orthogonal. To verify this, simply observe that

$$(u(i), v(j)) = Eu(i)\overline{v(j)} = Eu(i)E\overline{v(j)} = 0$$

for all i and j. Therefore $u(n)$ and $v(n)$ are orthogonal.

11.1 Random Vectors

Recall that E denotes the expectation. In particular, Eg is the mean of the random variable g. Let $\mathcal{K}$ be the Hilbert space generated by the set of all random variables g such that $E|g|^2$ is finite. Throughout we always assume that all of our random variables are in $\mathcal{K}$. The inner product on $\mathcal{K}$ is determined by the expectation, that is, $(f, g) = Ef\overline{g}$ where f and g are in $\mathcal{K}$. We say that f is a *random vector* with values in $\mathbb{C}^k$ if f is a vector of the form $f = \begin{bmatrix} f_1 & f_2 & \cdots & f_k \end{bmatrix}^{tr}$ where $\{f_j\}_1^k$ are all random variables. (Recall that tr denotes the transpose.) In this case, Ef is the vector in $\mathbb{C}^k$ defined by $Ef = \begin{bmatrix} Ef_1 & Ef_2 & \cdots & Ef_k \end{bmatrix}^{tr}$. The *correlation matrix* $\mathbf{R}_f$ is the matrix on $\mathbb{C}^k$ defined by $\mathbf{R}_f = Eff^*$. To be precise,

$$\mathbf{R}_f = Eff^* = \begin{bmatrix} Ef_1\overline{f}_1 & Ef_1\overline{f}_2 & \cdots & Ef_1\overline{f}_k \\ Ef_2\overline{f}_1 & Ef_2\overline{f}_2 & \cdots & Ef_2\overline{f}_k \\ \vdots & \vdots & \ddots & \vdots \\ Ef_k\overline{f}_1 & Ef_k\overline{f}_2 & \cdots & Ef_k\overline{f}_k \end{bmatrix}. \tag{11.1.1}$$

Notice that the j-k entry of $\mathbf{R}_f$ is given by $(\mathbf{R}_f)_{jk} = Ef_j\overline{f}_k$. Finally, it is noted that $\mathbf{R}_f$ is the Gram matrix determined by $\{f_j\}_1^k$. The following result shows that $\mathbf{R}_f$ is positive.

Theorem 11.1.1. *Let $f = \begin{bmatrix} f_1 & f_2 & \cdots & f_k \end{bmatrix}^{tr}$ be a random vector with values in $\mathbb{C}^k$. Then $\mathbf{R}_f$ is a positive matrix on $\mathbb{C}^k$. Moreover, $\mathbf{R}_f$ is strictly positive ($\mathbf{R}_f > 0$) if and only if the random variables $\{f_j\}_1^k$ are linearly independent.*

Proof. Let $\alpha = \begin{bmatrix} \alpha_1 & \alpha_2 & \cdots & \alpha_k \end{bmatrix}^{tr}$ be any vector in $\mathbb{C}^k$. Then

$$(\mathbf{R}_f\alpha, \alpha) = (Eff^*\alpha, \alpha) = E(ff^*\alpha, \alpha) = E\|f^*\alpha\|^2 = E\left|\sum_{j=1}^k \overline{f}_j\alpha_j\right|^2 \geq 0. \tag{11.1.2}$$

Hence $(\mathbf{R}_f\alpha, \alpha) \geq 0$ for all α in $\mathbb{C}^k$. Therefore $\mathbf{R}_f$ is positive.

Equation (11.1.2) shows that

$$(\mathbf{R}_f\alpha, \alpha) = E\left|\sum_{j=1}^k \overline{f}_j\alpha_j\right|^2.$$

Recall that if g is a random variable, then $E|g|^2 = 0$ if and only if $g = 0$. So $(\mathbf{R}_f\alpha, \alpha) = 0$ if and only if $\sum_1^k f_j\overline{\alpha}_j = 0$. By definition the positive matrix $\mathbf{R}_f$ is strictly positive if $\alpha = 0$ is the only solution to $(\mathbf{R}_f\alpha, \alpha) = 0$. On the other hand, $\{f_j\}_1^k$ are linearly independent if the only solution to $\sum_1^k f_j\beta_j = 0$ is $\beta_\nu = 0$ for all integers $1 \leq \nu \leq k$. Therefore $\mathbf{R}_f$ is strictly positive if and only if $\{f_j\}_1^k$ are linearly independent. □

11.2 Least Squares Estimation of a Random Vector

Recall that $\mathcal{K}$ is the Hilbert space generated by the set of all random variables g such that $\|g\|^2 = E|g|^2$ is finite. Let $f = \begin{bmatrix} f_1 & f_2 & \cdots & f_k \end{bmatrix}^{tr}$ be a random vector with values in $\mathbb{C}^k$, that is, $\{f_j\}_1^k$ are all random variables in $\mathcal{K}$. Let $\mathcal{H}$ be a subspace in $\mathcal{K}$ and $P_\mathcal{H}$ be the orthogonal projection onto $\mathcal{H}$. Then the orthogonal projection of f onto $\mathcal{H}$ is the random vector $P_\mathcal{H}f$ with values in $\mathbb{C}^k$ defined by

$$P_\mathcal{H}f = \begin{bmatrix} P_\mathcal{H}f_1 \\ P_\mathcal{H}f_2 \\ \vdots \\ P_\mathcal{H}f_k \end{bmatrix}. \tag{11.2.1}$$

It is noted that $P_\mathcal{H}Af = AP_\mathcal{H}f$ where A is a constant matrix of the appropriate size. In this section we will present the classical least squares method to compute $P_\mathcal{H}f$ when $\mathcal{H}$ is a finite dimensional subspace.

As before, let $f = \begin{bmatrix} f_1 & f_2 & \cdots & f_k \end{bmatrix}^{tr}$ be a random vector with values in $\mathbb{C}^k$. Recall that $\mathbf{R}_f$ is the positive matrix on $\mathbb{C}^k$ defined by $\mathbf{R}_f = Eff^*$. Moreover, $\mathbf{R}_f$ is strictly positive if and only if the random variables $\{f_j\}_1^k$ are linearly independent in $\mathcal{K}$. Let $g = \begin{bmatrix} g_1 & g_2 & \cdots & g_m \end{bmatrix}^{tr}$ be a random vector with values in $\mathbb{C}^m$. Obviously, $\mathbf{R}_g$ is a positive matrix on $\mathbb{C}^m$. We say that $\mathcal{H}$ is the subspace generated by g if $\mathcal{H}$ equals the linear span of $\{g_j\}_1^m$. Let us use the notation $\bigvee g$ to denote the linear span of $\{g_j\}_1^m$. Clearly, $\mathcal{H}$ is a subspace of $\mathcal{K}$. The *correlation matrix* $\mathbf{R}_{fg}$ is the matrix from $\mathbb{C}^m$ into $\mathbb{C}^k$ defined by $\mathbf{R}_{fg} = Efg^*$. The j-ν entry of $\mathbf{R}_{fg}$ is $Ef_j\overline{g}_\nu$. By taking the adjoint, it follows that $\mathbf{R}_{fg}^* = \mathbf{R}_{gf}$. Notice that the subspace $\bigvee f$ generated by f is orthogonal to $\mathcal{H}$ if and only if $Ef_j\overline{g}_\nu = 0$ for all $j = 1, 2, \ldots, k$ and $\nu = 1, 2, \ldots, m$. In other words, $\bigvee f$ is orthogonal to $\bigvee g$ if and only if $\mathbf{R}_{fg} = 0$. Motivated by this we say that the random vectors f and g are *orthogonal* if $Efg^* = 0$. It is emphasized that Efg^* is in general a constant matrix and not an inner product. The following classical result provides a formula for computing $P_\mathcal{H}f$.

Theorem 11.2.1. *Let f be a random vector in $\mathbb{C}^k$, and g a random vector in $\mathbb{C}^m$. Let $P_\mathcal{H}$ be the orthogonal projection onto the subspace $\mathcal{H}$ generated by g, set $\widehat{f} = P_\mathcal{H}f$ and the error $\widetilde{f} = f - \widehat{f}$. Then the following holds.*

(i) *If M is any matrix from $\mathbb{C}^m$ into $\mathbb{C}^k$, then*

$$\mathbf{R}_{\widetilde{f}} = E(f - \widehat{f})(f - \widehat{f})^* \leq E(f - Mg)(f - Mg)^*. \tag{11.2.2}$$

(ii) *The equality $\mathbf{R}_{\widetilde{f}} = E(f - Mg)(f - Mg)^*$ holds if and only if $Mg = \widehat{f}$, or equivalently, $\mathbf{R}_{fg} = M\mathbf{R}_g$. In this case, the estimation error is determined by*

$$E(f - \widehat{f})(f - \widehat{f})^* = \mathbf{R}_f - M\mathbf{R}_g M^* = \mathbf{R}_f - \mathbf{R}_{fg}M^*. \tag{11.2.3}$$

(iii) *If $\mathbf{R}_g$ is invertible, or equivalently, the set $\{g_j\}_1^m$ is linearly independent, then*

$$P_{\mathcal{H}}f = \mathbf{R}_{fg}\mathbf{R}_g^{-1}g. \tag{11.2.4}$$

(iv) *If $\mathbf{R}_g$ is invertible, then the covariance for the error $\widetilde{f} = f - \widehat{f}$ is given by*

$$\mathbf{R}_{\widetilde{f}} = E(f - \widehat{f})(f - \widehat{f})^* = \mathbf{R}_f - E\widehat{f}\widehat{f}^* = \mathbf{R}_f - \mathbf{R}_{fg}\mathbf{R}_g^{-1}\mathbf{R}_{gf}. \tag{11.2.5}$$

Proof. Let M be any matrix mapping $\mathbb{C}^m$ into $\mathbb{C}^k$. Since the components of $\widehat{f}$ and Mg are both in $\mathcal{H}$, it follows that the components of $\widehat{f} - Mg$ are also in $\mathcal{H}$. According to the projection theorem, each component of $f - \widehat{f}$ is orthogonal to $\mathcal{H}$. In other words, $f - \widehat{f}$ is orthogonal to $\widehat{f} - Mg$. Hence

$$\begin{aligned}
E(f - Mg)(f - Mg)^* &= E\left(f - \widehat{f} + \widehat{f} - Mg\right)\left(f - \widehat{f} + \widehat{f} - Mg\right)^* \\
&= E(f - \widehat{f})(f - \widehat{f})^* + E(f - \widehat{f})(\widehat{f} - Mg)^* \\
&\quad + E(\widehat{f} - Mg)(f - \widehat{f})^* + E(\widehat{f} - Mg)(\widehat{f} - Mg)^* \\
&= E(f - \widehat{f})(f - \widehat{f})^* + E(\widehat{f} - Mg)(\widehat{f} - Mg)^* \\
&= \mathbf{R}_{\widetilde{f}} + E(\widehat{f} - Mg)(\widehat{f} - Mg)^* \geq \mathbf{R}_{\widetilde{f}}.
\end{aligned}$$

The third equality follows from the fact that $f - \widehat{f}$ is orthogonal to $\mathcal{H}$, that is, $f - \widehat{f}$ is orthogonal to any vector whose components are in $\mathcal{H}$. This readily yields the inequality in (11.2.2).

To verify Part *(ii)* notice that the previous calculation also shows that

$$E(f - Mg)(f - Mg)^* = \mathbf{R}_{\widetilde{f}} + E(\widehat{f} - Mg)(\widehat{f} - Mg)^*.$$

So we have $E(f - Mg)(f - Mg)^* = \mathbf{R}_{\widetilde{f}}$ if and only if $E(\widehat{f} - Mg)(\widehat{f} - Mg)^*$ is zero, or equivalently, $\widehat{f} = Mg$. By the projection theorem, $\widehat{f} = Mg$ if and only if each component of $f - Mg$ is orthogonal to $\mathcal{H}$, or equivalently, $E(f - Mg)g^* = 0$. Therefore $E(f - Mg)(f - Mg)^* = \mathbf{R}_{\widetilde{f}}$ if and only if $\mathbf{R}_{fg} = M\mathbf{R}_g$.

Now let us establish the error formulas in (11.2.3). By the projection theorem $\widehat{f}$ is orthogonal to $\widetilde{f} = f - \widehat{f}$. Since $f = \widehat{f} + \widetilde{f}$, we obtain

$$\mathbf{R}_f = Eff^* = E(\widehat{f} + \widetilde{f})(\widehat{f} + \widetilde{f})^* = E\widehat{f}\widehat{f}^* + E\widetilde{f}\widetilde{f}^* = E\widehat{f}\widehat{f}^* + \mathbf{R}_{\widetilde{f}}.$$

Hence $\mathbf{R}_f = E\widehat{f}\widehat{f}^* + \mathbf{R}_{\widetilde{f}}$. Using $\widehat{f} = Mg$ and $\mathbf{R}_{fg} = M\mathbf{R}_g$, we have

$$\mathbf{R}_{\widetilde{f}} = \mathbf{R}_f - E\widehat{f}\widehat{f}^* = \mathbf{R}_f - MEgg^*M^* = \mathbf{R}_f - M\mathbf{R}_gM^* = \mathbf{R}_f - \mathbf{R}_{fg}M^*.$$

Therefore (11.2.3) holds.

Clearly, Part *(iii)* follows by inverting $\mathbf{R}_g$ in Part *(ii)*. Let us directly prove Part *(iii)* from the projection theorem. Since $\mathcal{H}$ equals the span of $\{g_j\}_1^m$, equation (11.2.1) shows that $\widehat{f} = P_{\mathcal{H}}f = Ag$ where A is a matrix from $\mathbb{C}^m$ into $\mathbb{C}^k$. According to the projection theorem, $f - Ag$ is orthogonal to g, that is, the subspace generated by the span of all the components of $f - Ag$ is orthogonal to the subspace generated by the span of all the components of g. In other words, the matrix $E(f - Ag)g^* = 0$. Hence $\mathbf{R}_{fg} = AEgg^* = A\mathbf{R}_g$. This readily implies that $A = \mathbf{R}_{fg}\mathbf{R}_g^{-1}$. Therefore $\widehat{f} = Ag = \mathbf{R}_{fg}\mathbf{R}_g^{-1}g$ is given by (11.2.4).

To complete the proof it remains to establish the error formulas in (11.2.5). By employing $M = \mathbf{R}_{fg}\mathbf{R}_g^{-1}$ and $\mathbf{R}_{gf} = \mathbf{R}_{fg}^*$ in (11.2.3), we obtain

$$\mathbf{R}_{\widetilde{f}} = \mathbf{R}_f - M\mathbf{R}_gM^* = \mathbf{R}_f - \mathbf{R}_{fg}\mathbf{R}_g^{-1}\mathbf{R}_g\mathbf{R}_g^{-1}\mathbf{R}_{gf} = \mathbf{R}_f - \mathbf{R}_{fg}\mathbf{R}_g^{-1}\mathbf{R}_{gf}.$$

This yields (11.2.5). □

If f in $\mathbb{C}^k$ and g in $\mathbb{C}^m$ are random vectors, then $f \bigvee g$ is the subspace of $\mathcal{K}$ generated by the span of all the components of both f and g. If $\mathcal{H}_1$ and $\mathcal{H}_2$ are two subspaces of random variables, then $\mathcal{H}_1 \bigvee \mathcal{H}_2$ is the subspace formed by the closed linear span of $\mathcal{H}_1$ and $\mathcal{H}_2$. If $\mathcal{M}$ is a subspace of $\mathcal{K}$, then $P_{\mathcal{M}}$ denotes the orthogonal projection onto $\mathcal{M}$. Recall that the notation $\mathcal{F} \oplus \mathcal{G}$ means that $\mathcal{F}$ and $\mathcal{G}$ are two orthogonal subspaces and $\mathcal{F} \oplus \mathcal{G} = \mathcal{F} \bigvee \mathcal{G}$. In this case, $P_{\mathcal{F}\oplus\mathcal{G}} = P_{\mathcal{F}} + P_{\mathcal{G}}$. The following result will be used in our derivation of the Kalman filter.

Lemma 11.2.2. *Let f be a random vector with values in $\mathbb{C}^k$. Let $\mathcal{M}$ be a subspace of random variables, and $\mathcal{Y}$ be the subspace generated by the random vector y in $\mathbb{C}^m$. Set $\mathcal{H} = \mathcal{M} \bigvee \mathcal{Y}$. Then $\mathcal{H} = \mathcal{M} \oplus \mathcal{E}$ where $\mathcal{E}$ is the subspace generated by the vector $\varphi = y - P_{\mathcal{M}}y$. Moreover, if $\mathbf{R}_\varphi$ is invertible, then*

$$P_{\mathcal{H}}f = P_{\mathcal{M}}f + \mathbf{R}_{f\varphi}\mathbf{R}_\varphi^{-1}\varphi. \tag{11.2.6}$$

This decomposition is orthogonal, that is, $P_{\mathcal{M}}f$ is orthogonal to $\mathbf{R}_{f\varphi}\mathbf{R}_\varphi^{-1}\varphi = P_{\mathcal{E}}f$. Finally, the error covariance for $f - P_{\mathcal{H}}f$ is given by

$$E(f - P_{\mathcal{H}}f)(f - P_{\mathcal{H}}f)^* = E(f - P_{\mathcal{M}}f)(f - P_{\mathcal{M}}f)^* - \mathbf{R}_{f\varphi}\mathbf{R}_\varphi^{-1}\mathbf{R}_{f\varphi}^*. \tag{11.2.7}$$

Proof. By the projection theorem, the vector $\varphi = y - P_{\mathcal{M}}y$ is orthogonal to $\mathcal{M}$. So the subspace $\mathcal{E}$ generated by φ is orthogonal to $\mathcal{M}$. Since the components of φ are contained in $\mathcal{M}\bigvee\mathcal{Y}$, we have $\mathcal{M}\oplus\mathcal{E}\subset\mathcal{H}$. Using $y = P_{\mathcal{M}}y+\varphi$, it follows that $\mathcal{Y}$ is contained in $\mathcal{M}\oplus\mathcal{E}$. Therefore $\{\mathcal{M},\mathcal{Y}\}$ and $\{\mathcal{M},\mathcal{E}\}$ span the same space. In particular, $\mathcal{H} = \mathcal{M}\oplus\mathcal{E}$. This readily implies that the orthogonal projection $P_{\mathcal{H}} = P_{\mathcal{M}} + P_{\mathcal{E}}$. Because φ generates the subspace $\mathcal{E}$, Part (iii) of Theorem 11.2.1 shows that if $\mathbf{R}_\varphi$ is invertible, then $P_{\mathcal{E}}f = \mathbf{R}_{f\varphi}\mathbf{R}_\varphi^{-1}\varphi$. Thus

$$P_{\mathcal{H}}f = P_{\mathcal{M}}f + P_{\mathcal{E}}f = P_{\mathcal{M}}f + \mathbf{R}_{f\varphi}\mathbf{R}_\varphi^{-1}\varphi.$$

Hence (11.2.6) holds. Since $P_{\mathcal{E}}f = \mathbf{R}_{f\varphi}\mathbf{R}_\varphi^{-1}\varphi$ and $\mathcal{M}$ is orthogonal to $\mathcal{E}$, the random vector $P_{\mathcal{M}}f$ is orthogonal to $\mathbf{R}_{f\varphi}\mathbf{R}_\varphi^{-1}\varphi$.

To verify that (11.2.7) holds, let us first establish the following:

$$\begin{aligned} Ef(P_{\mathcal{E}}f)^* &= E(P_{\mathcal{E}}f)(P_{\mathcal{E}}f)^* = \mathbf{R}_{f\varphi}\mathbf{R}_\varphi^{-1}\mathbf{R}_{f\varphi}^*,\\ E(P_{\mathcal{E}}f)f^* &= \mathbf{R}_{f\varphi}\mathbf{R}_\varphi^{-1}\mathbf{R}_{f\varphi}^*. \end{aligned} \tag{11.2.8}$$

To show this, according to the projection theorem $f - P_{\mathcal{E}}f$ is orthogonal to $P_{\mathcal{E}}f$. Using this with $P_{\mathcal{E}}f = \mathbf{R}_{f\varphi}\mathbf{R}_\varphi^{-1}\varphi$, we obtain

$$\begin{aligned} Ef(P_{\mathcal{E}}f)^* &= E(f - P_{\mathcal{E}}f + P_{\mathcal{E}}f)(P_{\mathcal{E}}f)^* = E(P_{\mathcal{E}}f)(P_{\mathcal{E}}f)^*\\ &= \mathbf{R}_{f\varphi}\mathbf{R}_\varphi^{-1}E\varphi\varphi^*\mathbf{R}_\varphi^{-1}\mathbf{R}_{f\varphi}^* = \mathbf{R}_{f\varphi}\mathbf{R}_\varphi^{-1}\mathbf{R}_{f\varphi}^*. \end{aligned}$$

So the first equation in (11.2.8) holds. The second part of (11.2.8) follows by taking the adjoint. Recall that $\mathcal{M}$ is orthogonal to $\mathcal{E}$. This with (11.2.8) implies that

$$\begin{aligned} E(f - P_{\mathcal{H}}f)(f - P_{\mathcal{H}}f)^* &= E((f - P_{\mathcal{M}}f) - P_{\mathcal{E}}f)((f - P_{\mathcal{M}}f) - P_{\mathcal{E}}f)^*\\ &= E(f - P_{\mathcal{M}}f)(f - P_{\mathcal{M}}f)^* - Ef(P_{\mathcal{E}}f)^*\\ &\quad - E(P_{\mathcal{E}}f)f^* + E(P_{\mathcal{E}}f)(P_{\mathcal{E}}f)^*\\ &= E(f - P_{\mathcal{M}}f)(f - P_{\mathcal{M}}f)^* - \mathbf{R}_{f\varphi}\mathbf{R}_\varphi^{-1}\mathbf{R}_{f\varphi}^*. \end{aligned}$$

This yields (11.2.7). □

11.3 Time Varying State Space Systems

In this section we will introduce discrete time varying systems which play a fundamental role in Kalman filtering. Consider the *time varying state space* system

$$x(n+1) = A(n)x(n) + B(n)u(n) \quad \text{and} \quad y = C(n)x(n) + D(n)v(n). \tag{11.3.1}$$

For every integer $n \geq 0$, the operator $A(n)$ is on $\mathcal{X}$ and $B(n)$ is an operator mapping $\mathcal{U}$ into $\mathcal{X}$ while $C(n)$ is an operator from $\mathcal{X}$ into $\mathcal{Y}$ and $D(n)$ is an operator mapping $\mathcal{V}$ into $\mathcal{Y}$. The state $x(n)$ is in $\mathcal{X}$, the input $u(n)$ is in $\mathcal{U}$, the input $v(n)$ is in $\mathcal{V}$ and the output $y(n)$ is in $\mathcal{Y}$ for all integers n. The initial condition $x(0) = x_0$. Throughout $\Psi(n,\nu)$ is the *state transition matrix* for $A(n)$ defined by

$$\begin{aligned}\Psi(n,\nu) &= A(n)A(n-1)\cdots A(\nu+1) && \text{if } n>\nu\\ &= I && \text{if } n=\nu.\end{aligned} \tag{11.3.2}$$

For example, $\Psi(4,1) = A(4)A(3)A(2)$ and $\Psi(3,-1) = A(3)A(2)A(1)A(0)$, while $\Psi(3,3) = I$. Finally, it is noted that $\Psi(n+1,\nu) = A(n+1)\Psi(n,\nu)$.

To obtain a solution to the time varying state space system in (11.3.1), observe that the state $x(1) = A(0)x_0 + B(0)u(0)$. By recursively solving for the state $x(n)$ in (11.3.1), we obtain

$$\begin{aligned}x(2) &= A(1)x(1) + B(1)u(1)\\ &= A(1)(A(0)x_0 + B(0)u(0)) + B(1)u(1)\\ &= \Psi(1,-1)x_0 + \Psi(1,0)B(0)u(0) + \Psi(1,1)B(1)u(1);\end{aligned}$$

$$\begin{aligned}x(3) &= A(2)x(2) + B(2)u(2)\\ &= A(2)A(1)A(0)x_0 + A(2)A(1)B(0)u(0) + A(2)B(1)u(1) + B(2)u(2)\\ &= \Psi(2,-1)x_0 + \Psi(2,0)B(0)u(0) + \Psi(2,1)B(1)u(1) + \Psi(2,2)B(2)u(2).\end{aligned}$$

By recursively solving for the state $x(n)$, it follows that the solution to the time varying state space system in (11.3.1) is given by

$$x(n) = \Psi(n-1,-1)x_0 + \sum_{j=0}^{n-1} \Psi(n-1,j)B(j)u(j), \tag{11.3.3}$$

$$y(n) = C(n)\Psi(n-1,-1)x_0 + \sum_{j=0}^{n-1} C(n)\Psi(n-1,j)B(j)u(j) + D(n)v(n).$$

Finally, it is noted that time varying systems play a basic role in linear systems.

11.4 The Kalman Filter

In this section we will present the discrete time Kalman filter. Recall that $w(n)$ is a discrete time random process with values in $\mathbb{C}^m$ if $w(n)$ is a random vector in $\mathbb{C}^m$ for all integers n. We say that $w(n)$ is a *mean zero process* if $Ew(n) = 0$ for all n. Finally, $w(n)$ is a *white noise process* if $w(n)$ is a mean zero random process and

$$Ew(j)w(k)^* = \delta_{j-k}I. \tag{11.4.1}$$

Here δ_j is the Kronecker delta. By definition, $\delta_0 = 1$ and $\delta_j = 0$ if $j \neq 0$.

Consider the time varying state space system given by

$$x(n+1) = Ax(n) + Bu(n) \qquad \text{and} \qquad y(n) = Cx(n) + Dv(n). \tag{11.4.2}$$

Here A is an operator on $\mathcal{X}$ and B maps $\mathcal{U}$ into $\mathcal{X}$ while C maps $\mathcal{X}$ into $\mathcal{Y}$ and D maps $\mathcal{V}$ into $\mathcal{Y}$ where $\mathcal{X}$, $\mathcal{U}$, $\mathcal{Y}$ and $\mathcal{V}$ are all $\mathbb{C}^k$ spaces of the appropriate size. It is emphasized that $\{A, B, C, D\}$ can be time varying matrices, that is, $A = A(n)$, $B = B(n)$, $C = C(n)$ and $D = D(n)$ for all integers n. However, the index n is suppressed in our development. The initial condition x_0 is a random vector with values in $\mathcal{X}$. The disturbance $u(n)$ and $v(n)$ are independent white noise random process. Moreover, we assume that the initial condition x_0, $u(n)$ and $v(m)$ are all independent random vectors for all integers n and m. In particular, this implies that x_0, $u(n)$ and $v(m)$ are orthogonal for all integers n and m. We also assume that the initial condition $\widehat{x}_0$ for the Kalman filter and the initial covariance $Q_0 = Ex(0)x(0)^*$ for the discrete time Riccati equation are known. Finally, it is noted that $u(n)$ is called the *disturbance* or *state noise*, while $v(n)$ is the *measurement noise*.

The Kalman filtering problem is to compute the best estimate $\widehat{x}(k)$ of the state $x(k)$ given the past output $\{y(j)\}_0^{k-1}$. The Kalman filter is an optimal state estimator. To be explicit, let $\mathcal{M}_n$ be the subspace generated by the random vectors $\{y(j)\}_0^n$, that is, $\mathcal{M}_n = \bigvee_{j=0}^n y(j)$. Let $P_{\mathcal{M}_n}$ be the orthogonal projection onto $\mathcal{M}_n$ for all integers $n \geq 0$. Then the best estimate $\widehat{x}(k)$ of the state $x(k)$ is given by the orthogonal projection $\widehat{x}(k) = P_{\mathcal{M}_{k-1}} x(k)$. Finally, we assume that $y(-1) = 0$, or equivalently, $\mathcal{M}_{-1} = \{0\}$. (If x_0, $u(n)$ and $v(n)$ are all jointly Gaussian, then $P_{\mathcal{M}_{k-1}} x(k)$ equals the conditional expectation $E(x(k)|y(0), y(1), \ldots, y(k-1))$.) The following result known as the Kalman filter yields a recursive algorithm to compute the optimal state estimate $\widehat{x}(k)$.

Theorem 11.4.1. *Consider the state space system*

$$x(n+1) = Ax(n) + Bu(n) \quad \textit{and} \quad y(n) = Cx(n) + Dv(n) \tag{11.4.3}$$

where $u(n)$ and $v(n)$ are independent white noise random processes, which are independent of the initial condition $x(0)$. Then the optimal estimate $\widehat{x}(k) = P_{\mathcal{M}_{k-1}} x(k)$ of the state $x(k)$ given the past $\{y(j)\}_0^{k-1}$ is recursively computed by

$$\widehat{x}(n+1) = A\widehat{x}(n) + \Lambda_n (y(n) - C\widehat{x}(n)), \tag{11.4.4}$$

$$\Lambda_n = AQ_nC^* \left(CQ_nC^* + DD^*\right)^{-1}. \tag{11.4.5}$$

The state covariance error $Q_k = E(x(k) - \widehat{x}(k))(x(k) - \widehat{x}(k))^*$ is recursively computed by solving the Riccati difference equation

$$Q_{n+1} = AQ_nA^* + BB^* - AQ_nC^* \left(CQ_nC^* + DD^*\right)^{-1} CQ_nA^*, \tag{11.4.6}$$

subject to the initial condition $Q_0 = Ex(0)x(0)^*$.

Remark 11.4.2. It is emphasized that when implementing the Kalman filter we always assume that the inverse of $CQ_nC^* + DD^*$ exists for all integers $n \geq 0$. If DD^* is invertible, then the positivity of Q_n implies that $CQ_nC^* + DD^*$ is strictly positive, and thus, invertible for all n.

Proof of the Kalman filtering Theorem 11.4.1. Let us give a proof of the Kalman filter by implementing Lemma 11.2.2 with $\mathcal{H} = \mathcal{M}_n = \mathcal{M}_{n-1} \bigvee \mathcal{Y}$. In our setting $\mathcal{Y}$ is the span of $y(n)$, the subspace $\mathcal{M} = \mathcal{M}_{n-1}$ and $\varphi = \varphi(n) = y(n) - P_{\mathcal{M}_{n-1}}y(n)$. Recall that the solution to the difference equation in (11.4.3) is given by

$$x(n) = \Psi(n-1,-1)x(0) + \sum_{j=0}^{n-1} \Psi(n-1,j)B(j)u(j), \tag{11.4.7}$$

$$y(n) = C(n)\Psi(n-1,-1)x(0) + \sum_{j=0}^{n-1} C(n)\Psi(n-1,j)B(j)u(j) + D(n)v(n).$$

Here $\Psi(n,\nu) = A(n)A(n-1)\cdots A(\nu+1)$ and $\Psi(k,k) = I$. This readily shows that

$$\mathcal{M}_n = \bigvee_{k=0}^{n} y(k) \subset \bigvee\{x(0), u(0), u(1), \ldots, u(n-1), v(0), v(1), \ldots, v(n)\}. \tag{11.4.8}$$

Because $u(k)$ and $v(k)$ are independent white noise processes and orthogonal to $x(0)$, the random vector $v(n)$ is orthogonal to $\mathcal{M}_{n-1}$. In particular, $P_{\mathcal{M}_{n-1}}v(n) = 0$. Recall that the optimal state estimate is given by $\widehat{x}(n) = P_{\mathcal{M}_{n-1}}x(n)$. Using this along with the fact that C and D are not random, we have

$$\begin{aligned}\varphi(n) &= y(n) - P_{\mathcal{M}_{n-1}}y(n)\\ &= y(n) - P_{\mathcal{M}_{n-1}}(Cx(n) + Dv(n))\\ &= y(n) - CP_{\mathcal{M}_{n-1}}x(n)\\ &= y(n) - C\widehat{x}(n).\end{aligned}$$

Hence $\varphi(n) = y(n) - C\widehat{x}(n)$. By definition the *state estimation error* $\widetilde{x}(n) = x(n) - \widehat{x}(n)$. Since $y(n) = Cx(n) + Dv(n)$, we have $\varphi(n) = C\widetilde{x}(n) + Dv(n)$. This yields the following two useful formulas,

$$\varphi(n) = y(n) - C\widehat{x}(n) = C\widetilde{x}(n) + Dv(n). \tag{11.4.9}$$

By consulting (11.4.7) and (11.4.8), we see that $v(n)$ is orthogonal to both $x(n)$ and $\mathcal{M}_{n-1}$. Hence $v(n)$ is orthogonal to $\widetilde{x}(n) = x(n) - \widehat{x}(n)$. This and $\varphi(n) = C\widetilde{x}(n) + Dv(n)$ implies that

$$\begin{aligned}E\varphi(n)\varphi(n)^* &= E\left(C\widetilde{x}(n) + Dv(n)\right)\left(C\widetilde{x}(n) + Dv(n)\right)^*\\ &= CE\widetilde{x}(n)\widetilde{x}(n)^*C^* + DD^*.\end{aligned}$$

By definition $Q_n = E\widetilde{x}(n)\widetilde{x}(n)^*$ is the error covariance. Therefore

$$\mathbf{R}_{\varphi(n)} = E\varphi(n)\varphi(n)^* = CQ_nC^* + DD^*. \tag{11.4.10}$$

Equation (11.4.8) shows that $u(n)$ is orthogonal to $\mathcal{M}_{n-1}$. In other words, $P_{\mathcal{M}_{n-1}}u(n) = 0$. By employing (11.2.6) in Lemma 11.2.2 with $\varphi = \varphi(n)$ and

$y = y(n)$ and $\mathcal{M}_n = \mathcal{M}_{n-1} \bigvee y(n)$, we obtain

$$\begin{aligned}\widehat{x}(n+1) &= P_{\mathcal{M}_n} x(n+1) = P_{\mathcal{M}_{n-1}} x(n+1) + \mathbf{R}_{x(n+1)\varphi(n)} \mathbf{R}_{\varphi(n)}^{-1} \varphi(n) \\ &= P_{\mathcal{M}_{n-1}} (Ax(n) + Bu(n)) + \mathbf{R}_{x(n+1)\varphi(n)} \mathbf{R}_{\varphi(n)}^{-1} \varphi(n) \\ &= A\widehat{x}(n) + \mathbf{R}_{x(n+1)\varphi(n)} \mathbf{R}_{\varphi(n)}^{-1} \varphi(n). \end{aligned} \tag{11.4.11}$$

We need an expression for $\mathbf{R}_{x(n+1)\varphi(n)}$. Since $\widehat{x}(n)$ is contained in $\mathcal{M}_{n-1}$, the random vector $\varphi(n) = y(n) - C\widehat{x}(n)$ is contained in $\mathcal{M}_n$. Hence $\varphi(n)$ is orthogonal to $u(n)$; see (11.4.8). Moreover, $v(n)$ is orthogonal to $x(n)$; see (11.4.7). Using $\varphi(n) = C\widetilde{x}(n) + Dv(n)$, we have

$$\begin{aligned} Ex(n+1)\varphi(n)^* &= E(Ax(n) + Bu(n))\varphi(n)^* = AEx(n)\varphi(n)^* \\ &= AEx(n)\left(C\widetilde{x}(n) + Dv(n)\right)^* = AEx(n)\widetilde{x}(n)^* C^* \\ &= AE(\widehat{x}(n) + \widetilde{x}(n))\widetilde{x}(n)^* C^* = AE\widetilde{x}(n)\widetilde{x}(n)^* C^* \\ &= AQ_n C^*. \end{aligned}$$

The second from the last equality follows from the fact that $\widehat{x}(n)$ is orthogonal to $\widetilde{x}(n)$. The previous calculation yields the result

$$\mathbf{R}_{x(n+1)\varphi(n)} = Ex(n+1)\varphi(n)^* = AQ_n C^*. \tag{11.4.12}$$

Substituting $\mathbf{R}_{x(n+1)\varphi(n)} = AQ_n C^*$ and the expression for $\mathbf{R}_{\varphi(n)}$ in (11.4.10) into (11.4.11) yields

$$\widehat{x}(n+1) = A\widehat{x}(n) + AQ_n C^*(CQ_n^* C^* + DD^*)^{-1} \varphi(n). \tag{11.4.13}$$

Finally, using $\varphi(n) = y(n) - C\widehat{x}(n)$ gives the state space formula for $\widehat{x}(n)$ in (11.4.4).

Now let us use equation (11.2.7) in Lemma 11.2.2 to derive the discrete time Riccati equation in (11.4.6). Recall that $u(n)$ is orthogonal to $\mathcal{M}_{n-1}$. Using $P_{\mathcal{M}_{n-1}} u(n) = 0$ along with the optimal state estimate $\widehat{x}(n) = P_{\mathcal{M}_{n-1}} x(n)$, we obtain

$$\begin{aligned} x(n+1) - P_{\mathcal{M}_{n-1}} x(n+1) &= Ax(n) + Bu(n) - P_{\mathcal{M}_{n-1}} (Ax(n) + Bu(n)) \\ &= Ax(n) + Bu(n) - A\widehat{x}(n) = A\widetilde{x}(n) + Bu(n). \end{aligned}$$

This readily implies that

$$x(n+1) - P_{\mathcal{M}_{n-1}} x(n+1) = A\widetilde{x}(n) + Bu(n). \tag{11.4.14}$$

By virtue of (11.4.7) we see that $u(n)$ is orthogonal to $x(n)$. Since the optimal estimate $\widehat{x}(n) = P_{\mathcal{M}_{n-1}} x(n)$ is a vector in $\mathcal{M}_{n-1}$, the random vector $u(n)$ is also orthogonal to $\widehat{x}(n)$; see (11.4.8). Hence $u(n)$ is orthogonal to the error $\widetilde{x}(n) =$

$x(n) - \widehat{x}(n)$. Using this fact in (11.4.14) along with $E\widetilde{x}(n)\widetilde{x}(n)^* = Q_n$, we arrive at

$$E\left(x(n+1) - P_{\mathcal{M}_{n-1}}x(n+1)\right)\left(x(n+1) - P_{\mathcal{M}_{n-1}}x(n+1)\right)^* = AQ_nA^* + BB^*.$$

Recall that $Ex(n+1)\varphi(n)^* = AQ_nC^*$; see (11.4.12). Finally, by employing equation (11.2.7) in Lemma 11.2.2 with the expression for $\mathbf{R}_{\varphi(n)}$ in (11.4.10), we have

$$\begin{aligned} Q_{n+1} &= E\left(x(n+1) - P_{\mathcal{M}_n}x(n+1)\right)\left(x(n+1) - P_{\mathcal{M}_n}x(n+1)\right)^* \\ &= E\left(x(n+1) - P_{\mathcal{M}_{n-1}}x(n+1)\right)\left(x(n+1) - P_{\mathcal{M}_{n-1}}x(n+1)\right)^* \\ &\quad - Ex(n+1)\varphi(n)^*\mathbf{R}_{\varphi(n)}^{-1}(Ex(n+1)\varphi(n)^*)^* \\ &= AQ_nA^* + BB^* - AQ_nC^*\left(CQ_nC^* + DD^*\right)^{-1}CQ_nA^*. \end{aligned}$$

This is the Riccati difference equation in (11.4.6). To obtain the initial condition, recall that $\mathcal{M}_{-1} = 0$, that is, $y(-1) = 0$. Hence $\widetilde{x}(0) = x(0) - P_{\mathcal{M}_{-1}}x(0) = x(0)$. Thus $Q_0 = E\widetilde{x}(0)\widetilde{x}(0)^* = Ex(0)x(0)^*$. □

Remark 11.4.3. As in Theorem 11.4.1, consider the state space system determined by

$$x(n+1) = Ax(n) + Bu(n) \quad \text{and} \quad y(n) = Cx(n) + Dv(n) \tag{11.4.15}$$

where $u(n)$ and $v(n)$ are independent white noise random processes, which are independent of the initial condition $x(0)$. The process $\varphi(n) = y(n) - C\widehat{x}(n)$ is called the *innovations process* for the Kalman filter. The innovations process is orthogonal, that is, $E\varphi(n)\varphi(k)^* = 0$ for all integers $n \neq k$. To see this, without loss of generality assume that $k < n$. In this case, $\varphi(k) = y(k) - P_{\mathcal{M}_{k-1}}y(k)$ is in $\mathcal{M}_k \subseteq \mathcal{M}_{n-1}$. According to the projection theorem, $\varphi(n) = y(n) - P_{\mathcal{M}_{n-1}}y(n)$ is orthogonal to $\mathcal{M}_{n-1}$. Hence $\varphi(n)$ is orthogonal to $\varphi(k)$, which proves our claim. So the innovations process $\{\varphi(n)\}_0^\infty$ can be viewed as a white noise process with variance $CQ_nC^* + DD^*$. Moreover, the Kalman filter admits an orthogonal decomposition of the form

$$\widehat{x}(n+1) = A\widehat{x}(n) + \Lambda_n\varphi(n)$$

where the optimal state estimate $\widehat{x}(n) \in \mathcal{M}_{n-1}$ and the innovations $\varphi(n) \in \mathcal{M}_{n-1}^\perp$ are orthogonal. Finally, it is noted that the Kalman filter in (11.4.4) can be used to recursively compute the innovations $\varphi(n)$ for the process $y(n)$. For further results on the innovations approach to Kalman filtering and stochastic processes see Kailath-Sayed-Hassibi [143].

Let us complete this section with the following result.

Proposition 11.4.4. *Let Q_n be the solution for the Riccati difference equation in (11.4.6) associated with $\{A, B, C, D\}$. Then Q_n is also a solution to the Riccati difference equation*

$$\begin{aligned} Q_{n+1} &= (A - \Lambda_nC)Q_n(A - \Lambda_nC)^* + BB^* + \Lambda_nDD^*\Lambda_n^*, \\ \Lambda_n &= AQ_nC^*\left(CQ_nC^* + DD^*\right)^{-1}. \end{aligned} \tag{11.4.16}$$

In particular, if the initial condition Q_0 is positive, then this also shows that Q_n is positive for all integers $n \geq 0$.

Proof. By consulting the form of the Riccati difference equation in (11.4.6), we obtain

$$\begin{aligned}
Q_{n+1} &= AQ_nA^* - AQ_nC^* \left(CQ_nC^* + DD^*\right)^{-1} CQ_nA^* + BB^* \\
&= AQ_nA^* - \Lambda_n CQ_nA^* + BB^* \\
&= (A - \Lambda_n C)Q_nA^* + BB^* \\
&= (A - \Lambda_n C)Q_n(A - \Lambda_n C)^* + (A - \Lambda_n C)Q_nC^*\Lambda_n^* + BB^* \\
&= (A - \Lambda_n C)Q_n(A - \Lambda_n C)^* + AQ_nC^*\Lambda_n^* \\
&\quad - \Lambda_n(CQ_nC^* + DD^*)\Lambda_n^* + \Lambda_n DD^*\Lambda_n^* + BB^* \\
&= (A - \Lambda_n C)Q_n(A - \Lambda_n C)^* + \Lambda_n DD^*\Lambda_n^* + BB^*.
\end{aligned}$$

This yields (11.4.16). □

11.5 The Steady State Kalman Filter

In this section we will present the steady state Kalman filter. Throughout we assume that the system in (11.4.2) is time invariant, that is,

$$x(n+1) = Ax(n) + Bu(n) \quad \text{and} \quad y(n) = Cx(n) + Dv(n) \tag{11.5.1}$$

where A is an operator on $\mathcal{X}$ and B maps $\mathcal{U}$ into $\mathcal{X}$ while C maps $\mathcal{X}$ into $\mathcal{Y}$ and D maps $\mathcal{V}$ into $\mathcal{Y}$. In other words, the operators $\{A, B, C, D\}$ are all fixed and do not depend upon n. We also assume that the operator D is onto $\mathcal{Y}$, or equivalently, DD^* is invertible. As before, let Q_n be the solution to the Riccati difference equation in (11.4.6) associated with $\{A, B, C, D\}$, that is,

$$Q_{n+1} = AQ_nA^* + BB^* - AQ_nC^* \left(CQ_nC^* + DD^*\right)^{-1} CQ_nA^*. \tag{11.5.2}$$

According to Proposition 11.4.4, this Riccati difference equation can also be rewritten as

$$\begin{aligned}
Q_{n+1} &= (A - \Lambda_n C)\, Q_n \,(A - \Lambda_n C)^* + BB^* + \Lambda_n DD^*\Lambda_n^*, \\
\Lambda_n &= AQ_nC^* \left(CQ_nC^* + DD^*\right)^{-1}.
\end{aligned} \tag{11.5.3}$$

Moreover, assume that the initial condition Q_0 is positive. The form for the Riccati difference equation in (11.5.3) shows that Q_n is positive for all integers $n \geq 0$. Since DD^* is invertible and Q_n is positive, $(CQ_nC^* + DD^*)$ is strictly positive. In particular, $(CQ_nC^* + DD^*)$ is invertible. Hence this Riccati difference equation is well defined for all n. The following result shows that if the initial condition $Q_0 = 0$, then the solution Q_n is increasing. Moreover, by replacing A by A^*, B by C^* and setting $R_{11} = BB^*$, $R_{22} = DD^*$ and $R_{21} = 0$ in Theorem 10.7.1 with the fact that $\{A, B\}$ is controllable if and only if $\{A, BB^*\}$ is controllable, we obtain the following result.

Theorem 11.5.1. *Consider the time invariant system $\{A, B, C, D\}$ where D is onto. Let Q_n be the solution for the Riccati difference equation in* (11.5.2) *where the initial condition $Q_0 = 0$. Then the following holds.*

(i) *The solution $\{Q_n\}_0^\infty$ forms an increasing sequence of positive operators. To be precise, $Q_n \leq Q_{n+1}$ for all integers $n \geq 0$.*

(ii) *If the pair $\{C, A\}$ is observable, then Q_n converges to a positive operator P as n tends to infinity, that is,*

$$P = \lim_{n\to\infty} Q_n. \tag{11.5.4}$$

In this case, P is a positive solution for the algebraic Riccati equation

$$P = APA^* + BB^* - APC^* \left(CPC^* + DD^*\right)^{-1} CPA^*. \tag{11.5.5}$$

(iii) *If $\{A, B, C, D\}$ is controllable and observable, and P is any positive solution to the algebraic Riccati equation in* (11.5.5)*, then P is strictly positive. In this case, $A - K_P C$ is stable where $K_P = APC^* \left(CPC^* + DD^*\right)^{-1}$.*

(iv) *If $\{A, B, C, D\}$ is controllable and observable, then there is only one positive solution P to the algebraic Riccati equation in* (11.5.5)*. In this case, P is strictly positive. Finally, this solution is given by $P = \lim_{n\to\infty} Q_n$.*

The steady state Kalman filter. Assume that $\{A, B, C, D\}$ is a controllable and observable system. Let P be the positive solution to the algebraic Riccati equation in (11.5.5) determined by (11.5.4). Notice that Λ_n converges to K_P and n tends to infinity. By passing to limits in the Kalman filter (11.4.4) and (11.4.5), we arrive at the *steady state Kalman filter* defined by

$$\begin{aligned} \zeta(n+1) &= (A - K_P C)\zeta(n) + K_P y(n), \\ K_P &= APC^* \left(CPC^* + DD^*\right)^{-1}. \end{aligned} \tag{11.5.6}$$

Theorem 11.5.1 guarantees that $A - K_P C$ is stable. The steady state Kalman filter provides an approximation $\zeta(n)$ for the optimal state estimate $\widehat{x}(n)$ for large n or once the system reaches steady state. The Kalman filter converges to the steady state Kalman filter. In other words, the steady state Kalman filter is an optimal state estimator in the limit.

11.6 Wide Sense Stationary Processes

In this section we introduce the notion of a wide sense stationary random process and its autocorrelation function. Recall that $y(n)$ is a *random process* if $y(n)$ is a random vector in some $\mathcal{Y} = \mathbb{C}^k$ space. A random process $y(n)$ with values in $\mathcal{Y}$ is called *wide sense stationary* if $Ey(n) = c$ is constant for all integers n, and

$Ey(n)y(m)^* = \tau(n-m)$ is a function of the time difference $n-m$ for all integers n and m. If $y(n)$ is wide sense stationary, then the *autocorrelation function* $R_y(n)$ for $y(n)$ is defined by $Ey(n)y(0)^* = R_y(n)$. Notice that $R_y(n)$ is an operator on $\mathcal{Y}$ and

$$Ey(n)y(m)^* = R_y(n-m)$$

for all n and m. Finally, it is noted that $Ey(n)y(m)^* = \tau(n-m)$ for all integers n and m if and only if $Ey(n+k)y(k)^* = \tau(n)$ is a function of just n for all integers n and k. So $y(n)$ is wide sense stationary if and only if $Ey(n) = c$ is constant for all n and $Ey(n+k)y(k)^* = \tau(n)$ is a function of n for all n and k. In this case, $R_y(n) = Ey(n+k)y(k)^*$.

If $y(n)$ is a wide sense stationary random process with values in $\mathcal{Y}$, then $R_y(n) = R_y(-n)^*$ for all integers n. To see this observe that

$$\begin{aligned} R_y(n) &= Ey(n+k)y(k)^* = (Ey(k)y(n+k)^*)^* \\ &= R_y(k-n-k)^* = R_y(-n)^*. \end{aligned}$$

Hence $R_y(n) = R_y(-n)^*$ for all n. In particular, if $y(n)$ is a scalar-valued wide sense stationary random process, then $R_y(n) = \overline{R}_y(-n)$.

Let $y(n)$ be a wide sense stationary random process with values in $\mathcal{Y}$. Let g be the random vector defined by

$$g = \begin{bmatrix} y(n) & y(n+1) & \cdots & y(n+\nu-1) \end{bmatrix}^{tr}.$$

Clearly, g is a random vector. Hence $T_{R_y,\nu} = Egg^*$ is a positive matrix. Using $Ey(n)y(m)^* = R_y(n-m)$, it follows that $T_{R_y,\nu}$ is a Toeplitz matrix of the form

$$T_{R_y,\nu} = \begin{bmatrix} R_y(0) & R_y(-1) & \cdots & R_y(1-\nu) \\ R_y(1) & R_y(0) & \cdots & R_y(2-\nu) \\ \vdots & \vdots & \ddots & \vdots \\ R_y(\nu-1) & R_y(\nu-2) & \cdots & R_y(0) \end{bmatrix} \text{ on } \begin{bmatrix} \mathcal{Y} \\ \mathcal{Y} \\ \vdots \\ \mathcal{Y} \end{bmatrix}. \tag{11.6.1}$$

Notice that the j-k entry of $T_{R_y,\nu}$ is given by $\{T_{R_y,\nu}\}_{jk} = R_y(j-k)$. The matrix $T_{R_y,\nu}$ in (11.6.1) is referred to as the Toeplitz matrix generated by $\{R_y(k)\}_0^{\nu-1}$. Therefore if $y(n)$ is wide sense stationary, its autocorrelation function uniquely determines a positive Toeplitz T_{R_y} defined by

$$T_{R_y} = \begin{bmatrix} R_y(0) & R_y(-1) & R_y(-2) & \cdots \\ R_y(1) & R_y(0) & R_y(-1) & \cdots \\ R_y(2) & R_y(1) & R_y(0) & \cdots \\ \vdots & \vdots & \vdots & \ddots \end{bmatrix}. \tag{11.6.2}$$

Finally, it is noted that T_{R_y} admits a unique controllable isometric representation $\{U \text{ on } \mathcal{K}, \Gamma\}$, that is, $R_y(-n) = \Gamma^* U^n \Gamma$ for all integers $n \geq 0$ where U is an isometry.

We say that two random processes $x(n)$ and $y(n)$ are *independent* if the random vectors $x(n)$ and $y(m)$ are independent for all integers n and m. The random processes $x(n)$ and $y(n)$ are *orthogonal* if the random vectors $x(n)$ and $y(m)$ are orthogonal for all n and m. Finally, it is noted that if $x(n)$ and $y(n)$ are two mean zero independent processes, then $x(n)$ and $y(n)$ are also orthogonal random processes. To see this observe that in this case $Ex(n)y(m)^* = Ex(n)Ey(m)^* = 0$. Here we used the fact that if two random variables f and g are independent, then $Efg = EfEg$. The following result is useful.

Proposition 11.6.1. *Assume that* $y(n) = \sum_{k=1}^{\mu} y_k(n)$ *where* $y_k(n)$ *are mutually orthogonal mean zero wide sense stationary random processes. Then* $y(n)$ *is also a mean zero wide sense stationary random process. Moreover, the autocorrelation function for* $y(n)$ *is given by*

$$R_y(n) = \sum_{k=1}^{\mu} R_{y_k}(n). \tag{11.6.3}$$

Proof. Since the mean of $y_k(n)$ is zero, and $y(n) = \sum_{k=1}^{\mu} y_k(n)$, it follows that $Ey(n) = 0$. If $k \neq r$, then $y_k(n)$ is orthogonal to $y_r(m)$ for all integers n and m. Using this orthogonality, we obtain, for all integers n and ν,

$$\begin{aligned}
Ey(n+\nu)y(\nu)^* &= \sum_{k=1}^{\mu}\sum_{r=1}^{\mu} Ey_k(n+\nu)y_r(\nu)^* \\
&= \sum_{k=1}^{\mu} Ey_k(n+\nu)y_k(\nu)^* = \sum_{k=1}^{\mu} R_{y_k}(n).
\end{aligned}$$

Therefore $Ey(n+\nu)y(\nu)^*$ is just a function of n for all n and ν. So $y(n)$ is wide sense stationary and its autocorrelation function $R_y(n)$ is determined by (11.6.3). □

11.6.1 A sinusoid process

For an example of a wide sense stationary process, let $\zeta(n)$ be the random process given by $\zeta(n) = a\cos(\omega n + \theta)$ where the amplitude a and the frequency ω are scalars while the phase θ is a uniform random variable over $[0, 2\pi]$. Recall that the probability density function f_θ for θ is given by

$$\begin{aligned}
f_\theta(\phi) &= \frac{1}{2\pi} \qquad \text{if } 0 \leq \phi \leq 2\pi \\
&= 0 \qquad \text{otherwise.}
\end{aligned} \tag{11.6.4}$$

We claim that $\zeta(n)$ is a mean zero wide sense stationary random process. Moreover, its autocorrelation function

$$R_\zeta(n) = \frac{1}{2}\,|a|^2\cos(\omega n).$$

To show that the mean of $\zeta(n)$ is zero simply notice that

$$E\zeta(n) = Ea\cos(\omega n + \theta) = \int_{-\infty}^{\infty} a\cos(\omega n + \phi) f_\theta(\phi)d\phi$$
$$= \frac{1}{2\pi}\int_0^{2\pi} a\cos(\omega n + \phi)d\phi = 0.$$

Hence $E\zeta(n) = 0$ for all integers n. Recall for any α and β, we have

$$\cos(\alpha)\cos(\beta) = \frac{1}{2}\cos(\alpha - \beta) + \frac{1}{2}\cos(\alpha + \beta).$$

If n and k are two integers, then this identity yields

$$E\zeta(n+k)\zeta(k)^* = |a|^2 E\cos(\omega(n+k) + \theta)\cos(\omega k + \theta)$$
$$= \frac{1}{2}|a|^2 E\cos(\omega n) + \frac{1}{2}|a|^2 E\cos(\omega(n+2k) + 2\theta)$$
$$= \frac{1}{2}|a|^2\cos(\omega n).$$

Clearly, $E\zeta(n+k)\zeta(k)^*$ is a function of only n for all integers n and k. Thus $\zeta(n)$ is wide sense stationary and $R_\zeta(n) = \frac{1}{2}|a|^2\cos(\omega n)$.

Consider the random process given by

$$y(n) = \sum_{k=1}^{\mu} a_k\cos(\omega_k n + \theta_k). \tag{11.6.5}$$

Here we assume that the amplitudes $\{a_k\}_1^\mu$ and the frequencies $\{\omega_k\}_1^\mu$ are distinct scalars while the phases $\{\theta_k\}_1^\mu$ are all independent uniform random variables over $[0, 2\pi]$. Then $y(n)$ is a mean zero wide sense stationary random process whose autocorrelation function is determined by

$$R_y(n) = \frac{1}{2}\sum_{k=1}^{\mu} |a_k|^2\cos(\omega_k n). \tag{11.6.6}$$

To verify this simply observe that

$$y(n) = \sum_{k=1}^{\mu} y_k(n)$$

where $y_k(n) = a_k\cos(\omega_k n + \theta_k)$ are mean zero wide sense stationary independent random processes for all $1 \le k \le \mu$. In particular, $y_k(n)$ for $k = 1, 2, \ldots, \mu$ are mean zero mutually orthogonal wide sense stationary random processes. To see this notice that for $k \ne j$, we have $Ey_k(n)y_j(m) = Ey_k(n)Ey_j(m) = 0$. Here we used the fact that if $f(x)$ and $g(z)$ are functions of two independent random variables

x and z, then $Ef(x)g(z) = Ef(x)Eg(z)$. Moreover, our previous analysis with $\zeta(n) = y_k(n)$ shows that $R_{y_k}(n) = |a_k|^2 \cos(\omega_k n)/2$. By consulting Proposition 11.6.1, it follows that $y(n)$ is a mean zero wide sense stationary process and its autocorrelation function is given by (11.6.6).

Let $\{V \text{ on } \mathbb{C}^{2\mu}, \Gamma\}$ be the controllable unitary representation determined by

$$V = \begin{bmatrix} e^{\imath\omega_1} & 0 & \cdots & 0 & 0 \\ 0 & e^{-\imath\omega_1} & \cdots & 0 & 0 \\ \vdots & \vdots & \ddots & \vdots & \vdots \\ 0 & 0 & \cdots & e^{\imath\omega_\mu} & 0 \\ 0 & 0 & \cdots & 0 & e^{-\imath\omega_\mu} \end{bmatrix} \quad \text{and} \quad \Gamma = \frac{1}{2}\begin{bmatrix} a_1 \\ a_1 \\ \vdots \\ a_\mu \\ a_\mu \end{bmatrix}.$$

Notice that V is a diagonal matrix with $\{e^{\imath\omega_1}, e^{-\imath\omega_1}, \ldots, e^{\imath\omega_\mu}, e^{-\imath\omega_\mu}\}$ appearing on the main diagonal. Using Euler's identity for the cosine, it follows that $\{V \text{ on } \mathbb{C}^{2\mu}, \Gamma\}$ is the controllable unitary representation for $\{R_y(n)\}$, that is, $R_y(-n) = \Gamma U^n \Gamma$ for all integers n. Finally, all controllable isometric representations for $\{R_y(n)\}$ are unitarily equivalent to $\{V, \Gamma\}$; see Theorem 5.1.1.

11.7 The Spectral Density

In this section we will introduce the spectral density. Let $y(n)$ be a wide sense stationary random process with values in $\mathcal{Y}$. Then the *spectral density* $\mathbf{S}_y$ for $y(n)$ is the Fourier transform of its autocorrelation function, that is,

$$\mathbf{S}_y(e^{\imath\omega}) = \mathcal{F}\{R_y(k)\}_{-\infty}^{\infty} = \sum_{k=-\infty}^{\infty} e^{-\imath\omega k} R_y(k),$$

$$R_y(k) = \frac{1}{2\pi}\int_0^{2\pi} e^{\imath\omega k} \mathbf{S}_y(e^{\imath\omega}) d\omega. \tag{11.7.1}$$

(Because we did not want to confuse the spectral density with the unilateral shift, we have used a bold face $\mathbf{S}_y$ to represent the spectral density.) Throughout this chapter, we assume that the spectral density is a well-defined integrable function. In fact, in all of our applications the spectral density is a rational function. Since $\{R_y(n)\}_0^\infty$ defines a positive Toeplitz matrix T_{R_y}, its spectral density $\mathbf{S}_y$ is almost everywhere a positive operator on $\mathcal{Y}$ with respect to the Lebesgue measure; see Bochner's Theorem 5.6.1. Finally, it is noted that $\operatorname{trace}(R_y(0))$ is referred to as the *energy* in the process y. So the energy in the process y is given by

$$\operatorname{trace} Ey(n)y(n)^* = \operatorname{trace}(R_y(0)) = \frac{1}{2\pi}\int_0^{2\pi} \operatorname{trace}(\mathbf{S}_y(e^{\imath\omega})) d\omega.$$

Let $y(n)$ be any random process with values in $\mathcal{Y}$ such that the mean is a constant. Then $\vec{y}$ is the infinite vector defined by

$$\vec{y} = \begin{bmatrix} \cdots & y(-2) & y(-1) & y(0) & y(1) & y(2) & \cdots \end{bmatrix}^{tr}. \tag{11.7.2}$$

Using this notation, it follows that the j-k entry of the matrix $E\vec{y}\vec{y}^*$ is given by $(E\vec{y}\vec{y}^*)_{j,k} = Ey(j)y(k)^*$. In particular, the process $y(n)$ is wide sense stationary if and only if $(E\vec{y}\vec{y}^*)_{j,k}$ is a function of $j-k$, or equivalently, $E\vec{y}\vec{y}^*$ is a Laurent matrix. In this case, $E\vec{y}\vec{y}^* = L_{\mathbf{S}_y}$ where $L_{\mathbf{S}_y}$ is the Laurent matrix determined by the spectral density $\mathbf{S}_y$. In particular, the j-k entry of $L_{\mathbf{S}_y}$ is given by $(L_{\mathbf{S}_y})_{jk} = R_y(j-k)$.

If $w(n)$ is a white noise process with values in $\mathcal{W}$, then its spectral density $\mathbf{S}_w(e^{\imath\omega}) = I$. Recall that the autocorrelation function for a white noise process is $R_w(n) = \delta_n I$ where δ_j is the Kronecker delta. Hence $\mathbf{S}_w(e^{\imath\omega}) = \mathcal{F}\{\delta_n I\} = I$.

Theorem 11.7.1. *Let $u(n)$ be a zero mean wide sense stationary process with values in $\mathcal{U}$, and assume that its spectral density $\mathbf{S}_u$ is a function in $L^\infty(\mathcal{U},\mathcal{U})$. Let $y(n)$ be the random process with values in $\mathcal{Y}$ determined by*

$$y(n) = \sum_{k=-\infty}^{\infty} G_{n-k}u(k) = \sum_{k=-\infty}^{\infty} G_k u(n-k) \tag{11.7.3}$$

where $G = \sum_{-\infty}^{\infty} e^{-\imath\omega k} G_k$ is a function in $L^2(\mathcal{U},\mathcal{Y})$. Then $y(n)$ is a mean zero wide sense stationary random process whose spectral density is given by

$$\mathbf{S}_y(e^{\imath\omega}) = G(e^{\imath\omega})\mathbf{S}_u(e^{\imath\omega})G(e^{\imath\omega})^*. \tag{11.7.4}$$

In particular, if $u(n)$ is white noise, then $\mathbf{S}_y(e^{\imath\omega}) = G(e^{\imath\omega})G(e^{\imath\omega})^*$.

Proof. Notice that the random process y is given by $\vec{y} = L_G\vec{u}$ where L_G is the Laurent matrix determined by G. Thus $E\vec{y} = L_G E\vec{u} = 0$. Hence the mean of $y(n)$ is zero for all integers n. By employing $\vec{y} = L_G\vec{u}$, we obtain

$$E\vec{y}\vec{y}^* = E(L_G\vec{u})(L_G\vec{u})^* = L_G E\vec{u}\vec{u}^* L_G^\sharp = L_G L_{\mathbf{S}_u} L_{G^*} = L_{G\mathbf{S}_u G^*}.$$

So $E\vec{y}\vec{y}^* = L_{G\mathbf{S}_u G^*}$ is the Laurent matrix determined by the symbol $G\mathbf{S}_u G^*$. In particular, $Ey(j)y(k)^* = (L_{G\mathbf{S}_u G^*})_{jk}$ is a function of $j-k$. Therefore $y(n)$ is a wide sense stationary random process. Finally, since $L_{\mathbf{S}_y} = E\vec{y}\vec{y}^* = L_{G\mathbf{S}_u G^*}$, we see that $\mathbf{S}_y = G\mathbf{S}_u G^*$. □

11.8 Jointly Wide Sense Stationary Processes

Let $x(n)$ be a random process with values in $\mathcal{X}$ and $y(n)$ be a random process with values in $\mathcal{Y}$ where $\mathcal{X}$ and $\mathcal{Y}$ are finite dimensional Euclidean spaces. Then we say that $x(n)$ and $y(n)$ are *jointly wide sense stationary* if the following three conditions hold:

(i) The process $x(n)$ is wide sense stationary.

(ii) The process $y(n)$ is wide sense stationary.

(iii) $Ex(n)y(m)^* = R_{xy}(n-m)$ is a function of $n-m$ for all integers n and m.

In this case, $R_{xy}(n)$ is called the *joint autocorrelation function* for x and y. Notice that $Ex(n)y(m)^* = R_{xy}(n-m)$ for all n and m if and only if $Ex(n+k)y(k)^* = R_{xy}(n)$ is just a function of n for all integers n and k. The processes x and y are jointly wide sense stationary if and only if the processes y and x are jointly wide sense stationary. In this case, $R_{xy}(n) = R_{yx}(-n)^*$. To see this simply observe that

$$\begin{aligned} R_{xy}(n) &= Ex(n+k)y(k)^* = (Ey(k)x(n+k)^*)^* \\ &= R_{yx}(k-n-k)^* = R_{yx}(-n)^*. \end{aligned}$$

Therefore we have $R_{xy}(n) = R_{yx}(-n)^*$. Finally, it is noted that $R_{xy}(n)$ is an operator from $\mathcal{Y}$ into $\mathcal{X}$ and $R_{yx}(n)$ is an operator from $\mathcal{X}$ into $\mathcal{Y}$ for all integers n.

Now assume that $x(n)$ and $y(n)$ are jointly wide sense stationary. Then the *joint spectral density* $\mathbf{S}_{xy}$ for x and y is the Fourier transform of their joint autocorrelation function R_{xy}, that is,

$$\mathbf{S}_{xy}(e^{\imath\omega}) = \mathcal{F}\{R_{xy}(k)\}_{-\infty}^{\infty} = \sum_{k=-\infty}^{\infty} e^{-\imath\omega k} R_{xy}(k).$$

For our purposes throughout this chapter, we assume that the joint spectral density $\mathbf{S}_{xy}$ is integrable. In this case, $\mathbf{S}_{xy}(e^{\imath\omega})$ is almost everywhere an operator from $\mathcal{Y}$ into $\mathcal{X}$. Finally, $\mathbf{S}_{xy}(e^{\imath\omega}) = \mathbf{S}_{yx}(e^{\imath\omega})^*$ almost everywhere. To verify this, notice that $R_{xy}(n) = R_{yx}(-n)^*$ yields

$$\begin{aligned} \mathbf{S}_{xy}(e^{\imath\omega})^* &= \sum_{k=-\infty}^{\infty} R_{xy}(k)^* e^{\imath\omega k} = \sum_{k=-\infty}^{\infty} R_{yx}(-k) e^{\imath\omega k} \\ &= \sum_{k=-\infty}^{\infty} R_{yx}(k) e^{-\imath\omega k} = \mathbf{S}_{yx}(e^{\imath\omega}). \end{aligned}$$

In other words, $\mathbf{S}_{xy}^* = \mathbf{S}_{yx}$.

Let $\xi(n)$ be the random process with values in $\mathcal{X} \oplus \mathcal{Y}$ determined by

$$\xi(n) = \begin{bmatrix} x(n) \\ y(n) \end{bmatrix}. \tag{11.8.1}$$

A simple calculation shows that for all integers n and k, we have

$$E\xi(n+k)\xi(k)^* = \begin{bmatrix} Ex(n+k)x(k)^* & Ex(n+k)y(k)^* \\ Ey(n+k)x(k)^* & Ey(n+k)y(k)^* \end{bmatrix}.$$

Notice that $\xi(n)$ is wide sense stationary if and only if $x(n)$ and $y(n)$ are jointly wide sense stationary. In this case, the autocorrelation function R_ξ for the process $\xi(n)$ is the operator matrix given by

$$R_\xi(n) = \begin{bmatrix} R_x(n) & R_{xy}(n) \\ R_{yx}(n) & R_y(n) \end{bmatrix}. \tag{11.8.2}$$

Moreover, the spectral density $\mathbf{S}_\xi$ for the process $\xi(n)$ is determined by

$$\mathbf{S}_\xi = \begin{bmatrix} \mathbf{S}_x & \mathbf{S}_{xy} \\ \mathbf{S}_{yx} & \mathbf{S}_y \end{bmatrix}. \tag{11.8.3}$$

Because the spectral density is a positive operator, $\mathbf{S}_\xi(e^{\imath\omega})$ is almost everywhere a positive operator on $\mathcal{X} \oplus \mathcal{Y}$. In particular, $\mathbf{S}_\xi(e^{\imath\omega})$ is almost everywhere a self-adjoint operator. This also shows that $\mathbf{S}_{xy}(e^{\imath\omega})^* = \mathbf{S}_{yx}(e^{\imath\omega})$.

Let x and y be two wide sense stationary processes. The j-k entry of the matrix $E\vec{x}\vec{y}^*$ is given by $(E\vec{x}\vec{y}^*)_{j,k} = Ex(j)y(k)^*$. In particular, the process $x(n)$ and $y(n)$ are jointly wide sense stationary if and only if $(E\vec{x}\vec{y}^*)_{j,k}$ is a function of $j-k$, or equivalently, $E\vec{x}\vec{y}^*$ is a Laurent matrix. In this case, $E\vec{x}\vec{y}^* = L_{\mathbf{S}_{xy}}$ where $L_{\mathbf{S}_{xy}}$ is the Laurent matrix determined by the joint spectral density $\mathbf{S}_{xy}$. In particular, the j-k entry of $L_{\mathbf{S}_{xy}}$ is given by $(L_{\mathbf{S}_{xy}})_{jk} = R_{xy}(j-k)$. The following is a generalization of Theorem 11.7.1.

Theorem 11.8.1. *Let $u(n)$ and $v(n)$ be jointly wide sense stationary zero mean processes with values in $\mathcal{U}$ and $\mathcal{V}$, respectively such that $\mathbf{S}_v$, $\mathbf{S}_u$ and $\mathbf{S}_{vu}$ are in the appropriate L^∞ space. Let $x(n)$ be the random process with values in $\mathcal{X}$ and $y(n)$ be the random process with values in $\mathcal{Y}$ determined by*

$$x(n) = \sum_{k=-\infty}^{\infty} H_{n-k}v(k) \quad \textit{and} \quad y(n) = \sum_{k=-\infty}^{\infty} G_{n-k}u(k). \tag{11.8.4}$$

Here $H = \sum_{-\infty}^{\infty} e^{-\imath\omega n}H_n$ is a function in $L^2(\mathcal{V},\mathcal{X})$ while $G = \sum_{-\infty}^{\infty} e^{-\imath\omega n}G_n$ is a function in $L^2(\mathcal{U},\mathcal{Y})$. Then $x(n)$ and $y(n)$ are mean zero jointly wide sense stationary random processes whose joint spectral density is given by

$$\mathbf{S}_{xy}(e^{\imath\omega}) = H(e^{\imath\omega})\mathbf{S}_{vu}(e^{\imath\omega})G(e^{\imath\omega})^*. \tag{11.8.5}$$

In particular, if $u(n) = v(n)$ is white noise, then $x(n)$ and $y(n)$ are mean zero jointly wide sense stationary and $\mathbf{S}_{xy}(e^{\imath\omega}) = H(e^{\imath\omega})G(e^{\imath\omega})^*$.

Proof. Observe that $\vec{x} = L_H\vec{v}$ and $\vec{y} = L_G\vec{u}$. Hence

$$E\vec{x}\vec{y}^* = E(L_H\vec{v})(L_G\vec{u})^* = L_H E\vec{v}\vec{u}^* L_G^\sharp = L_H L_{\mathbf{S}_{vu}} L_{G^*} = L_{H\mathbf{S}_{vu}G^*}.$$

So $E\vec{x}\vec{y}^* = L_{H\mathbf{S}_{vu}G^*}$ is the Laurent matrix determined by the symbol $H\mathbf{S}_{vu}G^*$. Therefore $x(n)$ and $y(n)$ are jointly wide sense stationary random process. Finally, since $L_{\mathbf{S}_{xy}} = E\vec{x}\vec{y}^* = L_{H\mathbf{S}_{vu}G^*}$, we obtain $\mathbf{S}_{xy} = H\mathbf{S}_{vu}G^*$. □

11.9 Wiener Filtering

In this section we will state and solve a classical Wiener filtering problem. To establish some general notation in systems theory, let L be a linear map sending an input sequence $u(n)$ with values in $\mathcal{U}$ into an output sequence $y(n)$ with values in $\mathcal{Y}$. Then L is *time invariant* if when the input $u(n)$ is shifted by k, then the output is also shifted by time k, that is, if $y = Lu$, then $y(n+k) = (Lv)(n)$ where $v(n) = u(n+k)$. In our terminology L is time invariant if L defines a Laurent matrix. The map L is *causal* if the output $y(n)$ depends only on the input $\{u(j)\}_{-\infty}^{n}$ up to time n. So L is causal if and only if L defines a lower triangular matrix. In particular, a linear time invariant causal map corresponds to a lower triangular Laurent matrix. There are several notions of stability. For our purposes, we say that a linear time invariant map L is *stable* if $L_G u\delta_{n-k}$ is in $\ell^2(\mathcal{Y})$ for all u in $\mathcal{U}$ where δ_j is the Kronecker delta. In other words, a linear time invariant map L is stable if and only if $L = L_G$ where L_G is the Laurent matrix determined by a function G in $L^2(\mathcal{U},\mathcal{Y})$. Finally, L is a stable linear time invariant causal map if and only if $L = L_G$ where G is a function in $H^2(\mathcal{U},\mathcal{Y})$.

Let us establish some notation. Let $\mathcal{Y}$ and $\mathcal{X}$ be finite dimensional Hilbert spaces. Then the inner product on $L^2(\mathcal{Y},\mathcal{X})$ is given by

$$(F,G) = \frac{1}{2\pi}\int_0^{2\pi} \operatorname{trace}(FG^*)d\omega = \frac{1}{2\pi}\int_0^{2\pi} \operatorname{trace}(G^*F)d\omega.$$

If F is a function in $L^2(\mathcal{Y},\mathcal{X})$, then the causal part of F is defined by $F_c = P_+F$ where P_+ is the orthogonal projection from $L^2(\mathcal{Y},\mathcal{X})$ onto $H^2(\mathcal{Y},\mathcal{X})$, that is, $F_c = \sum_0^\infty F_k e^{-\imath\omega k}$ where $F = \sum_{-\infty}^\infty F_k e^{-\imath\omega k}$. Moreover, the anti-causal part of F is given by $F_a = (I - P_+)F$ which is the orthogonal projection of F onto $L^2(\mathcal{Y},\mathcal{X}) \ominus H^2(\mathcal{Y},\mathcal{X})$. In other words, $F_a = F - F_c$. Finally, it is noted that

$$\|F\|^2 = \|F_c\|^2 + \|F_a\|^2 \qquad (F \in L^2(\mathcal{Y},\mathcal{X})). \tag{11.9.1}$$

Let $\mathbf{S}_y$ be the spectral density for a wide sense stationary random process $y(n)$. Then we say that $\mathbf{S}_y$ admits a *co-outer spectral factor* Θ if $\mathbf{S}_y = \Theta\Theta^*$ where Θ is a co-outer function. In this case, Θ is called a co-outer spectral factor for $\mathbf{S}_y$. (Recall that Θ is co-outer if $\widetilde{\Theta}(z) = \Theta(\overline{z})^*$ is outer.) Notice that Θ is a co-outer spectral factor for $\mathbf{S}_y$ if and only if $\widetilde{\Theta}$ is an outer spectral factor for $\mathbf{S}_y(e^{-\imath\omega})$. So the co-outer spectral factor is unique up to a unitary constant operator on the right.

Let $x(n)$ and $y(n)$ be two jointly wide sense stationary processes with values in $\mathcal{X} = \mathbb{C}^k$ and $\mathcal{Y} = \mathbb{C}^m$, respectively. The idea behind Wiener filtering is to find the best linear stable time invariant causal estimate of $x(n)$ given the past $\{y(j)\}_{-\infty}^{n}$ of y. In other words, one would like to find a function (or a filter in engineering terminology) H in $H^2(\mathcal{Y},\mathcal{X})$ such that the process determined by $L_H\vec{y}$ is the best estimate of $x(n)$ given the past $\{y(j)\}_{-\infty}^{n}$. Now let $H(z)$ be any transfer function in $H^2(\mathcal{Y},\mathcal{X})$, and consider the error process $\varepsilon(n)$ defined by $\vec{\varepsilon} = \vec{x} - L_H\vec{y}$, that is,

$$\varepsilon(n) = x(n) - \sum_{j=-\infty}^{n} H_{n-j}y(j) \tag{11.9.2}$$

where $H(z) = \sum_0^\infty z^{-j} H_j$. Notice that $\varepsilon(n)$ is a wide sense stationary process depending upon our choice of H. In fact, the spectral density for ε is given by

$$\mathbf{S}_\varepsilon = \mathbf{S}_x - \mathbf{S}_{xy} H^* - H\mathbf{S}_{yx} + H\mathbf{S}_y H^*. \tag{11.9.3}$$

To see this simply observe that

$$\begin{aligned} E\vec{\varepsilon}\vec{\varepsilon}^* &= E(\vec{x} - L_H\vec{y})(\vec{x} - L_H\vec{y})^* \\ &= E\vec{x}\vec{x}^* - E\vec{x}\vec{y}^* L_H^\sharp - L_H E\vec{y}\vec{x}^* + L_H E\vec{y}\vec{y}^* L_H^\sharp \\ &= L_{\mathbf{S}_x} - L_{\mathbf{S}_{xy}} L_{H^*} - L_H L_{\mathbf{S}_{yx}} + L_H L_{\mathbf{S}_y} L_{H^*}. \end{aligned}$$

Thus $E\vec{\varepsilon}\vec{\varepsilon}^*$ is the Laurent matrix whose symbol is given by $\mathbf{S}_\varepsilon$ in (11.9.3). In other words, ε is wide sense stationary.

The *energy* in the process $\varepsilon(n)$ is defined by

$$\operatorname{trace} E\varepsilon(n)\varepsilon(n)^* = \operatorname{trace} R_\varepsilon(0) = \frac{1}{2\pi}\int_0^{2\pi} \operatorname{trace}(\mathbf{S}_\varepsilon(e^{\imath\omega}))d\omega.$$

So the Wiener filtering problem is to find a function H in $H^2(\mathcal{Y}, \mathcal{X})$ to minimize the energy of the error process, or equivalently, minimize the area under the trace in the spectral density $\mathbf{S}_\varepsilon$. To be precise, the Wiener filtering problem is to find a causal function H which solves the optimization problem

$$\mu = \inf\{\frac{1}{2\pi}\int_0^{2\pi} \operatorname{trace}(\mathbf{S}_\varepsilon(e^{\imath\omega}))d\omega : H \in H^2(\mathcal{Y}, \mathcal{X}) \text{ where } \varepsilon(n) \text{ is in } (11.9.2)\}. \tag{11.9.4}$$

Here μ is the error in the Wiener filter.

Now assume that the spectral density $\mathbf{S}_y$ for $y(n)$ admits a spectral factorization of the form $\mathbf{S}_y = \Theta\Theta^*$ where Θ is an invertible co-outer function, that is, Θ and Θ^{-1} are both in $H^\infty(\mathcal{Y}, \mathcal{Y})$. Then the unique solution $\widehat{H}$ to the Wiener filtering problem is given by

$$\widehat{H}(z) = (\left[\mathbf{S}_{xy}\Theta^{-*}\right]_c)(z)\Theta(z)^{-1}. \tag{11.9.5}$$

Moreover, the error is determined by

$$\mu = \frac{1}{2\pi}\int_0^{2\pi} \mathit{trace}\, \mathbf{S}_\varepsilon(e^{\imath\omega})d\omega = \mathit{trace}\, R_x(0) - \|\left[\mathbf{S}_{xy}\Theta^{-*}\right]_c\|^2. \tag{11.9.6}$$

To verify this, we see by (11.9.3) that the spectral density for the error process is given by

$$\begin{aligned} \mathbf{S}_\varepsilon &= \mathbf{S}_x - H\mathbf{S}_{yx} - \mathbf{S}_{xy}H^* + H\mathbf{S}_y H^* \\ &= \mathbf{S}_x - H\mathbf{S}_{yx} - \mathbf{S}_{xy}H^* + H\Theta\Theta^* H^* \\ &= \mathbf{S}_x - \mathbf{S}_{xy}\Theta^{-*}\Theta^{-1}\mathbf{S}_{yx} + \mathbf{S}_{xy}\Theta^{-*}\Theta^{-1}\mathbf{S}_{yx} \\ &\quad - H\mathbf{S}_{yx} - \mathbf{S}_{xy}H^* + H\Theta\Theta^* H^* \\ &= \mathbf{S}_x - \mathbf{S}_{xy}\Theta^{-*}\Theta^{-1}\mathbf{S}_{yx} + \left(\mathbf{S}_{xy}\Theta^{-*} - H\Theta\right)\left(\mathbf{S}_{xy}\Theta^{-*} - H\Theta\right)^*. \end{aligned}$$

The last equality follows from the fact that $\mathbf{S}_{xy}^* = \mathbf{S}_{yx}$. By combining the previous expression for $\mathbf{S}_\varepsilon$ with (11.9.1), we obtain

$$\begin{aligned}
\operatorname{trace} R_\varepsilon(0) &= \frac{1}{2\pi}\int_0^{2\pi} \operatorname{trace}(\mathbf{S}_\varepsilon(e^{\imath\omega}))d\omega \\
&= \frac{1}{2\pi}\int_0^{2\pi} \operatorname{trace}(\mathbf{S}_x(e^{\imath\omega}))d\omega - \|\mathbf{S}_{xy}\Theta^{-*}\|^2 + \|\mathbf{S}_{xy}\Theta^{-*} - H\Theta\|^2 \\
&= \operatorname{trace} R_x(0) - \|\mathbf{S}_{xy}\Theta^{-*}\|^2 + \|\left[\mathbf{S}_{xy}\Theta^{-*}\right]_a + \left[\mathbf{S}_{xy}\Theta^{-*}\right]_c - H\Theta\|^2 \\
&= \operatorname{trace} R_x(0) - \|\mathbf{S}_{xy}\Theta^{-*}\|^2 + \|\left[\mathbf{S}_{xy}\Theta^{-*}\right]_a\|^2 + \|\left[\mathbf{S}_{xy}\Theta^{-*}\right]_c - H\Theta\|^2 \\
&= \operatorname{trace} R_x(0) - \|\left[\mathbf{S}_{xy}\Theta^{-*}\right]_c\|^2 + \|\left[\mathbf{S}_{xy}\Theta^{-*}\right]_c - H\Theta\|^2.
\end{aligned}$$

The fourth equality follows from the fact that $H\Theta$ is causal, and the last equality from the causal and anti-causal decomposition of $\mathbf{S}_{xy}\Theta^{-*}$. This readily yields the result

$$\operatorname{trace} R_\varepsilon(0) = \operatorname{trace} R_x(0) - \|\left[\mathbf{S}_{xy}\Theta^{-*}\right]_c\|^2 + \|\left[\mathbf{S}_{xy}\Theta^{-*}\right]_c - H\Theta\|^2. \tag{11.9.7}$$

Hence the causal filter $\widehat{H}$ in $H^2(\mathcal{Y}, \mathcal{X})$ which minimizes the energy $\operatorname{trace} R_\varepsilon(0)$ in the error process $\varepsilon(n)$ is the function $H = \widehat{H}$ which makes the last term $\left[\mathbf{S}_{xy}\Theta^{-*}\right]_c - H\Theta$ in (11.9.7) equal to zero. So the optimal Wiener filter $\widehat{H}$ is given by

$$\widehat{H}(z) = (\left[\mathbf{S}_{xy}\Theta^{-*}\right]_c)(z)\Theta(z)^{-1}.$$

Notice that $\widehat{H}$ is in $H^2(\mathcal{Y}, \mathcal{X})$ because $\left[\mathbf{S}_{xy}\Theta^{-*}\right]_c$ is in $H^2(\mathcal{Y}, \mathcal{X})$ and Θ^{-1} is in $H^\infty(\mathcal{Y}, \mathcal{Y})$. So their product is in $H^2(\mathcal{Y}, \mathcal{X})$. Finally, it is noted that our solution $\widehat{H}$ is unique because this $\widehat{H}$ is the only function in $H^2(\mathcal{Y}, \mathcal{X})$ which makes the last term $\left[\mathbf{S}_{xy}\Theta^{-*}\right]_c - H\Theta$ in (11.9.7) equal to zero. Finally, by setting $H = \widehat{H}$ in (11.9.7), we see that the error in estimation

$$\mu = \operatorname{trace} R_x(0) - \|\left[\mathbf{S}_{xy}\Theta^{-*}\right]_c\|^2.$$

This completes our derivation of the Wiener filter.

11.10 Steady State Kalman and Wiener Filtering

In this section we will show that the steady state Kalman filter is precisely the Wiener filter for the corresponding system.

11.10.1 State space systems with $x(-\infty) = 0$

Let us review some elementary facts concerning discrete time invariant systems starting in the infinite past. Consider the state space system given by

$$x(n+1) = Ax(n) + Bu(n) \quad \text{and} \quad y = Cx(n) + Dv(n). \tag{11.10.1}$$

Here $\{A \text{ on } \mathcal{X}, B, C, D\}$ are all constant operators acting between the appropriate spaces and the state space $\mathcal{X}$ is finite dimensional. The solution to (11.10.1) subject to the initial condition $x(m) = x_m$ is given by

$$x(n) = A^{n-m}x_m + \sum_{k=m}^{n-1} A^{n-k-1}Bu(k) \qquad (x(m) = x_m), \tag{11.10.2}$$

$$y(n) = CA^{n-m}x_m + Dv(n) + \sum_{k=m}^{n-1} CA^{n-k-1}Bu(k). \tag{11.10.3}$$

Recall that a finite dimensional operator A on $\mathcal{X}$ is *stable* if all the eigenvalues of A are contained in the open unit disc $\{z : |z| < 1\}$. Now assume that A is stable and $x_m = f$ is a fixed vector in $\mathcal{X}$. Then $A^{n-m}f$ converges to zero as m tends to minus infinity. Hence as m tends to minus infinity, the initial condition x_m does not play a role in the solution to the state space system (11.10.1); see (11.10.2) and (11.10.3). So without loss of generality, if the initial condition starts at minus infinity, then we can assume that $x(-\infty) = 0$. Moreover, if $x(-\infty) = 0$, then the solution to the state space system in (11.10.1) is given by

$$\begin{aligned} x(n) &= \sum_{k=-\infty}^{n-1} A^{n-k-1}Bu(k), \\ y(n) &= Dv(n) + \sum_{k=-\infty}^{n-1} CA^{n-k-1}Bu(k). \end{aligned} \tag{11.10.4}$$

In other words, if $\Omega = (zI - A)^{-1}B$ and $G = C(zI - A)^{-1}B$, then the state x and output y corresponding to $x(-\infty) = 0$, is given by

$$\vec{x} = L_\Omega \vec{u} \quad \text{and} \quad \vec{y} = L_D\vec{v} + L_G\vec{u}. \tag{11.10.5}$$

Because A is stable, L_Ω mapping $\ell^2(\mathcal{U})$ into $\ell^2(\mathcal{X})$ and L_G mapping $\ell^2(\mathcal{U})$ into $\ell^2(\mathcal{Y})$ are well-defined operators. Finally, L_D is the Laurent operator from $\ell^2(\mathcal{U})$ into $\ell^2(\mathcal{Y})$ formed by placing D on the diagonal and zeros elsewhere.

Assume that $\{A, B, C, D\}$ is a stable, controllable and observable system. Moreover, assume that $u(n)$ and $v(n)$ are independent white noise processes. Then the solution to the state space system in (11.10.1) with $x(-\infty) = 0$ is given by (11.10.4) or (11.10.5). In this case, $x(n)$ and $y(n)$ are jointly wide sense stationary processes. In fact,

$$\begin{aligned} \mathbf{S}_x(z) &= (zI - A)^{-1}BB^*(zI - A)^{-*}, \\ \mathbf{S}_y(z) &= C(zI - A)^{-1}BB^*(zI - A)^{-*}C^* + DD^* \qquad (|z| = 1), \\ \mathbf{S}_{xy}(z) &= (zI - A)^{-1}BB^*(zI - A)^{-*}C^*. \end{aligned} \tag{11.10.6}$$

To see this, Theorem 11.7.1 shows that $x(n)$ is a wide sense stationary process and the spectral density $\mathbf{S}_x = \Omega\Omega^*$. Because $u(n)$ and $v(n)$ are both mean zero processes and $\vec{y} = L_D\vec{v} + L_G\vec{u}$, the mean of $y(n)$ is also zero. Using the fact that u and v are orthogonal white noise processes ($E\vec{v}\vec{u}^* = 0$), we obtain

$$\begin{aligned} E\vec{y}\vec{y}^* &= E(L_D\vec{v} + L_G\vec{u})(L_D\vec{v} + L_G\vec{u})^* \\ &= L_D E\vec{v}\vec{v}^* L_D^* + L_G E\vec{u}\vec{u}^* L_G^* \\ &= L_D L_D^* + L_G L_G^* = L_{DD^*+GG^*}. \end{aligned}$$

Hence $E\vec{y}\vec{y}^* = L_{\mathbf{S}_y}$ equals the Laurent matrix determined by the symbol $DD^* + GG^*$. So y is wide sense stationary, and $\mathbf{S}_y = DD^* + GG^*$. To verify that x and y are jointly wide sense stationary, it remains to show that $E\vec{x}\vec{y}^*$ is a Laurent matrix. To this end, observe that

$$E\vec{x}\vec{y}^* = EL_\Omega\vec{u}(L_D\vec{v} + L_G\vec{u})^* = EL_\Omega\vec{u}\vec{u}^* L_G^* = L_\Omega L_{G^*} = L_{\Omega G^*}.$$

Therefore $E\vec{x}\vec{y}^* = L_{\mathbf{S}_{xy}}$ equals the Laurent matrix determined by the symbol ΩG^*. In other words, x and y are jointly wide sense stationary, and $\mathbf{S}_{xy} = \Omega G^*$.

Assume that DD^* is invertible. We claim that $\mathbf{S}_y$ admits an invertible co-outer spectral factorization, that is, $\mathbf{S}_y = \Theta\Theta^*$ where Θ is an invertible co-outer function in $H^\infty(\mathcal{Y}, \mathcal{Y})$. (A function is an invertible outer function if and only if it is an invertible co-outer function.) To compute Θ, let P be the unique positive solution to the algebraic Riccati equation determined by the steady state Kalman filter, that is,

$$P = APA^* + BB^* - APC^*\left(CPC^* + DD^*\right)^{-1} CPA^*. \tag{11.10.7}$$

Then the invertible co-outer spectral factor Θ for $\mathbf{S}_y$ is given by

$$\mathbf{S}_y = \Theta\Theta^* \quad \text{where} \quad \Theta(z) = C(zI - A)^{-1}APC^*N^{-1/2} + N^{1/2}. \tag{11.10.8}$$

Here $N = CPC^* + DD^*$ and $N^{1/2}$ is the positive square root of N.

To prove (11.10.8), observe that $\{A, APC^*N^{-1/2}, C, N^{1/2}\}$ is a realization for Θ. The inverse of Θ is determined by

$$\begin{aligned} \Theta(z)^{-1} &= N^{-1/2} - N^{-1/2}C(zI - J)^{-1}APC^*N^{-1}, \\ J &= A - APC^*\left(CPC^* + DD^*\right)^{-1} C; \end{aligned} \tag{11.10.9}$$

see Remark 14.2.1. According to Theorem 11.5.1, the feedback operator J is stable. Therefore Θ is an invertible co-outer function in $H^\infty(\mathcal{Y}, \mathcal{Y})$. To verify that $\mathbf{S}_y = \Theta\Theta^*$, set $\Phi = (zI - A)^{-1}$. Notice that

$$(zI - A)^{-1} = z^{-1}I + z^{-1}A(zI - A)^{-1}. \tag{11.10.10}$$

Clearly, $\Phi A = A\Phi$. Using the algebraic Riccati in (11.10.7) and $|z| = 1$, we have

$$\begin{aligned}
\mathbf{S}_y &= GG^* + DD^* \\
&= C\Phi BB^*\Phi^*C^* + DD^* \\
&= C\Phi\left(P - APA^* + APC^*N^{-1}CPA^*\right)\Phi^*C^* + DD^* \\
&= C\left(z^{-1}I + z^{-1}A\Phi\right)P\left(z^{-1}I + z^{-1}A\Phi\right)^*C^* \\
&\quad - C\Phi APA^*\Phi^*C^* + C\Phi APC^*N^{-1}CPA^*\Phi^*C^* + DD^* \\
&= CPC^* + CA\Phi PC^* + CP\Phi^*A^*C^* \\
&\quad + C\Phi APC^*N^{-1}CPA^*\Phi^*C^* + DD^* \\
&= C\Phi APC^* + CPA^*\Phi^*C^* + C\Phi APC^*N^{-1}CPA^*\Phi^*C^* + N \\
&= \left(C\Phi APC^*N^{-1/2} + N^{1/2}\right)\left(C\Phi APC^*N^{-1/2} + N^{1/2}\right)^*.
\end{aligned}$$

In other words, $\mathbf{S}_y = \Theta\Theta^*$ where $\Theta = C\Phi APC^*N^{-1/2} + N^{1/2}$ is the invertible co-outer spectral factorization for $\mathbf{S}_y$.

It is noted that one can use the results in Theorem 10.1.4 in Chapter 10 to compute the co-outer spectral factor Θ for $\mathbf{S}_y$. To see this observe that for $|z| = 1$, we have

$$\begin{aligned}
\sum_{n=-\infty}^{\infty} z^n R_y(n) = \mathbf{S}_y(\overline{z}) \qquad\qquad (11.10.11)\\
= DD^* + C(\overline{z}I - A)^{-1}BB^*(zI - A^*)^{-1}C \\
= DD^* + \sum_{j=0}^{\infty}\sum_{k=0}^{\infty} z^j\overline{z}^k CA^jBB^*A^{*k}C^*.
\end{aligned}$$

Let Q be the observability Gramian for the controllable pair $\{A, B\}$, that is, let Q be the unique solution to the Lyapunov equation

$$Q = AQA^* + BB^*. \qquad (11.10.12)$$

Recall that $Q = \sum_0^\infty A^jBB^*A^{*j}$. By matching like coefficients of z^ν in (11.10.11), we obtain

$$R_y(0) = DD^* + \sum_{j=0}^{\infty} CA^jBB^*A^{*j}C^* = DD^* + CQC^*,$$

$$R_y(-1) = \sum_{j=0}^{\infty} CA^jBB^*A^{*j}A^*C^* = CQA^*C^*,$$

$$\vdots$$

$$R_y(-n) = \sum_{j=0}^{\infty} CA^jBB^*A^{*j}A^{*n}C^* = CQA^{*n}C^*.$$

In other words,

$$R_y(0) = DD^* + CQC^* \quad \text{and} \quad R_y(-n) = CQA^*A^{*n-1}C^* \qquad (n \geq 1).$$

By consulting Theorem 10.1.4 with $\widehat{C} = CQA^*$ and replacing A by A^* and B by C^*, we see that the outer spectral factor Θ_o for $\mathbf{S}_y(\overline{z})$ is given by

$$\begin{aligned} \Theta_o(z) &= D_o + C_o(zI - A^*)^{-1}C^*, \\ D_o &= (R_y(0) - CP_oC^*)^{1/2} = (DD^* + C(Q - P_o)C^*)^{1/2}, \\ C_o &= D_o^{-*}(\widehat{C} - CP_oA^*) = D_o^{-*}(CQA^* - CP_oA^*). \end{aligned} \qquad (11.10.13)$$

Here P_o is the stabilizing solution to the following algebraic Riccati equation

$$P_o = AP_oA^* + (CQA^* - CP_oA^*)^*(R_y(0) - CP_oC^*)^{-1}(CQA^* - CP_oA).$$

Now let $P = Q - P_o$. Then subtracting this algebraic Riccati equation from the Lyapunov equation $Q = AQA^* + BB^*$, we arrive at the algebraic Riccati equation in (11.10.7). Using this P in (11.10.13), the outer spectral factor Θ_o for $\mathbf{S}_y(\overline{z})$ is given by

$$\Theta_o(z) = N^{1/2} + N^{-1/2}CPA^*(zI - A^*)^{-1}C^*.$$

Since $\Theta_o(z)$ is an outer spectral factor for $\mathbf{S}_y(\overline{z})$, it follows that $\Theta(z) = \Theta_o(\overline{z})^*$ is the co-outer spectral factor for $\mathbf{S}_y(z)$. In other words, the results in Chapter 10 can also be used to compute the co-outer spectral factor for $\mathbf{S}_y$.

11.10.2 The Wiener filter for $\{A, B, C, D\}$

As before, consider the stable controllable and observable state space system $\{A, B, C, D\}$ given by (11.10.1) where $u(n)$ and $v(n)$ are independent white noise processes subject to the initial condition $x(-\infty) = 0$. Moreover, we assume that DD^* is invertible. Then our Wiener filtering problem is to find the best causal estimate of $x(n+1)$ given the past $\{y(j)\}_{-\infty}^{n}$, that is, find the best function $\widehat{H}$ in $H^2(\mathcal{Y}, \mathcal{X})$ such that

$$\widehat{x}(n+1) = \sum_{j=-\infty}^{n} \widehat{H}_{n-j}y(j)$$

is the best estimate of $x(n+1)$ given the past $\{y(j)\}_{-\infty}^{n}$. The solution to this Wiener filtering problem is given by the state space system starting at $x(-\infty) = 0$,

$$\begin{aligned} \widehat{x}(n+1) &= (A - K_PC)\widehat{x}(n) + K_Py(n), \\ K_P &= APC^*\left(CPC^* + DD^*\right)^{-1}. \end{aligned} \qquad (11.10.14)$$

The optimal transfer function from $y(n)$ to $\widehat{x}(n+1)$ is given by

$$z(zI - (A - K_PC))^{-1}K_P.$$

In other words, the Wiener filter is precisely the steady state Kalman filter.

To set up this Wiener filtering problem and prove these results, let $\psi(n)$ be the wide sense stationary process determined by $\psi(n) = x(n+1)$. By consulting (11.10.4), we see that

$$\psi(n) = \sum_{k=-\infty}^{n} A^{n-k}Bu(k) \quad \text{or equivalently} \quad \vec{\psi} = L_{z\Omega}\vec{u}. \tag{11.10.15}$$

Recall that $\Omega = (zI - A)^{-1}B$. Notice that $\psi(n)$ and $y(n)$ are jointly wide sense stationary processes. In fact,

$$\begin{aligned} \mathbf{S}_{\psi} &= (zI - A)^{-1}BB^*(zI - A)^{-*}, \\ \mathbf{S}_{\psi y} &= z(zI - A)^{-1}BB^*(zI - A)^{-*}C^*. \end{aligned} \tag{11.10.16}$$

Theorem 11.7.1 shows that $\psi(n)$ is a wide sense stationary process and $\mathbf{S}_{\psi} = \Omega\Omega^*$. Recall that $y(n)$ is a wide sense stationary process. To verify that ψ and y are jointly wide sense stationary it remains to show that $E\vec{\psi}\vec{y}^*$ is a Laurent matrix. To this end, observe that

$$E\vec{\psi}\vec{y}^* = EL_{z\Omega}\vec{u}(L_D\vec{v} + L_G\vec{u})^* = EL_{z\Omega}\vec{u}\vec{u}^*L_G^* = L_{z\Omega}L_{G^*} = L_{z\Omega G^*}.$$

Therefore $E\vec{\psi}\vec{y}^* = L_{\mathbf{S}_{\psi y}}$ equals the Laurent matrix determined by the symbol $z\Omega G^*$. In other words, ψ and y are jointly wide sense stationary, and $\mathbf{S}_{\psi y} = z\Omega G^*$.

The optimal Wiener filter to estimate $\psi(n) = x(n+1)$ given the past $\{y(j)\}_{-\infty}^{n}$ is determined by

$$\widehat{H} = \left[\mathbf{S}_{\psi y}\Theta^{-*}\right]_c \Theta^{-1} \tag{11.10.17}$$

where Θ is the co-outer spectral factor for $\mathbf{S}_y$. To complete this section, it remains to show that this Wiener filter is given by the state space representation in (11.10.14).

Set $\Phi(z) = (zI - A)^{-1}$ and assume that $|z| = 1$. Using the Riccati equation in (11.10.7) with (11.10.10) and $\Phi A = A\Phi$, we arrive at

$$\begin{aligned} \Phi BB^*\Phi^* &= \Phi\left(P - APA^* + APC^*N^{-1}CPA^*\right)\Phi^* \\ &= \left(z^{-1}I + z^{-1}A\Phi\right)P\left(z^{-1}I + z^{-1}A\Phi\right)^* \\ &\quad - \Phi APA^*\Phi^* + \Phi APC^*N^{-1}CPA^*\Phi^* \\ &= P + PA^*\Phi^* + \Phi AP + \Phi APC^*N^{-1}CPA^*\Phi^*. \end{aligned}$$

This readily implies that

$$\Phi BB^*\Phi^* = P + PA^*\Phi^* + \Phi AP + \Phi APC^*N^{-1}CPA^*\Phi^*. \tag{11.10.18}$$

We claim that

$$\mathbf{S}_{\psi y} = zPC^* + zPA^*\Phi^*C^* + z\Phi APC^*N^{-1/2}\Theta^* \tag{11.10.19}$$

where Θ is the co-outer spectral factor for $\mathbf{S}_y$ given in (11.10.8). Using (11.10.18), we arrive at

$$\begin{aligned}
\mathbf{S}_{\psi y} &= z\Phi BB^*\Phi^*C^* \\
&= zPC^* + zPA^*\Phi^*C^* + z\Phi APC^* + z\Phi APC^*N^{-1}CPA^*\Phi^*C^* \\
&= zPC^* + zPA^*\Phi^*C^* + z\Phi APC^*N^{-1/2}\left(N^{1/2} + N^{-1/2}CPA^*\Phi^*C^*\right) \\
&= zPC^* + zPA^*\Phi^*C^* + z\Phi APC^*N^{-1/2}\Theta^*.
\end{aligned}$$

Therefore (11.10.19) holds. By employing (11.10.19), we arrive at

$$\begin{aligned}
\left[\mathbf{S}_{\psi y}\Theta^{-*}\right]_c &= \left[zPC^*\Theta^{-*}\right]_c + \left[zPA^*\Phi^*C^*\Theta^{-*}\right]_c + \left[z\Phi APC^*N^{-1/2}\right]_c \\
&= z\Phi APC^*N^{-1/2}.
\end{aligned}$$

Notice that $\Theta^{-*} = \sum_0^\infty z^n\Xi_n$. Thus $zPC^*\Theta^{-*}$ is anti-causal. The second term drops out because the product of $zPA^*\Phi^*C^*$ with Θ^{-*} is also anti-causal. Hence

$$\left[\mathbf{S}_{\psi y}\Theta^{-*}\right]_c = z\Phi APC^*N^{-1/2}. \tag{11.10.20}$$

Recall that the inverse of Θ is given by (11.10.9). Therefore the Wiener filter is given by

$$\begin{aligned}
\widehat{H} &= \left[\mathbf{S}_{\psi y}\Theta^{-*}\right]_c\Theta^{-1} = z\Phi APC^*N^{-1/2}\Theta^{-1} \\
&= z\Phi APC^*N^{-1/2}\left(N^{-1/2} - N^{-1/2}C(zI-J)^{-1}APC^*N^{-1}\right) \\
&= z\Phi APC^*N^{-1}\left(I - C(zI-J)^{-1}APC^*N^{-1}\right) \\
&= z\Phi\left(I - APC^*N^{-1}C(zI-J)^{-1}\right)APC^*N^{-1} \\
&= z\Phi\left(zI - J - APC^*N^{-1}C\right)(zI-J)^{-1}APC^*N^{-1} \\
&= z\Phi\left(zI - A\right)(zI-J)^{-1}APC^*N^{-1} \\
&= z(zI-J)^{-1}APC^*N^{-1} = z(zI-J)^{-1}K_P.
\end{aligned}$$

Since $K_P = APC^*N^{-1}$ and $J = A - K_PC$, the Wiener filter is determined by

$$\widehat{H}(z) = z(zI - (A - K_PC))^{-1}K_P.$$

Moreover, $\widehat{H}$ admits a Taylor series expansion of the form

$$\widehat{H}(z) = \sum_{k=0}^{\infty} z^{-k}\widehat{H}_k = \sum_{k=0}^{\infty} z^{-k}J^kK_P.$$

The Fourier coefficients $\widehat{H}_k$ for $\widehat{H}$ are determined by $\widehat{H}_k = J^kK_P$ for all integers $k \geq 0$. So the optimal estimate $\widehat{x}(n+1) = \widehat{\psi}(n)$ of $x(n+1) = \psi(n)$ given the past output $\{y(j)\}_{-\infty}^{n}$ is computed by

$$\widehat{x}(n+1) = \widehat{\psi}(n) = \sum_{k=-\infty}^{n} \widehat{H}_{n-k}y(k) = \sum_{k=-\infty}^{n} J^{n-k}K_Py(k).$$

In other words,

$$\begin{aligned}\widehat{x}(n+1) &= \sum_{k=-\infty}^{n} J^{n-k} K_P y(k) \\ &= K_P y(n) + J \sum_{k=-\infty}^{n-1} J^{n-1-k} K_P y(k) \\ &= J\widehat{x}(n) + K_P y(n).\end{aligned}$$

Therefore the state space equation for our Wiener filter is given by (11.10.14), which is precisely the steady state Kalman filter.

11.11 Notes

Kalman filtering started with the seminal paper of Kalman [144] and the continuous time version in Kalman-Bucy [148]. Our approach to Kalman filtering was taken from Luenberger [166]. Theorem 11.5.1 was taken from Section 3.5 in Caines [47]. For further results, historical comments, and a more detailed discussion of Kalman filtering; see Anderson-Moore [11], Davis [66], Caines [47], Kailath [138] and Kailath-Sayed-Hassibi [143]. The connection between the steady state Kalman filter and Wiener filtering is classical.

In general the Kalman filter does not require that the state noise $u(n)$ and $v(n)$ be independent. In this case, the state space set up for the Kalman filter is given by

$$x(n+1) = Ax(n) + u(n) \quad \text{and} \quad y(n) = Cx(n) + v(n) \tag{11.11.1}$$

where $u(n)$ and $v(n)$ are mean zero Gaussian random process which are independent of the initial condition $x(0)$. However, $u(n)$ and $v(n)$ may be correlated. As before, $A(n)$ and $C(n)$ can be function of the time index n and we suppress this index. Moreover, assume that

$$E \begin{bmatrix} u(n) \\ v(n) \end{bmatrix} \begin{bmatrix} u(m)^* & v(m)^* \end{bmatrix} = \begin{bmatrix} R_{11}(n) & R_{12}(n) \\ R_{21}(n) & R_{22}(n) \end{bmatrix} \delta_{n-m}.$$

Here δ_j is the Kronecker delta. It is emphasized that $R_{jk}(n)$ can be a function of n. Let $\mathcal{M}_n = \operatorname{span}\{y(k)\}_0^n$. As before, $\widehat{x}(n) = P_{\mathcal{M}_{n-1}} x(n)$ denotes the optimal state estimate. Finally, the error $\widetilde{x}(n) = x(n) - \widehat{x}(n)$, and

$$Q_n = E\widetilde{x}(n)\widetilde{x}(n)^* = E\left(x(n) - \widehat{x}(n)\right)\left(x(n) - \widehat{x}(n)\right)^*$$

is the error covariance.

In this setting the Kalman filter estimate for the optimal state is given by

$$\begin{aligned}\widehat{x}(n+1) &= A\widehat{x}(n) + \Lambda_n\left(y(n) - C\widehat{x}(n)\right)\\ &= (A - \Lambda_n C)\,\widehat{x}(n) + \Lambda_n y(n),\\ \Lambda_n &= (AQ_nC^* + R_{12})\,(CQ_nC^* + R_{22})^{-1}.\end{aligned} \tag{11.11.2}$$

The recursion for the error covariance is determined by

$$Q_{n+1} = AQ_nA^* + R_{11} - (AQ_nC^* + R_{12})\,(CQ_nC^* + R_{22})^{-1}\,(AQ_nC^* + R_{12})^*$$

where the initial condition $Q_0 = Ex(0)x(0)^*$. Another form for the Riccati difference equation is given by

$$Q_{n+1} = (A - \Lambda_n C)\,Q_n\,(A - \Lambda_n C)^* + \begin{bmatrix} I & -\Lambda_n \end{bmatrix} \begin{bmatrix} R_{11} & R_{12} \\ R_{21} & R_{22} \end{bmatrix} \begin{bmatrix} I \\ -\Lambda_n^* \end{bmatrix}.$$

The proof of this result is a minor modification of the classical Kalman filter and left as an exercise.

To obtain the steady state Kalman filter, assume that $\{C, A\}$ are time invariant. Moreover, assume that

$$R = \begin{bmatrix} R_{11} & R_{12} \\ R_{21} & R_{22} \end{bmatrix} \text{ on } \begin{bmatrix} \mathcal{X} \\ \mathcal{Y} \end{bmatrix} \tag{11.11.3}$$

is positive and R_{22} is strictly positive. Now consider the algebraic Riccati equation determined by

$$P = APA^* + R_{11} - (APC^* + R_{12})\,(CPC^* + R_{22})^{-1}\,(APC^* + R_{12})^*. \tag{11.11.4}$$

By taking the appropriate adjoint in Theorem 10.7.1, we obtain the following result which was taken from Section 3.5 in Caines [47].

Theorem 11.11.1. *Consider the pair $\{C, A \text{ on } \mathcal{X}\}$ where C maps $\mathcal{X}$ into $\mathcal{Y}$. Let Q_n be the solution for the Riccati difference equation*

$$Q_{n+1} = AQ_nA^* + R_{11} - (AQ_nC^* + R_{12})\,(CQ_nC^* + R_{22})^{-1}\,(AQ_nC^* + R_{12})^*$$

subject to the initial condition $Q_0 = 0$. Moreover, assume that R in (11.11.3) is positive and R_{22} is strictly positive. Finally, let $\Delta = R_{11} - R_{12}R_{22}^{-1}R_{21}$ be the Schur complement for R with respect to R_{11}. Then the following holds.

(i) *The solution $\{Q_n\}_0^\infty$ forms an increasing sequence of positive operators. To be precise, $Q_n \le Q_{n+1}$ for all integers $n \ge 0$.*

(ii) *If the pair $\{C, A\}$ is observable, then Q_n converges to a positive operator P as n tends to infinity, that is,*

$$P = \lim_{n\to\infty} Q_n. \tag{11.11.5}$$

In this case, P is a positive solution for the Riccati equation (11.11.4).

(iii) *If* $\{A - R_{12}R_{22}^{-1}C, \Delta\}$ *is controllable and* $\{C, A\}$ *is observable and* P *is any positive solution to the algebraic Riccati equation in* (11.11.4)*, then* P *is strictly positive. Moreover,* $A - K_P C$ *is stable where* K_P *is the operator defined by*

$$K_P = (APC^* + R_{12})(CPC^* + R_{22})^{-1}. \tag{11.11.6}$$

(iv) *If* $\{A - R_{12}R_{22}^{-1}C, \Delta\}$ *is controllable and* $\{C, A\}$ *is observable, then there is only one positive solution* P *to the algebraic Riccati equation in* (11.11.4)*. In this case,* P *is strictly positive. This solution is given by* $P = \lim_{n\to\infty} Q_n$.

Assume that $\{A - R_{12}R_{22}^{-1}C, \Delta\}$ is controllable and $\{C, A\}$ is observable. Let P be the positive solution to the algebraic Riccati equation in (11.11.4) determined by (11.11.5). By passing to limits in the Kalman filter (11.11.2), we arrive at the *steady state Kalman filter* defined by

$$\begin{aligned}\zeta(n+1) &= (A - K_P C)\zeta(n) + K_P y(n), \\ K_P &= (APC^* + R_{12})(CPC^* + R_{22})^{-1}.\end{aligned} \tag{11.11.7}$$

Theorem 11.11.1 guarantees that $A - K_P C$ is stable. The steady state Kalman filter provides an approximation $\zeta(n)$ for the optimal state estimate $\widehat{x}(n)$ for large n or once the system reaches steady state. The Kalman filter converges to the steady state Kalman filter. In other words, the steady state Kalman filter is an optimal state estimator in the limit. Finally, it is noted that under the appropriate assumptions, the steady state Kalman filter can be viewed as a Wiener filter.

Part IV

Interpolation Theory

Chapter 12

Tangential Nevanlinna-Pick Interpolation

In this chapter we will use the Naimark representation theorem, along with state space techniques to obtain a solution to a positive real tangential Nevanlinna-Pick interpolation problem.

12.1 An Introduction to Nevanlinna-Pick Interpolation

Recall that a function F is a *$\mathcal{L}(\mathcal{E},\mathcal{E})$-valued positive real function* if F is a $\mathcal{L}(\mathcal{E},\mathcal{E})$-valued analytic function in $\mathbb{D}_+ = \{z \in \mathbb{C} : |z| > 1\}$ and $\Re F(z) \geq 0$ for all z in $\mathbb{D}_+$. The classical Nevanlinna-Pick interpolation problem is: Given a distinct set of complex numbers $\{\alpha_j\}_1^\nu$ in $\mathbb{D}_+$ and a set of complex numbers $\{\gamma_j\}_1^\nu$, then find a positive real function f satisfying the conditions

$$f(\alpha_j) = \gamma_j \qquad (\text{for } j = 1, 2, \ldots, \nu). \tag{12.1.1}$$

Sz.-Nagy-Korányi [196] was the first to use operator techniques to solve the positive real Nevanlinna-Pick interpolation problem. The classical Nevanlinna-Pick interpolation problem in (12.1.1) is a special case of a more general tangential Nevanlinna-Pick interpolation problem.

To introduce our tangential Nevanlinna-Pick interpolation problem, let A be a stable operator acting on a finite dimensional space $\mathcal{X}$, and let C and $\widetilde{C}$ be two operators mapping $\mathcal{X}$ into $\mathcal{E}$. Throughout we assume that the range of C is onto $\mathcal{E}$, or equivalently, CC^* is invertible. We refer to a triple of operators $\{A, C, \widetilde{C}\}$ with these properties as a *data* set. Given a data set $\{A, C, \widetilde{C}\}$ our tangential Nevanlinna-Pick interpolation problem is to find a $\mathcal{L}(\mathcal{E},\mathcal{E})$-valued positive real function F such that

$$\sum_{n=0}^{\infty} F_n C A^n = \widetilde{C}. \tag{12.1.2}$$

Here $F(z) = \sum_0^\infty z^{-n} F_n$ is the Taylor series expansion of F at infinity. Because A is stable and F is analytic in $\mathbb{D}_+$, the infinite sum in (12.1.2) converges. Following some of the ideas in [84], Chapter 1, its value is denoted by $(FC)(A)_{right}$, that is,

$$(FC)(A)_{right} = \sum_{n=0}^{\infty} F_n C A^n \quad \text{where} \quad F(z) = \sum_{n=0}^{\infty} z^{-n} F_n. \tag{12.1.3}$$

If F is a $\mathcal{L}(\mathcal{E},\mathcal{E})$-valued positive real function and $(FC)(A)_{right} = \widetilde{C}$, then F is called an *interpolant* or a *solution* for the data set $\{A, C, \widetilde{C}\}$.

To show that this interpolation problem covers the classical Nevanlinna-Pick problem, let $\{\alpha_j\}_1^\nu$ be a finite set of distinct points in $\mathbb{D}_+$ and $\{\gamma_j\}_1^\nu$ a set of complex numbers. Let A be the diagonal matrix on $\mathbb{C}^\nu$ defined by $A = \operatorname{diag}[\{1/\alpha_j\}_1^\nu]$. Let C and $\widetilde{C}$ be the row vectors of length ν given by

$$C = \begin{bmatrix} 1 & 1 & \cdots & 1 \end{bmatrix} \quad \text{and} \quad \widetilde{C} = \begin{bmatrix} \gamma_1 & \gamma_2 & \cdots & \gamma_\nu \end{bmatrix}. \tag{12.1.4}$$

Obviously, C is onto $\mathbb{C}$. Now let f be an analytic function in the open unit disc. Then (12.1.1) is equivalent to $\sum_0^\infty f_n C A^n = \widetilde{C}$, where $f(z) = \sum_0^\infty z^{-n} f_n$. So the classical Nevanlinna-Pick interpolation problem of finding a positive real function f satisfying (12.1.1) is a special case of our tangential Nevanlinna-Pick interpolation problem.

In this chapter the Lyapunov equation

$$\Lambda = A^* \Lambda A + C^* \widetilde{C} + \widetilde{C}^* C \tag{12.1.5}$$

plays an important role in our solution of the tangential Nevanlinna-Pick interpolation problem for the data $\{A, C, \widetilde{C}\}$. Since A is stable, equation (12.1.5) has a unique solution Λ. In fact this solution is given by

$$\Lambda = \sum_{n=0}^{\infty} A^{*n} \left(C^* \widetilde{C} + \widetilde{C}^* C \right) A^n. \tag{12.1.6}$$

We shall prove that the Nevanlinna-Pick problem for the data $\{A, C, \widetilde{C}\}$ is solvable if and only if this unique solution Λ is positive.

To state one of our main results, we have to introduce some further notation. Assume that the unique solution Λ of (12.1.5) is positive. Then Λ admits a factorization of the form $\Lambda = M^* M$ where M is an operator from $\mathcal{X}$ onto $\mathcal{H}$, that is, $M\mathcal{X} = \mathcal{H}$. It is well known that $M = \Phi \Lambda^{1/2}$ where Φ is a unitary operator mapping $\Lambda^{1/2}\mathcal{X}$ onto $\mathcal{H}$. The Lyapunov equation in (12.1.5) implies that

$$\|Mx\| = \|MAx\| \qquad (x \in \ker C). \tag{12.1.7}$$

Now consider the spaces

$$\begin{aligned} &\mathcal{H} = M\mathcal{X}, \quad \mathcal{H}_1 = MA \ker C, \quad \mathcal{H}_2 = M \ker C, \\ &\mathcal{D}_1 = \mathcal{H} \ominus \mathcal{H}_1 \quad \text{and} \quad \mathcal{D}_2 = \mathcal{H} \ominus \mathcal{H}_2. \end{aligned} \tag{12.1.8}$$

Then (12.1.7) implies that there exists a partial isometry $T_\circ$ on $\mathcal{H}$ of the form

$$T_\circ = \begin{bmatrix} V & 0 \\ 0 & 0 \end{bmatrix} : \begin{bmatrix} \mathcal{H}_1 \\ \mathcal{D}_1 \end{bmatrix} \rightarrow \begin{bmatrix} \mathcal{H}_2 \\ \mathcal{D}_2 \end{bmatrix}, \tag{12.1.9}$$

where V is the unique unitary operator mapping $\mathcal{H}_1$ onto $\mathcal{H}_2$ such that

$$VMAx = Mx \qquad (x \in \ker C). \tag{12.1.10}$$

(To compute $T_\circ$ in Matlab, let $\Psi = \text{null}(C)$ be the unitary matrix computed from the Matlab command null, whose columns form a basis for the kernel of C. Then compute $T_\circ = M * \Psi * \text{pinv}(M * A * \Psi)$.) We shall refer to $T_\circ$ as the *partial isometry determined* by the unitary operator V. Finally, throughout this chapter N is the operator given by

$$N = C^*(CC^*)^{-1} : \mathcal{E} \rightarrow \mathcal{X}. \tag{12.1.11}$$

Recall that if G is analytic in $\mathbb{D}_+$, then $\widetilde{G}(z) = G(\overline{z})^*$. The proof of the following result will be given in Section 12.3.

Theorem 12.1.1. *The tangential Nevanlinna-Pick interpolation problem for the data set $\{A, C, \widetilde{C}\}$ is solvable if and only if the unique solution Λ of the Lyapunov equation* (12.1.5) *is positive. In this case:*

(i) *A special solution is given by*

$$\begin{aligned} G(z) = {} & \widetilde{C}N + N^*(A^*M^* - M^*T_\circ)MAN \\ & + N^*(M^*T_\circ - A^*M^*)(zI - T_\circ)^{-1}(M - T_\circ MA)N \end{aligned} \tag{12.1.12}$$

where M maps $\mathcal{X}$ onto $\mathcal{H}$ such that $\Lambda = M^*M$ and $T_\circ$ is the partial isometry determined by the unitary operator V in (12.1.10) *while $N = C^*(CC^*)^{-1}$.*

(ii) *The maximal outer spectral factor Θ for the Toeplitz matrix*

$$\Upsilon_G = \begin{bmatrix} G_0 + G_0^* & G_1 & G_2 & \cdots \\ G_1^* & G_0 + G_0^* & G_1 & \cdots \\ G_2^* & G_1^* & G_0 + G_0^* & \cdots \\ \vdots & \vdots & \vdots & \ddots \end{bmatrix} \quad \textit{with} \quad G = \sum_{n=0}^{\infty} z^{-n} G_n$$

is given by

$$\Theta(z) = \Pi_{\mathcal{D}_2} MN + \Pi_{\mathcal{D}_2}(zI - T_\circ^*)^{-1}(T_\circ^* M - MA)N \tag{12.1.13}$$

where $\Pi_{\mathcal{D}_2}^*$ is the canonical embedding of $\mathcal{D}_2$ into $\mathcal{H}$.

Definition 12.1.2. The positive real function G in (12.1.12) is called the *central interpolant* for the data set $\{A, C, \widetilde{C}\}$.

Recall that $\ell_+(\mathcal{E})$ is the linear space consisting of all unilateral sequences $u = \begin{bmatrix} u_0 & u_1 & u_2 & \cdots \end{bmatrix}^{tr}$ where $u_j \in \mathcal{E}$ for all $j \geq 0$. (The transpose is denoted by tr.) Furthermore, $\ell_+^c(\mathcal{E})$ denotes the $\mathcal{E}$-valued sequences in $\ell_+(\mathcal{E})$ with finite support. Obviously,

$$\ell_+^c(\mathcal{E}) \subset \ell_+^2(\mathcal{E}) \subset \ell_+(\mathcal{E}).$$

The usual inner product on $\ell_+^2(\mathcal{E})$ extends to a sesquilinear linear form between $\ell_+^c(\mathcal{E})$ and $\ell_+(\mathcal{E})$. Indeed, let

$$g = \begin{bmatrix} g_0 & g_1 & g_2 & \cdots \end{bmatrix}^{tr} \text{ and } h = \begin{bmatrix} h_0 & h_1 & h_2 & \cdots \end{bmatrix}^{tr}$$

be sequences with entries in $\mathcal{E}$. Then

$$(g, h) = \sum_{j=0}^{\infty} (g_j, h_j) \tag{12.1.14}$$

is well defined if g or h is a sequence of finite support (because in that case the sum is finite) or if g and h are both square summable.

Now, let F be a positive real function with values in $\mathcal{L}(\mathcal{E}, \mathcal{E})$. Define Υ_F to be the Toeplitz operator matrix given by

$$\Upsilon_F = \begin{bmatrix} F_0 + F_0^* & F_1 & F_2 & \cdots \\ F_1^* & F_0 + F_0^* & F_1 & \cdots \\ F_2^* & F_1^* & F_0 + F_0^* & \cdots \\ \vdots & \vdots & \vdots & \ddots \end{bmatrix}, \tag{12.1.15}$$

where $F(z) = \sum_{n=0}^{\infty} z^{-j} F_j$ is the Taylor series expansion for F at infinity. We emphasize that, for notational convenience, we placed $\{F_j\}$ in the first row of Υ_F and $\{F_j^*\}$ in the first column. Observe that $\Upsilon_F = T_{\widetilde{F}} + T_{\widetilde{F}}^{\sharp}$ where $\widetilde{F}(z) = F(\overline{z})^*$. Recall that F is positive real if and only if $\widetilde{F}$ is positive real. Therefore F is positive real if and only if Υ_F defines a positive Toeplitz matrix. Notice that Υ_F maps $\ell_+^c(\mathcal{E})$ into $\ell_+(\mathcal{E})$. So we may consider the optimization problem

$$\gamma(F, u) = \inf\{(\Upsilon_F g, g) : g = \begin{bmatrix} u & g_1 & g_2 & g_3 & \cdots \end{bmatrix}^{tr} \in \ell_+^c(\mathcal{E})\}, \tag{12.1.16}$$

where u is a vector in $\mathcal{E}$. We shall show that there exists a unique positive operator $\Delta(F)$ on $\mathcal{E}$ such that

$$\gamma(F, u) = (\Delta(F)u, u) \qquad (u \in \mathcal{E}).$$

The next theorem is one of our main results. The proof is given in Section 12.4.

Theorem 12.1.3. *Let $\{A, C, \widetilde{C}\}$ be a data set for a tangential Nevanlinna-Pick interpolation problem. Assume that the solution Λ to the Lyapunov equation in (12.1.5) is positive. Finally, let G be the central interpolant and F be an arbitrary interpolant for the data $\{A, C, \widetilde{C}\}$. Then $\Delta(G) \geq \Delta(F)$ with equality if and only if $F = G$.*

Remark 12.1.4. Assume that F is a solution to the tangential Nevanlinna-Pick interpolation problem with data $\{A, C, \widetilde{C}\}$. Let W_o be the observability Gramian for the pair $\{C, A\}$ defined by

$$W_o = \begin{bmatrix} C \\ CA \\ CA^2 \\ \vdots \end{bmatrix} : \mathcal{X} \to \ell_+^2(\mathcal{E}). \tag{12.1.17}$$

Then the solution Λ to the Lyapunov equation in (12.1.5) is given by $\Lambda = W_o^* \Upsilon_F W_o$. In particular, if the Nevanlinna-Pick interpolation problem is solvable, then Λ is a positive operator. (Theorem 12.1.1 shows that the converse is also true.)

To see this let, $\widetilde{W}_o$ be the observability Gramian for $\{\widetilde{C}, A\}$ obtained by replacing C by $\widetilde{C}$ in (12.1.17). Then using $\sum_0^\infty F_n C A^n = (FC)(A)_{right} = \widetilde{C}$, we see that $\widetilde{W}_o = T_{\widetilde{F}} W_o$. (Because A is stable and F is analytic in $\mathbb{D}_+$, the multiplication $T_{\widetilde{F}} W_o$ is well defined.) Hence

$$\begin{aligned} W_o^* \Upsilon_F W_o &= W_o^* \left(T_{\widetilde{F}} + T_{\widetilde{F}}^\sharp \right) W_o = W_o^* \widetilde{W}_o + \widetilde{W}_o^* W_o^* \\ &= \sum_{n=0}^\infty A^{*n} \left(C^* \widetilde{C} + \widetilde{C}^* C \right) A^n = \Lambda. \end{aligned}$$

Therefore Λ is a solution to the Lyapunov equation in (12.1.5). Since Υ_F is positive, $\Lambda = W_o^* \Upsilon_F W_o$ is also positive.

12.2 The First Main Result on Interpolation

The proofs of our results are based on the following theorem which is a restatement of Theorems 5.2.1 and 5.4.1.

Theorem 12.2.1. *Let Υ_F be the Toeplitz matrix in* (12.1.15) *determined by a $\mathcal{L}(\mathcal{E}, \mathcal{E})$-valued sequence of operators $\{F_k\}_0^\infty$. Let F be the function formally defined by $F(z) = \sum_{k=0}^\infty z^{-k} F_k$. Then the following statements are equivalent.*

(i) *The Toeplitz matrix Υ_F is positive.*

(ii) *Υ_F admits controllable isometric representation $\{U \text{ on } \mathcal{K}, \Gamma\}$.*

(iii) *The function F admits a controllable state space realization of the form*

$$F(z) = F_0 + \Gamma^* U (zI - U)^{-1} \Gamma$$

where U is an isometry and $F_0 + F_0^* = \Gamma^* \Gamma$.

(iv) *The function F is positive real.*

In this case, $F_0 = \Gamma^*\Gamma/2 + \Psi$ *where* Ψ *is an operator on* $\mathcal{E}$ *satisfying* $\Psi = -\Psi^*$. *Moreover, all controllable realizations* $\{U, \Gamma, \Gamma^* U, F_o\}$ *of* F *where* U *is an isometry and* $F_0 + F_0^* = \Gamma^*\Gamma$ *are unitarily equivalent. The maximal outer spectral factor for* Υ_F *is determined by*

$$\Theta(z) = z\Pi_{\mathcal{L}}(zI - U^*)^{-1}\Gamma \quad (z \in \mathbb{D}_+)$$

where $\mathcal{L} = \ker U^*$ *and* $\Pi_{\mathcal{L}} : \mathcal{K} \to \mathcal{L}$ *is the orthogonal projection from* $\mathcal{K}$ *onto* $\mathcal{L}$. *Finally, if* F *admits a finite dimensional stable realization, then* U *is a unilateral shift and* Θ *is the outer spectral factor for* Υ_F, *that is,* $T_\Theta^* T_\Theta = \Upsilon_F$.

Remark 12.2.2. Assume that F is a solution to the tangential Nevanlinna-Pick interpolation for the data set $\{A, C, \widetilde{C}\}$. Then F is a positive real function with values in $\mathcal{L}(\mathcal{E}, \mathcal{E})$ and

$$(FC)(A)_{right} = \sum_{j=0}^{\infty} F_j C A^j = \widetilde{C}, \tag{12.2.1}$$

where $F(z) = \sum_0^\infty z^{-j} F_j$. Since F is positive real, there exists a controllable isometric pair $\{U \text{ on } \mathcal{K}, \Gamma\}$ such that

$$F_0 + F_0^* = \Gamma^*\Gamma \quad \text{and} \quad F(z) = F_0 + \Gamma^* U (zI - U)^{-1}\Gamma. \tag{12.2.2}$$

Notice that $F_j = \Gamma^* U^j \Gamma$ for all integers $j \geq 1$. In terms of the isometric pair $\{U, \Gamma\}$, the interpolation condition (12.2.1) is equivalent to the requirement that

$$F_0 C + \sum_{j=1}^{\infty} \Gamma^* U^j \Gamma C A^j = \widetilde{C}. \tag{12.2.3}$$

Since $r_{\text{spec}}(A) < 1$ and U is an isometry, the series in (12.2.3) converges in the operator norm. Set $K = \sum_0^\infty U^j \Gamma C A^j$. Then K satisfies the Lyapunov equation

$$K = UKA + \Gamma C \quad \text{and} \quad \Gamma^* UKA = \widetilde{C} - F_0 C. \tag{12.2.4}$$

The second equality is a reformulation of (12.2.3). It follows that

$$\begin{aligned} K^*K &= (A^*K^*U^* + C^*\Gamma^*)(UKA + \Gamma C) \\ &= A^*K^*KA + C^*\Gamma^*UKA + A^*K^*U^*\Gamma C + C^*\Gamma^*\Gamma C \\ &= A^*K^*KA + C^*(\widetilde{C} - F_0 C) + (\widetilde{C}^* - C^* F_0^*)C + C^*(F_0 + F_0^*)C \\ &= A^*K^*KA + C^*\widetilde{C} + \widetilde{C}^* C. \end{aligned} \tag{12.2.5}$$

Thus $\Lambda = K^*K$ is a positive solution to the Lyapunov equation

$$\Lambda = A^*\Lambda A + C^*\widetilde{C} + \widetilde{C}^* C. \tag{12.2.6}$$

So if the tangential Nevanlinna-Pick problem with data $\{A, C, \widetilde{C}\}$ is solvable, then there exists a positive solution to the Lyapunov equation (12.2.6); see also Remark 12.1.4. We shall also prove the converse, that is, if Λ is positive, then there exists a solution.

Isometric extensions relative to a subspace. Let $\mathcal{H}$ be a Hilbert space. Let V be a unitary operator mapping $\mathcal{H}_1$ onto $\mathcal{H}_2$ where $\mathcal{H}_1$ and $\mathcal{H}_2$ are both subspaces of $\mathcal{H}$. We say that U on $\mathcal{K}$ is an *isometric extension of* V *relative to* $\mathcal{H}$ if U is an isometry on $\mathcal{K}$ such that

$$\mathcal{H} \subset \mathcal{K} \quad \text{and} \quad U|\mathcal{H}_1 = V. \tag{12.2.7}$$

An isometric extension U of V relative to $\mathcal{H}$ is called *minimal* if

$$\mathcal{K} = \bigvee_{k=0}^{\infty} U^k \mathcal{H}. \tag{12.2.8}$$

Finally, two isometric extensions U_1 on $\mathcal{K}_1$ and U_2 on $\mathcal{K}_2$, both relative to $\mathcal{H}$, are said to be *isomorphic* if there exists a unitary operator Φ from $\mathcal{K}_1$ onto $\mathcal{K}_2$ such that $\Phi|\mathcal{H} = I_{\mathcal{H}}$ and $U_2\Phi = \Phi U_1$.

Now let U on $\mathcal{K}$ be an isometric extension of V relative to $\mathcal{H}$. Let $\mathcal{D}_1$, $\mathcal{D}_2$ and $\mathcal{G}$ be the subspaces defined by

$$\mathcal{D}_1 = \mathcal{H} \ominus \mathcal{H}_1, \quad \mathcal{D}_2 = \mathcal{H} \ominus \mathcal{H}_2 \quad \text{and} \quad \mathcal{G} = \mathcal{K} \ominus \mathcal{H}. \tag{12.2.9}$$

Then U admits an operator matrix representation of the form

$$U = \begin{bmatrix} V & 0 & 0 \\ 0 & U_{22} & U_{23} \\ 0 & U_{32} & U_{33} \end{bmatrix} : \begin{bmatrix} \mathcal{H}_1 \\ \mathcal{D}_1 \\ \mathcal{G} \end{bmatrix} \to \begin{bmatrix} \mathcal{H}_2 \\ \mathcal{D}_2 \\ \mathcal{G} \end{bmatrix} \tag{12.2.10}$$

where the operator matrix

$$\begin{bmatrix} U_{22} & U_{23} \\ U_{32} & U_{33} \end{bmatrix} : \begin{bmatrix} \mathcal{D}_1 \\ \mathcal{G} \end{bmatrix} \to \begin{bmatrix} \mathcal{D}_2 \\ \mathcal{G} \end{bmatrix} \tag{12.2.11}$$

is an isometry. The partitions in (12.2.10) and (12.2.11) give us a hint on how to construct an isometric extension of V. Indeed, choose $\mathcal{G} = \ell_+^2(\mathcal{D}_1)$. Now consider the operator $U_\circ$ on $\mathcal{K}_\circ = \mathcal{H} \oplus \ell_+^2(\mathcal{D}_1)$ defined by

$$U_\circ = \begin{bmatrix} V & 0 & 0 \\ 0 & 0 & 0 \\ 0 & \tau & S \end{bmatrix} : \begin{bmatrix} \mathcal{H}_1 \\ \mathcal{D}_1 \\ \ell_+^2(\mathcal{D}_1) \end{bmatrix} \to \begin{bmatrix} \mathcal{H}_2 \\ \mathcal{D}_2 \\ \ell_+^2(\mathcal{D}_1) \end{bmatrix},$$
$$\tau = \begin{bmatrix} I & 0 & 0 & 0 & \cdots \end{bmatrix}^{tr} : \mathcal{D}_1 \to \ell_+^2(\mathcal{D}_1). \tag{12.2.12}$$

Here S is the unilateral shift on $\ell_+^2(\mathcal{D}_1)$ and τ is the isometry embedding $\mathcal{D}_1$ into the first component of $\ell_+^2(\mathcal{D}_1)$. In other words, $U_\circ$ is the isometry given by

$$U_\circ = \begin{bmatrix} V & 0 & 0 & 0 & \cdots \\ 0 & 0 & 0 & 0 & \cdots \\ 0 & I & 0 & 0 & \cdots \\ 0 & 0 & I & 0 & \cdots \\ \vdots & \vdots & \vdots & \vdots & \ddots \end{bmatrix} : \begin{bmatrix} \mathcal{H}_1 \\ \mathcal{D}_1 \\ \mathcal{D}_1 \\ \mathcal{D}_1 \\ \vdots \end{bmatrix} \to \begin{bmatrix} \mathcal{H}_2 \\ \mathcal{D}_2 \\ \mathcal{D}_1 \\ \mathcal{D}_1 \\ \vdots \end{bmatrix}. \tag{12.2.13}$$

Clearly, $U_\circ$ is an isometric extension of V relative to $\mathcal{H}$. Moreover, the closed linear span of $\{U_\circ^k \mathcal{H}\}_0^\infty$ equals $\mathcal{K}_\circ$. So $U_\circ$ is a minimal isometric extension of V. We will refer to $U_\circ$ as the *central extension* of V relative to $\mathcal{H}$.

Let us return to the tangential Nevanlinna-Pick interpolation problem with data $\{A, C, \widetilde{C}\}$. Recall that A is a stable operator on a finite dimensional space $\mathcal{X}$, while C and $\widetilde{C}$ are operators mapping $\mathcal{X}$ into $\mathcal{E}$. Moreover, we assume that C is onto, and hence CC^* is invertible.

Since A is stable, the Lyapunov equation in (12.2.6) has a unique solution Λ. Assume that this solution is positive. Then $\Lambda = M^*M$ where M maps $\mathcal{X}$ onto $\mathcal{H}$. Using $\Lambda = M^*M$, rewrite (12.2.6) in the equivalent form

$$\|Mx\|^2 = \|MAx\|^2 + 2\Re(Cx, \widetilde{C}x) \qquad (x \in \mathcal{X}). \tag{12.2.14}$$

Set

$$\mathcal{H} = M\mathcal{X}, \quad \mathcal{H}_1 = MA \ker C \quad \text{and} \quad \mathcal{H}_2 = M \ker C. \tag{12.2.15}$$

Formula (12.2.14) shows that $\|Mx\| = \|MAx\|$ for each x in $\ker C$. Hence there exists a unique unitary operator V mapping $\mathcal{H}_1$ onto $\mathcal{H}_2$ such that

$$VMAx = Mx \qquad (x \in \ker C). \tag{12.2.16}$$

We shall show that any minimal isometric extension of V relative to $\mathcal{H}$ produces an interpolant F for the data $\{A, C, \widetilde{C}\}$, that is, F is a positive real function and $(FC)(A)_{right} = \widetilde{C}$. Moreover, all interpolants are parameterized by the minimal isometric extension of V relative to $\mathcal{H}$.

Theorem 12.2.3. *Let $\{A, C, \widetilde{C}\}$ be a data set for a tangential Nevanlinna-Pick interpolation problem. Then this problem has a solution if and only if the unique solution Λ of the Lyapunov equation* (12.2.6) *is positive. In this case, all solutions are obtained in the following way: Let $\mathcal{H}$, $\mathcal{H}_1$ and $\mathcal{H}_2$ be the subspaces determined by* (12.2.15)*, and let V be the unitary operator mapping $\mathcal{H}_1$ onto $\mathcal{H}_2$ defined by* (12.2.16)*. Let U on $\mathcal{K}$ be a minimal isometric extension of V relative to $\mathcal{H}$, and set*

$$\begin{aligned} \Gamma &= (M - UMA)N : \mathcal{E} \to \mathcal{K}, \\ N &= C^*(CC^*)^{-1} : \mathcal{E} \to \mathcal{X}. \end{aligned} \tag{12.2.17}$$

Then $\{U, \Gamma\}$ is a controllable isometric pair, the function

$$F(z) = (\widetilde{C} - \Gamma^* UMA)N + \Gamma^* U(zI - U)^{-1}\Gamma \tag{12.2.18}$$

is an interpolant for $\{A, C, \widetilde{C}\}$, and $\{U, \Gamma, \Gamma^*U, F(\infty)\}$ is a controllable realization for F. Moreover, there is a one to one correspondence between the set of all interpolants for $\{A, C, \widetilde{C}\}$ and the (set of classes of unitarily equivalent) minimal isometric extensions of V relative to the subspace $\mathcal{H}$. Finally, it is noted that

$$\Upsilon_F = W^\sharp W \quad \textit{where} \quad W = \begin{bmatrix} \Gamma & U\Gamma & U^2\Gamma & \cdots \end{bmatrix}. \tag{12.2.19}$$

Proof. We split the proof into five parts. In the first three parts we assume that Λ is a positive solution to (12.2.6), and that U on $\mathcal{K}$ is a specified minimal isometric extension of V. In the two final parts we assume that F is an interpolant for the data $\{A, C, \widetilde{C}\}$.

Part 1. Assume that Λ is positive. Let V be the unitary operator mapping $\mathcal{H}_1$ onto $\mathcal{H}_2$ defined by (12.2.16). Let U on $\mathcal{K}$ be a minimal isometric extension of V relative to $\mathcal{H}$, and define Γ by (12.2.17). Notice that M maps $\mathcal{X}$ into $\mathcal{H} \subset \mathcal{K}$. Thus the product UM makes sense, and $M - UMA$ maps $\mathcal{X}$ into $\mathcal{K}$. Hence the operator Γ in (12.2.17) is well defined. Obviously, $\{U, \Gamma\}$ is an isometric pair. To prove that $\{U, \Gamma\}$ is controllable, we first show that

$$\Gamma C = M - UMA. \tag{12.2.20}$$

According to the definition of V we have

$$Mx - UMAx = Mx - VMAx = 0 \qquad (x \in \ker C). \tag{12.2.21}$$

Next, observe that $P = C^*(CC^*)^{-1}C = NC$ is the orthogonal projection onto the range of C^*. Hence $I - P$ is the orthogonal projection onto $\ker C$. Thus (12.2.21) implies that the operator product $(M - UMA)(I - P) = 0$. This readily implies that

$$\begin{aligned}\Gamma C &= (M - UMA)NC = (M - UMA)P \\ &= (M - UMA)\big(I - (I - P)\big) \\ &= M - UMA.\end{aligned}$$

In other words, (12.2.20) holds. From (12.2.20) we see that M is the solution to the Lyapunov equation $M = UMA + \Gamma C$. Thus for x in $\mathcal{X}$, we have

$$Mx = \sum_{k=0}^{\infty} U^k \Gamma C A^k x \quad \text{and hence} \quad \mathcal{H} = M\mathcal{X} \subset \bigvee_{k=0}^{\infty} U^k \Gamma \mathcal{E}. \tag{12.2.22}$$

Notice that $\bigvee_{k=0}^{\infty} U^k \Gamma \mathcal{E}$ is an invariant subspace for U. So for any integer $n \geq 0$, the subspace $U^n \mathcal{H}$ is also contained in $\bigvee_{k=0}^{\infty} U^k \Gamma \mathcal{E}$. Therefore $\bigvee_{k=0}^{\infty} U^k \mathcal{H} \subset \bigvee_{k=0}^{\infty} U^k \Gamma \mathcal{E}$. Since U is a minimal isometric extension of V relative to $\mathcal{H}$, we conclude that the pair $\{U, \Gamma\}$ is controllable.

Part 2. Let $\{U, \Gamma\}$ be as in the previous part. Now we will show that F in (12.2.18) is an interpolant for $\{A, C, \widetilde{C}\}$. Let H be the positive real function defined by

$$H(z) = \frac{1}{2}\Gamma^*\Gamma + \Gamma^* U (zI - U)^{-1}\Gamma,$$

and set $\widehat{C} = (HC)(A)_{right}$. Thus H is an interpolant or solution for the data $\{A, C, \widehat{C}\}$. According to (12.2.5) in Remark 12.2.2 with $\widehat{C}$ replacing $\widetilde{C}$, this implies that

$$K^*K = A^*(K^*K)A + C^*\widehat{C} + \widehat{C}^*C,$$

where $K = \sum_{j=0}^{\infty} U^j \Gamma C A^j$. However, from the first equality in (12.2.22) we know that $Kx = Mx$ for each $x \in \mathcal{X}$, and hence $K^*K = \Lambda$. So Λ is also a solution to the Lyapunov equation

$$\Lambda = A^*\Lambda A + C^*\widehat{C} + \widehat{C}^*C.$$

By subtracting this from the Lyapunov equation (12.2.6), we have $C^*\Delta + \Delta^*C = 0$ where $\Delta = \widetilde{C} - \widehat{C}$. Multiplying by C on the left and rearranging terms, yields $CC^*\Delta = -C\Delta^*C$. Thus $\Delta = \Psi C$ where $\Psi = -(CC^*)^{-1}C\Delta^*$ is an operator on $\mathcal{E}$. Substituting this into $C^*\Delta + \Delta^*C = 0$, yields $C^*(\Psi + \Psi^*)C = 0$. Since C is onto $\mathcal{E}$, we have $\Psi = -\Psi^*$. Therefore

$$\widetilde{C} = \widehat{C} + \Psi C \quad \text{where} \quad \Psi = -\Psi^*. \tag{12.2.23}$$

Clearly, $G = \Psi + H$ is a positive real function. Moreover, $G(\infty) = \frac{1}{2}\Gamma^*\Gamma + \Psi$. Furthermore, (12.2.23) implies that

$$(GC)(A)_{right} = (HC)(A)_{right} + \Psi C = \widehat{C} + \Psi C = \widetilde{C}.$$

In other words, G is a positive real function satisfying $(GC)(A)_{right} = \widetilde{C}$. Finally, it is noted that G admits a controllable realization of the form

$$G(z) = G_0 + \Gamma^*U(zI - U)^{-1}\Gamma \quad \text{where} \quad G_0 + G_0^* = \Gamma^*\Gamma \tag{12.2.24}$$

and $\{U, \Gamma\}$ is a controllable isometric pair.

Let $\{G_j\}_0^\infty$ be the Taylor coefficients of G at infinity, that is, $G = \sum_0^\infty z^{-n}G_n$. Since $G_j = \Gamma^*U^j\Gamma$ for all integers $j \geq 1$, we can use the first equality in (12.2.22) and (12.2.20) to show that

$$\begin{aligned}
\widetilde{C} &= \sum_{j=0}^{\infty} G_j C A^j = G_0 C + \sum_{j=1}^{\infty} \Gamma^* U^j \Gamma C A^j \\
&= G_0 C + \Gamma^* U \left(\sum_{j=0}^{\infty} U^j \Gamma C A^j \right) A \\
&= G_0 C + \Gamma^* U M A.
\end{aligned}$$

Thus $G_0 C = \widetilde{C} - \Gamma^*UMA$. By multiplying this identity on the right by $N = C^*(CC^*)^{-1}$, we see that $G_0 = F(\infty)$; see (12.2.18). Hence $G = F$, and F is an interpolant for $\{A, C, \widetilde{C}\}$.

In particular, if Λ is positive, then F is a solution to the Nevanlinna-Pick interpolation problem with data $\{A, C, \widetilde{C}\}$. Combining this with Remark 12.2.2 or Remark 12.1.4, we see that the Nevanlinna-Pick interpolation problem has a solution if and only if Λ is positive.

Part 3. Assume $\widehat{U}$ on $\widehat{\mathcal{K}}$ is an isometric extension of V that is isomorphic to U on $\mathcal{K}$, where U is the isometric extension of V considered in the two previous parts. Thus there exists a unitary operator Φ from $\mathcal{K}$ onto $\widehat{\mathcal{K}}$ such that

$$\Phi U = \widehat{U}\Phi \quad \text{and} \quad \Phi|\mathcal{H} = I_{\mathcal{H}}. \tag{12.2.25}$$

Let $\widehat{\Gamma}$ and $\widehat{F}(z)$ be defined as in (12.2.17) and (12.2.18) with $\widehat{U}$ in place of U. We claim that $\widehat{F} = F$. To prove this we first show that $\Phi\Gamma = \widehat{\Gamma}$. Recall that $\mathcal{H} = M\mathcal{X}$. Thus $\Phi|\mathcal{H} = I_{\mathcal{H}}$ shows that $\Phi Mx = Mx$ for each $x \in \mathcal{X}$, and

$$\begin{aligned}\Phi\Gamma &= (M - \Phi UMA)N = (M - \widehat{U}\Phi MA)N\\ &= (M - \widehat{U}MA)N = \widehat{\Gamma}.\end{aligned}$$

It follows that the pairs $\{U, \Gamma\}$ and $\{\widehat{U}, \widehat{\Gamma}\}$ are unitarily equivalent. To verify that $F = \widehat{F}$ it suffices to show that $F(\infty) = \widehat{F}(\infty)$. This follows from $\Phi^* = \Phi^{-1}$ and the fact that

$$\widehat{\Gamma}^*\widehat{U}M = \Gamma^*\Phi^*\widehat{U}M = \Gamma^*U\Phi^*M = \Gamma^*UM.$$

Part 4. Let F be an interpolant for the data $\{A, C, \widetilde{C}\}$. In this part, we prove that F can be represented as in (12.2.18). Choose a controllable isometric pair $\{U, \Gamma\}$ such that (12.2.2) holds. As we have seen in (12.2.5) of Remark 12.2.2, this implies that the Lyapunov equation in (12.2.6) has a positive solution, namely $\Lambda = K^*K$ where $K = \sum_{j=0}^{\infty} U^j\Gamma CA^j$. The two identities in (12.2.4) yield

$$\Gamma C = K - UKA \quad \text{and} \quad F_0 C = \widetilde{C} - \Gamma^*UKA. \tag{12.2.26}$$

By multiplying these equalities on the right by $N = C^*(CC^*)^{-1}$, we obtain

$$\Gamma = (K - UKA)N \quad \text{and} \quad F_0 = (\widetilde{C} - \Gamma^*UKA)N.$$

It follows that F is of the form (12.2.18) provided that we replace M by K in both (12.2.18) and (12.2.17).

Now set

$$\mathcal{H}_K = K\mathcal{X}, \quad \mathcal{H}_{1,K} = KA\ker C \quad \text{and} \quad \mathcal{H}_{2,K} = K\ker C.$$

The first identity in (12.2.26) also implies that $UKAx = Kx$ for each $x \in \ker C$. Thus there exists a unique unitary operator V_K mapping $\mathcal{H}_{1,K}$ onto $\mathcal{H}_{2,K}$ such that

$$V_K KAx = Kx \qquad (x \in \ker C). \tag{12.2.27}$$

In particular, U is an isometric extension of V_K relative to $\mathcal{H}_K$. Using $\Gamma C = K - UKA$ and the fact that C is onto, we see that $\Gamma\mathcal{E}$ is contained in $\mathcal{H}_K \bigvee U\mathcal{H}_K$. Thus $U^n\Gamma\mathcal{E}$ belongs to $\bigvee_{k=0}^{\infty} U^k\mathcal{H}_K$ for all n. Since the pair $\{U, \Gamma\}$ is controllable, this shows that U is a minimal isometric extension of V_K.

Recall that $\Lambda = K^*K$. So there exists a unitary operator Φ from $\mathcal{H} = M\mathcal{X}$ onto $\mathcal{H}_K$ such that $\Phi M = K$. In particular,

$$\Phi\mathcal{H}_1 = \mathcal{H}_{1,K}, \quad \Phi\mathcal{H}_2 = \mathcal{H}_{2K} \quad \text{and} \quad \Phi V = V_K\Phi|\mathcal{H}_1.$$

Hence without loss of generality we may assume that $K = M$ and $V_K = V$. In that case U is a minimal isometric extension of V, and F has the desired representation.

Part 5. Let U on $\mathcal{K}$ and $\widetilde{U}$ on $\widetilde{\mathcal{K}}$ be minimal isometric extensions of V relative to $\mathcal{H}$. Define Γ and $F(z)$ as in (12.2.17) and (12.2.18), respectively. Let $\widetilde{\Gamma}$ and $\widetilde{F}(z)$ be defined in the same way with $\widetilde{U}$ in place of U. Assume that $F = \widetilde{F}$. In this part we show that U and $\widetilde{U}$ are unitary equivalents as extensions of V relative to $\mathcal{H}$. From the results proved in Part one we know that $\{U, \Gamma\}$ and $\{\widetilde{U}, \widetilde{\Gamma}\}$ are controllable isometric pairs, and the realizations for F and $\widetilde{F}$ provided by (12.2.18) are controllable realizations. Since $F = \widetilde{F}$, it follows that $\{U, \Gamma\}$ and $\{\widetilde{U}, \widetilde{\Gamma}\}$ are unitarily equivalent, that is, there exists a unitary operator Φ from $\mathcal{K}$ onto $\widetilde{\mathcal{K}}$ such that $\Phi\Gamma = \widetilde{\Gamma}$ and $\Phi U = \widetilde{U}\Phi$. Recall that

$$Mx = \sum_{k=0}^{\infty} U^k\Gamma C A^k x \quad \text{and} \quad Mx = \sum_{k=0}^{\infty} \widetilde{U}^k\widetilde{\Gamma} C A^k x \qquad (x \in \mathcal{X}).$$

It follows that $\Phi Mx = Mx$ for each $x \in \mathcal{X}$. But then Φ acts as the identity operator on $\mathcal{H} = M\mathcal{X}$. Since $\Phi U = \widetilde{U}\Phi$, we conclude that U and $\widetilde{U}$ are unitarily equivalent isometric extensions of V relative to $\mathcal{H}$. $\square$

12.3 Proof of Theorem 12.1.1

This section is devoted to a proof of Theorem 12.1.1. First we use the central isometric extension in (12.2.13) to present a special solution to the tangential Nevanlinna-Pick interpolation problem for the data $\{A, C, \widetilde{C}\}$. Throughout we assume that the solution Λ to the Lyapunov equation in (12.2.6) is positive. Moreover, $\Lambda = M^*M$, where M maps $\mathcal{X}$ onto $\mathcal{H} = M\mathcal{X}$. As before, $\mathcal{H}_1 = MA\ker C$ and $\mathcal{H}_2 = M\ker C$. Recall that V is the unitary operator mapping $\mathcal{H}_1$ onto $\mathcal{H}_2$ determined by $VMAx = Mx$ where x is a vector in $\ker C$. Finally, the central extension $U_\circ$ of V relative to $\mathcal{H}$ is the isometry on $\mathcal{K}_\circ = \mathcal{H} \oplus \ell_+^2(\mathcal{D}_1)$ defined by

$$U_\circ = \begin{bmatrix} V & 0 & 0 & 0 & \cdots \\ 0 & 0 & 0 & 0 & \cdots \\ 0 & I & 0 & 0 & \cdots \\ 0 & 0 & I & 0 & \cdots \\ \vdots & \vdots & \vdots & \vdots & \ddots \end{bmatrix} : \begin{bmatrix} \mathcal{H}_1 \\ \mathcal{D}_1 \\ \mathcal{D}_1 \\ \mathcal{D}_1 \\ \vdots \end{bmatrix} \to \begin{bmatrix} \mathcal{H}_2 \\ \mathcal{D}_2 \\ \mathcal{D}_1 \\ \mathcal{D}_1 \\ \vdots \end{bmatrix}. \tag{12.3.1}$$

Here $\mathcal{D}_1 = \mathcal{H} \ominus \mathcal{H}_1$ and $\mathcal{D}_2 = \mathcal{H} \ominus \mathcal{H}_2$.

According to Theorem 12.2.3 the function G defined by

$$G(z) = (\widetilde{C} - \Gamma_\circ^* U_\circ MA)N + \Gamma_\circ^* U_\circ (zI - U_\circ)^{-1}\Gamma_\circ, \tag{12.3.2}$$
$$\Gamma_\circ = (M - U_\circ MA)N \tag{12.3.3}$$

is a positive real interpolant for the data $\{A, C, \widetilde{C}\}$. The function G in (12.3.2) is called the *central interpolant* for the data $\{A, C, \widetilde{C}\}$. Let $T_\circ = VP_{\mathcal{H}_1}$ be the partial isometry on $\mathcal{H}$ determined by V; see (12.1.9).

We claim that the central interpolant is also determined by equation (12.1.12) in Theorem 12.1.1, that is,

$$\begin{aligned} G(z) &= \widetilde{C}N + N^*(A^*M^* - M^*T_\circ)MAN \\ &\quad + N^*(M^*T_\circ - A^*M^*)(zI - T_\circ)^{-1}(M - T_\circ MA)N. \end{aligned} \tag{12.3.4}$$

Using $U_\circ^*\mathcal{H} \subset \mathcal{H}$ and $U_\circ^*|\mathcal{H} = T_\circ^*$ in the formula for $\Gamma_\circ$ in (12.3.3), it follows that $U_\circ^*\Gamma_\circ\mathcal{E}$ is contained in $\mathcal{H}$. Moreover,

$$\begin{aligned} U_\circ^*\Gamma_\circ &= (T_\circ^* M - MA)N, \\ \Gamma_\circ^* U_\circ &= N^*(M^*T_\circ - A^*M^*)P_{\mathcal{H}}. \end{aligned} \tag{12.3.5}$$

The second equality follows by taking the adjoint. As expected, $P_{\mathcal{H}}$ is the orthogonal projection onto $\mathcal{H}$. By employing $P_{\mathcal{H}}U_\circ = T_\circ P_{\mathcal{H}}$ in the definition of $\Gamma_\circ$, we obtain

$$P_{\mathcal{H}}\Gamma_\circ = (M - T_\circ MA)N. \tag{12.3.6}$$

Using (12.3.5), we have

$$(\widetilde{C} - \Gamma_\circ^* U_\circ MA)N = \widetilde{C}N - N^*(M^*T_\circ - A^*M^*)MAN. \tag{12.3.7}$$

Substituting (12.3.5), (12.3.6), (12.3.7) and $P_{\mathcal{H}}U_\circ = T_\circ P_{\mathcal{H}}$ into (12.3.2), we see that the central interpolant for the data $\{A, C, \widetilde{C}\}$ is given by (12.3.4).

Proof of Part (ii) *of Theorem* 12.1.1. Recall that $\{U_\circ, \Gamma_\circ, \Gamma_\circ^* U_\circ, G(\infty)\}$ is a controllable realization for the central solution G and $\{U_\circ, \Gamma_\circ\}$ is a controllable isometric pair. Notice that $\mathcal{D}_2 = \ker U_\circ^*$ where $\mathcal{D}_2 = \mathcal{H} \ominus M \ker C$. By consulting (12.2.19), we see that $\Upsilon_G = W^\sharp W$ where

$$W = \begin{bmatrix} \Gamma_\circ & U_\circ\Gamma_\circ & U_\circ^2\Gamma_\circ & \cdots \end{bmatrix}.$$

According to Theorem 5.2.1 in Chapter 5, the function

$$\Theta(z) = z\Pi_{\mathcal{D}_2}(zI - U_\circ^*)^{-1}\Gamma_\circ = \sum_{k=0}^{\infty} z^{-k}\Pi_{\mathcal{D}_2}U_\circ^{*k}\Gamma_\circ \qquad (z \in \mathbb{D}_+) \tag{12.3.8}$$

is the maximal outer spectral factor for Υ_G. Recall that $\Gamma_\circ = MN - U_\circ MAN$; see (12.3.3). Since $\Pi_{\mathcal{D}_2}U_\circ = 0$, this yields $\Pi_{\mathcal{D}_2}\Gamma_\circ = \Pi_{\mathcal{D}_2}MN$. Using $U_\circ^*\Gamma_\circ = (T_\circ^*M - MA)N$ with $k > 0$ and $U_\circ^*|\mathcal{H} = T_\circ^*$, we obtain

$$\Pi_{\mathcal{D}_2}U_\circ^{*k}\Gamma_\circ = \Pi_{\mathcal{D}_2}U_\circ^{*k-1}U_\circ^*\Gamma_\circ = \Pi_{\mathcal{D}_2}T_\circ^{*k-1}(T_\circ^*M - MA)N.$$

Substituting this into (12.3.8) with $\Pi_{\mathcal{D}_2}\Gamma_\circ = \Pi_{\mathcal{D}_2} MN$, yields

$$\begin{aligned}\Theta(z) &= \Pi_{\mathcal{D}_2}\Gamma_\circ + \sum_{k=1}^{\infty} z^{-k}\Pi_{\mathcal{D}_2} U_\circ^{*k}\Gamma_\circ \\ &= \Pi_{\mathcal{D}_2} MN + \sum_{k=1}^{\infty} z^{-k}\Pi_{\mathcal{D}_2} T_\circ^{*k-1}(T_\circ^* M - MA)N \\ &= \Pi_{\mathcal{D}_2} MN + \Pi_{\mathcal{D}_2}(zI - T_\circ^*)^{-1}(T_\circ^* M - MA)N.\end{aligned}$$

Therefore Θ in (12.1.13) is the maximal outer spectral factor for Υ_G. □

12.4 Proof of Theorem 12.1.3

In this section we will show that the central interpolant G in (12.1.12) satisfies a maximum principle. To this end, let F be a positive real function with values in $\mathcal{L}(\mathcal{E},\mathcal{E})$, and Υ_F the corresponding positive Toeplitz matrix defined in (12.1.15). Recall that $\gamma(F,u)$ is the cost in the optimization problem

$$\gamma(F,u) = \inf\{(\Upsilon_F g, g) : g = \begin{bmatrix} u & g_1 & g_2 & g_3 & \dots \end{bmatrix}^{tr} \in \ell_+^c(\mathcal{E})\}. \qquad (12.4.1)$$

Here u is a specified vector in $\mathcal{E}$ which is also viewed as the first component of g in $\ell_+^c(\mathcal{E})$.

Proposition 12.4.1. *Let F be a positive real function, and $\{U$ on $\mathcal{K},\Gamma,\Gamma^*U,F_0\}$ the controllable realization for F where $\{U,\Gamma\}$ is a controllable isometric pair. Let $\Delta(F)$ be the positive operator on $\mathcal{E}$ given by*

$$\Delta(F) = \Gamma^* P_{\mathcal{L}}\Gamma \qquad (12.4.2)$$

where $P_{\mathcal{L}}$ is the orthogonal projection onto the subspace $\mathcal{L} = \mathcal{K}\ominus U\mathcal{K}$. Then $(\Delta(F)u,u) = \gamma(F,u)$ for all u in $\mathcal{E}$.

Proof. Recall that $\Upsilon_F = W^\sharp W$ where W is the controllability matrix determined by the pair $\{U,\Gamma\}$; see (12.2.19). Let S be the forward shift on $\ell_+^c(\mathcal{E})$. Let $v = \begin{bmatrix} u & 0 & 0 & \dots \end{bmatrix}^{tr}$ where u is in $\mathcal{E}$, and f be any vector in $\ell_+^c(\mathcal{E})$. Using $WS = UW$ and $P_{\mathcal{L}} = I - UU^*$, we have

$$\begin{aligned}(\Upsilon_F(v-Sf),(v-Sf)) &= \|W(v-Sf)\|^2 \\ &= \|Wv\|^2 - 2\Re(Wv, WSf) + \|WSf\|^2 \\ &= \|\Gamma u\|^2 - 2\Re(\Gamma u, UWf) + \|UWf\|^2 \\ &= \|P_{\mathcal{L}}\Gamma u\|^2 + \|UU^*\Gamma u\|^2 - 2\Re(U^*\Gamma u, Wf) + \|Wf\|^2 \\ &= (\Gamma^* P_{\mathcal{L}}\Gamma u, u) + \|U^*\Gamma u - Wf\|^2 \geq (\Gamma^* P_{\mathcal{L}}\Gamma u, u).\end{aligned}$$

This readily shows that $\gamma(F,u) \geq (\Gamma^* P_{\mathcal{L}}\Gamma u, u)$. Because the span of $\{U^k\Gamma\mathcal{E}\}_0^\infty$ is dense in $\mathcal{K}$, we can choose a sequence $\{f_k\}$ in $\ell_+^c(\mathcal{E})$ such that $\|U^*\Gamma u - W f_k\|^2$ converges to zero as k tends to infinity. Therefore $\gamma(F,u) = (\Delta(F)u, u)$ where $\Delta(F) = \Gamma^* P_{\mathcal{L}}\Gamma$. □

The following result (which contains Theorem 12.1.3) shows that the central solution satisfies a maximum principle, or is the unique "maximal entropy solution" to the tangential Nevanlinna-Pick interpolation problem.

Theorem 12.4.2. *Let $\{A, C, \widetilde{C}\}$ be a data set for a tangential Nevanlinna-Pick interpolation problem. Assume that the solution Λ to the Lyapunov equation in* (12.2.6) *is positive. Finally, let G be the central interpolant and F be an arbitrary interpolant for the data $\{A, C, \widetilde{C}\}$. Then the following holds.*

(i) $\Delta(G) \geq \Delta(F)$ *with equality if and only if $F = G$;*

(ii) $\Delta(G) = N^*M^*P_{\mathcal{D}_2}MN$ *where $P_{\mathcal{D}_2}$ is the orthogonal projection onto the subspace $\mathcal{D}_2 = \mathcal{H} \ominus M \ker C$.*

Proof. Recall that $\{U_\circ \text{ on } \mathcal{K}_\circ, \Gamma_\circ, \Gamma_\circ^* U_\circ, G(\infty)\}$ is a controllable realization for G where $U_\circ$ is the isometry defined in (12.3.1). Let $\mathcal{E}_\circ = \mathcal{K}_\circ \ominus U_\circ\mathcal{K}_\circ$. By consulting the form of $U_\circ$ in (12.3.1), it follows that $\mathcal{L}_\circ$ is the subspace of $\mathcal{H}$ given by $\mathcal{L}_\circ = \mathcal{D}_2 = \mathcal{H} \ominus \mathcal{H}_2$ where $\mathcal{H}_2 = M\ker C$. Since $\mathcal{L}_\circ$ is orthogonal to the range of $U_\circ$, we have $P_{\mathcal{L}_\circ}U_\circ = 0$. Applying $P_{\mathcal{L}_\circ}$ on the left of the Lyapunov equation $M = U_\circ MA + \Gamma_\circ C$ (see (12.2.20) where $\{U_\circ, \Gamma_\circ\}$ replaces $\{U, \Gamma\}$), yields $P_{\mathcal{L}_\circ}M = P_{\mathcal{L}_\circ}\Gamma_\circ C$. Multiplying by $N = C^*(CC^*)^{-1}$ on the right, gives $P_{\mathcal{L}_\circ}\Gamma_\circ = P_{\mathcal{L}_\circ}MN$. According to Proposition 12.4.1, we have

$$\Delta(G) = \Gamma_\circ^* P_{\mathcal{L}_\circ}\Gamma_\circ = N^*M^*P_{\mathcal{D}_2}MN.$$

Therefore part *(ii)* holds.

Let F be any positive real function satisfying $(FC)(A)_{right} = \widetilde{C}$. Then F admits a controllable realization $\{U \text{ on } \mathcal{K}, \Gamma, \Gamma^*U, F_0\}$ where U is a minimal isometric extension of V relative to $\mathcal{H}$. Moreover, (12.2.20) shows that $M = UMA + \Gamma C$. Using $U|\mathcal{H}_1 = V$, it follows that $\mathcal{L} = \mathcal{K} \ominus U\mathcal{K}$ is orthogonal to $\mathcal{H}_2 = V\mathcal{H}_1$. Let $P_{\mathcal{L}}$ be the orthogonal projection onto $\mathcal{L}$. By applying $P_{\mathcal{L}}$ on the left of the Lyapunov equation $M = UMA + \Gamma C$, we obtain $P_{\mathcal{L}}M = P_{\mathcal{L}}\Gamma C$. Recall that $\mathcal{H} = \mathcal{H}_2 \oplus \mathcal{L}_\circ$. For any vector x in $\mathcal{X}$, Proposition 12.4.1 and the identity $P_{\mathcal{L}_\circ}M = P_{\mathcal{L}_\circ}\Gamma_\circ C$ gives

$$\begin{aligned}(\Delta(F)Cx, Cx) &= \|P_{\mathcal{L}}\Gamma Cx\|^2 = \|P_{\mathcal{L}}Mx\|^2 = \|P_{\mathcal{L}}(P_{\mathcal{H}_2} + P_{\mathcal{L}_\circ})Mx\|^2 \\ &= \|P_{\mathcal{L}}P_{\mathcal{L}_\circ}Mx\|^2 \leq \|P_{\mathcal{L}_\circ}Mx\|^2 = \|P_{\mathcal{L}_\circ}\Gamma_\circ Cx\|^2 \\ &= (\Delta(G)Cx, Cx).\end{aligned} \tag{12.4.3}$$

Because the range of C equals $\mathcal{E}$, we obtain $\Delta(F) \leq \Delta(G)$.

To complete the proof, it remains to show that if $\Delta(G) = \Delta(F)$, then $G = F$. To this end, recall that $\widetilde{U}$ on $\widetilde{\mathcal{K}}$ is a minimal isometric lifting for a contraction

T on $\mathcal{V}$ if $\mathcal{V} \subset \widetilde{\mathcal{K}}$ is an invariant subspace for $\widetilde{U}^*$ satisfying $\widetilde{U}^*|\mathcal{V} = T^*$ and the closed linear span of $\{\widetilde{U}^k\mathcal{V}\}_0^\infty$ equals $\widetilde{\mathcal{K}}$. Notice that $U_\circ$ on $\mathcal{K}_\circ$ is a minimal isometric lifting of $T_\circ$. Finally, it is noted that two minimal isometric liftings of the same contraction are unitarily equivalent; see Theorem 5.5.1. Now assume that $\Delta(G) = \Delta(F)$. As before, let $\{U, \Gamma, \Gamma^* U, F(\infty)\}$ be a controllable realization for F where U is an isometry. We claim that U is a minimal isometric lifting for $T_\circ$. For all x in $\mathcal{X}$, equation (12.4.3) implies that $\|P_\mathcal{L} P_{\mathcal{L}_\circ} Mx\| = \|P_{\mathcal{L}_\circ} Mx\|$. Hence $\mathcal{H} \ominus \mathcal{H}_2 = \mathcal{L}_\circ \subset \mathcal{L}$. In particular, $U^* P_{\mathcal{L}_\circ} = 0$. Since $U|\mathcal{H}_1 = V$, we have $U^*|\mathcal{H}_2 = V^*$. For h in $\mathcal{H}$, we obtain

$$U^* h = U^* P_{\mathcal{H}_2} h + U^* P_{\mathcal{L}_\circ} h = V^* P_{\mathcal{H}_2} h = T_\circ^* h \qquad (h \in \mathcal{H}).$$

In other words, $\mathcal{H}$ is an invariant subspace for U^* satisfying $U^*|\mathcal{H} = T_\circ^*$. Since U is a minimal isometric extension of V relative to $\mathcal{H}$, the space $\mathcal{K}$ equals the closed linear span of $\{U^k\mathcal{H}\}_0^\infty$. Therefore U is a minimal isometric lifting of $T_\circ$. Because $U_\circ$ is a minimal isometric lifting of $T_\circ$, the operators $U_\circ$ and U are unitarily equivalent. So without loss of generality we can assume that $U = U_\circ$. In this case, the corresponding Lyapunov equations yield

$$\Gamma_\circ C = M - U_\circ MA = M - UMA = \Gamma C.$$

Because C is onto $\mathcal{E}$, we have $\Gamma_\circ = \Gamma$.

Since $U_\circ = U$ and $\Gamma_\circ = \Gamma$, the function $F(z) = F_0 + \Gamma_\circ^* U_\circ (zI - U_\circ)^{-1} \Gamma_\circ$. By consulting the state space realization for G in (12.3.2), we see that $G = D + F$ where D is an operator on $\mathcal{E}$. Hence

$$\widetilde{C} = (GC)(A)_{right} = DC + (FC)(A)_{right} = DC + \widetilde{C}.$$

So $DC = 0$. Because C is onto, $D = 0$. Therefore $G = F$. □

12.5 The Case when Λ is Strictly Positive

In this section we will develop explicit state space formulas to solve the tangential Nevanlinna-Pick interpolation problem when Λ is strictly positive. If Λ is strictly positive, then we obtain the following result.

Theorem 12.5.1. *Let $\{A \text{ on } \mathcal{X}, C, \widetilde{C}\}$ be a data set for the tangential Nevanlinna-Pick interpolation problem. Assume that the solution Λ to the Lyapunov equation* (12.2.6) *is strictly positive. Then the following holds:*

(i) *The central interpolant* G *for* $\{A, C, \widetilde{C}\}$ *is given by the state space realization*

$$\begin{aligned} G(z) &= D - X_0^{-1} C\Lambda^{-1} A^*(zI - J)^{-1} B, \\ J &= A^* - C^* X_0^{-1} C\Lambda^{-1} A^*, \\ B &= (I - J\Lambda A\Lambda^{-1}) C^* X_0^{-1}, \\ D &= X_0^{-1} - X_0^{-1} C\Lambda^{-1} \widetilde{C}^*, \\ X_0 &= C\Lambda^{-1} C^*. \end{aligned} \tag{12.5.1}$$

(ii) *The operator* J *is stable and similar to* $T_\circ$.

(iii) *The operator* $\Delta(G) = X_0^{-1}$ *and* Υ_G *is a strictly positive Toeplitz operator on* $\ell_+^2(\mathcal{E})$.

(iv) *The outer spectral factor* Θ *for* Υ_G *is an invertible outer function in* $H^\infty(\mathcal{E}, \mathcal{E})$ *and given by*

$$\begin{aligned} \Theta(z) &= X_0^{-1/2} - X_0^{-1/2} C(zI - J^*)^{-1} A\Lambda^{-1} C^* X_0^{-1}, \\ \Theta(z)^{-1} &= zC\,(zI - A)^{-1}\,\Lambda^{-1} C^* X_0^{-1/2}. \end{aligned} \tag{12.5.2}$$

Finally, $T_\Theta^* T_\Theta = \Upsilon_G$.

Proof. Since $\Lambda = M^*M$ and Λ is invertible, the operator M is invertible. So without loss of generality we assume that $\mathcal{H} = \mathcal{X}$. We claim that $M^*T_\circ = JM^*$. In particular, $T_\circ$ is similar to J. To see this it is sufficient to show that $J^* = M^{-1} T_\circ^* M$ where

$$J^* = A - A\Lambda^{-1} C^* X_0^{-1} C \quad \text{and} \quad X_0 = C\Lambda^{-1} C^*. \tag{12.5.3}$$

First observe that $\mathcal{D}_2 = \mathcal{H} \ominus M \ker C = M^{-*} C^* \mathcal{E}$. By consulting (12.1.9), we see that $T_\circ^* M^{-*} C^* \mathcal{E} = 0$. So for J^* in (12.5.3), we obtain

$$\begin{aligned} (MJ^*M^{-1} - T_\circ^*) M^{-*} C^* &= M(A - A\Lambda^{-1} C^* X_0^{-1} C)\Lambda^{-1} C^* - 0 \\ &= MA\Lambda^{-1} C^* - MA\Lambda^{-1} C^* X_0^{-1} X_0 = 0. \end{aligned}$$

Recall that $V^*M|\ker C = MA|\ker C$. If v is in $\ker C$, then

$$(MJ^*M^{-1} - T_\circ^*) Mv = MJ^*v - T_\circ^* Mv = MAv - V^*Mv = 0.$$

Since $\mathcal{X} = M \ker C \oplus M^{-*} C^* \mathcal{E}$, this implies that $MJ^*M^{-1} = T_\circ^*$, where J^* is given by (12.5.3). In other words, $M^*T_\circ = JM^*$.

Now let us show that

$$P_{\mathcal{D}_2} = M^{-*} C^* X_0^{-1} C M^{-1} \tag{12.5.4}$$

where $P_{\mathcal{D}_2}$ is the orthogonal projection onto $\mathcal{D}_2 = \mathcal{H} \ominus \mathcal{H}_2$. Notice that $\mathcal{D}_2$ equals the range of E where E is the left invertible operator from $\mathcal{E}$ into $\mathcal{X}$ defined by

$E = M^{-*}C^*$. Recall that if T is any left invertible operator, then the orthogonal projection onto the range of T is given by $T(T^*T)^{-1}T^*$. Since $\mathcal{D}_2$ equals the range of E, we see that

$$\begin{aligned} P_{\mathcal{D}_2} &= E(E^*E)^{-1}E^* = M^{-*}C^*(CM^{-1}M^{-*}C^*)^{-1}CM^{-1} \\ &= M^{-*}C^*(C\Lambda^{-1}C^*)^{-1}CM^{-1} = M^{-*}C^*X_0^{-1}CM^{-1}. \end{aligned}$$

Hence $P_{\mathcal{D}_2} = M^{-*}C^*X_0^{-1}CM^{-1}$.

We claim that Λ is a solution to the following Lyapunov equation:

$$\Lambda = J\Lambda J^* + C^*X_0^{-1}C. \tag{12.5.5}$$

Because $T_\circ = VP_{\mathcal{H}_1}$ is a partial isometry on $\mathcal{H}$ whose range equals $\mathcal{H}_2$, we have $I - T_\circ T_\circ^* = P_{\mathcal{D}_2}$. Using $M^*T_\circ = JM^*$ with (12.5.4), we obtain

$$\begin{aligned} \Lambda - J\Lambda J^* &= M^*M - JM^*MJ^* \\ &= M^*M - M^*T_\circ T_\circ^* M \\ &= M^*(I - T_\circ T_\circ^*)M \\ &= M^*P_{\mathcal{D}_2}M = C^*X_\circ^{-1}C. \end{aligned}$$

Therefore the Lyapunov equation in (12.5.5) holds.

According to (12.1.12) the central solution is given by

$$\begin{aligned} G(z) = \widetilde{C}N &+ N^*(A^*M^* - M^*T_\circ)MAN \\ &+ N^*(M^*T_\circ - A^*M^*)(zI - T_\circ)^{-1}(M - T_\circ MA)N. \end{aligned} \tag{12.5.6}$$

Using $M^*T_\circ = JM^*$ and $J = A^* - C^*X_0^{-1}C\Lambda^{-1}A^*$, we arrive at

$$\begin{aligned} N^*(M^*T_\circ - A^*M^*)(zI - T_\circ)^{-1} &= N^*(J - A^*)M^*(zI - T_\circ)^{-1} \\ &= -X_0^{-1}C\Lambda^{-1}A^*(zI - J)^{-1}M^*. \end{aligned} \tag{12.5.7}$$

By employing $M^*M = \Lambda$ and $M^*T_\circ = JM^*$ with the Lyapunov equation in (12.5.5), we obtain

$$\begin{aligned} M^*(M - T_\circ MA)N &= (\Lambda - JM^*MA)N = (\Lambda - J\Lambda A)N \\ &= \left(\Lambda - J\Lambda(A - A\Lambda^{-1}C^*X_0^{-1}C) - J\Lambda A\Lambda^{-1}C^*X_0^{-1}C\right)N \\ &= \left(\Lambda - J\Lambda J^* - J\Lambda A\Lambda^{-1}C^*X_0^{-1}C\right)N \\ &= \left(C^*X_0^{-1}C - J\Lambda A\Lambda^{-1}C^*X_0^{-1}C\right)N \\ &= \left(I - J\Lambda A\Lambda^{-1}\right)C^*X_0^{-1} = B. \end{aligned} \tag{12.5.8}$$

The Lyapunov equation for Λ in (12.2.6), yields

$$\begin{aligned}\widetilde{C}N + N^*(A^*M^* - M^*T_\circ)MAN &= \widetilde{C}N + N^*(A^* - J)M^*MAN \qquad (12.5.9)\\ &= \widetilde{C}N + N^*C^*X_0^{-1}C\Lambda^{-1}A^*\Lambda AN\\ &= \widetilde{C}N + X_0^{-1}C\Lambda^{-1}(\Lambda - C^*\widetilde{C} - \widetilde{C}^*C)N\\ &= \widetilde{C}N + X_0^{-1}CN - \widetilde{C}N - X_0^{-1}C\Lambda^{-1}\widetilde{C}^*CN\\ &= X_0^{-1} - X_0^{-1}C\Lambda^{-1}\widetilde{C}^* = D.\end{aligned}$$

By combining this with (12.5.6), (12.5.7), (12.5.8) and (12.5.9), we obtain the form of the central interpolant in (12.5.1).

Now let us show that $\Delta(G) = X_0^{-1}$. According to Part (ii) in Theorem 12.4.2, we have

$$\Delta(G) = N^*M^*P_{\mathcal{D}_2}MN. \qquad (12.5.10)$$

Substituting $P_{\mathcal{D}_2} = M^{-*}C^*X_0^{-1}CM^{-1}$ into (12.5.10), yields $\Delta(G) = X_0^{-1}$.

Now let us show that J is stable. Since J is similar to $T_\circ$ it is sufficient to show that $T_\circ$ is stable. Because $T_\circ$ is a contraction, the eigenvalues of $T_\circ$ are contained in the closed unit disc. We claim that all the eigenvalues of $T_\circ$ are contained in the open unit disc. If λ is an eigenvalue on the unit circle with eigenvector f for $T_\circ$, then (12.1.9) and $T_\circ f = \lambda f$ imply that

$$\|f\| = \|\lambda f\| = \|T_\circ f\| = \|VP_{\mathcal{H}_1}f\| = \|P_{\mathcal{H}_1}f\| \leq \|f\|.$$

Thus $\|f\| = \|P_{\mathcal{H}_1}f\|$, or equivalently, $f = P_{\mathcal{H}_1}f$ is contained in $\mathcal{H}_1 = MA\ker C$. So $f = MAg$ for some nonzero vector g in $\ker C$. Hence

$$\lambda MAg = \lambda f = T_\circ f = T_\circ MAg = VMAg = Mg.$$

Using the fact that M is invertible, $\lambda Ag = g$, or equivalently, $Ag = \overline{\lambda}g$. In other words, $\overline{\lambda}$ is an eigenvalue of A. This contradicts the fact that A is stable. In other words, all the eigenvalues of $T_\circ$ are contained in the open unit disc and J is stable. Therefore Part (ii) holds.

For another proof to show that J is stable, recall that the pair $\{C, A\}$ is observable. Hence $\{A^*, C^*\}$ is controllable. Since J is an operator of the form $J = A^* - C^*Z$, this readily implies that the pair $\{J, C^*X_0^{-1/2}\}$ is also controllable. (One can also apply the Popov-Belevitch-Hautus test to show that $\{J, C^*X_0^{-1/2}\}$ is controllable.) Because Λ is a strictly positive solution to the Lyapunov equation (12.5.5) and $\{J, C^*X_0^{-1/2}\}$ is controllable, the operator J is stable. Hence the contraction $T_\circ$ is also stable. Finally, it is noted that zero is an eigenvalue for J. To see this observe that $C\Lambda^{-1}J = 0$. This also follows from the fact that J is similar to the partial isometry $T_\circ$ and $T_\circ$ has a nonzero kernel.

To obtain Part (iv), let Ψ be the operator from $\mathcal{E}$ into $\mathcal{X}$ defined by

$$\Psi = M^{-*}C^*X_0^{-1/2} \quad \text{where} \quad X_0 = C\Lambda^{-1}C^*. \qquad (12.5.11)$$

Notice that Ψ is an isometry whose range is $\mathcal{D}_2$. Hence Ψ^* is unitarily equivalent to $\Pi_{\mathcal{D}_2}$, that is, $\Psi^* = \Phi\Pi_{\mathcal{D}_2}$ where Φ is a unitary operator. Recall that the outer spectral factor is unique up to a constant unitary operator on the left. So without loss of generality, we can replace $\Pi_{\mathcal{D}_2}$ in the formula for Θ in (12.1.13) with Ψ^*. By employing (12.5.11) along with this state space formula for Θ, we arrive at

$$\begin{aligned}
\Theta(z) &= \Psi^* MN + \Psi^*(zI - T_\circ^*)^{-1}(T_\circ^* M - MA)N \\
&= X_0^{-1/2} C M^{-1} MN + X_0^{-1/2} C M^{-1}(zI - T_\circ^*)^{-1} M(J^* - A)N \\
&= X_0^{-1/2} CN + X_0^{-1/2} C(zI - J^*)^{-1}(J^* - A)N \\
&= X_0^{-1/2} - X_0^{-1/2} C(zI - J^*)^{-1} A\Lambda^{-1} C^* X_0^{-1} CN \\
&= X_0^{-1/2} - X_0^{-1/2} C(zI - J^*)^{-1} A\Lambda^{-1} C^* X_0^{-1}.
\end{aligned}$$

In other words, we obtain

$$\Theta(z) = X_0^{-1/2} - X_0^{-1/2} C(zI - J^*)^{-1} A\Lambda^{-1} C^* X_0^{-1}. \tag{12.5.12}$$

To complete the proof it remains to show that the inverse of Θ is given by (12.5.2). Recall that if $H(z)$ is given by the state space realization

$$H(z) = D + C_o(zI - Z)^{-1} B_o$$

where $\{Z, B_o, C_o, D\}$ are operators acting between the appropriate spaces and D is invertible, then $H(z)$ is invertible in some neighborhood of the origin and

$$H(z)^{-1} = D^{-1} - D^{-1} C_o \left(zI - (Z - B_o D^{-1} C_o)\right)^{-1} B_o D^{-1}.$$

Using this fact on the state space formula for Θ in (12.5.12), we obtain

$$\begin{aligned}
\Theta(z)^{-1} &= \left(X_0^{-1/2} - X_0^{-1/2} C(zI - J^*)^{-1} A\Lambda^{-1} C^* X_0^{-1}\right)^{-1} \\
&= X_0^{1/2} + C\,(zI - A)^{-1}\, A\Lambda^{-1} C^* X_0^{-1/2} \\
&= \left(C\Lambda^{-1} C^* + C\,(zI - A)^{-1}\, A\Lambda^{-1} C^*\right) X_0^{-1/2} \\
&= C\left(I + (zI - A)^{-1} A\right) \Lambda^{-1} C^* X_0^{-1/2} \\
&= zC\,(zI - A)^{-1}\, \Lambda^{-1} C^* X_0^{-1/2}.
\end{aligned}$$

Hence $\Theta(z)^{-1}$ is given by the state space formula in (12.5.2). Since A is stable, $\Theta(z)^{-1}$ has no poles in the closed unit disc. In particular, Θ is an invertible outer function. Therefore, Υ_G is a strictly positive Toeplitz operator on $\ell_+^2(\mathcal{E})$, and $T_\Theta^* T_\Theta = \Upsilon_G$. $\square$

12.6 The Carathéodory Interpolation Problem

Let $\{F_j\}_0^n$ be a set of operators with values in $\mathcal{L}(\mathcal{E},\mathcal{E})$. The *Carathéodory interpolation problem* is to find the set of all $\mathcal{L}(\mathcal{E},\mathcal{E})$-valued positive real functions F such that F admits a Taylor series expansion of the form

$$F(z) = \sum_{k=0}^{\infty} z^{-k} F_k. \tag{12.6.1}$$

Here $\{F_j\}_0^n$ are the first $n+1$ Taylor coefficients of F. To solve the Carathéodory interpolation problem, consider the data set $\{A, C, \widetilde{C}\}$ defined by

$$\begin{aligned} A &= \begin{bmatrix} 0 & I & \cdots & 0 & 0 \\ 0 & 0 & \cdots & 0 & 0 \\ \vdots & \vdots & \ddots & \vdots & \vdots \\ 0 & 0 & \cdots & 0 & I \\ 0 & 0 & \cdots & 0 & 0 \end{bmatrix} \quad \text{on } \mathcal{E}^{n+1}, \\ C &= \begin{bmatrix} I & 0 & \cdots & 0 & 0 \end{bmatrix} : \mathcal{E}^{n+1} \to \mathcal{E}, \\ \widetilde{C} &= \begin{bmatrix} F_0 & F_1 & \cdots & F_{n-2} & F_n \end{bmatrix} : \mathcal{E}^{n+1} \to \mathcal{E}. \end{aligned} \tag{12.6.2}$$

Observe that A is the upper shift on $\mathcal{E}^{n+1}$, that is, the identity I appears immediately above the main diagonal and zero's appear everywhere else. Clearly, A is stable. Let F be a $\mathcal{L}(\mathcal{E},\mathcal{E})$-valued analytic function in $\mathbb{D}_+$. Then it follows that $(FC)(A)_{right} = \widetilde{C}$ if and only if $\{F_j\}_0^n$ are the first $n+1$ Taylor coefficients in the power series expansion for $F(z) = \sum_0^\infty z^{-k} F_k$ for F about infinity. So F is a solution to the Carathéodory interpolation problem if and only if F is a solution to the Nevanlinna-Pick interpolation problem corresponding to the data $\{A, C, \widetilde{C}\}$ in (12.6.2).

As before, consider the data $\{A, C, \widetilde{C}\}$ in (12.6.2). Then a simple calculation shows that the Toeplitz matrix

$$\Lambda = \begin{bmatrix} F_0 + F_0^* & F_1 & \cdots & F_{n-1} & F_n \\ F_1^* & F_0 + F_0^* & \cdots & F_{n-2} & F_{n-1} \\ \vdots & \vdots & \ddots & \vdots & \vdots \\ F_{n-1}^* & F_{n-2}^* & \cdots & F_0 + F_0^* & F_1 \\ F_n^* & F_{n-1}^* & \cdots & F_1^* & F_0 + F_0^* \end{bmatrix} \tag{12.6.3}$$

is the unique solution to the Lyapunov equation $\Lambda = A^*\Lambda A + C^*\widetilde{C} + \widetilde{C}^*C$. By consulting Theorem 12.1.1, we see that there exists a solution to the Carathéodory interpolation problem if and only if the Toeplitz matrix Λ in (12.6.3) is positive. In this case, (12.1.12) provides the central solution to the Carathéodory interpolation problem. In particular, if Λ is strictly positive, then Theorem 12.5.1 gives a state space solution to the Carathéodory interpolation problem in terms of Λ^{-1}.

Assume that Λ is strictly positive. In this case, Theorem 12.5.1, yields the same solution as the Levinson algorithm in Theorem 7.5.1. To be more specific, let Λ be the strictly positive Toeplitz matrix given by

$$\Lambda = \begin{bmatrix} R_0 & R_1^* & \cdots & R_{n-1}^* & R_n^* \\ R_1 & R_0 & \cdots & R_{n-2}^* & R_{n-1}^* \\ \vdots & \vdots & \ddots & \vdots & \vdots \\ R_{n-1} & R_{n-2} & \cdots & R_0 & R_1^* \\ R_n & R_{n-1} & \cdots & R_1 & R_0 \end{bmatrix}. \tag{12.6.4}$$

Consider the Carathéodory interpolation problem corresponding to the data $F_0 = R_0/2$ and $F_k = R_k^*$ for $k = 1, 2, \ldots, n$. Let $\{A, C, \widetilde{C}\}$ be the data in (12.6.2). Notice that Λ is also determined by (12.6.3). Now let G be the central solution corresponding to the data $\{F_j\}_0^n$. According to Theorem 12.5.1, the outer spectral factor Θ for Υ_G is determined by

$$\Theta(z)^{-1} = zC\left(zI - A\right)^{-1}\Lambda^{-1}C^*\left(C\Lambda^{-1}C^*\right)^{-1/2}.$$

Observe that

$$zC(zI - A)^{-1} = \sum_{k=0}^{n} z^{-k}CA^k = \begin{bmatrix} I & z^{-1}I & z^{-2}I & \cdots & z^{-n}I \end{bmatrix}.$$

This readily implies that

$$\Theta(z)^{-1} = \begin{bmatrix} I & z^{-1}I & z^{-2}I & \cdots & z^{-n}I \end{bmatrix}\Lambda^{-1}C^*(C\Lambda^{-1}C^*)^{-1/2}. \tag{12.6.5}$$

Now consider the Levinson system of equations

$$\Lambda \begin{bmatrix} I \\ A_1 \\ \vdots \\ A_n \end{bmatrix} = \begin{bmatrix} R_0 & R_1^* & \cdots & R_n^* \\ R_1 & R_0 & \cdots & R_{n-1}^* \\ \vdots & \vdots & \ddots & \vdots \\ R_n & R_{n-1} & \cdots & R_0 \end{bmatrix} \begin{bmatrix} I \\ A_1 \\ \vdots \\ A_n \end{bmatrix} = \begin{bmatrix} \Delta_{n+1} \\ 0 \\ \vdots \\ 0 \end{bmatrix}. \tag{12.6.6}$$

Here $\{A_j\}_1^n$ and Δ_{n+1} are operators on $\mathcal{E}$. There is a unique solution to this system of equations. In fact, the unique solution to the Levinson system (12.6.6) is given by

$$\Delta_{n+1} = \left(C\Lambda^{-1}C^*\right)^{-1} \quad \text{and} \quad \begin{bmatrix} I & A_1 & A_2 & \cdots & A_n \end{bmatrix}^{tr} = \Lambda^{-1}C^*\Delta_{n+1}.$$

This readily implies that

$$\begin{bmatrix} I & A_1 & \cdots & A_n \end{bmatrix}^{tr}\Delta_{n+1}^{-1/2} = \Lambda^{-1}C^*(C\Lambda^{-1}C^*)^{-1/2}.$$

So by consulting (12.6.5), we arrive at

$$\Theta(z) = \Delta_{n+1}^{1/2}\left(I + z^{-1}A_1 + z^{-2}A_2 + \cdots + z^{-(n-1)}A_{n-1} + z^{-n}A_n\right)^{-1}.$$

This is precisely the outer spectral factor computed by the Levinson algorithm presented in Theorem 7.5.1. Moreover, $T_\Theta^* T_\Theta = \Upsilon_G$ where G is the central solution for the data $\{A, C, \widetilde{C}\}$. Finally, Theorem 7.5.1 shows that $\Lambda = \Pi_{\mathcal{E}^{n+1}} \Upsilon_G | \mathcal{E}^{n+1}$. This also follows from Remark 12.1.4 and the fact that for $\{C, A\}$ in (12.6.2) its corresponding observability Gramian is given by

$$W_o = \begin{bmatrix} C \\ CA \\ CA^2 \\ \vdots \end{bmatrix} = \begin{bmatrix} I \\ 0 \end{bmatrix} : \mathcal{E}^{n+1} \to \begin{bmatrix} \mathcal{E}^{n+1} \\ \ell_+^2(\mathcal{E}) \end{bmatrix}.$$

Therefore $\Lambda = W_o^* \Upsilon_G W_o = \Pi_{\mathcal{E}^{n+1}} \Upsilon_G | \mathcal{E}^{n+1}$.

12.7 Notes

Sz.-Nagy-Koranyi [196] were the first to use operator techniques to solve the Nevanlinna-Pick interpolation problem. Here we simply modified their techniques by adding state space theory and the Naimark dilation to solve the tangential Nevanlinna-Pick interpolation. The results in this chapter were taken from Frazho-Kaashoek [98]. It is also emphasized that Sz.-Nagy-Koranyi [196] used an isometric extension to solve the Nevanlinna-Pick interpolation problem. Isometric extensions have played a role in characterizing the set of all solutions to the commutant lifting theorem; see Arocena [13, 14, 15], Bakonyi-Constantinescu [22] and Foias-Frazho [82]. For some further applications of isometric extensions related to interpolation problems; see Agler-McCarthy [3, 4], Ball [23], Ball-Trent-Vinnikov [26] and Cotlar-Sadosky [61].

In this chapter, we only presented one solution to the tangential Nevanlinna-Pick interpolation. The set of all solutions to the tangential Nevanlinna-Pick interpolation is parameterized by the unit ball in some H^∞ space. The set of all solutions to the tangential Nevanlinna-Pick interpolation using the Naimark dilation is presented in Frazho-Kaashoek [98]. The band method is a very powerful theory for solving and characterizing all solutions to many interpolation problems; see Gohberg-Kaashoek-Woerdeman [115] and Gohberg-Goldberg-Kaashoek [114]. For a parameterization to the set of all solutions to the tangential Nevanlinna-Pick interpolation using the band method see Kaashoek-Zeinstra [137] and Frazho-Kaashoek [98]. A state space method for solving positive real interpolation problems is given in Ball-Gohberg-Rodman [24] and Georgiou [107, 108, 109]. A solution for a Nevanlinna-Pick interpolation problem involving a McMillan degree constraint is presented in Byrnes-Georgiou-Lindquist [46]. A solution to a bilinear or a certain nonlinear positive real interpolation problem is given in Desai [69], Frazho [96] and Popescu [177].

Chapter 13

Contractive Nevanlinna-Pick Interpolation

In this chapter, we will use isometric realizations to solve a contractive tangential Nevanlinna-Pick interpolation problem. This chapter can be viewed as a contractive version or dual of the positive real Nevanlinna-Pick interpolation problem discussed in Chapter 12.

13.1 Isometric Realizations

Recall that an operator T mapping $\mathcal{V}$ into $\mathcal{Y}$ is a *contraction* if $\|T\| \leq 1$. We say that Θ is a *contractive analytic function* if Θ is a function in $H^\infty(\mathcal{E}, \mathcal{Y})$ and $\|\Theta\|_\infty \leq 1$, or equivalently, its corresponding Toeplitz matrix T_Θ is a contraction.

Let Θ be a function in $H^\infty(\mathcal{E}, \mathcal{Y})$. Then ∇_Θ is the upper triangular Toeplitz operator defined by

$$\nabla_\Theta = \begin{bmatrix} \Theta_0 & \Theta_1 & \Theta_2 & \cdots \\ 0 & \Theta_0 & \Theta_1 & \ddots \\ 0 & 0 & \Theta_0 & \ddots \\ \vdots & \vdots & \vdots & \ddots \end{bmatrix} : \ell_+^2(\mathcal{E}) \to \ell_+^2(\mathcal{Y}),$$

$$\Theta(z) = \sum_{n=0}^{\infty} z^{-n}\Theta_n \tag{13.1.1}$$

In this case, Θ is called the *symbol* for ∇_Θ. Observe that $\nabla_\Theta = T_{\widetilde{\Theta}}^*$ where $T_{\widetilde{\Theta}}$ is the lower triangular Toeplitz matrix generated by $\widetilde{\Theta}(z) = \Theta(\overline{z})^*$. Since

$$\|T_{\widetilde{\Theta}}^*\| = \|T_{\widetilde{\Theta}}\| = \|\widetilde{\Theta}\|_\infty = \|\Theta\|_\infty,$$

we obtain $\|\nabla_\Theta\| = \|\Theta\|_\infty$. In particular, Θ is a contractive analytic function if and only if ∇_Θ is a contraction. Throughout, $S_\mathcal{U}$ is the unilateral shift on $\ell_+^2(\mathcal{U})$. Moreover, for any Θ in $H^\infty(\mathcal{E},\mathcal{Y})$, it follows that ∇_Θ intertwines $S_\mathcal{E}^*$ with $S_\mathcal{Y}^*$, that is, $S_\mathcal{Y}^*\nabla_\Theta = \nabla_\Theta S_\mathcal{E}^*$. Finally, let ∇ be an operator mapping $\ell_+^2(\mathcal{E})$ into $\ell_+^2(\mathcal{Y})$. Then $S_\mathcal{Y}^*\nabla = \nabla S_\mathcal{E}^*$ if and only if $\nabla = \nabla_\Theta$ for some Θ in $H^\infty(\mathcal{E},\mathcal{Y})$. In this case, $\|\nabla_\Theta\| = \|\Theta\|_\infty$. This follows by taking the appropriate adjoints in Theorem 2.6.1.

Recall that $\{A \text{ on } \mathcal{X}, B, C, D\}$ is a *state space realization* for a $\mathcal{L}(\mathcal{E},\mathcal{Y})$-valued transfer function Θ if

$$\Theta(z) = D + C(zI - A)^{-1}B.$$

The pair $\{A, B\}$ is *controllable* if $\mathcal{X} = \bigvee_0^\infty A^n B\mathcal{E}$. The pair $\{C, A\}$ is *observable* if $\mathcal{X} = \bigvee_0^\infty A^{*n}C^*\mathcal{Y}$. The pair $\{C, A\}$ is observable if and only if $\{A^*, C^*\}$ is controllable. Two state space realizations $\{A \text{ on } \mathcal{X}, B, C, D\}$ and $\{A_1 \text{ on } \mathcal{X}_1, B_1, C_1, D_1\}$ are *unitarily equivalent* if $D = D_1$ and there exists a unitary operator Φ mapping $\mathcal{X}$ onto $\mathcal{X}_1$ such that

$$\Phi A = A_1\Phi, \quad \Phi B = B_1 \quad \text{and} \quad C_1\Phi = C.$$

We say that the system $\{A, B, C, D\}$ is an *isometric realization* (respectively a unitary realization) if its system matrix

$$\Omega = \begin{bmatrix} D & C \\ B & A \end{bmatrix} : \begin{bmatrix} \mathcal{E} \\ \mathcal{X} \end{bmatrix} \to \begin{bmatrix} \mathcal{Y} \\ \mathcal{X} \end{bmatrix} \tag{13.1.2}$$

is an isometry (respectively a unitary operator). It is noted that $\{A, B, C, D\}$ is an isometric realization if and only if $\Omega^*\Omega = I$, or equivalently,

$$\begin{aligned} A^*A + C^*C &= I, \\ B^*A + D^*C &= 0, \\ B^*B + D^*D &= I. \end{aligned} \tag{13.1.3}$$

The following is a classical result in operator theory.

Theorem 13.1.1. *Let Θ be a $\mathcal{L}(\mathcal{E},\mathcal{Y})$-valued analytic function in $\mathbb{D}_+$. Then Θ is a contractive analytic function in $H^\infty(\mathcal{E},\mathcal{Y})$ if and only if Θ admits an isometric realization. In this case, all controllable isometric realizations of Θ are unitarily equivalent.*

Proof. In Section 7.8 we showed that if $\{A, B, C, D\}$ is a contractive ($\|\Omega\| \le 1$) realization, then its transfer function Θ is a contractive analytic function. So if Ω is an isometry, then clearly Ω is a contraction, and thus, Θ is a contractive analytic function.

Now assume that Θ is a contractive analytic function. Then ∇_Θ is a contraction. To construct a controllable isometric realization for Θ, let D_{∇_Θ} be the positive square root of $I - \nabla_\Theta^*\nabla_\Theta$, and $\mathcal{D}_{\nabla_\Theta}$ the closure of the range of D_{∇_Θ}. Let Z be the operator determined by the first row of ∇_Θ, that is,

$$Z = \begin{bmatrix} \Theta_0 & \Theta_1 & \Theta_2 & \cdots \end{bmatrix} : \ell_+^2(\mathcal{E}) \to \mathcal{Y}.$$

We claim that there exists an isometry Φ mapping $\mathcal{D}_{\nabla_\Theta}$ into $\mathcal{Y} \oplus \mathcal{D}_{\nabla_\Theta}$ such that

$$\Phi D_{\nabla_\Theta} = \begin{bmatrix} ZS_\mathcal{E} \\ D_{\nabla_\Theta} S_\mathcal{E} \end{bmatrix}. \tag{13.1.4}$$

Recall that $I - S_\mathcal{Y} S_\mathcal{Y}^*$ is the orthogonal projection onto the first component $\mathcal{Y}$ of $\ell_+^2(\mathcal{Y})$. Using $D_{\nabla_\Theta}^2 = I - \nabla_\Theta^* \nabla_\Theta$ and $S_\mathcal{Y}^* \nabla_\Theta = \nabla_\Theta S_\mathcal{E}^*$, with the fact that the unilateral shift is an isometry, for h in $\ell_+^2(\mathcal{E})$ we have

$$\begin{aligned} \|D_{\nabla_\Theta} S_\mathcal{E} h\|^2 &= \|S_\mathcal{E} h\|^2 - \|\nabla_\Theta S_\mathcal{E} h\|^2 \\ &= \|h\|^2 - \|(I - S_\mathcal{Y} S_\mathcal{Y}^*)\nabla_\Theta S_\mathcal{E} h\|^2 - \|S_\mathcal{Y} S_\mathcal{Y}^* \nabla_\Theta S_\mathcal{E} h\|^2 \\ &= \|h\|^2 - \|ZS_\mathcal{E} h\|^2 - \|\nabla_\Theta S_\mathcal{E}^* S_\mathcal{E} h\|^2 \\ &= \|D_{\nabla_\Theta} h\|^2 - \|ZS_\mathcal{E} h\|^2. \end{aligned}$$

By rearranging the terms, we obtain

$$\|D_{\nabla_\Theta} h\|^2 = \|ZS_\mathcal{E} h\|^2 + \|D_{\nabla_\Theta} S_\mathcal{E} h\|^2 = \left\| \begin{bmatrix} ZS_\mathcal{E} \\ D_{\nabla_\Theta} S_\mathcal{E} \end{bmatrix} h \right\|^2.$$

So there exists an isometry Φ such that (13.1.4) holds.

Now set $\mathcal{X} = \mathcal{D}_{\nabla_\Theta}$, and consider the state space system $\{A \text{ on } \mathcal{X}, B, C, D\}$ defined by

$$\begin{aligned} AD_{\nabla_\Theta} &= D_{\nabla_\Theta} S_\mathcal{E} \quad \text{and} \quad B = D_{\nabla_\Theta} \Pi_\mathcal{E}^*, \\ CD_{\nabla_\Theta} &= ZS_\mathcal{E} \qquad \text{and} \quad D = \Theta(\infty) = \Theta_0. \end{aligned} \tag{13.1.5}$$

Here $\Pi_\mathcal{E}^*$ is the natural embedding of $\mathcal{E}$ into the first component of $\ell_+^2(\mathcal{E})$. We claim that (13.1.5) defines a controllable isometric realization for Θ. Equation (13.1.4) guarantees that A and C are contractions. In fact, for h in $\ell_+^2(\mathcal{E})$ we have

$$\left\| \begin{bmatrix} C \\ A \end{bmatrix} D_{\nabla_\Theta} h \right\|^2 = \left\| \begin{bmatrix} ZS_\mathcal{E} \\ D_{\nabla_\Theta} S_\mathcal{E} \end{bmatrix} h \right\|^2 = \|D_{\nabla_\Theta} h\|^2.$$

By the definition of $\mathcal{X}$, the range of D_{∇_Θ} is dense in $\mathcal{X}$. It follows that $\begin{bmatrix} C & A \end{bmatrix}^{tr}$ is an isometry mapping $\mathcal{X}$ into $\mathcal{Y} \oplus \mathcal{X}$. In particular, A and C are contractions.

For v in $\mathcal{E}$, we obtain

$$\begin{aligned} \left\| \begin{bmatrix} D \\ B \end{bmatrix} v \right\|^2 &= \|\Theta_0 v\|^2 + \|D_{\nabla_\Theta} \Pi_\mathcal{E}^* v\|^2 \\ &= \|\Theta_0 v\|^2 + \|\Pi_\mathcal{E}^* v\|^2 - \|\nabla_\Theta \Pi_\mathcal{E}^* v\|^2 \\ &= \|\Theta_0 v\|^2 + \|v\|^2 - \|\Theta_0 v\|^2 = \|v\|^2. \end{aligned}$$

So $\begin{bmatrix} D & B \end{bmatrix}^{tr}$ is an isometry mapping $\mathcal{E}$ into $\mathcal{Y} \oplus \mathcal{X}$. In other words, the columns of the matrix Ω in (13.1.2) are both isometries.

To show that Ω is an isometry, it remains to verify that the columns of Ω are orthogonal. For v in $\mathcal{E}$ and h in $\ell_+^2(\mathcal{E})$, using $\nabla_\Theta^*\nabla_\Theta\Pi_\mathcal{E}^* = Z^*\Theta_0$, we have

$$\begin{aligned}\left(\begin{bmatrix} D \\ B\end{bmatrix} v, \begin{bmatrix} C \\ A\end{bmatrix} D_{\nabla_\Theta} h\right) &= (Dv, CD_{\nabla_\Theta}h) + (Bv, AD_{\nabla_\Theta}h)\\ &= (\Theta_0 v, ZS_\mathcal{E}h) + (D_{\nabla_\Theta}\Pi_\mathcal{E}^* v, D_{\nabla_\Theta}S_\mathcal{E}h)\\ &= (Z^*\Theta_0 v, S_\mathcal{E}h) + (D_{\nabla_\Theta}^2\Pi_\mathcal{E}^* v, S_\mathcal{E}h)\\ &= (Z^*\Theta_0 v, S_\mathcal{E}h) + ((I - \nabla_\Theta^*\nabla_\Theta)\Pi_\mathcal{E}^* v, S_\mathcal{E}h)\\ &= (\Pi_\mathcal{E}^* v, S_\mathcal{E}h) + (Z^*\Theta_0 v, S_\mathcal{E}h) - (Z^*\Theta_0 v, S_\mathcal{E}h)\\ &= (\Pi_\mathcal{E}^* v, S_\mathcal{E}h) = 0.\end{aligned}$$

Therefore the columns of Ω are orthogonal and Ω is an isometry.

Now let us show that $\{A, B, C, D\}$ is a realization for Θ. For any integer $n \geq 1$, we obtain

$$CA^{n-1}B \quad = \quad CA^{n-1}D_{\nabla_\Theta}\Pi_\mathcal{E}^* = CD_{\nabla_\Theta}S_\mathcal{E}^{n-1}\Pi_\mathcal{E}^* = ZS_\mathcal{E}^n\Pi_\mathcal{E}^* = \Theta_n.$$

Therefore $\{A, B, C, D\}$ is an isometric realization for Θ.

Notice that $\ell_+^2(\mathcal{E}) = \bigvee_0^\infty S_\mathcal{E}^n\Pi_\mathcal{E}^*\mathcal{E}$. Using this we obtain

$$\bigvee_{n=0}^{\infty} A^n B\mathcal{E} = \bigvee_{n=0}^{\infty} A^n D_{\nabla_\Theta}\Pi_\mathcal{E}^*\mathcal{E} = \bigvee_{n=0}^{\infty} D_{\nabla_\Theta}S_\mathcal{E}^n\Pi_\mathcal{E}^*\mathcal{E} = \bigvee D_{\nabla_\Theta}\ell_+^2(\mathcal{E}) = \mathcal{X}.$$

Hence $\{A, B\}$ is controllable. Lemma 13.1.3 below shows that two controllable isometric realizations of the same transfer function are unitarily equivalent. □

Remark 13.1.2. Let Θ be a contractive analytic function in $H^\infty(\mathcal{E}, \mathcal{Y})$. Then the system $\{A, B, C, D\}$ computed by (13.1.5) is a controllable isometric realization for Θ.

Recall that two minimal realizations of the same rational transfer function are similar. Theorem 13.1.1 shows that controllable isometric realizations of the same transfer function are unitarily equivalent. However, a controllable isometric realization for a rational transfer function is not necessarily minimal. In fact, the controllable isometric realization for a rational transfer function can even be infinite dimensional. For example, consider the contractive analytic function $\theta(z) = \gamma/z$ where γ is a nonzero scalar such that $|\gamma| < 1$. Notice that $\{0, 1, \gamma, 0\}$ is a minimal realization for θ. The McMillan degree for θ is 1. To construct a controllable isometric realization for θ, consider the operator A on $\mathcal{X} = \mathbb{C} \oplus \ell_+^2$ defined by

$$A = \begin{bmatrix} 0 & 0 \\ \varphi & S\end{bmatrix} \text{ on } \begin{bmatrix} \mathbb{C} \\ \ell_+^2\end{bmatrix} \quad \text{where} \quad \varphi \quad = \quad \begin{bmatrix} \sqrt{1-|\gamma|^2} \\ 0 \\ 0 \\ \vdots \end{bmatrix} : \mathbb{C} \to \ell_+^2$$

and S is the unilateral shift on ℓ^2_+. Now set

$$B = \begin{bmatrix} 1 \\ 0 \end{bmatrix} : \mathbb{C} \to \begin{bmatrix} \mathbb{C} \\ \ell^2_+ \end{bmatrix} \quad \text{and} \quad C = \begin{bmatrix} \gamma & 0 \end{bmatrix} : \begin{bmatrix} \mathbb{C} \\ \ell^2_+ \end{bmatrix} \to \mathbb{C}.$$

Then $\{A, B, C, 0\}$ is a controllable isometric realization for $\theta(z) = \gamma/z$. In this case, the controllability matrix for the pair $\{A, B\}$ admits a matrix representation of the form:

$$\begin{bmatrix} B & AB & A^2B & \cdots \end{bmatrix} = \begin{bmatrix} 1 & 0 \\ 0 & \sqrt{1 - |\gamma|^2 I} \end{bmatrix} \text{ on } \begin{bmatrix} \mathbb{C} \\ \ell^2_+ \end{bmatrix}.$$

Clearly, this controllability matrix defines an invertible operator on $\mathbb{C} \oplus \ell^2_+$. Hence the pair $\{A, B\}$ is controllable. According to Theorem 13.1.1, all controllable isometric realizations for θ are unitarily equivalent to $\{A, B, C, 0\}$. In particular, the state space dimension for any isometric realization of θ is infinite dimensional.

The following result was used to prove part of Theorem 13.1.1.

Lemma 13.1.3. *Let $\{A \text{ on } \mathcal{X}, B, C, D\}$ be an isometric realization for a transfer function Θ and W_c the controllability matrix defined by*

$$W_c = \begin{bmatrix} B & AB & A^2B & A^3B & \cdots \end{bmatrix}. \tag{13.1.6}$$

Then the following holds.

(i) *The matrix W_c determines a contraction mapping $\ell^2_+(\mathcal{E})$ into $\mathcal{X}$.*

(ii) *The operator $I - \nabla_\Theta^* \nabla_\Theta$ admits a factorization of the form*

$$I - \nabla_\Theta^* \nabla_\Theta = W_c^* W_c. \tag{13.1.7}$$

In particular, ∇_Θ is a contraction.

(iii) *All controllable isometric realizations of Θ are unitarily equivalent.*

Proof. First let us show that W_c is a bounded operator. Because Ω in (13.1.2) is an isometry, its last row $\Psi = \begin{bmatrix} B & A \end{bmatrix}$ is a contraction mapping $\mathcal{E} \oplus \mathcal{X}$ into $\mathcal{X}$. Hence $\Psi\Psi^* = BB^* + AA^*$ is a contraction on $\mathcal{X}$. In other words, $I \geq BB^* + AA^*$. This implies that

$$\begin{aligned} I &\geq BB^* + AA^* \geq BB^* + A(BB^* + AA^*)A^* \\ &= BB^* + ABB^*A^* + A^2A^{*2} \\ &\geq BB^* + ABB^*A^* + A^2(BB^* + AA^*)A^{*2} \\ &= \sum_{j=0}^{2} A^j BB^* A^{*j} + A^3 A^{*3}. \end{aligned}$$

By continuing in this fashion, we see that

$$I \geq \sum_{j=0}^{n-1} A^j BB^* A^{*j} + A^n A^{*n} \geq \sum_{j=0}^{n-1} A^j BB^* A^{*j}$$

for all integers $n \geq 1$. This readily implies that

$$\Xi = \begin{bmatrix} B^* \\ B^* A^* \\ B^* A^{*2} \\ \vdots \end{bmatrix} : \mathcal{X} \to \ell_+^2(\mathcal{E})$$

is a contraction. Since W_c is the adjoint of Ξ, it follows that W_c is also a contraction. Therefore Part (i) holds.

Let $\Theta = \sum_0^\infty z^{-n}\Theta_n$ be the Taylor series expansion for Θ. Recall that

$$\Theta_0 = D \quad \text{and} \quad \Theta_k = CA^{k-1}B \quad (k \geq 1). \tag{13.1.8}$$

Observe that the first row of ∇_Θ is given by

$$Z = \begin{bmatrix} D & CW_c \end{bmatrix} = \begin{bmatrix} D & CB & CAB & CA^2B & \cdots \end{bmatrix} : \ell_+^2(\mathcal{E}) \to \mathcal{Y}.$$

Moreover, the operator ∇_Θ admits a matrix decomposition of the form

$$\nabla_\Theta = \begin{bmatrix} Z \\ ZS_\mathcal{E}^* \\ ZS_\mathcal{E}^{*2} \\ \vdots \end{bmatrix} = \begin{bmatrix} \begin{bmatrix} D & CW_c \end{bmatrix} \\ \begin{bmatrix} D & CW_c \end{bmatrix} S_\mathcal{E}^* \\ \begin{bmatrix} D & CW_c \end{bmatrix} S_\mathcal{E}^{*2} \\ \vdots \end{bmatrix}. \tag{13.1.9}$$

Now let $f = \begin{bmatrix} f_0 & f_1 & f_2 & \cdots \end{bmatrix}^{tr}$ be any vector in $\ell_+^2(\mathcal{E})$. By employing (13.1.9) and (13.1.3), we obtain

$$\begin{aligned}
(I - \nabla_\Theta^* \nabla_\Theta)f, f) &= \|f\|^2 - \|\nabla_\Theta f\|^2 \\
&= \|f\| - \sum_{k=0}^\infty \| \begin{bmatrix} D & CW_c \end{bmatrix} S_\mathcal{E}^{*k} f\|^2 \\
&= \|f\| - \sum_{k=0}^\infty \|Df_k + CW_c S_\mathcal{E}^{*k+1} f\|^2 \\
&= \|f\| - \sum_{k=0}^\infty \big(\|Df_k\|^2 + 2\Re(Df_k, CW_c S_\mathcal{E}^{*k+1} f) \\
&\qquad + \|CW_c S_\mathcal{E}^{*k+1} f\|^2\big)
\end{aligned}$$

$$= \sum_{k=0}^{\infty} \left(\|Bf_k\|^2 + 2\Re(f_k, B^* A W_c S_{\mathcal{E}}^{*k+1} f)\right)$$
$$+ \sum_{k=0}^{\infty} \left(\|AW_c S_{\mathcal{E}}^{*k+1} f\|^2 - \|W_c S_{\mathcal{E}}^{*k+1} f\|^2\right)$$
$$= \sum_{k=0}^{\infty} \left(\|Bf_k + AW_c S_{\mathcal{E}}^{*k+1} f\|^2 - \|W_c S_{\mathcal{E}}^{*k+1} f\|^2\right)$$
$$= \sum_{k=0}^{\infty} \left(\|W_c S_{\mathcal{E}}^{*k} f\|^2 - \|W_c S_{\mathcal{E}}^{*k+1} f\|^2\right)$$
$$= \lim_{n\to\infty} \sum_{k=0}^{n} \left(\|W_c S_{\mathcal{E}}^{*k} f\|^2 - \|W_c S_{\mathcal{E}}^{*k+1} f\|^2\right)$$
$$= \|W_c f\|^2 - \lim_{n\to\infty} \|W_c S_{\mathcal{E}}^{*n+1} f\|^2 = \|W_c f\|^2.$$

In other words, $((I - \nabla_\Theta^* \nabla_\Theta) f, f) = (W_c^* W_c f, f)$. Therefore Part (ii) holds.

Assume that $\{A \text{ on } \mathcal{X}, B, C, D\}$ and $\{A_1 \text{ on } \mathcal{X}_1, B_1, C_1, D_1\}$ are two controllable isometric realizations of Θ. Let W_{c1} be the controllability matrix corresponding to the pair $\{A_1, B_1\}$, that is,

$$W_{c1} = \begin{bmatrix} B_1 & A_1 B_1 & A_1^2 B_1 & \cdots \end{bmatrix}. \tag{13.1.10}$$

Part (i) shows that W_{c1} is a contraction mapping $\ell_+^2(\mathcal{E})$ into $\mathcal{X}_1$. According to Part (ii), we have

$$W_c^* W_c = I - \nabla_\Theta^* \nabla_\Theta = W_{c1}^* W_{c1}.$$

Since $\{A, B\}$ and $\{A_1, B_1\}$ are both controllable, the range of W_c is dense in $\mathcal{X}$ and the range of W_{c1} is dense in $\mathcal{X}_1$. So there exits a unique unitary operator Φ mapping $\mathcal{X}$ onto $\mathcal{X}_1$ such that $\Phi W_c = W_{c1}$. In particular,

$$A_1 \Phi W_c = A_1 W_{c1} = W_{c1} S_{\mathcal{E}} = \Phi W_c S_{\mathcal{E}} = \Phi A W_c.$$

Hence $A_1 \Phi W_c = \Phi A W_c$. Because the range of W_c is dense in $\mathcal{X}$, it follows that $A_1 \Phi = \Phi A$. Now observe that

$$B_1 = W_{c1} \Pi_{\mathcal{E}}^* = \Phi W_c \Pi_{\mathcal{E}}^* = \Phi B.$$

In other words, $B_1 = \Phi B$.

Let $\Theta(z) = \sum_0^\infty z^{-n} \Theta_n$ be the Taylor series expansion for Θ. Using the fact that $\{A, B, C, D\}$ and $\{A_1, B_1, C_1, D_1\}$ are two controllable realizations of Θ, we have $CA^{n-1}B = \Theta_n = C_1 A_1^{n-1} B_1$ for all integers $n \geq 1$. By virtue of $\Phi A^j B = A_1^j B_1$ for all integers $j \geq 0$, we obtain

$$C_1 \Phi A^j B = C_1 A_1^j B_1 = C A^j B \qquad (j \geq 0).$$

Because the pair $\{A, B\}$ is controllable, $\{A^j B\mathcal{E}\}$ spans a dense set in $\mathcal{X}$. Therefore $C_1\Phi = C$. Finally, $D = \Theta_0 = D_1$. Therefore Φ is a unitary operator which intertwines $\{A, B, C, D\}$ with $\{A_1, B_1, C_1, D_1\}$. Hence Part (*iii*) holds. □

We say that an operator A on $\mathcal{X}$ is *strongly stable*, if A^n converges to zero in the strong operator topology as n tends to infinity. If $\mathcal{X}$ is finite dimensional, then A is strongly stable if and only if A is stable.

Proposition 13.1.4. *Let $\{A$ on $\mathcal{X}, B, C, D\}$ be a controllable isometric realization for a contractive analytic function Θ in $H^\infty(\mathcal{E}, \mathcal{Y})$. Then Θ is an inner function if and only if A is strongly stable. In this case, its observability operator*

$$W_o = \begin{bmatrix} C \\ CA \\ CA^2 \\ \vdots \end{bmatrix} : \mathcal{X} \to \ell_+^2(\mathcal{Y}) \tag{13.1.11}$$

is an isometry, and the pair $\{C, A\}$ is observable.

Proof. Assume that A is strongly stable. Then using $I = A^*A + C^*C$ recursively, we have

$$\begin{aligned} I &= C^*C + A^*(C^*C + A^*A)A \\ &= C^*C + A^*C^*CA + A^{*2}A^2 \\ &= C^*C + A^*C^*CA + A^{*2}(C^*C + A^*A)A^2. \end{aligned}$$

By continuing in this fashion, we obtain

$$I = \sum_{j=0}^{n-1} A^{*j}C^*CA^j + A^{*n}A^n$$

for all integers $n \geq 1$. Because A is strongly stable, $I = \sum_0^\infty A^{*j}C^*CA^j$. This readily implies that $I = W_o^*W_o$ and W_o is an isometry. In particular, the pair $\{C, A\}$ is observable.

Since $\{A, B, C, D\}$ is a realization for Θ, we have $\Theta_n = CA^{n-1}B$ for all $n \geq 1$ and $D = \Theta_0$ where $\Theta(z) = \sum_0^\infty z^{-n}\Theta_n$. Hence the Toeplitz operator T_Θ admits a matrix decomposition of the form

$$\begin{aligned} T_\Theta &= \begin{bmatrix} \Gamma & S_\mathcal{Y}\Gamma & S_\mathcal{Y}^2\Gamma & \cdots \end{bmatrix} : \ell_+^2(\mathcal{E}) \to \ell_+^2(\mathcal{Y}), \\ \Gamma &= \begin{bmatrix} D \\ W_oB \end{bmatrix} : \mathcal{E} \to \begin{bmatrix} \mathcal{Y} \\ \ell_+^2(\mathcal{Y}) \end{bmatrix} \end{aligned}$$

where $S_\mathcal{Y}$ is the unilateral shift on $\ell_+^2(\mathcal{Y})$. To verify that Θ is inner, it is sufficient to show that T_Θ is an isometry. Notice that T_Θ is an isometry if and only if all the columns of T_Θ are isometric and orthogonal to each other. Due to the Toeplitz structure, T_Θ is an isometry if and only if Γ is an isometry and $\Gamma\mathcal{E}$ is orthogonal to $S_\mathcal{Y}^n\Gamma\mathcal{E}$ for all integers $n \geq 1$. Using $D^*D + B^*B = I$ (see (13.1.3)), we obtain

$$\Gamma^*\Gamma = D^*D + B^*W_o^*W_oB = D^*D + B^*B = I.$$

In other words, Γ is an isometry. Observe that $S_\mathcal{Y}^*W_o = W_oA$. For $n \geq 1$, we have

$$\begin{aligned}(S_\mathcal{Y}^n\Gamma)^*\Gamma &= \Gamma^*S_\mathcal{Y}^{*n}\Gamma = \Gamma^*S_\mathcal{Y}^{*n-1}W_oB = \Gamma^*W_oA^{n-1}B \\ &= \begin{bmatrix} D^* & B^*W_o^* \end{bmatrix} \begin{bmatrix} CA^{n-1}B \\ W_oA^nB \end{bmatrix} \\ &= D^*CA^{n-1}B + B^*AA^{n-1}B = 0.\end{aligned}$$

The last equality follows from $D^*C + B^*A = 0$; see (13.1.3). So $S_\mathcal{Y}^n\Gamma\mathcal{E}$ is orthogonal to $\Gamma\mathcal{E}$ for all $n \geq 1$. Therefore T_Θ is an isometry and Θ is inner.

Assume that Θ is an inner function. Let G be the function in $L^\infty(\mathcal{E}, \mathcal{Y})$ defined by $G(e^{\imath\omega}) = \Theta(e^{-\imath\omega})$. Because Θ is inner, G is rigid, that is, $G(e^{\imath\omega})$ is almost everywhere an isometry. In particular, its Laurent operator L_G is an isometry mapping $\ell^2(\mathcal{E})$ into $\ell^2(\mathcal{Y})$. Notice that G admits a Fourier series expansion of the form $G(e^{\imath\omega}) = \sum_0^\infty e^{\imath\omega n}\Theta_n$ where $\{\Theta_n\}_0^\infty$ are the Fourier coefficients of Θ, that is, $\Theta(e^{\imath\omega}) = \sum_0^\infty e^{-\imath\omega n}\Theta_n$. Hence

$$L_G = \begin{bmatrix} \ddots & \ddots & \vdots & \vdots & \\ \ddots & \Theta_0 & \Theta_1 & \Theta_2 & \cdots \\ \cdots & 0 & \boxed{\Theta_0} & \Theta_1 & \cdots \\ \cdots & 0 & 0 & \Theta_0 & \ddots \\ & \vdots & \vdots & \ddots & \ddots \end{bmatrix} : \ell^2(\mathcal{E}) \to \ell^2(\mathcal{Y}) \qquad (13.1.12)$$

is an upper triangular isometric Laurent matrix. (The box around Θ_0 represents the $\{0,0\}$ entry.) The operator $\nabla_\Theta = \Pi_{\ell^2_+(\mathcal{Y})}L_G|\ell^2_+(\mathcal{E})$. For f in $\ell^2_+(\mathcal{E})$, we obtain

$$\begin{aligned}\|D_{\nabla_\Theta}f\|^2 &= \|f\|^2 - \|\nabla_\Theta f\|^2 = \|f\|^2 - \|\Pi_{\ell^2_+(\mathcal{Y})}L_Gf\|^2 \\ &= \|f\|^2 - \|L_Gf\|^2 + \|\Pi_{\ell^2_-(\mathcal{Y})}L_Gf\|^2 \\ &= \|\Pi_{\ell^2_-(\mathcal{Y})}L_Gf\|^2 = \|H_\Theta f\|^2. \end{aligned} \qquad (13.1.13)$$

Here H_Θ is the Hankel matrix obtained by rearranging the rows of $\Pi_{\ell^2_-(\mathcal{Y})}L_G|\ell^2_+(\mathcal{E})$ in the following way:

$$H_\Theta = \begin{bmatrix} \Theta_1 & \Theta_2 & \Theta_3 & \cdots \\ \Theta_2 & \Theta_3 & \Theta_4 & \cdots \\ \Theta_3 & \Theta_4 & \Theta_5 & \cdots \\ \vdots & \vdots & \vdots & \ddots \end{bmatrix} : \ell^2_+(\mathcal{E}) \to \ell^2_+(\mathcal{Y}). \qquad (13.1.14)$$

Clearly, $\|\Pi_{\ell^2_-(\mathcal{Y})} L_G f\| = \|H_\Theta f\|$. Since $\|D_{\nabla_\Theta} f\| = \|H_\Theta f\|$ for all f, there exists a unitary operator Ψ mapping $\mathcal{X} = \mathcal{D}_{\nabla_\Theta}$ onto $\mathcal{H}_r = \overline{\text{ran} H_\Theta}$ such that $\Psi D_{\nabla_\Theta} = H_\Theta$. It is easy to verify that $S^*_\mathcal{Y} H_\Theta = H_\Theta S_\mathcal{E}$. Therefore $\mathcal{H}_r$ is an invariant subspace for the backward shift operator $S^*_\mathcal{Y}$.

Consider the operator A_r on $\mathcal{H}_r$ defined by $A_r = \Psi A \Psi^*$. Using the definition of A in(13.1.5) with $S^*_\mathcal{Y} H_\Theta = H_\Theta S_\mathcal{E}$, we obtain

$$A_r H_\Theta = A_r \Psi D_{\nabla_\Theta} = \Psi A D_{\nabla_\Theta} = \Psi D_{\nabla_\Theta} S_\mathcal{E} = H_\Theta S_\mathcal{E} = S^*_\mathcal{Y} H_\Theta.$$

Hence $A_r = S^*_\mathcal{Y} | \mathcal{H}_r$. Since $\mathcal{H}_r$ is an invariant subspace for the backward shift and $S^*_\mathcal{Y}$ is strongly stable, A_r is strongly stable. Therefore A is also strongly stable.

Using (13.1.5), we obtain

$$W_o D_{\nabla_\Theta} = \begin{bmatrix} C D_{\nabla_\Theta} \\ C A D_{\nabla_\Theta} \\ C A^2 D_{\nabla_\Theta} \\ \vdots \end{bmatrix} = \begin{bmatrix} Z S_\mathcal{E} \\ Z S^2_\mathcal{E} \\ Z S^2_\mathcal{E} \\ \vdots \end{bmatrix} = \begin{bmatrix} \Theta_1 & \Theta_2 & \Theta_3 & \cdots \\ \Theta_2 & \Theta_3 & \Theta_4 & \cdots \\ \Theta_3 & \Theta_4 & \Theta_5 & \cdots \\ \vdots & \vdots & \vdots & \ddots \end{bmatrix} = H_\Theta = \Psi D_{\nabla_\Theta}.$$

Hence $W_o = \Psi$ is an isometry mapping $\mathcal{X}$ into $\ell^2_+(\mathcal{Y})$. In particular, the pair $\{C, A\}$ is observable. □

Remark 13.1.5. Recall that a minimal realization for a rational transfer function in $H^\infty(\mathcal{E}, \mathcal{Y})$ is stable. Let $\{A \text{ on } \mathcal{X}, B, C, D\}$ be a controllable isometric realization for a rational contractive analytic function Θ in $H^\infty(\mathcal{E}, \mathcal{Y})$. Proposition 13.1.4 shows that A is stable if and only if Θ is inner. In other words, $\{A, B, C, D\}$ is a minimal realization if and only if Θ is inner. In fact, if Θ is not inner, then $\mathcal{X}$ is infinite dimensional.

Remark 13.1.6. Assume that Θ is an inner function in $H^\infty(\mathcal{E}, \mathcal{Y})$. Consider the state space realization $\Sigma_r = \{A_r \text{ on } \mathcal{H}_r, B_r, C_r, \Theta_0\}$ defined by

$$A_r = S^*_\mathcal{Y} | \mathcal{H}_r, \quad B_r = \begin{bmatrix} \Theta_1 \\ \Theta_2 \\ \Theta_3 \\ \vdots \end{bmatrix} : \mathcal{E} \to \mathcal{H}_r \quad \text{and} \quad C_r = \Pi_\mathcal{Y} | \mathcal{H}_r : \mathcal{H}_r \to \mathcal{Y}.$$

As before, $\mathcal{H}_r = \overline{\text{ran} H_\Theta}$. Here $\Pi_\mathcal{Y}$ is the operator which picks out the first component of $\ell^2_+(\mathcal{Y})$. It is noted that Σ_r is the restricted backward realization for Θ; see Section 14.6.2. Then Σ_r is a controllable and observable isometric realization for Θ.

To see this, it is sufficient to show that the unitary operator Ψ intertwines $\Sigma = \{A, B, C, \Theta_0\}$ in (13.1.5) with Σ_r. We have all ready seen that $A_r = \Psi A \Psi^*$. Notice that

$$\Psi B = \Psi D_{\nabla_\Theta} \Pi^*_\mathcal{E} = H_\Theta \Pi^*_\mathcal{E} = B_r.$$

Hence $\Psi B = B_r$. Moreover, (13.1.5) yields

$$C_r \Psi D_{\nabla_\Theta} = C_r H_\Theta = \Pi_{\mathcal{Y}} H_\Theta = Z S_{\mathcal{E}} = C D_{\nabla_\Theta}.$$

Thus $C_r \Psi = C$. So Ψ intertwines Σ with Σ_r. In other words, Σ_r is a controllable and observable isometric realization for Θ. Finally, it is noted that one can also directly verify this fact.

Corollary 13.1.7. *Let Θ be a function in $H^\infty(\mathcal{E}, \mathcal{Y})$. Then Θ is a two-sided inner function if and only if Θ admits a controllable and observable unitary realization $\{A, B, C, D\}$ such that A and A^* are strongly stable. In this case, if $\{A, B, C, D\}$ is a controllable, isometric realization for Θ, then $\{A, B, C, D\}$ is an observable, unitary realization where A and A^* are strongly stable.*

Proof. If Θ admits a controllable and observable unitary realization $\{A, B, C, D\}$ such that A and A^* are strongly stable, then Proposition 13.1.4 shows that Θ is inner. Since $\{A^*, C^*, B^*, D^*\}$ is a strongly stable, isometric controllable realization for $\widetilde{\Theta}(z) = \Theta(\overline{z})^*$, it follows that $\widetilde{\Theta}$ is inner, and thus, Θ is two-sided inner.

Now assume that $\Sigma = \{A, B, C, D\}$ is a controllable, isometric realization for a two-sided inner function Θ. Proposition 13.1.4 shows that Σ is observable and A is strongly stable. Without loss of generality we can assume that Σ is determined by (13.1.5). We claim that

$$D_{\nabla_\Theta} = P_{\mathcal{X}} \quad \text{where} \quad \mathcal{D}_{\nabla_\Theta} = \mathcal{X} = \ell^2_+(\mathcal{E}) \ominus L_{\widetilde{\Theta}} \ell^2_+(\mathcal{Y}). \tag{13.1.15}$$

Let us first show that $\ker D_{\nabla_\Theta} = L_{\widetilde{\Theta}} \ell^2_+(\mathcal{Y})$. Recall that $G(e^{\imath\omega}) = \Theta(e^{-\imath\omega}) = \widetilde{\Theta}(e^{\imath\omega})^*$. Hence the Laurent operator $L_G = L^*_{\widetilde{\Theta}}$. Because Θ is two-sided inner, the Laurent operator $L_{\widetilde{\Theta}}$ is unitary. According to (13.1.13) there exists a unitary operator φ mapping the closure of $\Pi_{\ell^2_-(\mathcal{Y})} L^*_{\widetilde{\Theta}} \ell^2_+(\mathcal{E})$ onto $\mathcal{D}_{\nabla_\Theta}$ such that

$$D_{\nabla_\Theta} = \varphi \Pi_{\ell^2_-(\mathcal{Y})} L^*_{\widetilde{\Theta}} | \ell^2_+(\mathcal{E}).$$

Notice that $D_{\nabla_\Theta} h = 0$ if and only if $h \in \ell^2_+(\mathcal{E})$ and $\Pi_{\ell^2_-(\mathcal{Y})} L^*_{\widetilde{\Theta}} h = 0$, or equivalently, $h \in \ell^2_+(\mathcal{E})$ and $L^*_{\widetilde{\Theta}} h \in \ell^2_+(\mathcal{Y})$. Since $L_{\widetilde{\Theta}}$ maps $\ell^2_+(\mathcal{Y})$ into $\ell^2_+(\mathcal{E})$ and $L_{\widetilde{\Theta}}$ is unitary, $D_{\nabla_\Theta} h = 0$ if and only if h is in $L_{\widetilde{\Theta}} \ell^2_+(\mathcal{Y})$. Therefore $\ker D_{\nabla_\Theta} = L_{\widetilde{\Theta}} \ell^2_+(\mathcal{Y})$. In particular, the subspace $\mathcal{D}_{\nabla_\Theta}$ is determined by (13.1.15).

Now let us show that $D_{\nabla_\Theta} | \mathcal{X}$ is an isometry. A vector g is in $\mathcal{X}$, if and only if $g \in \ell^2_+(\mathcal{E})$ and g is orthogonal to $L_{\widetilde{\Theta}} \ell^2_+(\mathcal{Y})$, or equivalently, $g \in \ell^2_+(\mathcal{E})$ and $L^*_{\widetilde{\Theta}} g$ is orthogonal to $\ell^2_+(\mathcal{Y})$, which is equivalent to g being in $\ell^2_+(\mathcal{E})$ and $v = L^*_{\widetilde{\Theta}} g$ where v is in $\ell^2_-(\mathcal{Y})$. Hence g is in $\mathcal{X}$ if and only if g is in $\ell^2_+(\mathcal{E})$ and $g = L_{\widetilde{\Theta}} v$ where v is in $\ell^2_-(\mathcal{Y})$. Using this form for $g \in \mathcal{X}$, we obtain

$$\|D_{\nabla_\Theta} g\| = \|\Pi_{\ell^2_-(\mathcal{Y})} L^*_{\widetilde{\Theta}} g\| = \|\Pi_{\ell^2_-(\mathcal{Y})} L^*_{\widetilde{\Theta}} L_{\widetilde{\Theta}} v\| = \|v\| = \|g\|.$$

Therefore $D_{\nabla_\Theta} | \mathcal{X}$ is an isometry and $D_{\nabla_\Theta} | \mathcal{X}^\perp = 0$. Because D_{∇_Θ} is a positive operator, D_{∇_Θ} maps $\mathcal{X}$ into $\mathcal{X}$, and thus, D_{∇_Θ} admits a matrix representation of the form

$$D_{\nabla_\Theta} = \begin{bmatrix} \Phi & 0 \\ 0 & 0 \end{bmatrix} \text{ on } \begin{bmatrix} \mathcal{X} \\ \mathcal{X}^\perp \end{bmatrix}$$

where Φ is a positive isometry on $\mathcal{X}$. In particular, $\Phi = \Phi^*$ and $\Phi^2 = \Phi^*\Phi = I$. So the unique positive square root Φ of Φ^2 equals I. Therefore $D_{\nabla_\Theta} = P_\mathcal{X}$ is the orthogonal projection onto $\mathcal{X}$ and (13.1.15) holds.

We claim that $A = \Pi_\mathcal{X} S_\mathcal{E}|\mathcal{X}$. For x in $\mathcal{X}$, we have

$$Ax = AP_\mathcal{X}x = AD_{\nabla_\Theta}x = D_{\nabla_\Theta}S_\mathcal{E}x = \Pi_\mathcal{X}S_\mathcal{E}x.$$

Hence $A = \Pi_\mathcal{X} S_\mathcal{E}|\mathcal{X}$. Notice that $\mathcal{X}^\perp = L_{\widetilde{\Theta}}\ell_+^2(\mathcal{Y}) = T_{\widetilde{\Theta}}\ell_+^2(\mathcal{Y})$ is an invariant subspace for $S_\mathcal{E}$. So $\mathcal{X}$ is an invariant subspace for $S_\mathcal{E}^*$, and $A^* = S_\mathcal{E}^*|\mathcal{X}$. Since $S_\mathcal{E}^*$ is strongly stable, A^* is strongly stable.

Using $B = D_{\nabla_\Theta}\Pi_\mathcal{E}^* = \Pi_\mathcal{X}\Pi_\mathcal{E}^*$, we have

$$B^* = \Pi_\mathcal{E}\Pi_\mathcal{X}^* \quad \text{and} \quad C^* = S_\mathcal{E}^*Z^*.$$

To verify that $C^* = S_\mathcal{E}^*Z^*$ notice that

$$C = C\Pi_\mathcal{X}\Pi_\mathcal{X}^* = CD_{\nabla_\Theta}\Pi_\mathcal{X}^* = ZS_\mathcal{E}\Pi_\mathcal{X}^*.$$

By taking the adjoint $C^* = \Pi_\mathcal{X}S_\mathcal{E}^*Z^*$, we claim that the range of $S_\mathcal{E}^*Z^*$ is contained in $\mathcal{X}$. For ϕ in $\mathcal{Y}$ and y in $\ell_+^2(\mathcal{Y})$, we obtain

$$(S_\mathcal{E}^*Z^*\phi, L_{\widetilde{\Theta}}y) = (Z^*\phi, S_\mathcal{E}T_{\widetilde{\Theta}}y) = (T_{\widetilde{\Theta}}\Pi_\mathcal{Y}^*\phi, T_{\widetilde{\Theta}}S_\mathcal{Y}y) = (\Pi_\mathcal{Y}^*\phi, S_\mathcal{Y}y) = 0.$$

Here we used that fact that the Toeplitz operator $T_{\widetilde{\Theta}}$ is an isometry. Thus $S_\mathcal{E}^*Z^*\mathcal{Y}$ is orthogonal to $L_{\widetilde{\Theta}}\ell_+^2(\mathcal{Y})$. In other words, the range of $S_\mathcal{E}^*Z^*$ is a subspace of $\mathcal{X}$. In particular, $C^* = \Pi_\mathcal{X}S_\mathcal{E}^*Z^* = S_\mathcal{E}^*Z^*$.

Since $\{A, B, C, D\}$ is a strongly stable, controllable and observable isometric realization for Θ and A^* is stable, it follows that $\{A^*, C^*, B^*, D^*\}$ is a strongly stable, controllable and observable realization for $\widetilde{\Theta}$. To complete the proof, it remains to show that

$$\Omega^* = \begin{bmatrix} D^* & B^* \\ C^* & A^* \end{bmatrix} : \begin{bmatrix} \mathcal{Y} \\ \mathcal{X} \end{bmatrix} \to \begin{bmatrix} \mathcal{E} \\ \mathcal{X} \end{bmatrix}$$

is an isometry. Since Θ is two-sided inner, the first column Z^* of $T_{\widetilde{\Theta}}$ is an isometry. Hence

$$DD^* + CC^* = \Theta_0\Theta_0^* + ZS_\mathcal{E}S_\mathcal{E}^*Z^* = ZZ^* = I.$$

So the first column of Ω^* is an isometry. To show that the second column of Ω^* is an isometry, notice that

$$BB^* + AA^* = \Pi_\mathcal{X}\Pi_\mathcal{E}^*\Pi_\mathcal{E}\Pi_\mathcal{X}^* + \Pi_\mathcal{X}S_\mathcal{E}S_\mathcal{E}^*\Pi_\mathcal{X}^* = \Pi_\mathcal{X}\Pi_\mathcal{X}^* = I.$$

To verify that Ω^* is an isometry, it remains to show that the two columns of Ω^*, are orthogonal, that is,

$$DB^* + CA^* = \Theta_0 \Pi_{\mathcal{E}} \Pi_{\mathcal{X}}^* + Z S_{\mathcal{E}} S_{\mathcal{E}}^* \Pi_{\mathcal{X}}^* = Z \Pi_{\mathcal{X}}^* = 0.$$

The last equality follows from $\mathcal{X} = \ell_+^2(\mathcal{E}) \ominus L_{\tilde{\Theta}} \ell_+^2(\mathcal{Y})$ and the fact that the space $Z^* \mathcal{Y} = L_{\tilde{\Theta}} \Pi_{\mathcal{Y}}^* \mathcal{Y}$ is orthogonal to $\mathcal{X}$. So Ω^* is an isometry, and Ω is unitary. □

Finally, it is noted one can use the restricted backward shift realization in Remark 13.1.6 to prove Corollary 13.1.7.

13.1.1 Rational contractive realizations

In this section, we will develop a finite sections algorithm to compute a contractive minimal realization for a rational contractive analytic function Θ in $H^\infty(\mathcal{E}, \mathcal{Y})$. Recall that $\{A, B, C, D\}$ in (13.1.5) is a controllable isometric realization for Θ. Let $\mathcal{X}_c$ be the observability subspace defined by $\mathcal{X}_c = \bigvee_0^\infty A^{*k} C^* \mathcal{Y}$. Consider the state space system $\{A_c \text{ on } \mathcal{X}_c, B_c, C_c, D\}$ determined by

$$A_c = \Pi_{\mathcal{X}_c} A | \mathcal{X}_c, \quad B_c = \Pi_{\mathcal{X}_c} B \quad \text{and} \quad C_c = C | \mathcal{X}_c. \tag{13.1.16}$$

Because $\{A_c, B_c, C_c, D\}$ is obtained by extracting the observable part from the controllable realization $\{A, B, C, D\}$, it follows that $\{A_c, B_c, C_c, D\}$ is a controllable and observable realization for Θ. In particular, due to the fact that Θ is rational, A_c is stable. Moreover, $\{A_c, B_c, C_c, D\}$ is a contractive realization, that is, its systems matrix

$$\Omega_c = \begin{bmatrix} D & C_c \\ B_c & A_c \end{bmatrix} : \begin{bmatrix} \mathcal{E} \\ \mathcal{X}_c \end{bmatrix} \to \begin{bmatrix} \mathcal{Y} \\ \mathcal{X}_c \end{bmatrix} \tag{13.1.17}$$

is a contraction. To see this simply observe that

$$\Omega_c = \begin{bmatrix} I_{\mathcal{Y}} & 0 \\ 0 & \Pi_{\mathcal{X}_c} \end{bmatrix} \begin{bmatrix} D & C \\ B & A \end{bmatrix} \begin{bmatrix} I_{\mathcal{E}} & 0 \\ 0 & \Pi_{\mathcal{X}_c}^* \end{bmatrix}.$$

Since Ω in (13.1.2) is an isometry, Ω_c is a contraction. Therefore $\{A_c, B_c, C_c, D\}$ is a stable contractive minimal realization for Θ.

Now let us develop a finite sections method to compute the realization $\{A_c, B_c, C_c, D\}$. To this end, let $\mathcal{X}_n$ be the subspace of $\mathcal{X} = \mathcal{D}_{\nabla_\Theta}$ defined by $\mathcal{X}_n = D_{\nabla_\Theta} \mathcal{E}^n$. (Here $\mathcal{E}^n$ is embedded in the first n components of $\ell_+^2(\mathcal{E})$.) Notice that $\{\mathcal{X}_n\}_1^\infty$ is a sequence of increasing subspaces ($\mathcal{X}_n \subseteq \mathcal{X}_{n+1}$) whose closed linear span equals $\mathcal{X}$. So the orthogonal projections $P_{\mathcal{X}_n}$, are increasing ($P_{\mathcal{X}_n} \leq P_{\mathcal{X}_{n+1}}$) and $P_{\mathcal{X}_n}$ converges to $I_{\mathcal{X}}$ in the strong operator topology. By consulting (13.1.5), we see that $A \mathcal{X}_{n-1} \subseteq \mathcal{X}_n$. Consider the system $\{A_n \text{ on } \mathcal{X}_n, B_n, C_n, D\}$ defined by

$$A_n = A P_{\mathcal{X}_{n-1}} | \mathcal{X}_n, \quad B_n = B \quad \text{and} \quad C_n = C P_{\mathcal{X}_{n-1}} | \mathcal{X}_n.$$

The operators A_n, B_n and C_n respectively, converge to A, B and C in the strong operator topology as n tends to infinity. Using the fact that $A\mathcal{X}_{n-1} \subseteq \mathcal{X}_n$, it follows that

$$A_n^j B_n = A^j B \quad \text{for } j = 0, 1, \ldots, n-1.$$

Since $A^j B = D_{\nabla_\Phi} S_\mathcal{E}^j \Pi_\mathcal{E}^*$ and $\mathcal{E}^n = \oplus_0^{n-1} S_\mathcal{E}^j \Pi_\mathcal{E}^* \mathcal{E}$, the span of $\{A_n^j B_n \mathcal{E}\}_0^{n-1}$ equals $\mathcal{X}_n$. Hence the pair $\{A_n, B_n\}$ is controllable. Furthermore, $C_n A_n^{k-1} B_n = \Theta_k$ for $k = 1, 2, \ldots, n-1$ where $\Theta(z) = \sum_0^\infty z^{-\nu}\Theta_\nu$. Finally, it is noted that the systems matrix for $\{A_n, B_n, C_n, D\}$ is determined by

$$\Omega_n = \begin{bmatrix} D & C_n \\ B_n & A_n \end{bmatrix} = \begin{bmatrix} I_\mathcal{Y} & 0 \\ 0 & P_{\mathcal{X}_n} \end{bmatrix} \begin{bmatrix} D & C \\ B & A \end{bmatrix} \begin{bmatrix} I_\mathcal{E} & 0 \\ 0 & P_{\mathcal{X}_{n-1}}|\mathcal{X}_n \end{bmatrix}.$$

Because Ω is an isometry, Ω_n is a contraction. Therefore $\{A_n, B_n, C_n, D\}$ is a controllable contractive realization.

If $\|\Theta\|_\infty < 1$, then A_n is stable. To verify this, recall that $A_n = AP_{\mathcal{X}_{n-1}}|\mathcal{X}_n$ and A is a contraction. So A_n is also a contraction, and all the eigenvalues for A_n are contained in the closed unit disc. Now let us proceed by contradiction. Assume that λ is an eigenvalue for A_n on the unit circle and x is its corresponding eigenvector. Then $\lambda x = A_n x$ yields

$$\|x\| = \|\lambda x\| = \|A_n x\| = \|AP_{\mathcal{X}_{n-1}} x\| \le \|P_{\mathcal{X}_{n-1}} x\| \le \|x\|.$$

Thus $\|x\| = \|P_{\mathcal{X}_{n-1}} x\|$ and x is a vector in $\mathcal{X}_{n-1}$. In other words, there exists a vector f in $\mathcal{E}^{n-1}$ such that $x = D_{\nabla_\Theta} f$. Using (13.1.5), we obtain

$$D_{\nabla_\Theta} \lambda f = \lambda x = A_n x = A_n D_{\nabla_\Theta} f = A D_{\nabla_\Theta} f = D_{\nabla_\Theta} S_\mathcal{E} f.$$

Because $\|\Theta\|_\infty < 1$, the operator D_{∇_Θ} is invertible. Hence $\lambda f = S_\mathcal{E} f$, and λ is an eigenvalue for the unilateral shift $S_\mathcal{E}$. However, the unilateral shift has no eigenvalues, and f must be zero. So x is also zero, and this contradicts the fact that an eigenvector is nonzero. Therefore A_n has no eigenvalues on the unit circle and A_n is stable.

Now let us find an approximation for the contractive minimal realization $\{A_c, B_c, C_c, D\}$ of Θ. For for n sufficiently large, compute the singular value decomposition

$$U\Lambda V^* = \begin{bmatrix} C_n^* & A_n^* C_n^* & A_n^{*2} C_n^* & \cdots & A_n^{*\nu} C_n^* \end{bmatrix} \tag{13.1.18}$$

where $\nu > \delta(\Theta)$, the McMillan degree of Θ. Here U and V are unitary matrices and Λ is the diagonal matrix consisting of the singular values in decreasing order. Then Λ will contain $\mu = \delta(\Theta)$ significant singular values. Let $\Phi = U|\mathbb{C}^\mu$ be the isometry mapping $\mathbb{C}^\mu$ into $\mathcal{X}_n$ obtained by keeping the first μ columns of U. Then

$$A_c \approx \Phi^* A_n \Phi, \quad B_c \approx \Phi^* B_n \quad \text{and} \quad C_c \approx C_n \Phi. \tag{13.1.19}$$

If $\{A_c, B_c, C_c, D\}$ is not a realization for Θ, then one must choose a larger n.

To implement this algorithm, we need a computational method to compute the controllable realization $\{A_n, B_n, C_n, D\}$. To this end, let $\nabla_{\Theta,n}$ be the upper triangular block Toeplitz matrix mapping $\mathcal{E}^n$ into $\mathcal{Y}^n$ determined by $\nabla_{\Theta,n} = \nabla_\Theta|\mathcal{E}^n$, that is,

$$\nabla_{\Theta,n} = \begin{bmatrix} \Theta_0 & \Theta_1 & \cdots & \Theta_{n-2} & \Theta_{n-1} \\ 0 & \Theta_0 & \cdots & \Theta_{n-3} & \Theta_{n-2} \\ \vdots & \vdots & \ddots & \vdots & \vdots \\ 0 & 0 & \cdots & \Theta_0 & \Theta_1 \\ 0 & 0 & \cdots & 0 & \Theta_0 \end{bmatrix} : \mathcal{E}^n \to \mathcal{Y}^n.$$

We claim that there exists a unitary operator Ψ mapping $\mathcal{X}_n$ onto $\mathcal{D}_{\nabla_{\Theta,n}}$ satisfying

$$\Psi D_{\nabla_\Theta} h = D_{\nabla_{\Theta,n}} h \qquad (h \in \mathcal{E}^n). \tag{13.1.20}$$

(If M is any contraction, then D_M is the positive square root of $I - M^*M$ and $\mathcal{D}_M$ is the closure of the range of D_M.) To see this, simply observe that the upper triangular structure of ∇_Θ yields

$$\|D_{\nabla_\Theta} h\|^2 = \|h\|^2 - \|\nabla_\Theta h\|^2 = \|h\|^2 - \|\nabla_{\Theta,n} h\|^2 = \|D_{\nabla_{\Theta,n}} h\|^2.$$

Hence (13.1.20) holds.

Let J_n and Q_n be the operators defined by

$$J_n = \begin{bmatrix} I \\ 0 \end{bmatrix} : \mathcal{E}^{n-1} \to \begin{bmatrix} \mathcal{E}^{n-1} \\ \mathcal{E} \end{bmatrix} \quad \text{and} \quad Q_n = \begin{bmatrix} 0 \\ I \end{bmatrix} : \mathcal{E}^{n-1} \to \begin{bmatrix} \mathcal{E} \\ \mathcal{E}^{n-1} \end{bmatrix}. \tag{13.1.21}$$

By consulting (13.1.5) and (13.1.20), we see that up to the unitary operator Ψ, the operators A_n, B_n and C_n are determined by

$$\begin{aligned} A_n D_{\nabla_{\Theta,n}} J_n &= D_{\nabla_{\Theta,n}} Q_n \quad \text{and} \quad B_n = D_{\nabla_{\Theta,n}} \Pi_\mathcal{E}^*, \\ C D_{\nabla_{\Theta,n}} J_n &= \begin{bmatrix} \Theta_0 & \Theta_1 & \cdots & \Theta_{n-2} & \Theta_{n-1} \end{bmatrix} Q_n. \end{aligned} \tag{13.1.22}$$

Here $\Pi_\mathcal{E}^*$ is the natural embedding of $\mathcal{E}$ into the first component of $\mathcal{E}^n$. By taking the Moore-Penrose pseudo inverse (denoted by M^{-r}), we obtain

$$\begin{aligned} A_n &= D_{\nabla_{\Theta,n}} Q_n (D_{\nabla_{\Theta,n}} J_n)^{-r} \quad \text{and} \quad B_n = D_{\nabla_{\Theta,n}} \Pi_\mathcal{E}^*, \\ C_n &= \begin{bmatrix} \Theta_1 & \Theta_2 & \cdots & \Theta_{n-2} & \Theta_{n-1} \end{bmatrix} (D_{\nabla_{\Theta,n}} J_n)^{-r}. \end{aligned} \tag{13.1.23}$$

So to approximate $\{A_c, B_c, C_c, \Theta(\infty)\}$, compute $\{A_n, B_n, C_n\}$ in (13.1.23). Then compute the singular value decomposition in (13.1.18). Form the isometry Φ by keeping the appropriate number of significant singular values. Then the minimal contractive realization $\{A_c, B_c, C_c, \Theta(\infty)\}$ of Θ is given by (13.1.19). Finally, it is noted that this algorithm is more efficient when $\|\Theta\|_\infty < 1$.

Example. Let us demonstrate how this algorithm works on the example in Section 7.8.13. To this end, consider the transfer function in (7.8.13) given by

$$\theta(z) = \frac{-0.7165z^2 + 0.1796z - 0.0706}{z^3 - 0.2824z^2 - 0.0580z + 0.0003}. \tag{13.1.24}$$

A simple computation using the fast Fourier transform shows that $\|\theta\|_\infty = .92$. We ran Matlab on this example for $n = 100$. It is noted that one can use the fast Fourier transform or state space techniques to compute the Taylor coefficients $\{\theta_k\}$ for $\theta(z) = \sum_0^\infty z^{-k}\theta_k$. This immediately yields the upper triangular Toeplitz matrix $\nabla_{\Theta,n}$ (the Matlab command is "toeplitz"). Then we used "sqrtm" in Matlab to compute the positive square root $D_{\nabla_{\Theta,n}}$. Next we computed $\{A_n, B_n, C_n, D\}$ by using (13.1.23). The spectral radius for A_n is 0.7117. For $\nu = 10$ in (13.1.18), we discovered that the corresponding matrix has only three significant singular values $\{0.759, 0.1934, 0.0163\}$. So we formed the isometry Φ mapping $\mathbb{C}^3$ into $\mathbb{C}^{100}$ by keeping the first three columns of U. Finally, our approximation to $\{A_c, B_c, C_c, \theta(\infty)\}$ in (13.1.19) is given by

$$A_c = \begin{bmatrix} 0.0983 & -0.2506 & -0.0026 \\ -0.6529 & 0.1097 & -0.0825 \\ 0.0789 & 0.9595 & 0.0744 \end{bmatrix}, \quad B_c = -\begin{bmatrix} 0.9577 \\ 0.1721 \\ 0.2305 \end{bmatrix},$$
$$C_c = \begin{bmatrix} 0.7447 & 0.0226 & -0.0025 \end{bmatrix}.$$

Here $D = \theta(\infty) = 0$. Notice that our state space realization $\{A_c, B_c, C_c, 0\}$ is different from the state space realization of the same transfer function $\theta = g$ in Section 7.8.1; see (7.8.15). However, they both realize the same transfer function. In fact, using the fast Fourier transform in Matlab, $\|C_c(zI - A_c)^{-1}B_c - \theta\|_\infty = 1.0825 \times 10^{-15}$. Finally, $\|\Omega_c\| = 1$ where Ω_c is the systems matrix for $\{A_c, B_c, C_c, 0\}$ in (13.1.17).

13.2 Tangential Nevanlinna-Pick Interpolation

In this section, we will employ the method of Agler-McCarthy to solve the tangential Nevanlinna-Pick interpolation problem for contractive analytic functions.

The classical Nevanlinna-Pick interpolation problem for contractive analytic functions is: Given a distinct set of complex numbers $\{\alpha_j\}_1^n$ in $\mathbb{D}_+$ and a set of complex numbers $\{\gamma_j\}_1^n$, then find a contractive analytic function f in H^∞ satisfying the following conditions

$$f(\alpha_j) = \gamma_j \qquad (\text{for } j = 1, 2, \ldots, n). \tag{13.2.1}$$

The classical Nevanlinna-Pick interpolation problem in (13.2.1) is a special case of a more general tangential Nevanlinna-Pick interpolation problem. To introduce this Nevanlinna-Pick problem, let $\{E, \Lambda\}$ be an observable pair, where Λ is a strongly stable operator on $\mathcal{H}$, and E maps $\mathcal{H}$ into $\mathcal{E}$. By *strongly stable* we mean that Λ^n converges to zero in the strong operator topology as n approaches infinity. Let $\widetilde{E}$ be an operator from $\mathcal{H}$ into $\mathcal{Y}$. Let W and $\widetilde{W}$ be the observability operators defined by

$$W = \begin{bmatrix} E \\ E\Lambda \\ E\Lambda^2 \\ \vdots \end{bmatrix} : \mathcal{H} \to \ell_+^2(\mathcal{E}) \quad \text{and} \quad \widetilde{W} = \begin{bmatrix} \widetilde{E} \\ \widetilde{E}\Lambda \\ \widetilde{E}\Lambda^2 \\ \vdots \end{bmatrix} : \mathcal{H} \to \ell_+^2(\mathcal{Y}). \tag{13.2.2}$$

Throughout we assume that both W and $\widetilde{W}$ are operators, that is, bounded linear maps. In particular, if Λ is a stable operator and $\mathcal{H}$ is finite dimensional, then W and $\widetilde{W}$ are well-defined operators. Let Θ be a function in $H^\infty(\mathcal{E}, \mathcal{Y})$. As before, ∇_Θ is the upper triangular Toeplitz operator defined by (13.1.1). Recall that $\|\nabla_\Theta\| = \|\Theta\|_\infty$.

Definition 13.2.1. The tangential Nevanlinna-Pick interpolation problem for the data $\{\Lambda, E, \widetilde{E}\}$, is to find a contractive analytic function Θ in $H^\infty(\mathcal{E}, \mathcal{Y})$ such that $\nabla_\Theta W = \widetilde{W}$. In this case, Θ is called a contractive interpolant or solution for the data $\{\Lambda, E, \widetilde{E}\}$.

Assume that $\{\Lambda, E, \widetilde{E}\}$ a data set. Let $\Theta = \sum_0^\infty z^{-n}\Theta_n$ be any function in $H^\infty(\mathcal{E}, \mathcal{Y})$. Then it follows that

$$\nabla_\Theta W = \widetilde{W} \quad \text{if and only if} \quad \sum_{n=0}^{\infty} \Theta_n E \Lambda^n = \widetilde{E}. \tag{13.2.3}$$

So Θ is a solution to the tangential Nevanlinna-Pick interpolation problem if and only if Θ is a contractive analytic function in $H^\infty(\mathcal{E}, \mathcal{Y})$ satisfying $\sum_0^\infty \Theta_n E\Lambda^n = \widetilde{E}$.

To show that this interpolation problem covers the classical Nevanlinna-Pick problem, let $\{\alpha_j\}_1^n$ be a finite set of distinct points in $\mathbb{D}_+$ and $\{\gamma_j\}_1^n$ a set of complex numbers. Let Λ be the diagonal matrix on $\mathbb{C}^n$ defined by $\Lambda = \text{diag}[\{1/\alpha_j\}_1^n]$. Let E and $\widetilde{E}$ be the row vectors of length n given by

$$E = \begin{bmatrix} 1 & 1 & \cdots & 1 \end{bmatrix} \quad \text{and} \quad \widetilde{E} = \begin{bmatrix} \gamma_1 & \gamma_2 & \cdots & \gamma_n \end{bmatrix}. \tag{13.2.4}$$

Let f be an analytic function in $\mathbb{D}_+$. Then (13.2.1) is equivalent to $\sum_0^\infty f_n E\Lambda^n = \widetilde{E}$, where $f(z) = \sum_0^\infty z^{-n} f_n$. In other words, (13.2.1) is equivalent to $\nabla_f W = \widetilde{W}$; see (13.2.3). So the classical Nevanlinna-Pick interpolation problem of finding a contractive analytic function f satisfying (13.2.1) is a special case of our tangential Nevanlinna-Pick interpolation problem.

The tangential Nevanlinna-Pick interpolation problem also includes the classical Schur interpolation problem. To see this, let $\{\Theta_j\}_0^{n-1}$ be a specified set of operators with values in $\mathcal{L}(\mathcal{E},\mathcal{Y})$. The *Schur interpolation problem* is to find the set of all contractive analytic functions Θ in $H^\infty(\mathcal{E},\mathcal{Y})$ such that Θ admits a Taylor series expansion of the form

$$\Theta(z) = \sum_{k=0}^{\infty} z^{-k}\Theta_k. \tag{13.2.5}$$

Here $\{\Theta_j\}_0^{n-1}$ are the first n Taylor coefficients of Θ. To convert this to the tangential Nevanlinna-Pick problem, consider the data set $\{\Lambda, E, \widetilde{E}\}$ defined by

$$\begin{aligned}
\Lambda &= \begin{bmatrix} 0 & I & \cdots & 0 & 0 \\ 0 & 0 & \cdots & 0 & 0 \\ \vdots & \vdots & \ddots & \vdots & \vdots \\ 0 & 0 & \cdots & 0 & I \\ 0 & 0 & \cdots & 0 & 0 \end{bmatrix} \quad \text{on } \mathcal{E}^n, \\
E &= \begin{bmatrix} I & 0 & \cdots & 0 & 0 \end{bmatrix} : \mathcal{E}^n \to \mathcal{E}, \\
\widetilde{E} &= \begin{bmatrix} \Theta_0 & \Theta_1 & \cdots & \Theta_{n-2} & \Theta_{n-1} \end{bmatrix} : \mathcal{E}^n \to \mathcal{Y}.
\end{aligned} \tag{13.2.6}$$

Observe that Λ is the upper shift on $\mathcal{E}^n$, that is, the identity I appears immediately above the main diagonal and zero's appear everywhere else. Clearly, Λ is stable. Let Θ be a function in $H^\infty(\mathcal{E},\mathcal{Y})$. Then

$$\sum_{n=0}^{\infty} \Theta_n E \Lambda^n = \begin{bmatrix} \Theta_0 & \Theta_1 & \cdots & \Theta_{n-2} & \Theta_{n-1} \end{bmatrix}$$

where $\{\Theta_j\}_0^{n-1}$ are the first n Taylor coefficients of Θ. So according to (13.2.3), it follows that Θ is a solution for the Schur interpolation problem with data $\{\Theta_j\}_0^{n-1}$ if and only if Θ is a solution to the tangential Nevanlinna-Pick interpolation problem with data $\{\Lambda, E, \widetilde{E}\}$ determined by (13.2.6).

The fundamental Lyapunov equation. Now assume that Θ is a solution to the tangential Nevanlinna-Pick interpolation problem with data $\{\Lambda, E, \widetilde{E}\}$. Then we claim that there exists a positive solution Q to the Lyapunov equation

$$Q = \Lambda^* Q \Lambda + E^* E - \widetilde{E}^* \widetilde{E}. \tag{13.2.7}$$

Moreover, $Q = W^*W - \widetilde{W}^*\widetilde{W}$ is the unique positive solution to (13.2.7). Using the fact that ∇_Θ is a contraction, for all x in $\mathcal{H}$ we have

$$(\widetilde{W}^*\widetilde{W}x, x) = \|\widetilde{W}x\|^2 = \|\nabla_\Theta W x\|^2 \le \|Wx\|^2 = (W^*Wx, x).$$

Hence $Q = W^*W - \widetilde{W}^*\widetilde{W}$ is positive. Then using $S_{\mathcal{E}}^* W = W\Lambda$ and $S_{\mathcal{Y}}^* \widetilde{W} = \widetilde{W}\Lambda$, we obtain

$$\begin{aligned} Q - \Lambda^* Q \Lambda &= W^*W - \Lambda^* W^* W \Lambda - \widetilde{W}^*\widetilde{W} + \Lambda^* \widetilde{W}^* \widetilde{W} \Lambda \\ &= W^*(I - S_{\mathcal{E}} S_{\mathcal{E}}^*) W - \widetilde{W}^*(I - S_{\mathcal{Y}} S_{\mathcal{Y}}^*)\widetilde{W} \\ &= W^* P_{\mathcal{E}} W - \widetilde{W}^* P_{\mathcal{Y}} \widetilde{W} = E^*E - \widetilde{E}^*\widetilde{E}. \end{aligned}$$

(Here $P_{\mathcal{U}}$ is the orthogonal projection onto the first component $\mathcal{U}$ of $\ell_+^2(\mathcal{U})$.) Therefore Q is a positive solution to the Lyapunov equation in (13.2.7).

Notice that there is only one solution to the Lyapunov equation in (13.2.7). If Q_1 is another solution to (13.2.7), then subtracting $Q_1 = \Lambda^* Q_1 \Lambda + E^*E - \widetilde{E}^*\widetilde{E}$ from the Lyapunov equation in (13.2.7), shows that $\Delta = \Lambda^*\Delta\Lambda$ where $\Delta = Q - Q_1$. By recursively substituting $\Lambda^*\Delta\Lambda$ for Δ, we see that $\Delta = \Lambda^{*n}\Delta\Lambda^n$ for all integers $n \geq 0$. Because Λ^n converges to zero in the strong operator topology, $\Delta = 0$. In other words, $Q = Q_1$ and the solution to the Lyapunov equation in (13.2.7) is unique.

Contractive coupling. Now assume that Q is a positive solution to the Lyapunov equation in (13.2.7) with data $\{\Lambda, E, \widetilde{E}\}$. Let M be any operator from $\mathcal{H}$ into $\mathcal{M}$ such that $M^*M = Q$ and the range of M is dense in $\mathcal{M}$. Let Φ_1 and Φ_2 be the operators defined by

$$\Phi_1 = \begin{bmatrix} E \\ M\Lambda \end{bmatrix} : \mathcal{H} \to \begin{bmatrix} \mathcal{E} \\ \mathcal{M} \end{bmatrix} \quad \text{and} \quad \Phi_2 = \begin{bmatrix} \widetilde{E} \\ M \end{bmatrix} : \mathcal{H} \to \begin{bmatrix} \mathcal{Y} \\ \mathcal{M} \end{bmatrix}. \tag{13.2.8}$$

By employing the Lyapunov equation (13.2.7), we arrive at

$$\Phi_1^*\Phi_1 = \Lambda^* Q \Lambda + E^*E = Q + \widetilde{E}^*\widetilde{E} = \Phi_2^*\Phi_2.$$

This implies that there exists a unitary operator V from $\mathcal{H}_1$, the closure of the range of Φ_1, onto $\mathcal{H}_2$, the closure of the range of Φ_2, such that $V\Phi_1 = \Phi_2$. We say that U is a *contractive coupling* for the data set $\{\Lambda, E, \widetilde{E}\}$ if U is a contraction mapping $\mathcal{E} \oplus \mathcal{X}$ into $\mathcal{Y} \oplus \mathcal{X}$ such that $\mathcal{M} \subset \mathcal{X}$ and $U|\mathcal{H}_1 = V$. By choosing $\mathcal{X} = \mathcal{M}$, we see that $U = VP_{\mathcal{H}_1}$ mapping $\mathcal{E} \oplus \mathcal{M}$ into $\mathcal{Y} \oplus \mathcal{M}$ is a contractive coupling of $\{\Lambda, E, \widetilde{E}\}$. In other words, one can always construct a contractive coupling of $\{\Lambda, E, \widetilde{E}\}$.

Let U be any contractive coupling of $\{\Lambda, E, \widetilde{E}\}$. Then U admits a matrix representation of the form

$$U = \begin{bmatrix} D & C \\ B & A \end{bmatrix} : \begin{bmatrix} \mathcal{E} \\ \mathcal{X} \end{bmatrix} \to \begin{bmatrix} \mathcal{Y} \\ \mathcal{X} \end{bmatrix}. \tag{13.2.9}$$

The subspace $\mathcal{X}$ is referred to as the *state space* for the coupling U. We say the coupling U is *controllable*, if the pair $\{A, B\}$ is controllable. The coupling U is controllable if

$$\mathcal{X} = \bigvee_{n=0}^{\infty} (P_{\mathcal{X}} U P_{\mathcal{X}})^n P_{\mathcal{X}} U \mathcal{E}. \tag{13.2.10}$$

Finally, it is noted that U is a contractive coupling of $\{\Lambda, E, \widetilde{E}\}$, if and only if U admits a matrix representation of the form (13.2.9), the subspace $\mathcal{M} \subseteq \mathcal{X}$ and

$$\begin{bmatrix} D & C \\ B & A \end{bmatrix} \begin{bmatrix} E \\ M\Lambda \end{bmatrix} = \begin{bmatrix} \widetilde{E} \\ M \end{bmatrix}. \tag{13.2.11}$$

The central isometric coupling. Assume that there exists a positive solution Q to the Lyapunov equation (13.2.7) corresponding to the data $\{\Lambda, E, \widetilde{E}\}$. We will construct a special isometric coupling $U_\circ$ of $\{\Lambda, E, \widetilde{E}\}$ which plays a fundamental role in our solution to the tangential Nevanlinna-Pick interpolation problem. To this end, let $\mathcal{D}_1$ and $\mathcal{D}_2$ be the subspaces defined by

$$\mathcal{D}_1 = (\mathcal{E} \oplus \mathcal{M}) \ominus \mathcal{H}_1 \quad \text{and} \quad \mathcal{D}_2 = (\mathcal{Y} \oplus \mathcal{M}) \ominus \mathcal{H}_2. \tag{13.2.12}$$

Now consider the isometry defined by

$$U_\circ = \begin{bmatrix} V & 0 & 0 & 0 & 0 & \cdots \\ 0 & 0 & 0 & 0 & 0 & \cdots \\ 0 & I & 0 & 0 & 0 & \cdots \\ 0 & 0 & I & 0 & 0 & \cdots \\ 0 & 0 & 0 & I & 0 & \cdots \\ \vdots & \vdots & \vdots & \vdots & \ddots & \ddots \end{bmatrix} : \begin{bmatrix} \mathcal{H}_1 \\ \mathcal{D}_1 \\ \mathcal{D}_1 \\ \mathcal{D}_1 \\ \mathcal{D}_1 \\ \vdots \end{bmatrix} \to \begin{bmatrix} \mathcal{H}_2 \\ \mathcal{D}_2 \\ \mathcal{D}_1 \\ \mathcal{D}_1 \\ \mathcal{D}_1 \\ \vdots \end{bmatrix}. \tag{13.2.13}$$

Recall that $\mathcal{E} \oplus \mathcal{M} = \mathcal{H}_1 \oplus \mathcal{D}_1$ and $\mathcal{Y} \oplus \mathcal{M} = \mathcal{H}_2 \oplus \mathcal{D}_2$. So $U_\circ$ is an isometry mapping $\mathcal{K}_1 = \mathcal{E} \oplus \mathcal{M} \oplus \ell_+^2(\mathcal{D}_1)$ into $\mathcal{K}_2 = \mathcal{Y} \oplus \mathcal{M} \oplus \ell_+^2(\mathcal{D}_1)$. Since $U_\circ | \mathcal{H}_1 = V$, it follows that $U_\circ$ is an isometric coupling of $\{\Lambda, E, \widetilde{E}\}$. The isometry $U_\circ$ in (13.2.13) is called the *central coupling* for the data $\{\Lambda, E, \widetilde{E}\}$.

For another representation of the central isometric coupling, recall that $V P_{\mathcal{H}_1}$ mapping $\mathcal{E} \oplus \mathcal{M}$ into $\mathcal{Y} \oplus \mathcal{M}$ is a contractive coupling of $\{\Lambda, E, \widetilde{E}\}$. Notice that $V P_{\mathcal{H}_1}$ admits a matrix representation of the form

$$V P_{\mathcal{H}_1} = \begin{bmatrix} D_\circ & C_\circ \\ B_\circ & A_\circ \end{bmatrix} : \begin{bmatrix} \mathcal{E} \\ \mathcal{M} \end{bmatrix} \to \begin{bmatrix} \mathcal{Y} \\ \mathcal{M} \end{bmatrix}. \tag{13.2.14}$$

Then the central isometric coupling $U_\circ$ of $\{\Lambda, E, \widetilde{E}\}$ also admits a matrix representation of the form

$$U_\circ = \begin{bmatrix} D_\circ & C_\circ & 0 \\ B_\circ & A_\circ & 0 \\ B_1 & A_1 & S_{\mathcal{D}_1} \end{bmatrix} : \begin{bmatrix} \mathcal{E} \\ \mathcal{M} \\ \ell_+^2(\mathcal{D}_1) \end{bmatrix} \to \begin{bmatrix} \mathcal{Y} \\ \mathcal{M} \\ \ell_+^2(\mathcal{D}_1) \end{bmatrix}. \tag{13.2.15}$$

Here B_1 and A_1 are the operators defined by

$$B_1 = \begin{bmatrix} \Pi_{\mathcal{D}_1}|\mathcal{E} \\ 0 \\ 0 \\ \vdots \end{bmatrix} : \mathcal{E} \to \ell^2_+(\mathcal{D}_1) \quad \text{and} \quad A_1 = \begin{bmatrix} \Pi_{\mathcal{D}_1}|\mathcal{M} \\ 0 \\ 0 \\ \vdots \end{bmatrix} : \mathcal{M} \to \ell^2_+(\mathcal{D}_1).$$

Moreover, $\Pi_{\mathcal{D}_1}$ is the operator from $\mathcal{E} \oplus \mathcal{M}$ onto $\mathcal{D}_1$ defined by $\Pi_{\mathcal{D}_1} = P_{\mathcal{D}_1}$ where $P_{\mathcal{D}_1}$ is the orthogonal projection onto $\mathcal{D}_1$. In this case, the isometric realization $\{A, B, C, D\}$ for the central isometric coupling $U_\circ$ is determined by the matrix representation

$$\begin{aligned} U_\circ &= \begin{bmatrix} D_\circ & C \\ B & A \end{bmatrix} : \begin{bmatrix} \mathcal{E} \\ \begin{bmatrix} \mathcal{M} \\ \ell^2_+(\mathcal{D}_1) \end{bmatrix} \end{bmatrix} \to \begin{bmatrix} \mathcal{Y} \\ \begin{bmatrix} \mathcal{M} \\ \ell^2_+(\mathcal{D}_1) \end{bmatrix} \end{bmatrix}, \\ A &= \begin{bmatrix} A_\circ & 0 \\ A_1 & S_{\mathcal{D}_1} \end{bmatrix} \text{ on } \begin{bmatrix} \mathcal{M} \\ \ell^2_+(\mathcal{D}_1) \end{bmatrix}, \\ B &= \begin{bmatrix} B_\circ \\ B_1 \end{bmatrix} : \mathcal{E} \to \begin{bmatrix} \mathcal{M} \\ \ell^2_+(\mathcal{D}_1) \end{bmatrix}, \\ C &= \begin{bmatrix} C_\circ & 0 \end{bmatrix} : \begin{bmatrix} \mathcal{M} \\ \ell^2_+(\mathcal{D}_1) \end{bmatrix} \to \mathcal{Y}. \end{aligned} \tag{13.2.16}$$

In this case, the state space $\mathcal{X} = \mathcal{M} \oplus \ell^2_+(\mathcal{D}_1)$. We refer to this realization $\{A, B, C, D\}$ as the *central isometric realization* for the data $\{\Lambda, E, \widetilde{E}\}$. In a moment we will see that $\{A, B\}$ is controllable. Observe that both $\{A_\circ, B_\circ, C_\circ, D_\circ\}$ and $\{A, B, C, D\}$ have the same transfer function, that is,

$$\Theta(z) = D + C(zI - A)^{-1}B = D_\circ + C_\circ(zI - A_\circ)^{-1}B_\circ. \tag{13.2.17}$$

Finally, $\Theta(z) = D + C(zI - A)^{-1}B$ is called the *central interpolant* for $\{\Lambda, E, \widetilde{E}\}$. Because $U_\circ$ is a contraction, the central interpolant Θ is a contractive analytic function; see Theorem 13.1.1. Finally, Theorem 13.2.2 below shows that Θ is indeed a contractive interpolant for the data $\{\Lambda, E, \widetilde{E}\}$.

A computational algorithm in the finite dimensional case. As before, assume that there exists a positive solution Q to the Lyapunov equation (13.2.7). According to Theorem 13.2.2 below, the transfer function Θ in (13.2.17) is a solution for the data $\{\Lambda, E, \widetilde{E}\}$. Recall that $\{A \text{ on } \mathcal{X}, B, C, D_\circ\}$ is the isometric realization in (13.2.16), and $\{A_\circ \text{ on } \mathcal{M}, B_\circ, C_\circ, D_\circ\}$ is the contractive realization for Θ determined by (13.2.14). Clearly, $\dim \mathcal{M} \leq \dim \mathcal{X}$. In most cases, the state space $\mathcal{X}$ is infinite dimensional even when $\mathcal{H}$ is finite dimensional. Since $\dim \mathcal{M} \leq \dim \mathcal{H}$, it follows that $\{A_\circ, B_\circ, C_\circ, D_\circ\}$ is a finite dimensional realization when $\mathcal{H}$ is finite dimensional. If $\mathcal{H}$, is finite dimensional, then one can use Matlab to compute

$\{A_\circ, B_\circ, C_\circ, D_\circ\}$, and thus, the central interpolant Θ for $\{\Lambda, E, \widetilde{E}\}$. In the finite dimensional case, the pseudo-inverse in Matlab ("pinv"), yields

$$\begin{bmatrix} D_\circ & C_\circ \\ B_\circ & A_\circ \end{bmatrix} = \begin{bmatrix} \widetilde{E} \\ M \end{bmatrix} \begin{bmatrix} E \\ M\Lambda \end{bmatrix}^{-r}.$$

Here $-r$ denotes the Moore-Penrose restricted inverse. Later we will show that the pair $\{A_\circ, B_\circ\}$ is controllable. If $\Phi_1^*\Phi_1 = Q + \Lambda^* Q \Lambda$ is invertible, then $P_{\mathcal{H}_1} = \Phi_1(\Phi_1^*\Phi_1)^{-1}\Phi_1^*$. In this case,

$$\begin{aligned} VP_{\mathcal{H}_1} &= V\Phi_1(\Phi_1^*\Phi_1)^{-1}\Phi_1^* = \Phi_2(\Phi_1^*\Phi_1)^{-1}\Phi_1^* \\ &= \begin{bmatrix} \widetilde{E} \\ M \end{bmatrix} (Q + \Lambda^* Q\Lambda)^{-1} \begin{bmatrix} E^* & \Lambda^* M^* \end{bmatrix} = \begin{bmatrix} D_\circ & C_\circ \\ B_\circ & A_\circ \end{bmatrix}. \end{aligned}$$

In other words, when $Q + \Lambda^* Q \Lambda$ is invertible:

$$\begin{aligned} A_\circ &= M\left(Q + \Lambda^* Q\Lambda\right)^{-1} \Lambda^* M^*, \\ B_\circ &= M\left(Q + \Lambda^* Q\Lambda\right)^{-1} E^*, \\ C_\circ &= \widetilde{E}\left(Q + \Lambda^* Q\Lambda\right)^{-1} \Lambda^* M^*, \\ D_\circ &= \widetilde{E}\left(Q + \Lambda^* Q\Lambda\right)^{-1} E^*. \end{aligned} \tag{13.2.18}$$

In particular, $\{A_\circ, B_\circ, C_\circ, D_\circ\}$ is a contractive controllable realization of the central interpolant Θ for the data $\{\Lambda, E, \widetilde{E}\}$.

Now assume that Q is invertible. Then M is an invertible operator mapping $\mathcal{H}$ onto $\mathcal{M}$. Consider the system $\{A_c, B_c, C_c, D_\circ\}$ defined by $A_c = M^{-1}A_\circ M$, $B_c = M^{-1}B_\circ$ and $C_c = C_\circ M$. By consulting (13.2.18), we see that

$$\begin{aligned} A_c &= \left(Q + \Lambda^* Q\Lambda\right)^{-1} \Lambda^* Q, \\ B_c &= \left(Q + \Lambda^* Q\Lambda\right)^{-1} E^*, \\ C_c &= \widetilde{E}\left(Q + \Lambda^* Q\Lambda\right)^{-1} \Lambda^* Q, \\ D_\circ &= \widetilde{E}\left(Q + \Lambda^* Q\Lambda\right)^{-1} E^*. \end{aligned} \tag{13.2.19}$$

Then M is an invertible operator which intertwines the system $\{A_c, B_c, C_c, D_\circ\}$ with $\{A_\circ, B_\circ, C_\circ, D_\circ\}$. In particular, $\{A_c, B_c, C_c, D_\circ\}$ is a controllable realization for the central solution Θ for the data $\{\Lambda, E, \widetilde{E}\}$.

The basic coupling theorem. The following result describes the set of all contractive interpolants in terms of isometric couplings.

Theorem 13.2.2. *Let $\{\Lambda, E, \widetilde{E}\}$ be the data for the tangential Nevanlinna-Pick interpolation problem. Let Q be the solution to the Lyapunov equation*

$$Q = \Lambda^* Q \Lambda + E^* E - \widetilde{E}^* \widetilde{E}. \tag{13.2.20}$$

Then there exists a solution to this Nevanlinna-Pick interpolation problem if and only if Q is positive. In this case, the set of all contractive interpolants for $\{\Lambda, E, \widetilde{E}\}$ is given by

$$\Theta(z) = D + C(zI - A)^{-1}B, \tag{13.2.21}$$

where

$$U = \begin{bmatrix} D & C \\ B & A \end{bmatrix} : \begin{bmatrix} \mathcal{E} \\ \mathcal{X} \end{bmatrix} \to \begin{bmatrix} \mathcal{Y} \\ \mathcal{X} \end{bmatrix} \tag{13.2.22}$$

is any controllable isometric coupling of $\{\Lambda, E, \widetilde{E}\}$. (The isometric couplings are unique up to unitary equivalence.) In particular, the central transfer function Θ determined by (13.2.14) *and* (13.2.17) *is an interpolant for the data $\{\Lambda, E, \widetilde{E}\}$.*

Proof. We have already seen that if Θ is a solution to the tangential Nevanlinna-Pick interpolation problem, then $Q = W^*W - \widetilde{W}^*\widetilde{W}$ is the unique positive solution to the Lyapunov equation in (13.2.7).

Now assume that Q is a positive solution to the Lyapunov equation in (13.2.7). Let U be any contractive coupling of $\{\Lambda, E, \widetilde{E}\}$. Using the matrix representation for U in (13.2.22) with (13.2.11), we obtain

$$\begin{bmatrix} D & C \\ B & A \end{bmatrix} \begin{bmatrix} E \\ M\Lambda \end{bmatrix} = \begin{bmatrix} \widetilde{E} \\ M \end{bmatrix}. \tag{13.2.23}$$

Rewriting this equation yields

$$M = AM\Lambda + BE \quad \text{and} \quad \widetilde{E} = CM\Lambda + DE. \tag{13.2.24}$$

By recursively solving for M, we have

$$M = \sum_{j=0}^{n-1} A^j BE\Lambda^j + A^n M \Lambda^n. \tag{13.2.25}$$

Because U is a contraction, A is a contraction. Hence A^n is also a contraction for all integers $n \geq 1$. Using the fact that Λ is strongly stable, we see that $A^n M \Lambda^n$ converges to zero in the strong operator topology. By taking the limit in (13.2.25), we arrive at

$$M = \sum_{j=0}^{\infty} A^j BE\Lambda^j. \tag{13.2.26}$$

The sum converges in the strong operator topology. Now let Θ be the contraction given by $\Theta(z) = D + C(zI - A)^{-1}B$. Observe that Θ admits power series expansion of the form $\Theta = \sum_0^\infty z^{-n}\Theta_n$ where $\Theta_0 = D$ and $\Theta_n = CA^{n-1}B$ for all integers $n \geq 1$. By employing (13.2.26) with the second equation in (13.2.24), we obtain

$$\sum_{n=0}^{\infty} \Theta_n E \Lambda^n = \Theta_0 E + \sum_{n=0}^{\infty} CA^n BE\Lambda^n \Lambda = DE + CM\Lambda = \widetilde{E}.$$

Therefore Θ is a contractive interpolant for $\{\Lambda, E, \widetilde{E}\}$.

Assume that Θ is a contractive interpolant for $\{\Lambda, E, \widetilde{E}\}$. Then Θ admits a unique controllable isometric realization $\{A \text{ on } \mathcal{X}, B, C, D\}$ for Θ. According to Lemma 13.1.3, the controllability operator

$$W_c = \begin{bmatrix} B & AB & A^2B & \cdots \end{bmatrix} \tag{13.2.27}$$

is a contraction. Since W is an operator, we see that

$$N = W_c W = \sum_{n=0}^{\infty} A^n B E \Lambda^n$$

is a well-defined operator. Notice that $S_{\mathcal{E}}^* W = W\Lambda$ and $AW_c = W_c S_{\mathcal{E}}$. Using this along with the fact that $I - S_{\mathcal{E}} S_{\mathcal{E}}^* = P_{\mathcal{E}}$ is the orthogonal projection onto the first component $\mathcal{E}$ of $\ell_+^2(\mathcal{E})$, we obtain

$$\begin{aligned} N - AN\Lambda &= W_c W - AW_c W\Lambda = W_c W - W_c S_{\mathcal{E}} S_{\mathcal{E}}^* W \\ &= W_c \left(I - S_{\mathcal{E}} S_{\mathcal{E}}^*\right) W = W_c P_{\mathcal{E}} W = BE. \end{aligned}$$

In other words, N is a solution to the Lyapunov equation $N = AN\Lambda + BE$. By employing the fact that $\widetilde{E} = \sum_0^\infty \Theta_n E \Lambda^n$, we have

$$\widetilde{E} = DE + \sum_{n=0}^{\infty} CA^n BE\Lambda^n \Lambda = \Theta_0 E + CN\Lambda.$$

This readily implies that

$$\begin{bmatrix} D & C \\ B & A \end{bmatrix} \begin{bmatrix} E \\ N\Lambda \end{bmatrix} = \begin{bmatrix} \widetilde{E} \\ N \end{bmatrix}. \tag{13.2.28}$$

Because the 2×2 matrix is an isometry, we see that

$$\Lambda^* N^* N \Lambda + E^* E = N^* N + \widetilde{E}^* \widetilde{E}.$$

In other words, $N^*N = Q$ is the unique solution to the Lyapunov equation in (13.2.20). So without loss of generality we can assume that M is the operator mapping $\mathcal{H}$ into $\mathcal{M} = \overline{N\mathcal{H}}$ defined by $N = M$. By replacing N by M in (13.2.28), we see that U is an isometric coupling of $\{\Lambda, E, \widetilde{E}\}$. Since all controllable isometric realizations of Θ are unitarily equivalent, this completes the proof. $\square$

The controllability of the central solution. Let $\{\Lambda, E, \widetilde{E}\}$ be the data for a tangential Nevanlinna-Pick problem. Moreover, assume that the Lyapunov equation (13.2.20) admits a positive solution Q and $Q = M^*M$ where M is an operator mapping $\mathcal{H}$ into a dense set of $\mathcal{M}$. Then we claim that $\{A_\circ, B_\circ\}$ is controllable, where $\{A_\circ, B_\circ, C_\circ, D_\circ\}$ is the contractive realization for the central solution Θ determined by (13.2.14).

To see this, simply observe that by replacing $\{A, B\}$ with $\{A_\circ, B_\circ\}$ in (13.2.26), we have $M = \sum_0^\infty A_\circ^j B_\circ E \Lambda^j$. Hence $\mathcal{M}$ is a subspace of $\bigvee_0^\infty A_\circ^n B_\circ \mathcal{E}$. Because $A_\circ$ acts on $\mathcal{M}$, the pair $\{A_\circ, B_\circ\}$ is controllable.

Now let us show that $\{A, B\}$ is controllable, where $\{A \text{ on } \mathcal{X}, B, C, D_\circ\}$ is the isometric state space realization for central solution Θ; see (13.2.14) to (13.2.17). Recall that $\mathcal{X} = \mathcal{M} \oplus \ell_+^2(\mathcal{D}_1)$. According to (13.2.26), we have $M = \sum_0^\infty A^j B E \Lambda^j$. Because the range of M is dense in $\mathcal{M}$, we see that $\mathcal{M} \oplus 0$ is contained in $\mathcal{X}_c = \bigvee_0^\infty A^n B \mathcal{E}$. Since $\mathcal{X}_c$ is an invariant subspace for A, the subspace $\bigvee_0^\infty A^n \mathcal{M}$ is also contained in $\mathcal{X}_c$. By consulting the form of A and B in (13.2.16), we obtain

$$0 \oplus B_1 \mathcal{E} = (I - P_{\mathcal{M}}) B \mathcal{E} \subseteq B\mathcal{E} \bigvee \mathcal{M} \subseteq \mathcal{X}_c,$$
$$0 \oplus A_1 \mathcal{M} = (I - P_{\mathcal{M}}) A \mathcal{M} \subseteq A\mathcal{M} \bigvee \mathcal{M} \subseteq \mathcal{X}_c.$$

By taking the closed linear span of both of these spaces, we see that $0 \oplus \mathcal{D}_1 \oplus 0 \oplus 0 \cdots$ is contained in $\mathcal{X}_c$. Because $\mathcal{X}_c$ is an invariant subspace for A, we have

$$\mathcal{X}_c \supseteq \bigvee_{n=0}^{\infty} A^n (0 \oplus \mathcal{D}_1 \oplus 0 \oplus 0 \cdots) = \bigvee_{n=0}^{\infty} \begin{bmatrix} 0 \\ S_{\mathcal{D}_1}^n (\mathcal{D}_1 \oplus 0 \oplus 0 \cdots) \end{bmatrix} = \begin{bmatrix} 0 \\ \ell_+^2(\mathcal{D}_1) \end{bmatrix}.$$

So $\mathcal{X}_c$ contains both $\mathcal{M} \oplus 0$ and $0 \oplus \ell_+^2(\mathcal{D}_1)$. Therefore

$$\mathcal{X} = \mathcal{M} \oplus \ell_+^2(\mathcal{D}_1) \subseteq \mathcal{X}_c = \bigvee_0^\infty A^n B \mathcal{E}.$$

In other words, the pair $\{A, B\}$ is controllable.

13.3 Isometric Realizations Revisited

In this section we will use the central isometric coupling to show that any contractive analytic function Θ in $H^\infty(\mathcal{E}, \mathcal{Y})$ admits an isometric realization. This provides another proof of part of Theorem 13.1.1. To this end, let $\Lambda = S_{\mathcal{E}}^*$ be the backward shift on $\mathcal{X} = \ell_+^2(\mathcal{E})$. Let $E = \Pi_{\mathcal{E}}$ be the operator mapping $\ell_+^2(\mathcal{E})$ onto $\mathcal{E}$ which picks out the first component of a vector in $\ell_+^2(\mathcal{E})$. Let $\widetilde{E}$ be the operator mapping $\ell_+^2(\mathcal{E})$ into $\mathcal{Y}$ given by the first row of ∇_Θ. To be precise,

$$E = \begin{bmatrix} I & 0 & 0 & \cdots \end{bmatrix} : \ell_+^2(\mathcal{E}) \to \mathcal{E},$$
$$\widetilde{E} = \begin{bmatrix} \Theta_0 & \Theta_1 & \Theta_2 & \cdots \end{bmatrix} : \ell_+^2(\mathcal{E}) \to \mathcal{X} \qquad (13.3.1)$$

where $\Theta(z) = \sum_0^\infty z^{-n} \Theta_n$ is the power series expansion for Θ. In this case, $W = I$ and $\widetilde{W} = \nabla_\Theta$. We claim that $Q = I - \nabla_\Theta^* \nabla_\Theta$ is a positive solution to the Lyapunov equation

$$Q = \Lambda^* Q \Lambda + E^* E^* - \widetilde{E}^* \widetilde{E}. \qquad (13.3.2)$$

Using the fact that ∇_Θ intertwines $S_\mathcal{E}^*$ with $S_\mathcal{Y}^*$, we obtain

$$\begin{aligned} Q - \Lambda^* Q \Lambda &= I - \nabla_\Theta^* \nabla_\Theta - S_\mathcal{E} S_\mathcal{E}^* + S_\mathcal{E} \nabla_\Theta^* \nabla_\Theta S_\mathcal{E}^* \\ &= P_\mathcal{E} - \nabla_\Theta^* \left(I - S_\mathcal{Y} S_\mathcal{Y}^*\right) \nabla_\Theta \\ &= E^* E - \nabla_\Theta^* P_\mathcal{E} \nabla_\Theta = E^* E - \widetilde{E}^* \widetilde{E}. \end{aligned}$$

Because the Lyapunov equation in (13.3.2) admits a positive solution, there exists a solution to the tangential Nevanlinna-Pick interpolation problem with data $\{S_\mathcal{E}^*, E, \widetilde{E}\}$ in (13.3.1). To construct a solution, let

$$U_\circ = \begin{bmatrix} D & C \\ B & A \end{bmatrix} : \begin{bmatrix} \mathcal{E} \\ \begin{bmatrix} \mathcal{M} \\ \ell_+^2(\mathcal{D}_1) \end{bmatrix} \end{bmatrix} \to \begin{bmatrix} \mathcal{Y} \\ \begin{bmatrix} \mathcal{M} \\ \ell_+^2(\mathcal{D}_1) \end{bmatrix} \end{bmatrix} \tag{13.3.3}$$

be the cental isometric coupling for $\{\Lambda, E, \widetilde{E}\}$; see also (13.2.13). (To construct this coupling we did not use the fact that a contractive analytic function admits an isometric realization.) Then $\Psi = D + C(zI - A)^{-1}B$ is a solution to this Nevanlinna-Pick problem, that is, Ψ is a contractive analytic function satisfying $\nabla_\Psi W = \widetilde{W}$. Since $W = I$ and $\nabla_\Theta W = \widetilde{W}$, we must have $\Theta = \Psi$. Therefore $\{A, B, C, D\}$ is a controllable isometric realization for Θ. Combining this analysis with Lemma 13.1.3, we obtain another proof of Theorem 13.1.1. In other words, Θ *is a contractive analytic function in* $H^\infty(\mathcal{E}, \mathcal{Y})$ *if and only if* Θ *admits an isometric realization. In this case, all controllable isometric realizations of* Θ *are unitarily equivalent.* (Section 7.8 shows that if Θ admits an isometric realization, then Θ is a contractive analytic function.)

The previous analysis yields another method to construct an isometric realization for a contractive analytic function Θ. To this end, consider the data $\{S_\mathcal{E}^*, E, \widetilde{E}\}$ in (13.3.1). Since $W = I$, there is a unique solution to $\nabla_\Theta W = \widetilde{W}$. In other words, there is only one solution Θ to the Nevanlinna-Pick problem with data $\{S_\mathcal{E}^*, E, \widetilde{E}\}$. To construct a controllable isometric realization for Θ, let M be any operator mapping $\ell_+^2(\mathcal{E})$ into $\mathcal{M}$ such that $I - \nabla_\Theta^* \nabla_\Theta = M^* M$ and the range of M is dense in $\mathcal{M}$. Let $U_\circ$ be the central isometric coupling of $\{S_\mathcal{E}^*, E, \widetilde{E}\}$ in (13.3.3), obtained by constructing the corresponding matrix representation for $U_\circ$ in (13.2.13). Then $\{A, B, C, D\}$ is a controllable isometric realization for Θ.

13.4 The Maximum Principle

In this section we will show that the central solution satisfies a maximum principle. To this end, assume that the tangential Nevanlinna-Pick interpolation problem with data $\{\Lambda, E, \widetilde{E}\}$ admits a solution. In other words, there exists a positive solution Q to the Lyapunov equation in (13.2.20). Let U mapping $\mathcal{K} = \mathcal{E} \oplus \mathcal{X}$ into $\mathcal{K}_1 = \mathcal{Y} \oplus \mathcal{X}$ be any controllable isometric coupling for $\{\Lambda, E, \widetilde{E}\}$. Fix y in $\mathcal{Y}$ and consider the optimization problem

$$\delta(y,U) = \inf\{\|y - Ug\| : g \in \mathcal{K}\} \qquad (y \in \mathcal{Y}). \tag{13.4.1}$$

Let $\mathcal{L} = \ker U^*$. Then we claim that $\delta(y,U) = \|P_{\mathcal{L}}y\|$. To verify this, let $\mathcal{R} = U\mathcal{K}$ be the range of U. Recall that $\mathcal{L} = \ker U^* = \mathcal{K}_1 \ominus U\mathcal{K}$. Using the fact that $P_{\mathcal{L}} = I - P_{\mathcal{R}}$, we have

$$\delta(y,U) = \inf\{\|y - Ug\| : g \in \mathcal{K}\} = \|y - P_{\mathcal{R}}y\| = \|P_{\mathcal{L}}y\|.$$

Therefore $\delta(y,U) = \|P_{\mathcal{L}}y\|$ for all y in $\mathcal{Y}$.

Recall that $\mathcal{D}_2 = (\mathcal{Y} \oplus \mathcal{M}) \ominus \mathcal{H}_2$ where $\mathcal{H}_2$ is the closure of the range of Φ_2. We claim that $\delta(y,U) \le \|P_{\mathcal{D}_2}y\|$. Since $U|\mathcal{H}_1 = V$, the subspace $\mathcal{H}_2 \subseteq \mathcal{R}$. Hence

$$\begin{aligned}\delta(y,U) &= \inf\{\|y - Ug\| : g \in \mathcal{K}\} \le \inf\{\|y - Vg\| : g \in \mathcal{H}_1\}\\ &= \inf\{\|y - g\| : g \in \mathcal{H}_2\} = \|y - P_{\mathcal{H}_2}y\| = \|P_{\mathcal{D}_2}y\|.\end{aligned}$$

Hence $\delta(y,U) \le \|P_{\mathcal{D}_2}y\|$. Motivated by this we say that U is a *maximal coupling* of $\{\Lambda, E, \widetilde{E}\}$ if U is an isometric controllable coupling for $\{\Lambda, E, \widetilde{E}\}$ and $\delta(y,U) = \|P_{\mathcal{D}_2}y\|$ for all y in $\mathcal{Y}$.

Let $U_\circ$ be the central isometric coupling of $\{\Lambda, E, \widetilde{E}\}$ determined by (13.2.13) or (13.2.15). Then $U_\circ$ is a maximal coupling of $\{\Lambda, E, \widetilde{E}\}$. To see this, observe that the kernel of $U_\circ^*$ equals $\mathcal{D}_2$, that is,

$$\ker U_\circ^* = 0 \oplus \mathcal{D}_2 \oplus 0 \oplus 0 \cdots .$$

Since $U_\circ$ maps $(\mathcal{H}_1 \oplus \mathcal{D}_1) \oplus \ell_+^2(\mathcal{D}_1)$ into $(\mathcal{H}_2 \oplus \mathcal{D}_2) \oplus \ell_+^2(\mathcal{D}_1)$ and $\mathcal{H}_2 \oplus \mathcal{D}_2$ is just another orthogonal decomposition of $\mathcal{Y} \oplus \mathcal{M}$, we see that

$$\delta(y,U_\circ) = \|P_{\ker U_\circ^*}y\| = \|P_{\mathcal{D}_2}y\|.$$

Therefore the central coupling $U_\circ$ is a maximal coupling of $\{\Lambda, E, \widetilde{E}\}$. The following result shows that $U_\circ$ in (13.2.13) is the only maximal controllable coupling of $\{\Lambda, E, \widetilde{E}\}$.

Theorem 13.4.1. *Assume that the tangential Nevanlinna-Pick interpolation problem with data $\{\Lambda, E, \widetilde{E}\}$ admits a solution. Then the central coupling $U_\circ$ in (13.2.13) is the only maximal controllable coupling of $\{\Lambda, E, \widetilde{E}\}$. In other words, all maximal controllable couplings of $\{\Lambda, E, \widetilde{E}\}$ are unitarily equivalent to the central isometric coupling $U_\circ$.*

Proof. Assume that U is an isometric controllable coupling of $\{\Lambda, E, \widetilde{E}\}$ such that

$$\|P_{\mathcal{L}}y\| = \delta(y,U) = \|P_{\mathcal{D}_2}y\|$$

for all y in $\mathcal{Y}$ where $\mathcal{L} = \ker U^*$. We claim that $P_{\mathcal{D}_2}|\mathcal{Y} = P_{\mathcal{L}}|\mathcal{Y}$. By employing $\|P_{\mathcal{L}}y\| = \|P_{\mathcal{D}_2}y\|$ with y in $\mathcal{Y}$, we obtain

$$\|P_{\mathcal{H}_2}y\|^2 = \|y\|^2 - \|P_{\mathcal{D}_2}y\|^2 = \|y\|^2 - \|P_{\mathcal{L}}y\|^2 = \|P_{\mathcal{R}}y\|^2.$$

In other words, $\|P_{\mathcal{H}_2}y\| = \|P_{\mathcal{R}}y\|$. Since $\mathcal{H}_2$ is a subspace of $\mathcal{R} = \operatorname{ran} U$, we have $P_{\mathcal{H}_2}y = P_{\mathcal{R}}y$ for all y in $\mathcal{Y}$. Thus

$$P_{\mathcal{D}_2}y = (I - P_{\mathcal{H}_2})y = (I - P_{\mathcal{R}})y = P_{\mathcal{L}}y.$$

Therefore $P_{\mathcal{D}_2}|\mathcal{Y} = P_{\mathcal{L}}|\mathcal{Y}$.

We claim that $\mathcal{L} = \mathcal{D}_2$. To see this first observe that for h in $\mathcal{H}$, we have

$$P_{\mathcal{L}}\Phi_2 h = P_{\mathcal{L}}V\Phi_1 h = P_{\mathcal{L}}U\Phi_1 h = 0.$$

Hence $0 = P_{\mathcal{L}}|\mathcal{H}_2$. Clearly, $0 = P_{\mathcal{D}_2}\mathcal{H}_2$. By consulting (13.2.8), we see that $\mathcal{Y}\oplus\mathcal{M}$ equals the closed linear span of $\{\mathcal{Y}, \mathcal{H}_2\}$. Thus $P_{\mathcal{L}}(y\oplus m) = P_{\mathcal{D}_2}(y\oplus m)$ for all $y\oplus m$ in $\mathcal{Y}\oplus\mathcal{M}$. Since $\mathcal{D}_2$ is a subspace of $\mathcal{Y}\oplus\mathcal{M}$, it follows that $\mathcal{D}_2$ is a subspace of $\mathcal{L}$.

Recall that U admits a matrix representation of the form

$$U = \begin{bmatrix} D & C \\ B & A \end{bmatrix} : \begin{bmatrix} \mathcal{E} \\ \mathcal{X} \end{bmatrix} \to \begin{bmatrix} \mathcal{Y} \\ \mathcal{X} \end{bmatrix}.$$

Using the fact that $P_{\mathcal{L}}U = 0$ and $P_{\mathcal{D}_2}|\mathcal{Y} = P_{\mathcal{L}}|\mathcal{Y}$ with $\mathcal{D}_2 \subseteq \mathcal{L}$, for g in $\mathcal{E}$ we obtain

$$\begin{aligned} P_{\mathcal{L}}Bg &= P_{\mathcal{L}}P_{\mathcal{X}}Ug = P_{\mathcal{L}}Ug - P_{\mathcal{L}}P_{\mathcal{Y}}Ug \\ &= -P_{\mathcal{D}_2}P_{\mathcal{Y}}Ug = P_{\mathcal{D}_2}P_{\mathcal{X}}Ug - P_{\mathcal{D}_2}Ug \\ &= P_{\mathcal{D}_2}P_{\mathcal{X}}Ug - P_{\mathcal{D}_2}P_{\mathcal{L}}Ug = P_{\mathcal{D}_2}P_{\mathcal{X}}Ug \\ &= P_{\mathcal{D}_2}Bg. \end{aligned}$$

Thus $P_{\mathcal{L}}B = P_{\mathcal{D}_2}B$. For any integer $n \geq 1$, we have

$$\begin{aligned} P_{\mathcal{L}}A^nBg &= P_{\mathcal{L}}P_{\mathcal{X}}UA^{n-1}Bg \\ &= P_{\mathcal{L}}UA^{n-1}Bg - P_{\mathcal{L}}P_{\mathcal{Y}}UA^{n-1}Bg \\ &= -P_{\mathcal{D}_2}P_{\mathcal{Y}}UA^{n-1}Bg \\ &= P_{\mathcal{D}_2}P_{\mathcal{X}}UA^{n-1}Bg - P_{\mathcal{D}_2}P_{\mathcal{L}}UA^{n-1}Bg \\ &= P_{\mathcal{D}_2}P_{\mathcal{X}}UA^{n-1}Bg = P_{\mathcal{D}_2}A^nBg. \end{aligned}$$

Because the pair $\{A, B\}$ is controllable, $P_{\mathcal{L}}x = P_{\mathcal{D}_2}x$ for all $x \in \mathcal{X}$. Since $P_{\mathcal{L}}|\mathcal{Y} = P_{\mathcal{D}_2}|\mathcal{Y}$ and $\mathcal{Y}\oplus\mathcal{X}$ is the whole space, $P_{\mathcal{L}} = P_{\mathcal{D}_2}$. Therefore $\mathcal{L} = \mathcal{D}_2$.

Now let us show that U is unitarily equivalent to the central isometric coupling $U_\circ$ in (13.2.13). Since $\mathcal{L} = \mathcal{D}_2$, the isometry U admits a matrix representation of the form

$$U = \begin{bmatrix} V & 0 & 0 \\ 0 & 0 & 0 \\ 0 & U_{32} & U_+ \end{bmatrix} : \begin{bmatrix} \mathcal{H}_1 \\ \mathcal{D}_1 \\ \mathcal{X}\ominus\mathcal{M} \end{bmatrix} \to \begin{bmatrix} \mathcal{H}_2 \\ \mathcal{D}_2 \\ \mathcal{X}\ominus\mathcal{M} \end{bmatrix}. \tag{13.4.2}$$

In particular, $\mathcal{X}\ominus\mathcal{M}$ is an invariant subspace for U, and $U_+ = U|(\mathcal{X}\ominus\mathcal{M})$ is an isometry on $\mathcal{X}\ominus\mathcal{M}$.

Because U is an isometry, U_{32} is an isometry mapping $\mathcal{D}_1$ into $\mathcal{X} \ominus \mathcal{M}$. Moreover, $U_{32}\mathcal{D}_1$ is orthogonal to the range of U_+, or equivalently, $U_{32}\mathcal{D}_1$ is in the kernel of U_+^*. Hence $U_{32}\mathcal{D}_1$ is a wandering subspace for U_+ of dimension $\dim \mathcal{D}_1$. In this setting, A and B admit matrix representations of the form

$$A = \begin{bmatrix} A_\circ & 0 \\ U_{32}\Pi_{\mathcal{D}_1}|\mathcal{M} & U_+ \end{bmatrix} \text{ on } \begin{bmatrix} \mathcal{M} \\ \mathcal{X} \ominus \mathcal{M} \end{bmatrix},$$
$$B = \begin{bmatrix} B_\circ \\ U_{32}\Pi_{\mathcal{D}_1}|\mathcal{E} \end{bmatrix} : \mathcal{E} \to \begin{bmatrix} \mathcal{M} \\ \mathcal{X} \ominus \mathcal{M} \end{bmatrix}. \tag{13.4.3}$$

Since $\mathcal{X} \ominus \mathcal{M}$ is an invariant subspace for U and $A = \Pi_{\mathcal{X}} U|\mathcal{X}$, it follows that A admits a lower triangular matrix representation of the form in (13.4.3) where the lower right-hand corner is given by U_+. Using $UP_{\mathcal{H}_1} = VP_{\mathcal{H}_1}$, the lower left-hand corner of A is determined by

$$\Pi_{\mathcal{X}\ominus\mathcal{M}} U|\mathcal{M} \;=\; \Pi_{\mathcal{X}\ominus\mathcal{M}} U(P_{\mathcal{D}_1} + P_{\mathcal{H}_1})|\mathcal{M} = U_{32}\Pi_{\mathcal{D}_1}|\mathcal{M}.$$

The second component of B is given by

$$\Pi_{\mathcal{X}\ominus\mathcal{M}} U|\mathcal{E} \;=\; \Pi_{\mathcal{X}\ominus\mathcal{M}} U(P_{\mathcal{D}_1} + P_{\mathcal{H}_1})|\mathcal{E} = U_{32}\Pi_{\mathcal{D}_1}|\mathcal{E}.$$

Since $\{A, B\}$ is controllable, the form $\{A, B\}$ in (13.4.3) yields

$$\mathcal{X} \ominus \mathcal{M} = \bigvee_{n=0}^{\infty} \Pi_{\mathcal{X}\ominus\mathcal{M}} A^n B\mathcal{E} \subseteq \bigvee_{n=0}^{\infty} U_+^n U_{32}\mathcal{D}_1 \subseteq \mathcal{X} \ominus \mathcal{M}.$$

In other words, $\mathcal{X} \ominus \mathcal{M} = \oplus_0^\infty U_+^n U_{32}\mathcal{D}_1$, and $U_{32}\mathcal{D}_1$ is a cyclic wandering subspace for U_+. Because $U_{32}\mathcal{D}_1$ and $\mathcal{D}_1$ have the same dimension, U_+ is unitarily equivalent to the unilateral shift $S_{\mathcal{D}_1}$ on $\ell_+^2(\mathcal{D}_1)$. By consulting the (13.4.2) and (13.2.13), we see that U is unitarily equivalent to the central isometric coupling $U_\circ$. □

Computing the cost. Assume that Q is a strictly positive solution to the Lyapunov equation (13.2.20), corresponding to the Nevanlinna-Pick data $\{\Lambda, E, \widetilde{E}\}$. Recall that the cost in the optimization problem (13.4.1), associated with the central coupling $U_\circ$ is given by

$$\delta(y, U_\circ)^2 = \|P_{\mathcal{D}_2}\|^2 = (\Pi_{\mathcal{Y}} P_{\mathcal{D}_2} y, y) \qquad (y \in \mathcal{Y}).$$

This cost is determined by the positive operator $\Pi_{\mathcal{Y}} P_{\mathcal{D}_2}|\mathcal{Y}$. If Q is invertible, then

$$\Pi_{\mathcal{Y}} P_{\mathcal{D}_2}|\mathcal{Y} = (I + \widetilde{E} Q^{-1} \widetilde{E}^*)^{-1}. \tag{13.4.4}$$

To verify this notice that $\mathcal{H}_2$ equals the range of Φ_2, and thus,

$$P_{\mathcal{H}_2} = \Phi_2(\Phi_2^*\Phi_2)^{-1}\Phi_2^* = \Phi_2(\widetilde{E}^*\widetilde{E} + Q)^{-1}\Phi_2^*;$$

see (13.2.8). This implies that

$$\begin{aligned}\Pi_{\mathcal{Y}} P_{\mathcal{D}_2}|\mathcal{Y} &= I - \Pi_{\mathcal{Y}} P_{\mathcal{H}_2}|\mathcal{Y} = I - \widetilde{E}(\widetilde{E}^*\widetilde{E} + Q)^{-1}\widetilde{E}^* \\ &= I - \widetilde{E}(Q^{-1}\widetilde{E}^*\widetilde{E} + I)^{-1}Q^{-1}\widetilde{E}^* \\ &= I - \widetilde{E}Q^{-1}\widetilde{E}^*(\widetilde{E}Q^{-1}\widetilde{E}^* + I)^{-1} \\ &= (\widetilde{E}Q^{-1}\widetilde{E}^* + I)^{-1}.\end{aligned}$$

Therefore (13.4.4) holds.

13.5 Notes

For completeness we added this short chapter on contractive Nevanlinna-Pick interpolation. We also wanted to develop a connection between Nevanlinna-Pick interpolation, isometric realizations and contractive analytic functions. The main ideas behind the results in this chapter were taken from Agler-McCarthy [3, 4]. It is emphasized that Agler-McCarthy [3, 4] developed a contractive interpolation theory for complex functions of several variables; see Agler-McCarthy [4] for a history of this subject area, and Ball [23], Ball-Trent-Vinnikov [26] and Cotlar-Sadosky [61] for further results in this direction.

Theorem 13.1.1 is a classical result in operator theory. Our approach to constructing a contractive isometric realization is a minor modification of the de Branges-Rovnyak model theory [40, 41]. For some further results on this model theory see de Branges [38, 39]. The theory of unitary systems started with Livšic [163, 164]. Then using dilation theory Sz.-Nagy-Foias developed the characteristic function; see [198]. The characteristic function plays a fundamental role in operator theory. The Sz.-Nagy-Foias characteristic function can be used to study the spectrum and invariant subspaces of contractions. For further results on unitary systems; see Brodskii [43, 44], Chapter 28 in Gohberg-Goldberg-Kaashoek [114], Arocena [15] and Arov [19, 20].

The literature on H^∞ interpolation theory is massive. Here we only presented one solution, that is, the central solution to the tangential Nevanlinna-Pick interpolation problem. The set of all solutions is parameterized by the unit ball in some $H^\infty(\cdot,\cdot)$ space. For example, see Ball-Gohberg-Rodman [24] or Foias-Frazho-Gohberg-Kaashoek [84]. The Sz.-Nagy-Foias [197] commutant lifting theorem is a general theorem for solving interpolation problems, including Nevanlinna-Pick, Nehari, Sarason and many other H^∞ interpolation problems. The commutant lifting theorem was motivated by the interpolation results of Sarason [190]. For further results and a history of the commutant lifting theorem see Foias-Frazho [82], Foias-Frazho-Gohberg-Kaashoek [84] and Rosenblum-Rovnyak [182]. For some nice results on the Nehari interpolation problem see Adamjan-Arov-Krein [1, 2]. The band method is also a very powerful theory for solving many interpolation problems; see Gohberg-Kaashoek-Woerdeman [115] and Gohberg-Goldberg-Kaashoek

[114]. The state space method for solving H^∞ interpolations problems started with Glover [111]. For further results on state space methods in H^∞ interpolation theory; see Ball-Gohberg-Rodman [24], Green-Limebeer [123] and Zhou-Doyle-Glover [204].

Part V

Appendices

Chapter 14

A Review of State Space

In this chapter we will review some of the state space methods and terminology which will be used throughout the monograph. All the results in this chapter are classical.

14.1 State Space Realization Theory

Let us introduce some standard terminology and results from systems theory; see [60, 140, 189] for further details. The discrete time state space system is defined by

$$x(n+1) = Ax(n) + Bu(n) \quad \text{and} \quad y(n) = Cx(n) + Du(n). \tag{14.1.1}$$

Here A is an operator on a space $\mathcal{X}$ and B is an operator mapping $\mathcal{E}$ into $\mathcal{X}$ while C is an operator from $\mathcal{X}$ into $\mathcal{Y}$ and D is an operator mapping $\mathcal{E}$ into $\mathcal{Y}$. The *state* $x(n)$ is in $\mathcal{X}$, the *input* $u(n)$ is in $\mathcal{E}$ and the *output* $y(n)$ is in $\mathcal{Y}$ for all integers $n \geq 0$. The initial condition is $x(0) = x_0$. The space $\mathcal{X}$ is called the *state space,* while $\mathcal{E}$ is the *input space* and $\mathcal{Y}$ is the *output space.* Finally, the state space system in (14.1.1) is denoted by $\{A, B, C, D\}$.

By recursively solving for the state $x(n)$ and the output $y(n)$ in the discrete time system (14.1.1), we see that

$$x(n) = A^n x_0 + \sum_{k=0}^{n-1} A^{n-k-1} Bu(k) \qquad (x(0) = x_0), \tag{14.1.2}$$

$$y(n) = CA^n x_0 + Du(n) + \sum_{k=0}^{n-1} CA^{n-k-1} Bu(k). \tag{14.1.3}$$

Let $\{F_n\}_0^\infty$ be the sequence of operators defined by

$$F_0 = D \quad \text{and} \quad F_n = CA^{n-1}B \quad (\text{if } n \geq 1). \tag{14.1.4}$$

Then (14.1.3) shows that

$$\begin{bmatrix} y(0) \\ y(1) \\ y(2) \\ \vdots \end{bmatrix} = \begin{bmatrix} C \\ CA \\ CA^2 \\ \vdots \end{bmatrix} x(0) + \begin{bmatrix} F_0 & 0 & 0 & \cdots \\ F_1 & F_0 & 0 & \cdots \\ F_2 & F_1 & F_0 & \cdots \\ \vdots & \vdots & \vdots & \ddots \end{bmatrix} \begin{bmatrix} u(0) \\ u(1) \\ u(2) \\ \vdots \end{bmatrix}. \tag{14.1.5}$$

Let T_F be the lower triangular Toeplitz matrix corresponding to $\{F_n\}_0^\infty$ and W_o the observability matrix defined by

$$T_F = \begin{bmatrix} F_0 & 0 & 0 & \cdots \\ F_1 & F_0 & 0 & \cdots \\ F_2 & F_1 & F_0 & \cdots \\ \vdots & \vdots & \vdots & \ddots \end{bmatrix} \quad \text{and} \quad W_o = \begin{bmatrix} C \\ CA \\ CA^2 \\ \vdots \end{bmatrix}. \tag{14.1.6}$$

Then (14.1.5) shows that

$$\vec{y} = W_o x(0) + T_F \vec{u} \tag{14.1.7}$$

where $\vec{u} = \begin{bmatrix} u(0) & u(1) & u(2) & \cdots \end{bmatrix}^{tr}$ and $\vec{y} = \begin{bmatrix} y(0) & y(1) & y(2) & \cdots \end{bmatrix}^{tr}$ are the input and output vectors. Finally, it is noted that T_F and W_o are not necessarily operators, that is, bounded linear maps. However, because T_F is lower triangular, all the multiplications are well defined.

The *transfer function* for the state space system in (14.1.1) is defined by

$$F(z) = \sum_{n=0}^{\infty} z^{-n} F_n.$$

By employing (14.1.4), we obtain

$$\begin{aligned} F(z) = \sum_{n=0}^{\infty} z^{-n} F_n &= D + \sum_{n=1}^{\infty} \frac{1}{z^n} CA^{n-1} B \\ &= D + z^{-1} C (I - z^{-1} A)^{-1} B \\ &= D + C(zI - A)^{-1} B. \end{aligned}$$

In other words, the transfer function for $\{A, B, C, D\}$ is given by

$$F(z) = D + C(zI - A)^{-1} B.$$

Finally, it is noted that $F(z)$ is analytic in some neighborhood of infinity, that is, $F(1/z)$ is analytic in some neighborhood of the origin.

To provide some further insight into the notion of a transfer function, let $U(z)$ and $Y(z)$ be the functions formally defined by taking the following z transforms of the input and output sequences $U(z) = \sum_0^\infty z^{-n} u_n$ and $Y(z) = \sum_0^\infty z^{-n} y_n$. The input sequence $\{u_n\}_0^\infty$ and $U(z)$ (respectively $\{y_n\}_0^\infty$ and $Y(z)$) uniquely determine each other. In systems theory, the transfer function is the unique function F determined by $Y(z) = F(z)U(z)$ where all the initial conditions are set equal to zero. By employing (14.1.3) or (14.1.5), we arrive at

$$Y(z) = C(zI - A)^{-1}x_0 + \big(D + C(zI - A)^{-1}B\big)\, U(z).$$

If the initial condition $x_0 = 0$, then the output $Y(z) = F(z)U(z)$ where the function $F(z) = D + C(zI - A)^{-1}B$. In other words, the transfer function F for $\{A, B, C, D\}$ is given by $F(z) = D + C(zI - A)^{-1}B$.

By a slight abuse of terminology we say that a function $G(z)$ is *analytic in some neighborhood of infinity* if $G(1/z)$ is analytic in some neighborhood of the origin. Notice that G is analytic in some neighborhood of infinity if and only if G admits a Taylor series expansion of the form $G = \sum_0^\infty z^{-n}G_n$ for some $|z| > r > 0$. For example, z is not analytic in some neighborhood of infinity. Let G be a rational function, that is, assume that $G = N(z)/d(z)$ where N is a $\mathcal{L}(\mathcal{E}, \mathcal{Y})$-valued polynomial and d is a scalar-valued polynomial. (It is emphasized that N and d are polynomials in z and not $1/z$.) We say that G is a *proper rational function* if the degree of N is less than or equal to the degree of d. Finally, it is noted that a rational function G is analytic in some neighborhood of infinity if and only if G is a proper rational function.

We say that $\{A \text{ on } \mathcal{X}, B, C, D\}$ is a *realization* for a function F with values in $\mathcal{L}(\mathcal{E}, \mathcal{Y})$ if

$$F(z) = D + C(zI - A)^{-1}B \tag{14.1.8}$$

in some neighborhood of infinity. Here A is an operator on $\mathcal{X}$, the operator B maps $\mathcal{E}$ into $\mathcal{X}$, while the operator C maps $\mathcal{X}$ into $\mathcal{Y}$ and D is an operator from $\mathcal{E}$ into $\mathcal{Y}$. The *state space* is $\mathcal{X}$, while $\mathcal{E}$ is called the *input space* and $\mathcal{Y}$ the *output space*. In systems theory F is referred to as the *transfer function* for $\{A, B, C, D\}$. Because $(zI - A)^{-1}$ is analytic in some neighborhood of infinity, F is also analytic in some neighborhood of infinity. Motivated by this, we say that F is a *transfer function* if F is a $\mathcal{L}(\mathcal{E}, \mathcal{Y})$-valued analytic function in some neighborhood of infinity. The realization $\{A, B, C, D\}$ is *finite dimensional* if the dimension of the state space $\mathcal{X}$ is finite. *Finally, throughout this chapter, we assume that the input space $\mathcal{E}$ and output space $\mathcal{Y}$ are both finite dimensional.*

Assume that $\{A, B, C, D\}$ is a realization for some transfer function F. Let

$$F = \sum_{n=0}^{\infty} z^{-n}F_n \tag{14.1.9}$$

be the Taylor series expansion for F. Clearly, $\|z^{-1}A\| < 1$ for all z such that $\|A\| < |z|$. Observe that $(zI - A)^{-1} = z^{-1}(I - z^{-1}A)^{-1}$. Hence $(zI - A)^{-1}$ admits a Taylor series expansion of the form

$$(zI - A)^{-1} = \sum_{n=0}^{\infty} \frac{1}{z^{n+1}}A^n \qquad (|z| > \|A\|). \tag{14.1.10}$$

Combining this with (14.1.8), we obtain

$$\sum_{n=0}^{\infty} z^{-n}F_n = F(z) = D + C(zI - A)^{-1}B = D + \sum_{n=1}^{\infty} \frac{1}{z^n} CA^{n-1}B.$$

So by matching like coefficients of z^{-n}, we see that

$$F_0 = D \quad \text{and} \quad F_n = CA^{n-1}B \quad (\text{for all } n \geq 1). \tag{14.1.11}$$

On the other hand, if $F(z) = \sum_0^\infty z^{-n}F_n$ is analytic in some neighborhood of infinity, and $\{A, B, C, D\}$ is a state space system such that (14.1.11) holds, then $\{A, B, C, D\}$ is a realization for F. Therefore $\{A, B, C, D\}$ is a realization for F if and only if (14.1.11) holds.

We say that Φ is an operator *intertwining*

$$\{A \text{ on } \mathcal{X}, B, C, D\} \text{ with } \{A_1 \text{ on } \mathcal{X}_1, B_1, C_1, D_1\}$$

if Φ is an operator mapping $\mathcal{X}$ into $\mathcal{X}_1$ such that

$$\Phi A = A_1\Phi, \quad \Phi B = B_1, \quad C_1\Phi = C \quad \text{and} \quad D = D_1. \tag{14.1.12}$$

We say that $\Sigma = \{A, B, C, D\}$ is *similar to* (respectively *unitarily equivalent to*) $\Sigma_1 = \{A_1, B_1, C_1, D_1\}$ if there exists a similarity (respectively unitary) transformation intertwining Σ with Σ_1. Notice that Σ is similar to (respectively unitarily equivalent to) Σ_1 if and only if Σ_1 is similar to (respectively unitarily equivalent to) Σ. Finally, it is noted that similar realizations generate the same transfer function.

14.2 Controllability and Observability

The pair $\{A \text{ on } \mathcal{X}, B\}$ is *controllable* if $\mathcal{X}$ equals the closed linear span of $\{A^n B\mathcal{E}\}_0^\infty$. The pair $\{C, A\}$ is *observable* if $CA^n x = 0$ for all integers $n \geq 0$, then $x = 0$. Controllability is the dual of observability, that is, the pair $\{C, A\}$ is observable if and only if $\{A^*, C^*\}$ is controllable.

Let W_c be the controllability matrix defined by

$$W_c = \begin{bmatrix} B & AB & A^2B & \cdots \end{bmatrix}. \tag{14.2.1}$$

The pair $\{A, B\}$ is controllable if and only if $W_c\ell_+^c(\mathcal{E})$ is dense in $\mathcal{X}$. If the state space is finite dimensional, then due to the Cayley-Hamilton Theorem, the pair $\{A, B\}$ is controllable if and only if

$$\mathcal{X} = \operatorname{ran} \begin{bmatrix} B & AB & A^2B & \cdots & A^{\nu-1}B \end{bmatrix}$$

where ν is the dimension of $\mathcal{X}$. Let W_o be the observability matrix defined by

$$W_o = \begin{bmatrix} C \\ CA \\ CA^2 \\ \vdots \end{bmatrix}. \tag{14.2.2}$$

The pair $\{C, A\}$ is observable if and only if W_o is one to one. If the state space is finite dimensional, then due to the Cayley-Hamilton Theorem, $\{C, A\}$ is observable if and only if

$$\{0\} = \ker \begin{bmatrix} C & CA & CA^2 & \cdots & CA^{\nu-1} \end{bmatrix}^{tr}.$$

Finally, it is noted that W_c and W_o are matrices and not necessarily operators.

For the moment assume that the state space $\mathcal{X}$ is finite dimensional. The *Popov-Belevitch-Hautus controllability test* says that the pair $\{A, B\}$ is controllable if and only if

$$\dim \mathcal{X} = \operatorname{rank} \begin{bmatrix} A - \lambda I & B \end{bmatrix} \qquad (\text{for all } \lambda \in \mathbb{C}).$$

In a similar fashion, the *Popov-Belevitch-Hautus observability test* states that the pair $\{C, A\}$ is observable if and only if

$$\ker \begin{bmatrix} A - \lambda I \\ C \end{bmatrix} = \{0\} \qquad (\text{for all } \lambda \in \mathbb{C}).$$

We say that the realization $\{A, B, C, D\}$ is controllable (respectively observable) if the pair $\{A, B\}$ is controllable (respectively the pair $\{C, A\}$ is observable). Assume that $\{A, B, C, D\}$ is a realization for F. Let H be the Hankel matrix defined by

$$H = \begin{bmatrix} F_1 & F_2 & F_3 & \cdots \\ F_2 & F_3 & F_4 & \cdots \\ F_3 & F_4 & F_5 & \cdots \\ \vdots & \vdots & \vdots & \ddots \end{bmatrix}. \tag{14.2.3}$$

Here $F(z) = \sum_0^\infty z^{-n} F_n$ is the Taylor series expansion for F. Moreover, H is called the *Hankel matrix generated* by $\{F_j\}_1^\infty$ or F. Using the fact that $F_n = CA^{n-1}B$ for all integers $n \geq 1$, it follows that $H = W_o W_c$. In particular, rank$H \leq \dim \mathcal{X}$. So if the state space is finite dimensional, then the rank of H is also finite. It is well known that a transfer function F admits a finite dimensional realization if and only if the rank of H is finite. Recall that the rank of the Hankel matrix H in (14.2.3) is finite if and only if F is a rational function. So a transfer function F is rational if and only if F admits a finite dimensional realization.

We say that $\{A \text{ on } \mathcal{X}, B, C, D\}$ is a *minimal realization* for F if the dimension of $\mathcal{X}$ is less than or equal to the state dimension of any realization for F, that is, if $\{A_1 \text{ on } \mathcal{X}_1, B_1, C_1, D_1\}$ is any realization for F, then $\dim \mathcal{X} \leq \dim \mathcal{X}_1$. Now assume that F admits a finite dimensional realization, or equivalently, F is a rational transfer function. Then it is well known that $\{A, B, C, D\}$ is a minimal realization for F if and only if $\{A, B, C, D\}$ is a controllable and observable realization for F. Moreover, in this case, all minimal realizations of F are similar. The dimension of the state space $\mathcal{X}$ for a minimal realization $\{A, B, C, D\}$ of F is called the *McMillan degree of* F, denoted by $\delta(F)$. Finally, it is noted that the McMillan degree for F equals the rank of its corresponding Hankel matrix H in (14.2.3).

Let us conclude this section with the following useful results.

Remark 14.2.1. Assume that $\{A, B, C, D\}$ is a realization for a transfer function F with values in $\mathcal{L}(\mathcal{E}, \mathcal{Y})$. If D is invertible, then $F(z)$ is invertible in some neighborhood of infinity. Moreover,

$$\Sigma = \{A - BD^{-1}C, BD^{-1}, -D^{-1}C, D^{-1}\}$$

is a realization for the inverse of F, that is,

$$F(z)^{-1} = D^{-1} - D^{-1}C\left(zI - (A - BD^{-1}C)\right)^{-1} BD^{-1}. \tag{14.2.4}$$

To show this simply verify that $F(z)F(z)^{-1} = I$ where $F(z)^{-1}$ is given by (14.2.4). Finally, $\{A, B, C, D\}$ is respectively, controllable, observable, minimal if and only if Σ is respectively, controllable, observable, minimal.

For a state space derivation of $F(z)^{-1}$, recall that $F(z)$ is the transfer function for the state space system

$$x(n+1) = Ax(n) + Bu(n) \quad \text{and} \quad y(n) = Cx(n) + Du(n). \tag{14.2.5}$$

Since D is invertible, the input $u(n) = D^{-1}y(n) - D^{-1}Cx(n)$. Substituting this into (14.2.5) yields the state space system

$$\begin{aligned} x(n+1) &= (A - BD^{-1}C)x(n) + BD^{-1}y(n), \\ u(n) &= -D^{-1}Cx(n) + D^{-1}y(n). \end{aligned} \tag{14.2.6}$$

It is emphasized that $F(z)^{-1}$ in (14.2.4) is the transfer function for this state space system. The state space system in (14.2.5) maps $u(n)$ into $y(n)$, while the system in (14.2.6) maps $y(n)$ into $u(n)$. In other words, the state space system in (14.2.6) is the inverse of the system in (14.2.5). Therefore F^{-1} in (14.2.4) must be the inverse of F the transfer function for (14.2.5).

Remark 14.2.2. Let $\Sigma_1 = \{A_1 \text{ on } \mathcal{X}_1, B_1, C_1, D_1\}$ be a realization for a $\mathcal{L}(\mathcal{V}, \mathcal{Y})$-valued transfer function F_1, and $\Sigma_2 = \{A_2 \text{ on } \mathcal{X}_2, B_2, C_2, D_2\}$ a realization for a $\mathcal{L}(\mathcal{E}, \mathcal{V})$-valued transfer function F_2. Observe that the output space $\mathcal{V}$ for F_2 equals the input space for F_1. The product F_1F_2 is also a transfer function. Moreover, a realization $\{A, B, C, D\}$ for F_1F_2 is given by

$$\begin{aligned} A &= \begin{bmatrix} A_1 & B_1C_2 \\ 0 & A_2 \end{bmatrix} \text{ on } \begin{bmatrix} \mathcal{X}_1 \\ \mathcal{X}_2 \end{bmatrix}, \quad B = \begin{bmatrix} B_1D_2 \\ B_2 \end{bmatrix} : \mathcal{E} \to \begin{bmatrix} \mathcal{X}_1 \\ \mathcal{X}_2 \end{bmatrix} \\ C &= \begin{bmatrix} C_1 & D_1C_2 \end{bmatrix} \begin{bmatrix} \mathcal{X}_1 \\ \mathcal{X}_2 \end{bmatrix} \to \mathcal{Y} \quad \text{and} \quad D = D_1D_2. \end{aligned} \tag{14.2.7}$$

To show that $\{A, B, C, D\}$ is a realization for F_1F_2 observe that

$$\begin{aligned}
F_1(z)F_2(z) &= \left(D_1 + C_1(zI - A_1)^{-1}B_1\right)\left(D_2 + C_2(zI - A_2)^{-1}B_2\right) \\
&= D_1D_2 + C_1(zI - A_1)^{-1}B_1D_2 + D_1C_2(zI - A_2)^{-1}B_2 \\
&\quad + C_1(zI - A_1)^{-1}B_1C_2(zI - A_2)^{-1}B_2 \\
&= D_1D_2 + \begin{bmatrix} C_1 & D_1C_2 \end{bmatrix} \\
&\quad \times \begin{bmatrix} (zI - A_1)^{-1} & (zI - A_1)^{-1}B_1C_2(zI - A_2)^{-1} \\ 0 & (zI - A_2)^{-1} \end{bmatrix} \begin{bmatrix} B_1D_2 \\ B_2 \end{bmatrix} \\
&= D_1D_2 + \begin{bmatrix} C_1 & D_1C_2 \end{bmatrix} \begin{bmatrix} zI - A_1 & -B_1C_2 \\ 0 & zI - A_2 \end{bmatrix}^{-1} \begin{bmatrix} B_1D_2 \\ B_2 \end{bmatrix} \\
&= D + C(zI - A)^{-1}B.
\end{aligned}$$

Therefore $\{A, B, C, D\}$ in (14.2.7) is a realization for F_1F_2.

For a more intuitive approach based on discrete time systems, observe that F_1 and F_2 are the transfer functions for the linear systems

$$\begin{aligned}
x(n+1) &= A_1x(n) + B_1v_1(n) \quad \text{and} \quad y(n) = C_1x(n) + D_1v_1(n), \\
\xi(n+1) &= A_2\xi(n) + B_2u(n) \quad \text{and} \quad v(n) = C_2\xi(n) + D_2u(n).
\end{aligned}$$

In other words, F_1 is the transfer function from $v_1(n)$ to $y(n)$, and F_2 is the transfer function from $u(n)$ to $v(n)$. By setting $v_1(n) = v(n)$, we see that F_1F_2 is the transfer function from $u(n)$ into $y(n)$. Substituting $v_1(n) = C_2\xi(n) + D_2u(n)$ into the first two equations yields

$$\begin{aligned}
\begin{bmatrix} x(n+1) \\ \xi(n+1) \end{bmatrix} &= \begin{bmatrix} A_1 & B_1C_2 \\ 0 & A_2 \end{bmatrix} \begin{bmatrix} x(n) \\ \xi(n) \end{bmatrix} + \begin{bmatrix} B_1D_2 \\ B_2 \end{bmatrix} u(n), \\
y(n) &= \begin{bmatrix} C_1 & D_1C_2 \end{bmatrix} \begin{bmatrix} x(n) \\ \xi(n) \end{bmatrix} + D_1D_2u(n).
\end{aligned} \tag{14.2.8}$$

So F_1F_2 is a transfer function for the discrete time system in (14.2.8). In other words, $\{A, B, C, D\}$ in (14.2.7) is a realization for F_1F_2.

To conclude this section, it is noted that the realization $\{A, B, C, D\}$ in (14.2.7) may not be minimal even if both Σ_1 and Σ_2 are minimal realizations. For example, $\{0, 1, -1, 1\}$ is a minimal realization for $F_1 = (z-1)/z$ and $\{1, 1, 1, 1\}$ is a minimal realization for $F_2 = z/(z-1)$. However, the corresponding realization $\{A \text{ on } \mathbb{C}^2, B, C, 1\}$ in (14.2.7) is not a minimal realization of $1 = F_1F_2$.

14.3 A Canonical Realization

Recall that a rational transfer function is a proper rational function. Any scalar-valued rational transfer function admits a representation of the form:

$$F(z) \quad = \quad \frac{b_0 + b_1z + b_2z^2 + \cdots + b_{\nu-1}z^{\nu-1}}{a_0 + a_1z + a_2z + \cdots + a_{\nu-1}z^{\nu-1} + z^\nu} + d. \tag{14.3.1}$$

Here $\{a_j\}_0^{\nu-1}$ and $\{b_j\}_0^{\nu-1}$ and d are all scalars. Furthermore, any $\mathcal{L}(\mathcal{E},\mathcal{Y})$-valued rational transfer function F admits a representation of the form

$$\begin{aligned} F(z) &= C_1\Omega(z)^{-1}\Gamma(z) + D \\ \Omega(z) &= A_0 + A_1 z + A_2 z^2 + \cdots + A_{\nu-1}z^{\nu-1} + z^\nu I \\ \Gamma(z) &= B_0 + B_1 z + B_3 z^2 + \cdots + B_{\nu-1}z^{\nu-1}. \end{aligned} \tag{14.3.2}$$

Here Ω is a polynomial with values in $\mathcal{L}(\mathcal{Y},\mathcal{Y})$, and Γ is a polynomial with values in $\mathcal{L}(\mathcal{E},\mathcal{Y})$. Moreover, C_1 is an operator on $\mathcal{Y}$, and D is an operator mapping $\mathcal{E}$ into $\mathcal{Y}$. By choosing $C_1 = I$ and $\Omega(z) = \alpha(z)I$ where α is a monic-valued polynomial, it follows that any rational transfer function has a representation of the form (14.3.2). (A *monic polynomial* is a scalar-valued polynomial $\alpha(z)$ of degree μ such that 1 is the coefficient of z^μ.)

To construct a realization $\{A, B, C, D\}$ for the transfer function F in (14.3.2), let A on $\mathcal{Y}^\nu$ and B mapping $\mathcal{E}$ into $\mathcal{Y}^\nu$ and C mapping $\mathcal{Y}^\nu$ into $\mathcal{Y}$ be the operators defined by

$$A = \begin{bmatrix} -A_{\nu-1} & I & 0 & \cdots & 0 & 0 \\ -A_{\nu-2} & 0 & I & \cdots & 0 & 0 \\ \vdots & \vdots & \vdots & \ddots & \vdots & \vdots \\ -A_2 & 0 & 0 & \cdots & I & 0 \\ -A_1 & 0 & 0 & \cdots & 0 & I \\ -A_0 & 0 & 0 & \cdots & 0 & 0 \end{bmatrix}, \quad B = \begin{bmatrix} B_{\nu-1} \\ B_{\nu-2} \\ \vdots \\ B_2 \\ B_1 \\ B_0 \end{bmatrix}, \quad \text{and}$$

$$C = \begin{bmatrix} C_1 & 0 & 0 & \cdots & 0 & 0 \end{bmatrix}. \tag{14.3.3}$$

(If $\mathcal{E}$ is a vector space, then $\mathcal{E}^\nu = \oplus_1^\nu \mathcal{E}$ denotes ν copies of $\mathcal{E}$.) The identity I on $\mathcal{Y}$ appears immediately above the main diagonal of A, the operators $\{-A_j\}_0^{\nu-1}$ appear in the first column of A in decreasing order, and all the other entries are zero. We claim that $\{A, B, C, D\}$ is a realization for F, that is,

$$F(z) = C(zI - A)^{-1}B + D. \tag{14.3.4}$$

Clearly, this realization is finite dimensional. Hence any rational transfer function admits a finite dimensional realization. Moreover, if C_1 is invertible, then the pair $\{C, A\}$ is observable. Furthermore, $\det[\Omega(z)]$ is the characteristic polynomial for A, that is,

$$\det[\Omega(z)] = \det[zI - A]. \tag{14.3.5}$$

(The determinant of any operator T on a finite dimensional space is denoted by $\det[T]$. To be precise, $\det[T]$ is the determinant of any matrix representation for T.) In particular, if Ω is a scalar-valued polynomial, then Ω equals the characteristic polynomial for A. Finally, it is noted that A is stable if and only if all the zeros of $\det[\Omega(z)]$ are contained in the open unit disc $\mathbb{D} = \{z : |z| < 1\}$. (An operator on a finite dimensional space is *stable* if all of its eigenvalues are contained in the open unit disc.)

To show that $\{A, B, C, D\}$ is a realization for $C_1\Omega^{-1}\Gamma + D$, first observe that

$$zI - A = \begin{bmatrix} zI + A_{\nu-1} & -I & 0 & \cdots & 0 & 0 \\ A_{\nu-2} & zI & -I & \cdots & 0 & 0 \\ \vdots & \vdots & \vdots & \ddots & \vdots & \vdots \\ A_2 & 0 & 0 & \cdots & -I & 0 \\ A_1 & 0 & 0 & \cdots & zI & -I \\ A_0 & 0 & 0 & \cdots & 0 & zI \end{bmatrix}. \tag{14.3.6}$$

Using this we obtain

$$\begin{bmatrix} z^{\nu-1}I & z^{\nu-2}I & \cdots & zI & I \end{bmatrix} (zI - A) = \begin{bmatrix} \Omega(z) & 0 & 0 & \cdots & 0 & 0 \end{bmatrix}.$$

Multiplying by $(zI - A)^{-1}$ on the right and Ω^{-1} on the left, we arrive at

$$\Omega(z)^{-1} \begin{bmatrix} z^{\nu-1}I & z^{\nu-2}I & \cdots & zI & I \end{bmatrix} = \begin{bmatrix} I & 0 & 0 & \cdots & 0 & 0 \end{bmatrix} (I - zA)^{-1}.$$

Multiplying by C_1 on the left and B on the right yields

$$C_1\Omega(z)^{-1}\Gamma(z) = C_1\Omega(z)^{-1} \sum_{j=0}^{\nu-1} z^j B_j = C(zI - A)^{-1}B.$$

In other words, $\{A, B, C, D\}$ is a realization for $C_1\Omega(z)^{-1}\Gamma + D$.

Now assume that the operator C_1 is invertible. To show that $\{C, A\}$ is observable, simply observe that

$$\begin{bmatrix} C \\ CA \\ CA^2 \\ \vdots \\ CA^{\nu-1} \end{bmatrix} = \begin{bmatrix} C_1 & 0 & 0 & \cdots & 0 & 0 \\ \star & C_1 & 0 & \cdots & 0 & 0 \\ \star & \star & C_1 & \cdots & 0 & 0 \\ \vdots & \vdots & \vdots & \ddots & \vdots & \vdots \\ \star & \star & \star & \cdots & C_1 & 0 \\ \star & \star & \star & \cdots & \star & C_1 \end{bmatrix}.$$

As expected, $\star$ denotes an unspecified entry. Because the matrix on the right-hand side is invertible, the pair $\{C, A\}$ is observable.

The scalar case. Let F be the scalar-valued transfer function presented in (14.3.1). Then an observable realization for F is given by $\{A, B, C, d\}$ where

$$A = \begin{bmatrix} -a_{\nu-1} & 1 & 0 & \cdots & 0 & 0 \\ -a_{\nu-2} & 0 & 1 & \cdots & 0 & 0 \\ \vdots & \vdots & \vdots & \ddots & \vdots & \vdots \\ -a_2 & 0 & 0 & \cdots & 1 & 0 \\ -a_1 & 0 & 0 & \cdots & 0 & 1 \\ -a_0 & 0 & 0 & \cdots & 0 & 0 \end{bmatrix}, \quad B = \begin{bmatrix} b_{\nu-1} \\ b_{\nu-2} \\ \vdots \\ b_2 \\ b_1 \\ b_0 \end{bmatrix}, \quad \text{and}$$

$$C = \begin{bmatrix} 1 & 0 & 0 & \cdots & 0 & 0 \end{bmatrix}. \tag{14.3.7}$$

In this case, $\det[zI - A] = a_0 + a_1 z + \cdots + a_{\nu-1} z^{\nu-1} + z^{\nu}$. Finally, it is noted that $\{A, B, C, d\}$ is a minimal realization for F if and only if there is no pole zero cancellation in F, that is, $a_0 + a_1 z + \cdots + a_{\nu-1} z^{\nu-1} + z^{\nu}$ and $b_0 + b_1 z + \cdots + b_{\nu-1} z^{\nu-1}$ have no common zeros.

14.3.1 A Schur decomposition perspective

Now let us use the 2×2 Schur matrix inversion formula to show that $\det[\Omega(z)] = \det[zI - A]$, and $\{A, B, C, D\}$ in (14.3.3) is a realization for $C_1 \Omega(z)^{-1} \Gamma(z) + D$. To this end, let T be a block operator matrix of the form

$$T = \begin{bmatrix} V & W \\ X & Y \end{bmatrix} \text{ on } \begin{bmatrix} \mathcal{V} \\ \mathcal{Y} \end{bmatrix}. \tag{14.3.8}$$

Assume that Y is invertible. Then the *Schur complement for T with respect to V* is defined by $\Delta = V - WY^{-1}X$. In this case, T is invertible if and only if its Schur complement Δ is invertible. Moreover,

$$T^{-1} = \begin{bmatrix} \Delta^{-1} & -\Delta^{-1} W Y^{-1} \\ -Y^{-1} X \Delta^{-1} & Y^{-1} + Y^{-1} X \Delta^{-1} W Y^{-1} \end{bmatrix} \tag{14.3.9}$$

and $\Delta = (\Pi_{\mathcal{V}} T^{-1} \Pi_{\mathcal{V}}^*)^{-1}$ where $\Pi_{\mathcal{V}} = \begin{bmatrix} I & 0 \end{bmatrix}$ maps $\mathcal{V} \oplus \mathcal{Y}$ onto $\mathcal{V}$.

To obtain the matrix inversion formula in (14.3.9), observe that T admits a factorization of the form

$$T = \begin{bmatrix} I & WY^{-1} \\ 0 & I \end{bmatrix} \begin{bmatrix} V - WY^{-1}X & 0 \\ 0 & Y \end{bmatrix} \begin{bmatrix} I & 0 \\ Y^{-1}X & I \end{bmatrix}. \tag{14.3.10}$$

Because Y is invertible, T is invertible if and only if $\Delta = V - WY^{-1}X$ is invertible. By taking the inverse of T in (14.3.10), we arrive at

$$T^{-1} = \begin{bmatrix} I & 0 \\ -Y^{-1}X & I \end{bmatrix} \begin{bmatrix} \Delta^{-1} & 0 \\ 0 & Y^{-1} \end{bmatrix} \begin{bmatrix} I & -WY^{-1} \\ 0 & I \end{bmatrix}.$$

By performing these matrix calculations, we obtain the form for the inverse of T in (14.3.9). Finally, it is noted that (14.3.10) also yields

$$\det[T] = \det[\Delta] \det[Y]. \tag{14.3.11}$$

Now let us use the Schur decomposition to show that $\det[\Omega(z)] = \det[zI - A]$. The matrix representation for $T = zI - A$ in (14.3.6), yields

$$zI - A = \begin{bmatrix} V & W \\ X & Y \end{bmatrix} \text{ on } \begin{bmatrix} \mathcal{Y} \\ \mathcal{Y}^{n-1} \end{bmatrix}.$$

Here $V = zI + A_{\nu-1}$ and $W = \begin{bmatrix} -I & 0 & 0 & \cdots & 0 & 0 \end{bmatrix}$ maps $\mathcal{Y}^{\nu-1}$ into $\mathcal{Y}$, while

$$X = \begin{bmatrix} A_{\nu-2} \\ A_{\nu-3} \\ A_{\nu-4} \\ \vdots \\ A_1 \\ A_0 \end{bmatrix} : \mathcal{Y} \to \mathcal{Y}^{\nu-1} \quad \text{and} \quad Y = \begin{bmatrix} zI & -I & 0 & \cdots & 0 & 0 \\ 0 & zI & -I & \cdots & 0 & 0 \\ 0 & 0 & zI & \cdots & 0 & 0 \\ \vdots & \vdots & \vdots & \ddots & \vdots & \vdots \\ 0 & 0 & 0 & \cdots & zI & -I \\ 0 & 0 & 0 & \cdots & 0 & zI \end{bmatrix} \text{ on } \mathcal{Y}^{\nu-1}.$$

The matrix Y has zI on the diagonal, $-I$ immediately above the main diagonal and zeros elsewhere. For nonzero z the inverse of Y is the upper triangular Toeplitz matrix given by

$$Y^{-1} = \frac{1}{z^{\nu-1}} \begin{bmatrix} z^{\nu-2}I & z^{\nu-3}I & \cdots & zI & I \\ 0 & z^{\nu-2}I & \cdots & z^2I & zI \\ \vdots & \vdots & \ddots & \vdots & \vdots \\ 0 & 0 & \cdots & z^{\nu-2}I & z^{\nu-1}I \\ 0 & 0 & \cdots & 0 & z^{\nu-2}I \end{bmatrix} \text{ on } \mathcal{Y}^{\nu-1}.$$

To verify this simply observe that $YY^{-1} = I$. The Schur complement $\Delta(z)$ for $V = zI + A_{\nu-1}$ is given by

$$\begin{aligned} \Delta(z) &= zI + A_{\nu-1} + \begin{bmatrix} I & 0 & \cdots & 0 & 0 \end{bmatrix} Y^{-1} X \\ &= zI + A_{\nu-1} + \frac{1}{z^{\nu-1}} \sum_{j=0}^{\nu-2} z^j A_j \\ &= \frac{1}{z^{\nu-1}} \Omega(z). \end{aligned}$$

In other words, $\Delta = \Omega / z^{\nu-1}$ where Δ is the Schur complement with respect to $zI + A_{\nu-1}$. Recall that Δ is an operator on $\mathcal{Y}$. Hence

$$\Delta(z) = \frac{1}{z^{\nu-1}} \Omega(z) \quad \text{and} \quad \det[\Delta] = \frac{\det[\Omega]}{(z^{\nu-1})^{\dim(\mathcal{Y})}} \qquad (z \neq 0).$$

Since $\det[Y] = (z^{\nu-1})^{\dim(\mathcal{Y})}$, equation (14.3.11) yields

$$\det[zI - A] = \det[\Delta] \det[Y] = \det[\Omega(z)]$$

for all nonzero z. Because $\det[\Omega(z)]$ and $\det[zI - A]$ are both polynomials which are equal for all nonzero z, it follows that $\det[\Omega(z)] = \det[zI - A]$ for all z.

To complete this section, let us show that $\{A, B, C, D\}$ is a realization for

$C_1\Omega^{-1}\Gamma + D$. The matrix inversion formula in (14.3.9) yields

$$C(zI-A)^{-1}B = C_1\begin{bmatrix}\Delta^{-1} & -\Delta^{-1}WY^{-1}\end{bmatrix}\begin{bmatrix} B_{\nu-1} \\ \begin{bmatrix} B_{\nu-2} \\ \vdots \\ B_0 \end{bmatrix}\end{bmatrix}$$

$$= z^{\nu-1}C_1\Omega^{-1}B_{\nu-1} + C_1\Delta^{-1}\begin{bmatrix} I & 0 & \cdots & 0 & 0\end{bmatrix} Y^{-1}\begin{bmatrix} B_{\nu-2} \\ B_{\nu-3} \\ \vdots \\ B_0 \end{bmatrix}$$

$$= C_1\Omega^{-1}\sum_{j=0}^{\nu-1} z^j B_j = C_1\Omega(z)^{-1}\Gamma(z).$$

Therefore $\{A, B, C, D\}$ is a realization for $C_1\Omega^{-1}\Gamma + D$.

14.4 The Controllability and Observability Gramian

We say that an operator A on $\mathcal{X}$ is *stable* if the spectrum of A is contained in some compact subset of the open unit disc $\mathbb{D}$. So a finite dimensional operator is stable if and only if all of its eigenvalues are contained in the open unit disc. We say that $\{A, B, C, D\}$ is a *stable realization* if A is a stable operator. If $\{A, B, C, D\}$ is a stable realization for a transfer function F, then F is analytic in $\{z : |z| > 1 - \epsilon\}$ for some $\epsilon > 0$, and thus, F is a function in $H^\infty(\mathcal{E}, \mathcal{Y})$. In other words, if F admits a stable realization, then F is in $H^\infty(\mathcal{E}, \mathcal{Y})$.

For the moment, assume that F is a rational transfer function. Then F admits a minimal finite dimensional realization $\{A, B, C, D\}$. Moreover, it is well known that α is a pole of F if and only if α is an eigenvalue of A; see Theorem 6.3.1 page 78 in [60]. Finally, A is stable if and only if all the poles of F are inside the open unit disc $\mathbb{D} = \{z : |z| < 1\}$, or equivalently, F is in $H^\infty(\mathcal{E}, \mathcal{Y})$.

Let $\{C, A\}$ be a stable pair. The *observability Gramian* P is the solution to the Lyapunov equation

$$P = A^*PA + C^*C. \tag{14.4.1}$$

The solution P to this equation is unique and given by

$$P = \sum_{n=0}^{\infty} A^{*n}C^*CA^n. \tag{14.4.2}$$

Notice that P is positive. Furthermore, if the state space is finite dimensional, then P is strictly positive if and only if the pair $\{C, A\}$ is observable. If A is stable, then the observability matrix

$$W_o = \begin{bmatrix} C \\ CA \\ CA^2 \\ \vdots \end{bmatrix} : \mathcal{X} \to \ell_+^2(\mathcal{Y}) \tag{14.4.3}$$

is a well-defined operator from $\mathcal{X}$ into $\ell_+^2(\mathcal{Y})$. Finally, it is noted that the observability Gramian $P = W_o^* W_o$.

Assume that $\{C, A\}$ is an observable pair and the state space is finite dimensional. Then there exists a strictly positive solution P to the Lyapunov equation $P = A^*PA + C^*C$ if and only if A is stable.

Let $\{A, B\}$ be a stable pair. The *controllability Gramian* Q is the solution to the Lyapunov equation

$$Q = AQA^* + BB^*. \tag{14.4.4}$$

The solution Q to this equation is unique and given by

$$Q = \sum_{n=0}^{\infty} A^n BB^* A^{*n}. \tag{14.4.5}$$

Notice that Q is positive. If the state space is finite dimensional, then Q is strictly positive if and only if the pair $\{A, B\}$ is controllable. If A is stable, then the controllability map

$$W_c = \begin{bmatrix} B & AB & A^2B & \cdots \end{bmatrix} : \ell_+^2(\mathcal{E}) \to \mathcal{X} \tag{14.4.6}$$

is a well-defined operator from $\ell_+^2(\mathcal{E})$ into $\mathcal{X}$. The controllability Gramian $Q = W_c W_c^*$.

Assume that $\{A, B\}$ is a controllable pair and the state space is finite dimensional. Then there exists a strictly positive solution to the Lyapunov equation $P = APA^* + BB^*$ if and only if A is stable.

Finite rank Hankel operators. Let H be the Hankel matrix defined in (14.2.3) where $\{F_j\}_1^\infty$ is a sequence of operators with values in $\mathcal{L}(\mathcal{E}, \mathcal{Y})$. Recall that H is a Hankel operator if and only if

$$H = \begin{bmatrix} F_1 & F_2 & F_3 & \cdots \\ F_2 & F_3 & F_4 & \cdots \\ F_3 & F_4 & F_5 & \cdots \\ \vdots & \vdots & \vdots & \ddots \end{bmatrix} : \ell_+^2(\mathcal{E}) \to \ell_+^2(\mathcal{Y}) \tag{14.4.7}$$

defines a bounded linear map. Now assume that the Hankel matrix H is finite rank, or equivalently, $F(z) = \sum_1^\infty z^{-n} F_n$ determines a rational transfer function. Let

$\{A, B, C, 0\}$ be a minimal realization for F. We claim that H is an operator if and only if A is stable. If H is an operator, then the first column of H is an operator from $\mathcal{E}$ into $\ell_+^2(\mathcal{Y})$, that is, $\sum_1^\infty F_n^* F_n < \infty$, or equivalently, F is in $H^2(\mathcal{E}, \mathcal{Y})$. Hence all the poles of F are in $\mathbb{D}$. Because $\{A, B, C, 0\}$ is a minimal realization for F, the operator A must be stable. On the other hand, if A is stable, then the controllability matrix W_c in (14.4.6) and observability matrix W_o in (14.4.3) are well-defined operators. Since $H = W_o W_c$, we see that H is also an operator.

Now assume that $\{A, B, C, 0\}$ is a stable, minimal realization for a transfer function and H is the corresponding Hankel operator. Let $P = W_o^* W_o$ be the observability Gramian for $\{C, A\}$ and $Q = W_c W_c^*$ be the controllability Gramian for $\{A, B\}$. Let $\sigma_1 \geq \sigma_2 \geq \cdots \geq \sigma_\nu$ be the nonzero singular values of H. Then we claim that $\{\sigma_j^2\}_1^\nu$ are precisely the eigenvalues of PQ. In particular, $\|H\|^2$ equals the largest eigenvalue of PQ. To see this, recall that $H = W_o W_c$. Hence $H^* H = W_c^* W_o^* W_o W_c = W_c^* P W_c$. So $H^* H$ and $W_c^* P W_c$ have the same eigenvalues. Since $\{\sigma_j^2\}_1^\nu$ are the nonzero eigenvalues of $H^* H$, we see that $\{\sigma_j^2\}_1^\nu$ are also the nonzero eigenvalues of $W_c^* P W_c$. Recall that if M and N are two operators acting between the appropriate spaces, then MN and NM have the same nonzero eigenvalues. Hence $\{\sigma_j^2\}_1^\nu$ are the nonzero eigenvalues of $P W_c W_c^* = PQ$.

14.5 The Kalman-Ho Algorithm

The Kalman-Ho algorithm allows us to compute a minimal realization for a rational transfer function from a finite section of its corresponding Hankel matrix. This algorithm also provides an effective method to compute a reduced order model for a rational transfer function.

Let $\{F_j\}_0^m$ be a sequence of operators with values in $\mathcal{L}(\mathcal{E}, \mathcal{Y})$. A state space system $\{A \text{ on } \mathcal{X}, B, C, D\}$ is called a *partial realization* for $\{F_j\}_0^m$ if

$$F_0 = D \quad \text{and} \quad F_j = C A^{j-1} B \quad (\text{for } 1 \leq j \leq m).$$

The system $\{A, B, C, D\}$ is a *minimal partial realization* of $\{F_j\}_0^m$ if it is a partial realization of $\{F_j\}_0^m$ of the lowest state dimension. Clearly, a minimal partial realization is controllable and observable. It is emphasized that not all minimal partial realizations of the same sequence $\{F_j\}_0^m$ are similar. For example, $\{0, 1, 1, 0\}$ and $\{1, 1, 1, 0\}$ are both minimal partial realizations of the sequence $\{0, 1\}$ and they are not similar. For another example, the minimal partial realizations

$$\left\{ \begin{bmatrix} 0 & 1 \\ 0 & 0 \end{bmatrix}, \begin{bmatrix} 0 \\ 1 \end{bmatrix}, \begin{bmatrix} 1 & 0 \end{bmatrix}, 0 \right\} \quad \text{and} \quad \left\{ \begin{bmatrix} 0 & 1 \\ 1 & 0 \end{bmatrix}, \begin{bmatrix} 0 \\ 1 \end{bmatrix}, \begin{bmatrix} 1 & 0 \end{bmatrix}, 0 \right\}$$

of the sequence $\{0, 0, 1\}$ are not similar.

Remark 14.5.1. Let $\Sigma_j = \{A_j \text{ on } \mathcal{X}_j, B_j, C_j, D_j\}$, for $j = 1, 2$, be two partial realizations of the sequence $\{F_j\}_0^{2k-1}$. Furthermore, assume that $k > \dim \mathcal{X}_i$ for $i = 1, 2$. Then Σ_1 and Σ_2 determine the same transfer function.

To see this, let $\{A \text{ on } \mathcal{X}_1 \oplus \mathcal{X}_2, B, C, 0\}$ be the state space system defined by

$$A = \begin{bmatrix} A_1 & 0 \\ 0 & A_2 \end{bmatrix}, \quad B = \begin{bmatrix} B_1 \\ B_2 \end{bmatrix} \quad \text{and} \quad C = \begin{bmatrix} C_1 & -C_2 \end{bmatrix}. \tag{14.5.1}$$

Because both Σ_1 and Σ_2 are partial realizations of $\{F_j\}_0^{2k-1}$, it follows that $CA^jB = 0$ for $0 \leq j \leq 2k-2$. However, the degree of the characteristic polynomial for A must be less than or equal to $2k-2$. By the Cayley-Hamilton Theorem, this readily implies that $CA^jB = 0$ for all integers $j \geq 0$. Hence, $C_1 A_1^j B_1 = C_2 A_2^j B_2$ for all $j \geq 0$. Obviously, $D_1 = F_0 = D_2$. Therefore, Σ_1 and Σ_2 determine the same transfer function.

Remark 14.5.1 readily yields the following result.

Remark 14.5.2. Let F be a rational transfer function of McMillan degree ν, and $\{F_j\}_0^{2k-1}$ the first $2k-1$ Taylor coefficients of $F = \sum_0^\infty z^{-j} F_j$ where k is any integer such that $k > \nu$. If $\{A, B, C, D\}$ is a ν dimensional partial realization for $\{F_j\}_0^{2k-1}$, then $\{A, B, C, D\}$ is a minimal realization for F.

Finite rank Hankel matrices. Let $\{A \text{ on } \mathcal{X}, B, C, D\}$ be a minimal realization for a $\mathcal{L}(\mathcal{E}, \mathcal{Y})$-valued rational transfer function F, and let $F = \sum_0^\infty z^{-j} F_j$ be its Taylor series expansion. Let H_k be the Hankel operator mapping $\mathcal{E}^k$ into $\mathcal{Y}^k$ contained in the upper $k \times k$ left-hand corner of the Hankel matrix H generated by $\{F_j\}_1^\infty$ in (14.2.3), that is,

$$H_k = \begin{bmatrix} F_1 & F_2 & F_3 & \cdots & F_k \\ F_2 & F_3 & F_4 & \cdots & F_{k+1} \\ F_3 & F_4 & F_5 & \cdots & F_{k+2} \\ \vdots & \vdots & \vdots & \ddots & \vdots \\ F_k & F_{k+1} & F_{k+2} & \cdots & F_{2k-1} \end{bmatrix} : \mathcal{E}^k \to \mathcal{Y}^k. \tag{14.5.2}$$

(As expected, $\mathcal{L}^k = \oplus_1^k \mathcal{L}$ consists of k orthogonal copies of a Hilbert space $\mathcal{L}$.) Let W_{ck} be the controllability operator and W_{ok} be the observability operator defined by

$$\begin{aligned} W_{ck} &= \begin{bmatrix} B & AB & A^2B & \cdots & A^{k-1}B \end{bmatrix} : \mathcal{E}^k \to \mathcal{X}, \\ W_{ok} &= \begin{bmatrix} C & CA & CA^2 & \cdots & CA^{k-1} \end{bmatrix}^{tr} : \mathcal{X} \to \mathcal{Y}^k. \end{aligned} \tag{14.5.3}$$

Using the fact that $F_j = CA^{j-1}B$ for all integers $j \geq 1$, we see that $H_k = W_{ok} W_{ck}$. Now let ν be the dimension of $\mathcal{X}$. By the Cayley-Hamilton Theorem, W_{ok} is one to one and W_{ck} is onto for all integers $k \geq \nu$. Hence $\operatorname{rank} H_k = \nu$ for all integers $k \geq \nu$. This fact allows us to work with finite Hankel matrices H_k to compute a minimal realization for F.

The Kalman-Ho Algorithm. Let H_k be the Hankel matrix generated by a sequence of operators $\{F_j\}_0^{2k-1}$ with values in $\mathcal{L}(\mathcal{E},\mathcal{Y})$; see (14.5.2). Assume that $\nu = \text{rank}H_k = \text{rank}H_{k-1}$. Using the singular-value decomposition find a factorization for H_k of the form $H_k = LR$ where $L = \begin{bmatrix} L_1 & L_2 & \cdots & L_k \end{bmatrix}^{tr}$ is a one to one operator mapping $\mathbb{C}^\nu$ into $\mathcal{Y}^k$ and $R = \begin{bmatrix} R_1 & R_2 & \cdots & R_k \end{bmatrix}$ is an operator mapping $\mathcal{E}^k$ onto $\mathbb{C}^\nu$, that is,

$$H_k = LR = \begin{bmatrix} L_1 \\ L_2 \\ \vdots \\ L_k \end{bmatrix} \begin{bmatrix} R_1 & R_2 & \cdots & R_k \end{bmatrix}. \tag{14.5.4}$$

As expected, ν is the number of significant singular values of H_k, that is, ν equals the rank of H_k computed numerically. (Notice that $F_{j+k-1} = L_j R_k$.) To be precise, let $U\Lambda V^* = H_k$ be the singular value decomposition of H_k, where U and V are unitary operators and Λ is a diagonal matrix consisting of the singular values of H_k in decreasing order. Then for example, one could choose

$$\begin{aligned} L &= U\Pi_{\mathbb{C}^\nu}^* && \text{and} \quad R = \Lambda_\nu \Pi_{\mathbb{C}^\nu} V^* \quad \text{or} \\ L &= U\Pi_{\mathbb{C}^\nu}^* \Lambda_\nu^{1/2} && \text{and} \quad R = \Lambda_\nu^{1/2}\Pi_{\mathbb{C}^\nu} V^* \end{aligned} \tag{14.5.5}$$

where Λ_ν is the diagonal matrix on $\mathbb{C}^\nu$ contained in the upper left hand corner of Λ.

A minimal partial realization $\{A \text{ on } \mathbb{C}^\nu, B, C, F_0\}$ for $\{F_j\}_0^{2k-1}$ is given by

$$\begin{aligned} A &= \left(\sum_{j=1}^{k-1} L_j^* L_j\right)^{-1} \left(\sum_{j=1}^{k-1} L_j^* L_{j+1}\right), \\ B &= R_1 \quad \text{and} \quad C = L_1. \end{aligned} \tag{14.5.6}$$

Finally, it is noted that if we take $L = U\Pi_{\mathbb{C}^\nu}^*$ to be an isometry, then

$$A = (I - L_k^* L_k)^{-1} \left(\sum_{j=1}^{k-1} L_j^* L_{j+1}\right).$$

In this case, if k is large and the minimal realization is stable, then $L_k \approx 0$ and state space operator $A \approx \sum_{j=1}^{k-1} L_j^* L_{j+1}$.

To derive the Kalman-Ho Algorithm, first observe that $H_{k-1} = \Pi_{k-1} H_k | \mathcal{E}^{k-1}$. Here Π_{k-1} is the orthogonal projection from $\mathcal{Y}^k$ onto $\mathcal{Y}^{k-1}$ which picks out the first $k-1$ components of $\mathcal{Y}^k$. So using $H_k = LR$, we obtain $H_{k-1} = \Pi_{k-1} LR|\mathcal{E}^{k-1}$. Since $\text{rank}H_k = \text{rank}H_{k-1}$, we can say of the operators that

$$\begin{aligned} \Pi_{k-1} L &= \begin{bmatrix} L_1 & L_2 & \cdots & L_{k-1} \end{bmatrix}^{tr} : \mathbb{C}^\nu \to \mathcal{Y}^{k-1} \text{ is one to one and} \\ R|\mathcal{E}^{k-1} &= \begin{bmatrix} R_1 & R_2 & \cdots & R_{k-1} \end{bmatrix} : \mathcal{E}^{k-1} \to \mathbb{C}^\nu \text{ is onto } \mathbb{C}^\nu. \end{aligned}$$

By exploiting the form of the Hankel matrix H_k, we arrive at

$$\begin{bmatrix} L_2 \\ L_3 \\ \vdots \\ L_k \end{bmatrix} \begin{bmatrix} R_1 & R_2 & \cdots & R_{k-1} \end{bmatrix} = \begin{bmatrix} F_2 & F_3 & \cdots & F_k \\ F_3 & F_4 & \cdots & F_{k+1} \\ \vdots & \vdots & \ddots & \vdots \\ F_k & F_{k+1} & \cdots & F_{2k-2} \end{bmatrix}$$
$$= \begin{bmatrix} L_1 \\ L_2 \\ \vdots \\ L_{k-1} \end{bmatrix} \begin{bmatrix} R_2 & R_3 & \cdots & R_k \end{bmatrix}. \qquad (14.5.7)$$

Because $R|\mathcal{E}^{k-1}$ is onto, the range of $\begin{bmatrix} L_2 & L_3 & \cdots & L_k \end{bmatrix}^{tr}$ must be contained in the range of $\begin{bmatrix} L_1 & L_2 & \cdots & L_{k-1} \end{bmatrix}^{tr}$. This implies that there exists an operator A on $\mathbb{C}^\nu$ such that

$$\begin{bmatrix} L_2 \\ L_3 \\ \vdots \\ L_k \end{bmatrix} = \begin{bmatrix} L_1 \\ L_2 \\ \vdots \\ L_{k-1} \end{bmatrix} A = \Pi_{k-1} L A. \qquad (14.5.8)$$

Recall that $\Pi_{k-1}L$ is one to one. By taking the pseudo inverse in (14.5.8), we arrive at the formula for A in (14.5.6). To be more explicit, multiplying on the left by $\begin{bmatrix} L_1^* & L_2^* & \cdots & L_{k-1}^* \end{bmatrix}$, we obtain

$$\sum_{j=1}^{k-1} L_j^* L_{j+1} = \left(\sum_{j=1}^{k-1} L_j^* L_j \right) A.$$

This readily yields the formula for A in (14.5.6).

Set $C = L_1$. Then (14.5.8) implies that $L_2 = L_1 A = CA$ and $L_3 = L_2 A = CA^2$. By continuing in this fashion, we see that $L_j = CA^{j-1}$ for $j = 1, 2, \ldots, k$. In other words, $L = W_{ok}$ where W_{ok} is the observability operator defined in (14.5.3). By substituting (14.5.8) into (14.5.7), we obtain

$$\Pi_{k-1} L A \begin{bmatrix} R_1 & R_2 & \cdots & R_{k-1} \end{bmatrix} = \Pi_{k-1} L \begin{bmatrix} R_2 & R_3 & \cdots & R_k \end{bmatrix}.$$

Since $\Pi_{k-1}L$ is one to one, this implies that

$$A \begin{bmatrix} R_1 & R_2 & \cdots & R_{k-1} \end{bmatrix} = \begin{bmatrix} R_2 & R_3 & \cdots & R_k \end{bmatrix}.$$

Setting $B = R_1$, we see that $R_2 = AR_1 = AB$ and $R_3 = AR_2 = A^2B$. By continuing in this fashion, we obtain $R_j = A^{j-1}B$ for $j = 1, 2, \ldots, k$. Thus $R = W_{ck}$ where W_{ck} is the controllability operator defined in (14.5.3). This readily implies that

$$H_k = LR = W_{ok} W_{ck}.$$

In other words, $\{A, B, C, F_0\}$ is a partial realization of $\{F_j\}_0^{2k-1}$. Because $L = W_{ok}$, is one to one, the pair $\{C, A\}$ is observable. Since $R = W_{ck}$, is onto, the pair $\{A, B\}$ is controllable. Therefore $\{A, B, C, F_0\}$ is controllable and an observable partial realization of $\{F_j\}_0^{2k-1}$. This completes our derivation of the Kalman-Ho algorithm.

To compute a controllable and observable realization for a rational transfer function $F(z) = \sum_0^\infty z^{-j} F_j$, simply apply the Kalman-Ho algorithm to the Taylor coefficients $\{F_j\}_0^{2k-1}$ for F where $k > \nu$ the McMillan degree of F. Then the Kalman-Ho algorithm yields a realization $\{A \text{ on } \mathcal{X}, B, C, D\}$ for F; see Remark 14.5.2. In fact, the degree of $\mathcal{X}$ may be less than the McMillan degree of F. This can happen because the Hankel matrix determined by F may have several small nonzero singular values, and the Kalman-Ho algorithm will eliminate the insignificant singular values. So the Kalman-Ho algorithm can also be used to find a reduced order model for F or any realization. Ruffly speaking, a reduced order model of $\{A \text{ on } \mathcal{X}, B, C, D\}$ is any state space system $\{\widetilde{A} \text{ on } \widetilde{\mathcal{X}}, \widetilde{B}, \widetilde{C}, D\}$ such that $\dim \widetilde{\mathcal{X}} < \dim \mathcal{X}$ and these two systems have "approximately" the same transfer function.

The Kalman-Ho algorithm works especially well when the rational transfer function F is in $H^\infty(\mathcal{E}, \mathcal{Y})$, or equivalently, the corresponding Hankel matrix is bounded. In this case, one can use large data sets to compute a minimal realization. If H is not bounded, then $\|H_k\|$ approaches infinity as k tends to infinity, and this may cause numerical problems for large data sets.

The Kalman-Ho algorithm can be used on experimental data to find a state space realization. Since the singular value decomposition is very efficient, one can apply the Kalman-Ho algorithm on large data sets when the underlying system is stable.

Example. Let us show how the Kalman-Ho algorithm can be used to compute a reduced order model for a nonrational function. Consider the transfer function $f(z) = e^{1/z}$. Clearly, this transfer function is not rational, and thus, there is no finite dimensional realization for f. By applying the Kalman-Ho algorithm on the corresponding Hankel matrix of length 200, we arrived at the following second-order model:

$$A = \begin{bmatrix} 0.4349 & 0.2333 \\ -0.2333 & 0.0491 \end{bmatrix}, \quad B = \begin{bmatrix} -1.0079 \\ -0.1257 \end{bmatrix} \quad \text{and} \quad C = \begin{bmatrix} -1.0079 & 0.1257 \end{bmatrix}.$$

As expected, $D = f(\infty) = 1$. The transfer function for $\{A, B, C, D\}$ is given by

$$g(z) = D + C(zI - A)^{-1}B = \frac{z^2 + 0.5161z + 0.09196}{z^2 - 0.4839z + 0.07578}.$$

Clearly, $\{A, B, C, D\}$ is not a realization of f. However, it turns out that $\|f - g\|_\infty \approx 0.0014$ and $\|f\|_\infty = e \approx 2.7183$. So the realization $\{A, B, C, D\}$ is a "fairly accurate" reduced order model for $e^{1/z}$. Finally, it is noted that one can obtain a more accurate approximation of $e^{1/z}$ by choosing a higher order state space model.

14.6 The Restricted Backward Shift Realization

Recall that a rational transfer function is a proper rational function. Section 14.3 shows that any rational transfer function admits a finite dimensional realization. In this section, we will use the backward shift to construct a minimal realization for a rational transfer function F with values in $\mathcal{L}(\mathcal{E}, \mathcal{Y})$. To this end, let $S^\sharp$ be the backward shift given by

$$S^\sharp = \begin{bmatrix} 0 & I_\mathcal{Y} & 0 & 0 & \cdots \\ 0 & 0 & I_\mathcal{Y} & 0 & \cdots \\ 0 & 0 & 0 & I_\mathcal{Y} & \cdots \\ \vdots & \vdots & \vdots & \vdots & \ddots \end{bmatrix} \quad \text{on} \quad \begin{bmatrix} \mathcal{Y} \\ \mathcal{Y} \\ \mathcal{Y} \\ \vdots \end{bmatrix}. \tag{14.6.1}$$

Here $S^\sharp$ is viewed as a linear map acting on the space $\ell_+(\mathcal{Y})$ of all unilateral vectors of infinite length with values in $\mathcal{Y}$. We do not need a topology on $\ell_+(\mathcal{Y})$. However, if F is a function in $H^2(\mathcal{E}, \mathcal{Y})$, then we can take $S^* = S^\sharp$ to be the backward shift on $\ell_+^2(\mathcal{Y})$; see Section 14.6.2 below. Let $\mathcal{H}_r$ be the range of the Hankel matrix H in (14.2.3), that is,

$$\mathcal{H}_r = \text{span}\{S^{\sharp n}\Xi\mathcal{E} : n \geq 0\} \quad \text{where} \quad \Xi = \begin{bmatrix} F_1 & F_2 & F_3 & \cdots \end{bmatrix}^{tr}. \tag{14.6.2}$$

Here Ξ is the first column of H, and $F = \sum_0^\infty z^{-n}F_n$ is the Taylor series expansion for F. Notice that $\mathcal{H}_r$ is an invariant subspace for $S^\sharp$. Because F is rational, $\mathcal{H}_r$ is finite dimensional. Let A_r be the linear map on $\mathcal{H}_r$ defined by $A_r = S^\sharp|\mathcal{H}_r$, and B_r the linear map from $\mathcal{E}$ into $\mathcal{H}_r$ determined by $B_r = \Xi$, and C_r the linear map from $\mathcal{H}_r$ into $\mathcal{Y}$ given by

$$C_r \begin{bmatrix} x_0 & x_1 & x_2 & \cdots \end{bmatrix}^{tr} = x_0 \quad \text{where} \quad \begin{bmatrix} x_0 & x_1 & x_2 & \cdots \end{bmatrix}^{tr} \in \mathcal{H}_r. \tag{14.6.3}$$

We claim that $\{A_r, B_r, C_r, F_0\}$ is a controllable and observable realization for F. This realization is called the *algebraic restricted backward shift realization of* F.

To show that $\{A_r, B_r, C_r, F_0\}$ is a realization of F, observe that for any integer $n \geq 1$, we have

$$C_r A_r^{n-1} B_r = C_r S^{\sharp n-1}\Xi = C_r \begin{bmatrix} F_n & F_{n+1} & F_{n+2} & \cdots \end{bmatrix}^{tr} = F_n.$$

Hence (14.1.11) holds, and thus, $\{A_r, B_r, C_r, F_0\}$ is a realization for F. Because $\mathcal{H}_r$ is invariant for $S^\sharp$ and $A_r = S^\sharp|\mathcal{H}_r$, we have

$$\mathcal{H}_r = \text{span}\{S^{\sharp n}\Xi\mathcal{E} : n \geq 0\} = \text{span}\{A_r^n B_r \mathcal{E} : n \geq 0\}.$$

So the pair $\{A_r, B_r\}$ is controllable. Let $x = \begin{bmatrix} x_0 & x_1 & x_2 & \cdots \end{bmatrix}^{tr}$ be any vector in $\mathcal{H}_r$. Notice that $C_r A_r^n x = x_n$. So $C_r A_r^n x = 0$ for all integers $n \geq 0$ if and only if $x = 0$. Thus the pair $\{C_r, A_r\}$ is observable. Therefore $\{A_r, B_r, C_r, F_0\}$ is a controllable and observable realization for F.

It is well known that a finite dimensional realization is controllable and observable if and only if it is minimal. Let us directly show that the restricted backward shift realization is minimal. To this end, let $\{A \text{ on } \mathcal{X}, B, C, D\}$ be a finite dimensional realization for F. Let W_o be the observability matrix defined by

$$W_o = \begin{bmatrix} C & CA & CA^2 & \cdots \end{bmatrix}^{tr} : \mathcal{X} \to \ell_+(\mathcal{E}).$$

Notice that $S^\sharp W_o = W_o A$. Using $F_n = CA^{n-1}B$ for all integers $n \geq 1$, we see that $\Xi = W_o B$. Hence

$$\begin{aligned} \mathcal{H}_r &= \text{span}\{S^{\sharp n}\Xi\mathcal{E} : n \geq 0\} \\ &= \text{span}\{S^{\sharp n}W_o B\mathcal{E} : n \geq 0\} \\ &= \text{span}\{W_o A^n B\mathcal{E} : n \geq 0\} \\ &\subseteq W_o\mathcal{X}. \end{aligned} \tag{14.6.4}$$

In other words, $\mathcal{H}_r \subseteq W_o\mathcal{X}$. So $\dim \mathcal{H}_r \leq \dim \mathcal{X}$. Since $\mathcal{H}_r$ is the state space for $\{A_r, B_r, C_r, F_0\}$, the restricted backward shift realization is minimal.

Now let us show that all minimal realizations of the same rational transfer function are similar. Assume that $\{A, B, C, D\}$ is a minimal realization for F. Then $\dim \mathcal{H}_r = \dim \mathcal{X}$. This with $\mathcal{H}_r \subseteq W_o\mathcal{X}$ implies that W_o maps $\mathcal{X}$ one to one and onto $\mathcal{H}_r$. Let Φ be the similarity transformation mapping $\mathcal{X}$ onto $\mathcal{H}_r$ determined by $\Phi = W_o$. Then using $S^\sharp W_o = W_o A$, we obtain

$$\begin{aligned} A_r\Phi &= S^\sharp W_o = W_o A = \Phi A, \\ \Phi B &= W_o B = \Xi = B_r, \\ C_r\Phi &= C_r W_o = C. \end{aligned} \tag{14.6.5}$$

Therefore Φ is a similarity transformation intertwining $\{A, B, C, D\}$ with the restricted backward shift realization. So all minimal realizations of the same transfer function are similar to the restricted backward shift realization. Because similar realizations are transitive, all minimal realizations of the same transfer function are similar.

Now let us show that a finite dimensional realization is controllable and observable if and only if it is minimal. A minimal realization is similar to the restricted backward shift realization. Because the restricted backward shift realization is controllable and observable and controllability and observability are preserved under a similarity transformation, a minimal realization must be controllable and observable. On the other hand, if $\{A, B, C, D\}$ is a controllable and observable realization of F, then observability shows that W_o is one to one. By consulting (14.6.4), controllability implies that $\mathcal{H}_r = W_o\mathcal{X}$. In other words, W_o maps $\mathcal{X}$ one to one and onto $\mathcal{H}_r$. Therefore $\mathcal{X}$ and $\mathcal{H}_r$ have the same dimension. Because the restricted backward shift realization is minimal, $\{A, B, C, D\}$ is also minimal. So any controllable and observable realization is minimal. In other words, a finite dimensional realization is controllable and observable if and only if it is minimal. In this case, all minimal realizations of the same transfer function are similar.

14.6.1 Rational functions in $H^2(\mathcal{E}, \mathcal{Y})$

Throughout we assume that $\mathcal{E}$ and $\mathcal{Y}$ are finite dimensional spaces. Recall $H^2(\mathcal{E}, \mathcal{Y})$ is the Hilbert space formed by set of all $\mathcal{L}(\mathcal{E}, \mathcal{Y})$-valued analytic functions F in $\mathbb{D}_+$ such that the trace $F^*(e^{\imath\omega})F(e^{\imath\omega})$ is summable. In this case,

$$\|F\|_2^2 = \text{trace} \sum_{n=0}^{\infty} F_n^* F_n \quad \text{where} \quad F(z) = \sum_{n=0}^{\infty} z^{-n} F_n \tag{14.6.6}$$

is the Taylor series expansion for F. Finally, it is noted that if F is a rational function in $H^2(\mathcal{E}, \mathcal{Y})$, then F must be a proper rational function.

Remark 14.6.1. Let $F(z)$ be a proper rational function with values in $\mathcal{L}(\mathcal{E}, \mathcal{Y})$. Then the following holds.

(i) The transfer function F is in $H^2(\mathcal{E}, \mathcal{Y})$ if and only if all the poles of F are contained in the open unit disc $\mathbb{D}$. In particular, F is in $H^2(\mathcal{E}, \mathcal{Y})$ if and only if F is also in $H^\infty(\mathcal{E}, \mathcal{Y})$.

(ii) The function F is in $H^\infty(\mathcal{E}, \mathcal{Y})$ if and only if F admits a stable, finite dimensional state space realization $\{A, B, C, D\}$. In this case,

$$\|F\|_2^2 = \text{trace}\,(D^*D + B^*PB) \quad \text{where} \quad P = A^*PA + C^*C. \tag{14.6.7}$$

Part (i) follows from the partial fraction expansion for a rational function, and is left as an exercise. According to Part (i) a rational function F is in $H^\infty(\mathcal{E}, \mathcal{Y})$ if and only if all the poles of F are contained in $\mathbb{D}$. So F is a rational function in $H^\infty(\mathcal{E}, \mathcal{Y})$ if and only if F admits a stable, finite dimensional realization $\{A, B, C, D\}$, that is,

$$F(z) = D + C(zI - A)^{-1}B$$

where A is a stable operator on $\mathcal{X}$. In particular, this implies that the Taylor coefficients $\{F_n\}_0^\infty$ of F are given by

$$F_0 = D \quad \text{and} \quad F_n = CA^{n-1}B \qquad (\text{for all } n \geq 1). \tag{14.6.8}$$

To obtain the formula for the H^2 norm of F in (14.6.7), recall that the observability Gramian P is the unique solution to the Lyapunov equation in (14.6.7). Moreover, $P = \sum_0^\infty A^{*n}C^*CA^n$. This and (14.6.8) yield

$$\sum_{n=0}^{\infty} F_n^* F_n = D^*D + \sum_{n=1}^{\infty} B^*A^{*n-1}C^*CA^{n-1}B = D^*D + B^*PB.$$

Therefore H^2 norm of F is given in (14.6.7).

14.6.2 The restricted backward shift realization in H^2

Previously we presented the algebraic restricted backward shift realization. If F is a function in $H^2(\mathcal{E}, \mathcal{Y})$, then we can introduce a topological structure on this realization. To this end, let F be a transfer function in $H^2(\mathcal{E}, \mathcal{Y})$, and $F(z) = \sum_0^\infty z^{-n} F_n$ its Taylor series expansion. Let Ξ be the operator defined by

$$\Xi = \begin{bmatrix} F_1 & F_2 & F_3 & \cdots \end{bmatrix}^{tr} : \mathcal{E} \to \ell_+^2(\mathcal{Y}).$$

Let S be the unilateral shift on $\ell_+^2(\mathcal{Y})$, and $\mathcal{H}_r$ the invariant subspace for S^* determined by

$$\mathcal{H}_r = \bigvee_{n=0}^{\infty} S^{*n} \Xi \mathcal{E}. \tag{14.6.9}$$

Let $\Pi_\mathcal{Y}$ be the orthogonal projection mapping $\ell_+^2(\mathcal{Y})$ onto $\mathcal{Y}$ which picks out the first component of $\ell_+^2(\mathcal{Y})$, that is,

$$\Pi_\mathcal{Y} = \begin{bmatrix} I & 0 & 0 & \cdots \end{bmatrix} : \ell_+^2(\mathcal{Y}) \to \mathcal{Y}.$$

The restricted *backward shift realization* of F is the controllable and observable realization $\{A_r \text{ on } \mathcal{H}_r, B_r, C_r, F_0\}$ defined by

$$\begin{aligned} A_r &= S^* | \mathcal{H}_r \text{ on } \mathcal{H}_r, \\ B_r &= \Xi : \mathcal{E} \to \mathcal{H}_r, \\ C_r &= \Pi_\mathcal{Y} | \mathcal{H}_r : \mathcal{H}_r \to \mathcal{Y}. \end{aligned} \tag{14.6.10}$$

It is left as a simple exercise to verify that $\{A_r, B_r, C_r, F_0\}$ is indeed a controllable and observable realization for F. Finally, it is noted that F is a rational function if and only if $\mathcal{H}_r$ is finite dimensional. In this case, A_r is stable. (Since S^{*n} converges strongly to zero, and $A_r = S^* | \mathcal{H}_r$, we see that A_r^n converges to zero. Because $\mathcal{H}_r$ is finite dimensional, A_r is stable.) The dimension of $\mathcal{H}_r$ is the McMillan degree of F.

Now assume that $\{A \text{ on } \mathcal{X}, B, C, D\}$ is a minimal realization for a rational transfer function F in $H^2(\mathcal{E}, \mathcal{Y})$. Then the operator

$$\Phi = \begin{bmatrix} C \\ CA \\ CA^2 \\ \vdots \end{bmatrix} : \mathcal{X} \to \mathcal{H}_r \tag{14.6.11}$$

is a similarity transformation intertwining $\{A, B, C, D\}$ with the restricted backward shift realization $\{A_r, B_r, C_r, D_r\}$ for F.

14.7 Notes

All the results in this chapter are classical; see [53, 60, 67, 68, 140, 149, 189, 193] for further results on state space and realization theory. For a nice paper on the Kalman-Ho algorithm see Damen-Van den Hof-Hajdasinski [63]. The restricted backward shift realization is now classical. Our approach to the backward shift realization was taken from Helton [132] and Fuhrmann [105]. For a shift realization approach to certain nonlinear systems see Frazho [92], Rugh [188] and Wong [201].

Chapter 15

The Levinson Algorithm

In this chapter we will develop the Levinson algorithm. We will also present the Gohberg-Semencul-Heinig inversion formula for strictly positive block Toeplitz matrices.

15.1 The Levinson Recursion

To present the Levinson algorithm, consider the strictly positive Toeplitz matrix Υ_n on $\mathcal{E}^n$ defined by

$$\Upsilon_n = \begin{bmatrix} R_0 & R_1^* & \cdots & R_{n-1}^* \\ R_1 & R_0 & \cdots & R_{n-2}^* \\ \vdots & \vdots & \ddots & \vdots \\ R_{n-1} & R_{n-2} & \cdots & R_0 \end{bmatrix} \text{ on } \mathcal{E}^n. \tag{15.1.1}$$

The *Levinson system of equations* is defined by

$$\Upsilon_n \begin{bmatrix} I \\ A_{n,1} \\ A_{n,2} \\ \vdots \\ A_{n,n-1} \end{bmatrix} = \begin{bmatrix} \Delta_n \\ 0 \\ 0 \\ \vdots \\ 0 \end{bmatrix} \text{ and } \Upsilon_n \begin{bmatrix} B_{n,0} \\ B_{n,1} \\ \vdots \\ B_{n,n-2} \\ I \end{bmatrix} = \begin{bmatrix} 0 \\ 0 \\ \vdots \\ 0 \\ \Lambda_n \end{bmatrix}. \tag{15.1.2}$$

Here $A_{n,j}$, $B_{n,j}$, Δ_n and Λ_n are all operators on $\mathcal{E}$. Moreover, we set $A_{n,0} = I$ and $B_{n,n-1} = I$. To simplify some notation, let us also set

$$\begin{aligned} A_n &= \begin{bmatrix} A_{n,1} & A_{n,2} & \cdots & A_{n,n-1} \end{bmatrix}^{tr} : \mathcal{E} \to \mathcal{E}^{n-1}, \\ B_n &= \begin{bmatrix} B_{n,0} & B_{n,1} & \cdots & B_{n,n-2} \end{bmatrix}^{tr} : \mathcal{E} \to \mathcal{E}^{n-1}, \\ X_{n-1} &= \begin{bmatrix} R_1 & R_2 & \cdots & R_{n-1} \end{bmatrix}^{tr} : \mathcal{E} \to \mathcal{E}^{n-1}, \\ Y_{n-1} &= \begin{bmatrix} R_{n-1} & R_{n-2} & \cdots & R_1 \end{bmatrix} : \mathcal{E}^{n-1} \to \mathcal{E}. \end{aligned} \tag{15.1.3}$$

Using this notation, we see that Υ_n admits two different matrix decompositions of the form

$$\Upsilon_n = \begin{bmatrix} R_0 & X_{n-1}^* \\ X_{n-1} & \Upsilon_{n-1} \end{bmatrix} \text{ on } \begin{bmatrix} \mathcal{E} \\ \mathcal{E}^{n-1} \end{bmatrix} \quad \text{and} \quad \begin{bmatrix} \Upsilon_{n-1} & Y_{n-1}^* \\ Y_{n-1} & R_0 \end{bmatrix} \text{ on } \begin{bmatrix} \mathcal{E}^{n-1} \\ \mathcal{E} \end{bmatrix}. \tag{15.1.4}$$

Moreover, the Levinson system in (15.1.2) is equivalent to

$$\begin{bmatrix} R_0 & X_{n-1}^* \\ X_{n-1} & \Upsilon_{n-1} \end{bmatrix} \begin{bmatrix} I \\ A_n \end{bmatrix} = \begin{bmatrix} \Delta_n \\ 0 \end{bmatrix} \quad \text{and} \quad \begin{bmatrix} \Upsilon_{n-1} & Y_{n-1}^* \\ Y_{n-1} & R_0 \end{bmatrix} \begin{bmatrix} B_n \\ I \end{bmatrix} = \begin{bmatrix} 0 \\ \Lambda_n \end{bmatrix}. \tag{15.1.5}$$

Remark 15.1.1. Assume that Υ_{n-1} is strictly positive. Then the solution to the Levinson system in (15.1.2) or (15.1.5) is unique and given by

$$\begin{aligned} \Delta_n &= R_0 - X_{n-1}^* \Upsilon_{n-1}^{-1} X_{n-1} \quad \text{and} \quad A_n = -\Upsilon_{n-1}^{-1} X_{n-1}, \\ \Lambda_n &= R_0 - Y_{n-1} \Upsilon_{n-1}^{-1} Y_{n-1}^* \quad \text{and} \quad B_n = -\Upsilon_{n-1}^{-1} Y_{n-1}^*. \end{aligned} \tag{15.1.6}$$

Furthermore, Υ_n is strictly positive if and only if Δ_n is strictly positive, or equivalently, Λ_n is strictly positive. Finally, Δ_n is the Schur complement for Υ_n with respect to R_0 in the upper left-hand corner of Υ_n.

By consulting the Levinson system in (15.1.5), we obtain $\Upsilon_{n-1} A_n = -X_{n-1}$ and $\Upsilon_{n-1} B_n = -Y_{n-1}^*$. Because Υ_{n-1} is strictly positive, both of these equations have a unique solution. Moreover, these solutions are given by $A_n = -\Upsilon_{n-1}^{-1} X_{n-1}$ and $B_n = -\Upsilon_{n-1}^{-1} Y_{n-1}^*$. This yields the second and fourth equation in (15.1.6). Substituting $A_n = -\Upsilon_{n-1}^{-1} X_{n-1}$ into $R_0 + X_{n-1}^* A_n = \Delta_n$, we arrive at the first equation in (15.1.6). Finally, substituting $B_n = -\Upsilon_{n-1}^{-1} Y_{n-1}^*$ into $R_0 + Y_{n-1} B_n = \Lambda_n$ yields the third equation in (15.1.6). Because A_n and B_n are uniquely determined, Δ_n and Λ_n must also be unique. Therefore the solution to the Levinson system in (15.1.2) is unique and given by (15.1.6).

Now let us show that Υ_n is strictly positive if and only if Δ_n is strictly positive. According to (15.1.6), the operator Δ_n is the Schur complement for the first matrix decomposition of Υ_n given in (15.1.4). By Lemma 7.2.1, the operator Υ_n is strictly positive if and only if its Schur complement Δ_n is strictly positive. A similar argument shows that Υ_n is strictly positive if and only if Λ_n is strictly positive.

The Levinson algorithm. Let Υ_n be a strictly positive Toeplitz matrix on $\mathcal{E}^n$. Then the solution to the Levinson system in (15.1.2) is recursively computed by

$$\begin{bmatrix} A_{n+1,1} \\ \vdots \\ A_{n+1,n-1} \\ A_{n+1,n} \end{bmatrix} = \begin{bmatrix} A_{n,1} \\ \vdots \\ A_{n,n-1} \\ 0 \end{bmatrix} - \begin{bmatrix} B_{n,0} \\ \vdots \\ B_{n,n-2} \\ I \end{bmatrix} \Lambda_n^{-1}\Omega_n,$$

$$\begin{bmatrix} B_{n+1,0} \\ B_{n+1,1} \\ \vdots \\ B_{n+1,n-1} \end{bmatrix} = \begin{bmatrix} 0 \\ B_{n,0} \\ \vdots \\ B_{n,n-2} \end{bmatrix} - \begin{bmatrix} I \\ A_{n,1} \\ \vdots \\ A_{n,n-1} \end{bmatrix} \Delta_n^{-1}\Omega_n^*, \tag{15.1.7}$$

$$\Delta_{n+1} = \Delta_n - \Omega_n^*\Lambda_n^{-1}\Omega_n,$$

$$\Lambda_{n+1} = \Lambda_n - \Omega_n\Delta_n^{-1}\Omega_n^*,$$

$$\Omega_n = R_n + \sum_{j=1}^{n-1} R_{n-j}A_{n,j}.$$

The initial conditions for the Levinson algorithm are given by

$$\begin{aligned} \Delta_2 &= R_0 - R_1^*R_0^{-1}R_1 \quad \text{and} \quad A_{2,1} = -R_0^{-1}R_1, \\ \Lambda_2 &= R_0 - R_1R_0^{-1}R_1^* \quad \text{and} \quad B_{2,0} = -R_0^{-1}R_1^*. \end{aligned} \tag{15.1.8}$$

To derive the Levinson algorithm observe that (15.1.5) yields

$$\Upsilon_n \begin{bmatrix} I \\ A_n \end{bmatrix} = \begin{bmatrix} \Delta_n \\ 0 \end{bmatrix} \quad \text{and} \quad \Upsilon_n \begin{bmatrix} B_n \\ I \end{bmatrix} = \begin{bmatrix} 0 \\ \Lambda_n \end{bmatrix}. \tag{15.1.9}$$

Using the structure of the Toeplitz matrix Υ_n in (15.1.4), we see that Υ_{n+1} admits a matrix decomposition of the form

$$\Upsilon_{n+1} = \begin{bmatrix} R_0 & X_{n-1}^* & R_n^* \\ X_{n-1} & \Upsilon_{n-1} & Y_{n-1}^* \\ R_n & Y_{n-1} & R_0 \end{bmatrix} \quad \text{on} \quad \begin{bmatrix} \mathcal{E} \\ \mathcal{E}^{n-1} \\ \mathcal{E} \end{bmatrix}. \tag{15.1.10}$$

By employing (15.1.3) and (15.1.9), we arrive at

$$\begin{bmatrix} R_0 & X_{n-1}^* & R_n^* \\ X_{n-1} & \Upsilon_{n-1} & Y_{n-1}^* \\ R_n & Y_{n-1} & R_0 \end{bmatrix} \begin{bmatrix} I \\ A_n \\ 0 \end{bmatrix} = \begin{bmatrix} \Delta_n \\ 0 \\ \Omega_n \end{bmatrix},$$

$$\begin{bmatrix} R_0 & X_{n-1}^* & R_n^* \\ X_{n-1} & \Upsilon_{n-1} & Y_{n-1}^* \\ R_n & Y_{n-1} & R_0 \end{bmatrix} \begin{bmatrix} 0 \\ B_n \\ I \end{bmatrix} = \begin{bmatrix} \Phi_n \\ 0 \\ \Lambda_n \end{bmatrix}.$$

Here Ω_n and Φ_n are the operators on $\mathcal{E}$ determined by

$$\Omega_n = R_n + Y_{n-1}A_n \quad \text{and} \quad \Phi_n = R_n^* + X_{n-1}^*B_n.$$

Notice that $\Omega_n = R_n + \sum_{j=1}^{n-1} R_{n-j}A_{n,j}$ is precisely the last equation (15.1.7). Now observe that

$$\Upsilon_{n+1}\left[\begin{bmatrix} I \\ A_n \\ 0 \end{bmatrix} - \begin{bmatrix} 0 \\ B_n \\ I \end{bmatrix} \Lambda_n^{-1}\Omega_n\right] = \begin{bmatrix} \Delta_n - \Phi_n\Lambda_n^{-1}\Omega_n \\ 0 \\ 0 \end{bmatrix},$$

$$\Upsilon_{n+1}\left[\begin{bmatrix} 0 \\ B_n \\ I \end{bmatrix} - \begin{bmatrix} I \\ A_n \\ 0 \end{bmatrix} \Delta_n^{-1}\Phi_n\right] = \begin{bmatrix} 0 \\ 0 \\ \Lambda_n - \Omega_n\Delta_n^{-1}\Phi_n \end{bmatrix}.$$

To obtain the next step in the Levinson algorithm, set

$$\begin{aligned} \begin{bmatrix} I \\ A_{n+1} \end{bmatrix} &= \begin{bmatrix} I \\ \begin{bmatrix} A_n \\ 0 \end{bmatrix} \end{bmatrix} - \begin{bmatrix} 0 \\ \begin{bmatrix} B_n \\ I \end{bmatrix} \end{bmatrix} \Lambda_n^{-1}\Omega_n, \\ \begin{bmatrix} B_{n+1} \\ I \end{bmatrix} &= \begin{bmatrix} \begin{bmatrix} 0 \\ B_n \end{bmatrix} \\ I \end{bmatrix} - \begin{bmatrix} \begin{bmatrix} I \\ A_n \end{bmatrix} \\ 0 \end{bmatrix} \Delta_n^{-1}\Phi_n, \\ \Delta_{n+1} &= \Delta_n - \Phi_n\Lambda_n^{-1}\Omega_n, \\ \Lambda_{n+1} &= \Lambda_n - \Omega_n\Delta_n^{-1}\Phi_n. \end{aligned} \tag{15.1.11}$$

Then equation (15.1.9) holds for $n+1$. In a moment we will show that $\Phi_n = \Omega_n^*$. This with (15.1.11) yields the Levinson recursion (15.1.7).

To complete the proof, it remains to show that $\Phi_n = \Omega_n^*$. Recall that $Y_{n-1}^* = -\Upsilon_{n-1}B_n$ and $X_{n-1} = -\Upsilon_{n-1}A_n$. Using $Y_{n-1} = -B_n^*\Upsilon_{n-1}$, we obtain

$$\begin{aligned} \Omega_n &= R_n + Y_{n-1}A_n = R_n - B_n^*\Upsilon_{n-1}A_n \\ &= R_n + B_n^*X_{n-1} = \left(R_n^* + X_{n-1}^*B_n\right)^* \\ &= \Phi_n^*. \end{aligned}$$

Therefore $\Phi_n = \Omega_n^*$.

Remark 15.1.2. Since Δ_j is the Schur complement for Υ_j, the operator Υ_n is strictly positive if and only if Δ_j is strictly positive for $j = 1, 2, \ldots, n$ where $\Delta_1 = R_0$; see Lemma 7.3.1. So one can use the Levinson algorithm to determine whether or not Υ_n is strictly positive. In this case, let L_n be the lower triangular matrix defined by

$$L_n = \begin{bmatrix} \widehat{A}_{n,0} & 0 & 0 & \cdots & 0 & 0 \\ \widehat{A}_{n,1} & \widehat{A}_{n-1,0} & 0 & \cdots & 0 & 0 \\ \widehat{A}_{n,2} & \widehat{A}_{n-1,1} & \widehat{A}_{n-2,0} & \cdots & 0 & 0 \\ \vdots & \vdots & \vdots & \ddots & \vdots & 0 \\ \widehat{A}_{n,n-2} & \widehat{A}_{n-1,n-3} & \widehat{A}_{n-2,n-4} & \cdots & \widehat{A}_{2,0} & 0 \\ \widehat{A}_{n,n-1} & \widehat{A}_{n-1,n-2} & \widehat{A}_{n-2,n-3} & \cdots & \widehat{A}_{2,1} & \widehat{A}_{1,0} \end{bmatrix}. \tag{15.1.12}$$

Here $\widehat{A}_{n,j} = A_{n,j}\Delta_n^{-1/2}$ for all $j = 0, 1, 2, \ldots, n-1$ while $\widehat{A}_{1,0} = R_0^{-1/2}$ and $\widehat{A}_{k,0} = \Delta_k^{-1/2}$. Finally, L_n is a lower triangular factorization for Υ_n^{-1}, that is,

$$\Upsilon_n^{-1} = L_n L_n^*. \tag{15.1.13}$$

The operator Υ_n is strictly positive if and only if Λ_j is strictly positive for $j = 1, 2, \ldots, n$ where $\Lambda_1 = R_0$. Now consider the upper triangular matrix determined by

$$U_n = \begin{bmatrix} \widehat{B}_{n,0} & \widehat{B}_{n-1,0} & \cdots & \widehat{B}_{3,0} & \widehat{B}_{2,0} & \widehat{B}_{1,0} \\ \widehat{B}_{n,1} & \widehat{B}_{n-1,1} & \cdots & \widehat{B}_{3,1} & \widehat{B}_{2,1} & 0 \\ \widehat{B}_{n,2} & \widehat{B}_{n-1,2} & \cdots & \widehat{B}_{3,2} & 0 & 0 \\ \vdots & \vdots & \vdots & \vdots & \vdots & \vdots \\ \widehat{B}_{n,n-2} & \widehat{B}_{n-1,n-2} & \cdots & \cdots & 0 & 0 \\ \widehat{B}_{n,n-1} & 0 & \cdots & \cdots & 0 & 0 \end{bmatrix}. \tag{15.1.14}$$

Here $\widehat{B}_{n,j} = B_{n,j}\Lambda_n^{-1/2}$ for all $j = 0, 1, 2, \ldots, n-1$ while $\widehat{B}_{1,0} = R_0^{-1/2}$ and $\widehat{B}_{k,k-1} = \Lambda_k^{-1/2}$. Finally, U_n is an upper triangular factorization for Υ_n^{-1}, that is,

$$\Upsilon_n^{-1} = U_n U_n^*. \tag{15.1.15}$$

Because L_n is a lower triangular matrix with diagonal entries $\widehat{A}_{k,0} = \Delta_k^{-1/2}$ for $k = 1, 2, \ldots, n$, the operator L_n is invertible. By inverting L_n and L_n^*, we see that $\Upsilon_n^{-1} = L_n L_n^*$ if and only if $L_n^{-1}\Upsilon_n^{-1}L_n^{-*} = I$. By taking the inverse, $\Upsilon_n^{-1} = L_n L_n^*$ if and only if $L_n^*\Upsilon_n L_n = I$. Now let us use induction to show that $L_n^*\Upsilon_n L_n = I$. Since $\Upsilon_1 = R_0$ and $L_1 = R_0^{-1/2}$, we have $L_1^*\Upsilon_1 L_1 = I$. To complete the induction argument, assume that $L_{n-1}^*\Upsilon_{n-1}L_{n-1} = I$. The operator L_n admits a matrix decomposition of the form

$$L_n = \begin{bmatrix} \Delta_n^{-1/2} & 0 \\ A_n\Delta_n^{-1/2} & L_{n-1} \end{bmatrix} \text{ on } \begin{bmatrix} \mathcal{E} \\ \mathcal{E}^{n-1} \end{bmatrix}. \tag{15.1.16}$$

Using the first Levinson system in (15.1.5) with $(X_{n-1} + \Upsilon_{n-1}A_n)^* = 0$, we obtain

$$\begin{aligned} L_n^*\Upsilon_n L_n &= L_n^* \begin{bmatrix} R_0 & X_{n-1}^* \\ X_{n-1} & \Upsilon_{n-1} \end{bmatrix} \begin{bmatrix} \Delta_n^{-1/2} & 0 \\ A_n\Delta_n^{-1/2} & L_{n-1} \end{bmatrix} \\ &= \begin{bmatrix} \Delta_n^{-1/2} & \Delta_n^{-1/2}A_n^* \\ 0 & L_{n-1}^* \end{bmatrix} \begin{bmatrix} \Delta_n^{1/2} & X_{n-1}^*L_{n-1} \\ 0 & \Upsilon_{n-1}L_{n-1} \end{bmatrix} \\ &= \begin{bmatrix} I & \Delta_n^{-1/2}\left(X_{n-1}^* + A_n^*\Upsilon_{n-1}\right)L_{n-1} \\ 0 & L_{n-1}^*\Upsilon_{n-1}L_{n-1} \end{bmatrix} \\ &= \begin{bmatrix} I & 0 \\ 0 & I \end{bmatrix}. \end{aligned}$$

Therefore $L_n^*\Upsilon_n L_n = I$, or equivalently, $\Upsilon_n^{-1} = L_n L_n^*$.

To complete the proof, observe that $\Upsilon_n^{-1} = U_n U_n^*$ if and only if $U_n^*\Upsilon_n U_n = I$. As before, let us use an inductive argument. Since $\Upsilon_1 = R_0$ and $U_1 = R_0^{-1/2}$, we have $U_1^*\Upsilon_1 U_1 = I$. Assume that $U_{n-1}^*\Upsilon_{n-1}U_{n-1} = I$. The operator U_n admits a matrix decomposition of the form

$$U_n = \begin{bmatrix} B_n\Lambda_n^{-1/2} & U_{n-1} \\ \Lambda_n^{-1/2} & 0 \end{bmatrix} \text{ on } \begin{bmatrix} \mathcal{E}^{n-1} \\ \mathcal{E} \end{bmatrix}. \tag{15.1.17}$$

Using the second Levinson system in (15.1.5), we obtain

$$\begin{aligned} U_n^*\Upsilon_n U_n &= U_n^* \begin{bmatrix} \Upsilon_{n-1} & Y_{n-1}^* \\ Y_{n-1} & R_0 \end{bmatrix} \begin{bmatrix} B_n\Lambda_n^{-1/2} & U_{n-1} \\ \Lambda_n^{-1/2} & 0 \end{bmatrix} \\ &= \begin{bmatrix} \Lambda_n^{-1/2}B_n^* & \Lambda_n^{-1/2} \\ U_{n-1}^* & 0 \end{bmatrix} \begin{bmatrix} 0 & \Upsilon_{n-1}U_{n-1} \\ \Lambda_n^{1/2} & Y_{n-1}U_{n-1} \end{bmatrix} \\ &= \begin{bmatrix} I & \Lambda_n^{-1/2}\left(B_n^*\Upsilon_{n-1} + Y_{n-1}\right)U_{n-1} \\ 0 & U_{n-1}^*\Upsilon_{n-1}U_{n-1} \end{bmatrix} \\ &= \begin{bmatrix} I & 0 \\ 0 & I \end{bmatrix}. \end{aligned}$$

Therefore $U_n^*\Upsilon_n U_n = I$, or equivalently, $\Upsilon_n^{-1} = U_n U_n^*$.

Assume that Υ_n is strictly positive. By mimicking the computations in Remark 7.1.3, or by a direct computation, we see that the solution to the Levinson system is unique and also given by

$$\begin{aligned} \Delta_n &= \left(\Pi_1\Upsilon_n^{-1}\Pi_1^*\right)^{-1}, \\ \Lambda_n &= \left(\Pi_n\Upsilon_n^{-1}\Pi_n^*\right)^{-1}, \\ \Upsilon_n^{-1}\Pi_1^*\Delta_n &= \begin{bmatrix} I & A_{n,1} & A_{n,2} & \cdots & A_{n,n-1} \end{bmatrix}^{tr}, \\ \Upsilon_n^{-1}\Pi_n^*\Lambda_n &= \begin{bmatrix} B_{n,0} & B_{n,1} & \cdots & B_{n,n-2} & I \end{bmatrix}^{tr}, \\ \Pi_1 &= \begin{bmatrix} I & 0 & \cdots & 0 & 0 \end{bmatrix} : \mathcal{E}^n \to \mathcal{E}, \\ \Pi_n &= \begin{bmatrix} 0 & 0 & \cdots & 0 & I \end{bmatrix} : \mathcal{E}^n \to \mathcal{E}. \end{aligned} \tag{15.1.18}$$

Finally, it is noted that the operators Π_1 and Π_n depend on n.

15.2 The Scalar Case

Now assume that Υ_n is a strictly positive Toeplitz matrix on $\mathbb{C}^n$. In this case, the entries of Υ_n are all complex numbers. In other words, Υ_n is a matrix of the form

$$\Upsilon_n = \begin{bmatrix} r_0 & \overline{r}_1 & \cdots & \overline{r}_{n-1} \\ r_1 & r_0 & \cdots & \overline{r}_{n-2} \\ \vdots & \vdots & \ddots & \vdots \\ r_{n-1} & r_{n-2} & \cdots & r_0 \end{bmatrix} \quad \text{on } \mathbb{C}^n. \tag{15.2.1}$$

In the scalar setting, the Levinson system of equations is defined by

$$\Upsilon_n \begin{bmatrix} 1 \\ a_{n,1} \\ a_{n,2} \\ \vdots \\ a_{n,n-1} \end{bmatrix} = \begin{bmatrix} \varepsilon_n \\ 0 \\ 0 \\ \vdots \\ 0 \end{bmatrix}. \tag{15.2.2}$$

Here $a_{n,j}$ and ε_n are all scalars. (We set $a_{n,j} = A_{n,j}$ and $\varepsilon_n = \Delta_n$ and $\varphi_n = \Omega_n$.) As expected, $a_{n,0} = 1$. In this case, the Levinson system in (15.2.2) is recursively computed by

$$\begin{bmatrix} a_{n+1,1} \\ \vdots \\ a_{n+1,n-1} \\ a_{n+1,n} \end{bmatrix} = \begin{bmatrix} a_{n,1} \\ \vdots \\ a_{n,n-1} \\ 0 \end{bmatrix} - \frac{\varphi_n}{\varepsilon_n} \begin{bmatrix} \overline{a}_{n,n-1} \\ \vdots \\ \overline{a}_{n,1} \\ 1 \end{bmatrix},$$

$$\varepsilon_{n+1} = \varepsilon_n - \frac{|\varphi_n|^2}{\varepsilon_n},$$

$$\varphi_n = r_n + \sum_{j=1}^{n-1} r_{n-j} a_{n,j}. \tag{15.2.3}$$

The initial conditions for the Levinson algorithm are given by

$$\varepsilon_2 = r_0 - \frac{|r_1|^2}{r_0} \quad \text{and} \quad a_{2,1} = -\frac{r_1}{r_0}. \tag{15.2.4}$$

The Toeplitz matrix Υ_n on $\mathbb{C}^n$ is strictly positive if and only if $\varepsilon_j > 0$ for $j = 1, 2, \ldots, n$ where $\varepsilon_1 = r_0$.

To prove this recursion it is sufficient to show that $\Lambda_n = \varepsilon_n$ and

$$\begin{bmatrix} B_{n,0} & B_{n,1} & \cdots & B_{n,n-2} & 1 \end{bmatrix}^{tr} = \begin{bmatrix} \overline{a}_{n,n-1} & \overline{a}_{n,n-2} & \cdots & \overline{a}_{n,1} & 1 \end{bmatrix}^{tr} \tag{15.2.5}$$

solves the second Levinson equation in (15.1.2), or equivalently, (15.1.9). To this end, let J be the unitary operator on $\mathbb{C}^n$ with 1's on the off diagonal and zeros elsewhere, that is,

$$J = \begin{bmatrix} 0 & 0 & \cdots & 0 & 1 \\ 0 & 0 & \cdots & 1 & 0 \\ \vdots & \vdots & \vdots & \vdots & \vdots \\ 0 & 1 & \cdots & 0 & 0 \\ 1 & 0 & \cdots & 0 & 0 \end{bmatrix} \quad \text{on } \mathbb{C}^n. \tag{15.2.6}$$

A simple calculation shows that $J\Upsilon_n J = \overline{\Upsilon}_n = \Upsilon_n^{tr}$. As expected, $\overline{\Upsilon}_n$ is the matrix formed by taking the complex conjugate of the entries of Υ_n. Using this with (15.2.2) and $J^2 = I$, we arrive at

$$\begin{bmatrix} 0 \\ 0 \\ \vdots \\ 0 \\ \varepsilon_n \end{bmatrix} = J \begin{bmatrix} \varepsilon_n \\ 0 \\ \vdots \\ 0 \\ 0 \end{bmatrix} = J\Upsilon_n JJ \begin{bmatrix} 1 \\ a_{n,1} \\ \vdots \\ a_{n,n-2} \\ a_{n,n-1} \end{bmatrix} = \Upsilon_n^{tr} \begin{bmatrix} a_{n,n-1} \\ a_{n,n-2} \\ \vdots \\ a_{n,1} \\ 1 \end{bmatrix}.$$

By taking the complex conjugate, we obtain the second equation in the Levinson system (15.1.2), that is,

$$\Upsilon_n \begin{bmatrix} a_{n,n-1} & a_{n,n-2} & \cdots & a_{n,1} & 1 \end{bmatrix}^* = \begin{bmatrix} 0 & 0 & \cdots & 0 & \varepsilon_n \end{bmatrix}^*.$$

(The $\{B_{n,j}\}$ are defined in (15.2.5) and $\Lambda_n = \varepsilon_n$.) This with the Levinson recursion in (15.1.7) yields the Levinson recursion for the scalar case in (15.2.3).

The following is a scalar version of Remark 15.2.1.

Remark 15.2.1. Let Υ_n on $\mathbb{C}^n$ be the Toeplitz matrix in (15.2.1). Then Υ_n is strictly positive if and only if ε_j is strictly positive for $j = 1, 2, \ldots, n$ where $\varepsilon_1 = r_0$. So one can use the Levinson algorithm in (15.2.3) to determine whether or not Υ_n is strictly positive. In this case, let L_n be the lower triangular matrix defined by

$$L_n = \begin{bmatrix} \widehat{a}_{n,0} & 0 & 0 & \cdots & 0 & 0 \\ \widehat{a}_{n,1} & \widehat{a}_{n-1,0} & 0 & \cdots & 0 & 0 \\ \widehat{a}_{n,2} & \widehat{a}_{n-1,1} & \widehat{a}_{n-2,0} & \cdots & 0 & 0 \\ \vdots & \vdots & \vdots & \ddots & \vdots & 0 \\ \widehat{a}_{n,n-2} & \widehat{a}_{n-1,n-3} & \widehat{a}_{n-2,n-4} & \cdots & \widehat{a}_{2,0} & 0 \\ \widehat{a}_{n,n-1} & \widehat{a}_{n-1,n-2} & \widehat{a}_{n-2,n-3} & \cdots & \widehat{a}_{2,1} & \widehat{a}_{1,0} \end{bmatrix}. \tag{15.2.7}$$

Here $\widehat{a}_{n,j} = a_{n,j}/\sqrt{\varepsilon_n}$ for all $j = 0, 1, 2, \ldots, n-1$ while $\widehat{a}_{1,0} = r_0^{-1/2}$ and $\widehat{a}_{k,0} = 1/\sqrt{\varepsilon_k}$. Finally, L_n is a lower triangular factorization for Υ_n^{-1}, that is, $\Upsilon_n^{-1} = L_n L_n^*$.

Now consider the upper triangular matrix determined by

$$U_n = \begin{bmatrix} \widehat{b}_{n,0} & \widehat{b}_{n-1,0} & \cdots & \widehat{b}_{3,0} & \widehat{b}_{2,0} & \widehat{b}_{1,0} \\ \widehat{b}_{n,1} & \widehat{b}_{n-1,1} & \cdots & \widehat{b}_{3,1} & \widehat{b}_{2,1} & 0 \\ \widehat{b}_{n,2} & \widehat{b}_{n-1,2} & \cdots & \widehat{b}_{3,2} & 0 & 0 \\ \vdots & \vdots & \vdots & \vdots & \vdots & \vdots \\ \widehat{b}_{n,n-2} & \widehat{b}_{n-1,n-2} & \cdots & \cdots & 0 & 0 \\ \widehat{b}_{n,n-1} & 0 & \cdots & \cdots & 0 & 0 \end{bmatrix}. \tag{15.2.8}$$

The entries of U_n are formed by taking the complex conjugate and reversing the order of the $\{\widehat{a}_{k,j}\}$, that is,

$$\begin{bmatrix}\widehat{b}_{k,0} & \widehat{b}_{k,1} & \cdots & \widehat{b}_{k,k-2} & \widehat{b}_{k,k-1}\end{bmatrix}^{tr} = \frac{1}{\sqrt{\varepsilon_k}}\begin{bmatrix}\overline{a}_{k,k-1} & \overline{a}_{k,k-2} & \cdots & \overline{a}_{k,1} & 1\end{bmatrix}^{tr}. \tag{15.2.9}$$

Finally, U_n is an upper triangular factorization for Υ_n^{-1}, that is, $\Upsilon_n^{-1} = U_n U_n^*$.

15.3 The Gohberg-Semencul-Heinig Inversion Formula

In this section we will present the Gohberg-Semencul-Heinig formula for inverting a strictly positive Toeplitz matrix. To this end, let Υ_n be the strictly positive Toeplitz matrix on $\mathcal{E}^n$ presented in (15.1.1). Let $\{A_{n,j}\}_0^{n-1}$ and $\{B_{n,j}\}_0^{n-1}$ be the operators on $\mathcal{E}$ in (15.1.2) where $A_{n,0} = I$ and $B_{n,n-1} = I$, or equivalently, the solution to the Levinson system in (15.1.7). Let $\{\widehat{A}_{n,j}\}_0^{n-1}$ and $\{\widehat{B}_{n,j}\}_0^{n-1}$ be the normalized Levinson operators on $\mathcal{E}$ defined by

$$\widehat{A}_n = \Upsilon_n^{-1}\Pi_1^*\Delta_n^{1/2} = \begin{bmatrix} I \\ A_{n,1} \\ A_{n,2} \\ \vdots \\ A_{n,n-1}\end{bmatrix}\Delta_n^{-1/2} = \begin{bmatrix} \widehat{A}_{n,0} \\ \widehat{A}_{n,1} \\ \widehat{A}_{n,2} \\ \vdots \\ \widehat{A}_{n,n-1}\end{bmatrix},$$

$$\widehat{B}_n = \Upsilon_n^{-1}\Pi_n^*\Lambda_n^{1/2} = \begin{bmatrix} B_{n,0} \\ B_{n,1} \\ \vdots \\ B_{n,n-2} \\ I\end{bmatrix}\Lambda_n^{-1/2} = \begin{bmatrix} \widehat{B}_{n,0} \\ \widehat{B}_{n,1} \\ \vdots \\ \widehat{B}_{n,n-2} \\ \widehat{B}_{n,n-1}\end{bmatrix}, \tag{15.3.1}$$

$$\Delta_n = (\Pi_1\Upsilon_n^{-1}\Pi_1^*)^{-1} \quad \text{and} \quad \Lambda_n = (\Pi_n\Upsilon_n^{-1}\Pi_n^*)^{-1}.$$

Notice that $\widehat{A}_n$ and $\widehat{B}_n$ are operators mapping $\mathcal{E}$ into $\mathcal{E}^n$, while Δ_n and Λ_n are operators on $\mathcal{E}$. One can compute $\widehat{A}_n$ and $\widehat{B}_n$ directly without using the Levinson algorithm. However, the Levinson algorithm is numerically more efficient. As before, Π_1 is the operator which picks out the first component of $\mathcal{E}^n$, and Π_n is the operator which picks out the last component of $\mathcal{E}^n$; see (15.1.18). Consider the lower triangular block Toeplitz matrices on $\mathcal{E}^n$ determined by

$$L_a = \begin{bmatrix} \widehat{A}_{n,0} & 0 & 0 & \cdots & 0 \\ \widehat{A}_{n,1} & \widehat{A}_{n,0} & 0 & \cdots & 0 \\ \widehat{A}_{n,2} & \widehat{A}_{n,1} & \widehat{A}_{n,0} & \cdots & 0 \\ \vdots & \vdots & \vdots & \ddots & \vdots \\ \widehat{A}_{n,n-1} & \widehat{A}_{n,n-2} & \widehat{A}_{n,n-3} & \cdots & \widehat{A}_{n,0}\end{bmatrix},$$

$$L_b = \begin{bmatrix} 0 & 0 & \cdots & 0 & 0 \\ \widehat{B}_{n,0} & 0 & \cdots & 0 & 0 \\ \widehat{B}_{n,1} & \widehat{B}_{n,0} & \cdots & 0 & 0 \\ \vdots & \vdots & \ddots & \vdots & \vdots \\ \widehat{B}_{n,n-2} & \widehat{B}_{n,n-3} & \cdots & \widehat{B}_{n,0} & 0 \end{bmatrix}.$$

Notice that $\widehat{A}_{n,0}$ appears on the main diagonal of L_a, while the main diagonal of L_b is zero and $\widehat{B}_{n,0}$ appears immediately below the main diagonal. Consider the upper triangular block Toeplitz on $\mathcal{E}^n$ matrices defined by

$$U_b = \begin{bmatrix} \widehat{B}_{n,n-1} & \widehat{B}_{n,n-2} & \widehat{B}_{n,n-3} & \cdots & \widehat{B}_{n,0} \\ 0 & \widehat{B}_{n,n-1} & \widehat{B}_{n,n-2} & \cdots & \widehat{B}_{n,1} \\ 0 & 0 & \widehat{B}_{n,n-1} & \cdots & \widehat{B}_{n,2} \\ \vdots & \vdots & \vdots & \ddots & \vdots \\ 0 & 0 & 0 & \cdots & \widehat{B}_{n,n-1} \end{bmatrix},$$

$$U_a = \begin{bmatrix} 0 & \widehat{A}_{n,n-1} & \widehat{A}_{n,n-2} & \cdots & \widehat{A}_{n,1} \\ 0 & 0 & \widehat{A}_{n,n-1} & \cdots & \widehat{A}_{n,2} \\ \vdots & \vdots & \vdots & \ddots & \vdots \\ 0 & 0 & 0 & \cdots & \widehat{A}_{n,n-1} \\ 0 & 0 & 0 & \cdots & 0 \end{bmatrix}.$$

Here the entries $\{\widehat{B}_{n,j}\}$ of U_b and $\{\widehat{A}_{n,j}\}$ of U_a appear in reverse order. Moreover, $\widehat{B}_{n,n-1}$ is on the main diagonal of U_b, while the main diagonal of U_a is zero and $\widehat{A}_{n,n-1}$ appears immediately above the main diagonal. It is emphasized that L_a, L_b, U_a and U_b all depend upon n. The inverse of Υ_n is given by

$$\Upsilon_n^{-1} = L_a L_a^* - L_b L_b^* = U_b U_b^* - U_a U_a^*. \tag{15.3.2}$$

In particular, if Υ_n is a strictly positive Toeplitz matrix on $\mathbb{C}^n$, then $\Delta_n = \Lambda_n$ and $\widehat{B}_{n,j} = \widehat{A}_{n,n-j-1}$ for $j = 0, 1, 2, \ldots, n-1$. In this case, $U_b = L_a^*$ and $U_a = L_b^*$; see (15.2.9). The inverse of Υ_n in (15.3.2) is referred to as the Gohberg-Semencul-Heinig inversion formula. Finally, it is noted that the Levinson with the Gohberg-Semencul-Heinig inversion formula provides a numerically efficient method to compute the inverse of Toeplitz matrices.

To prove the Gohberg-Semencul-Heinig inversion formula, let $\Pi_{\mathcal{L}}$ be the operator which picks out the last $n-1$ components of $\mathcal{E}^n$, that is,

$$\Pi_{\mathcal{L}} \quad = \quad \begin{bmatrix} 0 & I \end{bmatrix} : \begin{bmatrix} \mathcal{E} \\ \mathcal{E}^{n-1} \end{bmatrix} \to \mathcal{E}^{n-1}.$$

Notice that Υ_n admits a Schur type factorization on $\mathcal{E} \oplus \mathcal{E}^{n-1}$ of the form

$$\begin{aligned}\Upsilon_n &= \begin{bmatrix} R_0 & X_{n-1}^* \\ X_n & \Upsilon_{n-1} \end{bmatrix} \\ &= \begin{bmatrix} I & X_{n-1}^* \\ 0 & \Upsilon_{n-1} \end{bmatrix} \begin{bmatrix} \Delta_n & 0 \\ 0 & \Upsilon_{n-1}^{-1} \end{bmatrix} \begin{bmatrix} I & 0 \\ X_{n-1} & \Upsilon_{n-1} \end{bmatrix} \\ &= \begin{bmatrix} I \\ 0 \end{bmatrix} \Delta_n \begin{bmatrix} I & 0 \end{bmatrix} + \begin{bmatrix} X_{n-1}^* \\ \Upsilon_{n-1} \end{bmatrix} \Upsilon_{n-1}^{-1} \begin{bmatrix} X_{n-1} & \Upsilon_{n-1} \end{bmatrix} \\ &= \Pi_1^* \Delta_n \Pi_1 + \Upsilon_n \Pi_{\mathcal{L}}^* \Upsilon_{n-1}^{-1} \Pi_{\mathcal{L}} \Upsilon_n. \end{aligned} \tag{15.3.3}$$

Recall that X_{n-1} is defined in (15.1.3), and Δ_n is the Schur complement for Υ_n with respect to R_0, that is,

$$\Delta_n = R_0 - X_{n-1}^* \Upsilon_{n-1}^{-1} X_{n-1}.$$

It is noted that $\Delta_n^{-1} = \Pi_1 \Upsilon_n^{-1} \Pi_1^*$. Multiplying the last equation in (15.3.3) by Υ_n^{-1} on the right and left with (15.3.1), we arrive at

$$\begin{aligned}\Upsilon_n^{-1} &= \Upsilon_n^{-1} \Pi_1^* \Delta_n \Pi_1 \Upsilon_n^{-1} + \Pi_{\mathcal{L}}^* \Upsilon_{n-1}^{-1} \Pi_{\mathcal{L}} \\ &= \widehat{A}_n \widehat{A}_n^* + \Pi_{\mathcal{L}}^* \Upsilon_{n-1}^{-1} \Pi_{\mathcal{L}}. \end{aligned} \tag{15.3.4}$$

Now let $\Pi_{\mathcal{F}}$ be the operator which picks out the first $n-1$ components of $\mathcal{E}^n$, that is,

$$\Pi_{\mathcal{F}} = \begin{bmatrix} I & 0 \end{bmatrix} : \begin{bmatrix} \mathcal{E}^{n-1} \\ \mathcal{E} \end{bmatrix} \to \mathcal{E}^{n-1}.$$

The operator Υ_n admits another Schur type factorization on $\mathcal{E}^{n-1} \oplus \mathcal{E}$ of the form

$$\begin{aligned}\Upsilon_n &= \begin{bmatrix} \Upsilon_{n-1} & Y_{n-1}^* \\ Y_{n-1} & R_0 \end{bmatrix} \\ &= \begin{bmatrix} \Upsilon_{n-1} & 0 \\ Y_{n-1} & I \end{bmatrix} \begin{bmatrix} \Upsilon_{n-1}^{-1} & 0 \\ 0 & \Lambda_n \end{bmatrix} \begin{bmatrix} \Upsilon_{n-1} & Y_{n-1}^* \\ 0 & I \end{bmatrix} \\ &= \Pi_n^* \Lambda_n \Pi_n + \Upsilon_n \Pi_{\mathcal{F}}^* \Upsilon_{n-1}^{-1} \Pi_{\mathcal{F}} \Upsilon_n. \end{aligned} \tag{15.3.5}$$

Recall that Y_{n-1} is defined in (15.1.3), and Λ_n is the other Schur complement for Υ_n, that is,

$$\Lambda_n = R_0 - Y_{n-1} \Upsilon_{n-1}^{-1} Y_{n-1}^*. \tag{15.3.6}$$

Furthermore, $\Lambda_n^{-1} = \Pi_n \Upsilon_n^{-1} \Pi_n^*$. Multiplying the last equation in (15.3.5) by Υ_n^{-1} on the right and left with (15.3.1), we obtain

$$\begin{aligned}\Upsilon_n^{-1} &= \Upsilon_n^{-1} \Pi_n^* \Lambda_n \Pi_n \Upsilon_n^{-1} + \Pi_{\mathcal{F}}^* \Upsilon_{n-1}^{-1} \Pi_{\mathcal{F}} \\ &= \widehat{B}_n \widehat{B}_n^* + \Pi_{\mathcal{F}}^* \Upsilon_{n-1}^{-1} \Pi_{\mathcal{F}}. \end{aligned} \tag{15.3.7}$$

By combining (15.3.4) and (15.3.7), we arrive at the following two formulas for the inverse of Υ_n:

$$\begin{aligned}\Upsilon_n^{-1} &= \widehat{A}_n \widehat{A}_n^* + \Pi_{\mathcal{L}}^* \Upsilon_{n-1}^{-1} \Pi_{\mathcal{L}} \\ &= \widehat{B}_n \widehat{B}_n^* + \Pi_{\mathcal{F}}^* \Upsilon_{n-1}^{-1} \Pi_{\mathcal{F}}.\end{aligned} \tag{15.3.8}$$

Let Z_n be the lower shift on $\mathcal{E}^n$, that is,

$$Z_n = \begin{bmatrix} 0 & 0 & \cdots & 0 & 0 \\ I & 0 & \cdots & 0 & 0 \\ \vdots & \vdots & \ddots & \vdots & \vdots \\ 0 & 0 & \cdots & 0 & 0 \\ 0 & 0 & \cdots & I & 0 \end{bmatrix} \quad \text{on } \mathcal{E}^n.$$

The identity appears immediately below the main diagonal and zeros elsewhere. Now observe that

$$\Pi_{\mathcal{L}} Z_n = \Pi_{\mathcal{F}} \quad \text{and} \quad \Pi_{\mathcal{F}} Z_n^* = \Pi_{\mathcal{L}}.$$

In particular,

$$\begin{aligned}\Pi_{\mathcal{F}}^* \Upsilon_{n-1}^{-1} \Pi_{\mathcal{F}} &= Z_n^* \Pi_{\mathcal{L}}^* \Upsilon_{n-1}^{-1} \Pi_{\mathcal{L}} Z_n, \\ \Pi_{\mathcal{L}}^* \Upsilon_{n-1}^{-1} \Pi_{\mathcal{L}} &= Z_n \Pi_{\mathcal{F}}^* \Upsilon_{n-1}^{-1} \Pi_{\mathcal{F}} Z_n^*.\end{aligned}$$

Using this with (15.3.8), we obtain

$$\begin{aligned}\Upsilon_n^{-1} &= \widehat{A}_n \widehat{A}_n^* + \Pi_{\mathcal{L}}^* \Upsilon_{n-1}^{-1} \Pi_{\mathcal{L}} \\ &= \widehat{A}_n \widehat{A}_n^* + Z_n \Pi_{\mathcal{F}}^* \Upsilon_{n-1}^{-1} \Pi_{\mathcal{F}} Z_n^* \\ &= \widehat{A}_n \widehat{A}_n^* + Z_n \left(\Upsilon_n^{-1} - \widehat{B}_n \widehat{B}_n^* \right) Z_n^*.\end{aligned}$$

The yields the following Lyapunov equation for the inverse of Υ_n:

$$\begin{aligned}\Upsilon_n^{-1} &= Z_n \Upsilon_n^{-1} Z_n^* + \widehat{A}_n \widehat{A}_n^* - Z_n \widehat{B}_n \widehat{B}_n^* Z_n^*, \\ \Upsilon_n^{-1} &= Z_n^* \Upsilon_n^{-1} Z_n + \widehat{B}_n \widehat{B}_n^* - Z_n^* \widehat{A}_n \widehat{A}_n^* Z_n.\end{aligned} \tag{15.3.9}$$

The second equation follows by a similar calculation involving (15.3.8).

Because $Z_n^n = 0$, the solution to the first Lyapunov equation in (15.3.9) is given by

$$\begin{aligned}
\Upsilon_n^{-1} &= \sum_{j=0}^{n-1} Z_n^j \left(\widehat{A}_n \widehat{A}_n^* - Z_n \widehat{B}_n \widehat{B}_n^* Z_n^*\right) Z_n^{*j} \\
&= \begin{bmatrix} \widehat{A}_n & Z_n\widehat{A}_n & \cdots & Z_n^{n-1}\widehat{A}_n \end{bmatrix} \begin{bmatrix} \widehat{A}_n^* \\ \widehat{A}_n^* Z_n^* \\ \vdots \\ \widehat{A}_n^* Z_n^{*n-1} \end{bmatrix} \\
&\quad - Z_n \begin{bmatrix} \widehat{B}_n & Z_n\widehat{B}_n & \cdots & Z_n^{n-1}\widehat{B}_n \end{bmatrix} \begin{bmatrix} \widehat{B}_n^* \\ \widehat{B}_n^* Z_n^* \\ \vdots \\ \widehat{B}_n^* Z_n^{*n-1} \end{bmatrix} Z_n^* \\
&= L_a L_a^* - L_b L_b^*.
\end{aligned}$$

This yields the first Gohberg-Semencul-Heinig inversion formula in (15.3.2).

To obtain the second Gohberg-Semencul-Heinig inversion formula, observe that the solution to the second Lyapunov equation in (15.3.9) is given by

$$\begin{aligned}
\Upsilon_n^{-1} &= \sum_{j=0}^{n-1} Z_n^{*j} \left(\widehat{B}_n \widehat{B}_n^* - Z_n^* \widehat{A}_n \widehat{A}_n^* Z_n\right) Z_n^j \\
&= \begin{bmatrix} Z_n^{*n-1}\widehat{B}_n & Z_n^{*n-2}\widehat{B}_n & \cdots & \widehat{B}_n \end{bmatrix} \begin{bmatrix} \widehat{B}_n^* Z_n^{n-1} \\ \widehat{B}_n^* Z_n^{n-2} \\ \vdots \\ \widehat{B}_n^* \end{bmatrix} \\
&\quad - Z_n^* \begin{bmatrix} Z_n^{*n-1}\widehat{A}_n & Z_n^{*n-2}\widehat{A}_n & \cdots & \widehat{A}_n \end{bmatrix} \begin{bmatrix} \widehat{A}_n^* Z_n^{n-1} \\ \widehat{A}_n^* Z_n^{n-2} \\ \vdots \\ \widehat{A}_n^* \end{bmatrix} Z_n \\
&= U_b U_b^* - U_a U_a^*.
\end{aligned}$$

This yields the second Gohberg-Semencul-Heinig inversion formula.

15.4 Notes

The Levinson algorithm is classical and due to Levinson [156]. For further results on the Levinson algorithm and historical comments; see Caines [47], Kailath [138] and Kailath-Sayed-Hassibi [143]. For some applications of the Levinson algorithm to signal processing see Kailath [141], Marple [167] and Stoica and R. Moses [195]. The Levinson algorithm also plays a basic role in geophysics; see Claerbout [57], Foias-Frazho [82] and Robinson-Treitel [181]. The Gohberg-Semencul-Heinig inversion formula is now a classical and widely used result. The seminal papers are Gohberg-Semencul [120] and Gohberg-Heinig [118, 119]. This inversion formula also works for Toeplitz matrices which are not positive. In this case, one needs four different operators mapping $\mathcal{E}$ into $\mathcal{E}^n$ to compute the inverse. Our approach was taken from Gohberg-Kaashoek-van Schagen [121]; see also Frazho-Kaashoek [99]. For some applications of the Gohberg-Semencul-Heinig inversion formula see Kailath-Kung-Morf [142] and Constantinescu-Sayed-Kailath [58].

Bibliography

[1] V.A. Adamjan, D.Z. Arov and M.G. Krein, Analytic properties of Schmidt pairs for a Hankel operator and the generalized Schur-Takagi problem, *Math USSR Sbornick*, **15** (1971), 31–73.

[2] V.A. Adamjan, D.Z. Arov and M.G. Krein, Infinite Hankel block matrices and related extension problems, *Izv. Akad. Nauk. Armjan SSR, Matematika*, **6** (1971), 87–112, English Translation: *Amer. Math. Soc. Transl.*, **III** (1978), 133–156.

[3] J. Agler and J.E. McCarthy, Nevanlinna-Pick interpolation on the bidisk, *J. reine angew. Math.* **506** (1999), 191–204.

[4] J. Agler and J.E. McCarthy, *Pick interpolation and Hilbert function spaces,* American Mathematical Society, 2002.

[5] M.R. Allen and L.A. Smith, Optimal filtering in singular spectrum analysis, *Physics Letters* **234** (6), 419–428.

[6] D. Alpay, J.A. Ball and Y. Peretz, System theory and scattering: The time varying case, *J. Operator Theory* **47** (2002), 245–286.

[7] D. Alpay, V. Bolotnikov, and Y. Peretz, On the tangential interpolation problem for H_2 functions. *Trans. Amer. Math. Soc.* **347** (1995), 675–686.

[8] B.D.O. Anderson, A system theory criterion for positive real matrices, *SIAM J. Control,* **5** (2) (1967), 171–182.

[9] B.D.O. Anderson, An algebraic solution to the spectral factorization problem, *IEEE Transactions on Control,* **12** (4) (1967), 410–414.

[10] B.D.O. Anderson and J.B. Moore, *Linear Optimal Control*, Prentice Hall, Englewood Cliffs, New Jersey, 1971.

[11] B.D.O. Anderson and J.B. Moore, *Optimal Filtering*, Prentice Hall, New Jersey, 1979.

[12] A. Antoniou, *Digital Filters: Analysis, Design, and Applications,* McGraw-Hill, New York, 1993.

[13] R. Arocena, Generalized Toeplitz kernels and dilations of intertwining operators, *Integral Equations and Operator Theory,* **6** (1983), 759–778.

[14] R. Arocena, Unitary extensions of isometries and contractive intertwining dilations, in: *The Gohberg Anniversary Collection II,* Operator Theory: Advances and Applications, Vol. **41**, Birkäuser Verlag, Basel, 1989, pp. 13–23.

[15] R. Arocena, Unitary colligations and parametrization formulas, *Ukrainskii Mat. Zhurnal,* **46** (3) (1994), 147–154.

[16] N. Aronszajn, Theory of Reproducing Kernels, *Transactions of the American Mathematical Society,* **68** (3) (1950), 337–404.

[17] D. Arov, On the Darlington method in the theory of dissipative systems, *Dokl. Akad. Nauk USSR,* **201** (1971), 559–562.

[18] D. Arov, Darlingtons method for dissipative systems, *Soviet Physics Doklady,* **16** (1972), 954–956.

[19] D.Z. Arov, On unitary coupling with losses (scattering theory with losses), *Functional Analysis Application,* **8** (4) (1974), 280–294.

[20] D.Z. Arov, Passive linear steady-state dynamical systems, *Sibirsk. Mat. Zh.,* **20** (2) (1979), 211–228 (Russian).

[21] Gr. Arsene, Z. Ceausescu, and C. Foias, On intertwining dilations VIII, *J. Operator Theory* **4** (1980), 55–91.

[22] M. Bakonyi and T. Constantinescu, *Schur's algorithm and several applications,* Putman Series **261**, Longman, Harlow, 1992.

[23] J.A. Ball, Linear systems, operator model theory and scattering: multivariable generalizations, in: *Operator Theory and its Applications,* A.G. Ramm et al (eds.), Fields Inst. Commun. Vol. **25**, Amer. Math. Soc., 2000, pp. 151–178.

[24] J.A. Ball, I. Gohberg and L. Rodman, *Interpolation for Rational Matrix Functions,* Operator Theory: Advances and Applications, Vol. **45**, Birkhäuser Verlag, Basel, 1990.

[25] J.A. Ball and T.T. Trent, Unitary colligations, reproducing kernel Hilbert spaces, and Nevanlinna-Pick interpolation in several variables, *J. Functional Analysis* **157** (1998), 1–61.

[26] J.A. Ball, T.T. Trent and V. Vinnikov, Interpolation and commutant lifting for multipliers on reproducing kernel Hilbert spaces, in: *Operator Theory and Analysis*, H. Bart et al (eds.), Operator Theory: Advances and Applications, Vol.**122**, Birkhäuser Verlag, Basel, 2002, pp. 89–138.

[27] H. Bart, I. Gohberg and M. A. Kaashoek, *Minimal Factorization of Matrix and Operator Functions,* Operator Theory: Advances and Applications **1**, Birkhäuser-Verlag, Basel, 1979.

[28] H. Bart, I. Gohberg, M. A. Kaashoek and A.C.M. Ran, *A State Space Approach to Canonical Factorization with Applications*, Operator Theory: Advances and Applications, Vol. 200, Birkhäuser Verlag, Basel, 2010, to appear.

[29] V. Belevitch, *Classical Network Theory,* Holden Day, San Francisco, 1968.

[30] H. Bercovici, *Operator Theory and Arithmetic in* H^∞, American Mathematical Society, Providence Rhode Island, 1988.

[31] A. Beurling, On two problems concerning linear transformations in Hilbert space, *Acta Math.* **81** (1949), 239–255.

[32] W. Bhosri, *Geometric Approach to Prediction and Perodic Systems*, Ph.D. Thesis, School of Aeronautics and Astronautics, Purdue University, 2008.

[33] W. Bhosri, D. Du and A.E. Frazho, Mixed norm low order multirate filterbank design: relaxed commutant lifting approach, *Allerton Conference on Communication, Control, and Computing*, 2008.

[34] W. Bhosri, A.E. Frazho and B. Yagci, Maximal Nevanlinna-Pick expansion for points in the open unit disc, in: *Recent Advances in Operator Theory and its Applications*, M.A. Kaashoek et al (eds.), Operator Theory: Advances and Applications, Vol. **160**, Birkhäuser Verlag, Basel, 2005, pp. 41–52.

[35] A. Biswas, C. Foias and A.E. Frazho, Weighted Commutant Lifting, *Acta Sci. Math. (Szeged)* **65** (1999), 657–686.

[36] A. Böttcher and B. Silbermann, *Introduction to Large Truncated Toeplitz Matrices*, Springer Verlag, New York, 1998.

[37] S. Boyd, L. El Ghaoui, E. Feron, and V. Balakrishnan, *Linear Matrix Inequalities in System and Control Theory*, SIAM, Philadelphia, 1994.

[38] L. de Branges, Factorization and invariant subspaces, *J. Math. Anal. Appl.* **29** (1970), 163–200.

[39] L. de Branges, Unitary linear systems whose transfer functions are Riemann mapping functions, in: *Operator Theory and Systems*, H. Bart et al (eds.), Operator Theory: Advances and Applications, Vol. **19**, Birkhäuser Verlag, Basel, 1986, pp. 105–124.

[40] L. de Branges and J. Rovnyak, Canonical models in quantum scattering theory, in; *Perturbation Theory and its Applications in Quantum Mechanics*, C.H. Wilcox (ed.), Wiley, New York, 1966, pp. 295–392.

[41] L. de Branges and J. Rovnyak, *Square Summable Power Series*, Holt Rinehart and Winston, New York, 1966.

[42] R.W. Brockett, *Finite Dimensional Linear Systems*, John Wiley and Sons, New York, 1970.

[43] M.S. Brodskii, *Triangular and Jordan Representations of Linear Operators,* Transl. Math. Monographs, Vol. **32**, American Mathematical Society, Providence, R.I., 1970.

[44] M.S. Brodskii, Unitary operator colligations and their characteristic functions, *Uspekhi Math. Nauk,* **33** (4) (1978), 141–178 (Russian); English Transl., *Russian Math. Surveys* **33** (4) (1987), 159–191.

[45] A. Brown and P.R. Halmos, Algebraic properties of Toeplitz operators, *J. Reine Angew. Math.* **231** (1963), 89–102.

[46] C.I. Byrnes, T.T. Georgiou and A. Lindquist, A generalized entropy criterion for Nevanlinna-Pick interpolation with degree constraints, *IEEE Transactions on Automatic Control* **46** (2001), 822–839.

[47] P.E. Caines, *Linear Stochastic Systems,* Wiley, New York, 1988.

[48] F.M. Callier and C.A. Desoer, *Multivariable Feedback Systems,* Springer Verlag, New York, 1982.

[49] J. Capon, High-resolution frequency-wave number spectrum analysis, *Proceedings of the IEEE* **57** (1969), 1408–1418.

[50] C. Carathéodory, Über den Variabilitätsbereich der Koeffizienten von Potenzreihen, die gegebene Werte nicht annehmen, *Math. Ann.* **64** (1907), 95–115.

[51] C. Carathéodory, Über den Variabilitätsbereich der Fourierschen Konstanten von positiven harmonischen Funktionen, *Rend. Circ. Mat. Palermo* **32** (1911), 193–217.

[52] H.J. Carlin, Darlington synthesis revisted, *IEEE Transactions on Circuts and Systems-I:Fundamental Theory and Applications* **46** (1) (1999), 14–21.

[53] C.T. Chen, *Linear Systems Theory and Design,* Oxford University Press, New York, 1998.

[54] H. Chen and J. Maciejowski, An Improved Subspace Identification Method for Bilinear Systems, in: *Proceedings of the 39th IEEE conference on decision and control,* vol. 2, Sydney, Australia, December 2000, pp. 1573–1578.

[55] C.K. Chui and G. Chen, *Signal Processing and Systems Theory,* Springer Verlag, Berlin, 1992.

[56] J.A. Cima and W.T. Ross *The backward shift on the Hardy Space,* American Mathematical Society, Providence, R.I., 2000.

[57] J.F. Claerbout, *Fundamentals of Geophysical Data Processing,* McGraw-Hill, New York, 1976.

[58] T. Constantinescu, A.H. Sayed, and T. Kailath, Displacement structure and completion problems, *Siam J. Matrix Anal. Appl.* **16** (1995), 58–78.

[59] J.B. Conway, *A Course in Functional Analysis*, Springer, New York, 1990.

[60] M.J. Corless and A.E. Frazho, *Linear systems and control: An operator perspective*, Marcel Dekker, New York, 2003.

[61] M. Cotlar and C. Sadosky, On the Helson-Szegö theorem and a related class of modified Toeplitz kernels, *Proc. Symp. Pure Math. AMS* **35** (1) (1979), 383–407.

[62] F. Cucker and S. Smale, On the Mathematical Foundations of Learning, *Bulletin of the American Mathematical Society* **39** (1) (2001), 1–49.

[63] A.A.H. Damen, P.M.J. Van den Hof and A.K. Hajdasinski, Approximate realization based upon an alternative to the Hankel matrix: the Page matrix, *Systems & Control Letters* **2** (1982), 202–208.

[64] S. Darlington, Synthesis of reactance 4-poles which produce prescribed insertion loss characteristics, *J. Math. Phys.* **18** (4) (1939), 257–353.

[65] G. Daryanani, *Principles of Active Network Synthesis and Design,* John Wiley & Sons, New York, 1976.

[66] M.H.A. Davis, *Linear Estimation and Stochastic Control*, Chapman and Hall, New York, 1977.

[67] R.A. DeCarlo, *Linear Systems*, Prentice Hall, Englewood Cliffs, New Jersey, 1989.

[68] D.F. Delchamps, *State Space and Input-Output Linear Systems*, Springer Verlag, New York, 1988.

[69] U.B. Desai, Realization of stochastic bilinear systems, *IEEE Transactions on Automatic Control* **AC-31** (1986), 189–192.

[70] P. Dewilde, The Geometry of Darlington synthesis, *Int. J. Appl. Math. Comput. Sci.* **11** (6) (2001), 1379–1386.

[71] J.L. Doob, *Stochastic Processes,* John Wiley & Sons, New York, 1953.

[72] R.G. Douglas and J.W. Helton, Inner dilations of analytic matrix functions and Darlington synthesis, *Acta Sci. Math. (Szeged)* **34** (1973), 301–310.

[73] R.G. Douglas, H.S. Shapiro and A.L. Shields, Cyclic vectors and invariant subspaces for the backward shift, *Ann. Inst. Fourier Grenoble* **20** (1971), 37–76.

[74] J.C. Doyle, B.A. Francis and A.R. Tannenbaum, *Feedback control theory,* Macmillan, New York, 1992.

[75] J.C. Doyle, K. Glover, P.P. Khargonekar and B.A. Francis, State-space solutions to standard H_2 and H_∞ control problems, *IEEE Trans. Automatic Contrtrol* **34** (1988), 831–847.

[76] P. Duren, *Theory of H^p-Spaces*, Academic Press, New York, 1970.

[77] P.L. Faurre, Realisations Markoviennes de processus stationaires, *IRIA Rapport de Recherche* **13** (March 1973).

[78] P.L. Faurre, Stochastic realization algorithms, in: *System Identification: Advances and case studies*, R.K. Mehra and D.F. Lainiotis (eds.), Academic Press, New York, 1976.

[79] A. Feintuch, Linear Time-Varying Darlington Synthesis, in: *15th International Symposium on Mathematical Theory of Networks and Systems*, South Bend, Indiana, August 2002, web address:
`http://www.nd.edu/~mtns/papers/2447_2.pdf`

[80] P.A. Fillmore, *Notes on Operator Theory*, Van Nostrand Reinhold, New York, 1968.

[81] C. Foias and A.E. Frazho, Markov subspaces and minimal unitary dilations, *Acta Sci. Math. (Szeged)* **45** (1983), 156–175.

[82] C. Foias and A.E. Frazho, *The Commutant Lifting Approach to Interpolation Problems,* Operator Theory: Advances and Applications, Vol. **44**, Birkhäuser Verlag, Basel, 1990.

[83] C. Foias, A.E. Frazho and I. Gohberg, Central intertwining lifting, maximum entropy and their permanence, *Integral Equations and Operator Theory* **18** (1994), 166–201.

[84] C. Foias, A.E. Frazho, I. Gohberg and M. A. Kaashoek, *Metric Constrained Interpolation, Commutant Lifting and Systems,* Operator Theory: Advances and Applications, Vol. **100**, Birkhäuser Verlag, Basel, 1998.

[85] C. Foias, A.E. Frazho and M.A. Kaashoek, Relaxation of metric constrained interpolation and a new lifting theorem, *Integral Equations and Operator Theory* **42** (2002), 253–310.

[86] C. Foias, A.E. Frazho and M.A. Kaashoek, Contractive liftings and the commutator, *C. R. Acad. Sci. Paris, Ser. I* **335** (2002), 431–436.

[87] C. Foias, A.E. Frazho and M.A. Kaashoek, The distance to intertwining operators, contractive liftings and a related optimality result. *Integral Equations and Operator Theory* **47** (2003), 71–89.

[88] C. Foias, A.E. Frazho and P.J. Sherman, A geometric approach to the maximum likelihood spectral estimator for sinusoids in noise, *IEEE Transactions on Information Theory* **34** (1988), 1066–1070.

[89] C. Foias, A.E. Frazho and P.J. Sherman, A new approach for determining the spectral data of multichannel harmonic signals in noise, *Mathematics of Control, Signals, and Systems* **3** (1990), 31–43.

[90] C. Foias, H. Ozbay, and A.R. Tannenbaum, *Robust Control of Infinite Dimensional Systems*, Springer Verlag, London, 1996.

[91] B.A. Francis, *A Course in Control Theory*, Lecture Notes in Control and Information Science, Springer, New York, 1987.

[92] A.E. Frazho, A shift operator approach to bilinear systems theory, *SIAM J. Control* 18 (1980), 640–658.

[93] A.E. Frazho, Models for noncommutating operators, *J. Functional Analysis* **48** (1982), 1–11.

[94] A.E. Frazho, On stochastic realization theory, *Stochastics* **7** (1982), 1–27.

[95] A.E. Frazho, Complements to models for noncommutating operators, *J. Functional Analysis* **59** (1984), 445–461.

[96] A.E. Frazho, On stochastic bilinear realization theory, in: *Modeling and applications of stochastic processes,* U.B. Desai (ed.), Kluwer Academic, Boston, 1988, pp. 215–241.

[97] A.E. Frazho and M.A. Kaashoek, A Band method approach to a positive expansion problem in a unitary dilation setting, *Integral Equations and Operator Theory* **42** (2002), 311–371.

[98] A.E. Frazho and M.A. Kaashoek, A Naimark dilation perspective of Nevanlinna-Pick interpolation, *Integral Equations and Operator Theory* **49** (2004), 323–378.

[99] A.E. Frazho and M.A. Kaashoek, A contractive operator view on an inversion formula of Gohberg-Heinig, in: *Topics in Operator Theory. Vol. 1: Operators, Matrices and Analytic Functions*, J.A. Ball et al (eds.), Operator Theory: Advances and Applications, Vol. **201**, 2010, to appear.

[100] A.E. Frazho, M.A. Kaashoek and A.C.M. Ran, The non-symmetric discrete algebraic Riccati equation and canonical factorization of rational matrix functions on the unit circle, *Integral Equations and Operator theory,* to appear.

[101] A.E. Frazho and P.J. Sherman, On the minimum variance spectral estimator in non-stationary noise, *IEEE Transactions in Information Theory* **37** (5) (1991), 1957–1959.

[102] A.E. Frazho, S. ter Horst, and M.A. Kaashoek, Coupling and relaxed commutant lifting, *Integral Equations and Operator theory* **54** (2006), 33–67.

[103] A.E. Frazho, S. ter Horst, and M.A. Kaashoek, All solutions to the relaxed commutant lifting problem, *Acta Sci. Math. (Szeged)* **72** (2006), 299–318.

[104] A.E. Frazho, B. Yagci and H. Sumali, On sinusoid estimation in nonstationary noise, *IEEE Transaction on Automatic Control* **49** (2004), 777–781.

[105] P.A. Fuhrmann, *Linear Systems and Operators in Hilbert Space*, McGraw-Hill, New York, 1981.

[106] J. Garnett, *Bounded Analytic Functions*, Academic Press, New York, 1981.

[107] T.T. Georgiou, Spectral estimation via selective harmonic amplification, *IEEE Transactions on Automatic Control* **46** (2001), 29–42.

[108] T.T. Georgiou, Spectral analysis based on the state covariance: the maximum entropy spectrum and linear fractional parameterization, *IEEE Transactions on Automatic Control* **47** (2002), 1811–1823.

[109] T.T. Georgiou, The structure of state covariances and its relation to the power spectrum of the input, *IEEE Transactions on Automatic Control* **47** (2002), 1056–1066.

[110] L. Ya. Geronimus, *Orthogonal Polynomials*, Consultants Bureau, New York, 1961.

[111] K. Glover, All optimal Hankel-norm approximations of linear multivariable systems and their L_∞-error bounds, *International J. Control* **39** (1984), 1115–1193.

[112] I. Gohberg, S. Goldberg and M.A. Kaashoek, *Basic Classes of Linear Operators*, Birkhäuser Verlag, Basel, 2003.

[113] I. Gohberg, S. Goldberg and M.A. Kaashoek, *Classes of Linear Operators*, Vol. I, Operator Theory: Advances and Applications, Vol. **49**, Birkhäuser Verlag, Basel, 1990.

[114] I. Gohberg, S. Goldberg and M.A. Kaashoek, *Classes of Linear Operators*, Vol. II, Operator Theory: Advances and Applications, Vol. **63**, Birkhäuser Verlag, Basel, 1993.

[115] I. Gohberg, M.A. Kaashoek and H.J. Woerdeman, The band method for positive and strictly contractive extension problems: an alternative version and new applications, *Integral Equations and Operator Theory* **12** (1989), 343–382.

[116] I. Gohberg, M.A. Kaashoek, and H.J. Woerdeman, A maximum entropy principle in the general framework of the band method, *J. Functional Analysis* **95** (2) (1991), 231–254.

[117] I. Gohberg, M.A. Kaashoek and F. van Schagen, On inversion of finite Toeplitz matrices with elements in an algebraic ring, *Linear Algebra and its Applications* **385** (2004), 381–389.

[118] I.C. Gohberg and G. Heinig, On the inversion of finite Toeplitz matrices, *Math. Issled* **8** (3) (1973), 151–155 (in Russian).

[119] I.C. Gohberg and G. Heinig, The inversion of finite Toeplitz matrices consisting of elements of a non-commutative algebra, *Rev. Roum. Math. Pures et Appl.* **20** (1974), 623–663 (in Russian).

[120] I.C. Gohberg and A.A. Semencul, On the invertibility of finite Toeplitz matrices and their continuous analogues, *Math. Issled* **7** (2) (1972), 151–155 (in Russian).

[121] I. Gohberg, M.A. Kaashoek, and F. van Schagen, On inversion of finite Toeplitz matrices with elements in an algebraic ring, *Lin. Alg. Appl.* **385** (2004), 381–389.

[122] G.H. Golub and C.F. Van Loan, *Matrix Computations*, The Johns Hopkins University Press, Baltimore, 1996.

[123] M. Green and D.J.N. Limebeer, *Linear Robust Control* , Prentice-Hall, Inc., New Jersey, 1995.

[124] P.R. Halmos, Shifts on Hilbert space, *J. reine angew. Math.* **208** (1961), 102–112.

[125] P.R. Halmos, *Finite Dimensional Vector Spaces,* Springer Verlag, New York, 1970.

[126] P.R. Halmos, *A Hilbert Space Problem Book,* Springer Verlag, New York, 1982.

[127] M.H. Hayes, *Digital Signal Processing, Schaum's Outline Series,* McGraw-Hill, New York, 1999.

[128] D. Hazony, *Elements of Network Synthesis,* Reinhold, New York, 1963.

[129] H. Helson, *Lectures on Invariant Subspaces,* Academic Press, New York, 1964.

[130] H. Helson and D. Lowdenslager, Prediction theory and Fourier series in several variables, *Acta Math.* **99** (1958), 165–202.

[131] H. Helson and D. Lowdenslager, Prediction theory and Fourier series in several variables II, *Acta Math.* **106** (1961), 175–213.

[132] J.W. Helton, Discrete time systems, operator models and scattering theory, *J. Functional Analysis* **16** (1974), 15–38.

[133] L. Hitz and B.D.O. Anderson, Discrete time positive real functions and their applications to system stability, Proc. IEEE **116** (1) (1969), 153–155.

[134] K. Hoffman, *Banach Spaces of Analytic Functions,* Prentice Hall, New Jersey, 1962.

[135] Vlad Ionescu, C. Oară and M. Weiss, *Generalized Riccati Theory and Robust Control. A Popov Function Approach,* John Wiley, West Sussex, 1999.

[136] C.R. Johnson and L. Rodman, Completion of partial matrices to contractions, *J. Functional Analysis* **69** (1986), 260–267.

[137] M.A. Kaashoek and C.G. Zeinstra, The band method and generalized Carathéodory-Toeplitz interpolation at operator points, *Integral Equations and Operator Theory* **33** (1999), 175–210.

[138] T. Kailath, A view of three decades of linear filtering, *IEEE Transactions on Information Theory* **20** (2) (1974), 146–184.

[139] T. Kailath, *Lectures on Linear Least-Squares Estimation*, CISM Courses and Lectures, **140**, Springer Verlag, New York, 1978.

[140] T. Kailath, *Linear Systems*, Prentice Hall, New Jersey, 1980.

[141] T. Kailath, *Modern Signal Processing*, Taylor and Francis, London, 1986.

[142] T. Kailath, S.-Y. Kung, and M. Morf, Displacement ranks of matrices and linear equations, *J. Math. Anal. Appl.* **68** (1979), 395–407.

[143] T. Kailath, A.H. Sayed and B. Hassibi, *Linear Estimation*, Prentice Hall, New Jersey, 2000.

[144] R.E. Kalman, A new approach to linear filtering and prediction problems, *Trans. ASME J. Basic Engineering* **82** (1960), 34–45.

[145] R.E. Kalman, Contribution to the theory of optimal control, *Bol. Soc. Matem. Mex.* **5** (1960), 102–119.

[146] R.E. Kalman, Mathematical description of linear dynamical systems, *SIAM J. Control* **1** (1963), 152–192.

[147] R.E. Kalman, Irreducible realizations and the degree of a rational matrix, *SIAM J. Applied Math.* **13** (1965), 520–544.

[148] R.E. Kalman and R.S. Bucy, New results in linear filtering and prediction theory, *ASME Trans. Part D* **83** (1961), 95–108.

[149] R.E. Kalman, P.L. Falb and M.A. Arbib, *Topics in Mathematical System Theory*, McGraw-Hill, New York, 1969.

[150] R.E. Kalman, Realization of covariance sequences, *Proc. Toeplitz Memorial Conference*, Tel Aviv Israel, 1981.

[151] P. Koosis, *Introduction to H_p Spaces*, 2nd ed., Cambridge University Pres, Cambridge, England, 1998.

[152] H. Kwakernaak and R. Sivan, *Linear Optimal Control Systems*, Wiley Interscience, New York, 1972.

[153] P. Lancaster and L. Rodman *Algebraic Riccati Equations,* Oxford Unversity press, New York, 1995.

[154] P. Lax, Translation invariant subspaces, *Acta Math.* **101** (1959), 163–178.

[155] P. Lax, Translation invariant subspaces, in: *Proceedings International Symposium, Jerusalem* (1960), pp. 299–306 (New York, 1961).

[156] N. Levinson, The Wiener RMS (root mean square) error criterion in filter design and prediction, *J. Math. Physics* **25** (1947), 261–278.

[157] M. Lindner, *Infinite Matrices and their Finite Sections: An Introduction to the Limit Operator Method*, Birkäuser Verlag, 2006.

[158] A. Lindquist and G. Picci, On the structure of minimal splitting subspaces in stochastic realization theory, in: *Proceedings of the 1977 Conference on Decision and Control,* New Orleans (1977), pp. 42–48.

[159] A. Lindquist and G. Picci, A Hardy space approach to the stochastic realization problem, in: *Proceedings of the 1978 Conference on Decision and Cnotrol,* San Diego (1978), pp. 933–939.

[160] A. Lindquist and G. Picci, On the stochastic realization problem, *SIAM J. Control* **17** (3) (1979), 365–389.

[161] A. Lindquist and G. Picci, On minimal splitting subspaces and Markovian representations, *Math. Systems Theory* **12** (1979), 271–279.

[162] A. Lindquist and G. Picci, State space models for Gaussian stochastic processes, in: *Stochastic Systems: The Mathematics of Filtering and Identification and Applications,* M. Hazewinkel and J.C. Willems (eds.), Reidel, Dordrecht, 1981, pp. 169–204.

[163] M.S. Livšic, On a class of linear operators in a Hilbert space, *Mat. Sbornik* **19** (61) (1946), 239–262 (Russian); English Transl., *Amer. Math. Soc. Transl.* (Series 2) **13** (1960), 1–63.

[164] M.S. Livšic, On the spectral resolution of linear non-selfadjoint operators, *Mat. Sbornik* **34** (76) (1954), 145–199 (Russian); English Transl., *Amer. Math. Soc. Transl.* (Series 2) **5** (1957), 67–114.

[165] M.S. Livšic, N. Kravitsky, A.S. Markus and V. Vinnikov, *Theory of commuting nonselfadjoint operators,* Mathematics and its Applications, Vol. **332**, Kluwer, Dordrecht, 1995.

[166] D.G. Luenberger, *Opimization by Vector Space Methods*, Wiley, New York, 1969.

[167] S.L. Marple, *Digital Spectral Analysis with Applications*, Prentice Hall, New Jersey, 1987.

[168] N.K. Nikolskii, *Treatise on the Shift Operator*, Springer Verlag, New York, 1986.

[169] A.V. Oppenheim, R.W. Schafer and J.R. Buck, *Discrete-Time Signal Processing,* Prentice-Hall, New Jersey, 1999.

[170] S. Parrott, On a quotient norm and the Sz.-Nagy-Foias lifting theorem, *J. Functional Analysis* **30** (1978), 311–328.

[171] E. Parzen, *Time Series Analysis Papers,* Holden-Day, San Francisco, 1967.

[172] A. Peled and B. Liu, *Digital Signal Processing: Theory, Design and Implementation,* Krieger Publishing Company, Krieger Pub Co, Malabar, FL, 1987.

[173] S.U. Pillai and T.I. Schim, *Spectrum Estimation and System Identification,* Springer Verlag, New York, 1993.

[174] G. Popescu, Multi-analytic operators and some factorization theorems, *Indiana University Math. J.* **38** (1989), 693–710.

[175] G. Popescu, On intertwining dilations for sequences of noncommutating operators, *J. Math. Anal. Appl.* **167** (1992), 382–402.

[176] G. Popescu, Multi-analytic operators on Fock spaces, *Math. Ann. Appl.* **303** (1995), 31–46.

[177] G. Popescu, Structure and entropy for positive-definite Toeplotz kernels on free semigroups, *J. Math. Anal. Appl.* **254** (2001), 191–218.

[178] G. Popescu, Multivariable Nelari problem and interpolation, *J. Functional Analysis* **200** (2003), 536–581.

[179] G. Popescu, Entropy and multivariable interpolation, Memoirs of the AMS, Vol. **184**, 2006.

[180] J. Rissanen and L. Barbosa, Properties of infinite covariance matrices and stability of optimum predictors, *Information Sciences* **1** (3) (1969), 221–236.

[181] E.A. Robison and S. Treitel, *Geophysical Signal Analysis,* Prentice-Hall, New Jersey, 1980.

[182] M. Rosenblum and J. Rovnyak, *Hardy Classes and Operator Thoery*, Oxford University Press, New York, 1985.

[183] H.L. Royden, *Real Analysis*, Prentice Hall, New Jersey, 1988.

[184] G. Ruckebusch, Representátions Markoviennes de processus Gaussiens sataionnaires, *CR Acad. Sci. (Paris), Ser. A.* **282** (1976), 649–651.

[185] G. Ruckebusch, *Théorie géométrique de la represéntations Markovienne,* Thèse de doctorate d'etat, Université de Paris VI, 1980.

[186] G. Ruckebusch, Théorie géométrique de la represéntations Markovienne, *Ann. Inst. Henri Poincaré* **XVI** (1980), 225–297.

[187] W. Rudin, *Real and Complex Analysis,* McGraw-Hill Series in Higher Mathematics, New York, 1987

[188] W.J. Rugh, *Nonlinear Systems Theory, The Volterra Wiener Approach,* The Johns Hopkins University Press, Baltimore, Maryland, 1981.

[189] W.J. Rugh, *Linear System Theory,* Prentice Hall, New Jersey, 1993.

[190] D. Sarason, Generalized interpolating in H^∞, *Trans. American Math. Soc.* **127** (1967), 179–203.

[191] R.W. Schafer and A.V. Oppenheim, *Digital Signal Processing,* Prentice-Hall, New Jersery, 2002.

[192] R. Schaumann and M. E. Van Valkenburg, *Design of Analog Filters,* Oxford University Press, New York, 2001.

[193] R.E. Skelton, *Dynamic Systems Control,* John Wiley and Sons, New York, 1988.

[194] P. Stoica and R. Moses, *Spectral Analysis of Signals,* Prentice Hall, Upper Saddle River, New Jersey, 2005.

[195] P. Stoica and R. Moses, *Introduction to Spectral Analysis,* Prentice Hall, New Jersey, 1997.

[196] B. Sz.-Nagy and A. Koranyi, Relations d'un problème de Nevanlinna et Pick avec la theorie des opérateurs de l'espace Hilbertien, *Acta Math.* **100** (1958), 171–202.

[197] B. Sz.-Nagy and C. Foias, Dilation des commutants d'opérateurs, *C. R. Acad. Sci. Paris, série A* **266** (1968), 493–495.

[198] B. Sz.-Nagy and C. Foias, *Harmonic Analysis of Operators on Hilbert Space,* North Holland Publishing Co., Amsterdam-Budapest, 1970.

[199] J.C. Willems, Least squares stationary optimal control and the algebraic Riccati equation, *IEEE Trans. Automatic Contrtrol* **16** (1971), 621–634.

[200] Wolovich, W.A., *Linear Multivariable Systems*, Springer Verlag, New York, 1974.

[201] W.S. Wong, *Operator Theoretic methods in nonlinear systems,* Ph.D thesis, Harvard University, Cambridge, MA, 1980.

[202] N. Young, *An Introduction to Hilbert Space*, Cambridge University Press, Cambridge, England, 1988.

[203] L.A. Zadeh, and C.A. Desoer, *Linear System Theory,* McGraw-Hill, New York, 1963.

[204] K. Zhou, J.C. Doyle, and K. Glover, *Robust and Optimal Control,* Prentice-Hall, Inc., 1996.

Index

MIX
Papier aus verantwortungsvollen Quellen
Paper from responsible sources
FSC® C105338

If you have any concerns about our products,
you can contact us on
ProductSafety@springernature.com

In case Publisher is established outside the EU,
the EU authorized representative is:
Springer Nature Customer Service Center GmbH
Europaplatz 3, 69115 Heidelberg, Germany

Printed by Libri Plureos GmbH
in Hamburg, Germany